The World Encyclopedia of Animals

The World Encyclopedia of Animals

General Editor: Maurice Burton

FUNK & WAGNALLS
New York

CONTENTS

A bird's-eye view of the animal kingdom

The classification of animals

One and a half million species of walking, jumping, crawling, flying, burrowing, or swimming animals inhabit the earth. The study of the structure and behavior of all these forms of animal life and the determination of their mutual relationships is the domain of zoology, the scientific study of animals. In such a study, it is necessary to order the great diversity of material as efficiently as possible and to bring all the phenomena within the confines of a system in which every species is assigned a logical and reasonable place.

An ideal system would be one that ordered all the animals in accordance with their relationships, taking into account not only their external and internal structures but also their manner of life, their development, and their evolution. But there are so many threads to be traced—evident, probable, and possible— that it is impossible to construct a system in which they all play their proper role. One could approximate the natural relationships, to some extent, only by a complicated spatial model; and such a three-dimensional system would no doubt be extremely difficult to use. But if one mixes the natural elements with a (rather arbitrary) quantity of artificial devices, one obtains a system which may well be somewhat simplistic, but which does at least have the advantage of practicality. (For an example, see illustration on page 11.)

In this way one arrives at an initial division of the animal kingdom into a number of large groups called phyla (singular, phylum), each of which is distinguished by one or more striking characteristics. There are many differences between an elephant, a canary, a viper, a salamander, and a goldfish; yet all these animals share a number of important characteristics: they all have a skeleton composed of bone and cartilage, and a hollow, tubular central nervous system contained in the brain and the spinal cord. At some stage in their development they all possess a remarkable rod of cells enclosed by a tough membrane. The support offered by these cells makes the whole structure, the notochord, a rather firm "temporary skeleton." The notochord is so typical of these animals that one includes all species which are equipped with it, either temporarily or permanently, in the phylum of the chordates (Chordata). In all "vertebrate" animals, the notochord is quickly replaced during development by the vertebrae; it remains only in a few primitive "jawless" fish. Invertebrates such as the lancelet and the Tunicata (Urochorda) are also entitled, thanks to their notochord, to claim a modest place in the lowest portion of the chordate phylum.

Within the phylum one distinguishes a number of classes, which are in turn subdivided into ever-smaller groupings as follows: orders, families, genera (plural of genus), species, and races (varieties). But this breakdown is not yet sufficient for the system builders; they have furnished the spiderweb of the system with a number of even finer connective meshes.

The binomial nomenclature

Once Aristotle had catalogued the animal world of his day, only a few attempts were made to improve his system. The fundamental principles which had been enunciated by a few scholars of genius at the dawn of science remained unassailed

The zebrafinch (p. 2) is a well-known aviary bird. The orange spot on the cheek is characteristic of the male, which can thus be easily distinguished from the female.

The galago, or bush baby (facing page), is related to the loris but has a longer tail and larger ears (which it folds up when going to sleep). The galago has keen senses and is a skillful insect-catcher. It makes a screeching noise that resembles the crying of nurslings; hence its alternate name "bush baby."

SUBDIVISIONS OF THE ANIMAL KINGDOM

The process of subdividing the animal kingdom usually begins with a division into subkingdoms, which are then further broken down into phyla, subphyla, classes, orders, families, genera, species, and races. Actually, scientists use even finer subdivisions in order to put each animal in its proper place. Below is a chart of the subdivisions of the animal kingdom. This chart is not complete—some phyla of less interest to the reader have been left out entirely, and the finer division into classes and orders is applied only when it helps to clarify the text of this introduction.

KINGDOM	SUB-KINGDOM	PHYLUM	SUBPHYLUM	CLASS	ORDER
	Protozoa	Protozoa (one-celled animals)		Rhizopoda (rhizopods)	
				Flagellata (flagellates)	
				Ciliata (ciliates)	
				Sporozoa (sporozoans)	
				Suctoria (suctorians)	
	Mesozoa				
	Parazoa	Porifera (sponges)			
		Coelenterata (coelenterates)	Acnidaria	Ctenophora (comb jellies)	
			Cnidaria	Hydrozoa (hydroids)	
				Scyphozoa (jellyfish)	
				Anthozoa (sea anemones, corals)	
		Platyhelminthes (flatworms)		Turbellaria (free-living flatworms)	
				Trematoda (flukes)	
				Cestoda (tapeworms)	
		Nemathelminthes (roundworms)		Nematoda (roundworms)	
				Rotatoria (wheel animalculates)	
		Annelida (segmented worms)		Chaetopoda (chaetopods)	
				Hirundinea (leeches)	

ANIMALIA (ANIMAL)

Metazoa

Echinodermata (echinoderms)

- Crinoidea (sea lilies)
- Holothuroidea (sea cucumbers)
- Echinoidea (sea urchins)
- Ophiuroidea (brittle stars)
- Asteroidea (starfish, sea stars)

Mollusca (mollusks)

- Amphineura (chitons)
- Lamellibranchiata (bivalves)
- Scaphopoda (elephant's tusks)
- Gastropoda (snails)
- Cephalopoda (squids and allies)

Arthropoda (arthropods)

- Insecta (insects)
- Myriapoda (centipedes, millipedes)
- Arachnida (spiders)

Chordata (chordates)

- Tunicata (tunicates)
- Acrania (lancelets)

Vertebrata (vertebrates)

- Cyclostomata (round-mouthed fish)
- Chondrichthyes (cartilaginous fish) — Sharks, rays, chimaeras
- Teleostomi (bony fish), *inter alia* — Sturgeons, morays, catfish, flatfish, herrings, perch, lungfish
- Amphibia (amphibians) — Caecilians, salamanders, frogs and toads
- Reptilia (reptiles) — Turtles, crocodilians, lizards, snakes
- Aves (birds), *inter alia* — Penguins, ostriches, rheas, grebes, storks, ducks, birds of prey, chickens, cranes, plovers, pigeons, parrots, cuckoos, owls, woodpeckers, songbirds
- Mammalia (mammals), *inter alia* — Primates, carnivores, seals, whales, even-toed hoofed animals, odd-toed hoofed animals, elephants, rodents, insectivores, bats, anteaters, pangolins, aardvarks, marsupials

THE MANNER OF WRITING ANIMAL NAMES

In normal usage, which is followed in this book, the names of animals begin with a small letter in their English form (insects, bivalves, lice).

The manner of writing Latin names diverges; in this language the substantive (the only or the first word of the name) begins with a capital letter, and subsequent adjectives with a small letter. Moreover, the names of genus and species are written in italics in Latin. Thus we have Hyaenidae (the Latin name of the family to which hyenas belong), but *Proteles cristatus* (the Latin name of the aardwolf).

Multicellular animals, which consist of tissues and organs (Metazoa), have a characteristic process of development. The fertilized egg cell divides into two identical cells. Repeated new divisions lead to formation of a clump of cells called a morula. A process of folding then leads to the emergence of the gastrula, formed of an exterior layer, interior layer, and future digestive cavity.

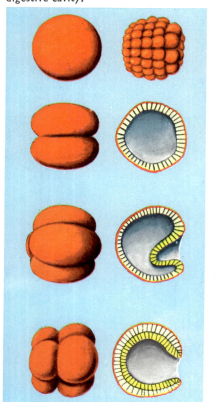

for centuries. It was not until the 18th century that scientists dared to break with the tradition and authority of the great pioneers. Independent observation and experimentation led to new concepts which were often completely at variance with the entrenched ideas dating from earlier times.

At that time, systematization was exercised with an enthusiasm bordering on fanaticism. Building on the lifework of John Ray, who described no fewer than 18,000 plants and combined them into a more or less natural system, the Swedish botanist Carolus Linnaeus (1707–1778) wrote his famous *Systema naturae* ("System of Nature"), surveying minerals, plants, and animals—stones, which grow; plants, which grow and live; and animals, which grow, live, and perceive. Linnaeus gave a short diagnosis of thousands of species. In the process he broke definitively with the prolixity of his predecessors and made consistent use of the system of binomial (two-part) nomenclature: every living creature was assigned a Latin name designating the genus and the species, a kind of scientific label from which one can deduce the relationship between an organism and other species. Although Linnaeus's system rests, it is true, upon an almost unbelievable knowledge of various species, it cannot be acquitted of the charge of a great onesidedness; it does not rise above the dryness and stiffness of a catalogue. The system is too artificial; it misses the natural connections among life forms that are the result of a protracted evolutionary history. The great systematizer had no doubts about the immutability of species; his scheme was based on the premise that, since the first moment of creation, all species have maintained themselves unchanged.

The great phyla of the animal kingdom

Every system is merely a human creation. The choice of the names and determination of the order remain a strongly personal matter, for what one person considers a primary and principal characteristic is viewed by another as merely an incidental feature. One usual division of the animal kingdom establishes four "subkingdoms" as the point of departure: Protozoa, Parazoa, Mesozoa, and Metazoa. A number of "phyla" are then placed within one or another of the subkingdoms. The members of all phyla but one are considered "lower animals"; the rest, constituting only one phylum, are assigned the status of "higher animals." The subkingdom of the Protozoa is very large, and its exact borders are difficult to trace. All its members are included in only one phylum, also called Protozoa. The name conveys the presupposition that these minute creatures were the first inhabitants of the earth. There are forms that display quite simple structures, as well as others that are surprisingly complex; yet they are all united by the attribute of being unicellular. Some investigators prefer to speak of "noncellular" organisms, since these correspond to a single cell of a higher organism only from a morphological point of view.

The remarkable phylum of the sponges (Porifera) is the only representative of the small Parazoa subkingdom. Its members consist of specialized types of cells, but the cells are able to shift to other activities if need be, since they have not yet coalesced to form tissues and organs.

A subkingdom of still more modest extent is that of the Mesozoa, the members of which consist of a few cells, usually arranged in two layers. The representatives of this class, of interest only to specialists, are peculiar creatures which move about by means of cilia and are parasitic on squids and other lower marine animals.

The very large subkingdom of the Metazoa includes the genuinely multicellular animals, constructed of tissues and organs. A tissue consists of a number of similar cells which are strongly "differentiated" from others in order to carry out

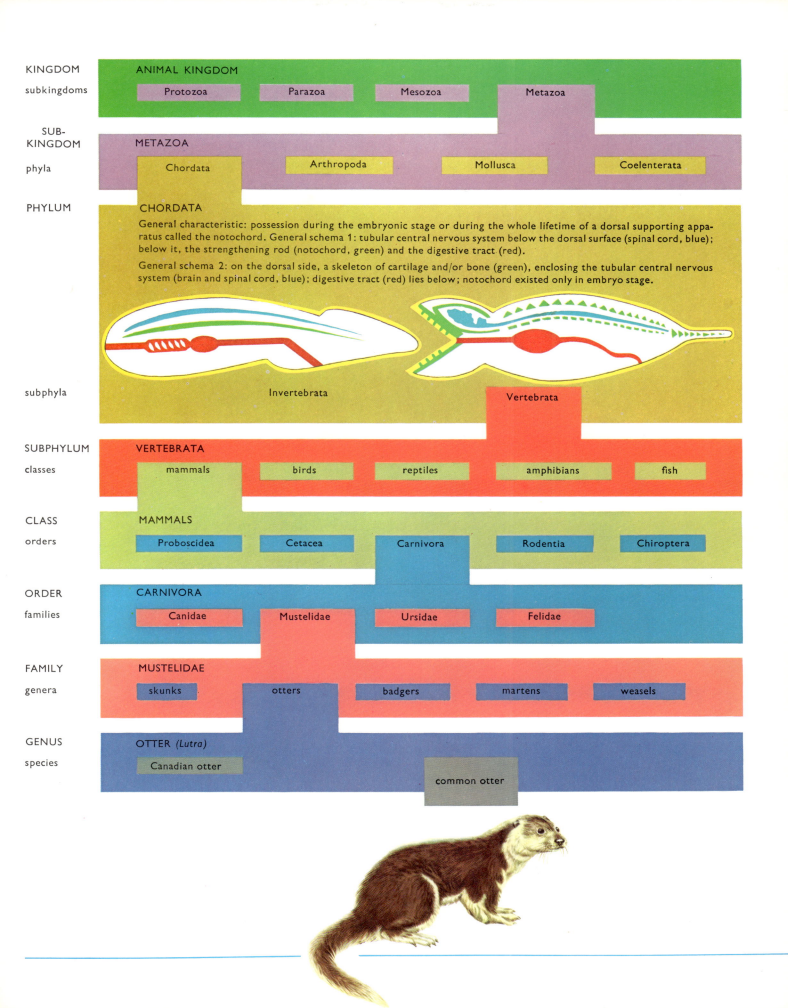

KINGDOM
subkingdoms

ANIMAL KINGDOM

Protozoa | Parazoa | Mesozoa | Metazoa

SUB-KINGDOM
phyla

METAZOA

Chordata | Arthropoda | Mollusca | Coelenterata

PHYLUM

CHORDATA

General characteristic: possession during the embryonic stage or during the whole lifetime of a dorsal supporting apparatus called the notochord. General schema 1: tubular central nervous system below the dorsal surface (spinal cord, blue); below it, the strengthening rod (notochord, green) and the digestive tract (red).

General schema 2: on the dorsal side, a skeleton of cartilage and/or bone (green), enclosing the tubular central nervous system (brain and spinal cord, blue); digestive tract (red) lies below; notochord existed only in embryo stage.

subphyla

Invertebrata | Vertebrata

SUBPHYLUM
classes

VERTEBRATA

mammals | birds | reptiles | amphibians | fish

CLASS
orders

MAMMALS

Proboscidea | Cetacea | Carnivora | Rodentia | Chiroptera

ORDER
families

CARNIVORA

Canidae | Mustelidae | Ursidae | Felidae

FAMILY
genera

MUSTELIDAE

skunks | otters | badgers | martens | weasels

GENUS
species

OTTER *(Lutra)*

Canadian otter | common otter

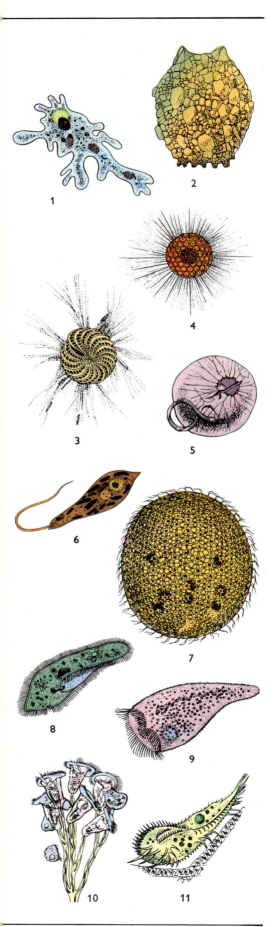

a certain vital function. During an early stage of development they consist of two cell layers, or germ layers: the ectoderm, which develops into the skin, and the endoderm, from which the digestive organs develop. At this stage the young gastrula generally resembles a kind of double-walled cup, the form of which can be compared with that of a strongly indented rubber ball. The opening of this cuplike organism is called the stomodaeum (primitive mouth), and the later mouth generally develops from it (see illustration on page 10).

The German naturalist Ernst Haeckel, who contributed to making Darwin's doctrine of evolution the central point in the 19th-century mechanistic philosophy of nature, saw this gastrula stage appear in most of the Metazoa, either as a development stage or—in the case of polyps, jellyfish, and corals—as the final stage. According to Haeckel, each animal in the course of its development displays, as it were, the evolution of its group in a more rapid tempo. This hypothesis, in a greatly exaggerated form, became a "biogenetic law." In the meantime, biologists have certainly become more cautious. In the fact that many development stages of higher animals, including man, correspond to the final stages of lower forms they see a genetic parallel which one cannot fail to recognize. The development of the individual and that of the species, as time elapses, clearly run parallel courses; but that fact does not prove they are identical.

Thus the gastrula consists of an outer and an inner layer. The former serves to protect the organism and to receive stimuli, and sometimes to make locomotion possible. The inner layer collects food and oxygen from the liquids entering the primitive mouth. All members of the phylum Coelenterata (coelenterates) experience this stage, which is to be compared to a gastrula.

In the case of the higher animals, a third germ layer develops between the other two. This third layer, called the mesoderm, provides *inter alia* for a firmer structure and for the movement of substances from the innermost germ layer to the outermost one. In the vertebrates, this primitive function is replaced by more specialized functions. Tissues and organs develop from the mesoderm. Connective tissue, cartilage, and bone impart solidity to the body; muscles make movement possible; and the blood in the heart and the blood vessels transports dissolved substances, while the kidneys purify the blood and maintain it at the proper concentration and acidity. Generally, a body cavity called a coelom develops in the mesoderm to shelter the respiratory and digestive organs, the heart, and the kidneys.

One-celled organisms: plant or animal?

In many cases the assignment of simple life forms to the plant or the animal kingdom is more or less a matter of personal preference, for the boundary between plant and animal is far from clearly marked in the lowest regions of life.

Higher plants evince certain peculiarities in structure and functions which distinguish them strongly from animals. Plants live more on the surface; they are more dependent than animals upon simple external organs for the exchange of substances with the environment, whereas animals have a more marked interior life in this respect. Moreover, the possibility of executing movements and of receiving various stimuli is much greater for animals than for plants. In addition, most higher plants are equipped with colored cell parts which serve the purposes of assimilation. They need to take in only simple inorganic substances, from which they make sugar and other organic material by using solar energy. The food of human beings and other animals must contain both organic and inorganic substances.

But if one makes this difference a rigid criterion for distinguishing between animal and plant, he immediately encounters difficulties as soon as he comes to the tiny cells that swarm in a drop of ordinary ditch water. Some of these one-celled organisms may behave as either a plant or an animal, depending upon the circumstances. One also finds, within the same family, species which acquire their food in a plantlike way alongside others which are fully entitled to be considered animals. Thus in the large class of the Flagellata (flagellates), we find flagellate plants which swim and assimilate along with non-green flagellate animals that chase prey or live as parasites. The trypanosomes, the notorious causative agents of sleeping sickness, are well-known examples of this last category.

The phylum Protozoa: primitive forms of animal life

Despite their tiny size, the one-celled animals make up a phylum of such diverse forms that a further division is desirable. Any representative of this phylum can be placed rather easily in one of the following classes.

Class Rhizopoda

These are naked, simply constructed protoplasts which move about lazily by means of pseudopodia ("false feet"). One also finds species floating in water, often strengthened by a tiny skeleton. The simplest animal organisms, the amoebas (see illustration 1 on page 12), and allied species that have a protective house (*Arcella*, *Difflugia*; illustration 2) live in fresh water. Among the striking rhizopods are the foraminifers (illustration 3), which live in the sea and have an external calcareous skeleton through which a delicate network of pseudopodia protrudes, and the even more handsome radiolarians (illustration 4).

Class Flagellata (Mastigophora)

These creatures move about by means of long, whiplike appendages called flagella. Both in fresh and in salt water, these delicate microorganisms constitute a very important part of the passively drifting marine life called plankton. Numerous species can with equal justification be classified as animal plankton (zooplankton) or as plant plankton (phytoplankton). Well-known species in this class include the dinoflagellates *Noctiluca* (illustration 5), which produce light and are responsible for much of the luminous appearance of the sea; *Euglena* (illustration 6), which can turn the water green; and animals forming spherical colonies, such as *Pandorina* and *Volvox* (illustration 7).

Class Sporozoa

This class owes its name to the fact that its life cycle includes propagation by means of spores as well as by sexual reproduction. Sporozoans live as parasites. The malaria parasite *Plasmodium malariae*, among others, is well known and notorious.

Class Ciliata

These animals, discovered by Anton van Leeuwenhoek, are called infusorians (Infusoria). For unicellular animals, they have an especially complicated structure. A noteworthy characteristic is their way of moving and catching prey by means of extremely delicate, rhythmically moving hairlike processes called cilia. A very common species, which lives in fresh water, is the paramecium (illustration 8). *Stylonichia* (illustration 11) has strong appendages called cirri on the "stomach" side, which enable it to walk over marine plants. Some species, including the bell-

ANIMAL EYES

The eyes of all vertebrate animals have, by and large, the same structure as those of human beings. Fish and aquatic mammals have a spherical lens, while the lens of land animals is more elliptical. In land vertebrates, accommodation results from a change in the shape of the lens itself; whereas in birds and reptiles the shape is changed by the sphincter of the iris. In the horse and cat families, the eyes shine brightly in dim light because of a special layer in the retina which reflects the light in such a way that the rays impinge twice upon the retina. The Leporidae (rabbits and hares) have a field of vision of almost 360° without moving their head. Dogs have a field of vision of 250°; humans, 160°.

like *Vorticella* (illustration 10) and *Stentor* (illustration 9), can form branched colonies.

Class Suctoria

These organisms are sessile for the most part. They use peculiar contractile, sucking tentacles to grasp a small prey, liquefy it, and take it up into the cell plasma. Some suctorians live as parasites.

Phylum Porifera (sponges)

Sponges, which are sessile and found primarily in the sea, have a body structure that is unique in more than one respect. They consist of more or less jellylike cups, the walls of which are permeated by canals. Numerous pores give access to this system of channels and hence to the flagellated chambers. These small chambers are lined with collar cells (choanocytes), each of which is provided with a plasma collar and a long flagellum. Coordinated action of the flagella maintains a continuous stream of water through the canals. The collars catch the edible particles, which consist of organic refuse or detritus and plankton; hence the collar cells work in the same way as in some unicellular organisms. The used water is expelled through openings called oscules.

A kind of skeleton supports the sometimes jellylike, sometimes leathery, and sometimes even hard body of the sponge. This consists of spicules of calcium carbonate or silicic acid. The subdivision of sponges into orders is based primarily on the form of these spicules. The skeleton can also consist wholly or partially of organic spongin fibers, chemically related to silk. Genuine tissues or organs are not yet present. There are no respiratory or excretory organs, no heart, and no blood vessels. Digestion and transportation of food are carried out by mobile amoebalike cells.

Phylum Coelenterata

The queer sea anemones and other fascinating showpieces in our large aquariums are—contrary to a common misapprehension—certainly not carnivorous plants nor plantlike animals; rather, they are complete animals which have only a superficial resemblance to flowers. These anthozoans ("flower animals"), together with jellyfish and corals, are representatives of a large animal phylum, the coelenterates (Coelenterata). In principle, a coelenterate has a simple structure. The more or less vase-shaped body is a double-walled cup, within which there is a cavity (coelenteron) that, *inter alia*, functions as a stomach. We may compare the two cell layers with the first two germ layers; a coelenterate advances no further, as it were, than the gastrula stage.

Coelenterates reproduce both sexually and asexually. The free-swimming jellyfish constitute a sexual stage and produce eggs and sperm. The fertilized eggs give rise to the sessile, asexual polyps. New jellyfish develop by means of division or budding (i.e., asexually). There are, however, many variations on this simple theme.

Nematocysts

One of the most striking peculiarities of most subphyla of the coelenterates is the possession of explosive stinging cells called nematocysts. These poison-filled cells, absent only in the Ctenophora, are used to attach the prey, cripple it, or kill it.

Sponges are stationary, cup-shaped animals which suck in water through numerous pores in their exterior walls. In small flagellated chambers, oxygen and microscopically small particles of plankton are removed from the water. Then the water is pumped back out of the chambers and expelled through one or more oscules. A large sponge can process 330 imperial gallons (1,500 liters) of water every 24 hours. Illustration 1 shows a schematic cross section of a sponge, with (a) pore, (b) flagellated chamber, (c) oscule. Illustration 2 shows a group of living sponges with their oscules.

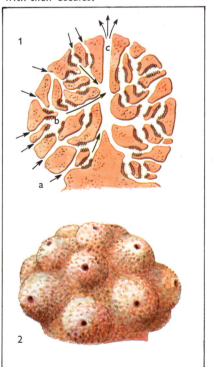

The nematocysts may be clearly seen in the freshwater polyp *Hydra*. These small animals, $\frac{2}{5}$ to $1\frac{1}{5}$ in. long (1 to 3 cm), had been described by Leeuwenhoek, but they became generally known in the 17th century through the discoveries of Trembley, who originally came from Geneva. In the Netherlands he studied the polyps in a small pond and investigated their locomotion, propagation, and ability to recover completely from serious wounds. Even when it was turned inside out, such a polyp appeared to recover quickly.

These coelenterates owe their official name *Hydra* to this regenerative ability; the fabulous monster known as the Hydra of Lerna grew two new heads for each one chopped off by Hercules.

A freshwater polyp demonstrates clearly the characteristic structure of polyps in general. The elongated, vase-shaped little animal has a number of tentacles around the mouth and an exterior layer containing the shiny little bodies known as nematocysts (or cnidocysts). These cells can throw out a hollow thread which functions as a lasso to hold a small prey or as an injection apparatus to cripple the prey by means of poison. The slightest touch on a *Hydra* nematocyst causes the small lid of the cell to spring open, after which the thread shoots out like a finger extended from a glove.

The classification of the coelenterates

In accordance with the presence or absence of nematocysts, the phylum Coelenterata can be divided into the subphylum Cnidaria, whose nematocysts make them a sort of animal stinging nettle, and the subphylum Acnidaria, the members of which have no nematocysts. The first subphylum includes the classes Hydrozoa (hydroids), Scyphozoa (jellyfish), and Anthozoa (sea anemones and corals). The glasslike sea gooseberries (Pleurobrachiidae, a family of the comb jellies), often found in enormous quantities along the high-water mark, lack the characteristic weapons of the coelenterates. To replace them, they have two long, retractile tentacles containing sticky cells known as colloblasts which enable the animal to capture plankton. These oval creatures have quite an odd manner of swimming; they move by beating their comb plates, which are such striking structures that they have caused the animals to be known as comb jellies.

In the jellyfish polyps, new individuals develop by budding, in the same way as in the case of the freshwater polyp. For the most part, the new animals do not separate themselves, but rather form a colony together with the mother polyp. The stomachs of the individuals retain open connections with each other. This form of social life often passes over into a higher state—a cooperation of individuals which display differences and are specialized for the performance of certain vital functions. Some are adapted to capture and digest prey, while others carry out the functions of defense or propagation. This type of colony evinces a division of labor.

The sexual form, the jellyfish or medusa, arises by budding or fission. A kind of mouth tube, called the manubrium or proboscis, hangs in the center of the young bell-shaped animal and serves as a stomach. This stalk, which has attached the young medusa to the polyp colony, pinches off, and the new generation swims away. Swimming results from rhythmic movement of circular and radial muscles.

True jellyfish

The large jellyfish which we can find on beaches are representatives of a class of coelenterates (Scyphozoa) in which the sexual stage is most in evidence. For the most part, we find numerous tentacles along the edge of the bell (umbrella). In the notorious red jellyfish (*Cyanea capillata*) these appendages, thickly inlaid with nematocysts, can reach a length of nearly 100 ft (30 m), and large specimens can

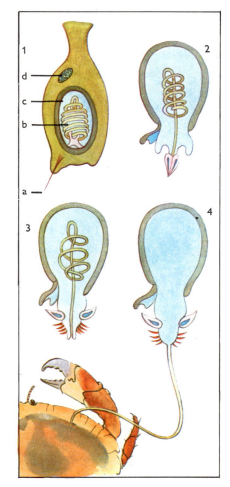

Most coelenterates are equipped with small but powerful poisonous weapons, the nematocysts, which explode at the slightest contact. 1, An unexploded nematocyst with (a) the sensitive process called a cnidocil (to be compared to the contact horns of a naval mine); (b) the rolled-up thread; (c) poison vesicle; (d) nucleus. 2–4, An exploding nematocyst. The thread is turned inside out and expelled from the cell with great rapidity. The poison is pumped into the skin of the prey or enemy through the hollow thread.

MUSCLE POWER

In the 18th century James Watt, the inventor of the steam engine, conducted experiments to determine how much a strong dray horse could pull. Since that time the unit of power, called a horsepower, has been defined as 550 footpounds per second (or, in the metric system, the ability to raise 75 kg one meter per second).

It is true that heavy dray horses have enormously developed muscles. History tells us that one horse once moved 16 railway cars loaded with 55 tons a total of 19 mi (30 km). But there are even stronger animals. With his trunk, which contains some 40,000 muscles, an adult elephant can lift a tree trunk weighing 4 tons from the ground. Among particularly strong mammals we may mention sperm whales, buffaloes, bears, tigers, gorillas, and moles, which can easily lift an object 40 times their own weight. It is indeed a striking fact that the muscular strength of small animals is often greater than that of a horse or an elephant. When we see a weasel dragging away a hare, it is as if a lion were to move a full-grown elephant at a speed of 9 mph (15 kph). The jaws of an adult elephant exercise a force of 1,650 lb (750 kg); but in the case of even the largest crocodile, the muscles that open the mouth are so weak that it is possible to hold the mouth closed with one hand.

Large birds of prey are also powerfully muscled, but there is probably not a single bird that is able, while flying, to carry a load heavier than its own weight. Since an eagle weighs no more than 16½ lb (7,5 kg), it is impossible for it to carry off such heavy loads as some stories have claimed.

Even mollusks, in certain circumstances, can give surprising displays of muscle power. There are snails which are able to attach themselves so firmly to rocks in the surf that a force of 88 lb (40 kg) must be exerted to pull them loose. The abalone, and above all the giant clam of the coral reefs, can clamp a diver's arm or leg as if in a steel vise.

attain a bell diameter of 6 ft (2 m). The crippled or dead prey is brought to the mouth by the curled mouth arms. The large central stomach is connected with the ring canal by a system of radial canals. This canal system distributes the food equally to all parts of the organism.

Most jellyfish are undaunted by catching a rather large prey, which they ingest in its entirety. Other jellyfish eat only small food particles. In place of a mouth to carry out this function, the common jellyfish *Rhizostoma* has a large number of pores, especially in the curved parts of the arms. Swimming compresses and extends the stomach, and this pump apparatus conducts organic debris and microplankton to the stomach.

Anthozoa

The sessile sea anemones and corals, known as Anthozoa, constitute a group of very ancient lineage. Anthozoa were living as early as the beginning of the Paleozoic era, some 1½ billion years ago. Anthozoa (polyps) are characterized by possession of a mouth tube (stomodaeum) which can be closed from above by muscles and which leads to the stomach, or coelenteron. Alternating with the tentacles the coelenteron contains a number of radially arranged membranous partitions called mesenteries, richly endowed with muscles. The edges of these partitions are packed with glandular cells and nematocysts. Small food particles, primarily of animal nature, are taken up in their entirety by special digestive cells and completely digested within them.

The mesenteries of the stomach also contain the sex organs—ovaries in the case of the females, testes in the males. Most Anthozoa are differentiated by sex. The larvae, which emerge from the fertilized eggs, swim out through the mouth of the mother, attach themselves to something after a few days, and grow into polyps. Asexual reproduction by budding is also frequently found. This process can give rise to very extensive colonies. The sea anemones attach themselves to the bottom by means of their pedal disc; this weak base also enables them to move about slowly, but unaided. Sea anemones include both particle eaters and genuine predators. Such a predator is the cosmopolitan beadlet anemone (*Actinia equina*), generally found along the North Sea coasts. This inhabitant of tidal regions is found in numerous color variations. As soon as several tentacles come in contact with a prey that has been crippled by poison from the nematocysts, they cling to it. Sensory cells, sensitive to chemical stimuli, send indications to a delicate and "diffuse" nervous system. With this guidance, the tentacles bring the prey to the mouth opening.

Coral animals: Anthozoa with a skeleton

Coral animals have the same body structure as sea anemones; they are also Anthozoa but are equipped with a skeleton built of a calcareous or horny material.

For the most part, a large number of individuals form a collective, a colony. The stomachs of the individual members are connected with each other by canals. If a portion of the colony is overgrown by other corals or by seaweed, the immobilized polyps can still be maintained for some time by their more fortunate associates. But if the distance between them and the still-free polyps becomes too great, they die off. Hence a coral reef lives and grows primarily at its outer rim. Corals require a great amount of light for their welfare, for they host numerous guests in their tissues. These microscopic unicellular plants, which produce organic substances for their hosts, cannot perform assimilation without light. The coral polyps feed on small animal plankton, and only during the night.

Besides asexual propagation, coral animals also display sexual reproduction.

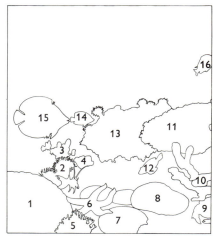

The coral reef, the result of centuries of activity by myriad coral animals, is a colorful community. A setting of fanciful coral forms teems with the most improbably shaped and colored fish, crabs, and echinoderms. Even worms and snails, with their fanlike gill plumage, are not absent. The corals themselves are the calcareous or horny skeletons and, at the same time, the hiding places of colony-dwelling Anthozoa, which can live only in oxygen-rich clear water with a temperature of around 77°F (25°C). Most corals live with unicellular seaweeds, which provide the necessary oxygen.

1, brain coral (Heliastraea); 2, hermit crab (Eupagurus) with sea anemone; 3, Cynthia; 4, Turbinaria; 5, stag's horn coral (Acropora); 6, starfish (sea star) Asterias; 7, coral Fungia; 8, sponge Euspongia; 9, crab (Cancer); 10, blood coral (Corallium rubrum); 11, organ-pipe coral (Tubipora); 12, Hypoplectrus unicolor; 13, sea anemone (Tealia); 14, angelfish (Pomacanthus arcuatus); 15, angelfish (Holacanthus); 16, porkfish (Anisotremus virginicus).

Larvae, equipped with cilia, develop from the fertilized egg cells. Most of these planulae do not succeed in finding an appropriate place to which to anchor themselves, but if an individual does succeed, it develops a mouth and tentacles. An initial layer of calcareous material is laid down. This skeletal plate soon grows outward in all directions, while the animal begins the formation of a new colony by budding.

Between the Tropic of Cancer and the Tropic of Capricorn, in areas in which the water is clear enough, no deeper than 135 ft (40 m), supplied with sufficient oxygen and salts, and no colder than 68°F (20°C)—corals are particularly demanding animals—one finds an extensive belt of fairylike underwater gardens. These coral reefs—the famed Great Barrier Reef off the northeast coast of Australia is about 1,200 mi (1,900 km) long and 20 mi (30 km) wide—constitute an ideal environment for the most diverse species of animals: fish of fantastic colors and forms, lobsters and crabs, mussels and snails, almost all as variegated in color as the corals themselves and often of incredibly bizarre form. Coral terraces or coastal (fringing) reefs can develop along rocky tropical coasts. Barrier reefs are parallel to the coast and are separated from it by a lagoon. These reefs run steeply down into the sea. A third reef form is the atoll, a barrier reef that is more or less round, with an outer rim plunging steeply into very deep water. A lagoon some 65 to 165 ft deep (20 to 50 m) lies within.

Phylum Echinodermata

This very old animal group, which includes the radially built starfish and sea urchins, constitutes a phylum which has never ventured outside fresh water. Most Echinodermata (echinoderms) are equipped with an armor of calcareous pieces and spines. The calcareous pieces constitute a complete shield, except in the case of the sea cucumbers. Some of the spines act as tiny tweezers and pincers; these "pedicellariae" keep the body surface clean, dispel dirt and sand corals, squeeze small skin parasites to death, and also catch edible morsels.

A starfish walks in a rather odd manner over the sea floor and over banks of mussels and oysters. The animal glides slowly along, generally keeping the tips of its arms somewhat elevated. On the lower side of the tops there is a red, simple eyelet which can distinguish between light and dark. The arms are equipped with mobile projections called tube feet, each of which has a suction cup at the end. The suction cups enable the animal to remain in one position against the action of the currents or to move slowly forward; but they also allow the sea star to take a tenacious grip on the shells of bivalves. When the muscles holding the mollusk closed finally cede, the starfish shows another side of its odd character —the animal eats out, as it were. The pressure of the body fluid forces the stomach out through the mouth, and after the stomach juices have digested a part of the tender mollusk meat, the stomach is returned to the interior.

Reproduction is simple. In the spring the starfish lays about 2 million eggs in a few hours. After fertilization by a sperm cell, the tiny egg develops into a larva furnished with cilia, from which in turn a young starfish develops in a rather complicated manner.

Asexual reproduction by fission is also found in some echinoderms; moreover, the animals have an impressive capacity for regeneration. Even a single broken-off arm can develop into a complete starfish.

Thanks to their calcareous armor, Echinodermata fossilize easily. The weal and woe of this group over more than a thousand million years no longer has many secrets for paleontologists. The Echinodermata have proved to be a quite conservative group; their evolutionary accomplishments are not sensational.

Worms—a heterogeneous society

In the ground, as well as in fresh and salt water, one finds many thousands of animal species which have more or less the elongated cylindrical exterior form of an earthworm. A closer look at these animals, which include a number of microscopic species, quickly shows that one cannot lump them all together.

Earthworms, marine clam worms, lugworms, and leeches are all included in the phylum of the segmented worms (Annelida); these animals are segmented both externally and internally.

In the phylum of the roundworms (Nemathelminthes), which includes more than 80,000 known species, segmentation is lacking. Well-known parasites such as roundworms, hookworms, and trichina worms belong to this phylum, as do an enormous number of other species parasitic on animals and plants. There are also free-living species to be found in almost every conceivable environment.

A strongly flattened structure is displayed by the free-living flatworms (Turbellaria) and the parasitic flukes (Trematoda) and tapeworms (Cestoda), three classes of which are united in the phylum Platyhelminthes (flatworms).

The exterior and interior of a worm

If I take a worm in my hand, I note that this cold, slippery and moist animal has a certain solidity. It attempts to escape by making twisting movements and by repeatedly lengthening, then shortening, its body segments. With some difficulty I can also distinguish the front and the rear end, not so much by the direc-

A worm is divided by partitions called septa into a series of compartments termed somites (or metameres), each of which contains virtually the same organs. Only the reproductive organs are limited to a few segments. Among the parts of a worm are: (a) mouth; (b) intestine; (c) septum; (d) closed blood-vessel system containing red blood; (e) ladderlike nervous system or nerve cord; (f) excretory organs, through which wastes are removed from the body cavity (coelom) and the blood.

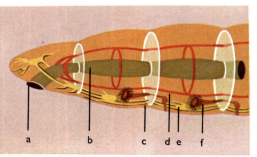

tion of movement (a worm can crawl backward just as easily as forward,) but rather by its mouth with a fleshy lip, which I can distinctly see with a magnifying glass on the lower side of the "head." At the other end I see the anus.

If I pull a worm through my fingers, I feel that the skin is not completely smooth; small, stiff bristles project from it. A set of muscles can turn these bristles in a direction opposed to that of the movement of the animal.

Two layers of muscles lie directly under the epidermis. Pressure from this "skin muscle sac" on the fluid filling the compartments of the body imparts a certain rigidity to the animal. Moreover, the combination of the muscles and the body compartments enables the worm to execute its typical movements. If the circular muscles contract while the longitudinal muscles in the same segment relax, the segment becomes elongated and thin. If the longitudinal muscles contract, it becomes short and thick. During this interplay of circular and longitudinal muscles, which is called peristalsis, the bristles are positioned so as to prevent the animal from sliding backward. If it had no bristles, the peristaltic movements would make the worm longer and shorter, but it would not advance.

A worm uses the same technique to crawl through the ground. First it pushes its thin front or rear end into the sand. Then the longitudinal muscles of this portion of the worm contract, making it thicker and the hole larger. In this way the worm can burrow through a layer of earth about 3 ft thick (1 m). And though the oxygen supply in deeper layers is not especially good, there is not much danger of suffocation. The reddish color of an earthworm is caused by the blood circulating in innumerable extremely fine capillaries close to the skin. This blood contains the same red pigment as the blood of all vertebrates; this hemoglobin is a respiratory pigment that draws oxygen from the environment with special ease. There is, of course, also much oxygen dissolved in the body fluids; the amount is sufficient in normal circumstances, and the hemoglobin is intended more for special conditions or emergencies.

Phylum Mollusca (mollusks)

The possession of a shell is the most striking characteristic of the Mollusca, a very old phylum of the animal kingdom with a large number of species. The shell is produced by the mantle, a fold of skin that covers the body like a roof.

In the class of the snails (Gastropoda), the shell develops in the form of an asymmetrical turret. Some members of the present-day Cephalopoda, which include squids and cuttlefish, also inhabit such a snail shell. But in this case it is divided into chambers filled with gas, which lower the specific gravity. The last and largest chamber is the animal's lair. This structure was usual as early as the Cambrian period. In the course of time, however, these animals performed strange experiments with their shells. The oldest forms still had straight shells; the *Endoceras* lived in the front chamber of a shell that was about 15 ft long (4.50 m). But the squids and cuttlefish quickly developed into rapid swimmers, and in the process the long shells gave way to the much handier snail shells. Moreover, the gas chambers came to lie more above the center of gravity of the animal—a change which considerably increased stability and floating ability. In other species the portion divided into chambers gradually became internal; the air chambers disappeared, and the internal shell came into being, as we encounter it in the most recent species. The familiar white pieces of cuttlebone (*Sepia*), of which we often find great quantities along the high-water mark, are such internal cuttlefish shells.

Mussels have a dwelling composed of two joined valves; chitons (Polyplacophora) have eight overlapping calcareous shell plates.

The diversity of form and color among the mollusks is unbelievably great. In the

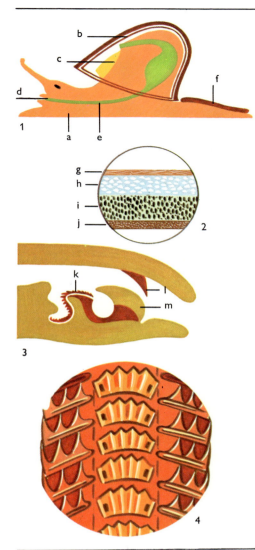

Snails are mollusks that have a spirally coiled shell. The snail (1) creeps along on its muscular foot (a). In a fold of skin called the mantle (b) lies the respiratory cavity, which may contain one or two gills (c). Other parts include (d) mouth, (e) intestine, and (f) operculum, which serves to seal off the shell. The shell (2) is composed of several layers: (g) organic membrane; (h) prism-shaped calcium carbonate crystals; (i) hard mother-of-pearl layer of leaf-shaped calcium carbonate crystals; (j) mantle. The cross section of the head (3) shows (k) radula, (l) chitinous "jaw," (m) tongue. The rasping organ, or radula (4), of a snail is covered by thousands of microscopically small teeth.

18th century the collecting of exotic shells became a real mania, and improbably high prices were paid for rare specimens. Cowrie shells were used as money in Africa, among other places, until the beginning of this century; and in New Guinea the Papuans still recognize shells as means of payment and exchange. Some cowrie shells (*Cypraea* species) are especially prized.

Snails

Gastropoda is the official name for the class of the snails. These animals glide slowly forward by using a "foot" which has a muscular interior and a rather slimy exterior. If a snail creeps over a glass surface, the gentle wavy movements of the foot muscles—about 50 waves per minute—can be clearly seen. The foot glides easily over a thin layer of mucus that is secreted by a gland under the head; the animal lays down a carpet of mucus while creeping along. The foot muscle withdraws immediately from sharp objects; a snail can crawl over a razor-sharp knife blade without risk.

During its often nocturnal expeditions, a snail orients itself primarily by the senses of touch and smell, which are located in a pair of short tentacles on the head. The eye stalks, very sensitive to touch, are well-known features. The rather primitive eyes on top do not allow the animal to do much more than distinguish light intensity and direction. Snails are equipped with an ingenious apparatus for acquiring food; the tongue is covered by a rasp, called the radula, containing thousands of tiny teeth.

Not all snails are herbivorous. This class of mollusks also includes predators and scavengers. The carnivorous species use the middle row of teeth of the radula to tear off small pieces of meat, and the outer rows to ingest them. The quite round little holes we often find in shells washed ashore are caused by the patient rasping and boring operations of a predatory snail called the moon shell (*Natica*). The entire rasp apparatus, together with an acid-secreting gland, can be inserted by means of proboscides through the hole bored in the shell, and the tender meat can easily be torn loose.

Snails cannot leave their shells; they are irrevocably bound to them. If we see what appears to be a snail without a shell, we are dealing with what is called a slug. Well-known naked land slugs include some that are very harmful and large black or reddish-orange types (*Agriolimax* and *Arion*).

Bivalves (class Lamellibranchia, or Pelecypoda)

Most of the shells we find on the beach are the products of laterally flattened mollusks that breathe through gills. This class is called bivalves (Lamellibranchia, or Pelecypoda). The shell valves, joined at the hinge by an elastic ligament, open mechanically, but can be closed quickly by one or two adductor muscles. The latter run transversely through the body and are attached to the inner side of the shell valves. In an empty shell one can clearly see the marks left by these muscles. When a bank of bivalves is left dry by low water, muscles hold the valves of the shell together. These muscles act slowly but are very economical, doing their work with sufficient strength and very low energy consumption. The underside of the mantle allows the foot (the creeping and burrowing tool) to protrude, and at the rear it has two openings, sometimes located at the end of long tubes called siphons.

Through the lower opening, water rich in oxygen and plankton is led past the gill bars by the beating of thousands of fine cilia; through the upper opening the used water is expelled. The cilia transport all the edible particles to the mouth.

The edge of the mantle enlarges the shells by continuous deposition of calcareous substances, and most bivalve shells show evident growth lines.

The word "mussel" is sometimes used in a narrow sense to indicate the well-known and prized representatives of the genus *Mytilus* (true mussels). Although these bivalves occur naturally everywhere along rocky coasts and breakwaters, they are also raised commercially in large quantities. The mussel farmer begins by harvesting young mussels and then distributes this "seed" over appropriate mussel areas. The young mussels then begin to "spin"; that is, by means of tough threads produced by the byssus gland, they attach themselves to each other and to the sea floor. When the mussels are half-grown, about 2 in. (5 cm) long, they are transferred to food-rich areas where they mature to become a remarkable product, a food rich in albumen, minerals, and vitamins. The whole growth process takes $1\frac{1}{2}$ to $2\frac{1}{2}$ years.

The oyster (*Ostrea edulis*) takes much longer, and oyster farming is more complicated. In contrast to most of the bivalves, the oyster is hermaphroditic. When grown, it behaves alternately as male and female; one summer it produces sperm, and the following summer eggs. The tens of thousands of tiny eggs remain for a time in the gills and are fertilized there by the sperm cells, which find their way into the gill cavity with the water incurrent. The larvae also remain for a time in the gills, where they do not lack for oxygen and plankton food. After about a week, the oyster "coughs" its brood out through the excurrent siphon. The delicate larvae swim around for a while by using their cilia, then sink and attach themselves. This sinking of the brood is a most important event for the oyster farmer. To receive the brood, he has put out thousands of whitewashed tiles in rows; enormous quantities of empty mussel shells are also used as "collectors." In the spring the tiles with the young oysters are fished out and transferred to growing grounds. In the following season the shells are separated from the tiles so that the animals can be placed in real, staked-off growing areas covering $12\frac{1}{2}$ to 25 acres, where they will mature to become worthy edible oysters. Finally they are transferred to the well-known oyster ponds, wet warehouses from which the oyster farmers deliver their product to gourmets (from September to April).

Cephalopoda—highly developed mollusks

This remarkable class diverges strongly from its allies in many respects. These mobile swimmers, with their highly developed senses and rapid reactions, hardly remind us at all of the slow snails or the still slower bivalves. The eyes are very similar to those of vertebrate animals, and the bodies are supported here and there by a tissue that is surely unique among invertebrates—cartilage. On the other hand, there are so many internal resemblances to snails and bivalves that we are obliged to consider the Cephalopoda (squid, octopus, cuttlefish) as very highly developed mollusks.

The common European cuttlefish (*Sepia officinalis*)—found in the North and Mediterranean seas, among other areas—has an oblong, bag-shaped body, equipped with a fringe of skin that serves as a fin; a round head with large, glittering eyes that always have somewhat of a surprised look; and eight short arms and two longer tentacles. These arms, to which the class owes its name of Cephalopoda ("head-footed"), grasp the prey and bring it to the mouth, located in the center of the arms. The hard beak and the tongue radula—a genuine characteristic of the snails—then process the food.

The edge of the mantle can be pressed firmly against the body and held in place by two knobs which fit into elastic cuplets; these are genuine press studs. This permits the animal to expel the water in the gill cavity with force through a funnel pointed forward, so that the cuttlefish shoots backward. The Cephalopoda are the originators of the press stud and an inspiration for jet propulsion. Moreover, the cuttlefish successfully applies smoke-screen tactics. If the animal is strongly excited, it expels a dark-brown product of the ink gland through the funnel in a

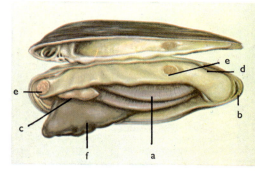

Platelike gills (a) hang in the mantle cavity between the two halves (valves) of the shell of a mussel; the latter is a member of the class of bivalves (also called Lamellibranchia or Pelecypoda). The gills take oxygen from the water that flows into the gill cavity through the incurrent opening (b). Moreover, food particles stick to a thin layer of slime which is brought past the labial palp (c) to the mouth by the beating of thousands of cilia. The stale water leaves the gill cavity through the excurrent opening (d). Two muscle bundles (e) can press the two valves of the shell against each other. (f) Foot.

1, Anatomy of a cephalopod of the group Dibranchia: (a) suckers, (b) mouth opening, (c) funnel, (d) fin, (e) mantle wall, (f) gill, (g) ink sac. 2, Cephalopods swim backward. The water is pressed forcefully out of the sac, spurts through the narrow opening of the funnel, and causes the animal to shoot backward through the water with a shock. The relaxed sac, which also contains the gills, then admits fresh water. Cephalopods are divided into Tetrabranchia, to which group only the nautilus (3, external shell of the *Nautilus pompilius*) belongs, and Dibranchia. The latter, in turn, are subdivided into Octopoda (4, the octopod *Octopus vulgaris*) and Decapoda, which have eight short arms and two long ones that rapidly grasp their prey.

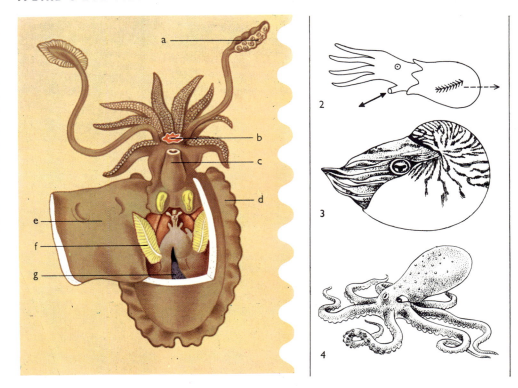

forward direction. The resulting jerky, backward swimming allows a cephalopod to escape its enemy. If it is in no hurry, it uses its fins for movement. In contrast to snails and bivalves, cephalopods are particularly intelligent. Their brain is at a remarkably high stage of evolution. In the case of most invertebrates, the nerve centers or ganglia are distributed over the whole body; but in the Cephalopoda, as among the vertebrates, they are centralized in the brain, an organization permitting a much more delicate coordination of movements and behavior. Their instincts are also equivalent to those of some fish. Researchers have even seen cephalopods use stones as tools with which to force open bivalve shells.

The evolution from primitive snail to cephalopod must certainly have been a very long process; hence one would surely not expect to find this supermollusk in very old Cambrian rocks. Millions of years before the vertebrates began to develop, the Cephalopoda—though by completely different paths—had reached the level of "almost higher animals."

Mollusks—A summary

Mollusks are animals possessing bilateral symmetry and are generally protected by a shell. Most species breathe through gills, but a few have lungs (the subclass of the pulmonates, Pulmonata). The phylum of the mollusks is divided into the following classes:

Amphineura, which includes the peculiar chitons (Polyplacophora), whose dorsal covering consists of eight shell plates.

Bivalvia (Lamellibranchiata), the mollusks which have a shell composed of two valves.

Scaphopoda (elephant's tusks), a class of which the genus *Dentalium* is best known.

Gastropoda, the class of the snails.

Cephalopoda (squids and allies), most of which have an internal shell.

In virtually all representatives of these classes the shell functions as a skeleton, offering support for the soft parts and an anchorage for the muscles. Perhaps the most primitive shell form is the conical type, which we encounter in snails of the genus *Patella*. In most classes we find a reduction of the shell, which becomes rudimentary.

The phylum Arthropoda (arthropods)

Crabs and lobsters, centipedes, spiders, beetles, flies, and butterflies all belong to the phylum Arthropoda, which has the greatest number of species of all the phyla. At first glance these highly developed animals certainly do not show any resemblance to an earthworm. Yet, when one considers them more closely, there are so many points of agreement as to justify the proposition that the Arthropoda must have had wormlike ancestors.

One of the most striking characteristics of the Arthropoda is their heavy armor. They have an exoskeleton composed of a light, but hard, substance called chitin (chemically a polysaccharide), still further strengthened in many species by the addition of calcareous substances. Arthropoda are "packed," as it were, within this framework, to the inner side of which the muscles are attached. The framework gives the animal the necessary support, so that functionally the armor corresponds to the skeletons of the higher animals which have them. Moreover, the thickness of the exoskeleton gives the animal especially effective protection against drought. This structure, however, has a serious disadvantage: it inhibits growth. The consequence is that during the growth period it must periodically be shed and replaced by a new one. Before the old armor splits open, a new, elastic layer of chitin in a soft state is laid down. As soon as the animal has successfully completed the rather drastic sloughing procedure, it dilates itself strongly by taking in air or water. It then remains in this pumped-up condition until the new armor has hardened sufficiently. Thus the animal, for the time being, again has enough room in which to grow. Insects slough only during the larval period. Once they have become adults, they do not grow any more, so that sloughing would be superfluous. Thus the dimensions of an adult insect have no relationship to its age; a small fly is not necessarily a young fly.

Movement of the body parts is possible only owing to the fact that they are articulated. The hard segments of the body, the limbs, the antennae, and other appendages are connected by flexible membranes, and joints are thereby created without danger of leakage.

The class Crustacea

The crustaceans (Crustacea), among which are crayfish and lobsters, have always chosen water as their milieu *par excellence*. They are, like fish, "primary" water animals that breathe through gills. This class does, to be sure, include certain descendants of stout-hearted pioneers who were able to conquer the land as a new environment. The gills did not disappear in the process, but they were rather drastically modified in order to function with sufficient efficiency out of the water. The well-known woodlice (sow bugs) are examples of such "dry-land crustaceans." But most crustaceans have remained true to the environment of their ancestors; the calcium-rich seawater is still *the* environment for an unbelievably large number of species. This milieu offers them easy access to sufficient calcium carbonate, which they ingest with their food. The blood brings the dissolved calcium carbonate to the skin, where it is transformed into very hard and insoluble calcium carbonate crystals, which turn the horny chitin plates into an especially solid "shell." This shell made it easy for the primitive crustaceans to fossilize; their armor served as casting molds that collected mud and fine sand, which then petrified. It is owing to this fortunate circumstance that we have extensive knowledge of the crustacean group. As early as the Cambrian period

THE NUMBER OF CEPHALOPODS

Nobody knows how many cephalopods there are living in the sea, but the number must certainly be very large, for many seals and nearly all toothed whales feed on squid. In some cases in which the sound waves from an echo-sounding device are unable to reach the sea floor, the reason is the fact that the waves are reflected by large schools of squid swimming close together at a depth of 650 ft (200 m).

ANIMAL LANGUAGE

Researchers have discovered that howling monkeys (howlers) have a command of 15 to 20 different cries, and chimpanzees of as many as 32. In the case of chickens, one can distinguish 10 different sounds with separate meanings, and 15 among rooks. There are remarkable differences in the sounds made by jackdaws in various regions—in England and on the European continent, for example. Thus these birds have "dialects."

Some animals can develop surprising speed. The record is held by the cheetah, but there are also other formidable racers. The table below gives the speeds, in mph (followed by the speed in kph, in parentheses), a few animals can attain.

African elephant	25	(40)
Rabbit	25	(40)
Grizzly bear	29	(46)
Virginia deer	29	(46)
Wild boar	29	(46)
Giraffe	31	(50)
Reindeer	31	(50)
Greyhound	38	(60)
Lion	38	(60)
Zebra	38	(60)
Coyote	44	(70)
Kangaroo	44	(70)
Jack rabbit (prairie hare)	44	(70)
Racehorse	44	(70)
Wapiti deer	44	(70)
Thomson's gazelle	50	(80)
Prongbuck	59	(95)
Mongolian gazelle	59	(95)
Springbok	59	(95)
Cheetah	69	(110)

Many fish are also known to be particularly fast. The following speeds have been recorded by means of a fish meter, a device which lets the fish unroll a long thin line while swimming.

Trout	25	(40)
Salmon	28	(45)
Blue shark	34	(55)
Tunny (tuna)	41	(66)
Marlin	50	(80)
Swordfish	50	(80)
Sailfish	59	(95)

—the oldest era from which clear evidence of life forms has come down to us—notable primitive crustaceans, the trilobites (three-lobed arthropods), were the most highly developed marine animals; and it is obvious that millions of years of unrecorded evolution must have preceded their emergence. Other crustacean orders were also represented as early as the Cambrian period; the eurypterids even included giants with a length of more than 6 ft (2 m).

The body of a crustacean has three clearly distinguishable parts: the head, the thorax, and the abdomen. In many species the thoracic rings and the head have fused to form a single, immobile region known as the cephalothorax. The body is richly equipped with articulated, more or less leglike appendages, some of which still clearly hark back to the primitive biramous (two-branched) form. This feature can, in turn, easily be traced back to the lateral projections which certain of the Annelida still have—a fact which constitutes one of the arguments in favor of a relationship between the Annelida and the Crustacea. Without entering upon a discussion of the terminology of all the crustacean appendages, we may note that in general the appendages of the head serve the purposes of detection and manipulation of prey (eyes on nonarticulated stalks, two pairs of antennae, one pair of upper mandibles and two pairs of lower mandibles), a task in which they are assisted by three pairs of appendages called maxillipeds ("foot jaws"). Some of the walking limbs are often provided with strong claws or pincers which enable the animal to grasp its prey, kill it, and bring it to the mouth; other limbs are exclusively for walking. In addition, the upper portion of these limbs is equipped with gills, concealed under a shieldlike fold of the integument called the carapace, which functions as a kind of "shell" around the thoracic parts and sometimes around the rest of the body. The segments of the abdomen of many crustaceans also carry appendages—swimming legs—which may also serve to hold fast the eggs and young. This procession of legs concludes with a broad, flat tail "fin." There are various signs indicating that each pair of appendages corresponds to an original body segment.

The development of a crustacean always proceeds by way of a series of larval stages. The series begins with an egg-shaped, still unsegmented nauplius, a playful little tumbler whose only appendages are the antennae and the upper mandibles; one small eye in the middle of the head completes its equipment. In most of the lower crustaceans (Entomostraca), this cyclopean eye remains throughout life. In the higher crustaceans (Malacostraca), it is often replaced by a set of much better organized compound eyes. At successive moltings, the larva becomes ever richer in jointed appendages. First the number of mouthparts increases, and then the number of legs. In this way, after each molting the larva moves closer to its final state.

Arthropods with eight legs—spiders and allies

The class of the spiders (Arachnida) is characterized particularly by the following:

a. A two-part body, consisting of a cephalothorax (joined head and thorax) and a nonsegmented abdomen joined by a thin bridge
b. No more, but no less, than eight legs, all on the cephalothorax
c. No wings
d. Silk-producing glands and spinnerets, fingerlike projections on the abdomen with which a thread can be spun.

Only a few of the 50,000 species of spiders and allies have acquired a modest measure of fame. For one thing, a spider does not possess much with which to impress the world; for the most part its dimensions are extremely modest, and

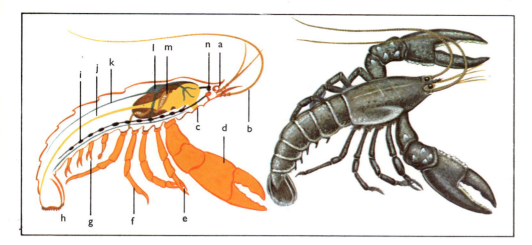

In the lobster, one of the crustaceans, the head and thorax have fused and are provided with numerous jointed appendages: antennae, mouthparts, and legs. Each segment of the abdomen also has a pair of "legs." (a) Eyes, (b) antennae, (c) mouth, (d) large pincer, (e) small pincer, (f) walking legs, (g) abdominal appendages, (h) swimming legs, (i) central nervous system, (j) intestinal tract, (k) blood-vessel system, (l) heart, (m) reproductive organ, (n) excretory organ.

most species lead a concealed existence in which everything seems to turn around spinning webs and catching insects. Still, a closer acquaintance with this class of Arthropoda offers more than one might expect, in many respects. Spiders are interesting animals—highly interesting even—not only because of their structure but also, and above all, because of the richly variegated life that the many species display.

Under the magnifying glass, the head of a spider will no doubt seem to be its most impressive part. Under the eight eyes are located two chelicerae, ending in needle-sharp poisonous claws. The pair of appendages behind the chelicerae are the palpi, which serve to grasp the prey. Male spiders use these appendages as organs of copulation. The hollow poison claws, connected with poison glands, function as injection needles; they easily bore through the chitinous armor of an insect, and the poison is quick and fatal. On the ventral surface of the abdomen we find the entrances to the oddly constructed, fan-shaped respiratory organs, the book lungs. Most spiders also have breathing tubes called tracheae, structured approximately like those of the insects.

The spinnerets are located on the ventral side of the abdomen, near the extreme rear of the body. Numerous small ducts from the silk-producing glands empty into them. The liquid silk, forced out through the spinnerets, immediately hardens in air to form hundreds of fine silk threads. These anchor the real spider's thread, which is generally composed of only a few thicker strands woven together.

Centipedes and millipedes (Myriapoda)

The name Myriapoda (centipedes and allies) is no longer used in modern systems of classification. Nevertheless, we shall continue to use the term, leaving aside the question of why this old class of animals has now been replaced by four new classes: Chilopoda (centipedes), Diplopoda (millipedes), Symphyla, and Pauropoda.

These light-shy, fast-moving animals, who live a hidden existence as predators and scavengers, constitute a rather varied company in which the variation in number of legs is not so great as the creatures' names would suggest. In general, these animals do have more than three pairs of legs, but there are species which do not have *many* more. There are others which have as many as 340 legs; and it is odd to note that the *milli*pede with the greatest number of legs (266) cannot compete with some *centi*pedes.

The heads of these animals, which are more closely related to insects than to spiders, are equipped with a pair of movable and threadlike antennae (sometimes composed of more than 400 parts) and a pair of strong, daggerlike claws connected with each other by poison glands.

BIRDS AND THEIR EGGS

The number of eggs in a clutch varies from species to species, from one to twelve, and the size of the eggs is also quite variable, even on a relative basis. Thus the egg of an ostrich weighs about 2.2 lb (1 kg), representing only 1/60 of the bird's body weight; but the egg of some hummingbirds weighs about 1 gram, corresponding to 1/5 of their body weight. A kiwi egg attains 1/4 of the bird's body weight. The full clutch of a golden-crested wren can weigh as much as 1½ times as much as her body. The incubation period varies from 8 days in hummingbirds to 80 days in the kiwi. Birds' eggs are very rich in yolk and have an air space that furnishes oxygen to the embryo. The space becomes enlarged during incubation and receives supplementary air through the somewhat porous shell.

Large tropical scolopendrine centipedes of the genus *Scolopendra*, which can grow as large as 10 in. (25 cm) long, can use these offensive and defensive weapons to give human beings an extremely painful poison injection.

The largest of all the animal classes—the insects (Insecta)

Experts on insects have reason to suppose that, up to the present, only one fourth of all existing species of insects (Insecta, formerly called Hexapoda) have become known. A number of unimposing insect orders have thus far been studied only superficially by these entomologists. An estimate of at least two million species of insects would certainly not be too bold.

Obviously these animals have had a particularly great measure of success in the struggle for existence. Among the factors that have contributed to this success are their small dimensions (there are insects smaller than the largest one-celled animals), their perfect internal organization, their rapid growth, and their great fertility. This enormous fertility constitutes a splendid counterweight to their relative defenselessness; on occasion, a single species can multiply to the point of becoming a real plague.

Metamorphosis is another advantage of this class of animals. An insect leads a double life: during the feeding and growing phase the larva lives in an environment that is appropriate for this purpose and is generally relatively safe; then, in the reproductive phase, its wings give the adult animal a large measure of mobility, enabling it to cover a large territory. The formation of social communities or states with extensive division of labor and the great adaptability of many species have also contributed to the creation of the immense and diverse world of the insects.

Although these generally small, generally winged, and always six-legged animals have never managed to conquer the sea as a habitat (with the exception of the water strider, *Halobates*), their evolution must still be called successful in every respect. With the exception of the sea, one can hardly name an environment in which they are lacking. Their history, compared with that of many other animal groups, does not go so far back into the earth's past; yet they were already present during the Carboniferous period. The remarkable forests of that period swarmed with cockroaches. Even more highly organized species date from that time, including even dragonflies with a wingspan of $2\frac{1}{2}$ ft ($\frac{3}{4}$ m). In the course of time, a number of animal groups displayed tendencies to giant growth, leading to excesses which as a rule preceded their dying out. The insects, however, were able to avoid this tendency. It is fortunate for the rest of the animal kingdom and for man that the insects, despite their enormous evolutionary possibilities, remained of modest dimensions.

Nonetheless, what these animals have attained since the Carboniferous period is no mean accomplishment. One may note, for example, the great multiplicity of mouthparts, variegated structures always based on the same plan: three pairs of mouthparts which move horizontally with respect to each other. They are the mandibles, the maxillae, and the lips, which arose from a fusion of parts. But how enormous are the differences in the working out of this structural plan in the butterflies and honeybees, the mosquito, the housefly, and the rove beetle! In every species we find a perfect harmony between structure and function. Predatory insects possess appropriate deadly weapons, daggerlike mouthparts that are sometimes hollowed out to allow the passage of poisonous saliva, which dissolves protein when injected into the prey. The aquatic larva of an ordinary dragonfly has a prehensile labium which can be extended with lightning speed to seize its prey. When a mosquito bites us, she—males do not bite—thrusts no fewer than five sharp projections through the skin, injects saliva to cause the blood to flow more readily, and begins to suck blood out by means of a micro-

INSECTS AS ENEMY NO. 1

One can say, somewhat romantically, that among land animals the struggle for existence is carried on primarily between insects and vertebrates—although, given the disparity in dimensions, the struggle is not very spectacular. There are numerous vertebrate animals which can feed exclusively or primarily on insects, but insects (leaving exceptions aside) which can overpower a vertebrate are great curiosities. On the other hand, it is precisely their smaller size that enables many insects to become parasitic upon vertebrates, and this phenomenon is very widespread. Moreover, certain insects appear in such numbers (tropical ants, locusts) that they can overwhelm the vertebrates. The continuous competition for food between the two groups is observable at all times and in nearly all places, but it is only seldom that it turns out to be of decisive importance for one of the two.

In recent times man has seen a chance in a number of places to use insect-killing chemical compounds to incline the power balance strongly to his side. However, numerous insects have developed a defensive mechanism called resistance, an insensitivity to certain substances, so that the struggle described above actually continues unabated.

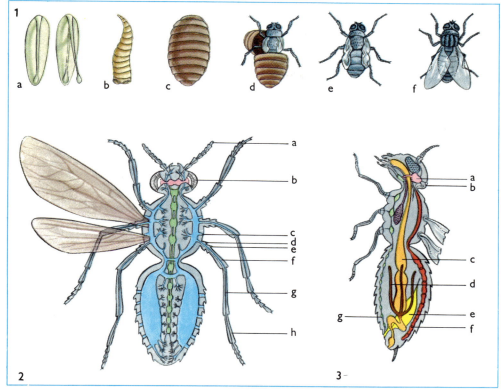

The often very drastic change of body form (metamorphosis) that occurs in many animal species (insects, amphibians) is among the most startling of natural phenomena. 1, The metamorphosis of a fly, with (a) egg; (b) larva; (c) pupa stage, which appears quiet, though actually within the puparium important changes are taking place under the influence of hormones; (d) almost full-grown fly, already equipped with legs and eyes but with rudimentary wings, creeps out of the puparium. The insect then quickly reaches the adult stage, called the imago (e,f).

2, Insects display all the characteristics of the phylum Arthropoda: a skin covered with a hard layer (cuticula), which also serves as an exoskeleton; a body which is clearly segmented; and jointed appendages. The body is divided into a head, thorax, and abdomen. The head has a pair of jointed antennae (a), in which the sense of smell is located, and three pairs of mouthparts. The thorax is provided with three pairs of legs and two pairs of wings. Each leg consists of a number of segments, namely, the coxa (d), trochanter (e), femur (f), tibia (g), and tarsus with five segments (h). On the sides of the head are large compound eyes composed of many facets (b), and on the forehead there are generally three simple eyes, or ocelli (see illustration 3b). The respiratory organ consists of a system of air tubes (tracheae), the finest branches of which penetrate to all the tissues. The openings (spiracles) of this system (c) are equipped with muscular rings which keep them closed as long as the animal is not active and needs only a small amount of oxygen; thus gas exchange and water loss are minimized.

3, The internal organs of an insect are located primarily in the abdomen. Nearest the back lies the tubelike heart (f), which consists of a number of segments separated by valves. Toward the front, it passes over into the aorta, the only part of the blood-vessel system that has its own wall (everywhere else the rather colorless blood flows freely among the organs). In the middle is the intestinal canal (c), consisting of the foregut, midgut, and hindgut. The long, tubular excretory organs (d) arise at the division between mid- and hindgut. The abdomen also contains the reproductive organs, such as the ovary (e). The nervous system consists of the brain (a) and the abdominal nerve cord (g), along which are strung clusters of nerve cell bodies called ganglia.

scopically small muscle pump. Nowhere else in nature is Darwin's "struggle for life" carried on with such vehemence; nowhere do we encounter such refined instruments for defense, attack, and killing, as in the world of those small, six-legged winged creatures, the insects. Their arsenal of weapons is comprehensive and greatly varied. It includes instruments for grasping, piercing, beating, biting, and sucking, with or without deadly poison glands, camouflage, and mimicry to allow the animal to escape an enemy or to attack its prey.

The following points characterize the insects:

1. The body, covered by a chitinous armor, consists of three parts: the head, the thorax, and the segmented abdomen.

2. Besides the above-mentioned mouthparts, the exterior of the head has a pair of antennae which bear numerous tactile and olfactory organs.

3. Most species also possess a pair of compound eyes, consisting of a small or very large number of identical facets called ommatidia. On the exterior of the eye these parts are visible as facets. In addition, many insects also have three small simple eyes called ocelli.

4. To the outside of the nonsegmented thorax are attached the means of locomotion: six jointed legs and one or two pairs of wings. There are primitive insects which have not yet developed wings and highly developed species which have discarded the wings of their ancestors, sometimes as an adaptation to a parasitic manner of life. The primary content of the thorax is the motors for locomotion; these muscles are considerably superior to those of man with respect to economy of energy consumption and reaction speed.

5. The abdomen, the exterior of which is composed of chitin rings, contains the principal organs of digestion, excretion, circulation, and reproduction.

Insects are complicated animals; moreover, most species do not follow the simplest path in the development from egg to adult individual. Before the final stage, the "imago," is reached, the insect passes through a series of developmental phases, separated by moltings.

Metamorphosis is not so perfect in all species. In general, the higher and more specialized the level of evolution attained by the insect, the more complicated is

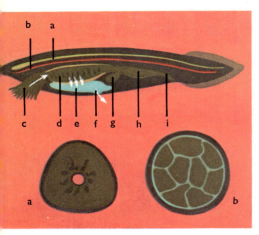

The lancelet, or Amphioxus (*Branchiostoma lanceolata*), is the celebrated link between nonvertebrate and vertebrate animals. This small marine animal has no skeleton, to be sure, but it is almost placed on the higher plane of the vertebrates by the presence of a spinal cord (a) and a supporting rod, the notochord (b), which all vertebrates possess, at least during development. The lancelet has no limbs, no heart, no eyes, and colorless blood. Besides spinal cord and notochord, the most important parts of the body are (c) mouth with processes propelling water to the pharynx; (d) pharynx with gill slits; (e) atrium; (f) atriopore, through which water is expelled; (g) liver; (h) midgut; (i) hindgut.

THE GIANT LUNGS OF A WHALE

We find the largest lungs, as is only natural, in the largest mammals. The lungs of a finback whale (rorqual) can hold 2,000 liters. This seems a surprisingly large amount; but actually, when we allow for the enormous weight of the body, it is only about half the lung capacity of land mammals. This statement would appear to be contradicted by the fact that whales can remain for a long time under water. However, in the first place, whales can breathe in and out very deeply (using up to 90% of the total volume of the lungs). Moreover, the blood absorbs much more oxygen; and finally, a diving whale takes a very large supply of oxygen along in its muscles. Hence fresh whale meat is of a notably dark-red color, and in the sperm whale it is almost black. The record-holding divers among the whales are the sperm whales and the Atlantic right whales, which can remain underwater for an hour and a half. In normal circumstances, however, they remain under for 50 minutes on the average and then swim for 10 minutes on the surface. During these 10 minutes, they change the air in their lungs about 60 times. Finbacks usually remain underwater no more than a quarter of an hour.

---

the process of development. Primitive "ur-insects" such as the well-known silverfish (*Lepisma saccharina*)—a quick, silvery little animal that lurks in old books and pantries—do not yet pass through any real metamorphosis. From the egg of a cricket or a locust there emerges a little animal which is not yet similar to an adult in all details (for one thing, it does not yet have wings) but which can nevertheless immediately be recognized as a little locust or cricket. It has exactly the same structure. At each molting the young become more perfect; they undergo a gradual metamorphosis.

Between invertebrates and vertebrates

The phylum of the chordates (Chordata) is divided into three subphyla, two of which—the tunicates (Tunicata) and the Acrania—constitute a transition to the third, the vertebrates (Vertebrata). Tunicata and Acrania are, to be sure, still nonvertebrate, but in many respects they already clearly display the vertebrate scheme. The vertebrates can be "derived" from these little animals. It is thought that in the distant past they could have originated from animals such as these. The Acrania, or "tube hearts," above all, have an internal organization which constitutes a preamble, so to speak, to that of the vertebrates. It is true that the lancelet (*Branchiostoma*), which belongs to the Acrania, has no skeleton, no limbs, no heart, and even no eyes; but it does have a simple blood-vessel system, a "genuine" spinal cord, and a notochord. This elastic dorsal rod, a tube filled with large turgid cells, is also characteristic of all vertebrate animals, at least during *some* stage of their development. In most vertebrates, however, the notochord is quickly displaced by vertebrae of cartilage or bone which completely enclose the spinal cord and, in front, develop into a container surrounding the brain, the skull.

Vertebrates (Vertebrata): animals with a skeleton

However great the differences may be among fish, frogs, lizards, birds, and mammals, all 80,000 species of vertebrates agree in their body plan. A small tropical fish hardly a centimeter long, a whale measuring 100 ft (30 m), a tiny sparrow, and a man all have a large number of body parts in common. The body consists of head, trunk, and tail (the latter being rudimentary in man); has two pairs of extremities for purposes of locomotion; and is covered by a series of cell layers, the skin. Ribs and two bony "girdles" (to which the limbs are connected by joints) are attached to the vertebral column. The spinal cord, an important component of the central nervous system, runs through the vertebral column. Segmentally arranged nerves branch out to the left and the right from this spinal cord. The most important part of the central nervous system, the brain, is enclosed by the skull. Blood circulates through a system of arteries, capillaries, and veins. The subphylum of the vertebrates is subdivided into seven classes: roundmouthed fishes (Cyclostomata), sharks and rays (Chondrichthyes), bony fishes (Teleostomi), amphibians (Amphibia), reptiles (Reptilia), birds (Aves), and mammals (Mammalia).

Round-mouthed fishes, sharks, and rays

The oldest vertebrates which have been discovered were jawless fishes. It is remarkable to note that at present there are still living fishes which can be regarded as direct descendants of these most primitive vertebrates. These roundmouthed fishes (Cyclostomata, lampreys and hagfishes) to a certain extent re-

VERTEBRATE ANIMALS

CLASS	SKIN	RESPIRATION	CIRCULATION	BODY TEMPERATURE	REPRODUCTION
Round-mouthed fishes	naked	gills	single-circuit heart: 1 atrium, 1 ventricle	variable	eggs laid in water, not incubated
Cartilaginous fishes	placoid scales	gills	singel-circuit heart: 1 atrium, 1 ventricle	variable	eggs laid in water, not incubated
Bony fishes	bony scales	gills	single-circuit heart: 1 atrium, 1 ventricle	variable	eggs laid in water, not incubated
Amphibians	naked	first—gills later—lungs	first—as in fish later—double heart—2 atria, 2 ventricles	variable	eggs laid in water, not incubated
Reptiles	horny scales	lungs	double	variable	eggs with horny or calcareous shell laid on land, not incubated
Birds	feathers (horn)	lungs	double	constant ca. 107.6°F	eggs with calcareous shell, incubated; young fed
Mammals	hair (horn)	lungs	double	constant ca. 98.6°F	development in uterus of ♀; young fed with milk

mind one of eels. They are parasites, which attach themselves to other fish by sucking.

The sharks and rays, the best-known representatives of the Chondrichthyes, directly follow the round-mouthed fishes in the system. The skin of a shark feels like sandpaper because of the numerous little bony teeth sitting on scales. The remarkable thing about these "placoid" (platelike) scales is that the little teeth consist of the same material as the teeth of all higher vertebrates. In the area of the snout, one finds clear transitions between these placoid scales and genuine teeth. From these facts the conclusion was drawn—and it must have seemed obvious—that teeth are modified placoid scales.

Bony fishes (fishes with skeletons)

One of the most characteristic properties of the bony fishes is their ability to withdraw dissolved oxygen from the water. When the oldest bony fishes came to feel at home in fresh water, they confronted the problem that the oxygen content of such water is not always so favorable as it is in the sea. In swampy areas and during droughts, above all, the gills did not always appear to be sufficient. A number of species were able to surmount this difficulty by developing a lung for emergency use. Gill-breathing fishes which have simple folds in the gullet as reserve lungs have not yet died out. The Australian lungfish (*Neoceratodus forsteri*) and the representatives of the genera *Protopterus* and *Lepidosiren* found in Africa and South America are such lungfish. From the systematic point of view, the lungfishes form a group of the bony fishes (Teleostomi), so called because of

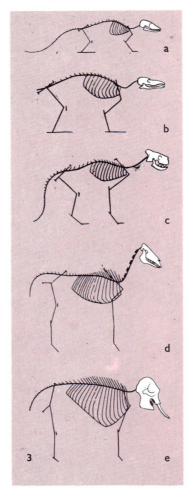

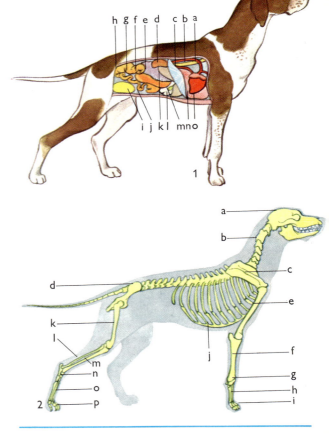

1, The internal organs of a mammal (dog): (a) anterior vena cava; (b) gullet (esophagus); (c) aorta; (d) kidney; (e) pancreas; (f) caecum; (g) duodenum; (h) rectum; (i) bladder; (j) small intestine; (k) liver; (l) stomach; (m) gall bladder; (n) heart; (o) left lung.

2, Skeleton of a mammal (dog): (a) skull; (b) cervical vertebrae; (c) shoulder blade (scapula); (d) pelvis; (e) humerus; (f) forearm (radius and ulna); (g) carpus; (h) metacarpus; (i) phalanges; (j) ribs; (k) femur; (l) fibula; (m) tibia; (n) tarsus; (o) metatarsus; (p) phalanges.

3, In all mammals the skeleton is composed of the same bones. There are, of course, differences in size, form, and orientation. (a) Insectivore, (b) rodent, (c) cat, (d) horse, (e) elephant.

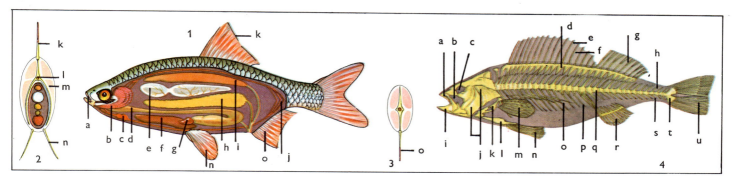

their more or less bony skeleton. In most bony fishes the lung remained present, but after some modification was given a completely different task. The organ came to be used for hydrostatic purposes. By regulating the volume of gas in this "swim bladder" (air bladder), a bony fish can make its specific gravity correspond accurately to that of its environment, so that the animal does not have to exert itself constantly to keep from sinking.

In 1938, off the east coast of South Africa, a rather large fish was caught which created a sensation in zoological circles. By its large scales and typically limblike fins the animal was immediately recognized as a member of the "fringe-finned fishes," which zoologists had thought extinct since the Mesozoic era. These fringe-finned fishes (Crossopterygii) were probably the ancestors of the lung-fishes and higher fishes, but also of the amphibians.

Animals that lead a double life

The enormous step from aquatic to terrestrial life was probably made in the Devonian period; we know fossil fishes and amphibians from that time which resemble each other very strongly.

The amphibians (Amphibia) have not yet freed themselves completely from aquatic life. Tailless frogs and toads (Anura) and tailed salamanders (Urodela) are the principal orders of this slippery class of animals. These animals are as cold to the touch as fish; they are "cold-blooded" (i.e., their body temperature is related to the temperature of the environment). The thin, slimy skin does not protect them against drying out, but rather makes it possible for them to breathe through the skin, both in the water and on land. Hence the lungs have a simple structure. One of the most noticeable characteristics of this class is the process of metamorphosis. Especially in the case of the tailless frogs and toads, an impressive modification of their structure is necessary in the transition from tadpole to four-legged land animal.

In reproduction the Amphibia are still completely dependent on the water. Their eggs are laid in the water. While the female frog is laying the eggs, the male sprays them with his sperm. The little ova are surrounded by a thin layer of jelly, which swells in the water. This change keeps the developing embryos sufficiently far from each other and gives them protection from the gluttony of ducks, fish, and other predators. A larva develops within a few days—if the temperature is not too low—by a rapid series of cell divisions. For a few days the little animal breathes through gill arches behind the head; these are then replaced by internal gills which, as in fish, lie at the rear of the mouth cavity. The water enters through the mouth, flows past the gills, and is then eliminated through an opening behind the head on the left side. The larva uses fine, horny teeth to scrape microscopically small bits of seaweed from aquatic plants; later it also feeds on dead animals. The hind legs appear after about two months, and then a month later the fore legs, which had developed in the gill cavity in the meantime. In the case of salamander larvae, the little legs break through shortly after the young animal leaves

Anatomy of a bony fish: 1, (a) mouth; (b) gills; (c) heart; (d) intestinal tract; (e) air bladder; (f) liver; (g) gall bladder; (h) ovary; (i) kidney; (j) capillaries. 2, Transverse cross section at about the middle of the body: (k) dorsal fin; (l) vertebral column; (m) muscle segments; (n) ventral fin. 3, Transverse cross section farther toward rear: (o) anal fin.
Skeleton of a bony fish: 4, (a) intermaxillary; (b) maxilla; (c) eye socket; (d) fin-ray support; (e) dermal rays; (f) dorsal fin; (g) soft dermal rays; (h) neural spines; (i) dentary; (j) opercular bones; (k) shoulder girdle; (l) pelvis; (m) pectoral fin; (n) pelvic (ventral) fin; (o) rib processes; (p) ribs; (q) vertebral column; (r) anal fin; (s) hemal spines; (t) fused tail vertebrae; (u) caudal fin.

the egg, so that they quickly come to resemble the adult animals.

Reptiles—vanished glory

Lizards, snakes, turtles, and crocodiles constitute the remainder of an animal group that has seen better days, a group that long ago brought forth the mighty of the earth—dinosaurs, pterosaurs, and other monsters. In a concise and objective way we may define reptiles (Reptilia) as follows: vertebrate animals which are covered with horny scales, breathe through lungs, are cold-blooded, and reproduce by means of eggs. Like fishes and amphibians, these animals lack the ability to regulate their body temperature. In a warm environment their metabolism is sufficient to maintain a reasonable level of activity; but if the temperature falls, the vital processes slow, the animal becomes more and more lethargic and has no need of food, and finally it becomes stiff. Lizards and turtles can remain in this condition for many months without damage. All kinds of structural peculiarities have a clear connection with this cold-bloodedness. Little oxygen is necessary for the low level of consumption, so that the lungs are smaller and of simpler construction than in the birds and mammals. The heart is also small (2% of the body weight, compared with 5% in mammals and up to 15% in birds) and shows some peculiarities correlated with the lower demands placed on the blood circulation. Since reptiles are thus particularly dependent upon the temperature of their environment, the great majority of the 5,000 species are found only in warm countries.

The name reptile (from the Latin *reptile* = creeping) is appropriate for only part of this class. Snakes, blindworms, and some other lizards do indeed slide on their bellies over the ground; but most reptiles, including the crocodiles, walk on four legs. One may even note that a few species walk on the hind legs only, using the tail as a balancing organ (e.g., the frilled lizard, *Chlamydosaurus kingi*). Many reptiles have also mastered the art of swimming, with varying degrees of success, and even the air is not forbidden to all species. The "flying lizards" of the genus *Draco* have winglike folds of skin, supported by elongated ribs, which they spread out in moving through the air. However, the accomplishments of these animals are far inferior to those of the prehistoric pterodactyls (Pterosauria), the largest of which had a wingspan of 40 ft (12 m). The small recent "flying lizards" can actually only glide.

Although most reptiles are predators, there are also purely herbivorous species. Thus the very large iguana species *Amblyrhynchus cristatus*, a representative of the very unusual fauna of the Galápagos Islands, feeds only on seaweed.

Crocodiles, lizards, and snakes

By their size, their more or less demoniacal appearance, and their armored skin, crocodiles give the impression of a prehistoric animal. But this only *appears* to be so. The crocodilian order of reptiles is indeed the most specialized of the reptilian orders, and hence the most recent. Thus the heart of a crocodilian is more modern and perfected than the heart of a lizard.

Lizards differ from snakes in a number of anatomical features, and they also have the remarkable gift of self-amputation. The tail breaks off easily, its areas of separation being weak spots in several tail vertebrae. The broken-off twisting tail diverts the attention of the attacker and gives the animal a chance to escape. A new tail then grows from the area of separation, though it contains no vertebrae and is much shorter than the original tail.

Snakes are "secondarily" limbless; they are descendants of limbed reptiles. Not only are there traces of a pelvis and hind limbs (so-called rudimentary organs) in some snakes, but also the nerves which branch off from the spinal cord, still in

four clearly delimited bundles, run to the places where front and hind limbs were located in earlier times.

Particularly characteristic of snakes is their ability to swallow surprisingly large prey whole. A small extra piece of bone between the skull and the lower jaw enables the mouth to open, proportionately, to an enormous width. The halves of the lower jaw are connected by an elastic ligament, the gullet and the skin are especially distensible, and the ribs are not attached to the breastbone, making it possible to turn them farther toward the outside. Moreover, the entrance to the windpipe lies forward in the mouth, so that respiration does not have to be interrupted during the long swallowing process. Another characteristic of snakes is the very large number of vertebrae (180 to 400), nearly all equipped with ribs, which help in locomotion. The long, forked tongue is not a stabbing apparatus, but rather serves the senses of touch and smell. The tongue brings odoriferous particles to Jacobson's organ, a kind of adjunct to the sense of smell.

Not all snakes possess poison fangs, and not all snakes with poison fangs are dangerous to human beings. Only those species with poison fangs located at the front of the snout can give a serious bite. A viper (*Vipera*) has only one functional poison fang, anchored to a hinged bone at the back of the mouth when not in use. When the mouth is opened, the fang is brought into a vertical position. The hollow or grooved poison fangs are connected with poison glands which produce a protein that destroys blood corpuscles or cripples the central nervous system.

The class of reptiles—which in the past had many mighty representatives, such as the brontosaurus—consists today only of snakes, lizards, crocodilians, turtles, and tuataras. They are all cold-blooded (i.e., their body temperature depends on the temperature of the environment).

1, sand lizard (*Lacerta agilis*) of north-central Europe
2, skeleton of a lizard
3, Nile crocodile (*Crocodilus vulgaris*)
4, skeleton of a crocodile
5, water, or grass, snake (*Natrix natrix*)
6, skeleton of a water snake
7, anatomy of a female water snake: (a) windpipe; (b) lung; (c) left aorta; (d) heart; (e) esophagus; (f) liver; (g) stomach; (h) fatty body; (i) gall bladder; (j) intestines; (k) ovary; (l) kidneys.
8, green turtle (*Chelonia mydas*)
9, skeleton of a turtle
10, ventral shell

1, The skeleton of a bird (pigeon) shows a great many adaptations to flight. It is light, yet strong. The eye sockets (a) are large. A special bone in the head, the quadratojugal (b), allows the beak to open very wide. The neck vertebrae (c) are movable. The rest of the vertebral column (d) comprises an almost rigid support for the thoracic skeleton. The shoulder girdle —consisting of shoulder blade (e), wishbone (f), and coracoid (g)—forms a strong pivot for the wing. The sternum has a large ridge (h) to which flight muscles are attached. The pelvis (i) is also solidly built, for both the legs and the steering mechanism (the tail, j) are attached to it. Flight feathers are attached to the forearm (k), the first finger (bastard wing, l), and the second finger (m).

2, Topography of a bird: (a) forehead; (b) auricular; (c) eye; (d) chin; (e) undertail coverts; (f) tail feathers; (g) upper tail coverts; (h) primary wing feathers; (i) secondary wing feathers; (j) wing coverts; (k) scapulars; (l) throat; (m) breast; (n) belly (abdomen); (o) rump; (p) back; (q) nape; (r) crown.

3, Anatomy of a bird (pigeon): (a) cere; (b) third eyelid; (c) esophagus; (d) crop (filled with grain kernels); (e) thumb wing; (f) lung; (g) liver; (h) heart; (i) small intestine; (j) vent; (k) gizzard; (l) left air sac; (m) trachea.

4–6, The forms and colors of birds have adapted to the environment in which the various species live.

Turtles

Turtles (also an order of reptiles) originated as land animals, living primarily on plants. They have a sharp horny beak in place of teeth. A number of species have specialized in aquatic living, and turtles are even to be found in the open sea.

When a certain type of animal is successful in competition with so many other forms over a period of some 200 million years, its structure certainly cannot be considered impractical. The turtle is such a special case. Its shell armor does not appear to have been a silly invention of nature's after all. Compared with an equally heavily armored lobster, the turtle has an important advantage: a lobster's shell does not grow along with the animal, so it must regularly slough it and is then defenseless for a time; the shell of a turtle, covered with horny shields, is part of the skeleton and hence grows along with the rest of the animal. With the exception of the Atlantic leatherback (*Dermochelys coriacea*)—which can be 8 ft (2½ m) long, weigh 1,100 lb (500 kg), and has a mosaic of plates of bone on its back—all species are equipped with a dorsal shield composed of fused vertebrae and bony plates. Thus a turtle is inseparably bound to its armor.

Birds

A number of anatomical peculiarities of the birds (class Aves) recall their distant ancestors, the reptiles. The famous primitive bird known as *Archaeopteryx*, the remains of which were found at Solnhofen, Bavaria, clearly points up this fact. This flying lizard was covered with feathers and had teeth in its beak, scales on its head, and claws on its wings; it was not much larger than a robust pigeon. One of the most characteristic peculiarities of the class of birds is the covering of feathers: downy feathers, which hold much air and hence counteract the cooling process (air is always a poor conductor); contour feathers, which provide streamlining, function as a recognition signal for members of the same species,

or serve as protection against enemies (camouflage); flight feathers for movement through the air; and tail feathers for steering. A feather is an extremely light, marvelously constructed epidermal horny structure. Old feathers are replaced by new ones from time to time.

Flight causes a bird's metabolism to be especially high; moreover, the bird must maintain a high body temperature—106° to 108°F (41°–42°C). Hence a bird must eat much and take in much oxygen. These requirements have produced a relatively large heart and a system of splendidly arranged blood vessels, appropriately arranged and fast-working digestive organs, and large lungs connected with air sacs.

The sensory life of birds is marked primarily by excellent vision, good hearing, and a fine sense of equilibrium. The eyes are particularly large in relation to the body. On the other hand, birds are microsmatic (i.e., have a poor sense of smell). Numerous species travel back and forth between their breeding grounds (summer quarters) and their wintering areas. This migratory behavior is one of the most surprising and intensively studied instinctive actions.

The highest class of animals

The mammals (Mammalia) undoubtedly constitute the crowning accomplishment of the evolutionary process. They may be defined as warm-blooded vertebrates which are covered with hair and breathe through lungs; the fertilized egg remains hidden in the womb or uterus, where it begins its embryonic development in close connection with the mother. After it is born, the young mammal is still fed for some time on a secretion of the mother's milk (mammary) glands and is also the center of special care. But there are exceptions to each of these rules. The primitive monotremes (echidnas and the platypus) lay eggs through the cloaca, a space into which both the intestine and the ducts from the kidneys and the reproductive organs empty. (This inelegant term is derived from the Latin

4, Bills have an appropriate structure for acquiring food that occurs in the bird's environment. The bill types are: (a) cruciform (crossbill), for eating pine-cone seeds; (b) conical (hawfinch), for seeds; (c) sawtoothed (goosander or merganser), for fish; (d) awl-shaped (European woodcock), for insects and slugs; (e) semicircular or hooked (eagle), for live prey; (f) downward-curved (tree creeper), for insects; (g) upward-curved (avocet), for small aquatic animals.

5, The form of the tail, like the form of the feet, can give important indications for determination of the species or type of bird. The tail types are: (a) squareended (starling); (b) cleft (linnet); (c) deeply forked (swallow); (d) with spiky feathers (woodpecker); (e) pintail (duck); (f) graduated (raven); (g) fantail (cuckoo); (h) elongated central rectrices (bee eater).

6, Foot types are: (a) grasping (swift: can hang in any position); (b) climbing (woodpecker: feeds while hanging); (c) walking (crow: walks on branches and the ground); (d) grallatorial (plover: walks in shallows); (e) wading (stork: stands in water); (f) lobed swimming foot (coot: dives for fish); (g) webbed swimming foot (duck: swims on the surface).

GESTATION PERIOD
OF A FEW MAMMALS

Mouse	21 days
Rabbit	30 days
Mole	30 days
Cat	55 days
Dog	63 days
Lion	110 days
Sheep	5 months
Bear	7 months
Ape	9 months
Cow	9 months
Finback	9–12 months
Sperm whale	16 months
Elephant	20–22 months

cloaca, meaning sewer or drain.) The duck-billed platypus (*Ornithorhynchus anatinus*) of Tasmania and southwestern Australia is a celebrated monotreme. It does indeed have a typically duck-shaped bill, but it is certainly not a transitional form between birds and mammals. The young uses an egg tooth to open the egg and is then fed with a jellylike substance from particularly primitive milk glands, which we may regard as modified sweat glands. The echidna, or spiny anteater (*Tachyglossus aculeatus*), a cave-dwelling land animal with a long cylindrical snout, goes a step farther. This strange creature lays only one egg, which it immediately places in a genuine pouch. The egg is hatched in the pouch, and the young are then carried around in it a few more weeks before being placed in a nest. Thus the echidna is an egg-laying marsupial mammal.

The genuine marsupials, who do not lay eggs but bring very imperfectly developed young into the world, stand on a somewhat higher level. The giant kangaroo (*Macropus giganteus*), which can be as much as 10 ft (3 m) long and weigh 330 lb (150 kg), has young that are no larger than those of a mouse. In contrast to the monotremes, the marsupials have good dentition, adapted for eating plants (herbivorous dentition), insects (insectivorous dentition), or meat (carnivorous dentition).

Another group of mammals, formerly classified in such orders as Edentata (sloths and allies) and Pholidota (pangolins), is also very old and primitive. The African and Asiatic pangolins, with their large scales, make one think of walking pine cones. Their enormously elongated tongue helps these animals ingest insects, the exoskeletons of which are then ground up in the stomach by means of stones the animals have swallowed. Thus they apply the same technique as the seed-eating birds.

The giant anteater, or antbear (*Myrmecophaga tridactyla*), also has a long sticky tongue. The powerful, inward-curving claws are particularly well adapted for

The structure of the dentition diverges widely among the various animal species, according to the type of food they consume (plant food requires a different type of dentition from animal food) and the evolutionary history of the species. In mammals one distinguishes a homodont dentition (1), in which there are many teeth, all of the same form; and a heterodont dentition (2–6), with various types of teeth. Some examples of skull and dentition types among mammals: (1) fish-eating (the homodont teeth are not used for rending the prey; e.g., dolphin); (2) omnivorous (for eating everything; e.g., gorilla); (3) carnivorous (meat-eating; e.g., cat); (4) herbivorous (plant-eating; e.g., cattle); (5) insectivorous (insect-eating; e.g., hedgehog); (6) frugivorous (fruit-eating, as the rodents; e.g. squirrel).

breaking open anthills and termite nests. The two-toed and three-toed sloths are arboreal animals that lead a literally upside-down life, hanging downward from tree branches.

Their accurately regulated internal central heating frees the mammals from dependence on a given habitat. In order to hold the body temperature constant, the blood temperature is subject to continuous close regulation by a temperature center in the brain. If the temperature falls, this center stimulates the vital processes; if it rises, oxidation is decreased and radiation of heat increased. The covering of hair offers extra protection against temperature influences.

While the mammalian body scheme is everywhere the same, in the skull, dentition, and limbs the skeleton shows clear adaptations to the type of life. The vertebral column is attached to ribs and two "girdles," with respect to which the limbs turn. The basic form of the limbs is the five-fingered type of slow plantigrades; and the bones of the upper arm and upper leg, forearm and lower leg, and hand and foot generally correspond. But many animals that were originally plantigrade (i.e., walking on the whole sole of the foot) have come to walk on the toes or on hoofs by a process involving the lifting up of the middle hand and middle foot, followed by the narrowing and firming of those parts and by the striving for a reduction of the number of fingers and toes. We see this final phase of the evolutionary process in solid-hoofed animals such as the horse: the hand and foot consist of only one finger or toe and a solid piece of bone in the middle hand or middle foot.

The many orders of the class of mammals display great differences in dentition, food processing, locomotion, and senses. Anatomically, the mammal group is much more richly variegated than the birds.

Apes and man

For many persons the apes, which belong to the order of the primates, are the most attractive representatives of the animal kingdom. The "human" body scheme of these highest mammals and the outstanding development of their central nervous system—and hence also of their psychic life—easily tempt one to liken their behavior to that of humans. But an ape is far from being a man, though he is undoubtedly a "personality" standing on a much higher level than flies, bees, or other insects, which are led primarily by instincts.

Not all primates give such a human impression. The primitive primates (Prosimiae) certainly stand below the level of the manlike primates (Anthropoidea), among which the apes (Pongidae) are the most highly developed family. Despite all the points of agreement between these apes and man, there are nevertheless many quantitative and qualitative differences to be pointed out. First, there are the differences in the skeleton. In the ape, the skull has a much smaller brain region but more powerful jaws. The human head is more mobile and is placed differently on the vertebral column. The shoulder and pelvic girdles are much more strongly developed in apes, and the long arms and relatively short legs are also striking in these animals. The form of the vertebral column does not permit a fully upright position of the body, a position which happens to be particularly characteristic of man—though after birth it takes about three years for the pelvis and the vertebral column to begin to approach their definitive form.

Moreover, apes have no outward-curving lips, and they must also go through life with no rump—or at least without that typically pronounced, muscular region which in man has a direct relationship with his upright stance. Apes also have no external nose, and their upper lip thus acquires great mobility. For the rest, only man possesses mental powers of such degree as to enable him to create science, art, and technology—the human reasoning powers that are expressed in language. Only man has been able to reach the dizzying heights of self-consciousness.

APES AND MEN

Researchers have reared chimpanzee babies and human babies at the same time. At first, the little apes are far ahead of the human babies, but that does not last long; in the case of the small chimpanzee, the ape soon begins to come more and more to the fore. Then the young demonstrates what it in fact really is—a strongly specialized arboreal animal—and the chasm becomes particularly noticeable as soon as the human child begins to speak. Apes also "talk," to be sure, but in *their* way. Their language is only an emotional one, which does enable them to express their feelings but which is inadequate to express a concept in a word symbol.

A

AARDVARK, *Orycteropus afer*, large African burrowing mammal with no close living relatives, placed by itself in the order Tubulidentata (the tube-toothed), so called because of the fine tubes radiating through each tooth. The teeth themselves are singular in having no roots or enamel.

The aardvark (the Afrikaans name for earth pig), has a sturdy body, 6 ft (180 cm) long including a 2 ft (60 cm) tail and stands 2 ft (60 cm) high at the shoulder. The tough gray skin looks almost naked but is in fact very sparsely covered with hair. The head is long and narrow with a snout bearing a round, piglike muzzle and small mouth. The ears are large, resembling those of a donkey. The limbs are very powerful, with four strong claws on the front feet and five on the hind feet.

Aardvarks excavate burrows 3-4 yd (3m) long with their powerful limbs and sharp claws, working with incredible speed.

In midsummer the female gives birth in her burrow to a single young, occasionally twins. After two weeks the young aardvark is strong enough to come out from the burrow and go foraging with its mother.

The aardvark feeds mainly on termites. It can rip a small hole in the rock-hard walls of the nests with its powerful claws, insert its muzzle and then pick out the swarming termites with its 18-in. (45-cm) long, slender, sticky tongue.

AARDWOLF, *Proteles cristatus*, African relative of the hyenas, in a separate family, the Protelidae. The aardwolf (the Afrikaans name for earth wolf) is somewhat larger than a fox, weighing 50-60 lb (22-27 kg) with a yellow-gray coat with black stripes and black legs below the knee. The muzzle is black and hairless, the tail bushy and black-tipped. The hair along the neck and back is long and may be erected when the animal is frightened. It is found throughout southern and eastern Africa as far north as Somalia, in sandy plains or bushy country. It is rarely seen, being solitary and nocturnal, spending the day in rock crevices or burrows.

It differs in form from true hyenas in having

five instead of four toes on the front feet, larger ears and a narrower muzzle with weaker jaws and teeth.

It feeds, therefore, almost entirely on termites, although, lacking strong claws to tear open the termite nests, it has to be satisfied with taking the insects from the surface or digging them out of soft soil. It sweeps up the termites in hundreds with amazing speed with its long sticky tongue.

After a gestation of 90-110 days, a single litter of two to four blind young ones is born each year in November or December in the southern part of its range.

ABALONE, *Haliotis*, the American name for the large Pacific ormer or ear shell, a gastropod mollusk. The single flat shell is very beautiful with iridescent colors on the inside and also on the outside when the surface is cleaned of encrustations. There are usually a number of perforations on the side of the shell through which the water emerges after passing the abalone's gills. Abalones are found along the California coasts, the largest being the red abalone, *H. rufescens*, with a shell up to 10 in. (25.4 cm) across. The green abalone, *H. fulgens*, is thin-shelled and smaller, up to 6 in. (15 cm), and the black abalone, *H. cracherodii*, 5 in. (13 cm) across, unlike the other two that cling

to rocks, lives among rocks in the surf and consequently has a clean and shining shell.

The flesh from the large muscular foot attached to the shell is regarded as a delicacy and eaten in many American cities along the Pacific coast, and the shell is used for decorative work.

ACCENTORS, sparrowlike birds with short rounded wings and fine pointed bills. They are generally brown or reddish-brown spotted or streaked above, and brown or gray below, usually with a reddish band on the breast or reddish streaks on the flanks. The twelve species are sufficiently alike to be included in the single genus *Prunella*. In most species, the nest is an untidy cup of grass, mosses and leaves, lined with wool or feathers, and placed low down or on the ground, in a small conifer, stunted shrub or crevice among boulders. The eggs are unmarked and vary in color from pale blue to blue-green, the usual clutch size being between three and five. All species feed mainly on the ground, hopping about stiffly with a mouselike action. Insects comprise a large part of the diet in spring and summer; in winter the food consists chiefly of small seeds and berries. Accentors are quiet, unobtrusive little birds, often extremely tame and confiding. They are usually rather solitary, but some species show a tendency to form flocks in winter. The song

The aardvark and (inset) crown view of tooth showing tubular structure.

◁ The bearded lizard (*Amphibolurus barbatus*), a well-known representative of the agamid family, in threat posture—mouth open and throat pouch inflated in order to display the "beards." Such subterfuges help many animals appear more dangerous to others than they really are. Yet agamids are sufficiently dangerous to iguanas to have driven out the latter wherever the two families were once found together.

is short and simple, generally soft, and delivered at a hurried pace, from a rock or low bush.

Two species, the alpine accentor, *Prunella collaris*, and the Himalayan accentor, *P. himalayana*, commonly breed at elevations of over 15,000 ft (4,570 m), and few accentors occur below 5,000 ft (1,520 m) even in winter.

The dunnock or hedge sparrow, *P. modularis*, of Europe is peculiar in that in some parts of its range it has descended from the coniferous-forest zone, and has become a characteristic bird of parks, gardens and hedgerows down to sea level. The Japanese accentor, *P. rubida*, a bird very similar in appearance to the dunnock, breeds in the zone of dwarf birches and pines in mountainous regions in Japan and winters in the undergrowth of dense forests at lower altitudes. More typical is the alpine accentor, which occurs in alpine meadows in mountainous regions from Spain and Morocco across Europe and Asia to Japan. This bird is commonly found around climbing huts and ski lifts in the Alps throughout the year. The Siberian accentor, *P. montanella*, breeds along the northern limits of the great coniferous forests of northern Asia from the Urals to Amurland, and is the only species that undertakes an extensive annual migration, its winter quarters being in northern China and Korea.

ACORN WORMS, sluggish, wormlike marine animals with the body in three clearly distinguishable parts: an anterior proboscis, shaped like a very elongate acorn, leading by a narrow neck to a "collar," commonly wider than the third and main part, the "trunk," which tapers gradually toward its rear end. The most obvious feature associating acorn worms with the true chordates is the paired gill slits or pores that open from the pharynx, the front part of the gut, through the body wall of the trunk just behind the base of the collar. These are numerous, and their number increases with age. Externally, these gill slits appear as small pores, but internally they have the elongate U-shaped

The aardwolf, although related to hyenas, lacks their powerful teeth and jaws.

slit with a tongue bar in the middle reminiscent of the gills of the primitive chordate, the lancelet (*Branchiostoma* or *Amphioxus*).

Most acorn worms are 4 in. (10 cm) in length, but *Balanoglossus gigas*, of Brazil, may be more than 3 ft (1 m) long. All are burrowers, and *B. gigas* makes a burrow several meters in length. Acorn worms live in soft sand or muddy sand between tidemarks or just below low tide level. For the most part they feed on sand, digesting the minute organisms living on or between the sand grains. They make a U-shaped burrow, eating the surface sand or mud, which is richest in nutrients. Like lugworms, they form coiled castings on the surface.

ACUSHIS, which are assigned to the genus *Myoprocta*, are a group of about five species of South American rodents belonging to the same family, Dasyproctidae, as the better-known agoutis. They are long-legged, agile

The dunnock, also called hedge-sparrow, feeding nestlings.

rodents, differing from agoutis in their smaller size (about 14 in. or 35 cm) and longer tails (about 2 in. or 5 cm). They live on the ground in wet forests throughout the northern half of South America. One of the better-known species is the green acushi, *Myoprocta pratti*, which is often seen in zoos.

ADDAX, *Addax nasomaculatus*, a medium-sized desert-living antelope, 42 in. (107 cm) high, with spiral horns up to 36 in. (90 cm) long; in males the horns may form 2½-3 spiral turns; in females, only 1½-2. Addax normally weigh 265-330 lb (120-150 kg) but a big male in good condition will weigh as much as 440 lb (200 kg). It has big, splayed hoofs; in summer it is white with a gray-brown tinge; in winter, gray-brown. Calves are red. Like the oryx, it has a remarkable ability to tolerate high temperature without significant water loss.

The addax is found in the whole of the desert region between the Nile and the Atlantic Ocean. Formerly its range extended to the Mediterranean coast in Libya and Egypt, and the Atlas range in the Mahgreb, and south into northern Nigeria and northern Cameroun.

In the summer rainy season (July to September), it goes south to the southern Sahara and the Sudan savanna; in the winter rains (November to March), it returns north. Only 5,000 addax remain in the wild; fortunately it breeds well in zoos. One young is born at a time, in winter or early spring.

Normally, addax are found in areas with large sand dunes and hard desert ground. They live in family groups of 5-15, of one male with several females and calves; other males live solitarily or associate with herds of addra gazelle, *Gazella dama*. During the cold weather, they dig small holes in the sand with their hoofs, for shelter against the wind.

ADDER, *Vipera berus*, or northern viper, is a representative of the large family of venomous snakes known as the Viperidae. Adders are variable in color and pattern but are most commonly cream, yellowish or reddish-brown with black or brown markings. They nearly always have a dark zigzag pattern down the middle of the back. These markings and the short, rather thick body, together with the copper-colored iris and vertical pupil, make this snake easy to recognize at close quarters. The adder occurs farther north than any other snake in Europe and Asia. It occurs within the Arctic Circle in Scandinavia and ranges well into southern Europe, where it tends to inhabit cooler mountainous areas. Dry, open moorlands and heaths, sunny hillsides and open woodland are its preferred habitats. Adders mate in April and May, and the young are born (usually free of the egg membrane) in August and September. They hibernate for much of the winter but may emerge early to bask in sunshine on warm days in February and March. In keeping with their northern distribution, they cannot tolerate very hot sun and retreat into shade during the hottest part of summer days. They hunt lizards and small mammals mainly in the evening and at night. Adder venom, like that of most of the Viperidae, is dangerously toxic to humans, especially small children.

AFRICAN EYE WORM, *Loa loa*, perhaps the most familiar migrant parasitic worm infecting man. It is one of a number of the roundworms that do not live permanently in one site in the body, but have the habit of wandering through various tissues and organs and in so doing may cause considerable damage and discomfort to the host. It is transmitted from person to person by the bite of a blood-sucking fly (*Chrysops*), which acts as an intermediate host in the life cycle. The adult female worms, living in the subcutaneous tissue of man, liberate microfilariae, which are carried around the body in the bloodstream. During the day, when the flies are actively feeding, the microfilariae are to be found in large numbers in the superficial blood vessels of the body, but at night they retreat to deeper vessels. Larvae are taken up by the flies as they suck blood and there undergo development in the body of the insect host. When infective to the human host, the larvae escape from the proboscis of the fly during feeding and enter the bite wound.

The adult worms are 1-3 in. (2.5-7.5 cm) long and live just below the surface of the skin, moving about from time to time. They cause irritating swellings on the skin, but these usually subside fairly quickly. More painful sensations arise when the worms are in the area of the eyes, their movement across the surface of the eyeball causing irritation, soreness, excessive watering and disturbed vision. However, in this position the worms are most obvious and may be removed surgically.

AGAMIDS, a large family of lizards living in the Old World. Agamids differ from most lizard families in their teeth, which are acrodont – that is, the teeth are fixed by their bases on the summit of the ridge of the jaw. The postorbital temporal line on the skull is complete. The tongue is short, thick and slightly forked. Agamids are small to medium-sized lizards with powerful claws. They live on the ground, in rocks or in trees. Their limbs are fully developed. The scaly skin very often consists of small spines that appear mainly on head and tail. Tail autotomy is completely absent in agamids. The ability to change color depending on temperature and emotional changes is well developed. Most agamids feed on insects and other small invertebrates. A few are omnivorous, others mainly herbivorous. Nearly all agamids lay eggs, only a few species of the genera *Phrynocephalus* and *Cophotis* being oviparous.

There are approximately 35 genera of agamids with 300 species. The focal point of their distribution is the oriental region, but from there they have spread to Africa, the Indo-Australian Archipelago and Australia. They did not reach Madagascar, which is inhabited by the iguanas that are otherwise confined to the New World. Nowhere in the world do agamids and iguanas occur together, as the former always drove out the latter. The agamids reached the borders of temperate zones only in the Old World: in Europe, the hardun, *Agama stellio*, is to be found in the southern part of the Balkan peninsula, and a few species of *Agama* and *Phrynocephalus* are found in the Central Asian steppe.

About 60 of the 300 species of *Agama* live in Southwest Asia and Africa. With their dorso-ventrally flattened heads and strong limbs, the medium-sized agamas look like typical lizards.

They live on the ground, among rocks or on thick tree trunks. Some of the rock dwellers have taken to living in the walls of houses, like the African *Agama agama*.

AGOUTI, almost tailless rodents, about the size of a large rabbit but with rather long, slender legs. They belong to the genus *Dasyprocta*, placed, with the pacas and acushis, in the family Dasyproctidae. They live in a variety of habitats, but are found especially in forest, from southern Mexico to southern Brazil, where they feed on leaves and fruit. Agoutis usually have rather speckled coats. The name "agouti" is also used by geneticists to describe the banded hairs of the wild form of house mouse in contrast to the mutant color varieties that are extensively used in genetical research.

AJOLOTE, *Bipes biporus*, is one of three species of worm lizard native to Mexico and southern Baja California. The soft skin of its body is folded into numerous rings, and this, with its cylindrical body, gives it a close resemblance to an earthworm. Its eyes are covered with skin, but it has two forelimbs each with five toes well equipped with claws. Few specimens were known to science until recently. Nothing is known of its breeding habits.

ALBATROSSES, 14 species of large, long-winged, gliding seabirds comprising the family Diomedeidae, one of the families of tube-nosed birds. They vary in length from 28-53 in. (71-135 cm) but in flight they seem much larger because of their long wings. The largest, the Wandering albatross *Diomedea exulans*, has the broadest

Adult acorn worm.

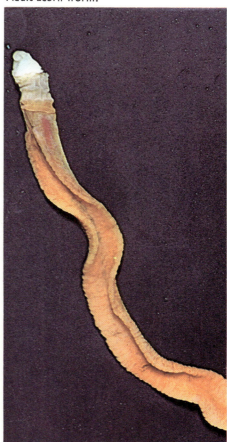

ALBATROSSES

Laysan albatross chick.

Wandering albatross at its breeding ground.

Black-browed albatross on nest.

wing-span of any living bird, up to 11½ ft (3.5 m).

Albatrosses are stoutly built with a white or brown plumage, often marked with darker brown on the wings, back and tail. The head is large and carries a strong, hooked bill, with nostrils opening through horny tubes, as in all species of the "tube-nosed" order. The legs are short, the hind toe is rudimentary or missing entirely, and the other three toes are webbed. The sexes are externally similar, except in the Wandering albatross.

There are only two generally accepted genera of albatrosses: *Diomedea* with 12 species, and *Phoebetria* the Sooty albatrosses with two. Most albatrosses live in the southern oceans, but the Galapagos or Waved albatross *D. irrorata*, breeds on the equator, and three species, the Short-tailed albatross *D. albatrus*, the Laysan albatross *D. immutabilis*, and the Black-footed albatross *D. nigripes*, breed on islands in the North Pacific.

When not on the breeding grounds they spend most of their time in flight. They take most kinds of marine animal which can be found at the surface.

The *Diomedea* albatrosses breed in colonies of many hundreds or thousands of individuals, mostly on remote oceanic islands. Nests vary from a simple scrape to a conical mound 1 ft (30 cm) or more high made of turf or soil and with a nest cup in the top lined with feathers and grasses. There is one egg, white often speckled with red-brown, particularly around the large end. The sexes share incubation which lasts from about 65 days in the smaller species such as the Laysan and Black-footed albatrosses to 70-80 days in the larger species.

Albatrosses have elaborate courtship displays involving mutual bowing and wingflapping and, in the various species, a wide variety of calls.

All the albatrosses except the Waved albatross lay their eggs between September and January, the southern spring.

The Waved albatross of the Galapagos, the only exclusively tropical species, lays during May and June.

The nestling period of the albatross is very

Alder fly *Sialis,* found near rivers and lakes in Europe, forms food for fishes.

American alligators in water.

ALLIGATOR SNAPPING TURTLE, *Macroclemys temmincki,* the largest freshwater turtle in the United States and one of the largest in the world. Reaching a weight in excess of 200 lb (90 kg), it is sluggish and heavily armored, frequenting the bottoms of lakes and rivers. It is unique in possessing a built-in fishing lure, a fleshy appendage on the floor of the mouth that resembles a twitching worm. Fishes, attracted by this lure, enter the mouth and are swallowed by the turtle. The ruse is enhanced by the dull coloration and rough shell, which is usually heavily covered with algae and serves to render the turtle invisible.

ALLIGATORS, an ancient, relict group of crocodilians comprising only two species out of a total of over 30 forms in the order Crocodylia. They are distinguished from crocodiles by the pattern and arrangement of the teeth and, not so reliably, by the shape of the snout. In the alligator the lower row of teeth project upward into a series of pockets in the upper jaw, so that when the mouth is closed, the only teeth exposed to view are the upper teeth. This gives the alligator an appearance of "smiling" when viewed from the side. The crocodile, on the other hand, normally has both rows of teeth exposed when the jaws are closed, and the teeth intermesh with one another. Particularly prominent is the enlarged tooth fourth from the front, which may even extend above the line of the upper jaw giving a constricted appearance immediately behind the nostrils. So the crocodile's "smile" resembles a toothy leer. Alligators are found only in two widely separated parts of the world: the upper Yangtse River valley in China and the southeastern United States. One theory is that the less aggressive alligator—which will run away from man—was once almost worldwide in distribution, but the more recent crocodiles have successfully weeded out the alligator throughout its range

Young alligators look almost benign.

to where it is now found only in two relict "islands."

Alligators in the wild construct underground dens, at least portions of which are filled with water. In fact, one method used by collectors for locating alligators is to probe with a long pole into the soft earth along a river bank. It is easy to tell when the pole enters a large, open cavern and contacts an alligator. At egg-laying time in the spring, a female American alligator constructs a large mound of mud and vegetation

extended, so much so that the larger species are not able to breed every year.

Both sexes feed the chick, by regurgitation, regularly for the first week or so, then with decreasing frequency, leaving the chick to its own devices for increasing periods.

The greatest danger the albatross chick has to face is predation by skuas when its parents are away. Unlike other albatrosses the *Phoebetria* species build their nests on inaccessible cliff shelves and edges.

The two species of *Phoebetria* also breed on oceanic islands but not in large colonies. They are found characteristically in the temperate subantarctic latitudes of the South Atlantic and Indian Oceans.

Clearly, that which makes albatrosses unique among birds, even among seabirds, is their faculty of sustained sailing, gliding flight. This is dependent on a high air speed, and is one of the primary reasons for albatrosses being found in the windiest oceanic regions. Albatrosses' wings have the highest aspect ratio (proportion of length to breadth) of any bird, from 20 to 25. The albatross in fact is admirably adapted for "dynamic soaring," which is the method of flight evolved for making use of the difference in speed between winds at sea-level and those at some height above the surface, normally up to 50 ft. In this way the albatross can travel for miles by a series of glides and climbs with never a wingbeat.

One further feature of albatrosses that must be mentioned is their habit, in common with other tube-nosed birds, of vomiting the oily contents of the stomach when molested, particularly when they are young.

The albatrosses, as a group, have suffered considerably over the years as a result of human activities, particularly in the 19th and 20th centuries. From early times man has used the bird for food, for both its flesh and eggs are palatable.

The Second World War was a great threat to many albatross populations in the Pacific. Hundred of thousands of birds were killed as a side-effect of normal warfare, and on one island at least the starving Japanese garrison ate the whole colony.

ALDER FLIES, four-winged insects of the suborder Megaloptera of the order Neuroptera. They are represented in Britain by only two species, *Sialis lutaria,* which occurs in ponds and lakes, and *Sialis fuliginosa,* which is usually found in running water. They are stout-bodied, brown in color, have long filamentous antennae and fold their wings along the back in the form of a ridged roof. The adults are commonly found among aquatic vegetation and reeds in May and early June. They are weak fliers and frequently fall onto the water surface, where they form a ready source of food for fishes. They are common in both stony and muddy-bottomed streams, and the fisherman's artificial fly, the "alder," is modeled on them. The eggs are laid in large batches on reeds and other emergent water vegetation. The larvae on hatching crawl or fall into the water, where they live at the bottom as active predators, feeding on worms, other insect larvae and small mollusks. The larva has a pair of long, bristled lateral appendages on the abdomen, which give it the appearance of being ten-legged. The larvae of the British species grow to about 1½ in. (4 cm) in length, but those of the American Dobson fly, *Corydalis,* are commonly over 3 in. (7½ cm) long and have enormous scissorlike mandibles retained by the male in adult life.

about 5-7 ft (1.5-2.5 m) wide at the base. 20-70 hardshelled eggs $3\frac{1}{3}$ in. (8.5 cm) by $2\frac{1}{2}$ in. (6.5 cm) are laid in a cavity in this, and are then covered with more debris and sealed in by the restless activity of the mother. The peeping of hatching young after approximately 10 weeks of natural incubation will encourage the female to tear open the nest and help release the 8-in. (20-cm), brightly colored black and yellow young, which will make for the nearest water and are then given some protection by the female for the next few months. They grow by about 12 in. (30 cm) a year during their first few years.

Alligators have been known to attain large sizes in the past, and some records claim individuals over 20 ft (6 m) in length (in which case, such individuals probably weighed well over half a ton). A 19 ft (5.8 m) specimen was shot in Louisiana a generation ago. Today, 10 ft (3 m) specimens are a rarity, and 8 ft (2.4 m) individuals are uncommonly hard to find in the wild.

ALPACA, *Lama pacos,* a camellike domesticated animal of the high Andes. It stands 3 ft (90 cm) at the shoulder, its long fleece, often reaching the ground, is of uniform black, reddish-brown or even white, as well as mixed colors, but piebalds are much rarer than in llamas. There is a tuft of hair on the forehead. The alpaca inhabits the high plateaus of Bolivia and, particularly, Peru. The principal center of breeding from time immemorial has been on the high plateau of Lake Titicaca in the departments of Puno, Cuzco and Arequipa. The alpaca thrives best at altitudes between 13,000 and 16,000 ft (4,300-5,300 m) with low air humidity, except in the rainy season. At lower altitudes the animals are often rachitic and their wool is poorer. The alpaca has very sensitive feet and prefers the soft moist ground with tender grasses of the "bofedales," where there are many pools and puddles in which they like to wallow. A lack of ground moisture is said to produce a fatal foot disease, and in rainless years the mortality rate is higher than at other times. Alpacas are more fastidious feeders than llamas.

The period of gestation is 11 months. The mares foal in the rainy season. From the age of seven onward alpacas are used only for meat production. At the present time the quality of

The alpaca, domesticated for its wool.

alpaca wool is only slightly lower than that of the vicuña and is being even further refined by crossing with vicuñas. The suri, a breed of alpaca with finer, thicker and longer hair, provides up to 11 lb (5 kg) of wool per annum, but shows a greater susceptibility to parasitic diseases. Shearing takes place every 2 years before the rainy season in November or December, when temperatures are fairly uniform. It is done with ordinary knives or with shears. The Indian women can divide the wool into seven classes just by feeling it. Black alpacas are particularly in demand owing to the heavy coat of hair. There are numerous large commercial farms, but most of the alpacas are owned by Indians. In Peru the number of alpacas is estimated at over 2 million, and in Bolivia at about 50,000.

AMBYSTOMATIDS, North American mole salamanders, are sturdily built, broadheaded, medium-sized salamanders. They include the marbled salamander, *Abystoma opacum,* which may attain a length of 5 in. (12.5 cm) and is terrestrial. It lives on hillsides near streams from New England to northern Florida and westward to Texas. Breeding takes place in autumn, fertilization is internal and the eggs are laid in shallow depressions on land, usually guarded by the female until the next heavy rain, when they hatch.

In most other ambystomatids, courtship takes place in water with an elaborate ritual.

The majority of salamanders, including the ambystomatids, are usually voiceless. A notable exception is the Pacific "giant" salamander, *Dicamptodon ensatus,* which occurs in the moist coastal forests from British Columbia to northern California and which may reach a length of 12 in. (30 cm). Almost all ambystomatids have well-developed lungs.

Neoteny (breeding in the larval state) is not uncommon in the Ambystomatidae, the best-known example being the axolotl, the permanent larva of *Ambystoma mexicanum.* It is found around Mexico City and keeps well in captivity. Some species of salamander are habitually

Tiger salamander, largest terrestrial salamander, one of the so-called mole salamanders.

neotenous in one part of their range and not in another. The major factor that contributes to the neoteny is a lack of iodine in the water.

The ambystomatids (family Ambystomatidae) together with the plethodontids (family Plethodontidae) form the suborder Ambystomatoidea, the largest group of tailed amphibians with some 27 genera and in excess of 200 species.

AMERICAN BUNTINGS, relatively small finchlike birds of North America belonging to the two subfamilies Pyrrhuloxiinae (=Richmondeninae) and Emberizinae of the family Emberizidae. Those Pyrrhuloxiinae called buntings comprise six species of small, highly colored and attractive birds in the genus *Passerina,* e.g., the indigo bunting, *P. cyanea,* the lazuli bunting, *P. amoena,* the painted bunting, *P. ciris,* and the varied (or beautiful) bunting, *P. versicolor.* Those species of North American Emberizinae popularly called bunting are the lark bunting, *Calamospiza melanocorys,* the snow bunting, *Plectrophenax nivalis,* its close relative McKay's bunting, *P. hyperbore,* and the rustic bunting, *Emberiza rustica,* a straggler from Asia. One species of North American longspur, the circumpolar Lapland longspur, *Calcarius lapponicus,* is known as the Lapland bunting in Europe. All the European Emberizinae are called buntings, so collectively the term may be applied to all species of American Emberizinae, including the juncos, the American "sparrows," towhees and brush finches. See also buntings, cardinals and song sparrows.

AMERICAN WOOD WARBLERS, a New World family of small perching birds, with nine primary wing feathers, which includes the tanagers, troupials, etc. There are about 113 species, varying from 4-7 in. (10-18 cm) long, found from Alaska and northern Canada south to southern South America.

They are brightly colored with striking patterns of orange, yellow, black and white plumage. Some, however, have dull, uniform gray or brown plumage although others have bright plumage with blue or red markings. Sexual and seasonal differences in plumage pattern are common in the more brightly colored of the northern species, but in many of the tropical ones females are also brightly colored and they keep their bright plumage all the year round.

The majority of the wood warblers inhabit areas of woodland or scrub, but some have adapted to marshes and swamps and even to open fields and the sparse vegetation of the edges of deserts.

Most of the species live in trees and vegetation and feed on insects caught among the foliage. Some that have developed wide, flat bills and pronounced bristles around the gape live by hawking flies and other airborne insects from a perch, as do the European flycatchers. A few species, including the North American black-and-white warbler, *Mniotilta varia*, creep up tree trunks and boughs, picking food from the bark like tree creepers.

Most wood warblers build cup-shaped or domed nests in bushes or trees, or on the ground concealed by vegetation. The oven bird, *Seiurus aurocapillus*, and some of the other terrestrial members of the family, build large dome-shaped nests on the ground with small side entrances.

They lay two to five eggs, incubated by the female alone, but both parents feed the young until after they fledge. The eggs of most species are white with brown or gray spots, but some tropical species have immaculate white eggs.

AMOEBAE, single-celled animals characterized by their mode of movement which is by means of pseudopodia—outpushings of cytoplasm. The amoebae have been recognized since 1755 but considerable confusion exists concerning the status of the various members of the group. The amoeba seen in 1755 by Rosel von Rosenhof was a large freshwater form. The best known of the large freshwater amoebae is *Amoeba proteus* which is the protozoan commonly studied in schools. *Amoeba proteus* lives in bodies of permanent water and occupying similar habitats are two other amoebae, *A. discoides* and *A. dubia*. All three of these amoebae possess a single nucleus. There are also a number of large amoebae which possess numerous nuclei and there are also a number of smaller amoebae which occur in ponds and streams.

An amoeba reproduces by dividing into two equal parts.

The amoeba feeds by phagocytosis, that is, it surrounds its food with its pseudopodia and engulfs it.

A quite separate group of amoebae is known collectively as the Limax amoebae. These are all very small and move by means of a single pseudopodium. They are found in fresh water and in the soil.

There are a number of true parasites which live in the intestines of invertebrates and vertebrates.

AMOEBAE, TESTATE, bottom-dwelling, single-celled animals found in fresh water and in damp places such as sphagnum moss. A few are marine. They divide asexually and form cysts under adverse conditions, which allow them to survive periods of drought. They possess a shell, consisting of a single chamber, which distinguishes it from the shells of the Foraminiferida, which typically have several chambers. Some shells consist of a pseudochitin secreted by the amoeba itself, as in *Arcella*, but

in others the substance produced by the amoeba may be impregnated with foreign substances such as sand grains or other similar particles, as in *Difflugia*. In one genus, *Euglypha*, the shell consists of siliceous plates secreted by the animal itself.

AMPHIPODA, with over 3,600 species from the seas and fresh waters of the world, the order is one of the most successful and dominant of the Crustacea.

The suborder Gammaridea contains over 3,000 species distributed over 57 families. They are regarded as the typical amphipods. The body, which is without a carapace, shows the usual division into head, thorax and abdomen. The legs hang vertically downwards and give no stability for a vertical stance. Thus amphipods lie on their sides and drag themselves over the substratum by the last five pairs of legs assisted in water by the pleopods of the abdomen.

The feeding methods of amphipods show great variety. They are scavengers, either vegetarian or omnivorous.

In breeding the male lies across the body of the female and passes sperm into the brood-pouch. The pair separate and the female immediately lays eggs into the pouch. The number varies from 2-750 and usually more than one brood is produced each year.

The Caprellidea contain the cyamids and the caprellids, often referred to as the Skeleton or Ghost shrimps.

The Ingolfiellidea are a small group of seven species within one genus.

The sandhopper *Orchestria gammarella*, shown nearly twice natural size, lives among small red seaweeds and tiny pebbles.

AMPHISBAENIDS, or worm lizards, represented by some 125 species in Africa, South America and Mexico. One species of the genus *Blanus* is found in Europe; one *Rhineura* in the United States and several in the West Indies. Three species of *Bipes*, found in Mexico and Lower California, are unique among the amphisbaenids in having diminutive but well-developed forelegs. With this exception, all worm lizards present essentially identical features. They closely resemble earthworms in their long and cylindrical body, with the integument arranged in rings separated by shallow grooves. Their movement, unlike that of snakes or legless lizards, is in a straight line, and the animal can move backward or forward with equal ease. There are no external ear openings, the eyes are covered with scales and,

when visible, appear as dark spots. Only the left lung is present. The exact relationships of worm lizards are not clear. Although long grouped with reptiles, there is considerable doubt whether these animals actually are lizards or, for that matter, reptiles.

Amphisbaenids are inoffensive and spend their lives in underground burrows, beneath forest litter or in the nests of ants and termites. Association with ant colonies is common, and, in parts of South America, the local peoples (who believe the animal to be a venomous snake) refer to it as *mai das saubas* or "mother of ants." Little is known of the feeding habits of these animals, examination of the stomach contents revealing little other than ants and termites.

Both oviparous and ovoviviparous species are known, but the majority of species probably lay eggs, live births being recorded for only a few African species.

AMPHIUMAS, elongated salamanders, which average about 24 in. (60 cm) and may reach 46 in. (116 cm). The tail, if undamaged, is approximately $\frac{1}{4}$ the total length and tapers to a point. The most distinctive anatomical feature is the presence of two pairs of small legs, with 1-3 toes according to species, which are totally useless for propulsion. The head is long, pointed and compressed from the top. The color is dark, uniformly brownish gray or slate gray above and lighter below. In contrast to the eel, there are no gill openings or fins behind the head, and no external gills as in the sirens. Vision is probably very poor, the eyes being very small.

Amphiumas are found in warm, weedy, quiet bodies of water in the lowlands of the southeastern and gulf-coastal plains of the United States and into the Mississippi Valley as far north as Missouri. They are mostly active at night, searching for small aquatic animals. Besides soft-bodied prey like worms and insect nymphs, their powerful jaws enable them to crush snails and subdue crayfish. An occasional frog or fish is caught with a surprisingly fast strike. They may occasionally come out on land on wet nights.

Fertilization is internal. In the two-toed amphiuma, fairly large elliptical eggs, $\frac{1}{3}$ in. (8 mm) in diameter, are extruded like a row of beads connected by a continuous gelatinous string. The eggs may number 48 or more and are sometimes guarded by the female, who stays with them in a sheltered hollow during the long incubation period of several months. The eggs hatch in about five months into 2-in. (5-cm) long larvae, which metamorphose and lose their external gills when 3 in. (7.5 cm). During growth the legs, which start off well developed, fail to keep pace with the rest of the body.

ANACONDA, *Eunectes murinus*, the largest of the nonvenomous snake family Boidae. One of the more aquatic boas, it inhabits swamps and slow-moving rivers in the northern parts of South America to the east of the Andes. It is the largest of living snakes, for although its length is a little less than that of the reticulated python of Asia, it is proportionately much thicker.

Its reputation as a man-eater is largely undeserved. A large anaconda may be capable of devouring a child, but such occurrences are rare. It generally shuns human habitations and preys chiefly on birds and small or medium-sized mammals such as rodents and peccaries. Fish and caimans are also included in its diet.

Like all boas the anaconda is ovoviviparous, the female giving birth to as many as 72 living young, each measuring about 3 ft (1 m) in length.

Even a relatively small snake seems almost unending as it crawls through undergrowth, so with a little embroidery about adventures in remote corners of the earth the record length claimed for an anaconda attained 140 ft (42 m). Such a record can easily be dismissed, but it is not easy to establish what is the greatest length to which an anaconda will grow. Experts on snakes give differing figures, for two reasons. They have different standards as to when a record can be considered reliable, one expert rejecting a record that another accepts, and, most important, it is very difficult to measure even a dead snake as its body stretches so easily that even pulling it into a straight line distorts it. The current record for anacondas seems to be 37½ ft (11.4 m), placing the anaconda just ahead of the reticulated python in record length.

ANCHOVIES, a family of usually small and silvery fishes allied to the herrings and found in temperate and tropical seas with a few species passing into fresh water or with permanent populations in rivers. Anchovies can be immediately distinguished from any small herring-like fishes by the pointed snout that overhangs the mouth and the long and slender lower jaw. The body is slender, more or less compressed depending on the species. There is a single soft-rayed dorsal and anal fin, and the tail is forked. The silvery scales are often easily shed. The majority of species are small, usually growing to 4-6 in. (10-15 cm), but a few species may reach 12 in. (30 cm).

The anchovies, of which about a hundred species are known, are essentially tropical fishes with a few species in temperate waters. The best known of the latter is the European anchovy, *Engraulis encrasicolus,* an elongated, round-bodied species that forms large shoals. It is found from Norway southward to the Mediterranean and the west coast of Africa and forms the basis for large fisheries.

ANGEL FISHES, members of the freshwater genus *Pterophyllum* of the family Cichlidae. These are highly compressed, deep-bodied fishes with slender, filamentous pelvic rays. When seen head-on through a growth of water plants, the extreme narrowness of the body makes the fish look like just another plant stem. The light brown flanks are marked with four darker vertical bars, so that even from the side the fish blends with its surroundings.

There are still problems concerning the numbers of species in the genus *Pterophyllum,* but three are usually recognized, *P. eimeki, P. altum* and *P. scalare.* Identification is difficult for the amateur and will become more so because the hybrids between *P. eimeki* and *P. scalare* are commonly sold. *P. altum* is imported less often than the other two species.

P. scalare is the largest of the three species in the Amazon, grows to a total length of 6 in. (15 cm) and has a body height (including fins) of 10¼ in. (26 cm). However, when kept in an

Freshwater angelfishes, popular with aquarists.

aquarium it does not grow so large. Breeding in captivity is not difficult, the main problem being to recognize the sexes, and this is best overcome by allowing the fishes to pair off themselves. They spawn on broad-leaved plants which have previously been cleaned by the fishes themselves. After the eggs have been deposited, the parents continually fan them, and at 86° F (30° C) the young hatch in about 30 hours. The parents assist the young to hatch by chewing at the eggs and spitting the young onto leaves. There they hang suspended from short threads until the parents remove them to a shallow depression or nest in the sand. After four to five days the young, who may number as many as a thousand, are able to swim, and the parents lead their brood out of the nest.

ANGEL SHARKS, or monkfishes, the large pectoral fins of which give them a "hybrid" appearance between a shark and a ray. Unlike the rays, the pectoral fins of angel sharks are not joined to the head. There are two dorsal fins, no anal fin, and the nostrils have two barbels that extend into the mouth. The latter is almost terminal and not underslung as in most sharks. Angel sharks are found principally in temperate waters, and they feed mainly on fishes. The monkfish, *Squatina squatina,* of European waters is the largest, reaching 8 ft (2.4 m) in length and weighing 160 lb (73 kg). It is found in British waters, chiefly off the south coast, and enters shallow waters in summer. Of the several species known, *S. dumeril* occurs off American Atlantic coasts and *S. californica* off American Pacific coasts.

ANGLERFISHES, a highly specialized group of marine fishes found in all oceans and at all depths, from shallow waters down to the abyssal trenches. The anglerfishes can be divided

into three main suborders: the Lophioidea, the Antennarioidea and the Ceratioidea. The three groups share certain anatomical features, the most outstanding and the one that has given them their common name being the tendency for the first ray of the dorsal fin to be long and to develop a lure at its tip with which the anglerfish "angles" for its prey. In addition, the pectoral fins in many species are borne on a fleshy limb, which adds to the bizarre appearance of the fish and gave rise to a former scientific name for the whole group, the Pediculati.

The first group of anglerfishes, the Lophioidea, are shallow-water fishes. The common anglerfish, *Lophius piscatorius,* sometimes known as the fishing frog, is found around European and American coasts. It is a greatly flattened fish, lying like a huge disc on the sea bottom. The mouth is enormous, and the jaws are lined with sharp, needlelike teeth. The eyes are on the top of the head, and the brown body is fringed with small flaps of skin so that the outline of the fish is broken and the fish itself rendered inconspicuous.

They are large fishes, reaching 4 ft (120 cm) in length. The spawn of the anglerfish is also peculiar. The eggs are contained in a long ribbon of mucus 1-3 ft (30-90 cm) wide and up to 50 ft (16 m) long. These long ribbons are often reported floating at the surface of the sea. The second group of anglers, the Antennarioidea, includes two important families. The family Antennariidae are flattened from side to side (not from top to bottom as in the lophioids). One of the best-known members is the Sargassum weed fish, *Histrio* spp. The second family is the Ogcocephalidae, often called the batfishes. They are flattened like the lophioids and have large heads and fairly small bodies. The most striking feature is the limblike pectoral fins, which are muscular and are used to crawl about the bottom with a slow, deliberate waddle. The lure is hidden in a tube above the mouth, and it can be projected out when the fish is hungry. Batfishes grow to about 15 in. (38 cm) and are found in both the Atlantic and Pacific oceans.

The third suborder of anglers, the Ceratioidea or deep-sea anglers, are among the most fascinating of all fishes. They are essentially midwater fishes of the deep seas. In this group, the body is rounded and not flattened, but the characteristic fishing lure is present. One of the giants of this group, *Ceratias holboelli*, grows to 36 in. (90 cm). Most of the ceratioids, however, are only a few inches long. They are usually dark brown or black, without scales and with a fragile, velvety skin, although some have warty projections on the body. The lure is luminescent. Male deep-sea anglerfishes are parasitic on the female.

ANGUID LIZARDS, a family of small to medium-sized lizards usually classified in seven genera with about 75 species. This family comprises standard lizardlike forms and limbless, snakelike ones. In America the anguids range from British Columbia in the north to Argentina in the south and the Alligator lizards form the most well known group. About 30 species of the Alligator lizard *Gerrhonotus* can be found in North and Central America. Their body structure is more primitive than that of the snakelike anguids. In *Gerrhonotus* the limbs are well developed, the head and body are protected by scales. Alligator lizards are ground dwellers, but some species are able climbers making use of their tails. Their food consists of insects which they stalk and seize, leaping at them from a short distance. Alligator lizards may be viviparous or oviparous.

ANGWANTIBO, *Arctocebus calabarensis,* a small, golden-brown mammal related to lorises and monkeys, with forward-directed, large eyes and a pointed snout. Its head and body measure 9 in. (22.5 cm) with a 2-in. (5-cm) tail almost hidden in the fur. Each hand has a widely opposed thumb, a very small nailless index finger and a short third finger. The big toe is widely opposed to the other toes. These grasping extremities are very powerful, and it is said that the angwantibo spends most of its time hanging upside-down from branches. Like the potto and the loris, it is very slow-moving and walks with a deliberate, clasping pace, its hands and feet acting like pincers.

The angwantibo is completely arboreal and lives high up in the trees in the tropical forest belt of West Africa from the Niger to the Congo. It is rarely seen, which explains why so little is known about it. It seems not to make a nest and probably sleeps clasped to a fork of a tree, like the potto. It is omnivorous, but the main item of diet is probably insects, supplemented by soft fruits, this being reflected in its dentition, which has long, pointed canines and well-developed sharp cusps on the cheek teeth (premolars and molars).

As with most of the primates, there is only one baby at a birth, after a gestation of about 4 months, and it climbs directly onto the mother's fur. Thereafter, it is carried around by the mother until reaching independence. There seems to be no well-defined breeding season, but there may be an annual rise and fall in the number of births. The angwantibo is apparently solitary or pair-living, so the baby probably moves off to establish its own home range or territory as soon as it is independent of the mother.

ANIS, three species of cuckoos, genus *Crotophaga* of southern U.S.A., Central and South America, inhabiting clearings and open country of the tropical lowlands. Of medium size, they have a mainly black plumage, slightly glossed with bronze, green or blue. Their tails are long and appear to be loosely joined to the body, while the legs and wings are relatively short. The birds often look awkward and ungainly. The most conspicuous feature is the bill, which, on the smooth-billed ani, *Crotophaga ani*, is very deep and laterally flattened, appearing large and arched in outline from the side, narrow and bladelike from in front. In the groove-billed ani, *C. sulcirostris*, it is less deep and has several longitudinal grooves on each side, while in the greater ani, *C. major*, the narrow flattened ridge is much reduced and confined to the basal part of the bill.

Anis are highly sociable and usually keep together and move about in small groups within a territory. Groups of smooth-billed anis will defend their territory against intruders of their own species. They feed extensively on the ground, eating insects, particularly grasshoppers, but they will take small lizards, and also berries.

A group of anis will usually nest communally, although at times a single pair may nest on their own. The nest, built in a tree, is a cup made of sticks, with an inner cup of finer material, the whole structure being very bulky if a number of birds are involved. Green leaves are used for the final lining, and both twigs and leaves may be added while incubation is in progress. Several females will lay in one nest. A single bird lays four or five eggs, and up to 29 have been found in one nest. In such a pile of eggs those at the bottom are not incubated sufficiently and fail to hatch. The eggs are blue-shelled, but the blue is covered with a white layer, thin enough in places for a blue tint to be apparent and become scratched during incubation. The several females in the group share both incubation and the care of the nestlings.

The sociable habits of the anis extend to their huddling together on a branch to roost at night and in the early morning. They also perch in the sun with their wings extended, and both habits suggest that they get chilled easily. This is supported by temperature measurements made on two captive smooth-billed anis. Their body temperatures fell from 41° C in the evening to 33-34° C by 4 a.m. the next morning.

An anglerfish *Antennarius* lies in wait on the sea-bed to catch passing prey.

Their temperatures slowly returned to normal by the late morning.

ANOA, the name of two small species of buffalo on the island of Celebes differing from the water buffalo and tamarao in their small size, rounded skulls, slender build and straight horns. They are placed in the subgenus *Anoa* of the genus *Bubalus.* The horns in both species are back-pointing in line with the forehead. The Lowland anoa, *Bubalus (Anoa) depressicornis,* is 34 in. (86 cm) high with horns 12 in. (30 cm) long in males, 9 in. (21.7 cm) in females. Both sexes are jet black with sparse hair, and the legs are white below the knee and hocks. The horns in adults are strong ridged and triangular in section at the base. The mountain anoa, *B. (A.) quarlesi,* is only 28 in. (69 cm) high with horns 6½ in. (16 cm) long in both sexes. The horns are simple, smooth and conical even in adults. Both sexes are usually black with thick woolly hair, and the white on the legs is restricted to a pair of spots above the hoofs. Calves of both species are golden brown; adult mountain anoas are sometimes brown.

ANOLE, American lizardlike reptiles of the family Iguanidae. There are 165 species, of which the best known is the green anole, *Anolis carolinensis.* Anoles, which are also known as American chameleons from their ability to change color rapidly, range from 5-19 in. (12.5-47.5 cm) in length. The toes of anoles are armed with small sharp claws and have adhesive pads of minute transverse ridges that enable them to cling to rough and smooth surfaces alike. The green anole is active in daylight, and moves continuously and rapidly, searching for insects in bushes and trees. Male anoles have a reddish throat sac which is extended to deter rival males from encroaching on another's territory.

ANTARCTIC CODS, fishes belonging to the family Nototheniidae, but in no way related to the true cods. They are placed in a suborder that includes the icefishes. The Antarctic cods are confined to the ocean surrounding Antarctica and live in waters that are permanently only just above freezing.

They are generally sluggish, bottom-living forms with large heads and jaws and show their relationship to other perchlike fishes by the presence of spines on the first part of the dorsal fin.

ANT BIRDS, a large diverse family of small American perching birds, sometimes called "ant thrushes."

There are 223 species, all less than 12 in. (30 cm) long. The plumage is in shades of black, gray, brown, rufous, chestnut, olive and white, and the females often differ from the males in having black replaced by brown, rufous or chestnut. There is often a short crest, and the feathers of the back, rump and flanks are long, loose and silky. The wings are short and rounded, the bill stout and strong with a hooked upper mandible and the feet and legs strong and well developed. The voice is generally harsh and unmusical.

The family is confined to wooded areas, essentially lowland and mountain forest, in Central

Female angwantibo carries her infant on her back.

and South America, from southern Mexico to Argentina, Bolivia and Peru. It reaches Trinidad and Tobago, but not the Antilles, and is absent from Chile and the treeless areas south of La Plata, extending farther south on the eastern side of the Andes than it does on the western side.

The nest may be a woven cup suspended by the rim from a forked twig or may be a normal nest supported in a fork, but some species are said to nest in holes or on the ground. Two white eggs with blotches, spots or scrawls of brown, purple, lavender, or reddish hues, often as a zone at the larger end, seem to be the normal clutch.

The incubation period is between 14 and 17 days, the nestling period between 9 and 13. The sexes share duties at the nest, and the female broods at night in those few species adequately studied.

ANT LIONS, name given to the larvae of a group of insects belonging to the Neuroptera and related to the lacewing flies. The larval ant lion digs itself a small conical pit in sand or soft soil and buries itself at the bottom of it. Ants and other small insects wandering across the ground fall into the pit and are seized by the ant lion before they are able to escape up the loose soil of the pit walls. Ant lions, *Myrmeleon* spp. are common in southern Europe, and the family is very widely distributed in tropical and subtropical regions.

The adult ant lion fly bears a superficial resemblance to the thin-bodied dragonflies, from which it is easily distinguished, however, by its long, clubbed antennae and two prominent parallel veins in the wings, which have no cross veinlets between them. Some of the tropical species may have a wingspan of up to 4 or 5 in. (10-12½ cm). Their wings are usually of equal size, translucent with mottled patches. Some tropical species have enormously elongated hind wings drawn out at the ends so that they are shaped roughly like a squash racquet. Ant lions are nocturnal, and their weak flight, achieved with their seemingly oversized wings, makes them conspicuous as they fly around lights.

The larvae generally have pear-shaped bodies with enormous curved mandibles. Some species in Southeast Asia have a greatly elongated prothorax, which gives the impression of a very long neck and therefore a rather bizarre appearance.

ANTS, small insects of which 3,500 species have been described from the tropics and the temperate zones. All are social and live in colonies, which may consist of a few individuals, as in the Ponerinae, or as many as 100,000 individuals, in, for example, the wood ant, *Formica rufa.* An ant is recognizable by its "waist" or petiole formed by a narrow segment or segments between abdomen and thorax. Females and males are winged when they leave the nest but are wingless at other times. Winged ants, which fly slowly in great clouds at certain times of the year, are not a separate species but the reproductive members of colonies that have left to swarm.

Ants are very clearly polymorphic, with worker, male and female castes. Males of all species are rather alike, being winged and having well-developed eyes and long antennae. They are

Wood ants' nests may be more than 3 ft (1 m) high.

usually to be found only at certain times of the year, for they do not survive mating for long and are not readmitted to the nest after the nuptial flight.

With very few exceptions, all species of ants have a clearly recognizable worker caste consisting of sterile females whose function is to forage, build the nest and care for the young. They are wingless, and often their eyes are small. In many species the workers do not produce eggs, but in some they lay eggs which are used to feed young larvae. These eggs, as they are not fertilized, could only give rise to males. Workers vary in size, and typically

Larva of ant lion *Myrmeleon alternatus.*

those hatched from the first eggs laid by a queen establishing a new nest are smaller than those forming the bulk of the population. However, there may be a spread of size at all times in the colony. The smaller ones then seem to pass most of their time within the nest while the larger ones protect the nest and forage. But there may be two very clearly defined kinds of workers, and in this case the larger ones, known as the soldiers, have very large, strongly chitinized heads and strong mandibles. One of their functions is to fight to protect the nest, but in addition they may help the smaller workers break up a large piece of food before it is transported to the nest.

The ant queen is solitary and not accompanied by a male when she establishes a nest. She is usually larger than males or workers but has fully functional mouthparts. After fertilization, which generally takes place in the air, the queen lands and pulls off her wings with her jaws or rubs them off against a solid object. She then begins to excavate a small chamber within which she remains until the following year. Very soon she lays a few eggs, which will develop into workers. She tends these eggs, and when the larvae hatch she feeds them upon salivary secretions, depending, herself, solely upon nutriment from her fat body and from her flight muscles, which degenerate during this period. These first workers show the effect of their reduced food supply in being small, but they can nevertheless break out of the chamber to forage and bring back food both for the queen and for the later young. Many species of ants found colonies in this way. A queen may live for as long as 15 years and throughout this time is capable of laying eggs that are fertilized by sperm deposited in her receptaculum seminis during the mating flight.

The eggs of ants are white and not more than $\frac{1}{50}$ in. (0.5 mm) long. The "ants' eggs" sold as fish food are the cocoons of ants and not the true eggs. Workers carry the eggs about as conditions within the nest change, always maintaining them in that part of the nest where the conditions are optimum. They are licked by the workers, and this keeps them free of fungal infection. The larvae are grublike with a head and about 13 segments. They are legless but seem, in some species, to solicit food from workers by side-to-side movements of the front end of their bodies. They are kept in piles of equal size and of roughly equal age.

On the whole, the larvae are fed upon regurgitated liquid material. The foragers transfer food to other workers, and the same kinds of trophallactic relationships exist among ants

Army ant soldier with workers.

as among honeybees. In some primitive ants (Ponerinae), insect prey is given to the larvae, which are active enough to tear it apart, and dehusked seeds are given to the larvae of harvest ants. In due course the larvae pupate, often first spinning a cocoon around themselves. Workers aid them in breaking out of this covering when they have changed to the adult form. The castle of an ant is determined by the amount of food that is fed to it as a larva. Those destined to be reproductives are given a high-protein diet, whereas the future workers receive a largely carbohydrate diet.

Colonies of ants live in a great variety of structures, usually made by themselves. Many make galleries in the soil with chambers scattered through the depth of the nest. In these chambers broods are kept, or seeds stored as food, or fungus grown upon beds of macerated leaves. These fungus gardens are typical of the leaf-cutter ants of the tropics (Attinae), which strip nearby trees to obtain the material upon which to grow their fungus. Their young are fed upon the bromatia, bodies

Busy scene at entrance to an ant's nest.

produced by the fungus only in this underground situation.

Other species make mound nests. The wood ant of Europe digs down a short distance and makes part of its nest below ground, but above it piles pine needles, small twigs and the like, between which the galleries of the nest penetrate. The entrances to the passages are closed when it is necessary to conserve heat within the mound and are reopened if the internal temperature rises too high. Some of the smaller ants, like *Leptothorax*, which have colonies of relatively few individuals, may live under the bark of sticks lying on the ground. "Paper" made from wood chewed by the insects themselves is another building material, and some tropical species occupy chambers within plants. The plants' response to this is often to produce gall-like formations which become riddled with the ants' galleries. One unusual type of nest is built by the tailor ant, *Oecophylla smaragdina*. Workers hold two leaves together, grasping one with their mandibles while gripping the other with their hind legs. Other workers, holding larvae in their jaws, "sew" the leaves together by moving the heads of the larvae to touch first one leaf and then the other. The larvae produce silk which holds the leaves firmly together. The army ants (Dorylinae) do not construct a nest. They move in long columns over the countryside, clearing it of other insects and even small birds and mammals as they go. Each night they bivouac below a log, or in a similar situation, and march on the next morning. Periodically they settle for a longer period in a hole or within a hollow log. These phases occur just after the queen has laid her eggs and when there are no larvae in the colony.

The habit of looking after aphids to obtain food from them is widespread. Sometimes ants will protect such greenflies by building shelters over

them or, in the case of *Lasius flavus*, excavating "stables" around the rootlets upon which the aphids feed. The honeydew collected by the ants is a food surplus to the greenfly's requirements that is exuded from its anus.

APHIDS, greenflies or plant lice, sap-feeding insects of the superfamily Aphidoidea, order Homoptera. This order also contains cicadas, scale insects, spittle bugs and whiteflies.

There are about 4,000 known species of aphids. Some are restricted to specific climatic or vegetational zones, while others, such as the peach-potato aphid, *Myzus persicae*, and the cabbage aphid, *Brevicoryne brassicae*, are cosmopolitan.

Aphids are small, soft-bodied insects with piercing mouthparts, and the tubular proboscis or rostrum usually has four evident segments. They feed by piercing the tissues of plants with their proboscis to take up the sap. The legs, usually long and slender, end in a bisegmented tarsus and a pair of claws. The four wings, when present, are transparent, and the hind wings are narrower and smaller than the forewings. The antennae have three to six segments. Most species have paired cylindrical tubes or cones, the cornicles or siphunculi, projecting from the upper surface of the fifth or sixth abdominal segment. The last abdominal tergite is usually prolonged over the anus into a cauda.

Aphids are among the most important insect pests of plants, not only causing direct damage by feeding but indirect damage by transmitting viruses. Some, such as the peach-potato aphid, feed on a number of different plants, but other species, such as the sycamore aphid, *Drepanosiphum platanoidis*, feed only on one kind of plant. Many species feed exclusively on leaves and young shoots, others on the branches of woody trees and shrubs, others below ground on roots, and some live in galls.

Adult males and females may be winged or wingless, and the females may produce eggs or living young. Females may be sexual, requiring fertilization, or they may reproduce parthenogenetically.

Many species have a complex life cycle, the sexual generation living on one kind of plant, the primary host, and the parthenogenetic generations living on others, the secondary hosts. In some species, however, the sexual and parthenogenetic generations occur on the same host. By contrast, some species have no sexual phase and reproduce only by parthenogenesis. Species that alternate between hosts usually spend the winter as eggs laid on the primary host, often a woody plant. An example is the black bean aphid, *Aphis fabae*. Its eggs are laid on the spindle tree and on the sterile guelder rose. They hatch in spring to produce wingless females, which reproduce parthenogenetically and viviparously. Their progeny reproduce in the same way, but in the third or later generations winged parthenogenetic viviparous females are produced. These migrate to herbaceous secondary hosts, such as beans, sugar beet and poppies, during May and June, and establish colonies. Throughout the summer new generations of winged or wingless parthenogenetic viviparous females are produced on these plants, and the winged forms fly to establish colonies on other plants. Toward

the end of summer or during early autumn, winged males and winged viviparous females are produced, which return to the winter hosts and reproduce parthenogenetically to produce wingless sexual females, which mate, and eggs are laid.

Others live continuously on one kind of plant, such as trees or grasses. For example, the sycamore aphid lays eggs on the sycamore tree, *Acer pseudoplatanus*, and on related species.

Aphids are very weak fliers, and the distances they migrate depend on the strength and direction of the air currents they enter. The average duration of single flights seems to be between one and three hours; distances traveled may range from a few yards to hundreds of miles.

Aphids and related insects, such as scale insects and cicadas, excrete a clear fluid called honey-dew. From very early times deposits of honey-dew on or near plants have attracted interest, especially because of their large sugar content. The "mannas" collected and consumed by peasants in the Middle East are mainly the accumulated excreta of aphids and scale insects, the deposits hardening and crystallizing in the sun.

Ant attendance. Many aphid species are attended by ants which collect and feed on their honeydew, the aphids often being referred to as "ant-cows." Some species are more or less adapted to live with ants and do not thrive without them. This association is termed myrmecophily. Other species are merely visited by ants on their host plants, when the aphids benefit by the removal of the sticky honeydew and some by the protection afforded by the ants against parasites and predators.

Many of the subterranean aphids feed on the roots of plants that penetrate ant nests, and it is in this situation that aphids receive the best care and protection.

The main insect predators are ladybirds, hoverflies, lacewings and anthocorid bugs. Only the larvae of hoverflies and lacewings, but both larval and adult ladybirds and anthocorids feed on aphids.

APOSTLEBIRD, *Struthidea cinerea*, a babbler-like Australian species, one of three in the family of mud-nestbuilders. These fluffy gray birds, with short thick bills, are highly sociable, living in groups, and are named after the 12 apostles. When resting, they huddle together and preen each other. They live in dry areas of eastern Australia, feeding on a mixed diet of insects and seeds. They run clumsily, climb with short leaps and fly weakly. The sexes are alike. The nest is a basin-shaped mud structure reinforced with grass, built on a horizontal limb, in which the group nests communally. Another name for these birds is happy families, from the way they keep together. The name "apostlebirds" seems to be from the way they travel about in flocks of a dozen.

ARACARIS, small to medium-sized toucans of the forests of Central and South America, belonging to two genera, *Andigena* and *Ptero-glossus*. The plumage is usually mainly black dorsally. It can, however, be red, yellow and black. The long bill has some toothlike serrations. The diet is fruit, with some insects.

Wood ants "milking" black aphids.

Arapaima is the largest freshwater fish.

Aracaris are highly sociable and usually found in small flocks. They roost communally in cavities in trees, single groups apparently using a number of alternative roosts. The nest is an unlined hole in a tree, and two to four rounded white eggs are laid. The young develop very slowly, taking about one and a half months to fledge. They may be fed by a group of adults.

ARAPAIMA, *Arapaima gigas*, a large and rather primitive fish of the Amazon basin. The arapaima is the largest member of the family Osteoglossidae or bonytongues and is one of the largest of all freshwater fishes, reaching about 9 ft (2.8 m) in length (reports of fishes reaching 15 ft [4.3 m] are probably exaggerated). It has a long, sinuous body, with the dorsal and anal fins set far back, the scales on the body thick and large, and the mouth not protrusible. The family to which the arapaima belongs is of considerable scientific interest because it appears to have been fairly widespread in Eocene times but is now represented by species isolated in South America, Africa, Australia and parts of the Indo-Australian archipelago.

There are two species of osteoglossids in South America, the arapaima and the related arawana, *Osteoglossum bicirrhosum*. Curiously enough, the most suitable bait for the arapaima seems to be the arawana. The arapaima is an avid fish-eater, and even the armored catfishes are readily taken. The front part of the body is a bronze-green, but nearer the tail red patches appear, and the tail itself can be mottled orange and green. Arapaima live in murky waters, and the swimbladder has been adapted for breathing atmospheric oxygen, the adult fishes coming to the surface about once every 12 minutes to breathe.

Male arapaima guard the eggs, which are laid in holes dug out of the soft bottom of the riverbed. An interesting method has been evolved to prevent the newly hatched larvae from straying too far and thus being snapped up by predators. For three months the young fishes stay with the father, and during that time they remain close to his head. What keeps them there is a substance secreted by the male from glands opening from the back of the head. This was formerly thought to be a kind of "milk" on which the young fed, but it is now known to be merely a substance that attracts the young. Should the male be killed while looking after the young, the latter will disperse until they encounter another male and will join his brood. The males apparently do not eat the young of their own species.

The arapaima has a rapid growth, and a 6-in. (15-cm) specimen at the London Zoo grew to nearly 6 ft (1.8 m) in a period of six years.

ARCHERFISH, *Toxotes jaculatus*, and related species, perciform fishes from Southeast Asian freshwaters that shoot down insects with droplets of water. The archerfishes are small, rarely reaching more than 7 in. (18 cm) in length in the wild. They have fairly deep bodies, and the dorsal and anal fins are set far back near the tail. The body is generally silvery, with three or four broad dark bars on the flanks.

The archerfishes live in muddy water and swim just below the surface searching for insects that rest on leaves overhanging the water. When a suitable insect is spotted, the fish pushes its snout out of the water and squirts droplets at the insect until it falls into the water and can be eaten. A small archerfish can shoot only a few inches, but an adult can hit a fly up to 3 ft (90 cm) away. Their aim is remarkable when it is considered that their line of sight is diffracted as it passes from water to air. It has been found, however, that before "shooting" an archerfish maneuvers to place its body in as nearly a vertical line as possible, to minimize refraction. The jet of water is squirted between the tongue and the roof of the mouth. Along the tongue there is a ridge and above it a groove along the palate; water is forced through the narrow channel between the two and emerges as a fine stream.

Archerfishes will eat other foods if no insects are available, but they retain their shooting habits in the aquarium and can be trained to shoot for their food. They usually hit with the first shot but will alter their position in the water and try again if they fail. The act of shooting seems to be induced by hunger since a well-fed fish will not shoot whereas a hungry one will do so, even aiming at blemishes on the glass of the aquarium above the water line. The force of the jet cannot be controlled with any accuracy. On occasions, if the jet is too powerful, the insect will be knocked out of reach so that it would seem that instinct rather than learned skill plays a major part in their shooting abilities. Other fish, perhaps erroneously, have also been credited with shooting abilities.

ARGENTINES, living in the midwaters of the North Atlantic. They resemble salmon and trout in having a small adipose fin behind the dorsal fin. Two species are occasionally trawled off European coasts, *Argentina silus* and *A. sphyraena*. The former is the larger of the two and reaches 2 ft (60 cm) in length. Its flesh is very palatable and it is known to live in shoals off the west coast of Ireland where it may in the future form the basis for an important fishery. The name "argentine" comes from the intensely silver coloring of these fishes, particularly in the form of a wide band down the flanks.

ARGUS, a name given to certain species of the marine perchlike fishes known as grouper. The blue-spotted argus, *Cephalopholis argus*, is perhaps the best known. It is found in the tropical Pacific and reaches a maximum length of 18 in. (46 cm). The body is brown or olive, passing into a deep blue on the fins with many small light blue spots edged with darker blue over both body and fins. There are also four to six whitish circular bands around the body, but these fishes are renowned for their quick color changes and the bands will appear and disappear in rapid succession. The blue-spotted argus has now been introduced into Hawaiian waters where it was previously unknown.

ARMADILLO, the only armored mammal, related to the sloths and anteaters.

Armadillo armor is remarkably modified skin in the form of horny bands and plates connected by flexible tissue that enables the animal to move and bend its body. Narrow bands across the back break the rigidity of the armor there. The underparts are covered with soft-haired skin, and in many armadillos hair projects either between the bands or from under the side plates and legs. The hairs are grayish brown to white, and the armor is brown to pinkish. Most armadillos can draw their legs and feet beneath the shell, and a few can roll into a ball. The three-banded armadillo, *Tolypeutes*, of the Argentine pampas, however, is the only one that can completely close up into a perfect sphere, snapping the shell shut like a steel trap. As a result of the armor, armadillos are heavy animals. They are reputed to be able to walk on the bottom of ponds and other bodies of water, but they are also good swimmers.

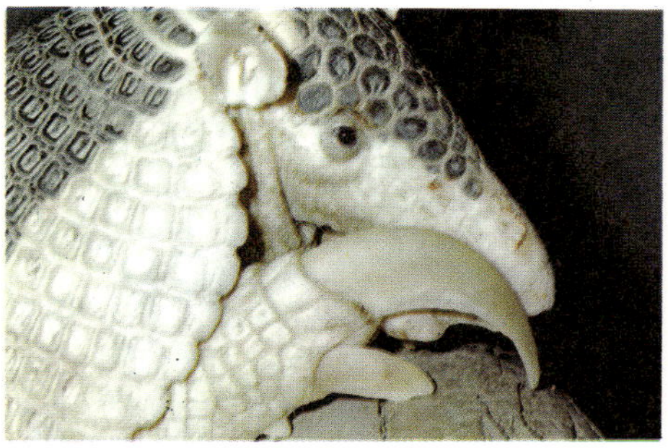

Armadillo's digging claw.

Armadillos have small ears, an extensible tongue, relatively good senses of sight, smell and hearing, and strong claws on the forefeet. They have normally 14 to 18 molars in each jaw, but the giant armadillo may have a combined total of 80 to 100. Armadillo teeth are among the most primitive found in mammals, being simply peg-shaped blocks of dentine without enamel, and they grow continuously at the roots.

The numerous species of armadillos range from southern Kansas and Florida to Patagonia. Most live in Brazil, Bolivia and Argentina. Only one, the nine-banded armadillo, *Dasypus*, reaches North America.

In the armadillos polyembryony (the production of two or more identical offspring from the division of a single fertilized ovum) commonly occurs, with as many as 12 young from a single fertilized egg, but the usual number per litter is two to four, and often only one embryo survives. The nine-banded armadillo, *Dasypus novemcinctus*, typically gives birth to quadruplets.

Armadillo young at birth are covered with soft leathery skin that gradually hardens into an armor.

Powerful diggers and scratchers, armadillos forage and root for a variety of food: insects and other arthropods, worms, eggs, small reptiles and amphibians, fruits, leaves and shoots, and even carrion. They generally walk on the tips of the claws and the entire soles of the feet. Most can run rapidly. They may be nocturnal or diurnal, and their home is an underground burrow dug with their claws or one abandoned by other animals.

ARROW WORMS, a genus of small wormlike creatures, most of which spend their lives drifting in the surface waters of the sea, therefore holoplanktonic. Three species are carried in deep oceanic currents and are bathypelagic, the largest of which, *Sagitta maxima*, is only 2 in. (5 cm) in length. All are transported long distances, in the characteristic fashion of planktonic animals, as their swimming speed is exceeded by the movements of the environment in which they live. Apart from the head and the posterior reproductive region, the arrow worm is largely transparent. On either side of the mouth there are hooklike bristles that are used as jaws; hence the name of the phylum to which it belongs, Chaetognatha (*L. chaeta*, bristle; Gk. *gnathos*, jaw). The body is flattened with thin lateral extensions or fins and a horizontal tail fin. The transparent body permits an all-around vision through spherical eyes, befitting an animal living in suspension and surrounded by both food and predators. Arrow worms feed on other animals of the plankton by suddenly darting toward them with an up-and-down flip of the tail. Animals of similar size are seized with the jaws and swallowed whole. At certain times of the year arrow worms are responsible for mortality in developing fish fry. They are, in their turn, consumed by the larger predators of the plankton, including jellyfishes and comb-jellies, together with plankton-eating fishes such as the herring.

Arrow worms are hermaphrodite, the ovary lying in front of the testes. These are connected with a bulge in the body surface behind the lateral fins where sperm can be stored. The male gonads mature first. Either cross-fertilization with another individual occurs or the sperm is stored until the eggs have matured when self-fertilization takes place. In the deep-

water *Eukronia hamata* the lateral fins form a brood pouch under the body. First the eggs and then the young are carried in it. Surface arrow worms of the Antarctic move into deeper water to breed.

ASCARIS, *Ascaris lumbricoides*, a roundworm of man, it is one of the largest and most widely distributed of human parasites. Infections are common in many parts of the world and the incidence of parasitism in populations may exceed 70%; children are particularly vulnerable to infection. The adult parasites are stout, creamy colored worms that may reach a length of 1 ft (30 cm) or more. They live in the small intestine, lying freely in the cavity of the gut and maintaining their position against peristalsis (the rhythmic contraction of the intestine) by active muscular movements.

The adult female has a tremendous reproductive capacity, and it has been estimated that one individual can lay 200,000 eggs per day. These have thick, protective shells, do not develop until after they have been passed out of the intestine, and for successful development of the infective larvae a warm, humid environment is necessary. The infective larva can survive in moist soil for a considerable period of time (perhaps years), protected by the shell. Man becomes infected by accidentally swallowing such eggs, often with contaminated food or from unclean hands. The eggs hatch in the small intestine, and the larvae undergo an involved and an as yet unexplained migration around the body before returning to the intestine to mature. After penetrating the wall of the intestine the larvae enter a blood vessel and are carried in the bloodstream to the liver and thence to the heart and lungs. In the latter they break out of the blood capillaries, move through the lungs to the bronchi, are carried up the trachea, swallowed and thus return to the alimentary canal. During this migration, which may take about a week, the larvae molt twice. The final molt is completed in the intestine, and the worms become mature in about two months.

An infected person may harbor one or two adults only and, as the worms feed largely on the food present in the intestine, will not be greatly troubled, unless the worms move from the intestine into other parts of the body. Large numbers of adults, however, give rise to a number of symptoms and may physically block the intestine. As in trichiniasis, migration of the larvae around the body is a dangerous phase in the life cycle, and, where large numbers of eggs are swallowed, severe and possibly fatal damage to the liver and lungs may result. Chronic infection, particularly in children, may retard mental and physical development. Medication is often effective in removing the adult worms, although surgical removal may be necessary when the worms have moved into other organs. In regions where infection is common, hygienic measures, particularly with regard to treatment and disposal of night soil, are essential in order to prevent reinfection.

Ascaris lumbricoides has probably evolved from the pig roundworm, *Asuum*, which is similar in many respects. Cross-infection between man and pig has been shown to take place from time to time, but the disparities between the organisms limit its extent.

ASITYS, two species of a family long isolated in Madagascar. The family also includes two species of false sunbird, *Neodrepanis*, which were classified as true sunbirds (Nectariniidae) until anatomical studies linked them with the asitys.

Asitys are plump, rather long-legged, arboreal birds 5-6 in. (13-15 cm) long. Very little is known about Schlegel's asity, *Philepitta schlegeli*, which inhabits the humid forests of western Madagascar, and only a little more about the velvet asity, *P. castanea*, which replaces it in the forests of eastern Madagascar. The male velvet asity is black with yellow fringes to its feathers when freshly molted, the male Schlegel's asity mainly yellowish with a black crown. Males of both species have a large greenish wattle, which surmounts the eye in the former species and surrounds it in the latter. The females and young of both species are duller and mainly greenish. The velvet asity builds a suspended, pear-shaped nest of moss and fibers in which it lays three white eggs. Both species feed mainly on small fruits and are quiet, solitary birds, but occasionally associate with flocks of other species.

ASS, a horselike animal but with longer ears and differing in color and size. All wild asses, *Equus asinus*, are very similar in appearance. The wild ass existed in three subspecies in the deserts and semideesrts of northern Africa, but has been largely exterminated by man. Their range has been restricted to the Danakil region

The African wild ass.

of Ethiopia, where the total population numbers an estimated two to three hundred Nubian wild asses, *E. a. africanus*, and to northern Somalia, the Somali wild ass, *E. a. somalicus*. There may also be some asses left in the Tibesti mountains of the central Sahara. It has not yet been ascertained that these asses are pure-blooded wild asses; they may well have interbred with escaped domestic donkeys. A few breeding herds of African wild asses are kept in zoological gardens.

African wild asses are gray or reddish-brown and have a dark cross on their withers, one line running down the back, the other at right angles down to the shoulders. They often have dark, horizontal stripes on their legs, especially on the outer side, possibly a vestigial, zebralike stripe pattern. They have a shoulder height of 48 in. (1.2 m).

The Asiatic wild asses are better known under

the vernacular names of their subspecies: the onager, *E. hemionus onager*, of Iran, Afghanistan, Turkmenia and Rann of Kutch, western India; the kulan, *E. h. hemionus*, of Mongolia, and the kiang, *E. h. kiang*, of Tibet and the Himalayas. A further subspecies, the Syrian wild ass, *E. h. hemippus*, is now extinct. The onager, shoulder height 48 in. (1.2 m), is light yellowish-brown in summer, but darker in winter. Its underparts, legs and muzzle are white. The kulan, shoulder height 50 in. (1.35 m), is darker and more reddish-brown. The kiang, shoulder height 54 in. (1.4 m), is the tallest of the Asiatic wild asses. Its coat is pale chestnut brown in summer and more reddish in winter.

Originally, the Asiatic wild asses had a range from Arabia, Syria and Turkey to northwest India and from the southern European USSR to Tibet and Mongolia. During the last few centuries they have been persecuted by man, being competitors of domestic stock for food and water, and their present range is now restricted to a few isolated areas. All subspecies can be considered to be in danger of extinction, and of the onager alone there is a population of several hundred in a wildlife reserve, Badchys, in Turkmenistan, USSR. Another population of 800-900 head lives in the Little Rann of Kutch in India, where it is protected due to the vegetarian habits of the surrounding human population. Small numbers live in Iran and Afghanistan. The kulan is apparently limited to central Mongolia; no recent information is available on its numbers and status, and there are probably not more than a few hundred survivors of this race. The same is true for the kiang.

All wild asses are inhabitants of steppes, semi-deserts and even deserts, but they depend on surface water as they need to drink every two or three days. The Asiatic wild asses live in groups of up to 15 head, which, during seasonal migrations, may join up to form larger herds. As in other equids, there is no set breeding time and foals are born in any month of summer.

Virtually nothing is known of the African wild ass's habits in the wild. Their social structure seems to resemble that of the Plains zebra. Little is known of the social structure of the Asian wild ass, but it is known that the groups consist of a stallion, a few mares, and their young, but these units do not seem to be permanent: mares have been observed to separate from their group when foaling. Normally, the stallion leads the group, but when they are attacked, the stallions keep to the rear and defend their mares and foals. During the rut, the young males are chased from the group by the stallion.

ASSASSIN BUGS, or kissing bugs, blood-sucking conenoses or masked hunters, predacious bugs usually living on the blood of other insects, but also attacking the higher animals and man. The 3,000 species are rather flattened, roughly oval-shaped insects, with small heads bearing protruding eyes and long antennae and with a prominent snout. The wings are folded flat on the back, and the six long legs give the insect a superficially spider-like appearance. Many species found in human dwellings may be from $\frac{1}{2}$-$1\frac{1}{2}$ in. (1.3-3.8 cm)

Avocet, a handsome wading bird with a turned-up bill and the victim of marsh drainage.

long. The mouthparts, as with all members of Hemiptera, are tubular and adapted for piercing the host's body and sucking its blood. When the insect is not feeding, the mouthparts, which are termed collectively the rostrum, are carried out of sight underneath the head and body. The edges of the abdomen are flattened as thin plates visible at the sides of the closed wings, and in some species the plates are brightly colored with red, yellow or pink, although more generally the whole insect is brownish.

The majority of the assassin bugs are found in the New World, especially Central America and northern South America. Other species occur in Europe, Africa (including Madagascar) and southern Asia.

Assassin bugs are active runners and good fliers. Most species live in the nests or burrows of rodents, armadillos and opossums. A few live in close association with man, spending the daytime in crevices in roofs and walls and feeding at night by sucking the blood of sleeping humans or domestic animals.

AUSTRALIAN MAGPIES, a family (Cracticidae) of birds peculiar to Australasia, having mainly black, gray and white plumage and a generalized resemblance to crows or large shrikes. The name is sometimes restricted to species of the genus *Gymnorhina*. All have loud and often melodious calls. They occur in a wide variety of habitats wherever some trees are present, being tree nesters. They make a typical cup-shaped nest built mainly of sticks with a lining of finer material. The eggs have a brown, green or bluish ground color, patterned with dark blotches or spots.

Species of the genus *Gymnorhina* are sometimes known as piping crows or bell magpies, more often simply as magpies, and are the most crowlike in size, shape and movement, but have strikingly pied plumage, the head, wings and tail being black and normally most of the body being white. The pattern varies with locality and with age, and from one to three different species are recognized. They feed mainly on the ground, walking about like

crows, and take mainly insects but also any other small creatures that are available. Instead of nesting in pairs the birds usually belong to a larger social group that defends a communal territory against other similar groups. The individual females make their own nests within the territory, caring for the eggs and young, the males helping when the young fledge. They are generally aggressive birds. The groups indulge in loud melodious calling in chorus.

AUSTRALIAN TREE CREEPERS, six species of finch-sized, slender-billed, strong-footed birds that are unrelated to other tree creepers but resemble them in taking insects from crevices while spirally ascending vertical tree trunks or branches. However, they also feed on the ground. When climbing, their tails are not pressed to the tree but held away from it, and they perch crossways on twigs like normal perching birds. The plumage is gray or brown with streaked areas, and light buff patches are visible on spread wings in flight. The nest is a cup built in a hole or hollow branch, and the eggs are heavily spotted.

AUSTRALIAN WRENS, a subfamily of small, slender-billed insectivorous birds usually regarded as warblers although some may be related to babblers. Their habitats range from forests to deserts with sparse vegetation. They build domed nests. The best known, the blue wrens, *Malurus* spp., have vividly colored males, and additional birds assist nesting pairs.

AVADAVAT, two species of Asian weaver finch frequently kept in captivity by aviculturists. The term is usually restricted to the red avadavat, *Estrilda amandava*, also known as the tiger finch, strawberry finch or red waxbill. The avadavat is small, about the size of the wren, and differs from other members of its family in having both breeding and non-breeding plumages. In the breeding season the male is basically a bright coppery or crimson red.

AVOCETS, wading birds of the genus *Recurvirostra*, which, with the stilts, make up the family Recurvirostridae. The four species are characterized by a long, slender, markedly upcurved bill and by long legs in proportion to body size; the legs trail behind the tail in flight. The plumage is chiefly black and white. The avocet, *R. avosetta*, breeds in the palearctic and Ethiopian faunal regions, chiefly on the coasts of the North Sea, the southern Baltic and around the Mediterranean, but also inland, in the Balkans and from the Black Sea south to Iraq and east to Mongolia. A few breed also in South Africa. Thus the temperatures tolerated by this species at its breeding grounds vary from about 60-90°F (15-32°C). The palearctic populations winter chiefly in Africa, mainly on the salt and alkaline lakes of the East African Rift Valley, and in Southeast Asia. Those wintering in South Africa may be only local birds. In recent years some have wintered on estuaries in the southwest of the British Isles.

Avocets nest in colonies on muddy or sandy islands, sometimes covered with vegetation. They use little or no nest material on the sandy sites, but make a lining of dead plants on muddy surfaces. They lay a single clutch of usually four eggs, though replacement clutches are often laid if the first clutch or young nestlings are lost.

Avocets frequent shallow lakes, marshes or mudflats at all times of year, with a preference for saline or brackish conditions, particularly in the breeding season. They feed on small crustaceans, by sweeping the slightly open bill from side to side, sifting the first few inches of water or the liquid top surface of muddy deposits. Sometimes they take insects from the surface, but usually they wade deep into the water, which may reach well above their intertarsal (knee) joint.

AXOLOTL, *Ambystoma mexicanum*, neotenous larva of the mole salamander, is confined to certain lakes around Mexico City. It measures 4-7 in. (10-17.5 cm) in length and is usually black or dark brown with black spots; but white or pale pink albino forms are common. The legs and feet are relatively small and weak, but the tail is long. A fin runs from the back of the head along the animal's back to the tail and then under the tail. Breathing is by three feathery gills on either side just behind the head and immediately in front of the forelimbs. During mating, the male attracts the female by an elaborate courtship display as he secretes certain chemicals from abdominal glands and makes violent movements of the tail, thought to serve in dispersing the chemicals. Sperm is deposited in a packet or spermatophore which the female picks up from the bottom with her cloaca. Fertilization is therefore internal. The

female lays 200-600 eggs in April-May and the young hatch 2-3 weeks later, reaching their maximum length by the winter, when they hibernate. They eventually become sexually mature and are able to breed while still in a larval condition. That is, they are neotenous. The axolotl was assigned to a separate genus, *Siredon*, until it was discovered that under certain conditions it could lose its external gills and larval characteristics and metamorphose into an adult salamander. The neoteny of the axolotl seems to be due to a deficiency of iodine in the water.

In fact the word "axolotl" is Mexican for "water sport," and the animal is eaten as a delicacy.

In some instances the term "axolotl" may be used in a more general way for the neotenous larvae of other salamanders. For example, the larvae of the tiger salamander, *Ambystoma tigrinum*, which develop in cold water at high altitudes, fail to metamorphose and are often referred to as the "axolotl" of the Rocky Mountains.

AYE-AYE, *Daubentonia madagascariensis*, a nocturnal mammal, about the size of a cat, with coarse black hair, large membranous ears, big forward-directed eyes and a bushy tail. The head and body length is 16 in. (40 cm) and the tail is 22 in. (55 cm). This is perhaps the most peculiar member of the order of primates. Its appearance and nocturnal habits make it eerie enough, and the presence of an elongated third finger on each hand adds a bizarre touch. It is therefore not surprising that villagers in Madagascar regard this animal as an evil omen. If one is found near a village, it is either killed, or the village is abandoned.

The aye-aye is restricted to Madagascar like all other lemurs. It is a rain-forest form, and its natural distribution range is the northeastern coastal forest. However, destruction of the forest habitat of the aye-aye has led to the virtual extinction of this remarkable species. Recently, a dozen aye-ayes have been introduced on the island reserve of Nosy Mangabe in the Bay of Antongil (northeast Madagascar), and this may well be the only chance of preserving the species.

Aye-ayes typically occur high up on the vertical trunks and larger branches of trees in the rain forest. They have a mixed diet of insect and plant food, but most of their energy is spent in hunting insects. Here, the two main peculiarities of the aye-aye are brought into play. Unlike other lemurs, *Daubentonia* has two continuously growing incisors in both upper and lower jaws. The canines are missing, and there is a gap (diastema) between the incisors

and the premolars. This typical gnawing type of dentition is combined with the long, thin middle finger of each hand in the search for woodboring insect larvae. The aye-aye seems to use both its sense of smell and its large, mobile ears to locate the larvae beneath the bark of dead branches. The incisor teeth are then used in a rapid, intense fashion to make a hole, and the thin middle finger of one hand is used to hook out the larva. An early report suggested that the aye-aye uses its elongated finger to tap on the bark when searching for larvae, but this is probably untrue.

Aye-ayes build complicated nests in bowllike forks of trees, about 40 ft (12 m) up and consisting of a framework of thin branches interwoven with leaves. Within a given home range, several nests are found, and it is likely that the aye-aye moves from one to the other for varying periods.

Like most lemurs, the aye-aye has one offspring at each birth, and it seems that this is carried around on the mother's back in the manner typical of the larger lemurs.

The aye-aye is nocturnal and therefore difficult to photograph in its natural surroundings. The author succeeded, after much waiting, in photographing one in Madagascar while it was eating fruit.

AYU, the common name for *Plecoglossus altivelis*, a peculiar salmonlike fish found in Japan. Most of the salmonid fishes have pointed teeth but the ayu, among other anatomical peculiarities, has platelike teeth. Because of this, it is placed in a family of its own, the Plecoglossidae.

The ayu, which grows to about 12 in. (30 cm), migrates into fresh water to spawn. During the upstream migration, Japanese fishermen bring their trained cormorants at night to the rivers, attach rings round the throats of the birds and then release them into the water. The ring prevents the cormorant from swallowing the ayu, and on the return of the bird the fisherman takes the fish from the cormorant and sends it off on another foray. In this way up to 50 ayu can be caught by one cormorant in a night. Cormorant fishing is now being replaced by other methods and is little more than a tourist spectacle.

The axolotl can be made to turn into a typical salamander.

B

BABBLERS are noisy birds, but otherwise inconspicuous inhabitants of forest and thick scrub. The majority occur in flocks, often within the mixed species flocks that are so characteristic of tropical forests. For the most part, babblers are insectivorous, though a number of species, particularly among the song babblers, supplement their diet with fruit outside the breeding season.

The jungle babblers are rather nondescript brownish species, which can be divided into two distinct ecological groups. For example, species of *Pellorneum* and *Trichastoma* are mainly terrestrial and have short wings and tails but long legs and strong feet. On the other hand, species of *Malacopteron* are more arboreal, and consequently have longer wings and tails but shorter legs. Jungle babblers occur in both the Ethiopian and oriental regions, extending as far eastwards as the Philippines and Celebes.

Scimitar babblers and wren babblers, though closely related, differ considerably in appearance. Typical scimitar babblers, such as species of *Pomatorhinus*, have long curved bills, long tails and short legs, while typical wren babblers, such as species of *Ptilocichla* and *Napothera*, have short straight bills, short tails and long legs. Scimitar babblers forage acrobatically around branches and creepers, probing with their long bills and using their long tails as a balance. By contrast, wren babblers are terrestrial, as their long legs and short tail would suggest. They occur from the Himalayas to Australia.

The tit babblers and tree babblers, as exemplified by the genera *Macronus* and *Stachyris*, are more uniform in size and proportions than the previous groups. They are mainly arboreal and forage in an acrobatic titlike manner. They occur mainly in the oriental region, where they range from India to the Philippines, Borneo and Java, though four species of *Neomixis* are confined to Madagascar.

The song babblers range in size from about 4-12 in. (10-30 cm). Species of *Turdoides* are characteristic of the Middle East and the drier areas of the Ethiopian and oriental regions, where they occur in almost any kind of thick scrub. Most other species occur in forest, by far the majority of them in the oriental region. They include the laughing thrushes, *Garrulax*, most of which are large and noisy, while many are beautifully patterned. A few sing particularly well and are much prized as cage birds in China. Species of *Pteruthius*, *Myzornis* and *Leiothrix* are also prettily marked, and one of the latter,

the red-billed leiothrix, *L. lutea*, is a very popular cage bird more commonly known as the Pekin robin.

The rockfowl were formerly thought to be related to starlings (Sturnidae) but are currently thought to be large and specialized babblers. The two species have heads completely devoid of feathers, hence their alternative name of bald crows. In one species, the gray-necked rockfowl, *Picathartes gymnocephalus*, the bald head is bright yellow; in the other, the white-necked rockfowl, *P. oreas*, it is pink. Both species have short, rounded wings, a long tail and long legs and are said to move gracefully on the ground by means of enormous hops. They live in the forests of West Africa, but only in areas in which large moss-covered boulders, cliffs and caves abound.

The ground babblers are another particularly diverse group. The group includes among its 20 or so species the rail babblers, *Eupetes*, and quail thrushes, *Cinclosoma*. The ground babblers are confined to the Australasian region, with the exception of a rail babbler, *E. macrocercus*, which lives in the forests of Malaysia.

Stachyris striata, one of the tit-babblers at its nest among reeds.

The jungle babblers, scimitar babblers and wren babblers, tit babblers and tree babblers, and a very few of the song babblers and ground babblers, build domed ball-like nests of moss and dead leaves, which are well hidden on or close to the ground. The majority of the song babblers build open cup-shaped nests, which are hidden in low trees, bushes and creepers, while

a few of the ground babblers, such as the quail thrushes, build open cup-shaped nests on the ground. The rockfowl differ from other babblers by nesting in small groups and by building cup-shaped nests of mud, plastered to rock faces in caves or under overhangs.

BABIRUSA, *Babyrousa babyrossa*, a large almost hairless hog with a brownish gray rough skin. The head and body length is 34-42 in. (86.4-106.7 cm); height at shoulder is 25-31 in. (63.5-78.7 cm). The tail is 10.8-12.5 in. (27.4-31.7 cm). The nose is typically hoglike with a cartilaginous mobile disc. There are four toes on each foot with the center two functional and the outer two forming dew claws. The dental formula is as follows: incisors $\frac{2}{3}$; canines $\frac{1}{1}$; premolars $\frac{2}{2}$; molars $\frac{3}{3} \times 2 = 34$. Unlike other members of the family, the upper tusks in males go up through the top of the muzzle and curve slightly backwards. The lower tusks do not touch the upper tusks and extend outward.

The babirusa is limited to the Celebes and Molucca Islands including Toguin, Buru and the Sulu Islands.

Information on breeding is scarce. In the London Zoo captive females have given birth to single young in January, March and April following a 125-150-day gestation period.

Babirusa inhabit dense, damp forests and bamboo thickets. Like most other hogs, they are social animals and travel in small groups. They feed mostly on fruits, green vegetation, roots and tubers.

BABOONS, typical open-country monkeys of Africa, distinguished by their long muzzle, especially in the male, and their large size. The most widespread species, *Papio hamadryas*, is distributed all over the savanna, semidesert and lightly forested regions of Africa south of the Sahara. There are also two species in the thick forests of the Cameroun region with very short tails, brightly colored buttocks and heavy ridges on either side of the nose. Savanna baboons are not highly colored, have no strong ridges, and their tails curve up, back and down in an arc.

It was once thought there were five species of baboon in the savanna areas, each confined to a portion of the range; but recent studies have demonstrated the existence of intergradation between them, with hybrid troops and clines of color and hair patterns through the range. To the north they are more stocky, the males have big manes, the face juts and the pointed nose extends beyond the end of the muzzle. They are light in color, the hairs being banded dark and light. To the south, baboons are more long-legged and rangy, the male has little or no mane, the face is more bent downward, and the nose does not overhang the end of the muzzle. The color is darker and the hair less banded. Eight subspecies are recognized:

Hamadryas, mantled or sacred baboon, *P. hamadryas hamadryas*. Rather small, 2½ ft (76 cm) long, tail 2 ft (61 cm), weight up to 40 lb (18 kg); females brown, males gray; males with a huge mane which ends in middle of back, very short hair on rest of body. Face pinkish, ischial callosities red. Northern Somalia, Eritrea and southwestern Arabia.

Olive baboon, *P. h. anubis*. Much larger, olive green; male with long mane. Face and ischial callosities black. Senegal east to Ethiopia and Kenya intergrades with hamadryas along Awash River, northern Ethiopia.

Guinea baboon, *P. h. papio*. Smaller, maned, reddish, with red ischial callosities. Restricted to small area of Senegal, Guinea and Gambia.

Yellow baboon, *P. h. cynocephalus*. Long-legged, no mane, yellowish. East Africa, from Amboseli (where it intergrades with olive baboon) to Zambezi.

Dwarf baboon, *P. h. kindae*. Like yellow baboon but much smaller and short-faced; Zambia and Katanga.

Kalahari chacma baboon, *P. h. ruacana*. Variegated, brownish, long face and long crest of hair on nape and withers. Southern Angola and Southwest Africa.

Gray-footed chacma, *P. h. griseipes*. Dark yellowish, larger than Kalahari chacma, otherwise similar. Rhodesia and Mozambique; intergrades with yellow baboon in eastern Zambia.

Black-footed chacma, *P. h. ursinus*. Blackish brown, with black hands and feet; largest of all baboons. South Africa, except for northern Transvaal.

Baboons, aggressive and dangerous, live in large troops whose composition and social behavior vary from area to area. In the savannas of East and South Africa the troops are smallish with 20–80 animals, and each troop stays together all the time. However, in the desert areas of Eritrea, northern Somalia and Southwest Africa, huge troops are found, as many as 150 together, but these split up during the day into one-male units and bachelor bands. This latter type of society, typical especially of hamadryas baboons, has a number of special features and must be described separately; the more typical

Baby baboon riding pick-a-back.

A babbler, the white-necked rockfowl *Picathartes oreas*.

savanna social organization will be described first.

Savanna baboons have a more or less hierarchical organization within the troop, a single male often being dominant in all respects. However, different individuals may appear to dominate in different situations, and sometimes tow or three of the less strong males may constantly associate together and take precedence in situations where a single stronger male would otherwise always dominate. Females are always subordinate to males, and have a kind of rank order among themselves, although this too may be broken: for example, a female with a young infant is treated with consideration by the rest of the troop.

Within the baboon troop the sex ratio is more or less equal; there are, however, two or three times as many adult females as there are adult males, since females mature at 4–5 years, males only after 7–10 years. The troop has a wide home range, with a core area consisting of a clump of trees. Around this, the troop wanders within a radius of a mile or more, and may travel as much as 6 miles in a day, foraging as they go. Baboons are omnivorous. They feed on seeds, tubers, grass, insects, even scorpions and snakes. Baboons have also been observed to

kill and eat young gazelles; they do not stalk them with intent as chimpanzees do, however, but probably merely stumble on a baby gazelle that has been left lying in the grass by its mother.

By day, a baboon troop will often associate with a herd of ungulates, especially impala. The association is mutually beneficial: the impala are alert and give warning of danger, while the baboons are powerful and fearsome and can offer protection to the antelopes. Leopards and cheetahs are not infrequently turned away by a group of male baboons. On the other hand, a lone baboon, even a male, may be killed and eaten by a leopard. On the Serengeti, baboons form 4% of the leopard's diet.

The baboon troop is very tightly knit; an individual very rarely changes troops. A very big, unwieldy troop will split into subgroups for foraging, and gradually two independent troops will result. Dominant males and females with infants travel in the center of a troop; younger and weaker males tend to move along the edges, and are the first to see and warn of danger. When this happens, the big adult males move forward. Like all Old World monkeys, a male baboon's canines are long, sharp and fearsome, with a razor edge up the back.

Gestation lasts six months; the infant is born black and is a focus of solicitous attention from other troop members. At first it rides on its mother's belly, later on her back. After four to six months the infant changes to the adult coloration and a little later is weaned.

Although olive baboons penetrate far into the forests of the northeastern Congo, the real forest baboons are the short-tailed drill, *Papio leucophaeus*, and the mandrill, *P. sphinx*. Mandrills are even larger than most baboons, dark brown with white cheek fringes, a yellow beard, and tuft of hair on the crown. The face is brilliantly colored in the adult male; the nose is red, the ridges on either side of it blue. In females and young the same colors are present, but much duller. The penis is red and the scrotum is blue; the ischial callosities are red, and the buttock hair blue and white. The maxillary ridges are diagonally grooved. The drill is smaller than the mandrill, more olive in color, with a black face and no grooves on the muzzle ridges; the chin is red; there is a white fringe all around the face.

The mandrill lives in Cameroun, Equatorial Guinea, Gabon and the Congo (Brazzaville), mainly near the coast. The drill is found in the same areas but prefers more inland forests, and extends onto Mt. Cameroun, the Cross River district of Nigeria, and the island of Fernando Po. Both live mainly on the forest floor but do not hesitate to climb trees, eating fruit and berries as well as bark and roots. The troops are large, up to 60 in number, but split into one-male parties at times (at least in the drill), which forage separately for several days. Intertroop location signals are given as they come back together again.

BACKSWIMMERS, common name for a group of aquatic bugs. They resemble water boatmen, family Corixidae, in general appearance, for their body is boat-shaped and streamlined and they swim by using their long hind legs as paddles. However, they differ from water boatmen in swimming on their backs and are vicious predators, often attacking and killing animals larger than themselves, including young fish. The sharp stylets in their mouthparts can inflict painful wounds, and they should be handled carefully.

BADGERS, medium-sized members of the weasel family Mustelidae, well known for their digging habits. Badgers comprise six genera, which are distributed over North America, Eurasia and Indonesia as far as Borneo. The South African ratel is sometimes included as a seventh genus. The name is derived from the French *bêcheur*, meaning digger or gardener. All badgers are heavily built animals, and their muscular legs appear even shorter than they are, due to the thickness of the coarse fur. The tail, usually shorter than in other mustelids, is 1-8 in. (2-20 cm) long. All badgers have potent anal scent glands, which appear to be particularly effective in the oriental stink badgers *Mydaus* and *Suillotaxus*. Considerable size and color variations exist. Three genera, *Meles*, *Taxidea* and *Melogale*, have distinct black-and-white facial "masks" contrasting with the inconspicuous salt-and-pepper gray of the back and the

Backswimmer, photographed from below. The irregular shape at the upper right is a surface reflection.

A pair of European badgers *Meles meles* leaving their set at nightfall.

black underparts and legs. Badgers usually excavate their burrows or sets where there is sufficient cover, in woodland or, in drier areas, brushland, and choose sandy limestone or clay. The European badger *Meles meles* is gregarious, and the sets may become, with successive generations, an extensive maze of tunnels up to 100 ft (30 m) long, leading to chambers lined with a "bedding" of leaves or grass. Sometimes up to ten entrances and exits can be discovered, but usually only two or three are in use at one time. Small ventilation holes air the tunnels closest to the surface. Dry bundles of bedding are collected, either pushed or carried backward into the set, clasped in the forearms, the badger shuffling along on elbows and hind feet. Bedding may also be nosed to the surface during the night and spread around the set to air, then

gathered again before daybreak. Not only do badgers keep the set clean underground but they also dig shallow pits nearby where dung is deposited.

The diet of this omnivorous mammal covers a wide range of vegetable and animal matter, showing great seasonal and local variations. Voles, hedgehogs and moles may be eaten as well as smaller frogs, slugs, beetles and, above all, earthworms. Fruits of every kind, maize, wheat and even fungi are also taken when available.

The European badger is found in Eurasia down to southern China. Males are slightly larger than females, measuring 27-40 in. (67-100 cm) and weighing 22-55 lb (10-25 kg), depending on the season. Large layers of fat are present in autumn. Hibernation occurs only in the colder

◁ The male olive baboon with cape-like mane.

areas of their range. Breeding usually occurs in spring or late summer, but there is delayed implantation varying from 2-10 months. The embryo develops for a few days, then lies dormant and is not implanted in the uterus until much later. The cubs are born in February or March. Numbering two to four, the blind cubs remain in a nesting chamber for several weeks. The American badger, *Taxidea taxus*, occurs throughout the drier regions of the United States down to southern Mexico. Stockier than the European badger, it appears to be even more "flattened." The black and white facial stripes are narrower and less distinct. Measuring 21-35 in. (52-87 cm) and weighing 8-22 lb (3.5-10 kg), they are usually smaller than the European badger. Unlike its European counterpart, the American badger is solitary most of the year and does not use the same burrow generation after generation. Breeding takes place in late summer and delayed implantation occurs, so that the cubs are not born until the following April. Weaned at six weeks, the young become independent by late autumn and set off on their own for the winter.

The hog badger, *Arctonyx collaris*, is found from India to Sumatra and up into China. More upstanding, albeit stout, this species nevertheless retains the same stocky appearance, with formidable claws. A large hog badger measures approximately 32 in. (80 cm) overall and weighs 25 lb (11 kg). The most distinguishing characteristic is the long, mobile snout, used in foraging for grubs and insects.

The Malayan stink badger, *Mydaus javanensis*, and the Palawan stink badger, *Suillo taxus marchei*, are the rarest members of this group. The Malayan species, which occurs also in Sumatra, Java and Borneo, has an overall dark brown or black coloration.

The Bornean ferret badger, *Melogale orientalis*, is widespread from Nepal to Borneo, and three species are currently recognized. This is the smallest badger, measuring 19-25 in. (48-63 cm), the bushy tail accounting for a third of the total length.

Bald eagle, America's national emblem.

BALD EAGLE, *Haliaetus leucocephalus*, a very large sea eagle confined to North American coasts, lakes and rivers, and the American national emblem. It feeds by fishing or by robbing other birds. Two races occur, one in central and southern United States, the other in Alaska and western Canada. The Florida population breeds in winter and migrates north in hot sum-

The bandicoot, which has earned a bad name by digging in lawns in search of insects.

mers. The Alaskan birds feed largely on dead and dying Pacific salmon for part of the year. Formerly common, it is now much reduced in central south U.S.A. through shooting (despite national status) and pesticide poisoning. It breeds in trees, making huge nests, laying two or three eggs and rearing one or two young per year.

BANDED ANTEATER, *Myrmecobius fasciatus*, or numbat, a pouchless, termite-eating marsupial that lives in the fallen hollow limbs of wandoo trees, *Eucalyptus redunca*, in southwestern Australia. The numbat shares common features of behavior, reproduction and chromosome cytology with the marsupial "cats" and their allies, to which it is undoubtedly closely related in spite of numerous aberrant features of dentition. Adult animals are rat-sized and weigh about 1 lb (454 gm).

The coat of the numbat is coarse, but it is nonetheless a remarkably beautiful animal, having bright rusty red fur broken by six or seven creamy white bars across the hind part of the body. The head is flat above, the nose of the adult animal long and pointed and the tail long and wandlike, especially when, as sometimes happens, it is held above the body with the hairs erect. There are five toes on the forefeet and four on the hindfeet. The skull is broad and expanded with four incisors, a single canine and three premolar teeth on each side of the upper jaw. As in all dasyurid marsupials, the lower jaw has only three incisor teeth and the same number of canines and premolars as in the upper jaw. Unlike most other marsupials, however, the numbat has more than four molar teeth in each jaw, and they are reduced in size and degenerate in accord with the insectivorous mode of feeding. There may be as many as six molars in each jaw, giving a total of about 50 teeth.

BANDICOOT, a rabbit-sized marsupial, insect and small animal eater with carnivorous dentition, as in native cats, but with the same foot structure as in kangaroos and wallabies (syndactylous). The bandicoots have many unique features and their earliest fossil history is un-

known so relationships with the other marsupials are problematical, although they are perhaps most nearly related to the dasyurids (native cats, etc.). They have pointed ears and tapering snouts.

The name "bandicoot" was apparently first applied to large rodents (genus *Bandicota*), which inhabit southern Asia. The word means "pig rat" in Telugu, a Dravidian language of India. The marsupial bandicoots are quadrupedal, but with the hindlimbs enlarged and carrying most of the body weight, while the forelimbs are used for scratching and digging. The insect food is mainly taken from the top few inches of soil or from rotting wood on the surface of the ground. Bandicoots show a remarkable amount of growth after sexual maturity, and the 20 or so species cannot be separated by size alone. The smallest bandicoot is *Microperoryctes* of west New Guinea Mountains, less than 1 ft (30 cm) long, with a hindfoot length of 1.2 in. (3 cm), while the largest forms of short-nosed and rabbit-eared bandicoots are up to 2½ ft (75 cm) long and have foot lengths three times greater. The tail is much shorter than the body length in all except rabbit-eared bandicoots. The predominant dentition of five upper and three lower incisor teeth in each jaw and the greatly elongated nasal region of the skull distinguish bandicoots from other marsupials. There are well-developed canines and three premolar and four molar teeth in each jaw; the deciduous premolar is minute and less than a third the size of the third premolar that replaces it.

BANTENG, *Bos javanicus*, species of wild ox in Southeast Asia, closely related to the gaur. Banteng are smaller than gaur, averaging a height of 5½ ft (170 cm), and longer-legged, with a smaller dorsal ridge. Like the gaur, the legs have white "stockings," but unlike them they also have a white patch on the buttocks. The horns are more angular, averaging 24 in. (60 cm) in length in bulls but only 12 in. (30 cm) or so in cows. The forehead between the horns is naked and heavily keratinized.

Banteng have a wide but discontinuous distribution. The typical race from Java, *B. j. javanicus*, has spreading horns, 25-30 in. (60-75 cm) across, and the bulls are shining blue black, while the cows are tawny. In the Borneo race, *B. j. lowi*, the bulls are chocolate brown with horns that are much more upwardly directed. It is also a much smaller animal. There are no banteng in Sumatra, and in Burma, Thailand, Cambodia and Vietnam the local race, *B. j. birmanicus*, is normally tawny in both sexes, but in Cambodia about 20% of bulls are blackish, and in peninsular Thailand most bulls are black. Bornean banteng are called temadau; in Burma they are called tsaine; and in Cambodia, ansong. There is some doubt whether truly wild banteng occur in Malaya although there are certainly 70 in Kedah, but these may be feral or recent immigrants from Thailand. Whatever their origin, they are sufficiently new to the Malays to be called "sapi utan" (wild ox).

Unlike gaur, banteng prefer flat or undulating ground, with light forest and glades of grass and bamboo. They are less timid than gaur, and more often enter cultivated fields. They live in herds of 10-30, but many bulls live solitary lives except in the rut. They are more or less nocturnal, feeding during the night and in the early morning, and lying up in the forest to chew the cud by day. In the monsoon season, they go up into the hills, ascending sometimes to 2,000 ft (600 m) to eat the young bamboo shoots. In the dry season they return to the grassy valleys. The time of the rut differs from place to place; in Burma and Manipur (their northernmost range) they mate in September and October, and calves are born in April and May. Calves are reddish, and cows mature in two years, bulls a little later.

In Bali, banteng have long been domesticated and form a characteristic breed, which is smaller than the true banteng with a more extensive dewlap and lower dorsal ridge; males are never quite black. In parts of Indonesia the Bali ox is extensively used as a draft animal and for milk. In Bali and Lombok some have run wild. In Java the domestication of the banteng is more recent, and domestic banteng are almost indistinguishable from wild ones.

BARBARY APE, *Macaca sylvana*, a large terrestrial macaque monkey, tailless and long-coated living in the forests of northwest Africa. A small colony has been maintained on Gibraltar for many years. Barbary apes roam the forests in large bands feeding on leaves, fruit, insects and scorpions, often raiding crops. They are the only species of macaque outside Asia, and it is now generally accepted that the species once ranged across the whole of north Africa and into Southwest Asia, and that all have died out except for the populations in Morocco and Algeria.

BARBARY SHEEP, *Ammotragus lervia*, one of the largest members of the "goat sheep" tribe as well as one of the most primitive, which shares characteristics exhibited by both sheep and goats.

It is a native of North Africa and inhabits the hot, arid mountains from the Red Sea to Morocco. It is a common resident of zoos and has been successfully introduced in southwest U.S.A. Barbary sheep have no preorbital and interdigital glands, but odoriferous glands on the naked underside of the long tail. In this they resemble goats. The long neck mane and cheek beards they share with the urials, the most primitive of sheep. Their horns are the same shape as those found in the most primitive of urials, in the "round horned goats," the west Caucasian ibex, *Capra cylindricornis*, and the bharal *Pseudois* from Tibet. Adults are rufous gray in color with lighter bellies, groins and rears. Barbary sheep have produced fertile and viable offspring with domestic goats but not with sheep, although their blood protein picture resembles that of sheep rather than goats. In their social adaptations they are much like the primitive sheep, but with a more generalized combat behavior. They clash in a similar way to mountain sheep and are better horn and shoulder wrestlers than these. In addition, they jab with their sharply pointed horns and can inflict severe wounds. In this way they resemble neither sheep nor goats but their ancestors, the goat antelopes, Rupicaprini. As in the latter, there is little difference in external appearance between adult males and females except in size. Four races of Barbary sheep are recognized.

BARBETS, stocky, powerfully built birds closely related to the honey guides. The largest is about 12 in. (30 cm) long, the smallest $3\frac{1}{2}$ in. (9 cm) long. The majority of barbets have relatively large heavy bills, sometimes with serrations or "teeth" along the cutting edges of the upper mandible. These "teeth" reach their greatest development in a number of African species, notably the double-toothed barbet, *Lybius bidentatus*, and presumably help in the plucking and manipulation of the fruits upon which these species feed. The feet of barbets are zygodactyl (with two toes pointing forwards, two backwards), as in woodpeckers and all other members of the order, and a number of species regularly clamber about tree trunks and branches in woodpecker fashion. Many barbets are brilliantly colored, notably the American species and the oriental barbets of the genus *Megalaima*, the latter being bright green, adorned with gaudy patches of blue, red, yellow and black around the head. The sexes are usually similar, though there is strong sexual dimorphism in some of the American species. Barbets are so named because the majority have well-developed chin and rictal bristles, or conspicuous tufts of feathers over their nostrils. Chin and rictal bristles reach their most extreme development in *Megalaima* and are often as long or longer than the bill. Normally these bristles are spread out in a fan around the bill, but they can be depressed so that they lie thickly

Barbary sheep, the only wild sheep of Africa, are found from Morocco to Upper Egypt and have goatlike horns.

along the bill, pointing forward. Until recently, the function of these bristles has been unknown, but now it seems likely that they are sense organs for measuring the size of fruits. Field observations on species of *Megalaima* have shown that when they feed on relatively large fruits the bristles are depressed and come into contact with each fruit as it is gripped by the bill. Some fruits are then rejected, others are swallowed. It is significant that such movements of the bristles are not usually seen when relatively small fruits are being eaten. While it is advantageous for specialized fruit eaters, such as barbets, to feed on fruits that are as large as possible, it seems that the sensory bristles prevent time being wasted in attempts to swallow fruits that are too large, or which might wedge in their throats.

The 76 species of barbets are found in the tropics of America, Africa and Asia, as far east as Borneo and Bali.

BARNACLES, crustaceans of the subclass Cirripedia. They are all marine and occur in countless millions on the shore line of almost every coast, and are found attached to almost all floating objects and to objects on the sea-bottom. Like most crustaceans, each individual animal goes through a series of juvenile stages before becoming an adult. Most barnacles have free-swimming larvae that pass through seven stages, six nauplius and a cypris stage, and in many places these form a notable part of the plankton at certain times of the year. Adult cirripedes are usually fixed to a solid support, and they then look quite unlike other crustaceans, so much so that they used to be classified as Mollusca. Then, in 1829, J. Vaughan Thompson found the larvae, which were nauplii, like the larvae of crabs.

Cirripedia divides into a number of orders. The first of these, the Thoracica is made up of the suborders Pedunculata and Operculata. In the Pedunculata, the most primitive, the animal is enclosed in a number of calcareous plates and borne on a stalk or peduncle. The Pedunculata include the familiar barnacle, often spoken of as the ships' barnacle.

The operculate or acorn barnacles are common in the sublittoral and on all rocky coasts, often completely covering wide stretches of rock. They have no stalk, and the calcareous plates have come together to form a chalky box containing the barnacle's body.

The Operculata or acorn barnacles are all very similar in construction and in their life cycle. Attention is largely confined here to these. *Balanus balanoides*, the common European species, is typical. Its soft parts are enclosed in a shell cemented securely to the substratum. The shell, composed largely of calcium carbonate, is made up of a basis, which in some species, including *B. balanoides*, is not calcified, and the compartments forming the walls. Opposite the basis, the shell opens to the exterior at the operculum, which consists of four valves on a flexible membrane under the control of the animal inside, and only when these are opened can the cirri be protruded and water flow in.

The body of the animal is attached underneath the opercular membrane around and beneath the large adductor muscle by which the two halves of the operculum are brought together. It consists of a thorax with a baglike prosoma

The great barracuda *Sphyraena barracuda*, of both sides of the Atlantic, has a reputation for ferocity.

from which arise the cirri. The abdomen is vestigial. The mouth is situated on the prosoma. It should perhaps be emphasized that living tissue is continuous around the whole of the interior of the shell. The cirri are pushed out and then swept downwards acting as a dragnet to capture food and at the same time driving a current of water through the mantle cavity, so bringing oxygen to the body surface through which it is absorbed. The early stages of the embryo are passed inside the parental body, at the base of the mantle cavity. Later, as larvae, they are expelled on the current of water induced by the movements of the cirri.

BARRACUDAS, tropical marine fishes related to the much more peaceable gray mullets (family Mugilidae). The barracudas are fierce predators, which, in some areas such as the West Indies, are more feared than sharks. The body is elongated and powerful, with two dorsal fins. The jaws are lined with sharp daggerlike teeth, which make a neat, clean bite. There are many records of barracudas attacking divers, and they appear to be attracted to anything that makes erratic movements or is highly colored. They feed on fishes and have been seen to herd shoals of fish, rather after the manner of sheepdogs, until they are ready to attack.

The smallest of the barracuda, *Sphyraena borealis*, grows to about 18 in. (46 cm) and is found along the North American Atlantic coast. The great barracuda, *Sphyraena barracuda*, which grows to 8 ft (2.4 m) in length, is found in the western Pacific and on both sides of the tropical Atlantic. A certain mystery surrounds its habits, for it is known to attack divers in the West Indies, but in the Pacific region, and particularly in Hawaii, it has the reputation of being harmless to man. In the Mediterranean there is a single species, *S. sphyraena*, which reaches 5 ft (1.5 m) in length.

BASILISKS, lizards of the genus *Basiliscus* containing several species inhabiting Central America and ranging as far north as central Mexico. They live along the banks of the smaller rivers or streams where they bask during the day or sleep at night on bushes that overhang the water. Basilisks are slender lizards with long, slim toes and tail, and the males are often adorned with crests.

Speed is the chief means of snatching up food (insects and small rodents or birds) and escaping

enemies. When attempting to escape, basilisks head for water and run across it. A fringe of scales along the lengthy rear toes provides support as they dash over the surface. Basilisks are known as "tetetereche" in some parts of their range, as this resembles the sound they make when running on water. Another name is the Jesus Cristo lizard for the ability to "walk on the water." As its speed slackens, however, the lizard begins to sink and must swim in a conventional manner like any other lizard.

The most colorful and largest of the basilisks is the rare green-crested basilisk, *Basiliscus plumbifrons* of Costa Rica. The male has a

The basilisk, South American reptile named after the legendary monster of Europe, is best known for running on water.

large, ornamental crest on its head, another along its back and one on the tail.

Basilisks hatch from $\frac{1}{2}$-$\frac{3}{4}$ in. (c. 2 cm), nearly round eggs that may be white or brown and are buried by the female in damp sand near stream banks. Hatching takes 18-30 days (normally 20-24 days), and the tiny youngsters are replicas of the adults except that they have no crests. Basilisks are normally some shade of brown with white or yellow bands or mottling.

BASKING SHARK, *Cetorhinus maximus*, second only to the whale shark in size and immediately recognizable by its very long gill clefts, which extend from the upper to the lower surface of the body. There are two dorsal fins, and one anal fin, very small teeth in the jaws, and the general body color is a gray-brown. The maximum size of these sharks is usually given as 45 ft (13.5 m) and certainly fishes of 30 ft (9 m) are not uncommon. Unlike most of its relatives, the basking shark is not carnivorous but feeds by straining plankton from the water. The gill arches are equipped with rows of fine rakers (up to 4 in. [10 cm] long and over 1,000 in each row) and these form a fine sieve through which the water is strained before leaving by the gill clefts. This system is clearly an efficient one since it can provide enough food for an animal that may weigh over 4 tons (4,000 kg). Basking sharks lacking gill rakers are sometimes found, and it is thought that the rakers may be shed in winter and regrown every spring. They derive their name from their habit of lying at the surface. They are not dangerous to man.

The basking shark is the only member of its family and appears to be found everywhere, but chiefly in temperate waters.

Bass of the coastal seas of northwest Europe are excellent angling fishes.

BASS, a term used in Europe for the sea perch, *Dicentrarchus labrax*, and its close relative the black-spotted bass, *D. punctatus* (both erroneously placed in the genus *Morone* in the older literature). The bass is considered by many to be one of the best of European angling fishes. It is a coastal fish that often enters estuaries and even ascends rivers. It is found in the Mediterranean and off the coasts of Spain and Portugal but reaches the southern coasts of the British Isles. It is found off shelving sand or shingle beaches and is often fished for in the breakers, where it feeds on fishes (sand eels, sprat and herring). Specimens of 18 lb (8 kg) have been caught, but fishes of 2-7 lb (0.9-3 kg) are more usual. There are two dorsal fins, the first spiny and separated from the second. The back is blue-green, the flanks silver with a black lateral line and a white belly. This fish is considered excellent eating. The black-spotted bass is a smaller fish, reaching 2 ft (60 cm) in length, and the body is speckled with black spots. This species does not reach as far north as the British Isles. See also Black bass.

BATFISHES, a common name given to two rather different groups of fishes. Members of the family Ogcocephalidae, a family of the anglerfishes, are sometimes referred to as batfishes although the term is more appropriately used for members of the Platacidae, a family containing marine perchlike fishes with greatly extended winglike dorsal and anal fins. Species of *Platax* have highly compressed, almost circular bodies, and their long fins give them a batlike appearance when swimming. They grow to about 2 ft (60 cm) and are found in the Indo-Pacific region. They are beautiful fishes with red-yellow coloring, and dark vertical bands in the young, which disappear with age. Such coloring would seem to make them more conspicuous, but in fact these fishes strongly resemble floating and yellowing leaves of the red mangrove. When chased by a predator in a mangrove swamp, the fishes stop swimming and drift motionless like leaves. They feed on small crustaceans and the general detritus of coastal and mangrove swamp areas.

BATH SPONGE, the fibrous skeleton of an animal used since time immemorial for a variety of purposes, not only for bathing but in a wide range of commercial uses. In life, the skeleton is covered with a yellow flesh bounded externally by a purplish black skin which may be pale yellow when the sponge has been growing in dim light. The bath, or commercial, sponge lives in warm seas, from shallow waters to not more than 600 ft (200 m) and may be one of a few species, the chief of which are *Spongia officinalis*, with a fine meshed skeleton, and *Hippospongia equina*, with a coarser skeleton. The main centers of sponge fisheries are the Mediterranean, especially the eastern half, and the Gulf of Mexico and Caribbean Sea, especially around the Bahamas, Florida and Hondu-

The batfish *Platax*, one of several kinds of remarkable fishes that have winglike fins.

ras. Bath sponges also occur, but in more limited numbers, in the Red Sea, Indian Ocean and on the Great Barrier Reef of Australia.

BATS are subdivided into two suborders—the Megachiroptera and the Microchiroptera. Although these names literally mean "big bats" and "small bats," there is a considerable overlap in their size ranges. The Megachiroptera vary in weight from about 1 oz (25-30 g) to 2 lb (900 g) with wingspans of 10 in. (25 cm) to over 5 ft (150 cm), while adult Microchiroptera range from about $\frac{1}{8}$ oz (3.5 g) to $6\frac{1}{2}$ oz (180 g) with wingspans of 6 in. (15 cm) to 3 ft (90 cm). The most obvious physical distinctions are that Megachiroptera have very large eyes and nearly always have a claw on the first finger as well as on the thumb. Because most of them have dog-like faces they are often called "flying foxes." The Megachiroptera are often called "fruit bats" but not all eat fruit; the Microchiroptera are often called "insectivorous bats" although their diets vary tremendously and some eat fruit.

The body form of bats is largely governed by the exacting requirements of flight. In this respect all bats have much in common, and they have solved the problems of flight in quite different ways from birds. In general, the skeleton is frail and light but the forelimbs are enormously developed. The upper arm is fairly short and strong with a powerful shoulder joint to bear the weight of the body. In the so-called advanced families, each arm has a double articulation with the shoulder blade, making a firm hinge for flapping movements. The forearm is very long with only a single bone, the radius, and bears a short, compact wrist in which many of the bones are fused together for further strength. From this projects a short thumb with a claw used for climbing or walking. In vampires, which are especially agile on the ground, the thumb is well developed as a "foot" and bears a fleshy pad that acts as a "sole." The fingers are all very long with extended shanks between the knuckles; the index and middle fingers are close together at the leading edge of the wing, but the ring and little fingers fan out behind to support the main wing surface. This whole arrangement gives fine control of the wing shape since all the joints can be flexed by tendons operated from muscles in the arm. Thus bats are by far the most maneuverable of aerial animals.

The whole of the forearm is webbed by a thin, double layer of skin that forms the wing membrane. This runs from the shoulder across the kink of the elbow to the wrist, between all the fingers (but rarely includes the thumb) and from the little finger backward along the side to the ankle of the back leg. It contains elastic strands and fine muscle fibers within it so that it collapses when the wing is folded and does not interfere with walking. In flight it also acts as a radiator as it contains a network of fine blood vessels in which the blood is cooled to prevent the bat becoming overheated by its exertions. When the bat lands, the blood supply to the wings is much reduced so that heat is then conserved.

The back legs of bats have thighs and shins of roughly equal length and a short, round foot with five small toes bearing sharp claws. The tendons in the foot prevent the toes from

straightening when the leg is extended so that bats can hang from a toehold without effort and indeed often die without falling off. The hips are also unusual, because the joints are permanently twisted so that the knees point upward over the back instead of forward. This position is the most convenient for hanging, and squirrels hold their legs this way when descending trees head first. In bats it also allows the legs to kick downward to assist the wingbeat when flying.

The tail and the flight membrane between the legs are very variable. The Megachiroptera have at most a simple flap along the inside leg, and the feet are held together in flight; the tail if present is short and free. In the Microchiroptera there is generally, but not always, a large tail membrane stretched between the legs, with the actual tail incorporated in it. The trailing edge of the tail membrane is strengthened by a whiplike cartilage called the calcar, which projects towards the midline from the ankle.

The breeding of bats is as unusual as their other characteristics. Being placental mammals, they have internal fertilization, nurture the fetus in the womb and give birth to well-developed young, usually singly, although twins are quite common in some species. Mating occurs in the daytime roost, and courtship appears to be perfunctory. Promiscuity is the rule, and both pair formation and paternal care are unknown. In temperate climates the breeding cycle of Microchiroptera is interrupted by hibernation, since mating occurs in the autumn. The sperm are then stored inside the female until ovulation, and fertilization occurs after she awakens in the spring. Although further matings may occur in the spring, it seems certain that delayed fertilization is responsible for the majority of conceptions. The young are born in all-female nursing colonies within a period of a few days during early summer. The mothers have two functional teats on the sides of the chest, but some bats also have a second, nonlactating pair in the groin, which the baby holds during flight. When the baby grows too heavy to be carried, it is left behind in the colony while the mother is hunting. Its limb bones rapidly grow to almost full length, long before the adult weight is achieved. This may be important, for the first flight is often critical. Bat roosts are generally inaccessible from the ground, and a failure to fly competently the first time would nearly always be fatal.

Tropical bats that live under constant climatic conditions appear to breed repeatedly throughout the year.

One of the most striking features of bats as a group is the wide range of their dietary specializations. In the Old World the Megachiroptera are completely vegetarian, mostly eating fruit, flowers or pollen. One small group in the Far East, with an isolated member in Africa, are almost exclusively nectar drinkers. They hover or perch close to night-opening flowers and sip the rich fluid with very long, rough-tipped tongues. The majority of the Old World Microchiroptera are insectivorous, but a few, notably the Megadermatidae, are "bats of prey," hunting small vertebrate animals such as mice, lizards and other bats. One or two species in Asia probably catch small fish with their back

A false vampire *Cartioderma cor*, characterized by extensive development of the ears.

feet as they fly low over open water.

In the New World, the closely related true vampires feed only on the blood of larger vertebrates, specializing either on birds or on mammals, which are attacked as they sleep. Not surprisingly, vampire bats possess a number of unusual adaptations for this mode of life; they have large, sharp incisor teeth to open a wound gently, a muscular, grooved tongue for sipping the blood and secretions in their saliva which prevent the blood from clotting. They can imbibe enormous quantities of blood and excrete the excess water rapidly, and are very agile on the ground in case their host should awaken when they are so engorged that they fly only with difficulty.

Hypsignathus monstrosus, being suckled.

Finally there are two specialized New World fishing bats, one in the Noctilionidae and a Mexican member of the Vespertilionidae. Both have very large back feet, which seize fish from the water during flight.

During the winter in temperate countries the food of bats becomes scarce. Unlike small insectivorous birds, bats seldom migrate for long distances. Instead, most of them hibernate. During the autumn they become very fat with accumulated food reserves and sink into a dormant state that is more extreme than daily sleep. Heart rate, breathing and all other bodily functions are greatly suppressed throughout the winter.

It is not true, however, that hibernation is un-interrupted, for bats are sometimes seen flying about in midwinter.

Only one kind of Megachiropteran is known to be able to fly safely in complete darkness. This is the genus *Rousettus* which has normal Megachiropteran eyes on which it relies whenever the light is adequate. When vision fails, these bats are able to use acoustic guidance by making short, high-pitched sounds and detecting echoes from nearby objects.

Flight guidance in the Microchiroptera is quite different. Nearly a quarter of the species, from all major families, have been investigated, and all are found to emit a constant train of sound pulses during flight or even when active on the ground. These pulses differ from those of *Rousettus* in being produced vocally from the modified voice box and in being almost entirely ultrasonic—that is, too high in pitch to be audible to man although their intensities may be very high. The actual sound frequencies range from about 15 KHz (just audible) to 150 KHz, having wavelengths of $\frac{5}{6}$ to $\frac{1}{12}$ in. (2.2-0.22 cm). The sound pulses also vary in duration from about $\frac{1}{4}$ millisecond to as much as 60 milliseconds. Insectivorous species have been shown to be capable of detecting and avoiding wires only 4 thousandths of an inch thick (0.1 mm) even in the presence of considerable background noise. Moreover they detect, locate and intercept their insect prey in midair by this acoustic method, a task that is apparently too difficult for even the best night vision.

BEACONFISH, *Hemigrammus ocellifer*, a small fish from the Amazon Basin and Guyana with the alternative and equally appropriate name of head-and-tail-light fish. It grows to

Flying fox *Rousettus aegypticus*, giving birth.

$1\frac{3}{4}$ in. (4.5 cm) and is one of the commonest aquarium fishes imported into Europe and the United States. The common name refers to the shining green-gold patch on the shoulder and a second patch at the base of the tail. The species is easy to breed, and the sexes can be more easily distinguished than in many other members of the family Characidae. The male is the slimmer of the two, and its swim bladder is quite clearly visible when the fish is viewed against a light; in the female the swim bladder is masked.

BEAKED WHALES, toothed whales that, with the bottle-nosed whales, make up the family Ziphiidae. Some of the family are reasonably well known, but a number are described only from a few or even single skulls. In general they are born toothless, but the males later develop a pair of teeth in the lower jaw, the females remaining essentially toothless. Rows of small vestigial teeth are, however, often found in both upper and lower jaws.

There are two species of bottle-nosed whales, a well-known northern form, *Hyperoodon rostratus*, and the little-known southern *H. planifrons*. The northern bottle-nosed whale is of truly whale proportions, males reaching 30 ft (10 m) and females 24 ft (8 m). The body is dark gray to black dorsally and light gray ventrally. The dorsal fin, as with all known members of the family, is well behind the middle of the body, virtually at the junction with the tail. The head has a prominent beak and above it a very rounded dome that contains an oil reservoir, big enough for the animal to be hunted in the late 19th century for the spermaceti in it. It is a fairly common species moving in small schools of 4-12 animals and feeding mainly on cuttlefish.

Similar in external appearance to the bottle-nosed whales are the two species of the genus *Berardius*, *B. bairdi* and *B. arnuxi*, the former from the northern Pacific and the latter from the southern hemisphere. They are so little known that they have no popular name. The northern animal grows to a size of over 40 ft (13 m), while the southern is some 10 ft (3 m) smaller. Cuvier's beaked whale, *Ziphius cavirostris*, is rather better known. The head has a much less pronounced dome. The color varies markedly. It grows to about 23 ft (7.5 m). In spite of being fairly common throughout the world, very little is known of its activities. The genus *Mesoplodon* is one of which nine species are described but only one, Sowerby's whale, *M. bidens*, is at all well known. It is found in the North Atlantic. It grows to a length of 15-16 ft (5 m) and has a slender, tapering head.

BEARS. Characterized by their heavy build, thick limbs, diminutive tail and small ears, members of the Ursidae comprise seven genera and nine species. Bears have a wide distribution, which covers the northern hemisphere and overlaps in a few places in the southern hemisphere. The fur is coarse and thick, and, with the exception of the polar bear, dark in color. The tuberculous molars indicate an omnivorous diet, whereas the plantigrade walk (the whole sole of the foot resting on the ground with each step) is an ancestral trait that has been retained, giving the bear a slow and ponderous gait. However, when pursuing or pursued it will gallop. Some species climb well.

The eyes and ears are small, and of all the senses smell is the sharpest and vision the least acute. Usually peaceful and timid creatures, bears may become formidable when wounded or suddenly disturbed, using their claws and teeth in conjunction with their great strength. Solitary and nocturnal, bears wander over their territory, the cubs remaining with the mother until half grown. Species living in colder areas may become lethargic during the winter, but true hibernation does not take place, only a winter dormancy, as physiological body functions remain the same.

The cubs are very small at birth, weighing approximately 1/350th of the adult weight. Gestation is between six to nine months, and delayed implantation occurs in most species, so the fragile young are born during "hibernation," emerging from the den two or three months later. Sexual maturity is reached at two years for the females and up to six years for the males. The total life span can attain 40 years under ideal captive conditions, but in the more arduous wild environment, it is less than half of that.

The European brown bear, *Ursus arctos*, the

A brown bear of the northern hemisphere.

most widespread species, has a typically short neck and large, doglike head. The grizzly, an American subspecies, is the largest terrestrial member of the Carnivora, some individuals reaching 113 in. (280 cm) in length and weighing over 1,700 lb (780 kg), but 80 in. (200 cm) and 550 lb (250 kg) are more usual measurements. Coat coloration varies from cream to blue-black. Distributed across Alaska and Canada down to the western portion of North America, it is also found in the mountains of Europe and Asia Minor, across to the Himalayas. Once known as *Ursus horribilis*, the grizzly has the reputation of being particularly ferocious.

The strength of the brown bear is proverbial, and it sometimes kills an adult cow with a swipe of the forepaw and drags the carcass back to the den. In certain areas fish are scooped out of the stream and form a large part of the diet. Fruit and grass are eaten in the spring as a laxative when the bears become active again after "hibernation."

The North American black bear, *Euarctos americanus*, is, however, smaller than the brown bear, rarely measuring more than 70 in. (180 cm) in length or weighing over 330 lb (150 kg). Different colorations exist: black, cinnamon and even white. The muzzle is often brown. The paws, which are less massive, make this species particularly agile, and full-grown adults can still climb trees. Breeding takes place in June or July, but the fertilized egg does not become implanted in the uterine wall until November. One to four cubs are then born in January or February and remain with the mother until the following autumn.

The spectacled bear, *Tremarctos ornatus*, is the only South American species, living both in low forests and up to 1,000 ft (300 m) in the Andean mountains. Measuring 60-70 in. (150-180 cm), adults weigh up to 300 lb (140 kg), but the shaggy coat contributes to the bulky appearance. Some individuals have uneven white streaks circling the eyes and extending to the muzzle, down the throat to the chest, whereas others are completely dark brown-black.

The polar bear, *Thalarctos maritimus*, is the only completely carnivorous bear. Its year-round cream color is perfect camouflage, blending with the background of ice floes in the Arctic. Among other characteristics, this large bear has a more slender and longer neck than the brown of black bears. The soles of the feet are thickly haired, providing both insulation and a secure grip on the ice. Furthermore, the forepaws are partially webbed, and the neck and shoulder muscles greatly developed, making the polar bear a tireless swimmer. Attaining occasionally 9 ft (200-280 cm) in length, the weight oscillates between 800 and 1,550 lb (400-700 kg) depending on the season. Nomadic in the extreme, polar bears are sometimes discovered resting on ice floes 200 miles (320 km) at sea.

Seals are the main prey. The bear waits for the seal to surface at its breathing hole in the ice then grabs it and hauls it out. 30 lb (13 kg) of seal blubber (rather than the flesh) can be eaten at one sitting. Mating occurs in mid-April or May, and while tracking the solitary females over great distances, the males become extremely irascible toward both other males and humans. In winter, the females excavate a shallow den under the snow, and there the cubs are born in January. Blind and helpless, the young are the size of a guinea pig at birth. The mother does not leave this makeshift shelter, living off her fat reserves, until the cubs can follow her about. When they emerge during the early summer months, the young polar bears are weaned on berries and Arctic hares, waiting until the following winter to be initiated to the seal-hunting technique. Because the cubs stay so long with the female, breeding takes place only every other year, which accounts for the low population numbers.

The Himalayan bear, *Selenarctos thibetanus*, occurs in the elevated forests of Baluchistan, Afghanistan, westward to the Himalayas and northward into China and Siberia. Its coat is black (or brown during the summer molt), and the white chevron fur pattern stands out clearly on the chest. The tufted ears are relatively larger and the muzzle shorter than in the European brown bear *Ursus arctos*, but they share common behavioral traits. "Moon" bears or "blue" bears, as they are also called, descend

A grizzly bear watches a puma feeding. Although they are basically vegetarian, grizzlies will readily eat flesh, either freshly killed or as carrion.

to the valleys to spend the winter and do not "hibernate," becoming intermittently dormant only during the height of a monsoon or a blizzard. Surprisingly enough, this heavy animal is an agile climber, building rough nests in low trees to sunbathe or nap. A den is preferred, though, when rearing cubs. Family parties have been observed foraging together in fruit trees, and it seems probable that the juveniles remain with the parents for at least a year, sometimes even after the next litter is born.

The Indian and Ceylonese sloth bear, *Melursus ursinus*, is quite different from other Asiatic bears. The muzzle is elongated into a snout, which shows great lateral mobility. Smaller and even shaggier than the Himalayan black bear, it has a mantle of long, coarse fur on its shoulders, giving it a humped appearance. There is also a wide, often semicircular-shaped yellowish expanse of fur on the chest, but the overall coloration varies from reddish-brown to black. It weighs 200-240 lb (90-110 kg) and measures 56-70 in. (140-180 cm) in length. Solitary and nocturnal, it spends most of the day sleeping in jungle caves. The outstanding behavioral trait of this bear is its method of feeding. Along with a long snout and extremely mobile lips, it has a gap in its front teeth, due to the absence of a pair of incisors in both upper and lower jaws. An observer in the field reports that after tearing a termite nest apart with its claws, this "vacuum-cleaner" bear places its muzzle near the hole and sucks in the grubs with such force that the noise can be located 200 yd (180 m) away. Another dental characteristic is the small molars, indicating a diet of fruit and insects. After a seven-month gestation, two or three cubs are born in the spring. Strangely enough, they are carried on the mother's back during her nightly rounds and when she climbs up a tree. Later, even when several months old, the young, if suddenly alarmed, scramble over each other to reach this safe vantage point.

The sun bear, *Helarctus malayanus*, of Malaya looks like a diminutive, short-haired version of the South American spectacled bear, but the chest and eye patches when present are usually of a tan color. This is the smallest member of the bear family, measuring 44-56 in. (110-140 cm) and weighing 50-140 lb (22-65 kg) at the most. It occurs in the forests of Burma down the Malaya Peninsula to Sumatra and Borneo. The claws are used, when feeding, to scatter fruit or tear open beehives and termite hills. Afterward, the bear licks up the insects among the debris with its long prehensile tongue. Small rodents, birds and eggs also complement the diet of this omnivore.

BEAVERS, largest rodents in the northern hemisphere, sometimes exceeding 88 lb (40 kg). The heavy body is covered with a thick fur making a waterproof coat. The hind paws are large and webbed to the tips of the five toes. The second toe has a double nail, similar to a bird's beak, which acts as a fine comb. The hindquarters are very powerful in contrast to the small forelimbs, which barely touch the ground when it is moving along. All orifices can be closed when the beaver is under water, including the cloaca, which is closed over the ano-urino genital organs. The tail is an indispensable counterpoise on land and a horizontal rudder when the animal is in the water with a heavy burden in its arms. The lips close behind the incisors, protecting the mucous membranes from water and from splinters of wood while the animal is working under water. The front paws, which are veritable hands, nimble and skillful, carry, push, pull, steer, scratch and groom. The beaver's respiratory system enables it to remain under water for up to 15 minutes. Scientists make a distinction between two kinds, the European *Castor fiber* and the American *C. canadensis*, but both types can be crossed and produce fertile hybrids.

The mating season is in February, when the winter ice is melting; the beavers pair under water, *more humano*, or even at the edge of the water. About 100 days later young are born, fluffy and resourceful, eyes already open. A litter averages three young. The mother keeps them in the burrow for two or three weeks, together with last year's young. The young are suckled during this period of confinement only, but the mother brings them tender young leaves a few days after birth.

The beaver is superior to all other mammals in the efficiency and technical skill used in organizing its domain. It prefers to settle on shallow lowland streams, where there is plenty of vegetation, and by means of dams converts them into a series of water levels, the value of which is obvious; they provide protection from enemies coming from the river banks and an easy and rapid means of getting around. Furthermore, by keeping the water at a constant level, they camouflage the underwater entrance to their burrow and insure access to their winter stores, even if the surface of the water is frozen. Although, during the summer months, the beavers live on all kinds of plant life, during the hard weather they have to live on the bark of the willows and poplars that they have collected in front of the entrance.

The beaver's shelter may be merely a cavity dug in the steep bank of the river, starting from a sloping gallery that begins just below the surface of the water. If the banks are too low, the beavers build a lodge, a wooden dome consolidated with mud and having a diameter of up to 18-20 ft (6-7 m). The largest lodges may have several rooms, each having an independent gallery. Inside, the nest may be seen, provided with a bedding of dry, shredded wood. Each family has several homes, inhabited successively apparently according to whim.

The most surprising piece of work is the dam, which may reach enormous widths, up to more than 1,000 yd (1,000 m). In spite of their size they are never the work of enormous colonies, at most two or three families living amicably together, since they are of the same blood. A family rarely exceeds ten members.

The communications network is almost entirely aquatic, except for a few short straight paths toward the felled trees. If it is necessary to go farther, the beavers can dig long canals, where they feel safer than on shore.

BEDBUGS, cosmopolitan blood-sucking insects associated with man and many animals, including birds and bats. They are true bugs, that is, they are members of the order Hemiptera. The common bedbug, *Cimex lectularius*, and the tropical bedbug, *Cimex rotundatus* (= *C. hemipterus*), commonly occur as parasites of man. The pigeon bug, *Cimex columbarius*, which lives closely associated with pigeons, doves and domestic fowl, is now usually considered to be merely a race of the common bedbug, and indeed, the two races are known to thrive on both man and birds. A similar but smaller species normally found in house martins' and swallows' nests will also bite man.

The adult bedbug is flat and roughly oval in shape, being about $\frac{1}{5}$ in. (5 mm) long, and $\frac{1}{8}$ in. (3 mm) broad. The body is covered with fine short hairs and is usually mahogany brown, but it may appear more reddish if it has recently fed or purple if an older meal is still present in its gut. Immature bedbugs (usually called nymphs) are paler in color than adults. When in need of a meal, bedbugs are paper-thin but appear much fatter, even almost globular, after a large meal. They have lost the ability to fly, and all that remains of the once functional wings are a pair of short flaps on the middle segment of the thorax.

The common bedbug is found throughout Europe, Russia, northern India, North Africa, North and South America and Australia. This species tends to be replaced in the tropics by the tropical bedbug, which is much better adapted to high temperatures but is otherwise very similar to the common bedbug. The fact that several species of bedbugs feed on birds and bats, which habitually breed in caves or rock clefts, suggests that man and the common bedbug may have commenced their "partnership" when man himself inhabited caves. But we shall probably never know for certain the original habitat of the bedbug. As it has been carried by commerce all over the world wherever suitable conditions occur, its primary center of distribution is also something of a mystery. Most authorities consider the Middle East as the most likely area of origin, especially on the eastern Mediterranean coasts. The insects were known to the Romans and Ancient Greeks, who regarded them as having wide medicinal properties when taken in a draft of water or wine. Bedbugs were recorded from the area now known as Germany in the 11th century, but apparently did not invade Britain until the early 16th century and Sweden until the 1800s.

After mating, the female bedbug lays up to 150-200 eggs at the rate of two or three per day. Each egg is creamy white, slightly curved and elongated in form, with a lid or operculum at one end, and is about $\frac{1}{25}$ in. (1 mm) long. They are deposited in crevices and cracks, behind skirting boards, wallpaper and similar sites, each being firmly fixed in place with a gluelike secretion produced by the accessory glands of the female reproductive apparatus. The eggs require a temperature of at least 55° F (13° C) to hatch. At 82° F (28° C) they will hatch in five or six days. When fully developed, the tiny nymph forces off the operculum and emerges as a miniature of the adult, about the size of a pinhead. It molts its skin and forms a new, larger skin to accommodate each of five increases in size between egg and adult. Each nymphal stage needs at least one full blood meal before proceeding to the next growth stage, but additional blood meals may occasionally be taken.

The bedbug sucks the blood of its human host using a rostrum or jointed beak beneath its head. When not in use the rostrum is carried pointing backward under the head, but it is swung forward and extended when the bug feeds. The rostrum is composed of two pairs of needlelike stylets, which are supported by a jointed lower lip (the labium). The two pairs of stylets when pressed together form two tubes, a larger one for sucking up blood and a smaller one through which saliva is pumped into the wound. A hungry bedbug carefully selects a suitable area of the skin with delicate probing movements of the rostrum and then penetrates the skin using the serrated tips of the stylets, so that these come to rest in a blood capillary. Saliva is pumped through the insect's mouthparts into the capillary from the salivary glands in the thorax. A mixture of blood and saliva is then sucked up the food canal in the stylets and passed into the gut. The saliva prevents coagulation of the host's blood, and hence clogging of the delicate stylet mechanism and the gut is avoided. The intense irritation that is well known to accompany bedbug bites is apparently caused by the saliva, the wound made by the stylets being a very minor puncture. Some people seem to be very much more sensitive to bug bites than others, and many tend to lose their sensitivity if bitten repeatedly. A feeding bedbug may take up to 12 minutes (frequently less) to gorge itself fully, depending, apparently, on the size of the blood capillary penetrated by the tip of the stylets. After feeding, the bug takes on an extremely bloated appearance, having imbibed up to six times its own weight of blood. Feeding normally occurs about once a week in summer, less frequently in cooler weather, and probably not at all in unheated premises in the winter in Britain.

Bedbugs are chiefly nocturnal and are probably only seen in broad daylight if very hungry. They are said to be most lively just before

Typical beaver pond and main dam, constructed of branches and logs plastered with mud.

Red-throated bee eater *Merops bulocki*.

Carmine bee eater *Merops nubicus*.

Odors play a large part in the economy of a honeybee colony. This honeybee is shown fanning its own scent into the air with its wings.

dawn, and it may be supposed that those seen walking in daylight are individuals that have failed to secure a suitable blood meal in the normal hours of activity. Bedbugs may be said to be truly bloodthirsty animals, as all the evidence indicates that they feed only on the blood of birds and mammals. Under laboratory conditions they can be induced to feed through a membrane simulating the host's skin. It is interesting to note that a bug steadfastly refuses to feed from otherwise suitable fluids unless skin or some other substitute membrane is present to be pierced initially.

BEE EATERS, a family of insect-eating, bright-plumaged attractive birds of the Old World; chiefly found in the tropics, with one species in Europe and one in Australia

The best-known species is the European bee eater, *Merops apiaster*. It is a summer visitor to the Mediterranean countries and western Asia, being found north of the Pyrenees, the Camargue, the Alps and Carpathians, 500 miles (800 km) north of the Black and Caspian seas and east to Kashmir. It spends the winter in Africa south of the Sahara, particularly in southern Africa.

Bee eaters live on flying insects, caught in graceful pursuit flights, and have pointed, rather long decurved beaks, pointed wings and very short legs. The sexes are similar. In some species the central tail feathers are elongated.

Bee eaters comprise a uniform group, and the 24 species are currently placed in three genera

of which the most important is *Merops*, with 21 species. The family Meropidae is classified in the order Coraciiformes, and although it is not particularly closely related to the other families, kingfishers, rollers, hoopoes, hornbills, etc., all of them share certain behavioral characteristics (e.g., hole nesting) and structural ones. In the American tropics, an unrelated family in another order, the jacamars (Galbulidae) fills exactly the same "niche" or place in nature as the bee eaters in Africa and Asia, and during their evolution the two families have converged in respect of diet and feeding habits, plumage, size, breeding biology and behavior.

Bee eaters are well named, for practically all the species that have been investigated in detail feed exclusively on airborne insects, with bees and their allies (Hymenoptera) comprising 80% or more of their diet. There are two principal ways of feeding; the smaller species keep watch for passing insects from a vantage point like a bush, fence post or telephone wire, and the larger species hunt on the wing. In either case, an insect is pursued with a fast and dextrous flight and snapped up in the bill. Generally the bird returns to its perch, where it beats the prey against the perch until it is inactive. Over much of the range of the family, the various bee eaters subsist largely on honeybees. The venomous workers of honeybees and other stinging Hymenoptera have their stings removed by an instinctive pattern of behavior, improved by the fledgling bee eater with experience. The insect

is held in the beak near the tip of its abdomen, which is rubbed against the perch so that the venom is discharged. Apart from bees, most other suitably sized flying insects are also preyed upon: demoiselle flies, termites, butterflies, bugs, beetles, grasshoppers, etc. After a rainstorm in Africa, flying ants and termites emerge in great profusion and are hunted by many kinds of birds; an excited flock of wheeling bee eaters is often in attendance.

BEES, honeybees. There are three species of honeybee. The most widespread is *Apis mellifera*. Its ability to regulate the temperature and humidity of its nest, together with its ability to survive on stored food during unfavorable periods, has undoubtedly played a major part in its successful colonization of most of the world.

Each honeybee colony contains three castes: the mother or fertile female known as the "queen," the infertile females known as "workers" and the fertile males known as "drones." Usually there is only one queen to a colony but up to 60,000 workers. In natural conditions the nest of a honeybee colony consists of a series of vertical wax combs in a hollow tree or cave, or under an overhanging rock or some such shelter. On either side of each wax comb are series of hexagonal cells in which the young are reared and the honey and pollen are stored. Most of the tasks of the colony are done by workers. Those born in the spring and summer generally undertake a series of duties that are dependent to some extent on the development of certain glands in their bodies. Thus, at first they clean cells and remove debris, but after a few days special brood-food glands in their heads have developed sufficiently for them to feed a protein-rich secretion to the larvae. A few days later the wax glands, which are located on the underside of the abdomen, have developed sufficiently for them to secrete wax and build a comb. Other tasks done inside the nest include packing pollen loads into cells, receiving nectar from foragers, converting it into honey and guarding the entrance to the nests. The task done by a particular worker depends not only upon its physiological condition but also on the requirements of the colony at the time. Indeed, a worker spends much of its time "patrolling" the combs and so becomes aware of current needs. A worker first leaves its nest when only a few days old to make orientation flights during which it learns the location of its home in relation to that of surrounding landmarks, and when it is about two to three weeks old it begins to forage. It continues to do so for the rest of its life, which may be only about four weeks in midsummer, but as long as six months for bees that emerged in late summer or autumn.

Like solitary bees and bumblebees, worker honeybees collect nectar and pollen, but they also, on occasion, collect propolis and water. Propolis is a sticky exudate of various buds, and the bees use it to cement and block gaps in the covering to their nest and to reduce the size of the entrance. Water is used for cooling the nest as well as to dilute honey. To cool the nest, the bees either spread it on the surface of the wax cells or manipulate it on their tongues to hasten its evaporation. They also aid cooling by fanning a current of air through the nest with their wings.

However, most foraging trips are for nectar or pollen or both. Although, unlike many solitary bees, honeybees do not restrict themselves to one or a few plant species, during any one foraging trip an individual bee tends to keep constant to one plant species only. Indeed they tend to become conditioned to make their flights at the time of day at which flowers of their particular species open or present pollen, at other times remaining inside their nest. Any fixation to a particular species is temporary only, and should a species cease flowering, or for some other reason fail to yield nectar or pollen, the bees concerned will soon forsake it for another. Although some bees may find a particular source of forage for themselves, more often than not they are informed about it by others. The ability of bees to communicate a favorable source of forage is one of the most remarkable biological discoveries of this century. On its return home a successful forager performs a "dance" on the surface of the comb, features of the dance indicating the direction of the food source in relation to the direction of the sun from the nest entrance and its distance from the nest. Bees that follow the dancing bee and receive the information are further helped to locate the food by the odor of the flowers that clings to the dancing bee's body.

The eagerness with which a dancing bee is followed, or with which the load of a successful forager is received, depends upon the extent of the colony's current need for the particular type of forage. Whereas nectar is welcome at any time, pollen demands are particularly high when brood rearing is extensive, and the amount collected is related to the size of the brood being reared.

In turn, egg laying and brood production are governed to a considerable extent by the amount of pollen a colony can collect. The queen is specialized for egg laying and, unlike the queen bumblebee, is incapable of doing the tasks of worker bees. Her egg production reaches its peak in late spring or early summer, when she may lay about 1,500 eggs per day. To maintain such a rate of egg production, a large protein intake is necessary, and the workers feed the queen entirely with glandular secretions to enable this to occur. The queen lays two types of egg, fertile eggs giving rise to workers or queens and unfertilized eggs giving rise to drones. The worker- and drone-producing eggs are laid one per cell, the drone-producing eggs in slightly larger cells than the worker-producing cells. The worker eggs hatch to larvae three days after they are laid, and after five days of feeding the larvae change into pupae and the tops of the cells are covered with a cap of wax. The larvae are visited and fed a great many times (progressive feeding) in contrast to the mass provisioning practiced by solitary bees. After 13 days as pupae, the soft downy adults emerge. Hence there are altogether 21 days from egg to adult. The larval and pupal stages of the drone are slightly longer, making a total of 24 days. Apart from their larger size, drone cells can be easily recognized during the pupal stage by their much more pronounced concave wax capping.

Drones are large and stocky with blunt abdomens. At first they are fed by the workers, but after about a week they are able to feed themselves from the honey stores of the colony. They are not usually produced until late spring.

A queen honeybee is attended by workers while laying eggs.

Their sole function is to fertilize any young queen, and to help them achieve this they have powerful thoracic flight muscles and large eyes. In warm afternoons they congregate and patrol in special areas, 30-50 ft (9-15 m) above ground, to which the queens are attracted and where mating occurs. The highly specialized genital apparatus of the drones includes an enormously developed penis which is torn from its base when it is everted into the vagina of the queen, and in the process of copulation the drone dies. However, although every queen mates more than once, only comparatively few drones can participate in the act, and when the summer's end is approaching and the usefulness of drones is at an end, they are left in a corner of the hive, starved of food and eventually dragged from the hive entrance.

New queens are reared in special cells that hang vertically downward from the comb. The queen larvae receive special food and attention, and their total period of development is only 16 days. They are either reared to replace the queen of the colony when she is dead or failing, or because the colony is about to reproduce by swarming and a new queen is needed to replace the mother queen that will leave with the swarm.

wattles. Their calls consist of far-carrying notes, like the tolling of a large bell or the clang of a hammer on an anvil, and may be monotonously repeated for long periods. These calls can be heard up to a mile away and may be intolerably loud at close quarters. The birds are forest-dwelling, fruit-eating species, and little is known of their life histories.

In Australia the name is applied to the crested bellbird, *Oreoica guttaralis*, a species of whistler (Pachycephalinae). This is a bird of dry regions, thrushlike and dull brown in color with a short erectile crest. The male has a mainly black head, and is white around the bill and throat. The bell-like call in this species consists of two slow notes followed by three quieter ones, all uttered in a muted ringing tone that makes the call sound like a distant cattle bell. It is ventriloquial in quality, the bird being difficult to locate from the sound which seems to come from much farther off.

In New Zealand the name is used for a species of honey eater, *Anthornis melanura*, Meliphagidae. This is again about the size of a thrush and a dull olive and yellowish green in color, the male plumage having some iridescent purple on the head and yellow tufts at the sides of the

no long hair on the cheeks, chin or neck. The horns are as large as those of the much larger Barbary sheep, reaching up to 31 in. (88 cm) in length with a basal circumference of $13\frac{1}{2}$ in. (34 cm). The body is stocky, and the legs are short. The ears are long, narrow and pointed, and the tail is long and naked on its underside. There are no preorbital, inguinal or interdigital glands. These are typical goat features. The skull is similar to that of the Barbary sheep, but more massive. Like all sheep or goats, the bharal has only two teats.

This is a strikingly colored little goat with white margins down hind and front legs, a white rump patch and belly, the latter bordered by a dark brown flank stripe. The front of the legs, as well as the nose, neck and chest are dark brown, or even blackish in old males. The body is brown-gray in autumn. The smooth horns tend to be dark olive in color.

Like sheep, bharal are social animals. Male and female groups range apart except in the rut, which begins sometimes in late October. The large horns and sturdy skulls suggest that bharal clash forcefully. Occasionally, they fight like goats. The gestation period has been suggested as lasting six months, which is probably correct. Twins are uncommon. The usual life expectancy of bharal is 12 to 15 years.

BICHIRS, *Polypterus,* a genus of primitive freshwater fishes of Africa comprising about ten species. The reedfish, *Erpetoichthys* (formerly known as *Calamoichthys*), is the only near relative of the primitive bichirs. The name *Polypterus* signifies "many fins," for when these fishes are alarmed or excited a row of 8-15 little finlets are erected along the back. These are not displayed during normal swimming. The body is covered by thick rhombic scales of a type known as ganoid (with a covering of ganoine as in certain extinct forms). A pair of spiracles (the vestigial first gill slits in most bony fishes) are conspicuous. In the larvae there are leaflike external gills such as are known in the South American and African (but not Australian) lungfishes and also in amphibians but in no other bony fishes. In some species of *Polypterus* these gills later disappear. The intestine has a spiral valve, which serves to increase the absorbent surface of the gut. This, too, is a primitive feature that is now found only in sharks and in such bony fishes as the sturgeons, lungfishes, the coelacanth and such fishes as the bowfin in the order Holostei. Bichirs can live out of water for a while breathing air into their lungs, which are large but not quite so efficient as those of the lungfishes. The pectoral fins are constructed in a way similar to those of the coelacanth and its fossil relatives, that is to say with the finrays arising from a fleshy lobe. *Erpetoichthys* is basically similar to *Polypterus* but has a more eellike body.

BINTURONG, *Arctictis binturong,* a relative of the palm civets but unlike them in appearance. It is long-bodied with short legs, $2-2\frac{1}{2}$ ft (60-76 cm) long with a bushy, slightly prehensile tail slightly shorter than this. Its coat is shaggy, black with brown or gray on the tips of the hairs, with long black ear tufts and with white on the ears, face and unusually long whiskers. The binturong is nocturnal and spends the day mainly curled up in the treetops, sometimes coming out to bask on a branch. It is known

Bearded bellbird *Procnias averano,* of South America and the island of Trinidad. The male has a metallic call which he makes from a fixed point throughout the day.

Bees that leave with the swarm each carry a supply of honey in their honey stomachs, and as soon as the swarm has settled in its new home some of this is rapidly converted into wax to start building the new comb.

BEIRA, *Dorcatragus megalotis,* a dwarf antelope 20-30 in (50-75 cm) high and weighing 20-24 lb (9-11 kg), has coarse hair, red-gray above, white below, with white eye rings and a dark flank stripe. It has big ears and long legs, and lives in rocky hills in Somalia. It is the only dwarf antelope to live in groups; four to seven are seen together.

BELLBIRD, a name used in various countries for birds with bell-like calls. In South America it is used for several species of cotingas of the genus *Procnias* (Cotingidae). These are about the size of large thrushes and may have a featherless throat or various types of vermiform

breast. The song, which may be heard at all times of the year, is a series of up to six notes, of a liquid quality, and at a distance the louder notes sound remarkably bell-like. The female also sings, but her song is shorter and weaker.

BELUGA, derived from the Russian for "white" is the name for two animals. The beluga *Huso huso* is the largest of the sturgeons and probably the largest of all freshwater fishes. It is found in the seas and rivers of the Soviet Union and reaches 29 ft (8.8 m). The beluga *Delphinapterus leucas*, or white whale, is a relative of the narwhal. It lives in northern seas and grows to 18 ft (5.5 m).

BHARAL, *Pseudois nayaur,* the least-known member of the goat-sheep tribe is not a large animal, the males rarely exceeding 150 lb (68 kg) in live weight. The head and horns resemble those of Barbary sheep and urials, but there is

Red plumed or Count Raggi's bird of paradise *Paradisea apoda raggiana*.

Prince Rudolph's blue bird of paradise *Paradisea rudolphi*, displaying, in an upside-down position.

to range from Burma to the Philippines, but it may possibly live farther to the north, in Assam, Nepal and southern China. Little is known of its biology except that it eats mainly fruit and green shoots, its teeth being blunter than those of typical carnivores, but it also hunts small mammals and birds.

BIRD-EATING SPIDERS, large spiders that sometimes catch and eat birds. Very small tropical birds are sometimes caught in the strong orb webs of *Nephila*, but even the giant theraphosid spiders like *Zasiodora* and *Grammostola* catch birds very rarely despite their being called mygale or bird spiders.

BIRDS OF PARADISE, strongly-built perching birds famous for the brilliantly colored and elaborately shaped plumes used in display. They are closely related to the bowerbirds. Birds of paradise range from 5-40 in. (13-100 cm) long. Their legs and feet are rather stout, and their tails vary from short and square to very long and wirelike. The bill may be fairly heavy, with or without a hooked tip, or long, thin and sickle-shaped. In some species the sexes are much alike, but others show such marked sexual dimorphism in plumage that each sex was originally described as a separate species. Birds of paradise are found in the forests of New Guinea and nearby islands, with a few species in northern Australia and the Moluccas. They feed mainly on fruit, but also on insects, spiders, tree frogs, lizards and other small animals.

The less ornate birds of paradise have black plumage with little or no decoration, as in the paradise crow or silky crow, *Lycocorax pyrrhopterus*, of the Moluccas, where both male and female are blackish with no ornamental feathers. In Princess Loria's bird of paradise, *Loria loria*, the sexes are different; the female dull olive-brown and the male black with iridescent patches on wings and face. In the more ornate species the female remains dull brown or gray, while the male has elaborate crests or long plumes on the body or tail. One of the more ornate species is the 12-wired bird of paradise, *Seleucidis melanoleuca*, where the male is black and yellow with yellow feathers on the flanks that end in 6-in. (15-cm) tips that bend forward. The king bird of paradise, the smallest species, has long curling, central tail feathers which are wirelike with "flags" at the tips. The body is red above, white below. Another remarkable species is the King of Saxony bird of paradise, *Pteridophora alberti*. Mainly black and yellow, the male has two long wirelike plumes set with small blue "flags" trailing from the head. These are 18 in. (45 cm) long, twice as long as the bird. Perhaps the most ornate, and bizarre, is the superb bird of paradise, *Lophorina superba*, which is black with a bright bronze, green or mauve gloss and has an enormous "cape" of long feathers on the nape, which are raised in display. The males of the plainly colored species, such as the trumpet birds or manucodes, *Manucodia*, chase the females through the trees and display to them, but the ornate males usually gather to display in particular places, sometimes with several males in one tree. The combination of displays and calls of several males is probably more successful in attracting females than the same number of isolated males. The males spread their plumes, flap the wings, sway and posture,

Bison bull, immediately recognizable by its heavy forequarters and low-slung head.

and even display hanging upside down, as in the greater bird of paradise, *Paradisea apoda*. The king bird of paradise hangs upside down with spread wings and walks along the undersides of branches.

The nests are cup-shaped and are built in thick creepers or in forked branches, but the king bird of paradise nests in tree cavities. The clutch is of two pale pink to brown eggs.

BISON, *Bison bison,* or buffalo, the largest land animal in America, the bulls weighing up to a ton or more (907 kg) and standing 5 to 6 ft (1.5-1.8 m) at the shoulder. The huge head, which appears extra large because of the long hair, the great hump on the shoulders, the dark brown woolly hair covering the forequarters, and the small naked hips are characteristic of the bull. Females are smaller and less striking. Both sexes have horns, but those of the bull are more massive. The calves are a light tawny color in contrast to their dark brown parents.

During the breeding season from July to September, the bulls leave their male herds and mingle with the cows and calves, the strongest bulls tending individual cows until copulation is completed. Cows are sexually mature at two and a half years, and normally produce a single calf after a gestation-period of about nine months.

The total population of bison was probably some 60 to 70 million animals when the white man first arrived in America. Despite their great abundance, the bison probably roamed mostly in groups of 20-30 individuals and only occasionally in herds of 100 or more, and the herds seem to have migrated only to the limited extent of perhaps 200 to 300 miles (300 to 500 km) rather than making the long northward

and southward treks described in early accounts.

BITTERLING, *Rhodeus sericeus,* a 3-in. (8-cm) carplike fish from lowland waters of Europe, the breeding cycle of which involves a freshwater mussel. The bitterling is normally silver, but in the breeding season the male develops violet and blue iridescence along the flanks and red on the belly, the fins becoming bright red edged with black. In color, the female is less spectacular, but develops a long pink ovipositor from the anal fin. It is this ovipositor, a 2-in. (5-cm) tube for depositing the eggs, that gives a clue to the extraordinary breeding biology of the bitterling. Most fishes of the family Cyprinidae merely scatter their eggs, but the female bitterling carefully deposits its eggs inside a freshwater mussel, hence the need for the long ovipositor. The male then sheds its milt, and this is drawn in by the inhalant siphon of the mussel and the eggs are fertilized inside. Normally, one has only to touch these mussels and the two halves of the shell are snapped shut. The female bitterling, however, conditions the mussel by repeatedly nudging it with its mouth. This remarkable nursery for the eggs is clearly of great value to fishes that would otherwise lose a large percentage of the eggs through predators. It is difficult to see how this association between the mussel and fish arose, but in return the mussel releases its larvae while the bitterling is laying its eggs. These fasten onto the skin of the bitterling, which carries them around until they change into young mussels and fall to the bottom.

BITTERNS, fairly large birds with a distinctive booming call, usually restricted to reedbeds and marshes. The cosmopolitan subfamily of bitterns comprises two genera, *Botaurus* and *Ixobrychus,* the genus *Botaurus* containing the Eurasian bittern, *B. stellaris,* the Australian *B. poiciloptilus,* the American *B. lentiginosus,* and the South American *B. pinnatus,* which replace each other and together form a superspecies. Of the genus *Ixobrychus,* the least bittern *I. exilis* of North America, the little bittern *I. minutus* of Europe, Asia, Africa and Australia and the Chinese little bittern *I. sinensis* form another superspecies.

The bittern *B. stellaris* is resident in British and European marshes, whereas the smaller American bittern is a rare vagrant to Europe, having been recorded in the Channel Islands, the Faeroes, Iceland, at least once in Germany and over 50 times in Britain.

The sexes of all the *Botaurus* species have similar plumage, that of the bittern *B. stellaris* being soft golden-brown and owllike, heavily mottled with black above, longitudinally streaked below. A mane of long feathers on the neck and throat can be erected at will. Their necks and legs are shorter than those of herons. The 3-in. (7.5-cm) bill is yellowish-green, the eyes yellow and the legs and feet pale green. They have powder-down patches on the breast and rump, one pair fewer than the herons, and the toothed middle claw is used to apply powder-down to the contour feathers in preening. The finely divided particles of the disintegrating filaments of powder-down coagulates slime, which coats the plumage when bitterns feed on eels. The bittern measures 30 in. (76 cm) and weighs about 3 lb (1.3 kg).

Distribution is restricted by nesting requirements to reedbeds and rank vegetation by sluggish water. The extensive marshes of North Jutland, the "plassen" of Zuid-Holland and Utrecht, Austria's Neusiedler See and the Danube marshes provide typically ideal haunts for many bitterns. In Europe bitterns breed from 60° N to the Mediterranean basin, and very large numbers are present in the USSR.

At any hour of the day or night from February

The bitterling *Rhodeus sericeus,* a carplike freshwater fish from Europe, inspecting a freshwater mussel.

Bittern, also an inhabitant of reed-beds throughout Europe, is famous for the booming call of the male during the breeding season. ▷

Little bittern looks like a miniature night heron. It haunts the reed-beds and scrub all over Europe.

to late June the male utters a resonant boom which may be repeated from three to six times and is notable for its carrying power, which is certainly over 1 mile (1.6 km).

The nest is built up above water level on matted roots in a reedbed and is constructed of reeds and sedge, lined with finer material. Three to six olive-brown eggs, without gloss, laid at intervals of two or three days, are incubated by the female alone. The young, too, are fed entirely by the female. Small mammals, birds (including young reedlings), water insects and crustaceans are all eaten, but the principal diet appears to be fish, mainly eels, and frogs.

The protective coloration is much enhanced by the bird's habit, when disturbed, of standing rigid with bill pointed skyward, presenting to the intruder the striped undersurface that blends so well with the surrounding reeds as to render the bird virtually invisible.

The American bittern, *B. lentiginosus*, is not so restricted to reedbeds as is the Eurasian species and is often observed feeding in open wet meadows. Its plumage lacks the black on the upperparts that are finely vermiculated, and the primaries are uniform gray-brown without barring. It is smaller, only 23 in. (58 cm) long.

The little bittern, *Ixobrychus minutus*, is the only European member of a genus containing a dozen species. Sluggish rivers, backwaters, even small ponds satisfy this bird's nesting requirements, and they nest practically throughout Europe, building sometimes just above water level and sometimes in willows up to 10 ft (3 m) above the water.

BLACK BUCK, *Antilope cervicapra,* the most common Indian antelope, closely related to the gazelles. Females and young are yellow-fawn with a white belly and white eye ring; after three years of age, the male begins to turn black. In the south of India males are usually dark brown; in all cases the true sable livery is assumed only during the rut, after the rains. Only the male has horns: these are ringed, closely spiraled, 25 in. (65 cm) long in the south, shorter in the north. Black buck stand

32 in. (80 cm) high and weigh 90 lb (40 kg). The black buck is found on all plains of peninsular India, south from Surat, and east into Bengal, west into Punjab and Rajputana. Herds number 20-30, and are led by a female; during the breeding season, the bucks split the herds up into harems; a buck struts about in front of his harem making short challenging grunts, with head thrown up so that the horns lie along the back, and the face glands widely open. The main rut occurs in February and March, but some breeding occurs in all seasons. Gestation is 180 days.

Once there were 4 million black buck in India; they were hunted by maharajahs with tame cheetah. Now there are some thousands, but they are thin on the ground. They occur in all plains country, entering both open forests and grasslands, as well as in cultivated areas. They graze in the morning, lie up during the heat of the day, and feed again in late afternoon. Eyesight is very keen. When alarmed, the herd moves off in light leaps and bounds, like gazelles, then breaks into a gallop.

BLACKFISH, *Centrolophus niger,* and its close relative the Cornish blackfish, *C. britannicus,* are marine stromateid fishes related to the barrelfish. These species are found in the eastern North Atlantic and Mediterranean. In *C. niger,* which reaches 3 ft (90 cm) in length, the body is elongated and purplish-black but paler on the head and belly. Little is known of the feeding habits of these fishes, but young specimens have been found which had fed on pollack.

In certain parts of the British Isles the name blackfish is also used for ripe female salmon.

BLACK GROUSE, two striking species of birds of the grouse family: *Tetrao tetrix* of northern Europe, including Britain, and *T. mlokosiewiczi* of Eurasia. The male black grouse, or blackcock, is 21 in. (53 cm) long and has a glossy blue-black plumage with an unusual lyre-shaped tail set off by white undertail coverts. There are also white markings on the wings and a bright red wattle above each eye. During the breeding season males from

adjacent territories gather at traditional display grounds, or "leks," for daily displays in which the tail is spread and raised, the wings partly opened and a variety of calls and posturings performed. The female black grouse or grayhen, 16 in. (40 cm) long, has the rufous-brown cryptic plumage typical of female ground birds. She becomes more interested in the lek behavior as spring progresses, eventually mating with one of the displaying birds and then leaving to rear the brood alone.

BLACK-HEADED GULL, *Larus ridibundus,* a smallish gull (family Laridae) found through much of Europe. It is 15 in. (38 cm) long and may be distinguished in flight by the white leading edge to the wings. It has a dark red bill and legs and, in summer, a chocolate brown head. In winter the head is white with a dark marking behind the eye. Young birds before their first winter are mottled gray-brown above, but have the pale leading edge to the wings. The black-headed gull is common in many areas, in town and cities as well as open country, and inland as well as on the coast. It nests in colonies on the ground in marshes or on islands, moorland, shingle beaches, sand dunes and other similar situations. Other features of its life history are very similar to those of other gulls.

BLACK WIDOW SPIDER, *Latrodectus mactans,* jet-black North American spider with a sinister reputation. Underneath the abdomen is a small hour-glass mark in vivid scarlet, which represents warning coloration. The bite of this spider causes intense pain accompanied by symptoms of nausea, partial paralysis and

Black-headed gull at nest on low-lying marsh.

difficulty in breathing, and the venom is said to be 15 times as potent as that of a rattlesnake. Fortunately the quantity injected is far less than that of a snake. Death is known, but this is not usual. A serum has been prepared in America, but in its absence doctors usually give intravenous injections of 10 cc of 10% calcium chloride or gluconate.

The black widow is not a large spider, being smaller than $\frac{3}{4}$ in. (18 mm). It lives a retiring existence and is not aggressive. Men are likely to be bitten more frequently than women because the cavity beneath the seat of earth privies is often selected by the spider to spin the strong coarse threads of its snare, and anything brushing against these threads is apt to be bitten.

Most members of the genus *Latrodectus* share the same evil reputation. These include the malmignatte of southern Europe, the redback of Australia, the katipo of New Zealand and others in Africa. These can usually be distinguished from the American black widow by the possession of scarlet spots or bands, but some have proved to be only subspecies. All have the globular abdomen typical of the family.

BLEAK, *Alburnus alburnus,* a small freshwater carplike fish found in slow-flowing waters and large lakes in Europe north of the Alps, but which has also been reported from brackish water in the Baltic. The bleak is gregarious and is frequently seen shoaling at the surface and catching insects; it will also browse on the bottom for aquatic larvae. During the breeding season of April to June the males develop a green-blue coloration on the back, and the fins become orange. The sticky eggs are laid between stones in shallow running water. Lake-dwelling bleak migrate up feeder streams to breed. Large numbers of bleak were formerly caught and their silvery scales used in the manufacture of artificial pearls. Since the adults are only about 8 in. (20 cm) long and are practically tasteless, bleak are rarely caught for any other purpose.

BLENNIES, a group of fairly small elongated marine and brackish water fishes comprising 15 families grouped in the suborder Blennioidei. The name "blenny" comes from the Latin *blennius* (the scientific name used for one of the principal genera) and indicates a worthless sea fish. Blennies have very long dorsal fins, and the pelvic fins, when present, are located in front of the pectoral fins under the head. Typically, the head is large, with a steeply rising forehead, and the body tapers evenly to the tail. A fleshy flap, the orbital tentacle, is often present just above the eyes. Blennies are carnivorous, bottom-living fishes, frequently well camouflaged for a life among rocks and in shallow waters. They are almost worldwide in their distribution.

The scaleless blennies (family Blenniidae) are among the most common shore fishes of the North Atlantic but are also found in most other seas. The butterfly blenny has a similar range but only reaches 6 in. (15 cm) in length. The first part of the dorsal fin is high and bears a black spot with a light border around it.

Klipfishes (family Clinidae) occur mostly in the southern hemisphere. These often highly colored fishes are live bearers, the male having an intromittent organ, formed from the spines of the anal fin, for introducing sperms into the female. *Heterostichus rostratus* from the Pacific coast of North America grows to 2 ft (60 cm) in length and is able to change its color to match its surroundings. *Neoclinus blanchardi* from the American Pacific has a greatly enlarged mouth with the lower jaw elongated resembling what Dr. Earl Herald terms a "vast scoop shovel." Some of the clinid fishes have fights while defending their territories, the jaws being opened wide and the gill covers extended to display two spots like eyes. After several sessions of display and aggression, one may bite the other or merely give way and swim off. Wolf fishes (family Anarhichadidae) from the

North Atlantic are the giants of the blennioid tribe. As a rule, all teeth in a fish's jaws are approximately the same shape, but in the wolf fishes the teeth at the front of the jaws are long and pointed whereas those at the back are flattened, crushing teeth. Two species of wolf fish are found in the northern Atlantic, and both are fished commercially. The wolf fish *Anarhichas lupus* is the more common in the North Sea and reaches 5 ft (1.5 m) in length. The spotted wolf fish, *A. minor,* grows to about 6 ft (1.8 m) long. Both species are sometimes known as cat fishes.

BLESBOK, *Damaliscus damaliscus phillipsi,* one of two subspecies of bastard hartebeest, the other being the bontebok. It is reddish, darker on the sides of the face, the neck, flanks and outside of the limbs as far as the knees and hocks. The rump, face-blaze and shanks, as

Common blenny or shanny out of water.

well as the underside, are white. The face-blaze is divided by a dark bar above the eyes, the rump patch is small and does not go around the tail-root, and the knees and hocks are not white. It stands approximately 36-48 in. (900-1,200 mm) high, and weighs 250-300 lb (114-136 kg).

Blesbok were found from the Cradock and Cathcart divisions of the Cape to the Orange Free State, southern Botswana and southern Transvaal; this area was separated by 200 miles from the nearest bontebok. Like its conspecific it was severely overhunted and survived only on farms; it has now been reintroduced over much of its former range, and has been introduced into northern Transvaal.

BLIND SNAKES or worm snakes are among the most primitive of living snakes. The 300 species are grouped into two families, the Typhlopidae, which contains the vast majority of the species, and the Anomalepididae. Blind snakes occur throughout the tropical and subtropical zones and extend also into South Africa and the southern region of Australia, but not Tasmania. They have reached numerous oceanic islands. All blind snakes are harmless; most are quite small and wormlike, though a few reach a length of 2 ft (61 cm) or more. Blind snakes have poorly developed teeth, on the mobile upper jaw only. Vestiges of the hind limbs are sometimes present.

Blind snakes feed on a variety of small soil animals, such as worms, termites, and ants and

their eggs and larvae. They are nocturnal and live underground, or under large flat stones, rotting logs or stumps, or in termite nests. The tail is very short and ends in a small, sharp spine, which is pushed into the soil and provides a purchase when the animal is moving forward. The eyes are small and very poorly developed, and each appears as a dark spot covered over by a transparent scale. Probably they can only distinguish between light and darkness.

BLISTER BEETLES, winged true beetles with characteristically soft bodies and long legs, comprising, with the flightless oil beetles, the family Meloidae of about 2,000 species.

The blister beetles are widespread and are particularly notable for the complex and hazardous life history of many species. For example, in one species the larvae hatch from eggs deposited in autumn near the nests of certain bees (*Anthophora* spp). In spring only a few of these minute larvae out of the 2-10,000 eggs laid by each female succeed in grasping the hairy bodies of the male bees, later transferring to female bees. The female bee constructs a nest of cells in the ground, each cell being stocked with honey, pollen and a single egg. When the bee deposits the egg, the beetle larva drops onto it and is then sealed up by the bee in the cell. The larva feeds first on the bee's egg and then on the food store of honey and pollen, and after a complicated development a mature beetle emerges from the cell.

Blister beetles contain the substance cantharidin in many of their tissues and structures. Perhaps the best known in this respect is the "Spanish fly," *Lytta vesicatoria,* a common insect in parts of southern Europe. In recent times the administration of cantharidin as an aphrodisiac by nonqualified persons has lead to deaths, and during the 19th century its use as a blistering agent for many kinds of ailments caused great misery.

BLUEBIRD, a name usually applied to three small North American thrushes of the genus *Sialia.* The male is blue in one species and blue and chestnut in the others. The females and young are brown with some blue on wings and tail. They are mainly insect feeders and are well known because they live around farms and orchards and nest in artificial nest boxes as well as natural holes. The name is also given to the fairy bluebird, *Irena puella,* which is a leafbird of the oriental region. The male is black and vivid glossy blue, but the female is dull blue.

BLUEBOTTLE or blowflies, names given to two similar species of true flies, *Calliphora erythrocephala* and *C. vomitoria,* which are a little under $\frac{1}{2}$ in. ($1\frac{1}{4}$ cm) long, stoutly built and of a metallic blue color. They sip sugary liquids, and the loud, buzzing, restless flight of females, which seems so frantic and is so irritating, is characteristic of their unceasing search for a suitable place to lay their eggs. They are often to be found in the kitchen, where their sense of smell directs them to any flesh food left unprotected. When no such site is to be found they fly backward and forward sometimes into the darkest corners and sometimes to bright lights. Out of doors they are found making their buzzing flight over decaying animal matter and dung, but generally they lay their eggs in the tissues of animals recently dead. Each female lays up to 600 eggs, which hatch in about a day. These larvae are the "gentles" used by anglers as bait, but they are commonly also called maggots. Under favorable conditions the larvae pupate in about a week and emerge as adults a fortnight later. Several other metallic blue flies occurring in the tropics are commonly called bluebottles. Among these are the screw worms. Greenbottle refers to a number of metallic green flies, many of them closely related to bluebottles. All of them have larvae that feed on decaying animal matter, but greenbottles are less prone to come indoors. Some of them lay their eggs in the wool of live sheep, and the larvae attack the flesh, causing a disease known as "sheep strike," which is often fatal.

Bluebottle fly *Calliphora*.

BLUEFISH, *Pomatomus saltatrix,* a marine perchlike species among the most savage and bloodthirsty of all fishes. It does not attack humans, however. Bluefishes are deep-chested, slender-tailed and derive their name from their color. They live in fast-moving shoals in all tropical and subtropical waters except the eastern Pacific, and reach 30 lb (13 kg) in weight.

The large shoals of bluefishes that move up and down the American Atlantic shores principally feed on the enormous shoals of menhaden, *Brevoortia,* species related to the shads. The bluefishes have the reputation for being animated chopping machines, the sole aim of which appears to be to cut to pieces, or otherwise mutilate, as many fish as possible in a short time. They have been seen to act like a pack of wolves, driving part of a shoal of menhaden into shallow coves from which they cannot escape. The menhaden apparently fling themselves onto the beach in an effort to escape from these savage predators. The sea is bloodstained and littered with pieces of fish after the bluefishes have eaten. Various impressive statistics have been cited to illustrate the destructive powers of the bluefishes. As many as 1 billion bluefishes may occur each summer season off American coasts, and if each eats only ten menhaden a day and the season lasts for 120 days, then the stock of menhaden must be depleted by 1,200,000,000,000 individuals during that time.

BLUE WHALE, or Sibbald's rorqual, *Balaenoptera musculus,* may reach 100 ft (33 m) in length and weigh up to 130 tons. It is dark blue-gray overall, with paler markings. Blue whales have been reported to have a sustained speed of 10-12 knots while submerged and can reach 15 knots. They are known to dive to over 1,500 ft (450 m) where the water pressure is 45 atmospheres, and can stay submerged for two hours. Blue whales seldom travel in schools and are usually seen singly or in pairs. They are found both north and south of the equator, and, although in general the groups remain apart, a few cross the equator.

During the winter blue whales move toward the warm waters for mating and calving, but in summer return to the richer krill feeding grounds of the Antarctic or Arctic.

A pygmy blue whale has recently been found, which may be a distinct species. It is, for the moment, classified as a subspecies *B. musculus brevicauda,* so named because of its relatively shorter tail. No individual captured had exceeded 80 ft (24.4 m); males generally reach 68 ft (20.7 m) and females about 72 ft (22 m). They tend to be found in different areas from the blue whale and only overlap somewhat on feeding ranges.

BOAS, nonvenomous snakes, most of which inhabit the warmer regions of the New World. Probably the best known is the boa constrictor, *Boa constrictor,* an inhabitant of Central America, tropical South America and the Lesser Antilles. Primarily a surface dweller, it may be found in a variety of habitats ranging from tropical rain forest to semidesert regions. It is divided into eight subspecies, each having its own area of distribution. Many stories have been written about the great size and prowess of the boa constrictor, but in reality it is only the fifth largest of living snakes, having a maximum length of a little over 18 ft (5.5 m). Its food consists mainly of small mammals and the occasional bird and lizard. Like most boas, its prey is killed by constriction before being swallowed whole.

This python clearly shows the notch in the upper lip through which the tongue is protruded.

The habitat of the boas is varied, ranging from arboreal to fossorial and semiaquatic. Some species, such as the boa constrictor and the rainbow boa, *Epicrates cenchria,* are surface dwellers. These tend to inhabit scrubland and wooded regions, where their blotched or reticulate pattern forms an excellent camouflage. The rainbow boa is primarily a surface dweller

Royal python curled into a ball, its defensive posture.

feeding on small rodents, but it has been known to climb trees and devour bats. It rarely exceeds 4 ft (122 cm) in length and is called the rainbow boa because of the bright green and blue iridescent sheen reflected by its scales.

The arboreal boas include the South American tree boas, *Corallus,* and the Malagasy tree boa, *Sanzinia madagascariensis.* These are often blotched like the surface dwellers, but two species, the emerald tree boa, *Corallus caninus,* of South America, and the adult of the Papuan tree python have acquired an effective camouflage in their bright green color with whitish markings. These two species are also alike inasmuch as the young differ from the adults in being yellow or pinkish brown with darker markings.

BOATBILL, *Cochlearius cochlearius,* or boatbilled heron, a small, gray heron with a black crown and long, black ornamental plumes on the back of the head. The bill is broad and scooplike. Some ornithologists consider the boatbill to be an aberrant night heron, although others place it in a family of its own—the Cochlearidae.

The boatbill is confined to Central America and northern South America, where it inhabits mangrove swamps. It feeds at night, as its large eyes suggest, and is usually inactive by day. Little is known of its diet or how it uses the oddly shaped bill. The clutch of two to four pale blue eggs are laid on a platform of dead sticks, built in a mangrove tree.

BOBCAT, *Lynx rufus,* the wide-ranging wildcat of America, which probably got its name from its short tail and a lolloping gait rather like that of a rabbit. Its total length is about 2½-3 ft (76-92 cm) with an average weight of 15-20 lb (7-9 kg), but it can be much larger. The color varies with the race and habitat but in general is brown spotted with gray or white. The short tail has a black bar on the upper side fringed with white hairs, and the ears are tipped with pointed tufts of hair that are said to improve the bobcat's hearing. Bobcats are found throughout most of the United States, Mexico and the south of Canada. Because of their small size and the variety of prey they feed on, ranging from small rodents to deer and domestic animals, they have largely survived the spread of agriculture. Usually two kits are born, at any time of the year, in a den or under logs.

A young bongo *Boocerus euryceros*. Bongos are probably the most handsome of antelopes.

BONGO, *Boocerus euryceros*, close relative of the eland but lacking the dewlap and the frontal tuft, and with open-spiraled horns more like the kudu group. The average height is 4 ft (120 cm), weight 480 lb (220 kg). There is an erect mane from shoulder to rump. The color is reddish, with 11-14 transverse white stripes, a white chevron between the eyes, white cheek spots, throat band, lips and chin, and a white stripe down the inner side of the limbs. The horns, which resemble those of a sitatunga, may be over 3 ft (1 m).

Bongos, essentially forest elands, inhabit the tropical forest belt of Africa from Sierra Leone to Togo. They occur again from Cameroun, south of the Sanaga River, across the Congo into the extreme southern Sudan, but are not found in Nigeria. East of the Congo, bongos are found only in isolated montane forests in Kenya. Bongos ascend to 7-10,000 ft (2,100-3,050 m). They live in the densest parts of the forest, where they move about by day. Bulls are often solitary, but some join with the herds of cows and calves, which may number as many as 20. They are fond of wallowing. They can move very fast through the forest, slipping under obstacles with the ease of a limbo dancer, with head held low and horns laid back. Bongos feed on leaves and shoots and can rear on their hind legs, planting the front hoofs on tree trunks, to browse. The horns are kept sharp by constant rubbing, but bongo appear to be as unaggressive and placid as eland. They bleat, like calves or like elands, rather than barking like bushbucks or kudus.

BONITOS, sometimes spelt bonitas in the United States, a name used for certain of the smaller tunalike fishes. In Europe, *Katsuwonus pelamis* is known as the oceanic bonito, but in the United States this species is referred to as the skipjack. The bonitos, in the broad sense of the term, are found in both the Atlantic and the Indo-Pacific region. They are highly stream-lined fishes, often with the fins folding into grooves, and the tail is crescentic. Their bodies are superbly adapted for an oceanic life. The pelamid or belted bonito, *Sarda sarda*, known in the United States as the Atlantic bonito, is found on both sides of the Atlantic as well as in the Mediterranean and sometimes reaches British coasts. It attains 3 ft (10 cm) in length, and its high-quality white meat is canned in the United States. The oceanic bonito has a similar distribution and reaches about the same size. It differs from the belted bonito in having bluish bands running horizontally along the lower part of the body (the bands run obliquely on the upper part of the body in the belted bonito). It has a remarkable turn of speed, about 25 mph (40 kph), which enables it to chase flying fishes, often leaping clear out of the water to do so. It is also said to circle shoals of fishes and will then charge into the middle of the shoal.

BONTEBOK, *Damaliscus damaliscus dorcas*, one of two subspecies of bastard hartebeest, the other being the blesbok. It is reddish, darker on the sides of the face, the neck, flank and outsides of the limbs as far as the knees and hocks. The rump, face-blaze and shanks, as well as the underside, are white. It differs from the blesbok in the face-blaze not being divided by a bar and in the white rump patch going all around the tail-root. Also, its knees and hocks are white.

The bontebok used to be found from the coast of the southwestern Cape from the Bot River (near Caledon) to Mossel Bay, restricted in the north by the Sondereind and Langeberge mountains; after severe reduction in numbers during the last century, it survived only on farms, but in spite of protection only 17 were counted in 1931. However, the numbers began to increase, and today there are about 750; the largest concentration is on Bowker's farm, E. Cape; the second largest, in the Bontebok National Park. Originally this park was in the Bredasdorp area; in 1961 it was moved to Swellendam because of the constant water inundation of the previous site and the consequent infestation of many animals with water-borne parasites.

Bontebok and blesbok are among the fastest antelope in all Africa. They run extended, low off the ground. When alarmed, they first stand in a stiff pose with head held high and neck up, the limbs held rigid and spread apart. The flight distance is 300 yards.

BOOBIES, fairly large seabirds of the genus *Sula* closely related to the gannets, with which they form the family Sulidae and which they replace in the tropical waters of the world. They resemble gannets in general appearance, physiology and many details of their life history, but are smaller and considerably lighter,

weighing from 2-4 lb (0.9-2 kg), and have a more extensive area of bare facial skin, which, together with the legs and feet, is more brightly colored in some species.

There are six species. The masked booby, *Sula dactylatra*, the red-footed booby, *Sula sula*, and the brown booby, *Sula leucogaster*, are all "pantropical" and occur widely in the three major oceans. The Peruvian booby, *Sula variegata*, and the blue-footed booby, *Sula nebouxii*, are confined to the eastern Pacific, the former occurring in the fish-rich, cold Humboldt current area off Peru and Chile (where it breeds on the famous "guano islands") and the latter occurring farther north as far as Baja California. Finally, the Abbott's booby, *Sula abbotti*, breeds only on Christmas Island in the Indian Ocean. The nestling periods of boobies are mostly longer than in gannets, lasting up to 20 weeks or more in the Abbott's booby. Unlike young gannets, the juvenile booby has no surplus fat but returns to the birth site for several weeks at least after fledging to be fed by its parents.

BOOKLOUSE, name given to over 1,000 very small to minute insects living in temperate and tropical regions. They have plump bodies, may be winged or wingless, and are characterized by modification of the maxillae of the mouthparts into a pair of chitinous rods or "picks" which, together with the mandibles, enable the animal to bite or rasp food from the surface of bark or leaves. They take their name from one wingless species, *Liposcelis divinatorius*, found especially in old books, feeding on the flour, size and glue of the bindings and on the molds that grow on old paper.

Most species are winged, but many are wingless

or short-winged, and these occur naturally under the bark of trees or in the nests of birds and mammals. Some have established themselves in human dwellings or in warehouses and ships' holds, thereby becoming cosmopolitan, and are classed as minor pests.

Also known as psocids, from their family name, they are among the few insects that spin silk when adult.

BOOMSLANG, *Dispholidus typus*, large African tree snake with three enlarged grooved fangs in the upper jaw below the eye. It is the most venomous of the back-fanged snakes, and its bite can prove fatal to man. Boomslang is Afrikaans for "tree snake."

The boomslang averages $4\frac{1}{2}$ ft (1.4 m) in length. It has a short head with very large eyes and a slender body and tail covered above with narrow, oblique, strongly keeled scales. It is common throughout the well-wooded parts of Africa south of the Sahara but is absent from the rain forest and semidesert regions.

Toward the end of the dry season the female boomslang lays 5-16 elongated eggs of about 1 in. (25 mm) in length. The newly hatched young are about 15 in. (38 cm) in length.

The boomslang hunts by day and may stay in a tree or group of trees for several days if food is plentiful. Its diet consists largely of chameleons, but during the nesting season many fledgling birds and eggs are eaten; adult birds are rarely caught. Lizards, frogs and rats are also devoured, but other snakes are rarely attacked. If disturbed, the boomslang will always try to escape at speed, but when cornered it inflates its throat with air, giving the impression of an enormous head, then makes savage lunges at its aggressor with gaping jaws. Its venom is

Blue-faced booby or masked booby, one of six species of seabirds closely related to gannets. This species is found in the Atlantic, Indian and Pacific Oceans.

extremely toxic, destroying the fibrinogen in the blood and causing extensive internal bleeding. Because the amount of venom produced by a boomslang is very small, the specific antivenom is in short supply.

BOT FLY, a fly, the larvae of which are better known than the fly and are parasitic in warm-blooded animals. The larvae of certain blow-flies can infest wounds and attack living flesh, as well as living in carrion, and bot flies represent a further step in evolution, the larvae having become completely parasitic. Since the larvae have a virtually unlimited supply of food, the adult flies have ceased to feed at all, and some have even lost their mouthparts. Adults are rarely seen, and some species are known only from specimens bred from larvae, which, in contrast, may be extremely numerous. The larvae are the "bots" and are generally less pointed than maggots and covered with rows of strong spines that help them to move about in the tissues of their host. The hind spiracles have many small pores, which make them less liable to become clogged.

The sheep nostril fly, *Oestrus ovis*, lives in the head sinuses of sheep, feeding mostly on mucus. *Cephenomyia* similarly infests deer, *Tracheomyia* the throats of kangaroos, *Cephalopina titillator* the head cavities of camels and *Pharyngobolus africanus* the gullet of African elephants. *Hypoderma bovis* and *H. lineatum* are the common warble flies of cattle in Europe and are more directly parasitic. Their larvae migrate through the body until they come to rest under the skin of the back where they form boils with an opening to the exterior for breathing purposes. *Oedemagena tarandi* is a warble fly of reindeer. Horses, zebras and

rhinoceroses are often heavily infested with stomach bots belonging to the family Gasterophilidae.

Bots do not normally attack man, but infrequently the eggs of *Oestrus* or *Hypoderma* may get into the eyes of shepherds or herdsmen, and the small first-stage larvae can cause irritation or even more serious damage to the eye.

BOWERBIRDS vary from 9-15 in. (23-38 cm) long and show a wide range of colors and patterns in their plumage. Bold patterns of green, orange, lavender, and yellow with gray or black are found in many species; some have a plain gray or brown plumage, and a few are spotted. In the more brightly colored species, the male bird is much brighter than the female, but the sexes are alike in the dull-colored forms. Some of the species have a crest of elongated feathers, the crest often being brilliantly colored. In a few species it forms an elaborate ruff or a mane hanging over the upper back. The bill is slightly hooked at the tip in all of the bowerbirds, but in some it is slightly down-curved, in others straight, in some it is thin and weak, in others again rather heavy, and in a few species the upper mandible has some small toothlike notches along its cutting edge.

Bowerbirds are most numerous in New Guinea. The stagemaker or tooth-billed bowerbird, *Scenopoeetes dentirostris*, is found only in Australia. It is an olive-brown bird, about 11 in. (28 cm) long, with prominent pale stripes on the underparts. The golden bowerbird, *P. newtoniana*, is found only in the mountain forests of northern Queensland, Australia. It is about 9 in. (23 cm) long, rather short-billed, with different plumages in the male and female birds. The male is bright olive-green with

Male satin bowerbird of the coastal area of eastern Australia, best known of the bowerbirds. It is decorating its bower with pebbles, shells and blue feathers.

yellow underparts, head, neck and tail, while the female is dull yellow-green with gray underparts. The satin bowerbird, *Ptilonorhynchus violaceus*, is the only member of its genus and is confined to eastern Australia. About 1 ft (30 cm) long, it is rather long-billed, the bill also being heavy and straight. The plumage is glossy black in the male and gray-green in the female. The bowerbirds are so called because of the complicated and often highly decorated structures that the males of some species use when they are displaying. These sometimes take the form of cleared areas the size of a table top containing a domed tunnel of sticks, decorated with brightly colored stones, fresh flowers, spiders' webs and colored insects' skeletons. Some of these bowers are so impressive that, when they were first discovered, the explorers believed that they could only be the product of human skill and artistry. The male defends a perch from which it sings, above a cleared patch of ground among slender young trees. It decorates its private lawn with large leaves laid upside down and a few snail shells. In display, the male flicks its wings open, bobs its head from side to side with its bill gaping, hops about erratically, fluffs its breast feathers and holds a leaf in the bill for long periods. The males only see the females at mating time and take no part in rearing the young. Each adult male defends a separate court and bower inside a traditional courtship area. The bower consists of a solid mat of small sticks with a wall of sticks on each side. The whole structure is painted with a paint made of vegetable juices and is brilliantly decorated by the male with shells, flowers, leaves and dead insects. In display, the male spends most of its time at the bower, uttering scraping, grating, cackling, churring and squeaking notes. It dances about with its tail raised over its back, jumps right over the bower, points the bill to the ground and becomes so excited that its eyes bulge outward. The female bird enters the bower and may crouch down as a signal to the male that she is ready to mate. The females appear to remain at the bower with the male for several days, but after this they separate, and the' female builds a nest and raises the young alone. It seems that the complexity of bowerbirds' plumage patterns bears a fairly close relationship to the complexity of the bower that is built. In the species that build no bowers, or only clear a ground court, the plumage is often colored with red, orange, yellow, green or black in varying combinations of brilliance, but in the species that build the most elaborate and brilliantly decorated bowers the birds are disappointingly dull in color, with the exception sometimes of the crest.

So far as is known, all of the bowerbirds live mainly on the fruits of trees and bushes, supplementing this diet with insects, larvae, spiders, and sometimes small snakes and lizards, tree frogs and seeds. Bowerbird nests vary from bulky cups built in bushes, to a bulky cup in a tree hole and a frail, shallow cup built in a bush. The female usually builds the nest alone, lays the clutch of from two to five eggs, incubates them and then feeds the young until they fledge and for a week or two afterward. Incubation periods of between 12-15 days have been recorded and approximate fledging periods of from 13-20 days, though some of these were probably inaccurately recorded.

BOWFIN, *Amia calva*, a member of one of the two surviving groups of Holostei, primitive ray-finned fishes that gave rise to all the modern bony fishes. Fossil bowfins have been found in Europe, but the only surviving species is now confined to the eastern side of North America. It is a cylindrical, solid-looking fish with a long dorsal fin and a heavy armor of scales. The body is dull brownish-green in color, lighter underneath, with several dark vertical stripes. A black spot is found near the base of the tail, margined in males with yellow. Underneath the lower jaw is a bony plate, the gular plate, a relict from its more primitive ancestors. In the intestine there are remnants of a spiral valve, a device that is found in many primitive fishes increasing considerably the digestive surface of the intestine. A spiral valve is also found in sharks. The swim bladder has a cellular structure that enables the bowfin to breathe atmospheric oxygen. Whereas most fishes swim by undulations of the body, the bowfin cruises majestically by a series of waves passing along its long dorsal fin. The normal method is adopted, however, for faster swimming.

Bowfins live in warm, sluggish waters, especially in shallow and weedy areas. In the breeding season in early summer the males make a round nest on sandy or gravelly bottoms or in clearings in weed patches. They then mate with

Another species of the boxfishes or trunkfishes, *Ostracion lentiginosus*.

Brain coral, named for the slight resemblance to the convolutions of the human brain.

several females, and after the eggs are laid guard them until the fry hatch and can swim well. They are carnivorous and seem to have a particular liking for game fishes. A large bowfin may reach almost 3 ft (100 cm) in length.

BOXFISHES or trunkfishes, fishes belonging to the genus *Ostracion*, the head and body of which are enclosed in a solid box of bony plates with only the fins, jaws and the end of the tail projecting and free to move. In cross section, the boxlike body is triangular, rectangular or pentagonal, the underside being flat. There are several species, growing to 20 in. (50 cm) found around the coral reefs and coasts of the Indo-Pacific area. Since the body is rigid, swimming can only be accomplished by sculling movements of the unpaired fins, with the pectoral fins helping to stabilize what would otherwise

be highly erratic movements. Many of the boxfishes are brightly colored with patterns and spots of red and blue, yellow and blue, blue with a red band, and so on. These bright colors probably serve to warn predatory fishes that the owner is not edible. When boxfishes are attacked, they secrete a virulent poison into the water that can kill other fishes.

BOX TURTLE, *Terrapene*, an animal like a European garden tortoise living in the eastern United States, where it is native, from New England to Florida and Texas. Its peculiarity is that the shell on the underside is hinged across the middle, and the two halves can be brought up in front and behind to close the shell completely after the head and legs have been withdrawn.

It is reported that some hounds used in deer

hunting and gundogs used for quail become addicted to box turtles. A hound may find a box turtle and carry it around instead of tracking deer, or a gundog may "point" a box turtle instead of a bird. Some dogs will habitually seek out the turtles, bring them home and bury them, but without doing them harm.

BRAIN CORAL, *Meandrina,* colonial polyps belonging to the true or stony corals. The surface of the massive, calcareous skeleton is marked by long curved depressions and bears a striking resemblance to the human brain. Each colony generally arises from a single planula larva, which settles onto a suitable object, such as a shell, and grows into a coral polyp. The polyp then starts to bud off other polyps. At an early stage in colony formation, each polyp is partly enclosed in a calcareous cup or theca, which bears ridges or sclerosepta radiating from the margin. But the budding of new polyps occurs very rapidly and in a manner termed intratentacular, a new mouth arising on the oral disc of the older polyp and inside the same ring of tentacles. In many corals, the skeleton becomes partitioned off into two cups to accommodate the old and new polyps. In brain corals, new mouths arise at a rate faster than the rate at which the skeleton is laid down, so that polyps share common walls of the cups. The polyps never become separated and share tentacles and internal mesenteries. The skeleton, which is the only part of the animal normally seen, shows these valleys which are lined with polyps in the living animal.

BREAMS, deep-bodied carplike fishes of European fresh waters. They are unrelated to the sea breams (family Sparidae). In England there are two species, the common bream, *Abramis brama,* and the silver bream, *Blicca bjoerkna,* which has been described elsewhere. The common bream has a compressed body with a very high back and short head. The upper parts are gray to black, the sides lighter, the belly silvery and the fins gray or blue-black. It is found chiefly in sluggish weedy waters throughout most of Europe north of the Pyrenees. Bream normally swim in shoals, each shoal made up of individuals of about the same size, usually near the bottom except in hot weather when they tend to lie still near the surface. They grow to over 12 lb (5.4 kg) in weight and are cunning

and difficult to catch. They feed on insect larvae, mollusks and worms, which they extract from great mouthfuls of mud sucked up from the bottom. The common bream often shoal with silver bream, and when small the two species are difficult to distinguish. The pharyngeal or throat teeth of the silver bream are in two rows, whereas those of the common bream are in a single row.

From the angler's point of view, the relative ease with which the bream will hybridize with the roach is a source of considerable annoyance. The hybrids strongly resemble the roach but grow to a much larger size, so that potentially record roach have, on closer examination, proved to be merely hybrids of the two species. From the roach, the hybrids can be distinguished fairly easily, having 15-19 branched rays in the anal fin (9-12 in the roach, but 23-29 in the bream).

BRENT GOOSE, *Branta bernicla,* a small dark goose with an arctic circumpolar distribution, breeding in the tundra. There are three recognizable races: the Russian or dark-bellied brent, *B. b. bernicla,* which has a black head, neck and breast, dark gray upper parts, a white rear and an incomplete white collar; the Atlantic or light-bellied brent, *B. b. hrota,* which is much paler beneath; and the Pacific brent or black brant, *B. b. orientalis,* which has a dark belly but light flanks. The Russian brent breeds in arctic Europe and Asia; the Atlantic brent on the islands and coasts of eastern Canada, northern Greenland, Spitzbergen and Franz Joseph Land; and the black brant on the islands and coasts of western Canada, northern Alaska and Siberia.

Brent geese are on the breeding grounds for some three months of the year from about mid-June. They nest on the ground in large colonies, often with eider ducks. Three to five eggs are laid and incubated by the female, with the male standing guard close by. The parents share in the care of the young, which are active soon after hatching and accompany the parents over the tundra, feeding on the tundra plants and any invertebrate animal food they can obtain.

After the breeding season, brent geese migrate south and winter on the shallow coastlines of the more temperate regions of the Atlantic and

Pacific oceans. During this period they take a wide variety of food, mostly plant material and including some algae, but on both the European and North American coasts of the Atlantic the preferred food has long been the greater eel grass, *Zostera maxima.* Around 1930 this plant was almost destroyed by disease, and the brent geese suffered drastic losses as a result. However, they were successful in adapting to other foods, and after 1940 their numbers built up again and the populations are now restored.

BRILL, *Scophthalmus rhombus,* a flatfish from the Mediterranean and eastern North Atlantic. It lies on its right side and is similar to the turbot, but is more oval in shape and has smooth scales with no tubercles. The general color is gray, brown or greenish with darker patches or mottlings and usually speckled with white spots. The brill, which is common around British coasts, lives on sandy bottoms at depths of 180-240 ft (54-72 m) and is often caught on the same grounds as the turbot. It grows to about 2 ft (60 cm) in length, and the flesh is considered most delicately flavored. In the North Sea, spawning takes place in spring and summer, the adult female producing about 800,000 eggs. After hatching, the young come inshore, but move back into deeper water after they have completed their metamorphosis and have attained the flat form of the adult.

BRISTLEMOUTHS, a name for small deep-sea fishes with thin fragile bodies, rarely more than 3 in. (7.5 cm) long, of the family Gonostomatidae. Although very abundant, they are rarely seen except by fish specialists. Their alternative name, lightfishes, expresses the fact that they have luminous organs, which in some genera are grouped together in glands although in others they are separate.

The genus *Cyclothone* has species all over the world, and many authors have voiced the opinion that they are among the most numerous fishes in the seas, but because of their unsubstantial bodies and the depths at which they live they are not likely ever to be of any economic importance. Some species take five years to reach 2 in. (5 cm) in length. Little is known of their breeding, but at least one species of the genus *Vinciguerra* has floating eggs and larvae that are difficult to distinguish from those of sardines.

Brent geese seldom come far inland. They normally feed in the beds of eel-grass in estuaries, and they usually roost at sea.

BRISTLETAILS

BRISTLETAILS, wingless insects believed to be closest to the ancestral type from which all present-day insects have evolved. About 400 species are known. Although they are found in all parts of the world they are rarely seen because with few exceptions they are less than 1 cm long. With the springtails and a few other primitive wingless types of insects they make up the subclass Apterygota ("without wings"), the rest of the insects being placed in a subclass Pterygota ("winged").

Bristletails are the only apterygote insects with compound eyes, like those found in the Pterygota. In most of them the mouthparts are of a rather unmodified type suitable only for biting and chewing living or dead plant material. The antennae are long and multisegmented. The abdominal segments carry a series of leglike appendages of unknown functions. At the end of the abdomen are three long, movable, antennalike appendages that give them their popular name. These are believed to be sense organs. Fertilization is internal. The young are like miniature adults in appearance, and, although they undergo several molts with increase in size before reaching adulthood, they show no metamorphosis.

What little is known of their feeding habits suggests that most bristletails feed on decomposing plant and animal material. Perhaps the best-known species is the silverfish, *Lepisma saccharina*, which is common in buildings all over the world, especially in damp places like kitchens and bathrooms. Its bright metallic color is due to a covering of shiny scales. The silverfish probably feeds on any sort of damp or decaying organic material and on the bacteria and fungi growing on it and is quite harmless. The gut contains symbiotic bacteria that enable the silverfish to digest cellulose, and experiments on related species have shown that adults can survive indefinitely on a diet of paper.

A species of bristletail that used to be much more common than it is now is the firebrat, *Thermobia domestica*. This is found in bakehouses and kitchens and other humid places where the exceptionally high temperatures the firebrat needs for the development of its young —up to 104° F (40° C)—are found. The firebrat probably feeds on food scraps dropped during cooking, and, since in recent years bakeries have become much cleaner places, the species has become quite rare.

BRITTLESTARS, mobile, star-shaped marine animals with the fivefold radial symmetry, endoskeleton of calcite plates and water vascular system characteristic of sea urchins, sea lilies, sea cucumbers and starfishes, to which they are related. They resemble the starfish in appearance but may be distinguished from them in details of structure, biochemistry, larval development and mode of life. The most obvious difference is that in brittlestars there is a clear distinction between the central disc that houses the vital organs and the flexible arms, whereas in starfishes this distinction is seldom marked.

The arms are long relative to the diameter of the central disc, narrow from base to tip, and somewhat bony in appearance because they are formed of little more than plates (ossicles) of calcite. Their snakelike form has given rise to the alternative popular name of "serpent

Bristletail *Petrobius*, showing the three movable appendages which give it its name.

Green broadbill *Calyptomena viridis* of Malaya.

The foliaceous coralline, a bryozoan commonly seen in European shores growing on rocks or on seaweeds.

stars" for the class Ophiuroidea (Gk *ophis*—snake). The arms break easily when handled, hence the name brittlestar.

The 2,000 species are divided into two main groups. The first is the Ophiurae or brittlestars proper. Their unbranched arms have a series of calcite plates embedded in the upper, lower and both side surfaces, with a central series of vertebralike ossicles that articulate by ball-and-socket joints, permitting lateral (sideways) movements only. The second, smaller group, is the Euryalae, including the basketstars. In these, disc and arms are covered with thick skin, sometimes with granules or tubercles. There is a central series of vertebral ossicles in each arm, and these articulate with broad hourglass-shaped surfaces, permitting all-around movement of the arms that may actually coil vertically. The arms may branch, and in some genera do so repeatedly. Those in which the arms branch right from the base are known as basketstars, for when the animals are alive and feeding the arms extend upward and outward to form an open mesh basket in which they catch their prey.

BROADBILLS, 14 species of small, squat tropical birds with short wings, outsize heads, short legs and strong feet. The bill is short and very broad, so that it appears triangular when viewed from above; the upper mandible overlaps the lower and terminates in a small hook. Broadbills are peculiar in a number of anatomical features, such as the number of neck vertebrae, 15 instead of the usual 14, and the musculature of the syrinx. In consequence they are regarded as the most primitive of the passerines, possibly allied to other primitive families like the Old World pittas and the New World cotingas. Many of the broadbills are brilliantly colored, and even the more somber of them with brown plumage usually have bright patches of color. In addition, several species have brightly hued eyes, bills or naked face patches. One species, *Eurylaimus steerii* of the Philippines, has wattles. Whitehead's broadbill, *Calyptomena whiteheadi*, is one of the most splendid members of the family. It is relatively large, 10 in. (25 cm) long, with brilliant green plumage marked boldly with black streaks and patches, which serve to break up the bird's outline and make it most inconspicuous as it sits among the dense evergreen vegetation in the mountain forests of Borneo.

Broadbills construct exquisite pendant nests usually slung from a branch overhanging a stream. The main body of the nest, below the suspending "rope," is pear-shaped, with an entrance hole at one side. A wide variety of fibrous nesting materials, roots, leaf midribs, lianas, etc., are used by different species. In many cases the outside is covered with moss or lichen, and strands of these materials are left hanging below the nest as a "beard." This undoubtedly helps to make the nest even more inconspicuous and confuses the nest robbers such as snakes, birds of prey, monkeys and other small mammals, which abound in the humid forests. The eggs are pale in color, usually two to four.

BRYOZOA, a phylum of quite common, though often inconspicuous aquatic animals, alternatively known as moss animals, which are present in fresh water but are especially

Budgerigars in the wild are seed-eaters and, like all seed-eating birds, are apt to be a menace to cereal crops. In Australia, where the photograph was taken, they live in flocks.

numerous in the sea. There are nearly 4,000 living species of Bryozoa and perhaps four times as many preserved as fossils. This number corresponds fairly closely to that of the echinoderms, for example; but whereas a starfish or sea urchin can be readily seen and admired, the beauty of bryozoans can generally be appreciated only by the use of a microscope.

They are colonial animals constructed from repeated units called zooids. While the zooids in a colony may be few, as in the Membranipora, which forms lacy patterns over the fronds of large seaweeds, they may number up to 2 million. Such a colony spreads over many square inches, but bryozoans are usually smaller than this. Incrusting species rarely cover more than 1 sq in. (6 sq cm), although the tufted and coralline types may be a little larger, reaching 2-3 in. (5-7½ cm) in height. One of these, the foliaceous coralline *Pentopora foliacea* of European waters, was once found in a clump 7 ft (2¼ m) in circumference.

BUDGERIGAR, *Melopsittacus undulatus*, small Australian parrot commonly kept as a cage bird. Wild budgerigars are small, 5½ in. (14 cm), long-tailed parrots with green underparts, a yellow back with closely spaced black barring, a bright yellow head with blue and black stripes on the cheeks, and a row of black spots on each side of the upper neck. Selective breeding in captivity has produced all-blue, all-white, all-yellow and all-green budgerigars, as well as patterned forms. Besides producing color variations, selective breeding has favored those with large heads and distinct face markings, and because of this many captive budgerigars bear little resemblance to the neatly proportioned wild ones.

Since budgerigars were first introduced into Britain in 1840 by John Gould, the famous bird artist, they have steadily increased in popularity and are now the commonest cage bird, considerably outnumbering the canary, which was formerly the favorite.

In Australia, budgerigars live in flocks throughout the arid interior regions and in dry coastal areas. They are most numerous in open country that is interspersed with belts of timber or patches of scrub. In prolonged droughts the flocks roam widely in search of food and water and are often seen in very large numbers, sometimes tens of thousands, at isolated water holes. Experiments on captive birds have, however, shown that they can live without water for up to 20 days, eating only dry seeds. This hardiness has probably been a key factor in the budgerigar's success as a household pet. Wild budgerigars feed mainly on small seeds from low-growing plants, running actively on the ground in search of food.

Like many other gregarious birds, budgerigars breed in colonies. No nest is built, and the clutches of 5-8 eggs are laid on the detritus in the bottom of a hole in a gnarled old tree. Several pairs often breed in one tree, so that large areas of old acacia trees are often occupied by breeding colonies. The oval or rounded eggs are incubated by the female alone, but the male helps to feed the nestlings, regurgitating partly digested seeds into their open bills. The young birds remain in the nest hole for about 20 days, and they are fed by the parents for a week or so after they fledge.

BULBULS, tropical tree-dwelling songbirds with few distinctive characters to immediately identify them. There are about 112 species (authorities differ in estimates), and they vary in size from sparrow- to thrush-size.

In general, bulbuls have the body plumage soft, dense and long, tending to be noticeably thick on the rump, with typically long, hairlike filoplumes on such parts as the nape or flanks. Another typical character appears to be the relative absence of feathering on the back of the short neck, which, when extended, shows a rather bare patch. The wings tend to be short and rounded at the tip; and the tail is usually relatively long, the shape of the tip varying from forked to rounded. The eyes are fairly large and dark. The bill is usually slender and of a type associated with insect eating, although for a great many of these species the most important food item is fruit, some insects being taken as well. The rictal bristles around the base of the bill are usually well developed. Compared with birds that move and feed in similar fashion, the bulbuls have legs and feet that are relatively weak and sometimes rather small.

The plumage is for the most part dull, olive-green, yellow or brown, and gray or black on one species, the black bulbul, *Microscelis madagascariensis*. Such plumage may, however, be varied by single patches of contrasting color, in white, red or yellow. A number of species, particularly those in Asia, have crests that vary from a loose, shaggy, rounded mass of elongated crown feathers to a slender, tapering, conical crest. The sexes are alike in size and color, and immature birds do not differ conspicuously from the adults.

Although nesting in scattered pairs, bulbuls are frequently gregarious at other times, traveling and feeding in small flocks, when they tend to be inquisitive and noisy. The call notes vary considerably, from the harsh or squeaky to the melodious, but many species also have loud and musical songs. These usually consist of brief notes or phrases, and at times some species will mimic calls of other birds. Bulbuls feed mainly on berries and other fruits, which they take in quantity, and where they occur in any number they are usually regarded as pests by fruit growers. In addition to fruit they also take insects and small invertebrates, especially when feeding their young, and also buds of plants and some nectar.

The largest and most widespread genus, *Pycnonotus*, has 47 species. These are mostly brown with black. Although some species of this genus are purely forest birds, many are birds of forest edge or scrub, and these have adapted themselves to the modified environment of cultivation and gardens. The red-whiskered bulbul, *P. jocosus* of southern Asia is one of these. It has red and white cheeks and a jaunty upcurved pointed crest. It is popular as a cage bird and has been introduced to other continents where similar changes in the environment through cultivation have created a niche for it. Another species of this group, the red-vented bulbul *P. cafer*, also of southern Asia, has become a tame species nesting on and around human habitation, while elsewhere to the south the yellow-vented bulbul, *P. goiavier* has become a bird of gardens.

There are a number of genera of bulbuls in Africa. These include a large number of small species with slender bills and rounded tails, mostly green and yellow and in many cases

Asia has also produced some bulbuls that have diverged from the typical form. The two species of finch-billed bulbuls of the genus *Spizixos*, for example, have short thick bills with arched culmens, more suggestive of finches than bulbuls. They live at the high altitude forest fringes in the Himalayas and China. *Microscelis* species are noisy, nervous and active with shaggy crests, slightly forked tails, longish bills and small legs. They are stronger on the wing than most species. The black bulbul, *M. madagascariensis*, has a variable plumage and very wide distribution, occurring from Madagascar and its adjacent islands, through India and Indochina to China.

BULLFINCH, *Pyrrhula pyrrhula*, a stocky little bird, about 6 in. (16 cm) long, with a short conical bill, black cap, bib, wings, and tail and a white rump. In the male, the breast is pinkish-red and the back bluish, but in the female both are brownish-gray. Bullfinches extend from the British Isles across Europe and Asia to Japan. Over most of this area, they live in conifer forests, but in western Europe they also extend south into deciduous woods, parks and gardens. For most of the year they eat a great variety of seeds from woody and herbaceous plants, but in spring they live mainly on buds; the young are reared on a mixture of seeds and insects. Compared to other finches, bullfinches eat many more seeds from fleshy fruits, like rowan and bramble.

The breeding season of the bullfinch is pro-

Budgerigars have usurped the popularity rating formerly held by canaries.

showing a brown or rufous tint on the rump or tail. They are, in the main, skulking species of thick forest and undergrowth and feed much more on insects than do other bulbuls. They are sometimes referred to as greenbuls, brownbuls and leafloves. Also in Africa there are three species of bristlebills, *Bleda* spp. These have a well-developed, compressed bill with a small hook at the tip and are sexually dimorphic, the females being smaller in body and bill. They are birds of forest undergrowth and one species at least, the green-tailed bristlebill, *B. eximia*, gets its food by raiding ant columns, either for the ants or for the prey they are carrying. These bristlebills have eggs heavily blotched with brown.

longed, up to three broods being raised each year. The nest is made of thin twigs and roots, and is placed in a thick bush in wood, hedge or garden. There are four to six eggs which are bluish white with red-brown spots. The species has several interesting courtship displays in which both sexes participate, the male puffing up the red abdominal feathers and turning his tail sideways toward the female.

BULLFROGS, large frogs the males of which have a call that has been likened to the bellowing of a bull. In different parts of the world the name refers to a particular species: the American bullfrog is *Rana catesbeiana;* the African bullfrog is *Pyxicephalus adspersus*, and the Indian bullfrog is *Rana tigrina*. These are only

Blackcrested yellow bulbul, one of over a hundred species of starling-sized birds spread over Africa and eastwards to the Philippines.

They are dormant during the dry season, emerging to breed at the beginning of the rains when they congregate in shallow pools. The calling of the males and the laying of eggs occurs in the daytime, and the frogs are not disturbed by intruders. In fact, they are aggressive and will jump at intruders with gaping jaws and bite viciously.

The Indian bullfrog is a similar olive color and reaches a length of about 6 in. (15 cm). It is shy and solitary and never found far from water, preferring ditches and marshes. In most areas it breeds at the beginning of the monsoon.

BUMBLEBEES, or humblebees, large, furry social bees, living in underground colonies founded by the mother or queen and containing her offspring, the sexually undeveloped females or workers. The queens are produced in the late summer. They hibernate and awake in the spring to begin feeding on the nectar of fruit trees until their ovaries develop. Each queen then searches for a nest site, which is frequently the disused burrow of a small mammal. Depending on the species, the site chosen will be approached by a long tunnel or may be on the surface of the ground but protected by tussocks of grass or moss. Within the nest, the queen prepares a small chamber lined with dried grass and moss and makes a shallow cup or egg cell from the wax she produces in thin sheets from glands situated between the segments of her abdomen. At this time, the queen begins foraging for pollen, which she stores in the egg cell until there is sufficient to feed her first offspring. 8-14 eggs are laid on the pollen in the cell, which is then sealed with a canopy of wax. Once the eggs have been laid, the queen makes a wax honeypot in which she stores some of the nectar she has collected and from which she feeds the growing larvae. This she does by making a hole in the egg cell and regurgitating the honey from her stomach into the cell.

When the maggotlike larvae have completed

related to one another in that they all belong to the family Ranidae or true frogs. Small frogs have a high-pitched call, large frogs have a low-pitched one, which is also louder.

The American bullfrog grows to 8 in. (20 cm), the females being larger than the males. It is robust and powerful, a greenish drab color with small tubercles on the skin, strictly aquatic, preferring still pools with shallows and plenty of driftwood or roots along the banks. The jumping ability of American bullfrogs is well known, and a contest is held every year in Calaveras to commemorate Mark Twain's famous tale "The Jumping Frog of Calaveras County." In fact, the bullfrog's jump, about 6 ft (2 m), is easily beaten by smaller, more athletic species of frogs from South Africa.

American bullfrogs emerge from hibernation in May, and breeding lasts until July, the males calling from the edges of lakes and ponds. The tadpole reaches a length of 6 in. (15 cm), and it is three years before it changes into a froglet. The adults eat anything of the right size, including mice, lizards, birds, fish, salamanders and other frogs, even those of the same species. Bullfrogs are easily caught by dangling a piece of cloth on a fishhook in front of them, and the meat of the large hindlimbs is considered a delicacy.

The African bullfrog is up to 9 in. (22.5 cm), the male being larger than the female, which is unusual. It is plump and olive-colored with many longitudinal folds in the skin. The mouth is enormous, reaching back to the shoulders, and there are three toothlike projections on the

lower jaw. African bullfrogs burrow, shuffling backward into the soil and remaining buried, with only the tip of the snout exposed. Like their American counterpart they will eat anything and are well known for their cannibalism.

The beautiful bullfinch has a bad reputation for stripping flower buds from shrubs and trees.

their feeding and are full grown, they spin silken cocoons within which they change into the adult form (pupation). After the young workers chew their way out, the vacated cocoon is used to store honey and pollen. The workers, the ovaries of which normally remain undeveloped, take over the foraging duties of the colony while the queen remains within the nest, making egg cells on top of the cocoons, laying eggs and incubating them by stretching herself over them like a broody hen.

When the colony is mature, the number of workers may vary from 100 to 400, although over 2,000 have been recorded in some tropical species. The considerable variation in size of workers results in a division of labor, with the larger bees performing the foraging duties and the smaller ones, or house bees, restricting themselves to duties within the nest.

BUNTINGS, the popular name for various groups of relatively small, finchlike birds, in particular the Old World representatives of the subfamily Emberizinae, which some authorities regard as a subdivision of the family Emberizidae, which includes tanagers and honey creepers. Other authorities do not recognize the Emberizidae as a distinct family and have split it up, placing the Emberizinae in the large finch family, the Fringillidae.

The three terms "bunting," "finch" and "sparrow," which are in everyday use, are of little taxonomic value because they have been applied so widely and indiscriminately. For instance, while Old World genera of the Emberizinae (e.g., *Emberiza*) are usually called "buntings," the more numerous New World genera are generally known as "sparrows," or "finches" and the name "bunting" is applied to genera in other groups, for example, to some members of the related subfamily Pyrrhuloxiinae (=Richmondeninae), which includes the cardinals. The "sparrows" of the Old World resemble buntings only superficially and are quite unrelated, most of them being closely related to the African weavers, although one familiar exception is the European "Hedge sparrow" or dunnock, *Prunella modularis*.

Structurally, the emberizine buntings and their allies are alike, mostly being 6 in. (15 cm) or less in length, although some may reach 8 in. (20 cm). Their bills, which cope with a mainly seed diet, are generally short, conical and attenuated. Some *Emberiza* species possess a distinctive bony knob on the palate for crushing the seeds. The legs are of medium length, although the feet tend to be rather large, being used to scratch for food in ground debris. The tail is often quite long, and in some is graduated or forked, frequently with the outer tail feathers partly or wholly white. Generally, the plumage is streaked or patterned in tones of brown, gray or olive, variously combined with bolder markings of black, white, yellow, green or chestnut. Although often bright, the emberizine buntings are never so brilliantly colored as some of the related pyrrhuloxine buntings, such as the indigo bunting, *Passerina cyanea*, and the painted bunting, *P. ciris* of North America. The males differ from the females in having a black head and bill and a contrasting white nape, mustachial stripe and breast.

The emberizine buntings occur in virtually all climatic zones of the world from the high Arctic

through temperate climates to the tropics. There are some 40 species of Old World buntings, most of which belong to the genus *Emberiza*. The exceptions are the snow bunting, *Plectrophenax nivalis*, and the Lapland bunting (or longspur), *Calcarius lapponicus*, both with a circumpolar distribution, and the distinctive crested bunting *Melophus lathani*, the only representative in tropical and subtropical Asia. The numerous *Emberiza* species are predominantly palearctic, although seven occur in Africa, and the rustic bunting, *E. rustica*, is a vagrant to North America. The genus is now represented in New Zealand following introductions of the cirl bunting, *E. cirlus*, and the yellowhammer, *E. citrinella*, by European settlers a century ago.

The origin of the emberizine buntings is believed to be within the New World, where, with around 150 species, the group is most abundant today. The precise relationships within the subfamily and with allied groups, such as the Pyrrhuloxiinae, are difficult to determine. Modern opinion recognizes the need for revision of the present classification of the American emberizines, for their division into over 50 genera, most of which contain only one or two species, hardly seems warranted, even allowing for their diversity. Characteristic of North America are the many genera of "sparrows," such as *Melospiza*, *Spizella* and *Zonotrichia*, as well as the juncos, *Junco* spp, and the

buds, or with small invertebrates, such as insect larvae, the main food of the nestlings. The cup-shaped nest is generally built on or quite near to the ground. From three to seven eggs may be laid, although the clutch is usually of four or five eggs. Most species are multibrooded. The coloring of the eggs varies, but they usually have a finely spotted or streaked pattern against a lighter background. They are, on the whole, monogamous, although for many years the relatively large corn bunting, *Emberiza calandra*, of Europe was believed to be polygamous, at least in some areas. This has recently been questioned by some authorities. Occasional bigamy apparently occurs in certain North American species, such as the song sparrow, *Melospiza melodia*, and the white-crowned sparrow, *Zonotrichia leucophrys*. A recent study of the reed bunting, *Emberiza schoeniclus*, in England has revealed an unusual situation in which individuals quite often change their mate during the course of the season (successive polygamy), while a few males may actually be mated to two or maybe even more females at the same time (harem polygamy). In most respects, however, the life cycle of the reed bunting is fairly typical of many other buntings.

Throughout the winter reed buntings scatter, singly or in small flocks, across open areas of stubble, weedy ground, etc., feeding on grain and other seeds. As early as January (in England) the males begin to leave these flocks and

A bumblebee on clover with its pollen baskets (the yellow protuberance on each hindleg) filled.

large and colorful towhees, *Pipilo* spp. Other genera are confined to South America, the largest of them being *Atapletes*, comprising numerous species of relatively large and heavily built buntings called brush finches.

The buntings are predominantly terrestrial, generally occupying relatively open terrain, such as scrubby grasslands and weedy areas, although some occur in denser scrub and woodland. They usually forage on or near to the ground, primarily feeding on seeds, supplemented with other vegetable matter, such as

arrive at the breeding grounds, staking claims to their territories around suitable areas of marsh or open water. Perching on small bushes or sedges, the males advertise their presence by their rather repetitive song ("tseek-tseek-tesek-tississisk"), successful birds maintaining their ground despite regular, often vociferous, skirmishes with other intruding males. Encouraged by periods of mild weather, this activity is soon suppressed during colder spells, when it is necessary to search for more food. The males and some females usually return to the same general area in successive years, the females arriving rather later. As the males become preoccupied with courtship and pair formation, their song intensity declines, usually until nest building and incubation commences in late April.

In England, four to five eggs are usually laid in a cup-shaped nest generally well hidden in ground vegetation. In northern Europe, where more young can apparently be reared due to the longer period of daylight, clutches are regularly of six or seven. The eggs hatch after about 13 days' incubation. The male often assists the female in incubating the eggs and after hatching. Although blind and naked and weighing only about $\frac{1}{12}$ oz (2 gm) when hatched, the young are alert and well feathered by the time they leave the nest 10-12 days later. They do not begin to fly, however, until they are nearly 20 days old, by which time their weight is about $\frac{1}{2}$-$\frac{3}{4}$ oz (17-19 gm). While reed buntings may successfully rear two broods during the season, most of the repeat clutches are replacements of earlier ones that were lost through predation.

Reed buntings are rather unusual among passerine birds in having a very marked distraction display to lure predators away from their nests and young. They feign injury, rustling along the ground with drooped and flapping wings and a widely spread tail, uttering their anxious alarm call "see."

By the end of the breeding season, which lasts from late April to August in England, the adult's plumage is very abraded, especially that of the female, and so there is a complete body molt between late June and early November. Continued breeding delays the onset of the molt, which is usually later in the females, for some of them remain in the breeding areas with late broods longer than their respective mates. Individual adults take 50-60 days to molt in Britain. Juveniles also molt in late summer, acquiring their adult plumages by a partial molt, usually confined to the head and body regions and a few areas on the wing.

Following the molt reed buntings become more gregarious, gathering together at communal roosts in reed beds overnight and feeding in loose flocks by day. In northern Europe, the breeding populations are entirely migratory, wintering in warmer climes to the south, while in Britain the species is only a partial migrant.

In northeast Asia, the reed bunting is replaced by a closely allied species, Pallas's reed bunting, *E. pallasi*, although in eastern Siberia the two species breed alongside each other. The song of *pallasi* is faster and higher-pitched than *schoeniclus*, more like that of the ortolan bunting, *E. hortulana*, while the upper parts are noticeably paler and the nape of the male is pale yellow, not white. Undoubtedly the two species originated from the same ancestral stock through geographical isolation. A similar pair consists of the ortolan and Cretzschmar's bunting, *E. caesia*. Both look very much alike, *hortulana* being widely distributed in open hilly country throughout much of the western Palearctic (but not Britain), while *caesia* only breeds in rocky and semiarid regions in parts of the eastern Mediterranean.

Many buntings occur in subspecific forms or races, but often the distinctions are not immediately obvious until the various types are

Baby thick-tailed bush-baby in nest.

critically compared in the hand. Thus, while yellowhammers in south and east Britain are typical of the continental race, *citrinella*, the plumage gradually becomes a little darker, more richly colored and more heavily streaked to the north and west, grading into the race *caliginosa*. Together with the corn bunting, the yellowhammer is a familiar species of arable areas throughout Europe, the bright lemon yellow head and underparts of the male being just as distinctive as its song, popularly rendered as "a-little-bit-of-bread-and-no-cheese." In eastern and central Asia, the yellowhammer is replaced by the pine bunting, which many authorities regard as a separate species, *E. leucocephalos*, although from central and western Siberia a large number of specimens with intermediate plumage are known, indicating that the two forms are "strong" subspecies which hybridize extensively in the area of overlap. The pine bunting lacks yellow coloration in its plumage, the male's head being boldly patterned in white and chestnut. Its preferred habitat is wooded or bushy steppe. In the Mediterranean climatic zone of Europe, the cirl bunting, *E. cirlus*, replaces the yellowhammer. The male can readily be distinguished from the male yellowhammer by its bold black and yellow face pattern and with practice by its more hurried, jingled song,

which lacks the terminal "cheese" phrase of the yellowhammer. But the two species are nevertheless very similar, and have probably also diverged from the same stock through geographical isolation. In Britain, where both species occur, *cirlus* is distinctly local and confined to the southern counties, while *citrinella* is widespread.

The black-headed bunting of the eastern Mediterranean and Iran and the red-headed bunting of Turkestan are two colorful *Emberiza* currently regarded as two "strong" races of *E. melanocephala*, although formerly they were thought to be distinct species (*E. melanocephala* and *E. bruniceps* respectively). Birds of intermediate plumage have been located in a small area of overlap, indicating that the two types are able to hybridize. True species do not interbreed. See also American buntings and Song sparrows.

BURBOT, *Lota lota,* the only freshwater member of the cod family. It derives its common name from the Latin for a beard, a reference to the barbel on its chin. The burbot lives in most European rivers except those in the extreme north and south. Its nearest relative is the ling. It has an elongated, subcylindrical body blotched with shades of brown. In recent times a variety of factors such as draining the land and other works have led to the burbot becoming very rare in England, and for several years there have been no reliable reports of specimens having been caught. A rather secretive fish, it lurks near the bottom among weeds during the day but comes out at night to feed on frogs and small fishes. The burbot is also found in Asia, and there are two subspecies in North America. It usually grows to 24 in. (60 cm) but sometimes to as much as 39 in. (98 cm).

BURYING BEETLES, also known as sexton beetles, are famous for burying the corpses of small birds and mammals, on which they lay their eggs, the larvae from these feeding on the decaying flesh. See Carrion beetles.

BUSH BABIES, nocturnal mammals with forward-pointing eyes and grasping hands and feet. The name refers to the soft, woolly fur and the large, staring eyes. The common bush baby, *Galago senegalensis*, which weighs just over 1 lb (500 gm), with head and body 17 in. (42 cm) and tail 10 in. (25 cm) long, is the best known and is often found in captivity in Europe and America. It has gray or grayish-brown fur, and the long tail is very bushy. The smallest of this group is the dwarf bush baby, *Galago demidovii*, which in weight, size and general habits is reminiscent of the mouse lemur of Madagascar. It weighs only 2 oz (60 gm); the head and body together measure only 6 in. (15 cm), and the tail is 8 in. (20 cm). The fur is a dark, rufous brown, and the pointed snout is very conspicuous. At the other end of the scale, there is the large thick-tailed bush baby, *Galago crassicaudatus*. With a head and body length of 13 in. (33 cm) and a tail 18 in. (45 cm) long, it is almost the size of a rabbit. There are also two less well known species, which are about the same size as the common bush baby. One is Allen's bush baby, *Galago alleni*, and the other is the needle-clawed bush baby, *Galago elegantulus*. The latter has a claw-like extension in the middle of every nail except those on the thumb and big toe, whereas all other bush babies have flat nails. The thumb and big toe are opposable in all species of bush baby and are very important in grasping the fine branches and trunks among which these animals live.

Bush babies are confined to Africa, and there is a distinction between rain-forest forms and steppe and savanna forms. The dwarf, Allen's and needle-clawed bush babies are rain-forest forms with dark brown fur and yellowish-brown underparts. They occur in central West Africa and may also extend into tropical deciduous forest in this area. On the other hand, the common and thick-tailed bush babies occur in steppe and savanna in Central and Southern Africa and have a more grayish fur. In general, these drier area forms are easier to keep in captivity.

Bush babies are omnivorous. Insects form the major part of the diet, with various other items of animal food and plant food such as fruits, gum and leaves making up the balance. However, each species has its own speciality.

BUSHBUCK, *Tragelaphus scriptus*, red-coated antelope with a variable number of white stripes and spots on the flanks. Also known in some areas as the harnessed antelope, it is related to the kudu and is found over most of Africa south of the Sahara.

The head of a young rough-legged buzzard, of northern Europe.

BUSTARDS, a well-defined family of 22 species related to the seriemas, cranes and rails, some being called "korhaans" in South Africa or "floricans" in India. They are birds of deserts, grassy plains and open savannas. They vary in size from about 14 in. (35 cm) long in the lesser florican, *Sypheotides indica*, of India, to more than 50 in. (130 cm) long in the kori bustard, *Ardeotis kori*, of Africa. Males are generally larger than females. The kori bustard and others of the same genus are among the largest

Female bushbuck or harnessed antelope, of Africa.

and heaviest of all flying birds, attaining weights of more than 30 lb (13.5 kg). They are close to the size limit above which flight is impossible and only fly reluctantly and for short distances. Smaller species fly strongly, many performing migrations of considerable length, but even so they depend more upon running and their cryptic coloring to escape predators. All bustards have long, strong legs and a long neck. Their three forward-pointing toes are short and broad, while the hind toe, or hallux, is absent as in many other cursorial species. The plumage is various shades of gray, brown and buff, beautifully vermiculated and spotted with black and white, and other shades of gray and brown. The majority of bustards have large white patches on their wings, which are usually visible only in flight. Several species have short crests, while the males of some, such as the great bustard, *Otis tarda*, of Eurasia and the Houbara bustard, *Chlamydotis undulata*, of northern Africa and southwestern Asia, have ornate bristles or plumes on the head and neck.

Bustards lay their eggs in a scrape on the ground, the clutch varying from one or two in the larger species to five in some of the smaller species. The female sits very tight while incubating and relies on her cryptic coloring to escape detection. The young leave the nest almost immediately.

Bustards are omnivorous, consuming a great variety of seeds, fruits, insects and small vertebrates.

BUTTERFISH, or gunnel, *Pholis gunnellus,* an elongated blennylike fish commonly found hiding under stones along European and North American coasts when the tide is out. Generally it is buff-colored with a row of black spots along the base of the long dorsal fin; the pectoral fin may be orange or yellowish. The body is covered with fine scales and is slimy; anybody trying to pick it up will be in no doubt why it is called the butterfish.

The breeding of this fish is quite unusual. The female lays the eggs in a small clump about 1 in. (2.5 cm) in diameter, the eggs being compacted into a ball by the female curving her body into a loop and laying the eggs within the circle. The

The typical outline of a soaring buzzard.

ball of eggs is then thrust into a hole in the rocks or into an empty shell. Both parents (which is a rare procedure in fishes) take turns in guarding the eggs until the young hatch after about four weeks. The young swim out to sea for several months and then return to the shore. Most fishes have paired ovaries, but in the butterfish there is a single long ovary. These fishes grow to about 10-12 in. (25-30 cm) in length.

BUTTONQUAILS, a family of small, skulking birds, very like quail in appearance but more closely related to the rails and cranes. They are terrestrial and have a generally brown plumage above, usually cryptically patterned and variegated in black, buff and white, while the underside is pale, often with a spotted pattern. In some species the female is distinguished by a bright chestnut breast patch. The bill is more slender and longer than that of a quail, but some Australian species have evolved heavy, blunt bills for seed-crushing. The eyes usually have pale irises and are staring. The feet are small, with no hind toes; the wings are rounded and the short tail mostly concealed by the wing coverts. Buttonquails are about 4-8 in. (10-20 cm) in length.

Buttonquails occur in grassland or low cover, both dry and swampy, and open woodland with some ground cover. They keep to cover and are difficult to see. The food is seeds, parts of plants, and insects. They have various whistling calls, and in the widely distributed striped buttonquail, *Turnix sylvatica,* and possibly some other species the female has a loud booming call.

The nest is a shallow scrape, lined with dry grasses and dead leaves, on the ground in a sheltered spot. Growing grasses may be pulled downward to conceal the nest. The clutch is usually of four eggs, rather rounded and glossy, grayish or buff with black specks and spots.

The young hatch as active downy chicks. The incubation period of 12-13 days is very short for birds of this type, and the young develop rapidly, becoming independent at about 2 weeks and adult in about 10 weeks.

BUZZARDS, large, broad-winged predatory birds renowned for their soaring flight, but spending much of the day perched on rocks, trees or telegraph posts. The body length is about 20 in. (50.8 cm) and the wing span 3-5 ft (91.4-152.4 cm). They live in open, wooded and mountainous habitats and feed principally on ground prey of mammals and reptiles, but also take carrion and birds. Most are dark brown and superficially resemble eagles, *Aquila* spp., but buzzards are smaller, have a less massive and less fierce appearance and lack the heavy bill and large claws of the eagles. There are 25 species of true buzzards included in the genus *Buteo,* with a further 10 associated genera of buzzardlike hawks that can conveniently be termed the subbuteonines.

Of the 25 species, the most widely distributed is the nominate race, the common buzzard *Buteo buteo,* originally described by Linnaeus in 1758. It breeds throughout the palearctic region, and several races have been described. Where suitable habitat exists it is found from sea level to 10,000 ft (3,000 m), and in the more northerly parts of its range it is migratory. Practically all its prey is taken on the ground, either by dropping onto it from a perch, or by diving onto it after circling or hovering briefly.

Breeding begins about February with striking aerial displays. These usually involve one or both birds circling together, interspersed with spectacular dives followed by upward swoops. The loud clear mewing note "peee-oo" is repeatedly uttered during the display. The nest is built by both birds of the pair in a tree or on a ledge, the clutch of from two to six oval eggs being laid in late March or early April. They are a dull white, sparingly marked with red or brown. Incubation begins between the first and second egg, and repeat clutches are rarely laid. Both sexes take part in the five-weeks incubation, and the young are in the nest for 40-50 days. In the early stages the female remains at

the nest with the young, while the male brings the food. After one or two weeks the female joins the male to help with the hunting. Breeding success appears to be related to food supply and varies from 1 to 3 per pair.

The common buzzard is replaced in the north by the rough-legged buzzard, *B. lagopus.* This Holarctic species breeds between latitudes 76°N and 61°N in Europe and America. It is highly migratory, moving south in winter to southern Europe, Asia Minor and the central United States. It is rather larger and much paler than *B. buteo* and, as its name implies, has a feathered tarsus.

In the New World the niche of *B. buteo* is filled by the red-tailed hawk, *B. jamaicensis,* of eastern North America and Swainson's hawk, *B. swainsonii,* of western North America. The migrations of Swainson's hawk are the most spectacular of any North American hawk, immense flocks passing each season along certain favored routes. Other North American species include the long-legged buzzard, *B. rufinus,* the upland buzzard, *B. hemilasius,* the ferruginous hawk, *B. regalis,* the red-shouldered hawk, *B. lineatus,* and the broad-winged hawk, *B. platypterus.*

BY-THE-WIND SAILOR, *Velella,* a kind of jellyfish, known as a siphonophore, living in the warmer oceans and the Mediterranean. It has a flat, almost oblong, float like a miniature raft, with a transparent gas-filled triangular sail on its upper surface, which the wind catches and drives the animal along the surface of the sea. The raft is deep blue and up to 2½ in. (63 mm) long and 1½ in. (38 mm) broad. On the undersurface is a central mouth surrounded by rings of reproductive bodies or gonophores and, outside these, by a ring of delicate mobile tentacles around the margin of the float. These are heavily charged with stinging cells for the capture of food and for the animal's protection. The gonophores bud off small male and female medusae, which lack a mouth. After release from the gonophores they produce eggs and sperm. They were named *Chrysomitra,* since they were originally thought to represent a separate genus.

By-the-wind sailor, an unusual jellyfish, of the kind known as siphonophores. Although typically oceanic, it is sometimes cast upon beaches in large numbers.

C

CACIQUE, the common name for tropical birds of the genus *Cacicus* (11 species) of the family of the New World orioles, which are confined to tropical Central and South America. The best known are the yellow-rumped cacique, *Cacicus cela*, which is black with yellow on the wings and a yellow rump and the red-rumped cacique, *Cacicus haemorrhous*, which is bluish black with a red rump. They live and nest in a colony, sometimes both species in mixed colonies which consist of a cluster of basket-like nests with the entrance right at the top and often around a wasp nest in trees and shrubbery. Nest building, incubation and feeding of the nestlings is done by the female only. Caciques feed on fruits and insects.

CADDIS FLIES, mothlike flies closely related to the moths and butterflies, with hairs instead of scales on the wings. The wings of caddis flies are held tentwise over the body and are never held vertically upward as are those of butterflies, and, as in moths, mandibles are absent. The antennae are slender, threadlike and sometimes exceptionally long, as much as three times the length of the body. Caddis flies are most often drably colored, browns and yellows predominating, although a number are blackish. Some are green or white, a few show iridescence and one is bright blue. They are mostly small to moderate in size, and many species are extremely small and often to be found running over the surfaces of stones or rocks adjacent to water. There are about 3,000 species. With only very few exceptions, the immature stages of caddis flies are aquatic so that the adult flies are to be found in the vicinity of water. In some parts of the world they exist in such large numbers that they assume pest proportions, causing jamming of air-conditioning plants and making outdoor activities impossible at the periods of maximum emergence. The eggs of caddis flies are laid either above the water or actually in it.

The larval stage of caddis flies is almost certainly better known than the adult, as the larvae of many species construct a case for themselves that is of such a size that when the larva is feeding or pulling itself along by its rather spidery legs, the abdomen is contained within it and, indeed, held fast to it by a pair of hooked appendages situated at the hind end. When alarmed or resting, the larva can retract itself completely within its case. The larval case is made of a wide variety of materials.

A number of unusual objects are often used in case construction. For example, *Limnephilus flavicornis* will utilize small water snails for its case whenever they are available. Often a caddis case is very delicately constructed, especially those made of small pebbles or sand grains, so the case becomes virtually invisible on the sandy bed of a stream or river.

Caddis larvae, by virtue of their large numbers, and in many cases large size, play an important role in freshwater environments by cutting up vegetable material, living or dead, for case construction and for food. In addition, they remove diatoms and algal felt from the substrate and feed on a wide range of freshwater animals, including species in their own order. On the other hand, larval, pupal and adult caddis form a not inconsiderable part of the diet of freshwater fishes, and in some cases, of birds and bats as well.

CAECILIANS, long-bodied, limbless amphibians without common names, which superficially resemble large earthworms. They are invariably blind, and the eyes are covered with opaque skin, or in some cases by the bones of the skull. There is a small sensory tentacle just in front of each eye, lying in a sac from which it can be protruded. The tentacle has two ducts, which communicate with Jacobson's organ, a sensory area adjacent to the nasal cavity. There is no tympanic membrane, and it is likely that caecilians "hear" by picking up vibrations in the ground via their lower jaw. Caecilians move, like snakes, by sinuous lateral undulations of the body. Many are able to burrow quite rapidly, and the majority spend most of their time below ground level. A number of species are, however, aquatic. Caecilians vary considerably in size. The largest, *Caecilia thompsoni*, from central Colombia, reaches a length of 55 in. (139.7 cm); the smallest is *Hypogeophis brevis*, from the Seychelles in the Indian Ocean, $4\frac{1}{2}$ in. (11.2 cm) long. The maximum diameter recorded for any caecilian is 1 in. (2.5 cm).

The body of caecilians is divided by a number of folds in the skin, which give a ringed appearance, as in an earthworm. Many species bear small scales in pockets just below the epidermis. The scales are small, rounded and composed of a large number of plates. Some species, such as *Ichthyophis*, lay eggs in cavities in the mud or among rocks close to water. The female usually coils around the developing eggs. The larvae develop within the egg capsule, and although gills are present these are resorbed before hatching. Others lay eggs that develop into free-swimming larvae with external gills. In other species the eggs are retained and develop within the mother, some species producing live

Black caiman, one of the eight species of crocodilian peculiar to South America.

Adult caddis fly *Stenophylax permistus*. Typically drab-colored, it is always found near water in which eggs are laid and larvae live.

young that have developed outside the egg capsule within the mother.

CAIMANS, tropical cousins of the alligator. They are smaller in size than the other crocodilians but possess the same general characteristics. The caiman's powerful tail, lashing from side to side, propels him through water at a rapid rate. The back has a tough hide reinforced by bony plates. The throat contains a fleshy flap that can be closed to permit breathing at the water's surface with only the tip of the nose showing. The nostrils have external valves, and the eyes are doubly protected by eyelids and a movable, clear membrane. The teeth of the lower jaw fit into pits in the upper jaw when the jaws are closed.

The smooth-fronted caiman, *Paleosuchus trigonatus*, and the dwarf caiman, *Paleosuchus palpebrosus*, from the Amazon, are the smallest. To protect their bodies from rapids in the swift waters they inhabit, the bony plates in the armor of these small creatures extend down to the belly.

Caimans occur only in the western hemisphere. South America contains most of the eight species of caiman. The huge black caiman, *Melanosuchus niger*, reaches a length of 15 ft (4.6 m) in the Amazon basin and Guiana region. Central America is the home of the dusky caiman, *Caiman crocodilus fuscus*.

The potential life span of the caiman can approach 40 years. Most caimans reach sexual maturity at a length of 5 or 6 ft (1.5-1.8 m). It is rare if one caiman reaches adult size from

an average hatch of two dozen young. Turtles, rodents, crocodilians and birds take a heavy toll of eggs and young. One month after conception the female builds a nest close to a stream, composed of decaying leaves and branches. The eggs are laid only a few inches deep in the mound. When the young in the eggs are ready to hatch, they respond to any activity on the mound by making "croaking" sounds. The female caiman hears the young and digs down to them. The young usually hatch the moment the rotting vegetation is removed and rush for the nearest water.

CAMEL, two species of large ruminants belonging to the Old World: the one-humped or Arabian camel, *Camelus dromedarius*, of North Africa and the Near East, and the two-humped or Bactrian camel, *C. bactrianus*, of Asia. The first is not known in the wild, the second survives in the wild in the Gobi Desert. The one-humped camel is commonly referred to as the dromedary, a name that strictly speaking should be reserved for a special breed used in riding.

Camels are up to 9 ft (3 m) long and stand nearly 7 ft (2.2 m) at the shoulder. The legs are long, the neck is long and curved, the ears are small, the eyes have long lashes and the nostrils can be closed as an additional protection against blown sand. The foot consists of two toes united at the sole by a web of skin. There are horny callosities on the chest and leg joints. Because of its fleshy lips and long papillae on the inside of the mouth a camel can eat

hard, thorny food. It will eat almost any kind of dry vegetation, and while food is abundant fat accumulates in the hump as an energy store.

The Arabian camel is more slender, with a shorter coat than the Bactrian, and the color of the coat is more variable. There are many skewbalds, but the white dromedaries are valued most. The Arabian camel was probably first domesticated by nomadic tribes from the interior of Arabia. However, it was first mentioned as a domestic animal in the 11th century B.C. in Palestine at the time of the Midianite invasions. The Arabian camel is found throughout the whole of the Sahara from Mauretania to Somaliland, across Arabia to Syria, Iran, Afghanistan and northern India, its southern limit of distribution in Africa being the latitude 13° N. Northward, it is spread beyond the Caspian Sea to Russian Turkestan. They were introduced later to southwest Africa and Australia where they are to be encountered at the present time in large, feral herds. In the cool season camels can travel across 620 miles (1,000 km) of waterless desert. They have a strong homing instinct and often escape from distant areas to their original home 300 miles (500 km) or even 620 miles (1,000 km) away. Gestation is 12-13 months, and lactation lasts 1-2 years. The mares foal every second year and go apart from the herd, to give birth either lying or standing. The foal can run after 2-3 hours, first of all in a mixture of stagger and amble.

The astonishing performances of the camel have given rise to many legends. The most famous is about storing water in the stomach. In 1954-55 the physiologists K. and B. Schmidt Nielsen proved that the "water cells" in the paunch, which Pliny long ago regarded as water reservoirs, cannot hold more than 11 pints (5 liters), and usually they contain only a wet chyme. In no case was stored drinking water found.

A camel can tolerate an extraordinary degree of dehydration, losing well over a third of its body weight, and because it is very economical with water, it can manage longer without drinking water than any other domestic animal. The water is drawn from the tissues and cell fluids, although the blood serum remains almost constant, and thus the circulation of the blood can continue without hindrance.

The animal can quench a moderate thirst (for a dromedary this is around 20 gallons [70-90 liters]) in one draft in barely ten minutes. When very thirsty it has to drink two or three times in several hours, in order to make good the water it has lost.

CANARY, *Serinus canaria*, a finch named after the islands from which it was first brought, which has now been domesticated to become a common cage bird. The wild bird is a sub-species of the European serin and is also found in the Azores and Madeira. The male is a streaky olive-green above and yellow-green below, although the female is duller and browner. The bird was apparently brought to Germany early in the 16th century, and the modern yellow varieties have been developed by selective breeding from chance mutations. During domestication, the sexual dimorphism of the wild bird has also been lost, and now the only way of telling the sexes apart is by the song of the male, which is also more elaborate than in the wild bird.

CANDIRU or carnero, a small South American catfish which becomes parasitic on larger fish. Habitually it lives in the gill cavities, and with its sharp teeth and the spines on its gill covers it induces a flow of blood on which it feeds. The best-known species and one that is greatly feared by the peoples of Brazil is *Vandellia cirrhosa*. This little fish enters the urinogenital apertures of men and women, particularly if they happen to urinate in the water. It seems likely that this is accidental, the fish mistaking the flow of urine for the exhalant stream of water from a fish's gills. Having penetrated, however, it is almost impossible to remove the candiru without surgery because of the erectile spines on the gill cover. The South American Indians often wear special sheaths of palm fibers to protect themselves. The candiru thus has the distinction of being the only vertebrate to parasitize man.

CAPERCAILLIE, *Tetrao urogallus*, a large game bird of the grouse family, found in forests, particularly of conifers, in northern Europe and Asia, with a relict population in the Pyrenees. The cock capercaillie is very striking, some 34 in. (86 cm) long, with a basically gray plumage, brown wing-coverts and dark glossy blue-green breast. The tail

Cock capercaillie, dandy among gamebirds. In his display attitude, he is challenging his rivals with song.

coverts and belly are marked with white and there is a white wing patch. The strong bill is dull white and there is a bright red wattle above each eye. The legs are feathered. The female, 24 in. (61 cm) long, is cryptically colored like other female game birds, mottled and barred with buff, gray and black.

Capercaillies prefer dense, shady forests with substantial underbrush interspersed with glades and boggy areas. They feed on a variety of plant materials, particularly the buds and needles of pine, spruce and larch. This makes them unpopular with foresters, but a moderate population does little real harm. The flesh of the capercaillie is palatable, and for centuries this species has been hunted for sport and food. This, with deforestation, brought about their extermination in Britain in the 16th century, but reintroductions from 1837 onward resulted in the species becoming reestablished.

Cock capercaillies have a particularly striking breeding display. In the spring, small numbers of cocks gather at the lek, or display ground, each beginning its display with a peculiar clicking utterance that develops to a crescendo and terminates with a loud "pop," followed by a short, wheezy, grating sound. The characteristic attitude during this song is with the neck stretched up to display beardlike feathers beneath the chin, with wings drooping to touch the ground and with the tail fanned and almost vertical. Noisy flapping of the wings and jumps into the air also take place. Rival males are charged with the wings drooping, but most of the attacks are ritualized.

The nest is usually on the ground, well hidden among the forest vegetation or under a fallen branch, frequently at the foot of a tree. 5-8 eggs are laid and incubated by the female for about 28 days.

CAPUCHIN, South American monkey so named because the hair on its head resembles the capuche or cowl of a Franciscan monk. It is the monkey formerly most favored by organ grinders.

White-throated capuchin searching for insects in the trunk of a palm tree. It also eats leaves, shoots and small birds and mammals.

CAPYBARA, similar in form to an overgrown guinea pig, it is the largest of all living rodents, being the size of a sheep or a large dog. A fully grown animal is about 4 ft (1.3 m) long with almost no tail and rather long legs. The unusually deep muzzle gives a very characteristic appearance and is correlated with the long, rootless, ever-growing cheek teeth, four in each row, that are adapted for grinding tough vegetation.

The best-known species, *Hydrochoerus hydrochaeris*, is common and widespread in South America, being found in woodland adjacent to rivers, lakes and swamps from the Parana River in Argentina northward. A smaller species, *H. isthmius*, occurs in Panama. Capybaras often live close to cultivated areas and sometimes damage crops of cereals and fruit, but their normal diet consists of aquatic vegetation and grasses.

They are sociable animals, and it is not unusual to find a party of 10-20 feeding together in a weed-filled waterway at dawn or dusk. In the water they behave rather like hippopotamuses, swimming with only the tops of their heads or even only the nostrils exposed and readily submerging if they are disturbed. When resting on land they sit on their haunches like dogs.

Capybaras have only one litter of young each year. Gestation is about four months, and usually about four or five rather well-developed young are born, which can swim competently from an early age.

CARACAL, *Caracal caracal*, usually known as a lynx, but in fact more closely related to the serval, must surely be classed as one of the most beautiful of the cats, if not the most beautiful. It is long-legged and is better adapted for running than most of the other cats, although it also shows remarkable agility and leaping ability. It is slightly smaller than the jungle cat, measuring about 27½ in. (70 cm), and is much more slender. The coloring is a regular reddish brown above, with no other markings except for the ears, which are black behind, a small dark spot above each eye. The underparts are white, with vague brownish red spots on the belly. The tail is short, and the ears have distinctive tufts or plumes at their tips.

The caracal is widely distributed throughout the drier areas of southern Africa and Asia and was formerly much more common than it is now. While the caracal prefers open country, it will also be found in mountainous areas, bush and desert. For the most part, it is semi-nocturnal and prefers to lie up during the day in hollows, under tree roots or even in a disused porcupine or aardvark den.

The main food source of the caracal is birds, which the cat hunts when they are roosting, although it has been observed to catch some prey when it was on the wing. In addition the caracal will take quite a wide range of small animals. The size of the litters varies between one and three, with a maximum of four and an average of two. The young resemble the adults at birth, and the majority of breeding occurs in July and August.

CARACARAS, large, long-legged carrion-eating birds of the subfamily Daptriinae, belonging to the falcon family, Falconidae, but looking quite unlike true falcons. Found commonly in parts of South and Central America, caracaras

associate with vultures at carcasses, but some are insectivorous or omnivorous. They build their own nests, unlike true falcons, but will rob other carrion-eaters of their food. They are well adapted for walking and running and live in forests, savanna, or more open country.

CARDINALS, a name given to some birds with red plumage and, by association, transferred to other similar or related forms.

In North America the name "cardinal" is usually applied to a large, heavy-billed bunting *Pyrrhuloxia (Richmondena) cardinalis*, also known at times as the "redbird." The male is almost entirely red save for a black patch on the throat, and has a small spiky erectile crest. The female is similar but mostly brown, with only a little red in the plumage. A related cardinal, *P. phoenicea*, a little less vivid in color, occurs in northwestern South America.

Young bull barren ground caribou, in Alaska.

These birds are members of the subfamily Pyrrhuloxinae, which are known as cardinals or cardinal grosbeaks.

The cardinal grosbeaks include the most colorful seed eaters in America, males being brightly colored although the females are usually brown. The males of the genus *Passerina* are variously deep blue, light blue and buff, blue, scarlet and green, purple, red and blue and blue, green and yellow. The male is deep blue and buff in the blue grosbeak, *Guiraca caerulea*, and black and white with a red breast patch in the rose-breasted grosbeak, *Pheucticus ludovicianus*. Other species are more drab. The pyrrhuloxia, *P. sinuata*, is gray and red with a slender crest and a heavy parrotlike bill. These heavy bills are used for seed eating, although insects and some fruit are taken.

CARIBOU, *Rangifer tarandus*, a large, rather ungainly looking deer of the tundra and northern forest regions. It has a wide distribution in both eastern and western hemispheres, being referred to as reindeer in the east and caribou in North America.

The bulls have a thick muzzle, maned neck and broad flat hoofs that are concave underneath and designed for traveling over snow and boggy terrain. The bulls are considerably larger than the cows, but both sexes of caribou are larger than the reindeer of northern Europe and Siberia. A full-grown caribou bull will vary in height at the shoulder from about 42 in. (107 cm) to just over 50 in. (127 cm) according to the subspecies, while the weight will vary from 200 lb (91 kg) to 600 lb (272 kg). A full-grown European reindeer bull will measure 44-45 in. (112-114 cm) at the shoulder, the cows being some 4-6 in. (10-15 cm) less. The largest animals would appear to come from north British Columbia.

The body color of the adult males is brown, with the neck a palish gray turning white in winter. There is more variation of color in the female, particularly in reindeer that have been domesticated. Females of both reindeer and caribou are unique in normally possessing a pair of antlers, similar in construction to, but considerably smaller than those of the males. The bulls generally have a palmated "shovel" close to the face on one of the antlers, although a shovel on each of the antlers is by no means a rarity. The upper points are also frequently palmated, and there is a back tine about halfway up the main beam. A common belief is that this shovel is used to dig through the snow to enable the deer to reach the lichen underneath. This is not true, for throughout most of the winter the older bulls will be without antlers and the forefeet are used to paw through the snow. The antlers on the older males are shed during the early part of the winter, but the new growth does not commence until the spring. The younger animals retain their antlers until the early part of the year, but not until late spring or early summer, at about the time the calves are being born, are the cows' antlers cast. A good pair of Scandinavian

The mirror carp, a domesticated variety of the common carp, is named for the large scales on its flanks.

antlers will measure about 50 in. (127 cm) in length, but North American heads may be up to 8 or 10 in. (20-25 cm) longer.

This deer is usually silent, and even during the rut, which takes place in late September-October, the bulls utter no special challenge call. The calves are born in late May and June.

CARPS, or carplike fishes, slender streamlined fishes almost entirely found in fresh water, although a few will occasionally go into brackish water, as in the Baltic Sea. Their main center of distribution is southern Asia whence they may have originally evolved. Carplike fishes are entirely absent from South America, Australia and also the island of Madagascar, which was isolated from the mainland before the carps entered Africa.

Typically, the carps are fairly slender, with silvery scales, a single dorsal fin set at about the midpoint of the body and a forked caudal fin. There are no teeth in the jaws, but these may develop a horny cutting edge for scraping algae or may bear disclike lips, which act as a sucker. One or two pairs of short barbels may be present at the corners of the mouth. Mastication of food, such as insects, plants, detritus, is archieved by a set of teeth in the throat, the pharyngeal teeth.

The common carp is now widespread in the United States, and its introduction dates from fishes imported into California from Germany in the 1870s. Conditions in America seem to suit it, and a magnificent specimen of 55 lb (25 kg) has been recorded from American waters. Wild goldfish, which escaped from ponds and aquaria, have now become established in some parts of America. In Japan, the common carp has for centuries been regarded as a symbol of fertility, but the earliest record of this fish is from China in 500 B.C. The first European record seems to be that of Theodoric (A.D. 475-526), king of the Ostrogoths, whose secretary Cassiderus was compelled to issue a circular to provincial rulers urging them to improve the supply of carp for the king's table. The so-called viviparous barb, *Barbus viviparus*, was at first thought to produce live young (like the tooth carps), but it is now known to reproduce like other cyprinids. Some species lay eggs in shallow water, some in deep, some

attach their eggs to aquatic plants, while yet others release their eggs into the water.

The common carp is a moderately deep, laterally compressed fish with a large dorsal fin, short anal fin and two pairs of barbels around the mouth. The scales in wild carp are large, there being about 35-37 along the lateral line, but in the domesticated varieties there are either just a few large scattered scales (mirror carp) or no scales at all (leather carp).

CASSOWARIES, large flightless birds of the genus *Casuarius*, living in the tropical rain forests of New Guinea and adjacent islands and in northern Australia.

Adults of the largest species, the double-wattled cassowary, *Casuarius casuarius*, stand about 6 ft (1.8 m) high when erect, but normally the head is held about 4 ft (1.2 m) above the ground. The body is shorter than that of the emu, and the bird appears stockier. The

Double crested cassowary, the commonest of the three species.

plumage is almost black and has a glossy texture. Each feather has two equal shafts, and as there are no barbules the barbs of the feathers do not link together to form a firm vane. The plumage hangs loosely from the body, looking more like hair than feathers. The head is embellished by a casque, or flattened horny crown, projecting up to 6 in. (15 cm) above the top of the skull and this often leans to the left. The skin of the head and neck is blue, and in two of the three species ornate wattles, or folds of loose skin colored red, orange and yellow, hang from the neck over the throat. The cassowaries have three toes, flattened beneath as in the emu, with a long sharp claw on the innermost toe of each foot. The bill is narrower than the emu's and perhaps a little longer. The wings are greatly reduced, and the wing quills, of which only the shaft is present, are horny spines, up to 15 in. (38 cm) long, which hang conspicuously at the bird's side as it stands.

The double-wattled cassowary lives mainly in the rain forests of New Guinea and the surrounding islands, as well as northeast Australia, as far south as Cardwell.

The single-wattled cassowary, *C. unappendiculatus*, stands 5 ft (1.6 m) tall and weighs up to 128 lb (58 kg). It lives in riverine and coastal swamp forests of New Guinea and adjacent islands.

The dwarf cassowary, *C. bennetti*, is considerably smaller than the others, standing about 3½ ft (1 m) high, and inhabits montane forests in most parts of New Guinea and the surrounding islands.

Cassowaries appear to live in pairs and family parties, each pair defending a territory during the breeding season. Eggs are laid from May onward, and have been found even in September in north Queensland. Incubation probably takes about seven weeks, and in captive birds the male undertakes the whole of the incubation, as in the emu. The clutches so far recorded range from three to eight eggs, and the nest is a scrape on the forest floor in the dense vegetation of the bird's normal habitat. The chicks are striped with a dark and light longitudinal pattern and probably resemble those of the emu. There is no information about the rate of growth, but it is clear that the juvenile plumage is brown and the ornamental wattles do not appear until the young bird matures, probably during its second or third year.

Wild cassowaries have been recorded eating the fruit of various native and cultivated plants throughout the year, and swallowing whole items as bulky as a large plum. Leaves have also been found in their gizzards, but it is likely that other plant, and probably animal food is taken at times when fruit is in short supply. Captive and cornered cassowaries have a reputation for attacking humans, and several fatalities have recently been reported from New Guinea. They apparently use the long claw on their innermost toe in these attacks, throwing both legs forward together as they jump at their opponent.

CATFISHES, widespread, chiefly freshwater fishes with barbels round the mouth. Some have completely naked, scaleless bodies, although others have a heavy armor of bony

Upside-down catfish, of Africa, lacks the silver belly usual in fishes that swim the right way up.

plates. There is a single dorsal fin followed by an adipose fin, which in some families (e.g., armored catfishes) is supported by a spine. The common name for these fishes refers to the barbels around the mouth, which look like whiskers and serve a sensory function in detecting food. The vast majority of catfishes are found in fresh waters, but two families are marine.

Europe has only two catfishes. The larger of the two, and one of the largest of the European freshwater fishes, is the wels, *Silurus glanis*, which is also found in Asiatic USSR. The second is *Parasilurus aristotelis*, an Asiatic species that is found in a few rivers in Greece. It was named after Aristotle, whose very accurate account of its biology was doubted for 2,000 years until it was realized that his descriptions referred to *Parasilurus* and not to the wels.

The African family Mochokidae contains one of the best known of all the catfishes, the famous upside-down catfish, *Synodontis nigriventris*. The body is naked, an adipose fin is present, and there is a strong spine in both dorsal and pectoral fins. This fish has the habit of swimming on its back, with the belly uppermost, a habit that is, however, shared with a few other members of this genus. In most fishes the back is much darker colored than the belly, but in *S. nigriventris*, as its Latin name suggests, the belly is darker than the back. This reversal of the normal pattern of counter-shading is a clear indication that this type of camouflage is of value to the fish.

Perhaps the most curious of the South American catfish families is the Trichomycteridae, which contains the parasitic forms. One of these, the candiru, is the only vertebrate parasite of man. *Stegophilus insidiosus* is a small, wormlike catfish that lives in the gill chambers of armored catfishes.

Two of the South American families of armored catfishes are very well known to aquarists. These are the Callichthyidae and the Loricariidae. The first is found throughout the tropical part of South America and in Trinidad. In these fishes the flanks are covered by two rows of bony plates and the adipose fin is preceded by a spine. They live in small shoals in slow-flowing water and grub around for food. The pectoral fins are provided with strong spines, which are used in some species to help them move overland when the ground is moist and the air humid.

CATS, thought to have been first domesticated by the Egyptians, are today familiar members of most households throughout the world. Although various small cats have been tamed since prehistoric times, the present-day short-haired domestic cat of Europe and many other parts of the world, which has been given the scientific name of *Felis catus,* seems to have been derived from the cafer cat or bush cat of Africa, *F. libyca*, perhaps with some admixture from the European wildcat, *F. sylvestris*. Given the somewhat derogatory name of alley cat or gutter cat in the United States, it nevertheless makes up about 99% of the domesticated cat population of the world. It is possible that most of the domesticated cats of India may have had a totally independent origin from those of Europe and that the Indian desert cat, *F. libyca ornata*, may have been its parent stock. The common occurrence of spotted cats in India, comparatively rare in Europe, suggests their origin from this spotted wildcat.

A typical short-haired domestic cat is about 2½ ft (76 cm) long, including a 9-in. (23-cm) long tail, which, unlike that of the wildcat, is held horizontally when walking. The weight varies considerably, but up to 21 lb (9 kg) has been recorded. It is a graceful animal with a well-knit, powerful body and a rounded face with a broad, well-whiskered muzzle. The whiskers are used to feel the way in the dark.

The cat uses its teeth to tear and chop meat. The legs are strong and well-boned, the feet small and neat with retractile claws, which are kept in good condition by scratching on a post or rough surface. They are a great help in climbing and enable a cat to shin swiftly up a tree to avoid a dog, or to rob birds' nests or just to lie on a branch basking in the sun.

The color of the coat varies considerably. The most common, the tabby, is of two kinds. The striped tabby has narrow vertical stripes on the body, similar to those of the bush cat and European wildcat. The blotched tabby is nearly the same color but has broad, mainly longitudinal dark lines and blotches on a light ground. In extreme cases the dark markings are relatively few, strongly drawn and standing out conspicuously against the lighter background. Such cats are recurrent mutants that parallel the king cheetah.

There are many color varieties of the ordinary domestic cat, including marmalade, ginger, tortoiseshell, blue, silver and black. All of the well-known color variations have arisen by mutation after the cat was domesticated. The pure white cat is either a total albino with pink eyes or a dominant white usually with blue eyes. Darwin was the first scientist to note that white cats with blue eyes are usually deaf. Subsequent research on kittens from deaf white parents has shown that, although the ears of ree kittens were normal for a few days after birth those of 75% soon degenerated to give deafness. Only a few of the remainder had normal hearing, some had hearing in one ear only.

There are many breeds of domestic cats, not all officially recognized in every country. American breeders in particular have been very enterprising in the production of new breeds and have introduced many new varieties. Some breeds are long-haired, but most are short-haired. The long-haired angora, said to come from Ankara in Turkey, and the smaller Persian, are now considered as one, owing to interbreeding. They have long silky fur with heavy ruffs of fur around the neck and thick tufts between the toes. The tail is long and bushy. The color varies from pure white to gray. Some authorities claim that they are descended from the manul or Pallas's cat found wild in Siberia, Tibet and Mongolia, but there is no direct evidence for this. The American Peke-faced Persian is a long-haired breed that has developed superficial characters similar to those of the Pekingese dogs. Although grotesque in features, it has nevertheless gained recognition and is popular in the United States. Of the short-haired varieties, one of the most popular breeds today is the Siamese, said to have descended from the sacred, royal or temple cat of Thailand. It was introduced into England in 1884 and into the United States in 1894. The color of the fur is remarkable in changing from pure white in the kitten to a pale fawn color in the adult, with the nose, mouth, ears, feet and tail dark brown. The eyes are a very bright blue, and the tail is sometimes kinked. The forelimbs are relatively short and the hindquarters high. The related Burmese is seal-brown with yellow or golden eyes, and it lacks the dark points of the Siamese.

The Abyssinian is nearest in looks to the cats of ancient Egypt. There seems no proof that it originated in Abyssinia. It has only recently been introduced into western Europe but is proving increasingly popular. It is a very lithe animal with large, yellow, hazel or green eyes but with a weak voice. Its coat is gray-brown to reddish. The Russian blue is also becoming popular. It has a soft seallike coat, bright blue in color and vivid green eyes.

One of the most familiar of the short-haired breeds is the Manx cat, remarkable for being tailless. It looks very like a lynx in outline and differs from the ordinary domestic cat in its short back, its higher hindquarters and its habit of walking with a kind of bobbing gait. Although not a good climber, it is by far the fastest on the ground of all domestic cats. In general, its coat is double, with long top hairs and shorter hairs forming a dense undercoat. Although it takes its name from the Isle of Man, it probably originated in the Far East, where many other cats have either very short tails or kinked tails.

A cat hunts by night by sound or sight, and accordingly its hearing and sight are very acute. Its hearing extends beyond the range of the human ear into the higher frequencies, and this is why a cat probably responds more readily to a woman's voice. It also probably enables it to hear rodents' voices when waiting beside a mousehole, when these are inaudible to the human ear.

Although a cat cannot see in total darkness, its eyes are so constructed as to make the fullest use of any light available when out hunting at night. During the day, its eyes are protected by an iris diaphragm, which helps to exclude the bright rays of full daylight, making the pupils become smaller until they are mere vertical slits. At night or in full shade of day, as in a room, the diaphragm opens fully, giving a rounded pupil to take advantage of all possible light. As with most nocturnal animals, the cat's eye has a layer of cells behind the retina called the tapetum, that causes the cat's eyes to glow, or shine in the dark, and to a lesser extent this can happen on a starlit night when there is no other illumination.

Domestic cats are sexually mature at 10 months or less, the earliest record being $3\frac{1}{2}$ months. The height of the breeding season is from late December to March, but the female may come in heat at intervals of 3-9 days from December to August. She is at her best for breeding purposes from 2-8 years. Males are at their highest potential from 3-8 years. Oestrus (heat) may last up to 21 days and is preceded by 2 or 3 days of excessive playfulness. Gestation is usually 65 days but may vary from 56-68 days. There may be up to 8 kittens in a litter, but 13 are known. Very young mothers have only one or two, and the number drops again as the female approaches 8 years of age.

Kittens are born blind, deaf and only lightly furred. The mother will often lift and move her young one by gently grasping its body in her mouth. The eyes open between 4-10 days. Milk teeth may appear at 4 days, but it may be 5 weeks before all have come through. Permanent teeth are cut between 4 and 7 months. Weaning begins at 2 months.

CATTLE, DOMESTIC, *Bos taurus,* large herbivorous mammals of the family Bovidae, exclusively living in some kind of association with man, their origins shrouded in archeological confusion and conflicting opinions. Many domestic cattle will successfully interbreed with their wild relatives thus obscuring their lineage, and the picture is further complicated by the zebu cattle, which is described as a separate species with an independent ancestor, yet breeds well with ordinary cattle. Moreover, as man has migrated to settle in new lands, he has spread his domestic animals far and wide at different

The British blue cat, one of the many breeds of domestic cat which, despite differences in color, are all of much the same build.

times and under a multitude of different conditions.

Despite all this, it seems fairly evident that modern cattle are descendants of the aurochs, the wild bovids that once roamed Europe and have become steadily more important with the continued development of human civilization. The first cattle were probably domesticated well over 6,000 years ago somewhere in western Asia and were later spread to Europe and elsewhere. Even this is not certain, since it is possible that it was the idea of domestication, rather than the cattle themselves, which was brought to Europe. Several distinctly recognizable breeds of cattle had been developed by 2500 B.C.

Modern Western Cattle. These stand about 5 ft (1.5 m) high and weigh 1,000-2,200 lb (450-1,000 kg) or more, the bulls being heavier than the cows. They are grazing animals, nibbling grass with their lower incisor teeth biting against the hard pad of the upper lip. Cattle eat about 150 lb (70 kg) of grass in an eight-hour day, and spend the rest of their time resting and chewing the cud.

Cows are mature when 18 months to three years old, depending on the breed and feeding. They are referred to as heifers before producing their first calf, and can continue breeding for over ten years. Usually only one calf is born,

Chartley bull, an ancient breed of park cattle.

Highland x Hereford cow.

after a gestation of about nine months. Twins sometimes occur. Male calves are usually killed early and sold for veal; only a few of the best are kept for breeding.

Cattle products include fat, glue, fertilizers, soap and leather, but their main uses in western society are as sources of meat and milk. Cows normally produce milk only when their calf is small, but lactation is prolonged, in certain specially developed breeds, for several months.

Milk production can also be extended by showing the cow its calf (a stuffed effigy is used by certain primitive herdsmen) and by the act of milking. Extended lactation allows the excess maternal efforts to be directed to the milk churn.

Notable meat producing breeds; Shorthorn. This originated in northeast England and is derived from a superior race of small-horned cattle clearly distinct as early as the 16th century. It is a stocky breed with a level back and a deep body boldly marked in reddish brown and white; the horns are very short and sharply curved inwards. Modern bulls weigh up to 2,200 lb (1,000 kg); the cows are smaller. Shorthorns have a mild temperament and are therefore easily handled. They are popular as fine, sturdy beef-producing animals which will also give a good milk yield compared with other beef cattle.

Hereford. This breed originated in Herefordshire (western England) and was formerly much used as a draft animal. Herefords are very solidly built and their markings (red-brown body, white head and belly) are highly characteristic. They thrive on very poor grazing land and are reared strictly for beef production, especially where it is uneconomical to give supplementary quality feed to boost production.

Aberdeen Angus. Angus cattle are short with a heavy build and straight back. They are black all over and hornless. They produce very high quality beef, well distributed on the carcass from a butcher's point of view. Angus cattle mature early and are therefore ready for sale sooner.

Notable breeds of Dairy Cattle. These are taller, more angular-looking than beef cattle. They are bred specially for the quantity or quality of their milk.

Ayrshire. A large brown and white blotched animal, which is able to produce plenty of milk, without needing the high quality fodder essential to the top dairy herds.

Friesian. Tall and robust (among the largest of western breeds) with variable markings, usually a pattern of black and white, which originated in the Netherlands. Friesians are famous for their generous milk production. They hold all the world records for sheer quantity of milk and butterfat. Many cows produce over 20,000 lb (9,000 kg) of milk (2,370 gall/9,000 lt) per year, and the record is 42,805 lb (19,262 kg) of milk (5,120 gall/ 19,454 lt), plus 1,246.4 lb (560.9 kg) of butterfat. The record production for a lifetime was an almost incredible 334,292 lb (150,431 kg) of milk plus 11,351 lb (5,108 kg) of butterfat. A good average yield is about 1,080 gall (4,100 lt) a year.

Jersey. This breed was developed on the island of Jersey in the 18th century. Strict legislation prevented the importation of other breeds to the island, so as to keep the local stock pure and maintain their excellent dairy qualities. Jerseys are light brown or fawn with a dark head. They are kept on good pastureland as producers of high-quality milk rather than large quantities. The milk is very fatty and Jersey cows thus hold the records for buttermilk production. Their milk appears a more creamy yellow than that of Friesians.

Guernsey. Fawn or brown and white cattle

developed in the Channel Islands. They are hardy and adaptable and can be used for quality production under severe conditions.

CAVY, a South American rodent with no tail, short legs, small round ears and a large head, of the genus *Cavia*, which includes the domesticated guinea pig. Wild cavies are represented by about 12 different species found throughout South America except in the extreme south. They range in size from a little smaller to almost twice as large as a guinea pig, i.e., from about 8-14 in. (20-35 cm) in length. They are fairly uniformly colored in various shades of brown.

Cavies live in burrows in a variety of habitats but most commonly in open country on mountains, savanna and swamps. They are especially common along the chain of the Andes and avoid lowland rain forest. They are sociable animals, and the burrows usually occur in groups, rather like a rabbit warren. Strictly vegetarians, they emerge at dusk to feed on grass and other fresh vegetation. The gestation is 63-71 days. The litter is small, usually two or three, and the young are very well developed and active at birth.

In the arid areas of northeastern Brazil is found a relative of the true cavies, the rock cavy, *Kerodon rupestris*, also a member of the family Caviidae. It differs from the true cavies in having longer legs and feet, with nails instead of claws. Rock cavies are exceedingly agile, climbing rocky cliffs or trees with equal facility. In its way of life and in its physical structure the rock cavy shows considerable convergence with the hyraxes of Africa although the latter are not rodents.

CENTIPEDES, swift-moving invertebrate predators with a long, thin segmented body, each segment bearing a single pair of legs. The first pair of legs are profoundly modified into a pair of robust pincerlike claws (hence Chilopoda—claw-footed) which meet each other horizontally beneath the head. The head carries a pair of threadlike antennae and three pairs of jaws, the mandibles and two pairs of maxillae.

The head consists of a single lentil-shaped capsule. The antennae are inserted on the forehead, and the groups of simple eyes or ocelli, if present, are located behind the antennae on the front and to the side of the head capsule. One order, the Geophilomorpha, have a constant number of segments (14) in the antennae but never possess eyes: the other three orders have a variable number of antennal segments and usually possess eyes.

The body is dorsoventrally flattened. The cuticle is like that of insects. The number of pairs of legs is always odd.

When walking, a wave of movement passes along the rows of legs, only one leg in eight being on the ground at any time. The legs of one side are in opposite phase to those of the other, thus staggering the points of support. The centipede is pushed along rather like a boat with one oarsman with only one oar that he uses first on one side then on the next, the resulting course being rather sinuous. The fastest centipede is *Scutigera*, which achieves a speed of 20 in. (50 cm) per second.

Centipedes are mainly nocturnal, resting by day in leaf litter, in the soil, under stones or loose bark and in crevices, and emerging at

Color varieties of guinea pigs, one of the easiest pets to keep and one of the most popular.

night to roam over the surface or climb trees and walls in search of their prey. Prey animals are immobilized by the injection of poison by the poison claws. The prey are usually invertebrates.

There are over 2,750 species of centipedes spread over all the world. Generally not much more than 1¼-1½ in. (3-4 cm) long in temperate regions, they may be larger in the tropics. *Scolopendra gigantea* from Brazil reaches 10½ in. (26.5 cm).

CHALCID WASPS, sometimes called chalcid flies, a family that includes some of the smallest of all insects with bodies only 0.25 mm in length. Chalcids can be distinguished immediately from typical wasps, for their body has no "waist" between the thorax and abdomen. Another distinguishing character of the chalcids can be found in the structure of the hind legs, in which the femur is greatly enlarged and, as a rule, very conspicuous. Many chalcids lay their eggs in the eggs of other insects, and when the larval wasp hatches it eats the tissues of its host. In this way, these wasps can inflict heavy mortality on the populations of insects injurious to man, and are used in the biological control of pests.

CHAMELEONS, Old World lizards especially noted for their adaptation to life in the branches of trees and shrubs. Most of the 90 chameleon species are to be found in tropical Africa and Madagascar.

The common chameleon, *Chamaeleo chamaeleon*, is the most northern representative of the family and can be found in North Africa, southern Spain and Portugal, Malta, Crete, in the north of the Arabian peninsula and in India and Ceylon. Most chameleons are medium-sized lizards of about 6-12 in. (15-30 cm) in length, but there are some tiny ones of just a few centimeters in length (e.g., species of the genus *Brookesia*) and some giants among the species of the genus *Chamaeleo* that reportedly reached a length of 32 in. (80 cm) in some forms in Madagascar and Africa.

The chameleon's body is usually flattened from side to side; the head often has a showy crest, horns or skinfolds. The eyelids are grown over to a circle of about the size of a pupil out of which the big eyeballs protrude on each side of the head. The eyes can be moved independently of each other, which especially fascinates the human contemplator. The chameleon's tongue is built like a catapult with a clublike tip and can be shot out at high speed to a length greater than the chameleon's total body length. The prey is hit by the tip of the tongue, glued to it and drawn back to the mouth at the same speed. The shooting out of the tongue

Female chameleon *Chamaeleo pardalis*.

greatly increases the radius within which food can be caught. The size of the prey depends on the size of the chameleon itself: small to medium-sized ones will catch mainly insects: the large ones will also capture other lizards, small mammals and birds.

The tongue bone is well developed in all chameleons and plays an important part in the catapult mechanism of the tongue. While at rest the muscular tongue is curled around

the tongue bone. Before shooting, the circular muscles at the back end of the tongue contract violently and try to push the tongue down from the pointed continuation of the tongue bone. This, however, is only possible after the longitudinal muscles relax and thus become inefficient in their role as adversaries to the circular muscles. As this relaxation is very sudden, the tongue shoots out of the mouth under the resulting pressure, rather as one would shoot an orange pip by squeezing it with the fingertips. It is withdrawn together with the glued-on prey through the elasticity of the tissue and a repeated contraction of the longitudinal muscles.

The limbs of a chameleon are long and thin and carry the body rather high. The toes are united into two opposing bundles on each foot: two toes on the outside and three on the inside on the front feet, and three on the outside and two on the inside on the hind feet. This has changed each foot into a pair of clasping tongs that enable the chameleon to get a firm grip on a perch. In addition, the species of the genus *Chamaeleo* have a prehensile tail, which even on its own could support the weight of the body.

The chameleon's ability to change its color is well known, although the layman often has a greatly exaggerated idea of this. The physiological color change in most chameleons is, however, a very noticeable and quick process. But the chameleon cannot always match its surroundings. The body coloration of the day-active arboreal chameleons is a good protection, for example, a green or bark-colored skin according to environment. Very often, however, coloration and markings play a part in disputes with another member of the same species: sometimes they just mirror a specific physiological condition. A surprising number of chameleons have a pale whitish sleeping coloration, and with the help of a torch can easily be found in the dark foliage where they would be completely protected by their color during the day. Male and sometimes also female chameleons hold territories that they will defend jealously against others. But only very rarely do real fights take place, although the horns and crests on the heads of a number of species would probably make good weapons. When the two rivals are at viewing distance they threaten each other by displaying their brilliant colors, always showing the side of the inflated body to the enemy in order to look more impressive. Characteristic swayings of the body emphasize the threat posture, and sometimes the mouth is opened wide to show the contrasting coloration of the mucous membrane.

Most chameleons lay eggs, which the female deposits in holes in the ground that she has previously dug herself. Digging holes into the ground and laying eggs in them is a highly dangerous affair for the arboreal chameleon. Some species, mainly in the subtropical climate of southern Africa and in some higher parts of the highlands, have become viviparous.

CHAMOIS, *Rupicapra rupicapra*, one of the so-called goat-antelopes, mountain-living and closely related to the goats; famous for its agility. 30-32 in. (75-80 cm) high, both sexes are of approximately equal size. Most individuals weigh 77-100 lb (35-45 kg), but members of the largest race, from the Carpathians, may weigh as much as 130 lb (60 kg). Chamois have stiff coarse hair, fawn or brown in summer and dark, nearly black, in winter; there is a black dorsal stripe and flank band, and the legs are black: the underparts are white and so is the face, except for a thick black line from the base of the horns through the eye to the muzzle (the gazelle face pattern). The hoof pad is somewhat elastic, giving the animal an enviable surefootedness. The horns are upright, thin and hooked back at the tip.

Chamois are found on all the major mountain ranges of Europe and Asia Minor: the Pyrenees, Cantabrian range, Chartreuse massif, Alps, Apennines, Tyrol, Tatra, Carpathians, Balkans, Pindus, Taurus and the Caucasus. Slightly different races are found on each range. They live in herds of 15-30, consisting of females and young; the males are solitary for most of the year, some being attached to the herds, following slowly behind as the herd moves along, single file. In the rutting season, August to October, big bisexual herds are formed, and the actual rut takes place at the end of October and beginning of November, when the males fight for possession of harems. Some harems are possessed by up to three males; most by only one. The male has glands behind the horns, which enlarge in the rutting season. The big herds re-form after the rut, breaking up again into the summer separate-sex pattern at the end of winter. Gestation is anything from 153 to 210 days. In the Caucasus, the young, usually single but sometimes two or three, are born in early or mid-May; in the Alps, one month later. Females mature at two years, males at three to four. A chamois may live as long as 22 years.

Chamois generally inhabit the alpine zone between the forests and the snowline; they have been recorded as high as 15,430 ft (4,750 m) on Mt. Blanc. In winter they descend to feed on pine shoots and moss in the forest. Each herd has a sentinel, which whistles and stamps when alarmed.

CHEECHAK, *Hemidactylus frenatus*, or common house gecko. It gets its name from the call it makes when prowling about at night for food. It prefers native huts and some city buildings as its habitat. Its origin was mainland Asia, but by hitching a ride on boats and in cargo, it has spread to most of the Pacific islands. In the daytime, the cheechak is dark brown to nearly black as it hides in crevices. At night, when on the prowl, it becomes almost ghostly white.

CHEETAH, *Acinonyx jubatus*, the "odd man out" of the family Felidae, it shows many dog-like characteristics that are not to be found in the other cats. It has long legs and is built for great turns of speed, from which it gets the reputation of being the fastest mammal in the world. The feet are of a very similar construction to the dog, in that they have hard pads with sharp edges, rather than the soft, elastic pads that are found in the rest of the cats. These pads and the blunt, non-retractile claws have a definite purpose in giving this very fast-moving animal the additional grip required for sudden stops and turns. The dew claws are worthy of note as they are more developed than in the other cats, and they play a big part in gripping prey. The head is small in comparison to the rest of the body, and the eyes are set high upon it, which helps the animal when peering over low cover and hillocks. The ears are small and rather flattened, reducing the silhouette still further. The nasal passages are larger than in the other cats to allow the intake of the extra oxygen required for the final sprint to the prey.

The cheetah is easily distinguished from the other cats, not only by its loose and rangy build, but also by the very distinctive markings. The ground color of the coat is a reddish shade of yellow and is broken by spots of solid black. The face is marked by the very noticeable "tear stripes" running from the corner of the eyes down the sides of the nose. There is another form, called the king cheetah, which has been given the suffix "rex," but this appears to be no more than a recurrent mutation. The male measures 7 ft (2.1 m) overall, of which the tail, a very effective aid to turning, measures 2½ ft (0.8 m). The height at the shoulder is about 2 ft 9 in. (0.83 m), and the total weight of the animal is about 130 lb (59 kg). The female is usually about three quarters of the size and weight of the male.

Cheetahs are distributed from Algeria and Morocco to the Transvaal, through Egypt, Ethiopia, Arabia, Syria, Persia and India, and throughout much of their range they have been captured and trained as hunting leopards, which is one of their names. When the cheetah is trained, it is used in very much the same way as a coursing greyhound. The only fossil remains are in Asia, and this suggests that this animal has migrated to Africa, probably through its association with man, although it is now more common in Africa than in Asia. In the wild, the cheetah will hunt either with a partner or as a member of a group, and it lives mainly on the smaller antelopes or the young of some of the larger species, although it has been observed taking quite small mammals. When the kill has been made, the cheetah prefers to eat the heart and kidneys first, and it also drinks the blood. After that, the head is eaten, and only then is the muscle meat attacked. It is not usual for a cheetah to return to a kill after the first feed. This difference in feeding habits is reflected in the teeth, which are neither as large nor as sharp as those of the leopard.

There appears to be no regular breeding season, and two to four kittens may arrive at any time of the year. At birth, the coat is a blue-gray color on the back, and the rest of the animal is brown with dark spots. The young are born blind, and the eyes open after about two weeks. The kittens, unlike the adult animals, are good climbers, and this fact, combined with other points of specialization, suggests that these are cats that are developing away from the general cat type.

The African race, *Acinonyx jubatus jubatus*, is still fairly plentiful, although it is not as widespread as it used to be, but the Asian race, *Acinonyx jubatus venaticus*, is now listed as decreasing and in danger of extinction. The cheetah is credited with being the fastest land animal, but it is very difficult to determine its maximum speed accurately. This is mainly due to the characteristic that the cheetah shares with other cats of being a sprinter. Prey is caught after a short burst of speed, and, if the prey eludes capture, the cheetah gives up.

CHIMAERAS or ratfishes (sometimes also called rabbitfishes), members of a subclass of cartilaginous fishes related to sharks and termed the Bradyodonti (or Holocephali). The chimaeras are characterized by the presence of a curious appendage or "clasper" on the head in front of the eyes. It is found only in the males, which suggests that it may serve some function in copulation. Like sharks, these fishes have a skeleton of cartilage and the males have one pair of claspers or more modified from the pelvic fins which are used for internal fertilization. They differ from sharks, however, in that the gill openings are covered by an operculum (resembling the gill cover of bony fishes); the primary upper jaw elements are fused to the skull (free from the skull in sharks but fused in the lungfishes); the anus does not discharge into a cloaca together with the urinary and genital products but has a distinct opening of its own (as in the bony fishes). Highly characteristic are the teeth of chimaeras, which are formed of three pairs of large flat plates, two above and one below, armed with hard points or "tritors" in some species but remaining beaklike in others. The majority of species belong to the genera *Chimaera* and *Hydrolagus* in which the snout is fairly blunt, the mouth ventral, the tail elongated and ratlike, and the first dorsal fin provided with a serrated spine capable of injecting a painful venom into wounds. *Chimaera monstrosa* is found along European shores and in the Mediterranean and grows to a length of 5 ft (1.5 m).

CHIMPANZEE, *Pan troglodytes*, the smaller of the two African apes and, with the gorilla, the species closest to man. Adult chimpanzees weigh 80-110 lb (36-50 kg); males are somewhat larger than females. The long, rather

Chimpanzees feed mainly on fruit, but also eat leaves, nuts, bark, ants, termites and even meat.

Infants sleep with their mothers.

members of the local group. These nests are made by bringing together, interlacing and patting down branches to make a platform with a rim, in which the chimpanzee lies with his legs drawn up and goes to sleep. Chimpanzees call loudly throughout the day and even sometimes during the night. Their calls can carry up to two miles, and the drumming that accompanies it carries even farther.

Within the local society there are no leaders, and there is very little trace of the "rank order" seen in monkey societies. But big males, especially old and graying ones, dominate other individuals and are given right of way, but they do not threaten or bully the others. Quarrels, which last only a few seconds and involve stamping and branch shaking as a kind of threat, often break out. Sudden attacks, never serious, may occur.

Chimpanzees constantly utilize natural objects for their own purposes. Not only do they catch termites with the aid of sticks, but they will also crumple up leaves to make a sponge to soak up water out of a hole in a tree, and use large sticks as weapons, either throwing them or actually rushing up to the enemy and giving it a resounding thwack.

sparse hair is black, and the skin changes from flesh-color in juveniles to bronze, then black in adults. The face is long and prognathous, and the lips thin and mobile. The brow ridges are marked, and the ears are very large, often remaining light colored long after the face has gone black.

Nearly half a chimpanzee's time is spent on the ground, walking quadrupedally on the backs of the knuckles. When it goes into trees it climbs and moves with all four limbs equally, as often as swinging from its arms alone.

Chimpanzees are found throughout the African lowland forest belt, from Senegal to northwestern Tanzania. They ascend mountains to quite high altitudes. Throughout their wide range, there is only one species of chimpanzee, but there are four well-marked subspecies, or geographic variants, which differ in size, color, head shape and development.

Chimpanzees are highly promiscuous. Females in oestrus develop huge pink genital swellings which serve as signals to males; a female at the height of oestrus may be served by half a dozen males in fairly rapid succession. The female's sexual cycle lasts 35 days: gestation is 225 days. Females in oestrus often leave the nursery group, even if they have infants, and join a bisexual band where they will mate.

An infant chimpanzee is carried on its mother's belly at first, clinging tight—a reflex which is well developed from birth. Sometimes she places the infant on her back and will give it food when it reaches out for some. While she is busy feeding or making her nest for the night, she may "hang up" her infant on a branch beside her. Infants scream and have temper tantrums, but generally they are ignored: indeed, the whole band of chimpanzees carries on with its activities as if the youngster were not there. Mothers and their infants continue to associate even after the young have

become independent, and up to a year after the birth of a second infant. The juveniles continue to build their nests near the mother's, but at this stage they are associating more with their peers, playing among themselves—wrestling, chasing and swinging around: as they play they make little panting noises—laughter.

Chimpanzees feed mainly on fruit, but also eat leaves, nuts, bark, ants, termites and even meat. They have been observed cooperatively stalking and killing prey, such as colobus monkeys or young antelope.

Chimpanzees live in large societies of as many as 60 or 80 individuals. During the day, they go around in four types of association: adult bisexual bands, all-male bands, "nursery groups" (females with their young) and mixed bands. Males sometimes move around on their own. The discovery by one of the wandering bands of a prolific food source is announced with hooting, screaming and drumming on trees, bringing other members of the local group to the spot.

Six to eight hours a day are spent foraging, the animals moving around the forest from one fruit tree to another. They may eat it on the spot, or they may carry off food, sometimes running along the ground bipedally with both hands full. At times when there is a sudden abundance of insects, such as termites, the chimpanzees spend a long time feeding on them. At other times they will break off a twig or grass stem, removing the side twigs to straighten it, and push it into a termite mound; termites crawl all over it, and the chimp then pulls it out and eats the termites. The same technique is used by forest chimpanzees to get honey from ground nests of wild bees. At night, chimpanzees make themselves nests in the trees wherever they happen to be, not returning to any special area or to other

Chimpanzees spend half their time in trees in which they make simple nests in which to spend the night.

CHINCHILLA, *Chinchilla laniger*, a rodent, rabbitlike in appearance but with a bushy tail and a body 10 in. (25 cm) long. It is best known for its long, soft, bluish-gray fur.

Wild chinchillas have been hunted almost to extinction since the value of their pelts was first appreciated in the 18th century. At one time, the Andes in Peru, Bolivia, Chile and Argentina supported thriving populations of chinchillas, but today they survive in the wild only high in the mountains in the northern parts of Chile. The South American governments saved the species from complete extinction by banning their hunting and export in the early 1900s. Later, government farms were set up, and, eventually, successful chinchilla farms were established outside South America, for example in California. As a result, the price of chinchilla fur dropped sufficiently to protect the remaining wild populations.

In their natural habitat on the slopes of the Andes the only food available is coarse grass and herbs. Water is scarce, and chinchillas must rely on dew and the moisture contained in the plants they eat.

Chinchillas live a communal life among rocks or in burrows in the mountains. Although they are nocturnal, they will bask in the morning and evening sun. Pairs mate for life, and the female, who is slightly the larger, dominates the male. Gestation varies according to the height at which the pair lives, being 115 days at 8,000 ft (2,500 m) and 125 days at 20,000 ft (6,000 m). The young are born fully furred and comparatively advanced in development, running about within hours of their birth, for example, and so are better able to withstand the cold of their environment. Although they begin to eat solid food within a week, they are not weaned until seven or eight weeks of age. They become sexually mature in less than a year, and there may be one to three litters born annually.

CHINESE WATERDEER, *Hydropotes inermis*, of which there are two subspecies, one in eastern and central China and the other in Korea, prefer reed swamps and grasslands close to a river to mountainous country. Standing about 20-21 in. (about 51-53 cm) high at the shoulder with an extremely short tail, the summer coat of the adult is reddish brown, which as winter advances, becomes a dull brown faintly flecked with gray. Twins and triplets are common and up to five or six fawns have been recorded. When fighting, the bucks often make a "chittering noise" which may be produced by clicking the tusks.

CHIPMUNK, in appearance and behavior it tends to be intermediate between the tree squirrel, with its long bushy tail, and the ground squirrel of the open plains, which usually has a rather small, short-haired tail. Chipmunks belong to two genera, *Tamias* and *Eutamias*, of the squirrel family, Sciuridae, and occur in the temperate forests of Siberia and North America. They live mainly on the ground but in woodland rather than on open plains, and they move with great agility on the ground and among shrubs, brushwood and small trees.

They are predominantly brown with five dark longitudinal stripes on the back alternating with pale stripes, both extending on to the sides of the face. They are about 4-6 in. (10-15 cm) long, excluding the tail, which is rather shorter than the head and body and only slightly bushy. There are about 16 species of chipmunks, all found in North America except for one species in Asia, the Siberian chipmunk, *Eutamias sibericus*, which is found throughout the coniferous forests from northern European Russia to eastern Siberia, Korea and Japan. Among the American species, one of the best known is the eastern chipmunk, *Tamias striatus*, which occurs throughout the eastern states of the U.S.A. and in the adjacent parts

of Canada. In the western states of the U.S.A., there is a much greater variety of chipmunks, some of them very local in their distribution. Chipmunks are diurnal, and this, combined with their striking pattern, nimble movements and chattering calls, makes them among the most familiar of all wild animals where they occur. They live in burrows in the ground, in stone walls or under rocks or fallen trees. They are inquisitive animals and, having taken flight, will quickly re-emerge from their retreat to keep a watchful eye on approaching danger. Chipmunks feed on all kinds of seeds and nuts, which they carry in their cheek pouches and store in great quantities for winter use. In Siberia, a record cache contained 13 lb (6 kg) of food including over 3,000 hazelnuts and about 3 lb of wheat. In Siberian markets, hazelnuts collected from chipmunk stores are reputed to fetch higher prices because the chipmunks are believed to choose only sound nuts. They hibernate more completely than tree squirrels but less so than the other ground squirrels, frequently awakening during the winter to feed on their stores of food.

Young chipmunks are born in spring in an underground nest and usually number between four and six. They are weaned by six weeks, and some species probably have a second litter later in the summer.

CHITAL, alternative name for the axis or Indian spotted deer, *Axis axis*.

CHITONS, about 600 species of primitive marine mollusks, deriving their name from the Greek word for a coat of mail, as the shell is formed from eight overlapping plates. They are generally oval in shape, mostly rather

Underside of a chiton, showing the foot.

small, between $\frac{1}{2}$-2 in. (1.5-5 cm) long, although some may reach 13 in. (34 cm) long. The eight overlapping shell plates run along the center of the back, surrounded by a fleshy girdle of tough, scaly or bristly mantle tissue, that secretes the shell.

Chitons are worldwide in distribution with particularly large numbers on the North American shores of the Pacific and in Australia. They are found particularly in intertidal regions, where they cling tightly to rocks.

The sexes are separate in nearly all chitons but are usually indistinguishable externally. Eggs and sperm are shed directly into the sea. Spawning is linked with tidal cycles in several species. For example, the Barrier Reef species, *Acanthopleura gemmata*, spawns once a month for several months at full moon, which almost coincides with the spring tide. The larvae develop in the plankton for periods varying from a little more than a day to almost two

weeks. They then settle low in the intertidal region and metamorphose into young chitons. Chitons move slowly, by muscular waves passing along the foot from front to back. If detached from a rock, they roll up, protecting the ventral surface like a woodlouse. Most chitons avoid light and move only at night.

In general they are herbivorous, feeding by rasping the surface of the rocks on which they live with the long radula.

CHOUGHS, certain species of predominantly black, crowlike birds of Europe and Asia or Australia. In Europe and Asia there are two species of chough of the crow family, Corvidae; in Australia there is one species of chough in the mud-nest-building family, Grallinidae.

The Cornish chough, more commonly known simply as the chough *Pyrrhocorax pyrrhocorax*, was formerly much more common in Britain than it is now. Today it is only found on the western edge of the British Isles, mainly in Ireland. Apart from its rarity in the Alps, its European distribution is the same as that of the alpine chough, *P. graculus*, that is, the mountainous regions from the Iberian Peninsula and the Atlas in the west to China in the east. Differences within this distribution are the absence of the alpine chough from maritime cliffs—which are commonly inhabited by the chough—and the absence of the chough from the higher altitudes inhabited by the alpine chough, the latter being found up to 27,000 ft (8,200 m) or more in the Himalayas.

Both species of *Pyrrhocorax* are about 15 in. (3.8 cm) long and have red legs, but the bill of the chough is long, curved and red, while that of the alpine chough is shorter, less curved and yellow. Both are aerial experts able to take full advantage of the updrafts and other air movements so typical of their habitat. They breed in cliffs, caves and rock crevices (including those in large stone buildings) and feed on insects and other invertebrates. Alpine choughs also take a considerable amount of garbage. Choughs, and probably alpine choughs, suffer from competition with jackdaws for nest sites and probably for food also. Neither species is migratory.

CHUCKWALLA, *Sauromalus obesus*, one of the iguanas, the New World equivalents of the Old World lizards (Lacertidae). The chuckwalla is a heavy-bodied reptile, which attains a length of $1\frac{1}{2}$ ft (46 cm). Its robust head, powerful legs, coat of small scales and long tail are typical of many iguanas. Predominantly herbivorous, the chuckwalla feeds on the flowers and fruits of cacti living in desert areas of the southwestern United States, northern Mexico and Lower California. It is notable for its ability to inflate its lungs so that the body becomes almost globular. When threatened, the chuckwalla runs into a rock crevice, inflates itself and is virtually impossible to remove by force.

CICADAS, large plant-sucking insects well known for their shrill, monotonous song. They are characterized by three beadlike simple eyes, ocelli, between the main eyes, and two pairs of membranous wings.

There are 1,500 known species, most of which live in the warmer regions of the world. One of the best-known species is the periodical cicada or 17-year locust, *Magicicada septen-*

decim, of the United States, remarkable for the long time required for development. The adults are about 1-1½ in. (25-40 mm) long. The female lays eggs in slits she makes in the branches of trees. The young emerge in about six weeks, drop to the ground and bury themselves. They then live below the surface, sucking sap from the fine roots of trees. They come to the surface again after 13 or 17 years, the time depending on the particular race.

Cicadas are famous for their sound-producing powers, the shrill noise some of them make in tropical forest being almost deafening. Only the males make sounds, the sound-producing organs consisting of a pair of shell-like drums at the base of the abdomen. These are vibrated by the action of powerful muscles, the sound produced varying greatly in note and intensity in different species. The function is almost certainly to attract females.

CICHLIDS, a family of perchlike fishes found chiefly in fresh waters (rarely brackish waters) in Africa and South America, but with one genus in India and Ceylon. There are over 600 species and more are being discovered every year. Typically, the cichlids have a fairly deep and compressed body, a single long dorsal fin (the first part having spines) and a shorter anal fin beginning with one spine or more. The body bears fairly large scales, which in some genera are smooth (cycloid) but in others are rough-edged (ctenoid). The cichlids have a single nostril on each side (paired nostrils are usual in other perchlike fishes).

Cichlid fishes are usually found in lakes or sluggish waters, although some are adapted to life in rivers and streams. Many are carnivorous, and all have a set of teeth in the throat, the pharyngeal teeth that are adapted to particular diets. In those that feed on fishes or large invertebrates, the pharyngeal teeth are fairly large and coarse, but in species which are filter feeders and live on algae or other small organisms, the teeth are fine and close set. In the mollusk eaters, the teeth of the pharyngeal tooth pads are flat for crushing and grinding. In many species, and not only the filter feeders, there is a microscopic series of little spines on the inner faces of all but the first gill arch. Each spine bears minute teeth. As the gills move over each other during breathing, the teeth probably comb out the single-cell organisms that collect on the mucus of the gills. A few species feed on higher plants, and there are also some that have teeth adapted for eating the scales of other fishes.

Most of the cichlids exhibit territorial behavior connected with the guarding of a nest in which the eggs are laid. In the nest-building forms, such as the species of *Tilapia*, the male usually excavates a circular hollow in the sandy or muddy bottom and goes through a fairly elaborate display of fin movements and swimming antics when a female arrives. When the female has deposited the eggs, the male sheds its sperm, and in the mouth-brooding species the female (occasionally the male) sucks the eggs into her mouth where they remain for the incubation period, and even after they hatch the fry remain there until they are able to fend for themselves. In other species, the eggs may stay where they are laid or be scooped

up and deposited at another site where they are aerated by fanning movements of the fins by one or both parents. In most species the male develops a characteristic breeding coloration, and this is thought to help keep the species separate when several are found on the same breeding ground.

Many cichlids are prettily colored, and because of this and their interesting breeding habits they are popular with aquarists.

CIVETS, little carnivores. From early times civets have been kept in captivity for the secretion that can be obtained from glands near their external reproductive organs. The secretion, of the consistency of honey, was used, until the invention of artificial substitutes, as a base for perfumes, and was "spooned" out of each civet at the rate of ⅛ oz per week. The secretion was known as either civet or musk. In *As You Like It*, Shakespeare writes, "Civet is of a baser birth than tar, the very uncleanly flux of a cat," but Topsell in the *History of Four-footed Beasts* says, "The musk cat is neither like a cat or a mouse." Nowadays, musk and civet refer to two different substances; the former from the musk deer, the latter from the civet. The active substances in each are known respectively as muscone and civetone. In 1926 they were involved in chemical history when it was shown that their molecules were made up of rings of more than six carbon atoms, until then considered an impossibility.

The civet's own use for its secretions is to scent-mark objects in its range for the same reason as a dog lifts its leg against a tree.

Civets, with the mongooses and genets, make

A typical cicada, a monotonous musician.

up the Viverridae, a family of carnivores placed between the weasel and the cat families. Like weasels, they are long-bodied with long tails and sharp muzzles. Their teeth represent an earlier form of carnivore dentition. They have about ten more teeth than the cats, and the face is generally longer. In habits and pattern of coat, however, they resemble the small cats. They are generally striped or spotted over a solid ground color and the claws are sometimes retractile, but this is not the rule. They differ from the cats by having a fifth toe on the hind feet. In common with the rest of the family, none of the civets has any great value as a fur-bearing animal, and because of this they have been largely left alone by man.

The African civet, *Civettictis civetta* of Africa south of the Sahara, is the best known of all civets. It is stout-bodied with short legs, some 30 in. (76 cm) in head and body length with another 18 in. (46 cm) of tail. It weighs 15-24 lb (5.4-10.2 kg). The fur is coarse with a ground color of ash gray, broken by black spots and stripes. The tail is ringed black and white. Another common species, the large Indian civet, *Viverra zibetha* of India, Burma and southern China, is very similar to the African civet in size and appearance, except for the ridge of erectile hairs down the spine, the distinctive black and white throat markings and the fact that the fur is thick and soft. The small Indian civet, *Viverricula indica*, which ranges through southern China to Ceylon and to Bali in the Malayan Archipelago, is similar but smaller.

All three are similar in habit. They are usually solitary and nocturnal and very seldom seen, as they keep in dense cover, in forests or undergrowth. They rest by day in abandoned burrows and come out to hunt only at night. They can climb and swim well. Their food is mainly animal, including insects, crabs, frogs, snakes and birds and their eggs. The African civet is said to assist reforestation. Although a carnivore it also eats a large amount of fruit and berries, the seeds from which are passed undigested in the excreta. There are usually two to three young in a litter born in a hole in the ground or in dense cover.

There are six species of palm civets in southern Asia and one in Africa. They differ from the true civets in several anatomical details. Their dentition shows that they are less carnivorous as the teeth are weaker and the carnassials, used for slicing flesh, are less well developed. Most species have naked soles and sharp curved claws, which aid their tree climbing. One of the best known and most widespread is the common Indian palm civet or toddy cat, *Paradoxurus hermaphroditus*, ranging from Ceylon and India through southern China and Southeast Asia to the Philippines.

CLICK BEETLES, or skipjacks, elongated, dull brown insects having the ability to spring upward into the air when placed on their backs. The leap, which is accompanied by a noticeable clicking sound, is achieved by a sudden movement of the joint between the first and second segments of the thorax. This old behavior pattern allows the beetle to right itself when it falls on its back and also to escape some of its enemies. Click beetles are important, as their larvae, the familiar wireworm, are

Click beetle, master acrobat.

among the most destructive agricultural pests. The genus to which they belong, *Agriotes*, is part of the family Elateridae, containing about 7,000 species.

CLIMBING PERCH, *Anabas testudinosus*, a labyrinth fish reputedly able to climb trees and suck their juices. It is found in Southeast Asia, and the first specimen to be seen by someone from the West, at the end of the 18th century, was found in a crack in a tree. Recent research has shown, however, that the tree-climbing legend is untrue, these fishes being most unwilling tree tenants, having usually arrived there after being dropped by birds. The climbing perch has an accessory breathing organ in the form of a series of plates in the upper part of the gill chamber richly supplied with blood vessels. The fish frequently moves from one pool to another, and it is probably during these overland journeys that it is seized by birds and later dropped. The gill covers are used as an extra pair of "legs" while it is moving overland. They are spiny, and are spread out to anchor the fish while the pectoral fins and tail push forward. The climbing perch can survive out of water for some time, and these fishes are carried about by local people as a fresh fish supply.

CLOTHES MOTHS, term including several species of small brown moths, ½ in. (1.5 cm) or a little more in wingspan, which appear periodically in houses, the most important of which is *Tineola bisselliella*. They are called clothes moths because their larvae, whitish creatures up to ½ in. (1.5 cm) long with pale brown heads, attack woolen fabrics, such as clothes, carpets and blankets. In a survey carried out in Britain in 1948 a quarter of the houses investigated had harbored clothes moths in the previous few months, and loss through moth damage was estimated at £1,500,000.

Clothes moth larvae are unusual in being able to digest keratin, the very resistant protein of hair and wool, but this alone is not adequate for their survival, and most infestations of clothes moths begin in parts that are soiled with sweat or urine. The soiling apparently adds traces of essential substances to the diet. But damage far exceeds the amount eaten because the larvae tend to bite through strands of wool, take a few bites and then move on elsewhere. Very often they spin tunnels of silk over themselves and construct cases in

which to molt and pupate. These are made of fragments of wool bound with silk.

Adult clothes moths cannot feed because their mouthparts are not fully developed. Hence their life is short, usually not much more than a week. The males fly fairly readily, but the females are not good fliers and mostly move by running from place to place. They may lay up to 100 tiny white eggs which hatch in about a month. The number of larval stages is very variable, depending on the adequacy of the diet, and from five to about 40 stages have been recorded.

CLOWNFISHES, small damselfishes from the Indo-Pacific region that have a remarkable association with species of sea anemone and for this reason are often called anemone fishes. About 12 species are known, all very brightly colored. *Amphiprion percula* grows to a few inches and has a bright orange body with three intensely white vertical bars, the fins being edged in black. Similar patterns are found in the other species of *Amphiprion*. They spend their lives among the waving tentacles of large sea anemones, which eat fish caught by means of the many hundreds of thousands of stinging cells along the tentacles, and the question arises how the clownfish manages to escape being stung. At first it was thought that the clownfish was agile enough to avoid the tentacles, but closer observation of members of the Australian clownfish genus *Stoichactis* has shown that these fishes actually rub against the tentacles to encourage the sea anemone to open. When inside, the fish often rests with its head poking out of the sea anemone's mouth. If danger threatens, the clownfish dives back into the sea anemone and will remain there even if the latter is removed from the water.

If the clownfish is deprived of its host it usually falls an easy prey to predators. It is difficult to see what the anemone gains from this curious association because the fish not only takes shelter but also may feed on the tentacles and on any food that the anemone traps. The anemone can, in fact, live quite well without its guest, but it has been suggested that the fish may have a beneficial aerating effect from entering the anemone's stomach, and one species of clownfish, *Amphiprion polymnus*, actually feeds the anemone by placing on its tentacles pieces of food from the sea floor.

COALFISH, *Pollachius virens*, also known as the coley or saithe, a codlike fish found from the Arctic to the Mediterranean. It resembles the cod but has only a rudimentary barbel under the chin, no speckling on the body and a dark green or almost black upper surface. It is a fairly deep-water fish usually found at 300-600 ft (100-200 m). One of the largest rod-caught specimens weighed nearly 24 lb (11 kg). Like many of the codlike fishes it undergoes extensive migrations.

COATI, social raccoonlike animals extending from Oklahoma through South America, dividing into three species along the way: *Nasua narica*, found in the southern parts of the United States, Mexico and Panama; *Nasua nelsoni* from Cozumel Island off the Yucatan coast; and *Nasua nasua*, which is distributed throughout the forested regions of South America. It is more slender than the raccoon, and the most conspicuous features of the coati are its long, ringed tail that is carried erect and the thin muzzle with the extremely mobile nose. The color ranges from red to black-brown. The males are larger than the females, measuring 38.6-47.2 in. (98-120 cm), the tail accounting for at least half the total length. Weights vary between a usual 12 lb (5.5 kg) to an extreme 23 lb (11.4 kg). Black or brown markings are present on the face with white patches below the eyes. The chin and chest may be light buff or whitish, and the coarse fur of the back has a reddish-brown hue. Coatis are diurnal and social, living in bands of 5-30 individuals.

Although small invertebrates nosed out of the litter are the major food items, coatis have powerful claws and easily dig out lizards, tarantulas and crabs from their burrows. When a tree bears fruit, coatis will make daily excursions, feeding until the supply is exhausted, ignoring more usual insect food. After a ten-week gestation, three to five young are born in a tree nest, where they remain for five weeks. The eyes and ears are closed at birth, opening a week later, and the cubs then begin to walk clumsily and hold their tails erect. They descend and join the band at six weeks, keeping close to their mother, who remains very protective.

The coati always holds its tail erect.

COBRAS, snakes capable of spreading a "hood" when alarmed: the long anterior ribs swing upward and forward, stretching the skin like the fabric of an umbrella.

When the Portuguese established themselves in India at the beginning of the 16th century, they encountered a snake that they called *cobra de capello* (snake with a hood), and the English subsequently adopted the name "cobra" for any African or Asian elapid snake that could spread a hood. The true cobras belong to the genus *Naja*, derived from the Sanskrit *naga*—snake. Africa, with five species,

is the center of distribution for the group, two more species occur in the Middle East and the Indian cobra is widespread in Southeast Asia.

The true cobras may reach 8-9 ft (c. 260 cm) in length. Their bodies are covered with smooth scales, which may be dull as in the Egyptian cobra or very shiny as in the forest cobra. The head is broad and flat, and the hollow fangs are set well forward in the upper jaw.

The Egyptian cobra, *Naja haje*, appears on ancient Egyptian royal headdresses, rearing up with hood spread. It has a wide distribution in Africa and Arabia, but is absent from rainforest and desert regions. It reaches a length of over 8 ft (2.4 m) and is a relatively thick and heavy snake. It is usually brown to black, often with lighter speckling. In southeastern Africa, a spectacular color variety is banded in black and yellow. Juveniles are uniform yellow.

The black-necked spitting cobra, *Naja nigricollis*, occurs in the savanna regions bordering the rain forests from West Africa south to Angola, Zambia, Malawi and Tanzania, with subspecies extending into the western Cape Province. Young cobras are dark gray, but adults are usually uniform black above. There is a single broad black band on the throat. The black-necked cobra reaches a length of 8 ft (2.4 m).

The Indian cobra, *Naja naja*, has a wide range from West Pakistan east to southern China and the Philippines. The body is usually brown, often with narrow light rings. On the back of the neck is often a dark-edged pale marking that may take the form of a pair of spectacles or a dark-centered ring.

Female cobras lay between 8 and 25 eggs in a hole. The newly hatched young are about 10 in. (25 cm) in length.

Cobras are active at night, when they do most of their hunting. During the day a cobra may often be found basking in the sun close to its refuge, which may be a termitarium, rodent burrow or a rock crevice. Cobras' diets include rodents, birds and their eggs, lizards, other snakes, amphibians, and fishes. Even locusts are occasionally devoured. Cobras are notorious poultry raiders, taking eggs and young birds. The forest cobra is the most accomplished catcher of fish, but the black-necked cobra is known to be a "fishing snake" in Victoria Nyanza.

A cobra will usually try to escape when disturbed by man, but if surprised or cornered, it will raise its head and neck, at the same time spreading the hood. If approached it will strike, or may even glide forward to attack. When a cobra bites successfully, it will often hang on and chew, injecting large quantities of venom. A full bite, if untreated, will cause death in man in about six hours through the action of the venom on the nervous system.

Four species of cobra have modified fangs that allow them to project a spray of venom into the eyes of an enemy, causing temporary blindness. The venom causes permanent damage to the eyes if not quickly washed out. The "spitting cobras" are the black-necked cobra, Mozambique cobra and ringhals of Africa and the southeast Asian populations of the Indian cobra.

The Indian cobra spreads its hood wide.

COCKATOOS, medium-sized to large parrots of the subfamily Cacatuinae, mostly short-tailed with white, gray or black plumage. The cockatoos are found from the Philippines to Australia and Tasmania, some species being commonly kept in captivity. Apart from some anatomical characters, cockatoos have no features to distinguish them from other parrots, but the combination of black, gray or white coloring, rather large size, square-ended tail and robust build serve to distinguish most of them.

There are about 17 species of cockatoo in five genera. In a genus of its own is the palm cockatoo, *Probosciger aterrimus*, a large, short-tailed parrot with dark gray plumage, and

The corella *Cacatua tenuirostris,* a cockatoo that stays near permanent water sources and is a waterguide to the bushman.

a crest of long feathers on the nape. It is a bird of woodland and jungle on the Cape York Peninsula of Australia and in the New Guinea region. It has a huge, curved bill, which enables it to crack open the palm nuts that form its principal food.

The four black cockatoos, *Calyptohynchus*, are rather similar and long-tailed. They are found in different parts of Australia and are distinguished from one another in the shape of the bill and the coloring of the tail. Except for the tail, the males of all species are completely black, but the females have narrow bars of pale yellow on their underparts. They are found in lightly wooded country, often associating in flocks. They feed on seeds, nuts, and insect grubs excavated from dead wood. In a genus of its own is the gang-gang cockatoo, *Callocephalon fimbriatum*, which is found only in wooded areas of eastern Australia. It is a medium-sized parrot with a short tail and gray plumage. The male has a crest of red-tipped feathers on the crown and the female has narrow pale yellow bars on its underparts. The bill is very stout, an adaptation to the diet of nuts and seeds.

The 10 species of *Cacatua* are distributed from Tasmania to the Philippines. Most species are predominantly white with erectile crests colored white, yellow, orange, red or pink, and some species have large pale yellow patches on the wing linings. The galah, *C. roseicapilla*, of Australia has gray upperparts, a pale pink cap on the head, and bright pink underparts. They are sociable, gregarious birds, which inhabit open or lightly wooded country. They feed mainly on nuts, fruit, seeds and the roots of low plants. With a few other cockatoos, those of the genus *Cacatua* differ from other parrots in that the eggs are incubated by both sexes rather than by the female alone.

The cokatiel, *Nymphicus hollandicus*, is a very distinctive small cockatoo with long, pointed tail feathers, gray plumage, a slender crest and yellow and orange markings on the cheeks. It is found in the arid interior regions of Australia, mainly in open country with scattered trees, where it feeds predominantly on small seeds collected on the ground. It seems likely that the slender build, long tail and distinctive plumage pattern and behavior of this peculiar cockatoo are adaptations to its open country habitat, ground feeding and diet of small seeds.

COCKCHAFER, common name for *Melolontha melolontha*, a scarabaeid beetle of economic importance in Europe because its larvae feed on the roots of potatoes, cereals and grasses. The adult is a relatively large beetle, coppery-brown in color, with prominent, club-shaped antennae. The wing covers are longitudinally striped, and the abdomen terminates behind in a narrow, cylindrical "tail-piece." Chafers can often be collected at night in large numbers around lights, particularly in early summer. The soil-dwelling larvae are known by the agriculturalist as "white grubs."

COCKLES, common marine bivalve mollusks belonging principally to the family Cardiidae. The best-known species is the intertidal common or edible cockle, *Cerastoderma edule*, formerly known as *Cardium edule*. It lives just

Sulphur-crested cockatoo *Cacatua galerita*, probably the best known of the cockatoos.

beneath the surface of slightly muddy sands, although it often occurs in coarse sands and sometimes also in gravels. The shell is whitish-yellow and globular with an obvious ligament and is crossed by 24-28 conspicuous ribs. When covered by the tide, the cockle filters plankton from the overlying water, using two very short siphons. When uncovered, however, the cockle retreats to about ¾ in. (2 cm) below the surface and often disturbs the black underlying deposits in the process. Buried cockles can be located by the discolored patches on the surface of the sand. Sometimes the population of cockles is so dense that the shells touch one another and growth is inhibited. More usually, however, a commercially exploited cockle bed may have several hundred cockles per sq yd (sq m). The method of collection varies. In some areas a large rake is used to bring the cockles to the surface, whereas elsewhere the deposits are passed through a coarse sieve, principally on the lower part of the shore where they are uncovered for a relatively short time. Cockles can also be dredged from a boat when the tide is high.

The fertilized egg of the common cockle give rise to a ciliated veliger larva, which spends two to three weeks in the plankton before coming to rest. Such larvae may be found in the plankton from March on through the summer.

COCKROACHES, largely nocturnal insects commonly thought of as domestic pests, but, although some occur in large numbers, the vast majority of the 3,500 species are independent of the domestic environment. They

are found mainly in tropical regions. In temperate countries, they usually occur indoors in places artificially heated: they used to be very common in bakeries, for instance. A few species occur naturally in temperate regions. The best-known domestic pests are *Periplaneta americana*, *Blatta orientalis*, sometimes known as the black beetle, and *Blattella germanica*.

The cockroaches are an ancient group of insects, very common during the Carboniferous period, 350-270 million years ago. Apart from *Blatta*, the commoner species have wings but rarely fly, unless it is very hot. The forewings, leathery in appearance, protect the large, delicate hindwings, while a large shield-like plate covers the front part of the body, often including the head. Cockroaches typically have very long antennae.

They are fast-running insects with soft, almost slippery bodies, which make them difficult and slightly unpleasant to handle, and these

Female of the bizarre cock of the rock.

features, added to the fact that they spoil food and have a peculiar smell, probably account for the general distaste with which they are viewed.

The eggs of cockroaches are coated with a secretion from special glands as they are laid, which is molded and hardens into a purselike structure or ootheca. This protects the eggs, especially from drying out. In *Periplaneta americana* there are usually about 20 eggs in two rows of ten in each purse, while *Blatta* has about 16, also in two rows.

The young emerging from the eggs escape from the ootheca when it splits open. They look like miniature versions of the adult except that they have no wings. The number of developmental stages varies, but in *Periplaneta* and *Blatta* there are normally 10-12, the insects getting bigger at each molt until the adult emerges.

COCKS OF THE ROCK, bizarre birds of the tropical American family Cotingidae. The bulk of our knowledge concerns the more common of the two species, the orange cock of the rock, *Rupicola rupicola*, the male of which is bright orange and the female brown. Some 12 in. (30 cm) long, with a short tail and rounded wings, and particularly strong, though short, feet and legs, these birds are unusual in some of their display features. The male has a peculiar crest in the form of a laterally compressed disc-shaped helmet, which, during the breeding displays, is erected in a fan that covers the top of the head and bill. The second species, the Peruvian cock of the rock, *R. peruviana*, is very similar but has red and gray plumage. The cock of the rock is highly terrestrial, spending most of its time near or on the forest floor where the displays are performed in small cleared areas or "courts." The displays are social affairs involving several males, but very little is known about the participation of the females. One unusual feature of the displays is the adoption of immobile, trancelike postures that may last several minutes. Prodigious leaps may also be performed. These displays, though little known in general, are familiar to the Jivaro Indians, who have based one of their dances on them.

R. rupicola breeds semisocially, several nests of mud and vegetable fibers being attached to a rock face within a few yards of one another. Two eggs seem to form the usual clutch.

CODS. The Atlantic cod is an olive green to brown with darker spots on the flanks and a silvery belly. It can be recognized by the small barbel under the chin, the pointed snout, the white lateral line and the presence of three dorsal fins. In the North Sea, cod grow to 8 in. (20 cm) in their first year, and a really large cod can reach 5 ft (1.5 m) in length and weigh up to 210 lb (95 kg). The cod is tolerant of a wide range of temperatures and is found in both temperate waters and the Arctic. It is omnivorous and voracious, feeding chiefly on crustaceans (especially Norway lobster, *Nephrops*, in European waters), mollusks, echinoderms, worms and fishes. The feeding of cod has been compared with the action of a vacuum cleaner, and among the curiosities recorded from their stomachs are a hare, a turnip and a bottle of whiskey! Cod are extremely prolific, and a large female may produce up to 7 million eggs, although only a very small

percentage ever reach maturity. The eggs and the newly hatched larvae are at first pelagic but after about ten weeks migrate to the bottom, and for the remainder of the summer and autumn the young fishes remain in shoals. Cod has long been a staple food in Europe, but the fishery really developed in the 16th century with the discovery of the Grand Banks of Newfoundland. The Basques and Portuguese were the first to make the voyage across the Atlantic to catch cod, which they did on long lines bearing up to 5,000 hooks. The cod were split down the middle, salted and brought home to be sold in the protein-hungry Mediterranean countries. Later, English and American fishing fleets exploited the fishing grounds, and trawling replaced long-lining. Dried or salt cod fed the growing cities of Europe and was used to provision ships and armies. Few parts of the fish were wasted, the swim bladders being turned into isinglass to clear beer and wine and the heads being boiled up for glue. The liver is rich in oil, and it became the most important part of the fish when it was used to combat rickets, a condition caused by vitamin D deficiency. In rickets the patient's bones are weakened, and it became common in large, smoky towns where the usual process of vitamin D formation by the action of sunlight on substances in the skin was inhibited. As early as 1820 it was found that cod liver was an antidote, but a century passed before its use was generally accepted and the importance of vitamin D was demonstrated.

COELACANTHS, primitive fishes belonging to the order Crossopterygii, once thought to be long extinct but now known to be represented

by a single living species, *Latimeria chalumnae*. They are related to the rhipidistian fishes, a group that gave rise to the first land vertebrates. Coelacanths first appear in the rocks of the Devonian period 400 million years ago, and the group finally disappears from the fossil record in the chalk of the Cretaceous period 90 million years ago. During all that time they altered remarkably little. They had heavily built bodies covered by thick, rough scales made up of four distinct layers and known as cosmoid scales (as found also in the fossil lungfishes). The vertebral column was poorly ossified and represented only by the cartilaginous notochord, and the fin spines were also hollow cartilage, hence the name "coelacanth" or "hollow spine." One of the most striking features of these fishes were the leglike lobes that supported the pectoral, anal, pelvic and second dorsal fins and the curious central lobe in the tail. The skull was hinged in the middle as in the earliest of the land vertebrates. In some coelacanths the swim bladder was ossified, for example in *Undina*, a feature not found in other fishes and difficult to account for.

On December 22, 1938, however, Miss Courtney Latimer of the East London Museum in South Africa went down to the quayside to inspect the fishes brought back by the trawlers. She recognized most of them, but lying on the pile was a large, heavily scaled fish that was 5 ft (1.5 m) long and weighed 127 lb (56 kg). It was deep blue in color and had curious fleshy lobes at the bases of the fins, and a central lobe in the tail. Miss Latimer made a sketch of the fish and sent it to Dr J. L. B. Smith, one of the most able ichthyologists in South Africa.

Common cockle, half buried in sand with siphon out for feeding.

Meanwhile, the specimen was sent to a taxidermist to be mounted, but the precious internal organs were lost. Dr Smith recognized the importance of the find, which represented a type thought to have been extinct for 90 million years. He named the fish *Latimeria chalumnae* in honor of Miss Latimer and to record its provenance off the mouth of the Chalumna River. Dr. Smith then launched a campaign to find more specimens of *Latimeria*. His patience was rewarded 14 years later in 1952, when a second specimen was caught off the Comoro Islands, north of Madagascar. A third coelacanth was caught nine months later, and since these exciting early days more have been found off the Comoro Islands and have been studied by French scientists led by Professor J. Millot.

COLLARED DOVE, *Streptopelia decaocto,* medium-sized bird, 11 in. (28 cm), of the family Columbidae, noteworthy for the speed with which it has spread northwestward across Europe from Asia Minor during the 20th century. The collared dove—sometimes called the collared turtledove—is grayish fawn above, paler beneath, with a pinkish flush on the breast. The tail is rather long, white-edged when seen from above and with the terminal half white when seen from below. There is a narrow black, white-edged half collar on the back of the neck, and the eyes are red.

This species has shown a remarkable increase in its range in a comparatively short time. Originally a native of India, it was introduced into China and Korea and possibly the Middle East, and from China into Japan, where it still occurs in a limited area. It has spread westward into some parts of Africa, but its main expansion has been into Europe, probably as

Collared dove, Europe's latest invader.

a result of the development of feeding habits that take advantage of human cultivation. It was probably a common bird in Constantinople during the 16th century, but it is not recorded farther east until 1835, when it reached Bulgaria. Around 1912 it appeared in Hungary, and about 1943 it was breeding near Vienna, near Venice and in Rumania. It was nesting in Germany in 1946 and had reached the Netherlands, Denmark and Sweden by 1949. In 1950 it bred in Poland and appeared in France, and by 1955 it was breeding in England and Switzerland. It first bred in Scotland in 1958 and now breeds regularly over much of the British Isles.

COLOBUS, the only leaf-eating monkeys in Africa, distinguished from their Asiatic relatives, the langurs, by their lack of a thumb—the

Mantled guereza, one of the best known of the black-and-white group of Colobus monkeys.

most that is present is a little nubbin. By and large, they are more compactly built and less rangy than langurs but rival them in their bright colors and variety of hair patterns. Colobus monkeys are 20-27 in. (50-70 cm) long in head and body, with a tail of about the same length.

Different authorities recognize different numbers of species of colubus, but all divide them into three groups, the black-and-white, the red and the olive colobus. Black-and-white colobus are found throughout the forest belt of Africa and in several of the forest "islands," both montane and lowland, of East Africa. Red colobus are just as widely distributed, but are found on fewer of the forest islands. Olive colobus are restricted to a belt of forest along the West African coast.

The mantled guereza, *Colobus guereza,* is the best known of the black-and-white group. It is deep shining black in color with a white ring round the face, a long white "veil" of hair on the flanks, which meets its fellow over the base of the tail, and a full white tuft on the tail. The pattern varies from one race to another, with the veil restricted to the lower flanks or spreading up onto the back, and the tail brush terminal only, or occupying nearly the whole of the tail. The light parts are white or creamy, varying geographically.

Guerezas live in troops of up to 20, with a single male in each. The surplus males live alone or in small bachelor bands. Each troop has its own territory, which is very small, only 20 acres on average. Normally the troop does not venture outside its territory, so that the borders are adjacent. The boundaries are marked by both visual and auditory means. A troop will sit on branches well exposed to view, with their tails hanging down, waving

their tails for their neighbors to see that they are in occupation of a territory. At dawn and dusk, and sometimes also in the middle of the night, the male roars in a territorial display; the sound can be heard a mile away, and as he roars—sometimes a long-drawn-out roar lasting as much as 20 minutes—he leaps from branch to branch very agitatedly, setting off the neighboring males who begin to roar also.

There is no special breeding season. A female will present to the male when she is in oestrus; she may also mate with the troop's subadults. The infants are born white and are thus very conspicuous. The other females crowd around the mother, and each takes turns in handling and holding the infant. This "passing-around" ceremony is found also among langurs. The king colobus, *C. polykomos,* is found in West Africa, from the Niger delta west to Gambia. It is also black, with whitish fringes on the cheeks and shoulders, and the whole tail is white, without any long "brush" of hair. The black colobus, *C. satanas,* is entirely black, but with long hair fringes on the cheeks and shoulders. It occurs in Cameroun, Gabon and parts of the Congo (Brazzaville). Finally, the Angolan colobus, *C. angolensis,* has long shoulder tufts but no white on the brows, and the white does not extend along the whole of the tail. It is found in the Congo (Kinshasa) and isolated forests in Rwanda, Tanzania and Uganda. The red colobus, *C. badius,* is more slenderly built and lacks the long hair tufts or the convex nose of the black-and-white. The color pattern is largely erythristic, with a dark back and lighter underside and limbs, but beyond this it is difficult to specify the colors.

Red colobus live in the tops of the trees, often in the emergent crowns. On Zanzibar the average troop size is said to be 15, but on the mainland 50 or more seems to be the rule. They are rather unwary.

The olive colobus differs profoundly from the other two species, and is now generally placed in a separate genus and called *Procolobus verus*. It is very small, with a head and body length under 20 in. (50 cm), olive green, the hairs being green-yellow at the base and black at the tip. The only adornment is a small crest on the crown. The head is small and rounded; the face is flesh-colored except around the eyes, where it is bluish. The nose is small and the face straight. It haunts the shrub and lower layers of the forest, not being usually found above 30 ft (10 m). It lives even in dry, scrubby areas where there are no other colobus. The troops of 10-15 constantly associate with troops of guenons.

COLORADO BEETLE, *Leptinotarsa decemlineata*, a leaf-eating beetle of the family Chrysomelidae, easily recognized, being about ⅜ in. (1 cm) long, very convex above and with longitudinal black and yellow stripes on the wing cases. The fat, humpbacked, red and black larvae are also easy to identify.

The Colorado beetle was originally native to the eastern slopes of the Rocky Mountains, where it fed on a weed called the buffalo bur,

Colorado beetle on a potato leaf.

Solanum rostratum. The cultivated potato, *Solanum tuberosum,* was first planted in the United States near the Atlantic seaboard, and as its cultivation spread westward it eventually reached the home of the Colorado beetle, which, in 1850, took to it and spread eastward across the potato fields, completing its conquest by about 1874. In the early 1920s the Colorado beetle became established in France and soon spread to more areas of western Europe, wherever potatoes were grown. The species has only been prevented from establishing itself in Britain by the extreme vigilance of the Ministry of Agriculture, the farming community and the population as a whole.

The female Colorado beetle lays batches of orange eggs on the early potato leaves, and these hatch in about one week into reddish larvae that feed on the foliage. When large numbers of larvae and adults are present, most of the foliage may be devoured so that the potato plants die, or the tuber yield is reduced.

COMB-FOOTED SPIDERS, small spiders mostly with a globular abdomen and slender short legs, which build irregular snares among vegetation. A leaf or a specially woven tent of silk and debris provides a retreat. The black widow, *Latrodectus,* is one example. Another is *Theridion tepidariorum,* native to the tropics, which has become a common inhabitant of hothouses in Europe.

Some threads in the snare are studded with gum globules, and the spider throws further threads over an entrapped insect. It will attack prey far larger than itself, first biting the legs and then waiting for the poison to take effect. The male courts the female by web vibrations, which are probably made distinctive by a stridulating apparatus between the abdomen and cephalothorax. Some females share insects with their young. In one European species, *Theridion sisyphium,* common in English gardens, the mother regurgitates a drop of fluid from her mouth from which the young spiders feed. In American species of *Anelosimus* the young never scatter but continue life in a communal web.

Species of the genus *Argyrodes* live in tropical and semitropical countries where they lead a commensal existence in the webs of much larger species of the family Argiopidae.

CONCH SHELLS, marine operculate snails found in the shallow waters of the tropics. The shells are widely collected as they are large, well colored and ornamental. They are spirally coiled, often armed with processes (as in the spider conch, *Pterocera lambis*) or with a huge reflected lip (as in the queen conch, *Strombus gigas*) or with both these features (as in the West Indian fighting conch, *Strombus pugilis*). The outside of the shell is often well marked,

and in some species, such as the queen conch, the inside of the reflected shell mouth is a delicate pink color. These animals are a staple food in the West Indies, and the shells are sold for the manufacture of cameo brooches. Conch shells were used as currency by the inhabitants of the islands of the western Pacific and also for medicinal purposes. The foot has not evolved in the same way as that of the majority of gastropod mollusks, for instead of being large and flat, it has a reduced sole and so is narrow and muscular, with the horny operculum adapted to form an anchor. The animal moves by extending the foot, digging in the operculum and then contracting the longitudinal muscle connecting the operculum to the shell (the columella muscle). This then pulls the heavy shell (about 3½ lb [1.5 kg] for the queen conch) backward along the ground. If the conch is attempting to avoid a predator— for example, the queen conch escaping from the tulip shell (a carnivorous operculate snail)—then the contraction of the columella muscle is very rapid, flinging the conch into the air, so being responsible for their leaping mode of progression.

Conchs have well-developed eyes, which are borne on long, movable eye stalks.

CONDORS, very large birds comprising two species of New World vultures: the California condor, *Gymnogyps californianus,* and the Andean condor, *Vultur gryphus.* The former is one of the rarest of living birds of prey, found only in a small area of California, and reduced to about 40 individuals. The Andean condor is more numerous and widespread, inhabiting the whole Andean chain from Venezuela to Patagonia, but even this bird is much reduced from former times. The California condor

West Indian conch shell, moving over a sandy beach.

Conger eel, a large marine specimen, sometimes swims at the surface on its side, with its body undulating.

The conger, like the freshwater eels, makes a spawning migration. The population of the North Atlantic migrates to an area between latitudes 30° and 40° N and spawns at depths of 4,800 ft (1,500 m). The Mediterranean population spawns in the deeper parts of that sea. The eggs float at that depth until the small, ribbonlike leptocephalus or larva hatches. It metamorphoses into the adult form near coasts but remains pinkish in color until it is about 12 in. (30 cm) long, after which it assumes the gray-brown of the adult. A large female may lay up to 8 million eggs.

Both the conger and the common freshwater eel lack pelvic fins, but in the conger eel the dorsal fin, which extends to the tip of the body, begins over the pectoral fins, whereas in the common eel the dorsal begins much farther back. There are no scales on the body. Congers are large, voracious fishes of nocturnal habits, feeding on mollusks, crabs and fishes and often haunting wrecks. Females grow larger than the males and may reach a length of 9 ft (2.7 m); one of the largest rod-caught specimens weighed 84 lb.

COOTS, small to medium-sized birds, the commoner species being about 14 in. (35 cm) long, with dark plumage and living in ponds, lakes or marshy areas.

They are members of the subfamily Fulicinae in the rail family Rallidae, and are like other rails in having a back-and-forth bobbing motion of the head when walking or swimming, but unlike them in commonly staying on open water where they are conspicuous. Other rails tend to be secretive, and skulk in waterside herbage; coots do this only when breeding.

Coots have a short neck and tail, and therefore appear rather squat. Their wings are proportionally short, but their legs are long. They are noteworthy for the large size of their feet and the lobes on the toes, which act as "mud boards" enabling the bird to walk over soft mud and floating vegetation, explaining the origin of one of its North American vernacular names—"mud hen." Another name given to the American coot is "blue peter," from its habit of pattering along the water surface before takeoff. Indeed, it has to do this in order to become airborne, but when in the air it is a strong flier. The lobes on the feet also act as swimming aids, and coots feed much in the water, diving with ease and bringing up various vegetable materials for sorting at the surface. This habit leads to a certain amount of attention being paid by ducks, and the coots are not uncommonly robbed of their food by gadwall, wigeon, and other ducks.

A distinctive feature of coots is the frontal shield, which continues the soft horny material of the bill up and over the front of the head. In the European species this is wholly whitish, like the bill, hence the phrase "as bald as a coot." In other species the shield may be wholly or partly pigmented. The function of this rather noticeable feature seems to be to assist the orientation of the young to the bill of the adult during feeding in the tangled undergrowth of the water's edge.

The European coot, *Fulica atra*, is found through much of Europe and in Asia and Australia. In New Guinea it is found up to 11,000 ft (3,350 m). The common coot of

probably never has been very numerous, but its main decline has occurred in the last century, and it appears to have been reduced by about a third since the early 1950s. Unless more effective conservation measures can be applied it must soon disappear. The males are as large as or larger than the females. California condors weigh 18-31 lb (8-14 kg), and recorded wingspans vary from 8 ft 2 in. to 9 ft 7 in. (2½-3 m). In the Andean condor males weigh about 25 lb (11 kg), females 17-22 lbs (7½-10 kg), and wingspans of up to 10½ ft (3.2 m) have been recorded. If wing area and weight are taken together, the Andean condor is the world's largest flying bird.

The California condor is mainly black, with white underwing coverts and edges to the secondaries. The bare skin of the head and neck is orange, sometimes yellowish or gray. The Andean condor is more variegated, mainly glossy black, but with white upper wing covers.

The tongue of the California condor is specialized, the distal 1½ in. (4 cm) being thick, with a U-shaped trough edged with hard, backward-pointing spines. In feeding, the tongue is worked to and fro, and the U-shaped trough can be partly closed so that soft flesh is gripped and pulled backward to the throat.

The wings of these huge birds are highly specialized for soaring. They cannot fly unless air currents or thermals are active, and their high wing loading also means that they have a wide turning circle. In wet weather they are more or less grounded, and later spread out their wings to dry. Both species inhabit mountainous country where strong air currents are normally prevalent and where they can find cliffs on which to roost and breed.

Both species lay single white eggs, very large, weighing about 10½ oz (280 g). The California condor does not always use the same nest cave, having several alternatives. Both sexes may incubate, either by day or night, and individuals stay on the nest for up to 20, even up to 46 hours. The incubation period is probably more than 50 days; 54-58 days is recorded for the Andean condor in captivity.

CONGER EEL, *Conger conger*, a very elongated fish with no pelvic fins. This is a large marine eel widely distributed in the North and South Atlantic, in the Mediterranean and in the Indo-Pacific region. It is not found off the west coast of America.

The European coot has a white blaze and lobed membranes on its toes.

North America, *F. americana*, is found from southern Canada to northern South America; it is very closely related to the European species and, like it, is widespread and common. Another closely related species is the crested coot, *F. cristata*, of South and East Africa, Madagascar, a relatively small area of northwest Africa, and a small part of southern Spain. This species is distinguished by two prominent red knobs on the head—one above each side of the bluish-white frontal shield. It is rather more secretive than the previous species and is found less often in open water.

The greatest development of the coot subfamily has been in South America, where the other seven species are found. One species, *F. caribea*, is found in the West Indies, and two species are found on the high lakes of the Andes. One of these is the large *F. gigantea* and the other the rare horned coot, *F. cornuta*, so called because of the large hornlike projection on the frontal shield. The latter species exists in numbers that are probably dangerously low. It is unlikely that there are more than 2,000 of them living around isolated lakes above 13,000 ft (4,000 m). It is also unusual in employing stones in the construction of its nest. Because there is little vegetation on the shores of the lakes, each pair of horned coots carry pebbles into the water to build a mound 3 ft (90 cm) in diameter and 2 ft (60 cm) high, and on top of this they make their nest from the sparse vegetation.

Coots nest around the edges of ponds and lakes, building large nests of reeds, sedges and other materials, sometimes floating among the aquatic vegetation. They usually lay up to ten eggs. The young have brightly colored head patches, presumably to show the adult where the chick's head is, to put food into its beak. After the eggs are hatched, extra nests may be built for brooding the young.

COPEPODA, small crustaceans, mostly microscopic, the larger ones being about the size

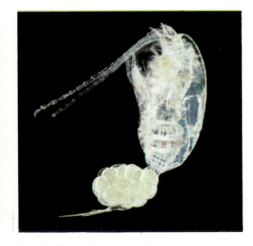

Pseudocalanus elongatus, a free-swimming copepod (side view of female carrying eggs).

of a pinhead, abundant in fresh water and the sea. Some are free-living in offshore waters, some are more sedentary, either on or near the bottom or live a semiparasitic life, whereas others are wholly parasitic and have become

The long, lobed toes of the coot enable it to walk on leaves and water-plants.

so changed that they are hardly recognizable as crustaceans, except when larvae. In all free-living copepods the body is elongated, torpedo- or pear-shaped, segmented, and with double-branched swimming feet (Gk *kope* = oar, *podos* = foot). There are paired appendages on the underside, not more than one pair to each segment, and in the center of the front end of the body is a single red eye made up of three units, or ocelli, each with retinal cells and its separate nerve from the brain. This single eye is not formed, as in many other crustaceans, from fusion of two eyes into a so-called compound eye.

Copepods lack the fold of skin or carapace familiar in crustaceans such as the lobster.

It has been suggested that there are numerically more copepods in the world than all other kinds of multicellular animals put together. If the vast numbers that can be taken from the open sea in a tow net are pictured as extending for thousands of miles in all directions through the oceans of the world, this suggestion is perhaps not greatly exaggerated.

COPPERHEAD, *Agkistrodon contortrix*, venomous snake related to the water moccasin and rattlesnakes. Adult copperheads are generally less than 1 yd (91 cm) long and are named from the coppery or reddish-brown head. The body is brownish with 15-25 rather irregular and darker brown cross bands. Copperheads often superficially resemble the common milk snake, *Lampropeltis doliata*, but may be distinguished from this harmless species by

the vertical elliptical pupil and the facial pit. The four subspecies of the copperhead are distributed over much of the central and eastern United States, from Kansas, Oklahoma and Texas to Connecticut and Massachusetts. Copperheads may be found in a variety of habitats, including woods, rocky country and semidesert. In the early spring they are active during daylight, but during the hotter summer months become mainly nocturnal and spend the day resting under logs and in crevices. Copperheads have frequently been reported as hibernating in assemblies, sometimes with other species of snakes. Mating of copperheads occurs in April and May, and two to ten young are born in August or September. Copperheads display a very catholic taste and have been reported as eating small mammals, birds, lizards, snakes, frogs, caterpillars and cicadas.

CORAL may be soft coral, horny coral, thorny coral or true or stony coral, the last being builders of reefs and atolls. They can be most easily thought of as polyps similar to sea anemones that lay down a skeleton, usually calcareous.

The typical coral polyp is small and lacks a pedal disc, the base of the animal sitting in a calcareous cup or theca which it secretes. The short column is smooth, and the oral disc bears tentacles arranged in rings and in multiples of six, each bearing a knob-shaped free end, which is packed with nematocysts or stinging cells, of two types, true nematocysts and spirocysts.

Corals display an inexhaustible variety of shapes and colors.

The strikingly patterned Arizona or Sonoran coral snake.

Most corals are colonial, and the polyps are connected to one another. These connections lie above the level of the calcareous skeleton, so that the colony generally sits above its massive skeleton. The skeleton of the colony is called the corallium, while that of each polyp is a corallite and consists of a cup or theca around the lower region of the polyp. The colonial corals range from solid spherical masses to those with delicate branches.

Even the largest coral colonies arise from a planula larva, which settles down on a solid object, develops into a polyp and then starts to bud off other polyps asexually. In branching forms each polyp is separated from the next by some distance, although connected by the extension from the body wall, so that the skeleton appears to be internal. In massive forms, polyps spread outward over the surface, so that the skeleton is underneath.

CORAL SNAKES, highly venomous snakes with hollow or grooved fangs fixed to the front of the upper jaw, related to the cobras of Africa and Asia, the mambas of Africa and the kraits of Asia. They are characterized by the head being only slightly defined from the neck, the slender body having an oval cross section and the eye having a vertical, elliptical pupil. The vertical pupil is associated with twilight or nocturnal activity and indicates that these snakes hunt their prey mostly at night.

Coral snakes are restricted to the New World, from the United States southward into the tropical forests of South America. The eastern coral snake, *Micrurus fulvius*, renowned as a snake eater and up to 3 ft (90 cm) long, is one of the few very dangerous snakes in the United States. The Arizona or Sonoran coral snake, *Micruroides euryxanthus*, of southeastern Arizona, western New Mexico and Mexico, is about half the size, but several other species, found in Mexico and the tropics of South America, may attain a length of 5 ft (150 cm). Coral snakes usually have a striking color pattern of broad, alternating orange or pink and black bands separated by narrow yellow rings. This is probably a warning coloration, indicating to other animals that the snake is dangerous and best left alone. In general, coral snakes are secretive burrowers and spend the day hidden in grass tussocks or other dense vegetation or in runs under stones and rocks. They are, in fact, rarely seen above ground during daylight, but may become active in the day after rain. The jaws of coral snakes are, unlike those of most snakes, only slightly distensible, so that they feed on small or slender prey such as other snakes and lizards.

CORMORANTS, long-necked, long-billed diving birds, sometimes known as "shags." There are some 30 species of cormorants—the exact status of some them, particularly the antarctic forms, is uncertain—distributed around the sea coasts and larger lakes and rivers of the world, excepting parts of northern Canada and northern Asia and certain islands in the Pacific. They vary in length from 19-40 in. (48-101 cm).

In most species the plumage is black, or black with a green or blue metallic sheen, hence the etymological development: *corvus marinus* ("sea crow"), *cor marin* (some French patois), cormorant. In the southern hemisphere a few species are gray and others white on the throat or underparts. The eyes, bill and facial skin are often brightly colored. In the breeding season many species have crests, or patches of white on the head, neck or flanks. The bill is long, though strong, and is hooked. The feet are large and fully webbed, and the legs are short and strong—all contributing to the cormorants, abilities as swimmers.

On the whole, cormorants enter the water only for fishing. This may be because the plumage becomes wet rather easily as a result of its general looseness. After fishing, cormorants stand for long periods with wings outstretched, presumably to dry them. The wings are used to a certain extent under water, particularly for steering and braking. Even

An Australian subspecies of the common cormorant of the North Atlantic.

Common cormorants with their nests (usually made of seaweed and dead vegetation) on a rocky cliff.

the Galápagos flightless cormorant, *Nannopterum harrisi*, has this habit.

Most cormorants are gregarious and breed colonially on rocky shores or cliff ledges or in trees. Some species are strictly marine, others restricted to fresh water. Some, however, may be found in either of these habitats, anywhere, in fact, where there is enough water to harbor the fishes, Crustacea or Amphibia on which they feed. The common cormorant, *Phalacrocorax carbo*, (the great cormorant of eastern North America, and the black cormorant of Australasia), for example, is almost cosmopolitan in its distribution, from the tundra to the tropics and from Canada eastward to Japan and Australasia. In many countries it is found on both fresh and salt water, and will nest in trees or on rocks. It takes a wide variety of animal food but feeds largely on fishes.

The sexes of cormorants are externally very similar, and there is a reduction in the dimorphism of their sexual display. There has also been some confusion on the part of observers as to the relative roles of the two sexes during display. In the family as a whole it would seem that breeding may be promiscuous, polygamous or polyandrous, with reversed copulation being quite common. The flightless cormorant, however, is monogamous, and other species probably are as well. Copulation normally takes place on the nest, or nest site.

The young cormorants are fed largely by regurgitation, both parents bringing partly digested fish into the mouth and gullet and allowing the young to insert their heads to feed. Sometimes the adults present food from the bill tip, and sometimes they disgorge onto the nest. In the double-crested cormorant, *P. auritus*, pebbles have been noted to be given incidentally with food and have been accumulated by the young until they were old enough to disgorge.

As cormorants are primarily fish eaters and exist in large numbers—colonies of the guanay, *P. bougainvillei*, of the western coast of South America, for example, may contain millions of individuals—they are potential or actual competitors for a certain proportion of man's food.

COTINGAS, a family comprising 90 species of heterogeneous New World birds. They occur from the southeastern part of the United States (Arizona and Texas) south through Central America to tropical South America. Cotingas vary in size from that of a sparrow to that of a crow. Practically all of them are birds of the forest, and for this reason the life histories of a great number, possibly the majority, remain wholly unknown. This even applies to quite common species. As they live in high trees in the rain forests, their nests are extremely difficult to locate. Some are dull-colored, but a great number have brilliant colors, with strangely modified wing feathers,

fleshy wattles around the bill or brightly colored naked skin on their heads.

COTTONTAIL. Seven species of rabbits are known by this common name because of the characteristic fluffy appearance of their tails. All belong to the genus *Sylvilagus* (Latin *sylva* = wood; Gk. *lagos* = hare). The eastern cottontail is the only true cottontail, but the other six species of the same genus are given this name by some authors.

The eastern cottontail, *Sylvilagus floridanus*, is the most abundant and best studied of the group. The upper parts of the body and tail are brownish or gray, and the underparts white; ears are dark grayish-tan bordered with black. Adults weigh up to $3\frac{1}{2}$ lb (1.6 kg). There are three molts annually. During the spring and summer months the pelage is replaced in spots and in autumn in sheets. All cottontails are confined to North America apart from the eastern cottontail, which has also invaded part of South America. Its wide geographic distribution is reflected in the large number of subspecies. It is found in almost all types of habitat except mature woodlands and is common where adequate shelter exists bordering grasslands or cultivated fields. It is not a particularly good runner, and freezes, stops, zigzags and seeks cover when alarmed.

The breeding season is from March to September, and four litters are produced annually averaging four kittens each. The litter size is smaller in more southerly latitudes. The

Cottontail, North American rabbit named for the cottony appearance of the tail hair.

Indian courser *Cursorius coromandelicus* with its two camouflaged eggs in the hole in the ground which serves as a nest.

tempo, volume, pitch and in cadence, often descending the scale during repetition. During calling the neck is inflated and the bill tip lowered.

The nest is a crudely made, domed structure, unlike the usual cup-shaped nests of non-parasitic cuckoos, with little evidence of an internal cup for the eggs. It may be on the ground, where it is concealed in long grass and very difficult to find or, if above ground, low down in a thick twiggy shrub. The eggs are more spherical than the typical birds' eggs, white in color with a mat surface. They may become stained by nest material.

The young have blackish skins and, unlike most nestling cuckoos, are not naked but have down, although this is sparse and remains in a sheath, appearing like distinct coarse white hairs. It is suggested that the black body color and these white hairs help to conceal the presence of the young coucals in a nest that, by reason of its terrestrial or near-terrestrial site, might otherwise be very vulnerable to attack by predators. In the nest the young coucals face away from the entrance and crouch motionless if the nest is disturbed. Coucals are general predators on any small animals that they encounter. Large insects are the principle item in the diet, but amphibians, reptiles, including snakes, and small mammals and birds may also be included. These cuckoos are known to take nestlings from the nests that they find and have even been known to carry off small birds that have fallen into long grass when shot by collectors. Although coucals are poor fliers, and will flutter up into trees only with apparent difficulty, they show more agility in planing down from perches and may at times descend on their prey in this manner.

COURSERS, wading birds, closely related to the pratincoles. Both are Old World groups, the coursers comprising ten species, one Australian, two Indian and the rest African. These include the Australian dotterel, *Peltohyas australis*, and the Egyptian plover, *Pluvianus aegyptius*, (neither of which is a true plover) and four species of each of the genera *Cursorius* and *Rhinoptilus*. Coursers resemble plovers in several respects, among them size, body shape, locomotory behavior and call, but they have pointed and sometimes down-curved bills (rather than blunt and straight) and longer legs in proportion to body size. All, apart from the Egyptian plover, are predominantly pale brown (though often with dark brown or black and white eye stripes), as befits their chosen habitat of semidesert and open ground; the leg color is also pale. The sexes are alike. The Egyptian plover, a bird of river margins and sandbanks, has gray upperparts, a characteristic head pattern of dark green crown and eye stripes and a black wing bar. It associates with basking crocodiles.

Of the primarily African species, the cream-colored courser, *C. cursorius*, which breeds on both borders of the Sahara desert belt (north and east to Persia, south and east to Kenya), is the only one that reaches Europe at all frequently. When approached, it runs swiftly for short distances with head lowered, then pauses to stand erect, in the same way as many plovers. Sometimes it crouches to escape detection, again like young plover chicks.

gestation period is 28 days. Kittens are dropped in a shallow excavation in the ground, lined with grass and the mother's fur and concealed from the top by dead grass. Newly born young are blind but have sparse hair over the body, differing in this respect from the European rabbit, which is born in an underground nest and is completely naked. The young are suckled for three weeks. Eyes open after seven days, and at ten days the young are capable of venturing outside the nest. The does are reported to carry their kittens in the mouth from one place to another.

COUCALS, a genus of 27 large, nonparasitic cuckoos. They are absent from the Americas but widespread in tropical regions elsewhere. They are heavily built, mainly terrestrial birds; pheasantlike, with heavy, slightly curved bills, large rounded heads, squat bodies with short rounded wings, long tails and long strong

legs. Their gait varies from a sedate stride to rapid running. They tend to be skulkers, keeping to low cover and occurring in grassland with tall grasses, in thick vegetation on the edges of swamps and forests, and in secondary growth in forests and clearings.

The plumage has a loose and hairy texture and is coarse and glossy, well able to withstand wear and abrasion. Immature birds, and hens of some species, may have a plumage cryptically colored in brown and black, whereas the adults tend to have boldly colored plumage in black, dark blue, chestnut and white, striated at times. Often there is a strong contrast between the dark head and tail and the chestnut and white body and wings.

The songs of coucals are very distinctive, being series of repetitions of low-pitched hollow notes that sound like air bubbling up through water. In different species the calls differ in

All these coursers must be able to withstand very high temperatures in their chosen breeding areas. For example, the Egyptian plover part-buries its eggs in the sand, which it may moisten from time to time with water, apparently by regurgitation. Other species lay their eggs on open ground and shade them from the sun, rather than incubate them. The air temperature is usually high enough, and extra heat can be provided if necessary by brief exposure to the sun. Courser chicks may also be shaded from the direct sun by their parents, and it has been reported that Egyptian plover chicks may be buried temporarily in the sand by their parents if danger approaches.

COWBIRDS, small to medium-sized perching birds of the American oriole family, remarkable for the habit of parasitizing other birds for the rearing of young, which is shown by all but one species of cowbird. Cowbirds are so-called because of their association with cattle.

It is very likely that the birds benefit considerably from the disturbance of insects and other small animals caused by the cattle as they walk through the vegetation. It has been shown, for example, that the groove-billed ani, *Crotophaga sulcirostris*, a tropical American cuckoo, feeds about three times as efficiently when with cattle. The brown-headed cowbird, *Molothrus ater*, was previously associated closely with bison and transferred this association to domestic cattle when the latter replaced bison as the most common large mammal.

The cowbirds are noteworthy for the wide range of stages of development of the parasitic habit that are shown among the various species of the group. The bay-winged cowbird, *Molothrus badius*, is nonparasitic and sometimes builds its own nest, though usually it appropriates the nest of another species in which to lay its eggs and rear its young. Some of the parasitic cowbirds are rather selective in their

"choice" of host. The screaming cowbird, *Molothrus rufoaxillaris*, apparently only parasitizes the bay-winged cowbird, and the giant cowbird, *Scaphidura oryzivorus*, limits itself to certain other icterids—the oropendolas and caciques. The brown-headed cowbird, on the other hand, is known to parasitize over 250 other species of birds.

COYOTE, *Canis latrans*, or prairie wolf, a North American wild dog closely related to the wolf, *Canis lupus*, but smaller in size. Weighing between 25-30 lb (11.5-13.5 kg), the coyote has grizzled black and yellow fur with black markings on the tail and shoulders. Although originally an inhabitant of the western plains of the United States, this species, unlike most carnivores, has expanded its range in recent years and is now found as far east as the New England states. The coyote's success is largely due to its cunning and opportunism. Whereas the wolf, a cooperative

Coypu, a rodent which can be a pest or profitable, according to circumstances.

pack hunter, cannot survive in the absence of large, hoofed mammals, the coyote takes advantage of numerous food sources, including young ungulates, small rodents and rabbits, carrion, insects, fruits, and even rubbish from garbage dumps. Its ability to hunt in packs composed of family groups, in pairs, or alone and to alter its habits as external conditions change is reminiscent of the jackals, which occupy a similar ecological niche in Africa and southern Asia. Both the coyotes and jackals are constantly hunted by man, who believes that they prey upon domestic livestock, but their cleverness in evading traps and snares has so far prevented their extermination. In the United States, recent attempts to reduce the coyote population with poisoned carcasses have not only killed members of this species but have also taken their toll of other wildlife as well.

Coyotes may pair for life; certainly the male aids in rearing the young, which are born in March or April after a 63-day gestation. While the cubs are helpless and need constant tending, the male does most of the hunting. Later, both parents hunt, and, during weaning, they regurgitate partly digested meat to the litter to ease the transition from a milk to meat diet. The family often remains together during the lean winter months, constantly on the move in search of carrion or larger mammals weakened by cold and hunger.

COYPU, *Myocastor coypus,* a South American rodent superficially resembling a large muskrat or small beaver.

Coypus have a head and body length of almost 2 ft (60 cm) when fully grown and weigh up to about 15 lb (7 kg). They are aquatic animals, but the tail, which is about two-thirds the length of the head and body, is quite round in section, not flattened as in most aquatic animals. The fur is yellowish-brown above and gray below. The underfur, especially on the underside of the body, is dense and soft, and coypu pelts, known in the trade as "nutria," are prepared by cutting along the upper surface of the body to leave the underside intact and by removing the longer bristly guard hairs. An unusual adaptation to an aquatic life is the very high position of the teats on the female's flanks, allowing the young to suckle while the mother is swimming.

Coypus are found throughout the southern half of South America wherever there is water. However, they have been kept all over the world on fur farms from which they have frequently escaped (and in the case of the USSR, been deliberately released) and have become established in the wild. Such colonies can be found in eastern England, many parts of Europe and the USSR and in much of North America. In some areas, e.g., in East Anglia in England, they are considered a pest because of damage to crops such as sugar beet, but in Russia wild coypu are considered a valuable asset. They are artificially fed in winter and systematically trapped for their fur. Although primarily herbivorous, coypus are known to vary their diet with shellfish. Breeding, at least in the northern hemisphere, appears to continue all year, and the young, usually about five in a litter, are very well developed at birth and are very soon accomplished swimmers. Depending on the habitat the nest may be in a burrow in a river bank or in a heap of vegetation built on the surface of a marsh.

CRABS, name given to a wide range of decapod crustaceans having a reduced abdomen that is carried tucked forward underneath a broad carapace, which is often wider than long, and the insertions of the legs are widely separated on the underside. When walking or running, many crabs move sideways or obliquely, but the most skilled runners can move forward, backward or sideways with great facility. A typical crab has a pair of well-developed pincers and four pairs of walking legs, making five pairs, or 10 legs in all (Gk: *deka*, ten; *pous, podos,* foot). One or two of the hind pairs of legs are reduced or even absent in certain families. Some of the fast-running crabs appear to use only two pairs of legs when traveling at high speed.

The eyes of crabs are borne on stalks, which vary greatly in length in different species. An extreme example is found in the genus *Podophthalmus,* where the eye stalks resemble matchsticks, and there is a joint just before the eye.

Crabs are often scavengers, eating whatever they can find, but they are also active predators, and their pincers are capable of killing small fish and breaking open the shells of various

A stone crab, *Lithodes maia,* in the Bergen Aquarium. It is named for its resemblance (when the legs are tucked into the body) to a stone.

mollusks. In addition to these generalized methods of feeding, there are specialized techniques in certain families.

Male crabs often have larger pincers than females, and the male abdomen is usually much narrower. When eggs are laid, they become attached to the female's abdominal appendages and form an egg mass between the abdomen and the underside of the thorax. Crabs measure only $\frac{1}{4}$ in. (6 mm) across the carapace, but the largest have a carapace over 1 ft (30 cm) in width, and the legs of the giant Japanese spider crab may span nearly 9 ft (3m). Most species are marine, living among rocks, or burrowing in sand and mud down to considerable depths in the sea. Others have invaded fresh water, particularly in the tropics. Yet a number of other crab species, on the other hand, have become semiterrestrial, often ranging far away from water, although they generally prefer moist places.

Demoiselle crane of Europe, Asia and Africa.

CRANES, large, long-legged, spectacular birds with a superficial resemblance to herons and storks, but hardly ever perching in trees. There are only 14 species, and some of these are in danger of extinction.

Cranes stand from 30-60 in. (76-150 cm) tall, of which the long legs and neck make up a considerable proportion. The plumage varies from white to slate-gray or brown, and the inner secondary feathers of the wing are curled or elongated to form a mass of plumes that droop over the hind end of the body when the wings are folded. The head is ornamented with plumes or has areas of red featherless skin, and the tail is rather short.

The bill is strong and straight; the legs and feet are also strong, with the toes connected at the base by a membrane. The hind toe is elevated, apparently playing little part in normal locomotion. The voice is loud and resonant—typically trumpeting but in others more whistling—aided by the unusual formation of the trachea, which in the males of most species is coiled upon itself. These tracheal coils occupy part of the sternum, and in some species even penetrate the breast muscles.

Cranes are worldwide, being absent only from South America, Madagascar, the Malayan archipelago, New Zealand and Polynesia. Typically they are found in open country: marshes, plains, prairies, lakes and seashores. In the breeding season they are particularly shy and nest in wild remote areas. Most species are migratory.

Outside the breeding season most cranes are strongly gregarious, migrating and wintering in large flocks. Cranes will take most kinds of small animal and perhaps even more vegetable material, particularly fruits and roots. They forage in cultivated fields whenever possible, and the understandable antagonism of farmers in an increasingly competitive and demanding market has been one of the reasons for the decline in numbers of cranes. In certain countries where cranes are partial pests there are governmental schemes for protecting crops and indemnifying farmers. In Canada, for

Spider crab *Stenorhynchus seticornis*, of the tropical Atlantic.

example, insurance against damage to crops by wildlife, including cranes, was instituted in Saskatchewan in 1953 and in Alberta in 1962. Also, exploders are used to scare cranes off the standing crops on which they feed, and alternative "lure crops" are provided.

One of the most outstanding characteristics of cranes is their habit of dancing. This may occur at any time of year, in winter as well as the breeding season, and may involve single birds or groups of either or both of the sexes. It is probably seen, however, in its most highly developed form in the dancing displays of a mated pair. Typically, the birds walk around

each other with rapid steps and partly spread wings, bowing, or bobbing their heads, and alternately jumping into the air in a manner which is graceful in spite of their bulk. From time to time they stop suddenly, erect. They also throw a stick or other pieces of vegetation into the air, either catching it or stabbing at it as it falls.

Cranes nest on the ground or in shallow water, and the nest varies from a shallow scrape in the soil to a bulky collection of vegetable material. Occasionally one, normally two and rarely three eggs are laid. They are incubated by both sexes.

Siberian white cranes, a species now much reduced in numbers.

Most living species of cranes are in the genus *Grus*. The present center of the group is eastern Asia, but there are 7 species spread across Eurasia. The common crane, *Grus grus*, breeds from northern and eastern Europe across most of Asia. It stands about 44 in. (112 cm) tall and is slate-gray with a white streak from the eye down the neck. There is a red patch on the back of the crown. It breeds in wet or damp open countryside such as swamps and marshes with or without scattered trees or scrub. In winter it prefers more open country, including dry grassland and steppe. In certain areas of the Siberian tundra it is replaced by the Siberian white crane, *G. leucogeranus*, though this species is now much depleted.

The other American species of crane, the whooping crane, *G. americana*, is far from being a nuisance because of the depredations of large flocks. On the contrary, this species, endemic to North America, has become famous among conservationists and ornithologists through the efforts made to save the few remaining birds from extinction.

The whooping crane stands some 4 ft high and has a wingspan of about 8 ft. It is white with black wing tips and has a large featherless area of red on the forehead and crown, on the front of the face and in a large mustachial streak. It would seem that this bird was not abundant even in the days of the first colonization of America by Europeans, though flocks of thousands were reported into the 19th century. By 1920, however, the population was down to 50 individuals, and the lowest level was reached in 1941 when there were only 23 individuals left, two of them in captivity.

The Manchurian crane, *G. japonensis*, breeds in a few areas of Manchuria and in Hokkaido, Japan. This is another largely white species with black wings and red crown, but it also has a large gray streak running down each side of the neck. It seems that most of the birds winter in Japan, where they are strictly protected, but the breeding population in Hokkaido in 1962 was no more than 200.

The white-necked crane, *G. vipio*, is gray with white throat and hind neck and white secondary wing feathers. It has a red face and forehead. It breeds in eastern Siberia and winters in Japan and also, like several other cranes, in parts of China and Korea.

The Sarus crane, *G. antigone*, is a very impressive bird, even for a crane, standing some 5 ft (1½ m) tall. It is blue-gray with a bare red head and upper neck and breeds in northern India and adjacent regions. The little-known black-necked crane, *G. nigricollis*, lives in central Siberia. The remaining species of the genus is the Australian crane, *G. rubicunda*, also called the "native companion." It is a silver-gray bird with a green crown and red on the face and back of the head and neck, and is found in southern New Guinea as well as Australia.

In eastern Europe and through much of central Asia to the south of the forest belt inhabited by the common crane is found the demoiselle crane, *Anthropoides virgo*. This bird is only about 3 ft (1 m) tall and has soft gray plumage with black flight feathers, face and front of neck, and white plumes curving back and down from behind the eye. It is very handsome

and is frequently kept in captivity. It winters in northeastern Africa and southern Asia.

The blue or Stanley crane, *Tetrapteryx paradisea*, is found only in South Africa. It is basically similar to the demoiselle but has black secondary wing feathers extending far beyond the tail.

Another South African species is the wattled crane, *Bugeranus carunculatus*, which is found in East Africa also. This is a large dark bird, gray above, black beneath and with the upper breast, neck and much of the head white. The face bears red warts, or carunculations, and below it hang two white feathered wattles, or lappets.

Finally, one of the most striking of all the cranes: the crowned crane, *Balearica pavonina*, which is a widely distributed species found over much of Africa from Abyssinia and the Sudan southward. It is 38 in. (96 cm) tall, and there are several races, one of which, from South Africa, is regarded by some authorities as being a separate species. The Ugandan race is particularly common and forms the Ugandan national emblem. The species is seen on the postage stamps of several African countries.

The crowned crane is basically a dark-plumaged bird, with rufous on the hind part and white on the wings, but its most striking feature is its head with the crown that gives it its name.

Sarus crane, the familiar crane of India and southeast Asia.

This crown is composed of a tuft of stiff straw-colored feathers standing up in a fan from the back of the head. The head is otherwise covered with short black plumes, and the bare cheeks are white. The bill is proportionally shorter than in other cranes.

The striking appearance of the crowned crane is enhanced by the dances that it performs, the crown being displayed to full effect in the bobbing and bowing. This is one of a number of species of birds that have provided models for the dances of local tribes in West Africa.

CRAYFISHES, freshwater decapod crustaceans, having five pairs of walking legs, with the first pair enlarged and modified to form characteristic stout pincers. The second and third pairs also bear small pincers. The body is elongated, and the tail is large, bearing a fan formed by an enlarged and flattened pair of abdominal appendages. A characteristic escape movement of a crayfish is to shoot backward, using a violent flexing of the fan.

The common European crayfish, *Astacus astacus*, lives in streams and ponds, where it takes a varied diet, including snails, insects and plants.

The crayfish respires by means of gills, which

are housed in a branchial cavity on either side of the thorax.

In November, the female of the common European crayfish lays a batch of relatively large eggs. These are attached to the underside of the abdomen and carried for six months until they hatch. The young emerging from the eggs have the same general structure as the adult, but the form is more rounded and the limbs are not fully developed. The European crayfish takes two or three years to reach maturity and may live for a total of 20 years. In winter, crayfishes tend to burrow into the banks of their streams, and the greater the likelihood of the water freezing the deeper they burrow. Some of the larger species do considerable damage to the banks of rivers, and some have extended their burrowing habits to other times of the year, becoming semi-terrestial.

CRESTED SWIFTS, or tree swifts, relatives of the true swifts (Apodidae), which they resemble. They differ from true swifts in that they perch and build their nests on tree branches and the males are brightly colored.

The three species of *Hemiprocne* are restricted to tropical Asia, where they live in open woodland, secondary forest and gardens. As with the true swifts, all their food is gathered on the wing and includes wasps and bees.

The nest is a tiny cup of saliva, bark and feathers glued to the top of a slender branch. The single egg is glued into the nest, which only just accommodates it. The mottled brown nestling sits still on the branch, protected by its general resemblance to a knot or piece of lichen.

CRICKETS, jumping insects of small to moderate size closely related to the bush crickets and belonging to the family Gryllidae. There are about 1,000 species, and although the group is mainly tropical there are many temperate species.

The antennae are threadlike and composed of more than 30 segments. Behind the head is a saddle-shaped structure, the pronotum, protecting the front part of the thorax. There are usually two pairs of fully developed wings, but in some species these are reduced in size or completely absent. The hind wings are membranous and fold up like a fan when at rest, forming two "spikes" that protrude from beneath the shorter and tougher fore wings. The females have a rod or needlelike egg-laying instrument (ovipositor) at the tip of the abdomen, and both sexes have a pair of long and conspicuous cerci.

The eggs are usually laid in the ground, but in the tree crickets of the subfamily Oecanthinae, which are exceptional in living in trees and shrubs, they are laid on bark, in the pith of twigs or in the stems of various plants. The young are similar to the adults in appearance, but lack wings. They reach maturity after molting from 8 to 11 or more times, an unusually large number for insects of this size. The young stages often pass through a resting phase during the winter or, in the tropics, the dry season, and so may take as long as a year to become adult.

Most male crickets can "sing" by rubbing the hing edge of the left fore wing against a row of teeth on the right fore wing. The sounds produced are often quite musical, and in some parts of the world crickets are sold in cages, to be kept as singing pets. The females are unable to sing, but both sexes have a hearing organ in each fore leg.

A few species of crickets have become associated with man and have been spread by him to many parts of the world. The best-known example is the European house-cricket, *Acheta domesticus*, which now occurs in indoor situations throughout the world and is sometimes sufficiently numerous to be a pest.

CROCODILES, 13 species belonging to the order Crocodylia, the others being known as alligators, caimans and gavials. They differ from alligators and caimans in that the large fourth tooth in the lower jaw that fits into a notch in the upper jaw remains visible even when the mouth is closed. Moreover, the teeth of upper and lower jaw are more or less in line, those of the lower jaw engaging between the teeth of the upper jaw.

The best-known species of true crocodiles is the Nile crocodile, *Crocodylus niloticus*, which occurs all over Africa south of the Sahara, as well as Madagascar and the Seychelles Islands. At one time this species reached lengths of up to 33 ft (10 m), but nowadays it is hard to find specimens more than 20 ft (6 m) long. In western and central Africa, as well as in Ujiji on Lake Tanganyika, lives the 13 ft (4 m) long-snouted crocodile, *Crocodylus cataphractus*, the snout of which is up to 3½ times as long as it is wide at the base. In contrast, the broad-nosed crocodile, *Osteolaemus tetraspis*, of western and central Africa, has an extremely short snout and is rarely more than 6 ft (1.8 m) long. It has brown eyes and is said to spend more time in the jungle than in water. It feeds mainly on freshwater crabs and soft turtles, using its comparatively blunt teeth.

In Asia lives the 13 ft (4 m) long mugger or swamp crocodile, *Crocodylus palustris*, mainly in India and Ceylon, perhaps also in Burma. It differs from the very similar Nile crocodile by its somewhat shorter snout and the less regular arrangement of the scutes on the back, the central longitudinal rows being somewhat broader than those at the sides. The Siam crocodile, *C. siamensis*, which has somewhat the same length, is restricted to Thailand and Indochina and the islands of Java and Borneo. It is characterized by a triangular raised portion in front of the eyes, with the apex of the triangle facing forward, and a longitudinally directed bony ridge between the eyes. Despite its name, the somewhat longer Sunda gavial, *Tomistoma schlegelii*, does not belong to the gavials (Gavialidae) but to the true crocodiles. Although its snout is extremely long, and up to five times as long as its width at the base, this does not stem from the main part of the skull in the manner of a beak but gradually merges with the skull. This species is found on the Malayan peninsula as well as on the islands of Sumatra and Borneo.

The estuarine crocodile, *Crocodylus porosus*, is spread over a vast territory stretching from Ceylon to the Fiji Islands, including northern Australia. It lacks the large postoccipital scutes, the scutes on the back overlie only small oval ossifications and on the snout there is a pair of scutelike bony ridges extending from the front corners of the eyes up to the nose. The estuarine crocodile reaches 33 ft (10 m) in length and lives at the mouths of rivers and in salt water. It can be seen swimming many miles from the coast in the open seas.

Similar bone ridges on the snout, even though shorter and less pronounced, are found in the New Guinea crocodile, *C. novaeguineae*, and the Philippine crocodile, *C. mindorensis*, which also occurs in the Sulu archipelago. Both forms, which differ from the estuarine crocodile

Crocodiles hatching in the Murchison Falls Park, Uganda. The mother opens the nest for the babies to crawl out.

by the well-built postoccipital scutes and the completely ossified scutes on the back, are perhaps only geographical races of a single species; their length is up to roughly 10 ft (3 m).

True crocodiles are also at home in the New World. The most northern species is the light olive-colored American or sharp-nosed crocodile, *C. acutus*, in the south of the Florida peninsula, on the islands of Cuba, Jamaica and Hispaniola, as well as from Mexico southward over the whole of Central America to Venezuela, Colombia, Ecuador and northern Peru. It is characterized by its narrow and long snout having a bulbous dome in front of the eyes. The scutes on the back are somewhat irregular and form only four to six interconnected longitudinal rows. Whereas this species reaches almost 12 ft (3.5 m), the bulbous crocodile, *C. moreletii*, found in the east of British Honduras and in Guatemala, is only a little more than 8 ft (2.5 m) long and is almost black when aged. Here, too, the snout carries a lump in front of the eyes but has a much broader and shorter effect than in the previous species. Restricted to Cuba there is the barely 7 ft (2 m) long Cuba crocodile, *C. rhombifer*, which has a triangular raised portion in front of the eyes similar to that in the Siam crocodile. The older individuals are almost black and exhibit light yellow spots on the back legs. The only wholly South American species is found in the region of the Orinoco River, namely the Orinoco crocodile, *C. intermedius*, which reaches the stately length of up to 15 ft (4.5 m). It differs clearly from all other American species of crocodile by its extremely narrow and very elongated snout.

CROSSBILLS, finches with peculiar crossed mandibles for extracting seeds from the conifer cones that provide almost their entire food. They also have large feet for gripping the cones and are adept at climbing among the branches, using feet and bill to pull themselves along in parrot fashion. They are found in northern regions throughout the world. The three species differ in body size, size of bill and preferred food. The largest, the parrot crossbill, *Loxia pytyopsittacus*, has a heavy bill and feeds from the hard cones of pine; the medium-sized common crossbill, *L. curvirostris*, has a medium bill and feeds from the softer cones of spruce; whereas the smallest, the two-barred crossbill, *L. leucoptera*, has a slender bill and feeds from the delicate cones of larch. This is true as a general tendency, but each sometimes takes cones of other types. In all three species, the males are colored brick-red, with brownish wings and tail, and the females and young are greenish; but the two-barred, as its name implies, also has white bars across its wings.

Crossbills nest high on the branches of conifers, laying three to four bluish-white spotted eggs. So that the young can be raised while seeds are still plentiful, all three species nest early in the year before the seeds have fallen. The main breeding season of the common crossbill lasts from January to March. In the winter of 1940-1941 some nests were found near Moscow when temperatures were down to 0°F (–18°C); but inside the nests, while the female brooded, they reached 100°F (38°C). It is perhaps because the young are often raised in cold

The common crossbill can pick seeds out of cones but cannot pick them up from the ground.

snowy weather that they develop more slowly than other finches and stay in the nest for up to 30 days.

The migrations of crossbills are also unusual, being adapted to their special food supply. Cone crops vary enormously in size, both from place to place in the same year and from year to year in the same place, a good crop usually being followed by a poor one. The birds therefore move around every summer, settling wherever cones are plentiful. Here they will remain for a year, but will move on at the end of the year in search of new areas rich in seed. Thus the crossbills differ from other migratory birds in making only a single movement each year. In addition, these movements sometimes develop into "irruptions," in which the birds leave their home range in enormous numbers and appear in regions quite unsuited to their needs. This is supposedly caused by widespread crop failure or "overpopulation."

CROWS are usually entirely black, or black with white, gray or brown. At close quarters the black plumage can usually be seen to have a blue, green or purple metallic sheen. The carrion crow and hooded crow, both probably *C. corone*, are found through most of the palearctic region. They are 18½ in. (47 cm) long, the hooded crow being distinguished by a gray back and belly. In North America these two species are replaced by the common crow, *C. brachyrhynchos*, a very similar species and also, to a certain extent, by the slightly smaller fish crow, *C. ossifragus*, which feeds particularly along rivers and tide lines. Another

species is the pied crow, *C. albus*, which is common through much of tropical and southern Africa and is the only corvid of any kind found in Madagascar. Stresemann's bush crow, *Zavattariornis stresemanni*, gray above, white below, with blue black wings and tail, is an inhabitant of the thorn bush country of southern Ethiopia and is about the size of a jackdaw. Relatively little is known about it.

CTENOPHORA, or comb jellies, exclusively marine and usually pelagic gelatinous animals, their distinguishing feature being eight rows of comb plates bearing fused cilia, 100,000 in each plate, which beat in a coordinated sequence and drive the animal through the water, mouth first.

The best known of the 80 species is the sea gooseberry, *Pleurobrachia*, ¾ in. (2 cm) long. Typically, ctenophores are egg-shaped, slightly flattened in one plane, thus appearing oval in cross section. At the pointed or oval end is a central mouth, while at the opposite, broader aboral end is a sense organ lying in a depression and composed of chalky particles supported on S-shaped cilia, as if on springs. If this sense organ is removed, the movements of the comb plates become uncoordinated and the animal loses the ability to remain vertical. On each side of a comb jelly and between two of the comb rows is a deep pouch from which a long retractile tentacle protrudes. This bears the lasso cells, with hemispherical heads containing adhesive granules that attach to passing prey. Small planktonic animals are caught on the outstretched tentacles, which contract and are wiped across the mouth so that food passes in.

Ctenophores are luminescent. They luminesce only after being in the dark for some while, and the way in which luminescence spreads when the comb jelly is touched is suggestive of nervous propagation.

CUCKOOS, a name that is, for most people, associated with birds that leave their eggs in the nests of other birds, which then rear the young. In fact, only 47 of the 127 species in this family are brood parasites.

Most cuckoos have zygodactyl feet, the outermost toe being reversed so that two are directed forward and two backward; relatively short legs, causing the bird to squat close to a perch; an elongated shape, a longish tail and a relatively large head and heavy bill. Some tropical species have bare and often conspicuously colored skin around the eyes, and some have noticeable eyelashes. Some have a small, usually tapering crest on the back of the head, and this can be erected to some extent. As a group they tend to be skulkers, keeping to the concealment of foliage, better known for their voices than for their appearance. The songs of cuckoos are usually a monotonous repetition of a single call or of several notes. Although the repetitive disyllabic call of the common cuckoo, *Cuculus canorus*, which has given the whole family its name, is hailed as a welcome sign of summer, the incessant calling of the brain-fever bird, *Cuculus varius*, of the Orient, combined with a hot climate, is reputed to drive men nearly mad.

The principle food of cuckoos ranges from insects and small invertebrates in the case of smaller species to snakes, lizards, small rodents and birds taken by the larger terrestrial forms. A few species also eat fruit. The cuckoos are among the few birds that regularly eat hairy caterpillars. The typical parasitic cuckoos occur throughout the Old World from Scandinavia and Siberia to New Zealand, but are absent from the Americas. They vary in size from tiny sparrow-sized glossy cuckoos to the giant channel-billed cuckoo, *Scythrops novaehollandiae*, of northern Australasia, which is over 2 ft (60 cm) long. In general, they have fairly long tails and long narrow wings and are strong fliers. Some species migrate, the common cuckoo of northern Europe traveling south to Africa in winter, while in the southern hemisphere the little shining cuckoo, *Chalcites lucidus*, breeds in Australia and migrates north over the Pacific Ocean as far as the Solomon Islands. The smallest species have plumages with vivid iridescent green tints and, in one case, a glossy violet color.

These cuckoos exploit a wide range of host species. The koel, *Eudynamys scolopacea*, parasitizes crows, the great spotted cuckoo, *Clamator glandarius*, uses magpies, other cuckoos of the genus *Clamator* use babblers, typical cuckoos of the genus *Cuculus* exploit a range of small songbirds, and the smallest of the glossy cuckoos, the violet cuckoo, *Chalcites xanthorhynchus*, lays eggs in the tiny pendent nests of sunbirds. These cuckoos lay an egg directly into the nest of the host. Where the nest is domed or partially inaccessible the cuckoo either forces its way in or else clings to the outside of the nest and lays the egg so that it falls into the nest. Only two or three seconds are required in which to lay the egg.

There is no good evidence that any other method is ever used. Usually only a single egg is laid in a nest by any one cuckoo, and one of the host's eggs is removed and often eaten. The cuckoo will have observed the nest of the host for a period prior to laying and will usually deposit its egg during the time the host is laying its own clutch. The cuckoo's egg has a shorter incubation period than that of the host and so hatches a little sooner. Subsequently, the young cuckoo grows much faster than the young of the host. The young cuckoo may simply supplant the other young by taking almost all of the available food, thus starving them, and forcing them out of the nest as it grows. The young cuckoos of the genus *Cuculus* have evolved a more certain method. For the first few days after hatching the young cuckoo appears to be physically irritated by the presence of an egg or nestling touching it and will wriggle underneath it until the egg or nestling rests on its back. The little cuckoo then rears up against the side of the nest and heaves the egg or nestling out, continuing this until it has the nest to itself. The host parents returning with food appear quite indifferent to their ejected nestlings. The young cuckoo may grow to such a size that it bursts the nest apart. The conspicuously colored gapes and begging calls of young cuckoos appear to be sufficient to stimulate their hosts to feed them.

In many instances, cuckoos lay eggs that closely mimic those of the host in color, pattern

and, to some extent, in size. The eggs of the black-and-white cuckoo, *Clamator jacobinus*, and some of those of the violet cuckoo may so closely resemble those of the host in whose nest they are laid that it is very difficult to distinguish them with certainty. The evolution of such a close resemblance seems to have been aided by the tendency for a female cuckoo to lay eggs in the nests of the same host species as that in which she was reared. Some young cuckoos reared in nests of crow

species that could harm intruding nestlings have a superficial resemblance to the young of their host. Although the advantage of this to the cuckoo is obvious, the fact that several adult cuckoos have a strong superficial resemblance to birds in other families is less obviously advantageous. Some cuckoos closely resemble hawks, and in some of these species the immature cuckoos also resemble the immature raptors. The drongo cuckoo, *Surniculus lugubris*, resembles the drongo, whose nest it parasitizes, both in its black plumage and its shape. The nonparasitic cuckoos are widespread, with a few species on each continent with the exception of Australia. They are slenderly built skulking birds of forest and thickets. They build shallow, cup-shaped nests in trees and incubate their own eggs and rear their own young. The oriental forms tend to have boldly colored bills, bare facial skin and long tails.

CURASSOWS, the common name for a family of forest-dwelling birds (Cracidae), ranging from southern Texas, throughout Central and South America to Uruguay and Argentina. There are 44 species, but only one, the plain chachalaca, *Ortalis vetula*, goes as far north as southern Texas. Curassows are most numerous in the tropical part of their range.

They are large birds, the largest being almost of the size of a turkey. The smallest are the chachalacas of the genus *Ortalis*, of which there are ten species. In many features they

Male common cuckoo; the female is browner. Both look like hawks, except for the bill and the longer neck.

resemble pheasants. Their plumage is mostly glossy and either black, brown or olive-brown, and in most of the species the sexes are alike. Some, such as members of the genus *Crax* with seven species, have a crest of curly feathers on the crown, while others like *Mitu* and *Pauxi* have bizarre-looking and colorful helmets on the base of their bills or skulls. Their bills are short, often with a brightly colored cere, their wings are rounded, the tails are long and flat and they have strong feet. Their flight is heavy but fast and direct even through a dense forest. All curassows are birds of the forest, where they live partly on the ground and partly in the trees. In trees they run fast over thick branches. Some of them are more or less gregarious. They feed on leaves and fruits.

The nests of curassows are small in comparison to the size of the birds. They make open cup nests of dead sticks, often lined with fresh leaves, in trees and shrubbery frequently quite near the ground and seldom at a great height. The eggs are white, and the clutch consists of only two or three eggs. The nestlings, which hatch with well-developed wing feathers, leave the nest soon after hatching.

CURLEWS, eight species of wading birds of the genus *Numenius* distinguished by long down-curved bills and characteristic call notes. Their plumage is largely cryptic, with mottled brown upperparts, but most have a white rump, conspicuous in flight. In the Palaearctic, the large curlew *N. arquata* breeds in wet meadows and on moorlands in the temperate zone, while the smaller whimbrel, *N. phaeopus*, replaces it at higher latitudes (and altitudes), where it has a discontinuous circumpolar distribution. Whimbrel are absent from north central Siberia, where the little or pygmy curlew, *N. minutus*, breeds, and from north-eastern Canada, where the now almost extinct Eskimo curlew, *N. borealis*, occurs. Farther south, in the temperate zone of the North American continent, the whimbrel is replaced by the ecological counterpart of the European curlew, the larger long-billed curlew, *N. americanus*. In the marshy steppes of Asia, the slender-billed curlew, *N. tenuirostris*, breeds. This species rarely reaches Europe except on passage.

Most species show strong migratory habits. The populations of the European curlew mostly move southwest in autumn, to winter in the British Isles and Iberia; some reach Africa and Southeast Asia. The whimbrel has a migration pattern that has been termed "leap-frog" migration, since it breeds to the north of the European curlew, e.g., in Iceland, but winters well to the south, on the coasts of western and southern Africa. Other breeding populations of whimbrel reach southern South America during the northern winter, and the pygmy curlew winters in Australasia. Curlews are birds chiefly of sandy and muddy (rarely rocky) shores of rivers and lakes outside the breeding season, but some, like the whimbrel, are largely confined to the sea coasts. The distribution of the curlew has altered considerably in some European countries in this century, as a result of changing agricultural practices. Reclamation and drainage of marshy fields and moorland, and afforestation of the latter, have led to local decreases, while conversion of forest to grassland in some parts of Scandinavia has led to increases there. The whimbrel's chosen habitat, marshy tundra with or without low shrub vegetation, has been affected much less by the hand of man, so the bird's numbers have probably altered little.

The curlews feed on a variety of small organisms living on or close to the surface of the soil, including insects and crustaceans, but also take many seeds and berries. The significance of the curved bill in obtaining food

is not known. Those species that stay to winter in the temperate zone may be forced to move to coastal areas in winter if their inland feeding grounds become frozen.

CUSCUS, name given to any of at least six species of marsupials of the genus *Phalanger*, resembling monkeys in appearance. The affinities of the cuscuses, which have a tropical distribution from the Celebes Islands east to the Solomon Islands, appear to be with the brush possums, their ecological counterpart in tropical and temperate regions of Australia and adjacent islands. Cuscuses are arboreal animals with rounded heads, small ears, large eyes, dense soft fur and a long prehensile tail, the terminal portion of which is without fur. The head and body of full-grown animals of the larger species is over 2 ft (60 cm) in length, and the tail is only a little shorter.

The bear cuscuses (*P. celebensis, ursinus*) are large, brown to black animals found only in the Celebes and adjacent islands. As in all the other phalangers, the cuscuses have three incisor teeth on each side of the upper jaw and the anterior lower pair of incisors enlarged and bent forward (procumbent). A pair of canine teeth are always present in the upper jaw. In the bear cuscuses, the third upper incisor is enlarged, and the canine is reduced in size, a dental condition approaching that of the larger kangaroos. No color differences between the sexes occur in the bear cuscuses, and white individuals are unknown.

The so-called gray cuscuses include a multiplicity of color phases ranging from pure white through various shades of gray, to chocolate brown or even distinctly reddish (*P. orientalis, gymnotis, vestitus*). The pure white phase is restricted to the male only. This is the most generalized and widest ranging group extending

from the Celebes and Timor, through New Guinea, eastward to most of the Solomon Islands and southward to Cape York on the Australian mainland.

The spotted cuscuses (*P. maculatus, atrimaculatus*) include forms in which both sexes may be pure white. Other forms are, by marsupial standards, brightly colored and spotted in a variable and irregular, nonsymmetrical fashion: a rarity amongst wild animals. The spotted cuscuses range from New Guinea to Cape York, the Moluccas and other islands near New Guinea but not to the Celebes or Solomon Islands. Except on certain islands northwest of New Guinea, female cuscuses appear to lack the spotted pattern.

CUSHION STAR. It is a small starfish, some 2 in. (5 cm) in diameter, of almost pentagonal shape, its body produced into five very short, stumpy arms. The upper surface is slightly swollen, the lower surface flat. The upper or dorsal surface is covered with overlapping scalelike plates, carrying small spines in groups of four to eight and also, but less abundantly, simple pedicellariae (pincerlike structures) between the spine groups. On the lower of oral surface, the central mouth is surrounded by five grooves that extend radially to the ends of the arms, each groove being flanked on either side by a series of suckered tube feet. Between these grooves the regular overlapping plates, which are imbedded in the skin, bear spines, fewer in number but slightly larger than those of the upper surface. The fairly thin edge of the animal is formed of small, mostly indistinct, marginal plates of calcite. In life, cushion stars vary from olive-green to yellowish or brownish-red.

This common cushion star is widely distributed in western European seas and the Mediterra-

Crested curassow *Crax alector* of the forests of the Amazon Basin.

nean, and it also extends down the west coast of Africa to the Azores. Although recorded down to depths of some 400 ft (125 m), it is mainly a littoral species that dwells on or under stones, but may be found on shell gravel in deeper water. It feeds mainly on mollusks but also eats worms and brittlestars.

When breeding, the female is generally accompanied by two or three males. The eggs are not shed at random into the surrounding water but deposited in small groups, typically attached to the undersurfaces of stones. They are very yolky and quite large (0.5 mm in

regions in water of 16-390 ft (5-120 m) depth. In summer they generally remain in water of 32-65 ft (10-20 m) depth. During the autumn and winter they migrate to deeper water, returning again to shallow waters in the spring. Along the length of each of the eight arms are four rows of stalked suckers. There are also two very long retractile tentacles, which when fully extended capture prey on a terminal pad bearing stalked suckers. The pad has several rows of suckers, the central row having five particularly large ones. These tentacles are normally retracted into two pockets, which lie

by means of an undulatory movement of the lateral fins. The current of water is also used to expel, through the funnel, the produce of the ink sac and the digestive, excretory and reproductive systems. All of these open into the mantle cavity.

Vision is very important during the courtship of the cuttlefish, as both the male and female display a zebralike pattern. The pattern, less brilliant in the female, is produced by the chromatophores or color cells of the skin. When the male approaches another cuttlefish the pattern intensities and the left fourth arm, which is more prominently marked, is extended toward the other. If the other cuttlefish is a female the display is not returned and the arm will touch. The animals then face each other with their arms interwoven. They remain in this position for at least two minutes during copulation. In this time the spermatophores, or packets of sperm, are transferred to the female by the fourth left arm of the male, which has a reduced number of suckers near the base. During copulation this region forms a groove into which a spermatophore is blown by a jet of water from the funnel.

The female lays her eggs a few days after mating. Each egg is blown out through the funnel, which is held between the fourth pair of arms. The female then "embraces" a clump of seaweed, or other suitable projection, with all her arms while the egg is attached to the object. It is not known how this is achieved, but a number of eggs are attached in this way. One female can lay 200-550 eggs, depositing them over a period of several days. She does not remain to care for the eggs as does the octopus. Each egg is elliptical, about $\frac{5}{16}$ in. (8 mm) in diameter, and contains sufficient yolk for the development of the embryo. Hatching occurs about 50 days after laying. The young cuttlefish measure about $\frac{1}{8}$ in. (10 mm) and immediately adopt a life on the bottom like the adults.

CYCLOPS, microscopic, somewhat pear-shaped freshwater crustaceans. While they all conform to the same general shape and structure, the detailed structure is variable, and this has given rise to much confusion in classification.

The body of a *Cyclops* has two clearly defined parts, a bulbous anterior cephalothorax and a narrow posterior abdomen. The first pair of antennae are no longer and often shorter than half the length of the body and have not more than 17 segments.

The eggs of *Cyclops* are carried in a pair of sacs, one on each side of the abdomen. Eggs are carried until they hatch, and an adult female needs to mate only once in her life. Thereafter she can lay a succession of fertile batches of eggs (up to about ten) with no male present. After hatching, *Cyclops* passes through a series of naupliar and copepodid stages, shedding its outer chitinous skin, increasing in size and adding on appendages and segments between each instar. The number of eggs laid can be as high as 150 but is often lower. Few of those that hatch survive to become adult. Life span from extruded egg to mature adult varies from two or three weeks to two or three years depending mainly on temperature and food.

Cuttlefish swimming. The animal assumes a protective striped color pattern when among seaweed and when courting.

diameter) and adhere by their surface membranes. The fertilized egg develops into an embryo that escapes from the egg membrane in a few days. Usually a starfish larva passes through two different stages of development in turn, named the bipinnaria and brachiolaria, but in *Asterina* the bipinnaria stage is always omitted, and the embryo either develops directly or, in some species, metamorphoses to the adult form after spending some ten days as a free swimming larva of brachiolaria type. In British seas, breeding usually takes place in May-June, but may occur earlier, and in April-May in the Mediterranean. It has been claimed that this cushion star behaves as a male when young and small and as a female when older and larger, but this lacks confirmation.

CUTTLEFISH, *Sepia officinalis,* marine mollusk related to the squid and octopus, having a well-defined head with two large eyes, and eight arms and two retractile tentacles encircling the mouth. The head is attached by a short neck to the body, which is supported internally by the cuttlebone. This structure not only acts as a support, it is also a buoyancy mechanism enabling the cuttlefish to remain on the bottom or to swim freely at any depth. The cuttlefish is a coastal animal and, like all known cephalopods, lives by preying on other marine animals.

About 100 species are known. They are found in tropical and temperate seas but rarely in colder waters. The cuttlefish inhabits coastal

close to the mouth between the base of the third and fourth arms. The head is broad and somewhat flattened. The eyes are situated on either side of the head so that the animal can see both in front and behind, with a horizontal visual field of 360°. The eyes are large and have a single lens that throws an image onto the retina, showing in this and in other ways considerable similarity to the vertebrate eye.

The mouth found at the center of the arms has an upper and a lower beak surrounded by a circular lip. The beaks are chitinous and resemble those of parrots, except that in the cuttlefish the upper one fits into the lower one. Inside the mouth is a tooth-bearing ribbon, the radula, used in the transport of food particles to the esophagus. This organ contains tiny projections which tear flesh into small pieces easy to digest.

The shape of the body of the cuttlefish is determined by the presence of the cuttlebone. Below it the visceral mass is enclosed by the muscular mantle. The mantle forms a cavity, open at the front, beneath the enclosed viscera. The gills are suspended within the mantle cavity, and a respiratory current of water is maintained by the mantle muscle. Water is drawn into the cavity and then expelled through a median funnel. Contraction of the mantle muscle forces a jet of water through the funnel, strong enough to bring about a propulsive movement. The cuttlefish uses this jet to move rapidly. It also swims more slowly

D

DAB, *Limanda limanda*, a small flatfish living in sandy bays along the coasts of northern Europe. The dab, which reaches 17 in. (43 cm) in length, can be recognized by the sharp curve of the lateral line near the pectoral fin and the spiny margins of the scales on the eyed side of the fish (these are smooth on the blind side). The general color of the eyed side is sandy brown, flecked with orange and black, and the blind side is white. Occasionally specimens are found that are colored on both sides, and the scales are spiny on the two sides as well. The dab is common around British shores and reaches as far north as Iceland. These fishes are inactive during the day but feed at night and are more easily caught then. The breeding season is from March to May, the eggs being pelagic and among the smallest of any of the European flatfishes (averaging 0.8 mm in diameter).

Dab, enormously abundant European flatfish.

DADDY-LONGLEGS, unspecialized flies with slender bodies, narrow wings and extremely long legs which easily break off. Thus Daddy-longlegs has become a common name. They tend to settle in cool damp places, under leaves and between grass stems. They are slow-flying and vulnerable to predators. Many of them bob up and down when they are perched, by bending and straightening their legs. Others hang upside-down holding on by the fore legs and rhythmically waving the hind legs. Some of these movements are probably involved in sexual attraction and are seen mainly in larger species. Some of the smaller gnatlike species form mating swarms of males, seen just before sunset or sunrise monotonously dancing up and down. At intervals a female enters the swarm and drops to the ground with a male.

Daddy-longlegs feed little as adults and take only nectar. It is the growing larvae which consume the most food. They are tough-skinned, known as leatherjackets, and have biting mouthparts. They live just under the surface of the soil, attacking roots and stems of a wide range of plants including crops of various kinds. In warm weather they will feed at night above ground and cut through the stems at

Daddy-longlegs or cranefly, female.

ground level. The larvae of *Tipula paludosa* are common farm pests and irregular bare patches in a field often betray their presence. This species lays eggs in the soil and the fully grown larvae pupate there in early summer. Adults emerge in two to three weeks. Other species have larvae which are semiaquatic, some live among moss and others in rotten logs. All of them are moisture loving and dry weather kills them in large numbers.

DAHLIA ANEMONE, *Tealia felina*, an anemone varying in color from crimson and green less commonly to yellow and dull orange, with a broad, short body that is rough due to the presence of scattered grayish warts. Although common between tide marks, the dahlia anemone is much less well known than either the beadlet or snakelocks anemone. It occurs in at least four varieties, three of which are offshore and deep-water forms. The variety found on the shore, generally at low-tide levels, is difficult to find because it covers itself with gravel that adheres to the very rough body. It is common around the coasts of Britain and extends north to Norway and the Faroes and south to the Atlantic coast of France. Two of the offshore varieties may be

seen rarely in very deep cracks where they are never exposed at low tide and are of variable color. A circumpolar variety is also known, which does not occur as far south as the British Isles. This lacks the rough body and is viviparous, in contrast to the other varieties in which fertilization of the ova and development of the young anemones occur in the sea.

DAMSELFLIES, dragonflies, which, although resembling other dragonflies structurally and ecologically in all important respects, can be recognized by their more delicate build and by their possession of similar fore and hind wings stalked at the base. The larvae, which are aquatic, are distinguished by having external gills. Compared with other dragonflies, they are weak fliers. Some of the commoner species in Britain (e.g., *Coenagrion puella*, *Enallagma cyathigerum*) look like pale-blue needles as they hover low among the vegetation along the edges of ponds and canals.

DARTERS, small freshwater perchlike fishes found in North America. They derive their common name from their habit of darting between stones. They occur only in the temperate parts of North America to the east of the Rocky Mountains, and about 95 species are known. They are bottom-living forms, and some have surprisingly bright colors compared with most freshwater fishes of temperate waters. They lack swim bladders and have two dorsal fins, the first spiny and the second soft-rayed. The darters are carnivorous, feeding on small insect larvae and tiny crustaceans. The entire body is scaled in most species. Their spawning habits vary widely: some bury their eggs in sand or gravel, others carefully guard the eggs and still others simply scatter the eggs and leave them. The Johnny darter, *Etheostoma nigrum*, spawns in the spring, the female depositing the eggs on the underside of stones and the male aerating them and guarding them in a most ferocious manner for a fish of only $2\frac{1}{2}$ in. (6.5 cm).

The eastern sand darter, *Ammocrypta pellucida*, has a row of scales only along the midline. It is sand-colored and translucent and like most darters is secretive, burying itself in the sandy beds of streams with only its eyes and snout protruding. It reaches 3 in. (7.5 cm) in length. Most darters live in clear shallow streams but *Etheostoma fusiforme* lives in murky, swampy waters and has been found in estuaries. The log perches are the largest of the darters. *Percina caprodes* grows to 6 in. (15 cm), and the flanks are marked by dark vertical bars that gave rise to the alternative name of zebrafish.

DARWIN'S FINCHES, a group of sparrow-like birds confined to the Galápagos archipelago (13 species) and Cocos Island 600 miles (960 km) northeast of the Galápagos (a single endemic species). Although the original stock is unknown, they probably arose by speciation within the islands after the colonization of the islands by a single species, probably a finch. The group provides the best avian example of adaptive radiation into ecological niches, as this radiation has taken place so recently that some of the intermediate forms are still extant.

The *Geospiza* species feed on the ground, and the different bills enable each species to deal efficiently with a different range of seeds so that on the larger islands, which have a range of different habitats, the various species can coexist. The exceptions are the two species of cactus finch, *G. scandens* and *G. conirostris*. These two species are restricted to islands with the prickly pear cactus *Opuntia*, and as only one species is found on any one island, the slight differences in the bills may be insufficient to prevent severe competition between them, so that one species will always oust the other. The other large group are the tree finches, *Camarhynchus* spp., the six species of which can be conveniently divided into three subgenera: *Camarhynchus* of three species that feed on largish insects and soft seeds; *Platyspiza* with a single species eating buds, leaves and fleshy fruits; and *Cactospiza* of two insect-eating species. Again there are marked differences in bill size and shape correlated with the different diets. The woodpecker and mangrove finches, *Cactospiza pallida* and *C. heliobates*, regularly use small twigs and cactus spines to prod out insects that they would otherwise be unable to reach. The two remaining finches, the warbler finch, *Certhidea olivacea*, and the cocos finch, *Pinarolaxias inornata*, have thin bills and eat mainly insects, although the latter species may also eat some nectar.

DASSIES, mammals about the size of a marmot yet showing definite links with such animals as the elephants, rhinoceroses and tapirs. One of the most remarkable features of these small animals is the feet. The fore feet have four functional toes and a rudimentary first toe, all of which have short nails in the manner of the rhinoceros. The hind feet have only three toes, of which the inner has a curved claw. The other toes are similar to those on the front feet. The alimentary canal is unusual in having a supplementary caecum, and the brain of the dassie is more akin to that of an ungulate than a rodent. The collarbones are not developed, and the tail is short. The incisors of the upper jaw grow from persistent pulps and are curved as in rodents. They do not follow the rodent plan of ending in a chisellike edge, but are prismatic and end in fine points. There is one pair of incisors in the top jaw, but those in the lower jaw number four and differ in being rooted. The outermost pair tend to lie flat and have trilobate crowns. The grinding or cheek teeth, separated from the incisors by a considerable gap, are seven in number and resemble those of a rhinoceros.

The dassie, rock hyrax, rock rabbit or coney, *Procavia capensis*, is about the size of a rabbit, has a blunt head with small ears and is cov-

The anhinga, an underwater javelin-thrower, has a trigger in its neck.

ered with soft brown fur. In the middle of the back is a yellowish white patch of hair that marks the site of the dorsal gland.

The dassie is found in Africa from Algeria and Libya through southern Egypt to the Cape Province of South Africa. It also occurs in southern Arabia and Syria. Dassies live in groups of 60 and more, making their homes in cliffs and on rocky hilltops.

They feed mainly on plants and fruits, but they will also eat lichens and seeds. Their feeding times are the early morning and the late afternoon. They also have the habit of using a communal latrine in their dens, and the excrement that collects here has a commercial value, as it contains an ingredient used in the manufacture of perfumes. They spend a large part of their day basking in the sun, but

Geospiza fortis, one of the ground finches among the 14 species of Darwin's finches.

they also have to keep at least one eye open for predators, such as leopards, hawks and eagles. When alarmed they either utter a whistling note or make a chattering noise. The litter of two or three young are born after a gestation period of $7\frac{1}{2}$ months (225 days). At the time of their birth the young are able to see, and very soon after they are born they can follow the mother. The Syrian hyrax, *Procavia syriaca*, is very similar to the rock hyrax or dassie. The tree dassie or bosdas, *Dendrohyrax arboreus*, is a tree-dwelling hyrax that differs slightly from the dassie in that the fur is often longer, the coat may vary from a rich brown to a grizzled gray and the soles of the feet are entirely naked, allowing it to move with ease in the trees. It is found from South Africa northward into the Congo, Tanzania and Kenya. In other respects it is very similar to the dassie. The cry differs in that it starts with a series of groans and culminates in a screaming wail.

DEAD MEN'S FINGERS, *Alcyonium digitatum*, a typical soft coral, drab yellowish-white in color and shaped rather like a human hand. It is a colony of octocoralline polyps that can be found in temperate waters, although most soft corals prefer warm seas. The colony is a mass of mesogloea in which the polyps are embedded. Each polyp bears eight feathery tentacles and is very much elongated, the cylindrical, tubelike body extending through the mesogloea to the base of the colony. It is connected to other polyps in the colony by tubelike extensions of the body wall that leave at varying places from the body. The skeleton consists of a mass of spicules, elongated spindles of calcium carbonate, which are secreted by ectodermal cells and generally distributed through the mesogloea. As typical members of the phylum Cnidaria, the polyps catch their prey by means of nematocysts on the feathery tentacles.

DESMAN, an aquatic insectivore of the mole family (Talpidae), resembling a very large shrew in appearance. Formerly they were widely distributed in Europe but are now represented by two very localized relict species: *Galemys pyrenaicus* in the Pyrenees and northern Portugal, and *Desmana moschata* far away in the rivers north of the Caspian Sea. The desman is highly adapted to its aquatic existence. Its hind feet are powerful and have webbed toes, the fore feet are also wholly or partly webbed. All four feet bear fringes of bristles to increase their effectiveness as paddles. The tail is scaly with sparse hairs and is used together with the feet for swimming. Desmans have long oily guard hairs that are water repellent and prevent the dense underfur from getting wet during swimming. This is a common arrangement in many groups of aquatic mammals but is not found in other members of the family Talpidae. The snout is probably the most characteristic feature of the desmans. It projects well beyond the teeth and lower lip and is so long that it may be bent back on itself and licked with the tongue. It is flattened, tubular, has two grooves along its length and is only sparsely covered with short bristly hairs. It is very mobile and is frequently waved about in scenting the air. It is also used to investigate food and other items of interest by touch and smell, rather as an elephant uses its trunk. The nostrils are

situated at the end of this snout, and, as a further adaptation to aquatic life, they may be closed by small skin flap valves. It has even been reported that the snout may be used as a kind of "snorkel," enabling the animal to breathe while its body is completely submerged. Eyes and ears are very tiny, and the desman relies very much on its snout and whiskers for investigating its environment.

Desmans live in burrows beside streams and rivers. An entrance tunnel is dug from water level up into the bank, often under tree roots. Their diet is predominantly aquatic. Insects and their larvae, crustaceans and fish are eaten, together with some worms and other terrestrial creatures. Desmans are active throughout the day and night, do not hibernate and usually lead solitary lives.

The Russian desman, *Desmana moschata*, is the largest member of the mole family with a head and body length of about 8 in. (20 cm). It is found beside water in southeastern Europe and central western Asia. It is reddish brown above, shading to gray below. The tail is flattened in the vertical plane forming an effective rudder and is about the same length as the head and body.

Strong scent glands around the base of the tail produce a powerful odor making the animal distasteful to most predators.

The Russian desman prefers slow-moving, muddy water around ponds and lakes, especially where there is plenty of rotting vegetation at the water's edge. Three to five young are produced in June.

The Pyrenean desman, *Galemys pyrenaicus*, with a head and body length of 4-5 in. (11-13 cm), lives in the Pyrenees in both France and Spain and in the mountains of northern Portugal. It prefers clean, fast-flowing water.

DEVIL RAYS, a family of large raylike cartilaginous fishes whose "devilish" reputation stems rather from their size and curious appearance than from their supposedly ferocious behavior. They have the winglike pectoral fins of the ordinary skate or ray and closely resemble the eagle rays except that the mouth is much wider and is provided with a pair of appendages on either side that are used as scoops in feeding. The devil rays are surface-living fishes.

They feed on small crustaceans and plankton. The mouth, which is terminal in most of the species, is kept open as the fish cruises along, and food is collected by fine, sievelike rakers before the water passes to the gills. Some species grow to an enormous size, and there is a record of a specimen that measured 22 ft (6.6 m) from tip to tip of the pectoral fins. There are many records of these fishes leaving the water and sailing through the air, the resounding noise of their return to water having been compared with the sound of a cannon.

DHOLE, *Cuon alpinus*, also called the Asiatic wild dog, known as a fierce hunter of great stamina. It weighs from 25-35 lb (11-15$\frac{1}{2}$ kg) and has a head and body length of 38 in. (96 cm) and a tail length of 18 in. (45 cm). It differs from most canids in having a reduced number of molar teeth in its lower jaw and a thick muzzle, both of which are adaptations to a strictly flesh-eating way of life. Distributed throughout the oriental tropical and mountain forests, in Nepal, India, Malaysia and Sumatra,

the dhole has rufous fur with a darker-colored tail ending in a black tail tip. It lives in groups of 5-20, although large packs of several families and their young may have up to 40 members.

Mating usually occurs during the late winter, and males pursue the females while uttering a soft squeak. Like domestic dogs, they "lock" or "tie" during mating, and the litter of between four and ten cubs is born about two months later, from January to March. Mothers rear their offspring in caves or holes which are dug before the birth. Other pack members participate in guarding the young and bring meat back to the litter, which is regurgitated to the pups after being chewed by the adults. Initially, the cubs are completely helpless and blind, but at two weeks their eyes open and weaning starts when they are five weeks old. When hunting, dholes display a certain degree of teamwork, with different individuals alternately chasing the prey and then resting. Favorite victims are wild pigs, chital, spotted deer, and muntjac, which are attacked mainly on the muzzle and face or rump region and then pulled down.

DIK-DIK, six species of small antelope (*Madoqua, Rhynchotragus*) found especially in eastern Africa from Ethiopia to northern Kenya with one species in southwest Africa. They are 21-27 in. (53-69 cm) in head and body length, stand 12-16 in. (30-40 cm) at the shoulder and weigh 7-11 1b (3-5 kg). The males have short horns often partly hidden by a tuft of hair on the top of the head. They live solitarily or in pairs, sometimes in small family parties, in dense undergrowth from which they race away, when flushed, on a more or less zigzag course.

DIPPERS, rather wrenlike songbirds, distributed in western North and South America, northwest Africa, Europe and the temperate parts of Asia. They resemble wrens in their short wings and tail, copious plumage and type of nest, but are specialized for feeding in or under running water, with the tarsus long and sturdy and with stout claws for gripping the riverbed.

The four species of dipper vary in size between a sparrow and a song thrush, have thin bills and are predominantly gray or brown. Although the sexes look alike (and both sing), the dipper *Cinclus cinclus* of Europe can be distinguished in the hand by the greater wing length of the male, 87-95 mm compared with 84-92 of the female.

Dippers are confined to hilly and mountainous regions. Several races of the white-capped dipper, *C. leucocephalus*, inhabit the Andes. The range of the American dipper, *C. mexicanus*, extends from western Panama to Alaska and the Aleutians. The European dipper ranges from Britain eastward to northwest Africa and central Asia reaching 16,000 ft (4,920 m) in the Himalayas and nearly 17,000 ft (5,230 m) in Tibet.

Nests are usually built on rock faces (including waterfalls), on man-made structures, for example on sluices, mills, culverts and bridges, and under overhanging banks and against tree trunks. The entrance is normally over water. Most nests are built at heights of 4-5 ft (1.2-1.5 m).

The large nest is like that of a wren and usually fits into a roofed cavity that determines its dimensions. The measurements of nests not limited in this way are 7-10 in. (18-25 cm) across, 5-9 in. (13-22 cm) deep, and 6-9 in. (16-23 cm) high. Nest building begins as early as February and proceeds slowly. Eggs may be laid in mid-March but more usually in April. Later nests may be built in as little as nine days. Four or five thin-shelled white eggs are laid, sometimes three, rarely as many as seven. They hatch in about 16 days, and the young are fed by both parents. Two broods are hatched in some years, perhaps those in which breeding starts earlier than usual.

Young dippers remain in the nest for about three weeks, and reach a weight of 2 oz (50-60 gm). From 14 days onward they have a disconcerting habit of leaping out of the nest into the water if disturbed, and swimming away downstream. Their first plumage, which they begin to molt at nine weeks, is more mottled than that of the parent birds, and the white feathers at chin, throat and upper breast are tipped with dark brown, giving a rather thrush-like appearance. They are fed by their parents until about five weeks old and then disperse. They can be distinguished from adults by the white tips of the greater wing coverts, which persist until the following April, although in some cases they become much abrabded.

Dipper feeding nestling.

DISCOGLOSSIDAE.

Fire-bellied toads occur in Europe and eastern Asia. When seen from above, the true fire-bellied toad, **Bombina** *bombina*, is a dull frog with a gray warty skin, but underneath it is vividly patterned with red. Fire bellies spend most of the time in water. The yellow-bellied toad, *B. variegata*, is the least discriminating and is found in any sort of pool or ditch throughout central and southern Europe. It hangs motionless in the water with only its snout and eyes above the surface and when disturbed can swim well, its toes being fully webbed.

The skin of fire-bellied toads is poisonous, producing a white frothy liquid which irritates the mucous membranes. Just looking into a bag of freshly caught specimens causes fits of sneezing and watering of the eyes. The bright coloration of their undersides is a warning associated with this poisonous nature. Since the colors are not normally visible from above, however, a fire-bellied toad, when

disturbed and unable to escape, arches its back and folds its limbs up over its body, revealing its bright colors to its attacker.

Fire-bellied toads leave the water at the approach of winter and hibernate in burrows. The breeding season is long, beginning in April or May and lasting for two or three months. About 100 large eggs are laid in water and are attached to stones or plants.

The single species of *Barbourula*, *B. busuangensis*, is similar to the fire-bellied toads except that it is even more aquatic in its habits. The fingers, as well as the toes, are webbed. It is found only in the Philippines.

The **painted frog**, *Discoglossus pictus*, a small, very active frog with a smooth skin that varies in its coloration, the olive-brown ground color being marked with dark bands or patches although a yellowish stripe down the middle of the back and reddish ones along the sides may be present. It lives in or near water and like the fire-bellied toads, but unlike most frogs, is able to swallow its food under water. Its breeding follows the pattern typical of most frogs; 300-1,000 small eggs are laid in water, where they lie in a mass on the bottom. The tadpoles are able to leave the water after two months.

The **midwife toad**, *Alytes obstetricans*, a frog with a very unusual method of breeding, the eggs being carried around attached to the hind limbs of the male until the tadpoles hatch out. Pairing occurs on land, and the female extends her hind limbs to form a receptacle for the eggs. These are large, and between one and five dozen are expelled in two long loops, each egg attached to the next by a tough elastic thread. The male fertilizes the eggs and then pushes his hind limbs through the loops until the eggs are firmly wound around him. The female moves away and has nothing further to do with the development, while the male withdraws to his hole until the next night when he comes out to feed. He is not particularly hampered by his load and may repeat the procedure to add a second lot of eggs to the first. After about three weeks the tadpoles break through the egg capsules while the male is in water and swim off to complete their development.

DIVERS, fish-eating diving birds of the family Garridae, also known as loons. They comprise a distinctive group of holarctic birds showing a very high degree of specialization to the aquatic mode of life. They are quite large, from the weight of a large duck to that of a small goose, with a long body and thick, strong neck. The tail is short, the legs and feet strong, with webs between the three front toes, and the bill is stout, long and pointed. These characteristics result in a very streamlined shape. This has come about as a progressive adaptation to underwater swimming, the loons being preeminent among birds in this environment. Particularly noteworthy is the structure of the hind limb. As in the grebes, the tibial, or central segment of the leg, like the femoral, or upper segment, is bound by its musculature to the pelvic region of the body. Thus the only externally visible part of the leg is the so-called tarsal, or lower segment, and the foot. The leg therefore emerges from the body in a position at the rear that allows great propulsive efficiency in water. But

this also means that loons have great difficulty in standing on land, and their mode of progression out of the water is by pushing themselves along on the belly.

The plumage of loons is of a hard, even harsh texture, except on the neck where it is soft. The young have two successive coats of down —that with which they are hatched, and a second coat that comes between this and the adult-type feathering of the first winter plumage. This replacement of the hatching down may be connected with the fact that loons leave the water only with difficulty and reluctance, and the young therefore are less likely to be brooded on dry land than the young of ducks and geese. And they are not carried on the back as much as young of grebes.

Adult loons are strikingly marked, with white stripes or spots contrasting with the basic color of black, gray or brown. The sexes are externally similar. They have a distinctive appearance in flight with head and legs hanging down somewhat and the feet projecting beyond the tail. In the red-throated diver, *G. stellata*, this attitude is seen in its most definite form giving the appearance of a flying banana! Because of their swimming adaptations, particularly the small wings, loons have difficulty in taking flight and need to taxi for a considerable distance to pick up speed before becoming airborne. Once aloft they fly strongly— indeed, they have to, because with their very high wing-loading they must maintain speed in order to stay aloft. Also, as hovering or slow speeds are impossible, they can only "land" by coming down in a shallow glide onto water, which slows them gradually.

Loons have a wide distribution throughout the Arctic, though the common loon is almost exclusively a nearctic breeder. During the breeding season all species are essentially freshwater birds, frequenting lakes, ponds, streams and slow-moving rivers. However, enormous tracts of land fulfilling these requirements in the Arctic are found near the sea, and loons in such areas may divide their time between fresh and salt water even in the breeding season. Outside that season their habitat is essentially the maritime one, particularly estuaries and inshore waters.

The distribution of these birds is largely determined by their food, consisting almost entirely of fishes, which they pursue and capture beneath the surface of the water. They dive effortlessly from the surface in a smooth forward plunge, swimming with the feet only, except in unusual circumstances, and stay beneath the surface for up to a minute or more, mostly in water of a depth of 6-18 ft (2-6 m). An unusual feature is the ability of loons to submerge only partially. They are thus able to remain with only the head above the surface — a valuable attribute in times of danger.

In winter loons migrate southwards to avoid the more extreme arctic conditions, largely following the coastlines and, particularly the common loon, traveling in flocks.

Loons' nests are placed at or near the water's edge, and they vary considerably in construction. Sometimes they are in the form of a simple depression in the vegetation covering the ground; at other times they may be substantial structures of plant material with the nest cup in the center. Two eggs are normally laid. They are rather more elongated than in most birds and are cryptically colored olive-brown or green with very dark spots. Both sexes take part in incubation, which lasts about four weeks, and share the care of the young.

The common loon or great northern diver breeds throughout arctic North America and in Greenland and Iceland and some smaller islands. It is one of the larger species, up to some 36 in. (91 cm) long. In breeding plumage the head and neck are black with a glossy green and mauve sheen, a horizontal line of vertical white stripes on the throat, and a more prominent collar of similar marks on the neck—incomplete at the back. The rest of the upperparts are black with white spots, the spots being largest in the scapular region. The underparts are white. Bill, feet and legs are black. In winter the plumage is brown above rather than black, without the spots, and the bill also is paler.

The yellow-billed loon (white-billed diver), *G. adamsii*, replaces the common loon in the far north and west of North America and also breeds around the European and Asian coasts of the Arctic Ocean.

The Arctic loon or black-throated diver has a circumpolar distribution in arctic and subarctic habitats. It is a smaller species some 24 in. (60 cm) long, separated by some into Eurasian and American species. Its breeding plumage is somewhat similar to that of the foregoing species, but the head and the back of the neck are gray and the white markings on the back form transverse bands. Also, at the sides of the neck and breast there is a beautiful pattern of fine vertical black and white lines. The bill and legs are as in the common loon, but slighter.

The red-throated loon or red-throated diver is slightly smaller than the black-throated species, though like all loons it varies considerably in size. It is another circumpolar species and is the most widely distributed of all the loons. Its breeding plumage is an almost uniform gray-brown above with a chestnut-red throat patch, and white beneath. The bill and legs are gray, and the former has a distinctly upturned appearance due to the upward sweep of the distal half of the lower mandible and a depression in the upper mandible in the region of the nostrils. In winter the upperparts may be seen to be flecked with white. This species is less extremely modified than the other loons and is able to take wing with greater ease.

DIVING BEETLES, large water beetles of the genus *Dytiscus*. They are carnivores and are unusual, as both adults and larvae live and hunt in water yet their pupae occur on the moist land surrounding the pond. The larvae are astoundingly voracious and can grow up to 2 in. (5 cm) long. They will feed on large freshwater animals such as newts, as well as smaller animals, pumping digestive enzymes into the carcass through their large hollow mandibles. The larvae breathe air through two posterior channels while hanging suspended from the surface and can then drag themselves below by crawling down the stems of water plants. When fully grown, the larvae pupate in the moist soil surrounding the pond, but this stage lasts only a few weeks.

The adults are large, up to 1¼ in. (3 cm), are active fliers and so can colonize new habitats such as waterbutts and fish ponds. They are also excellent swimmers, using their powerful, flattened hind legs for this purpose. They have to come to the surface to breathe, but dive with an air bubble trapped underneath their wing cases which is gradually used up during this activity.

DOG, a general term referring to several species in the family Canidae, including the wolf, *Canis lupus*, coyote, *Canis latrans*, domestic dog, *Canis familiaris*, and four species of jackal. Other members of this family, which arose about ten million years ago, are referred to as foxes although some, especially those from South America, cannot easily be ascribed to dogs or foxes. All canids have an acute sense of smell and hearing, but the dogs more than the foxes are specialized for a surface dwelling life and reliance on speed and endurance for hunting prey. The variability in size, shape and coloring both within and between species in the Canidae is unique among the mammals. For example, wolves from the same litter may dif-

Cape hunting dog, the wild dog of Africa.

fer radically in temperament and external appearance. This diversity is based on a vast gene pool. The success of this family with a worldwide distribution covering many types of habitats, is due to this polymorphism. Man

Female great diving beetle *Dytiscus marginalis*.

himself has used this diversity in selective breeding, and today there are more than 100 different breeds of domestic dogs.

The three possible ancestors of the domestic dog are the wolf, a wolf and golden jackal cross, and a now extinct wild canid resembling the pariah dogs and dingos. Most zoologists believe that the numerous breeds were developed from different subspecies of Asian and European wolf, but jackal and coyote blood could have been added by crossing already domesticated animals with wild ones.

The reasons for the taming and selective breeding of wild dogs are complex and probably depended on the needs of Mesolithic and Neolithic man in different environments. The initial motivating factor might have been the use of the dog as food. Many primitive peoples today eat dogs, as did some ancient peoples. Once an association developed between man and dog, the value of dogs in other spheres was no doubt soon recognized. Hunters and gatherers learned to exploit the keen senses of their tamed, but still wild-born captives in the pursuit of game. Wild dogs scavenging around human settlements in search of refuse not only had a sanitary function but also unconsciously acted as guards by howling and barking when unfamiliar animals or men approached. As selection for behavioral and physical characters continued, dogs assumed importance as sled or draft animals, for herding and protecting domestic livestock and for guarding homes. The dog's role as a home companion probably developed relatively late.

The classification of breeds differs from country to country and usually reflects the main dog interests of each nation. Breeds are normally grouped according to their function, but as this often varies with the locality, it is impossible to provide a classification that satisfies everyone. A recent one has attempted to classify dogs according to their probable origin, but this is unsatisfactory because we know the ancestry of only a few. Man has been trading and crossing dogs for centuries. Thus, most breeds are of untraceable and mixed origins. The six main groupings of the American Kennel Club are:

1) Sporting breeds. These include the retrievers, spaniels, setters and pointers, all of which were developed as bird or gun dogs. The spaniels are the oldest and most basic group, from which the others have been derived. Retrievers are of recent origin and have been bred mainly in England and North America.

2) Hounds. The hounds are divided into two groups, those that hunt by scent and those that use sight. In the first are the bloodhounds, beagles and foxhounds, used not only for tracking small carnivores like badgers and foxes, but also criminals. The sight hounds, including the greyhounds, Afghans, salukis and borzois, are an ancient group of different ancestry from the scent dogs. Indeed, the saluki is said to be the oldest purebred dog, having been kept in the Middle East for thousands of years. These breeds were developed in the open country and, being specialized for a cursorial life, they are unsurpassed for their speed and endurance. Also included among the hounds by the American Kennel Club are the dingo-type dogs, for example, the Rhodesian ridgeback, so-called

because of its shoulder patch of erect hair, and the barkless African basenji.

3) Working dogs. These are usually divided into herding, guard and sled dogs. Like the hounds, this group is of mixed ancestry. A distinction is normally made between the exclusively herding breeds like the collie, breeds like the Alsatian and Old English sheepdog, which both herd and guard, and finally those that guard livestock, like the St. Bernard and the Pyrenean mountain dog. The latter group also watch over property, as do the Alsatians and Doberman pinschers used widely in police work. The sled dogs of arctic regions include the Alaskan huskies, the malamute and the samoyed. Among the working dogs are found the largest breeds, the St. Bernard for example, weighs 150-200 lb (68-90 kg).

4) Terrier breeds. Terriers were initially used to flush badgers and foxes from their dens. Of recent origin, the majority of breeds have been developed in the British Isles, and are often named after the regions and cities where they were first bred. Terriers are known for their persistence and aggressiveness.

5) Toy dogs. Most members of this group are just smaller editions of a normal-sized breed and are valued as home companions or "lap dogs." Among them are the Pekingese, the Mexican hairless dog, the Maltese, the Yorkshire terrier and the griffon from Belgium. The smallest dog is the chihuahua, weighing only 3-5 lb (1.5-2.5 kg).

6) Non-sporting breeds. Like the toy dogs, this group contains mainly housebound breeds and show animals, mostly of different ancestry. Included are the Dalmatian, chow chow and poodle, originally a sporting dog.

Despite enormous differences among individual breeds in appearance and temperament, dogs readily recognize one another as conspecifics

when they meet. The reasons are varied, but two stand out. First, dogs have an excellent sense of smell, and no matter what the breed, a dog recognizes the scent of another dog. Second, all dogs have certain motor patterns easily interpreted by others, especially during social interactions. Most of these, like rapid tail-wagging to indicate friendliness and snarling or growling, which signify threat, are like the social behavior patterns of wolves. In fact, dogs and wolves, despite their radically different life styles today, share the same basic behavior repertoire. Both species are territorial and drive strangers from their homes. On unfamiliar ground, however, they are more tolerant, which explains the absence of aggression when two dogs meet in the street or a park. Wolves and dogs regularly deposit urine and feces on bushes, posts and rocks within their home range. Males always lift a hind leg before urinating, a response that appears at sexual maturity and is dependent on the presence of the male hormone testosterone. This tendency to distribute scent marks throughout the territory is strong, and careful training is necessary to

Dingo, the wild dog of Australia, is believed to be a feral domestic dog.

restrict the activities of housebound dogs. Individuals can almost certainly distinguish the scent marks of neighbors. A bitch in heat has a strongly scented urine that is readily recognized by males.

If allowed to run wild in groups, a dog litter will develop the same social structure as a wolf pack. Close relationships develop between individuals, and all members of the pack will be loyal to and obey the dominant animal. If a litter is split up early enough and the pups exposed to men, this loyalty is transferred to a human master. Dogs not given contact with man before 14 weeks old are almost untamable and very fearful, thus indicating that dogs have no special affinity with man despite thousands of years of close association.

Many wolflike traits arising from a cooperative pack hunting life have been exploited by man for his own needs. Dogs hunt both individually and in packs and will also retrieve game, a behavior recalling the wolf's tendency to carry prey back to the den. The tracking abilities of dogs have been employed in finding hidden criminals and discovering buried mines during wartime. Guiding the blind, which relies on the dog's loyalty and specialized senses is a recent and very worthwhile use of the dog.

As a result of domestication, the reproductive cycle of the dog has been altered in ways that increase its reproductive potential. Sexual maturity occurs earlier in domestic dogs than their wild counterparts, at one year instead of two years of age. Moreover, bitches come into heat twice a year at any season (except the basenji), and males are able to mate at any time, unlike the annually breeding wild dogs and foxes. The litter size of a domestic bitch is also larger than that of a wolf. It is probable that these changes have occurred through selective breeding, a constant supply of nutritious food, and reduced exposure to severe fluctuations in weather.

The gestation of the dog, like the wolf and coyote, is about 63 days. Occasionally when females have not been fertilized they will show all the characteristics of pregnancy, such as swelling of the abdomen and lactation, without being gravid, a condition known as pseudopregnancy or false pregnancy.

The maternal behavior of the bitch consists of grooming and suckling the young, retrieving them when they wander from the den, and keeping them warm. Puppies are completely helpless at birth, being both blind and deaf, and depend on the mother to fulfill all needs. At 14 days the eyes open.

DOGFISHES are characterized by having two dorsal fins, which lack a spine in front, and one anal fin. A spiracle is present, but there is no nictitating membrane or "third eyelid." The two most common European species are the greater spotted dogfish, *Scyliorhinus stellaris*, and the lesser spotted dogfish, *S. caniculus*. The former, which is also known as the nursehound or bullhuss, can be distinguished by the fact that its nostrils are farther apart and the nasal lobes or flaps of skin leading from the nostrils back toward the mouth are distinctly lobed. In both species the body is generally light brown with a fine speckling of black on the upper surfaces, the spots being larger in the greater spotted dogfish. The two are very common along all European coasts. They feed on worms, mollusks, crustaceans and echinoderms, and can be fished for by boat over sandy bottoms using lugworms (*Arenicola*) or small pieces of fish. There is an angling record of just over 20 lb (9 kg) for a larger spotted dogfish caught in British waters. The dogfishes are oviparous and produce rectangular egg cases with a spiraling tendril at each corner. The embryo is well supplied with yolk and does not hatch for seven months after fertilization.

DOLPHINS, a group of small toothed whales of which the best-known is the bottle-nosed dolphin, *Tursiops truncatus*, the highly intelligent and friendly creature now so popular in many large aquaria. The bottle-nosed dolphin is moderately large, up to 12 ft (4 m) in

The coyote or prairie wolf has suffered less than the timber wolf, largely because it is smaller.

length and has a medium-sized beak with 40 teeth in each jaw. It is usually gray on the back but may approach black; the belly, chest and throat are white or pale gray, but there is no sharp line between the two shades. It is found on both sides of the North Atlantic and into the Mediterranean. A very similar bottle-nosed dolphin, *T. aduncus*, is found in the Red Sea, Indian Ocean and around Australia, while *T. gilli* is the Pacific form.

The bottle-nosed dolphin is a favorite animal for study in captivity because it is relatively easy to catch, easy to handle in captivity and very tractable, becoming very attached to its human friends if properly treated. It is also easy to train, and, therefore, so long as it is not allowed to get bored, it will respond to experimental conditions in much the same way as it does for show purposes. It is highly intelligent and shows remarkable ability at developing ideas. For example, it will initiate games with an attendant. It will even use whatever material is available to serve its purpose. Had it hands it could well be an effective tool user.

Lesser spotted dogfish swimming in the aquarium at the Plymouth Biological Station.

There are said to be 25 species of long-beaked dolphins in the genus *Stenella*, but probably this should be reduced to about 10 true species. Even of these some are known only from skulls, and within the whole group the skulls tend to show similarities that make formal establishment of species difficult. The slender dolphin, *S. attenuata*, is found in warmer parts of the Atlantic, the spotted dolphin, *S. plagiodon*, is a deep-water species off the North American Atlantic coast, as is *S. euphrosine*. Other species are found in most of the warmer waters of the world. The rough-toothed dolphin, *Steno bredanensis*, is very closely related, but it has characteristically rough teeth and is found in the warmer parts of the Atlantic and the Indian Ocean.

Members of the genus *Sotalia* are small dolphins of only about 3½ ft (1 m) long. Four species are found in the Amazon and one in the harbor of Rio de Janeiro. The genus *Sousa*, also long-beaked, is found in tropical waters of the Eastern Atlantic, the Indian and Pacific Oceans. It includes *S. teuszii*, found in rivers in Senegal and the Cameroons, which was formerly thought to be the only whale to feed on vegetable material. *S. plumbea* is found off East Africa. The Chinese white dolphin, *S. sinensis*, is a truly white animal that is found up the Yangtze and in other Chinese waters. Very little is known of this genus.

The common dolphin, *Delphinus delphinus*, is found in large schools in temperate waters throughout the world. It is slender, up to 8 ft (2.4 m) in length and has a pronounced beak. Its back is dark gray to black and its belly white, with light gray or brownish stripes on the mouth. The dolphin pictures of classical times usually seem to be of common dolphins, but bottle-nosed dolphins also exist in the Mediterranean. It is a fast swimmer, probably the fastest of all dolphins, and is difficult to catch. When caught it has been found to be very nervous and difficult to keep in captivity as it needs to be with its fellows. At sea in freedom and company, however, it is apparently much more confident and will approach boats and even swimmers. The Pacific common dolphin, *D. bairdi*, is very similar and is probably a subspecies.

The Irrawaddy dolphin, *Orcaella brevirostris*, is a beakless dolphin about 7 ft (2.3 m) in length. It has a small dorsal fin, rather long broad flippers and an overall blue-gray color. It is found in the Irrawaddy over 900 miles (1,400 km) upriver. A rather different dolphin is found in the Bay of Bengal, around Malaya and off Thailand, but it is unlikely to be an entirely different species. The behavior of both is much the same. They are traditionally adopted by the local fishing communities and, although feeding exclusively on fish themselves, have the reputation for driving fish into the fishermen's nets.

There are two species of right whale dolphins, so called because they have no dorsal fin like the

Dolphin in a sea aquarium, a familiar sight to visitors over the last 30 years.

Pacific striped dolphin, found in the northern Pacific as far south as California in large, socially organized schools.

true right whales. *Lissodelphis peronii* is found in southern seas around New Zealand and southern Australia. It is a strikingly colored animal of some 6 ft (2 m) length, having the top of the head, back and flukes black and the rest of the body, including the flippers, white. The northern species, *L. borealis*, is found in the North Pacific. It is somewhat larger, being over 8 ft (2½ m), and the black extends downward to include the flippers, leaving white on the chest between the flippers and extending to the tail.

The genus *Cephalorhynchus* consists of about a dozen species, none of which is at all well known. Some are known only from skeletons and skulls. They are all found in southern seas, and those known in the flesh are all strikingly marked. At first sight they would appear to be porpoises rather than dolphins, having little or no beak and porpoiselike flippers and dorsal fins. Only four species are at all well known, but even this amounts to carcasses or occasional sea sighting.

Heaviside's dolphin, *C. heavisidei*, was named after Captain Haviside but corrupted by misunderstanding to Heaviside. It is a striking animal of about 4 ft (1.2 m), having a black back and a white or yellowish-white undersurface with lobes of white running up onto the side in front of and behind the black flippers and obliquely toward the tail. It is best known around South Africa.

Hector's dolphin, *C. hectori*, is similarly colored to Heaviside's but somewhat longer: 6 ft (1.8 m). The black undersides of the flippers extend across the chest, and a strip toward the head. It is best known around New Zealand.

Commerson's dolphin, *C. commersoni*, of southern South America and the Falkland Islands, is the most striking of all dolphins. It has a black head and flippers, vent, dorsal fin and tail and otherwise is white.

The white-bellied dolphin, *C. albiventris*, is also found in much the same waters though more rarely seen. It is about 4½ ft (1½ m) long and black with white on the throat, behind the flippers and belly.

Externally, the genus *Lagenorhynchus* can be compared with the above group, but there are differences in the bones of the skull and particularly the large number of vertebrae, which may be as many as 90. They have fairly large and characteristically dolphin-type dorsal fins with a concave posterior border. The genus is represented in the colder seas of both the northern and southern hemispheres, and one species in particularly well known, having been kept in captivity to a considerable extent. Wilson's hourglass dolphin, *L. wilsoni*, is seen only in the Antarctic along the edge of the pack ice. It is remarkable for its clear black and white coloring, which gives an hourglass effect, of white on the flanks and a white belly and tail to the black flukes.

The dusky dolphin, *L. obscurus*, is the best-known southern species. It is the commonest dolphin around New Zealand where it is found in large schools but extends across to the Falkland Islands and to South Africa. It is black and white with black back, mouth, eye, flippers and tail, but with white bands coming from the belly giving a marvelous effect. Lillie, in the "Terra Nova" Report on Cetacea, states, "This

Common dormouse, or hazel dormouse, seems most of the time to be asleep or about to fall asleep.

dolphin does not seem to occur further south than about Lat. 58° S, but when we were approaching, or leaving, the coast of New Zealand we invariably met large schools of Dusky dolphin which used to follow us and play around the bows of the ship, as though they were seeing us off or welcoming us back to the temperate lands." Several other southern species are known only from skeletal material. The three northern species are somewhat longer, being up to 10 ft (3 m) in length, as compared with 7 ft (2.1 m) for the dusky dolphin and 5-6 ft (1.7 m) for the others. They are all somewhat similar in appearance and behavior and probably are quite closely allied.

The white-sided dolphin, *L. acutus*, is black dorsally and white-bellied with a prominent white streak on its flank. It lives in the North Atlantic, where its range extends from Greenland in the north across to Norway – where it is the second most common dolphin – north British waters and Cape Cod, off America. It is highly gregarious, being found in schools of up to 1,500 animals.

The white-beaked dolphin, *L. albirostris*, is similar to the white-sided dolphin but has a characteristic white beak. The margin of the upper lip is gray or black and the lower lip may also show some color as well. It is also found in vast schools in the North Atlantic extending over much the same areas as the white-sided dolphin. The young of both species are born in the spring and early summer, and the mating season probably occurs at the same time of year.

The Pacific striped dolphin, *L. obliquidens*, is very similar in appearance to the white-sided dolphin and is found in the northern Pacific as far south as California. This animal is also found in large, socially organized schools. It is fairly easily caught and has proved a popular dolphin attraction in captivity. Hose's Sarawak dolphin, *Lagenodelphis hosei*, is known only from a single specimen sent as a skeleton to England in 1895 and was about 8 ft (2.4 m)

long. It appears to be midway between the *Lagenorhynchus* and *Delphinus* dolphins, but no others have been seen.

Risso's dolphin, *Grampus griseus*, is perhaps unfortunately named, because a common name for the killer whale is "grampus." There is, however, no similarity in appearance or behavior between the two animals. It is said that the word "grampus" is an abbreviation of *grand poisson*. Risso's dolphin is a longish, beakless dolphin of about 12-13 ft (4 m) in length. It is gray and may be black on the fin, flippers and tail while the underside is paler to lighter gray or even white. Often the skin shows white score marks, which are said to be healed tooth marks of other Risso's dolphins. It is very widely distributed in the North Atlantic and the Mediterranean and in the south around New Zealand and South Africa. It is fairly solitary, rarely being found in schools of more than a dozen. It feeds mainly on cuttlefish.

DORMICE, in general appearance as well as in their way of life tend to bridge the gap between squirrels and mice, and are represented by a variety of species throughout Africa and temperate Eurasia. With one exception they have bushy tails, but they are mostly mouse- rather than squirrel-sized and are nocturnal. Dormice are mostly agile climbers, living in trees or in the shrub layer of woodland. The fur is soft and dense. It varies in color from orange-brown to gray, and many species have dark marks around the eyes, accentuating the already large eyes. The feet are well adapted for climbing, with long, flexible toes, and the hind feet can be turned outward at right angles to the body enabling the animal to move confidently on slender twigs. The food is very varied, including both animal and vegetable material, although it is likely that seeds and berries predominate in the diet of most species. The dormice of the temperate region are noted for their hibernation, which begins in October. They become very fat in early autumn and during hibernation become quite torpid and cold.

Their reproductive rate is low, usually with only one or two litters per year, but the scarcity of their remains in the pellets of owls shows that their specialized habits allow them to escape the heavy predation suffered by most other kinds of mice. During hibernation, however, they are vulnerable to mammalian predators like weasels and martens, and it is probably at this time that they suffer the highest mortality.

Muscardinus avellanarius is one of the smallest and has a distinctive orange-brown coat. It lives in dense shrubby undergrowth where it constructs neat nests, often using the shredded bark of honeysuckle. The largest European species, the fat dormouse, *Glis glis*, is more arboreal, nesting in holes or in exposed positions quite high in the woodland canopy. It is also known as the edible dormouse, since the ancient Romans fattened it for the table in special jars. One of the most attractive of all dormice is the garden dormouse, *Eliomys quercinus*, a European species with a conspicuous black mask and a long tail, bushy only toward the tip. It frequents rocky hillsides as well as woodland and is especially common in the Mediterranean region. By far the most isolated species of this family, geographically speaking, is the Japanese dormouse, *Glirulus japonicus*, since dormice are quite absent from the mainland of temperate eastern Asia. The Japanese dormouse is small, lives in montane forests and is one of the very few genera of mammals endemic to Japan. A decidedly aberrant member of the family, and one of the least known, is the mouse-tailed dormouse, *Myomimus personatus*, found in Turkmenistan, Iran and Bulgaria. It has a long, almost naked tail and appears to be terrestrial rather than arboreal.

African dormice belong to the genus *Graphiurus* and are found throughout the forest and savanna zones of Africa. Although primarily arboreal, they are frequently found living in the thatch of houses.

DOUROUCOULIS, *Aotes,* also known as night monkeys, live in the forests of the Amazon northward into Central America. They are nocturnal, spend the day in holes in trees and move about by night in troops, taking a variety of food both plant and animal. The head and body length is 12 in. (30 cm) and the tail is slightly longer. The body is covered with long coarse hair, mottled gray or brown, and the forehead is whitish with three black stripes. The face is owllike, the eyes large, yellow and staring and the ears are small.

DRAGONETS, flattened, bottom-living fishes, members of *Callionymus* and related genera found along coasts in temperate regions. The dragonets rarely grow to more than 12 in. (30 cm) in length. They have flat, depressed heads and slender bodies, but many are so beautifully colored that they resemble some of the tropical reef fishes. There are no scales on the body. There is a sharp spine on the preopercular bone of the gill cover, and there are two dorsal fins of which the anterior is greatly elongated in the male.

Two species of dragonet are found along European shores, the common dragonet, *Callionymus lyra,* and the spotted dragonet,

C. maculatus. The former is common in the Mediterranean and reaches northward to Norway. The female is rather dull colored, but the male is a splendid fish, especially in the breeding season. The back is red-yellow with blue markings, and the flanks and lower part of the head are orange, again with blue spots and marks. There are two blue bands along the body, and the fins are marked with blue, yellow and green. The differences between the sexes are so striking that they were once thought to be quite different species, the "sordid" and the "gemmeous" dragonets. Spawning takes place in spring and summer, the male swimming around the female and displaying with gill covers and fins until the female is sufficiently stimulated. The two then swim together to the surface, close together and with the anal fins forming a gutter into which eggs and sperm are shed. The eggs have honeycomb markings on their surfaces.

The spotted dragonet is a smaller fish and can be distinguished by the three or four rows of ocellated spots instead of bands along the dorsal fin.

Common dragonets. The male is considerably larger than the female.

DRAGONFLIES, the robust, winged, carnivorous insects that comprise the order Odonata. Adult dragonflies are most often seen flying along the edges of ponds and rivers during sunny weather. They have two pairs of narrow, richly veined wings, a long slender abdomen and prominent eyes. Features that distinguish dragonflies from other insects are the skewed thorax and forwardly directed legs, the inconspicuous antennae and the existence of accessory genitalia on the second and third abdominal segments of the male. The larvae (sometimes referred to as nymphs) are fully aquatic, and have the labium (second maxillae) highly modified for seizing prey.

Dragonflies are primarily tropical insects but occur in temperate latitudes to the limit of trees. In the northern hemisphere this means that in parts of Canada, Alaska and Sweden

they reach the Arctic Circle. These hardy, northern forms are species of *Aeshna, Somatochlora* and *Leucorrhinia,* all genera that are represented in Britain and western Europe.

Dragonflies are able to reproduce in almost any kind of fresh water that is not too hot, acid or saline. They inhabit watercourses throughout their length, from highland streams to placid, mature rivers in the plains. There are species adapted to breeding in temporary ponds, rock pools, waterfalls, brackish marshes, and even the water that collects in the leaf bases of certain forest plants. In Hawaii lives the only species known to have a terrestrial larva, *Megalagrion oahuense.* It inhabits leaf litter on the forest floor, and its closest relatives live in forest streams from which their larvae occasionally wander.

Dragonfly larvae catch their prey in an unusual way. Typically, they remain motionless until a small creature comes near enough to be detected by sight or touch. When the creature comes within range, the larva suddenly extends the labium, at the tip of which are hooks that open to grasp the prey. The labium then draws the victim back to the mandibles. When in the resting position, the labium lies folded beneath the head and thorax, sometimes hiding the lower part of the face: for this reason it is sometimes called the "mask." The sudden extension of the labium is the result of a localized increase in blood pressure, a condition controlled by the abdominal diaphragm.

Adult dragonflies lay eggs in or near water. The aquatic larva casts its skin 8-15 times before completing development, passing through a corresponding number of intervening stages, or instars. The first instar (prolarva) is unlike the others in appearance and duration. It somewhat resembles the pupa of a wasp or beetle, the functionless appendages lying closely against the body. A few minutes after leaving the egg it molts to disclose the spiderlike second-instar larva, usually about $\frac{1}{25}$ in. (1 mm) long. From this stage on, the larva feeds actively on small water animals such as Protozoa and minute Crustacea and, later, beetles, mosquito larvae and even small fishes. At an early stage wing buds appear on the larva's body, and these become larger at each subsequent molt. There is no pupal stage. Near the end of the last instar the adult organs begin to form inside the larval skin. When these internal changes have been completed, the larva climbs out of the water and molts to disclose the adult, a process known as "emergence." This whole process, from leaving the water to being ready to fly, may take only 20 minutes in the tropics but several hours in temperate latitudes. As a rule, larger dragonflies leave the water after sunset and take to flight just before sunrise, so their emergence is seldom witnessed by man. Smaller species may be seen emerging during the daytime.

The newly emerged dragonfly is a pale, soft creature with glistening wings. Its first flight takes it away from water, and it spends the next few days (2-15 depending on the species and weather) feeding actively. During this time it is developing its full coloration and becoming sexually mature. When this maturation period is over, the adult seeks water and enters the reproductive period, which may last up to six

Adult dragonflies lay eggs in or near water. The aquatic larva casts its skin 8-15 times before completing development. At the end of the last stage the adult organs begin to form inside the larval skin. The first of these three pictures of the male *Aeshna cyanea* shows this stage, the nymph in water. It then climbs out of the water (second picture) and the adult dragonfly wriggles free of the ñymphal skin, then hangs on to it while its wings expand, harden and dry. Only then is it ready to take wing.

weeks. During the reproductive period, males assemble at the mating site, usually the pond or river where the eggs will be laid, and soon space themselves out along the shore, each coming to defend a well-defined area, or "territory." In several respects this activity resembles the swarming of other insects (e.g., mayflies and mosquitoes), the main difference being that adjacent dragonflies are farther apart. Large species have larger territories than do small species, but the size of territory tends to remain constant within a species.

DRONGOS, a family of medium-sized, arboreal songbirds with flycatcherlike habits. There are 20 species, showing relative similarity. They occur through Africa south of the Sahara, southern Asia from India to China, and south to Australasia. They vary from 7-15 in. (18-38 cm) in length, but this is due in part to a fairly long tail. Elaborate long outer tail feathers may, in some species, nearly double this length. Drongos show the typical modifications that occur in flycatching birds. The stance on a perch is upright, the legs are short, the wing long and tapering and the tail well-developed and forked. The bill is strong and stout, hooked at the tip and slightly notched to grip the prey, with long, strong bristles around the base. Drongos are agile and swift in flight and can maneuver quickly. The feet are strong and are used not only to seize large prey in the air but also to hold an insect and raise it toward the bill, parrot fashion.

The plumage is gray in two species, flight feathers are red-brown in another, but otherwise drongos are black, with a few showing some white on the underside, wing-coverts or head. The black plumage shows glosses of blue-green or purple to varying degrees, often confined to a part of the plumage. On some, such as the spangled drongo, *Dicrurus hottentottus*, the gloss is limited to spots at the

feather tips. The tips of the forked tail may be greatly extended, apparently as a decoration, and this may vary between forms of the same species living on different islands. Rarely, the outer pair of feathers are long, curved, bare barbs ending in racquets. Some birds show crests, varying in different species from small forehead tufts to full curly crests.

The nests are frail-looking, shallow cups suspended in forked twigs usually well up in trees. They are made of thin, flexible stems and roots, and bound together with spiders' webs. The two to four eggs vary from white to buff or pink: variably spotted, blotched or, rarely, streaked with brown, red or black. The female alone incubates, but the naked nestlings are cared for by both parents.

DUCKS, aquatic birds comprising of most of the smaller members of the family Anatidae, which also contains the geese and swans. The whistling ducks and the perching ducks commonly perch in trees and, like the goldeneyes and the buffleheads, even nest in trees, usually in holes. The sheld-ducks nest in holes in the ground, and some species, including the ubiquitous mallard, nest in a variety of situations, on or off the ground. The redhead (a pochard) and the black-headed duck (a stifftail) are brood parasites laying their eggs in the nests of other species of birds—not only ducks. Of the seven tribes of "true" ducks by far the most common and successful is the tribe Anatini containing those species that obtain all their food at or near the surface of the water or on damp ground or vegetation. This food may include a wide variety of edible materials, both plant and animal. Most of the species are birds of fresh water, though a few may nest near the seashore and a number regularly pass over the sea during migration. These migrations may be very extensive, totaling a few thousand miles in some species.

Nearly half the dabbling duck species show a strong sexual dimorphism, the male being brightly colored and the female dull and cryptic. This dimorphism is, as is usual in birds, connected with the breeding display of the male and the female's need for concealment, particularly while on the nest. Most of the other species are dull colored in both sexes, but a few, such as the Chiloe wigeon, *A. sibilatrix*, have bright plumage in both sexes. All dabbling ducks have two molts per year, which ensure the replacement of worn or damaged feathers. In the sexually dimorphic species the bright breeding plumage of the male is replaced in the post-breeding molt by an "eclipse" plumage similar to that of the female.

A further outstanding plumage feature of the dabbling ducks is the presence of a brightly colored patch of feathers on the wing. This is called the speculum and is seen in both sexes. Usually it involves the secondary wing feathers only, but other adjacent areas of feathering may play a part. Most specula are in metallic shades of green or blue, often combined with patches or stripes of white, and are very striking, although they may not be visible when the wing is folded. It is thought that the role of the speculum is one of communication within the species, and the display of the speculum plays an important part in the courtship ceremonies, probably helping to ensure mating between birds of the same species. It may also assist in the coordination of movements of bird flocks engaged in complex flight maneuvers, being a plainly visible feature of every extended wing.

The mallard group of species includes not only the well-known mallard duck itself, which, in various races, breeds .across most of the northern half of the world and has given rise to most of the domesticated forms of duck. In North America the canvasback, *Aythya valisi-*

neria, is a favorite quarry as is the redhead, *A. americana*. These birds are basically gray on the body, black on the tail and upper breast and chestnut-brown on the head in the male, the female being brown. The canvasback has a longer bill and is a paler gray. The European pochard, *A. ferina*, has the body shape of the redhead but in coloration and head shape is intermediate between the other two species. The food is largely vegetable, but the scaups take a larger proportion of animal food. The greater scaup, *A. marila*, is the sturdiest species and spends more time on the sea than the others. It ranges across the whole of the northern hemisphere. The lesser scaup, *A. affinis*, is restricted to America. The scaup males are gray-bodied with black rump and breast. The head and neck are basically black also, but the greater scaup has a green sheen on the head and the lesser is more purple. In America the scaups are also known as bluebills.

The pochard most commonly kept in captivity is the red-crested pochard, *Netta rufina*, the male of which has a striking plumage of black, pale gray and brown, with a light chestnut head and orange bill. It breeds sporadically in Europe but principally in western Asia. The rosybill, *N. peposaca*, is a striking South American pochard with a black, purple and gray plumage and a knobbed red bill.

The perching ducks and geese are somewhat peculiar species, which spend more time in trees than other ducks. They vary considerably in form, size and plumage pattern, from the tiny, pygmy geese of the genus *Nettapus* of which there are three species in Africa, southern Asia and Australia, to the very large, 32-in. (81-cm) spur-winged goose, *Plectropterus gambensis*, of Africa. Two species of perching ducks are very well known in aviculture: the mandarin, *Aix galericulata* of eastern Asia (also breeding as an escape in Britain), and the Carolina or wood duck, *Aix sponsa* of the eastern half of the United States and southern Canada. In each species the male has a striking multicolored plumage, including a very handsome backward-sweeping crest.

A further noteworthy species of the perching duck group is the muscovy, *Cairina moschata*, the ancestor of the domestic muscovy, which is a fairly common farm bird, particularly in America, although not as successful in domestication as the mallard.

The common sheld-duck, *Tadorna tadorna*, breeds from the coasts of western Europe eastward through much of central Asia, nesting usually in burrows in the ground. It has striking white, green-black and chestnut plumage, with red bill and legs. The sexes are similar although the female is smaller. In the sheldgeese of South America, which are larger and more gooselike, some, such as the ashy-headed goose, *Chleophaga poliocephala*, have similar sexes, while in others, such as the Magellan goose, *C. picta*, the female is basically brown and the male white.

The steamer ducks, genus *Tachyeres*, of South America and the Falkland Islands, are large,

Shoveler duck scoops small items of food from surface of water with its broad bill.

The mandarin, most showy of ornamental ducks.

powerfully swimming ducks of three species, two of which are flightless. The sea ducks include a variety of species that obtain their food by diving in both fresh and salt waters. They include the sawbills, or mergansers, of the genus *Mergus*, which have serrated bills for holding slippery fishes; the scoters, genus *Melanitta*, which are bulky and black and dive for shellfishes that are crushed in the especially muscular gizzard; the goldeneyes, genus *Clangula*, which nest in holes in trees, as do some of the mergansers to which they are closely related; and the harlequin and old squaw. The last two are striking arctic species, the harlequin *Histrionicus histrionicus* having a boldly marked blue, white and chestnut plumage and inhabiting fast-flowing streams, and the old squaw or long-tailed duck, *Clangula hyemalis*, having two different plumages, both basically black and white, one winter and

Black drongo or king crow, *Dicrurus macrocercus*, stoops like a falcon to scare predators attacking its nest.

one summer. The females are cryptically colored. There are four species of eider, three of them in the genus *Somateria*. The common eider, *S. mollissima*, has provided man with the best down for pillows and mattresses, and this important species has its own entry. Female eiders, like all the female Mergini, are cryptic, but the males are boldly marked in black and white and, in *Somateria*, have apple- or emerald-green on the head. Steller's eider, *Polysticta stelleri*, is smaller than the other species, and the male has a chestnut breast and belly. The male king eider, *S. spectabilis*, is an outstandingly handsome bird with soft gray on the head as well as green, and with an unusual helmeted forehead formed from a continuation of the orange horn of the bill.

Finally, the tribe of stifftails contains a number of dumpy little freshwater diving ducks, most of which are red-brown in color with blue bills. It includes the ruddy duck, *Oxyura jamaicensis*, which breeds, as a number of subspecies, from North to South America, and another South American species, the black-headed duck, *Heteronetta atricapilla*, which is parasitic, laying its eggs in the nests of other species of birds.

DUGONG, *Dugong dugon*, a large, totally aquatic, herbivorous mammal which lives in warm Indo-Pacific seas.

Male Reeves' muntjac, a deer also called barking deer.

Banded duiker or zebra antelope, of West Africa.

The dugong commonly grows to a length of about 10 ft (3 m), although lengths of 16 ft (5 m) have been noted. It is a heavily built animal of torpedo shape with a horizontally flattened whalelike tail. Like the whales, too, all trace of hind legs has disappeared in the course of evolutionary adaptation to aquatic life. The fore flippers are small and take no part in propulsion. All the bones are of exceptional density, and the skull is heavy and down-turned in front. The teeth are limited to a small number of crushing teeth, and there are horny pads in the front of each jaw. The young have five or six cusped molars on each side of each jaw, and these teeth fall out progressively from the front so that only two may remain in the adult. A pair of tusks is also present in the upper jaw, which, in the male, protrude without known function. The tusks of the female never become visible.

The gray skin is of immense thickness and toughness, but there is no substantial blubber layer beneath. The skin is nearly hairless but has individual hairs about $\frac{1}{4}$ in. (7 mm) long and about 2-3 in. (5-8 cm) apart. The bristly mask forms a flat and down-turned front to the head above the mouth. The nostrils are high on the head and, in adaptation to taking air from the surface, are equipped with powerful muscular valves. The eyes are small and circular, and the minute ear holes are only to be found by careful scrutiny.

The single pair of mammary glands are pectoral in position, almost in the armpit, as it were. The occasional suckling of a young one held vertically by a flipper must be one of the reasons for the mermaid myth.

One young is born at a time, and it probably accompanies its mother for a considerable period. Dugongs tend to live in groups and at times may be highly gregarious. Enormous herds moving together were recorded in Moreton Bay off Brisbane early in the last century. No evidence exists of regular migrations nor of movements across deep waters.

Their habit is to feed, especially at night, in the shallow areas where dense patches of dugong grass (*Zostera*, etc.) grow. The food is pulled off by the powerful lips, and it is dislodged vegetation floating to the surface that commonly informs the hunter of the creature's presence.

DUIKER, a group of small African antelopes of uncertain affinities. None is more than 30 in. (75 cm) tall, and many species are less than half that height. The head is conical with a convex nose and a small bare muzzle. The horns are smooth, somewhat keeled, and placed on a backwardly-projecting eminence of the frontal bone. In front of the eyes are face glands, which are quite different both in position and structure to those of other bovids. Instead of being close to the eye, they are halfway between it and the muzzle. There is a line of bare skin, lying at a slight angle to the eye-muzzle axis, which is studded with a line of glandular pores. As well as face glands, duikers have foot glands, and sometimes inguinal (groin) glands too. The duiker's fur is short and close, often rather harsh and longer on the rump than on the neck, and there is usually a long tuft between the horns.

Duikers occur all over sub-Saharan Africa. They inhabit both forest and savanna, but even the savanna forms are found mostly in places with thick ground cover. There is only one savanna species in the genus *Sylvicapra*; this has long legs, and the horns turn upward from the base. On the other hand, there are at least 11 species of the forest duikers, genus *Cephalophus*. Forest duikers have horns that point straight backward in a line with the face, and their build is wedge-shaped, with short fore legs and a high rounded rump—an obvious adaptation to pushing through thick undergrowth. The name comes from the Afrikaans word meaning "diver," referring to the way duikers plunge into the undergrowth when disturbed.

DUNG BEETLES, dark heavy-bodied insects of the family Scarabaeidae. The name refers to their habit of feeding on animal, especially mammalian, droppings. The adults of the larger, true dung beetles, such as *Copris* and *Canthon*, are dark brown or black with powerful spiny legs. The head often has a flat scooplike projection used to push balls of dung over the ground. The beetles fashion the balls by breaking off a portion from a large dung mass and tumbling it to and fro until it is spherical. The beetles may also roll dung balls by holding them between their hind legs. The spherical meal of dung is transported to a suitable retreat before being eaten. Some species of dung beetles are unique among insects in that a male and female may cooperate in their care of the young stages. They excavate two to seven chambers in the soil and stock each with a separate store of dung. A single egg is deposited in each chamber, and the female remains to guard the nest. In some species the female even remains long enough to tend her brood to maturity. The larvae develop into thick grubs with well-developed legs.

Giant scarab beetle *Helicopris standingeri*, 2 in. (5 cm) long, of Nigeria.

DWARF ANTELOPES, essentially small gazelles of the tribe Neotragini, the smallest living members of the Bovidae. It is uncertain whether it is a homogeneous group, or whether it is derived independently from larger gazelle-like antelopes. Only one species, the rhebok, *Pelea capreolus*, is distinctive enough to be of uncertain affinity. It may be a dwarfed reedbuck rather than a dwarfed gazelle. All appear to live in pairs.

EAGLE, name given to many powerful, rapacious, diurnal birds of prey. Eagles divide naturally into four main groups: the sea and fish eagles, the snake and serpent eagles, harpy and crested eagles and their relatives, and true "booted" or aquiline eagles, which differ from all the others in that the tarsus is feathered to the toes. They vary in size from the largest and most formidable of all birds of prey (and accordingly, the most formidable of all living birds) the harpy eagle, *Harpya harpya*, and Philippine monkey-eating eagle, *Pithecophaga jefferyi*, to relatively tiny species such as the Nias Island serpent eagle, *Spilornis cheela asturinus*, and Ayres hawk eagle, *Hieraetus dubius*, smaller than most buzzards.

The sea and fish eagles include two huge species, the European sea eagle or erne, *Haliaeetus albicilla*, and Steller's sea eagle, *Haliaeetus pelagicus*. The latter is perhaps the grandest and most impressive of all diurnal raptors—a huge bird, weighing 15 lb (6½ kg) or more, with a wingspan approaching 8 ft (2½ m), and capable of killing animals as large as a seal calf. No sea eagle or fish eagle is very small, the smallest of the genus *Haliaeetus* being the Madagascar fish eagle, *Haliaeetus vociferoides*.

Sea, fish and vulturine fish eagles share a number of features. They have bare tarsi, and adults show much white, especially on the head and tail; immature birds are brown. They are all more or less confined to aquatic habitats, large lakes, rivers or the seashore. Some, for example, the African fish eagle, *Haliaeetus vocifer*, are numerous and familiar in such habitats. Pallas' sea eagle, chiefly of interior Asia (and hence wrongly called a sea eagle) is perhaps the least aquatic of the Eurasian species, whereas the aberrant, frugivorous vulturine fish eagle occurs scattered through savannas lacking large bodies of water; it is nevertheless aquatic in, for instance, mangrove swamps. All these eagles except the vulturine fish eagle have spicules on the feet adapted to grasping fishes, but not to such an extreme degree as the osprey *Pandion*. Fish is important in the diet of all, even the vulturine fish eagle. Two species, the bald eagle, *H. leucocephalus*, and Steller's sea eagle, feed much on stranded and dying Pacific salmon. Besides fish, they eat carrion and some water birds, but seldom do they eat mammals, except for the young of aquatic species such as seals. The most unusual of the snake eagles is the African bateleur, *Terathopius ecaudatus*. Adults have exceptionally long wings and a very short tail. In immature birds the tail is longer. The

bateleur's relationship to other snake eagles is shown by similar development of the young, which is mainly brown, unlike the spectacular black, chestnut and white adult. Bateleurs appear to be specialized for high-speed gliding, and spend most of the day traversing the African skies, usually at 200-500 ft (60-150 m), at an airspeed of about 35-55 mph (56-88 kph). They hold their exceptionally long wings with a pronounced upward slant (dihedral) and, having little tail, apparently steer by canting from side to side—hence the name "bateleur" (derived from the French name for old-time tightrope walkers, who carried a long pole to aid balance). Bateleurs eat many snakes, but also take some mammals and ground birds and will eat carrion. They are also piratical, chasing and robbing vultures, a habit possibly derived from carrion feeding.

The harpy eagle is huge, females weighing about 15 lb (6½ kg) or more, males 8-10 lb (3½-4½ kg). It ranges widely in tropical South and Central American forests. Harpies are rather unlike buzzards in that they are apparently fierce and powerful predators, feeding upon large forest mammals such as monkeys, agoutis and sloths. It is probable that a female harpy eagle is about as large and heavy as an eagle can be while still being able to fly dexterously enough among trees to be an effective predator. However, so little is known about the field habits of these great birds that it is unwise to be dogmatic. A female harpy has legs almost as thick as a child's wrist and massive feet with huge curving talons.

Nearly as large as the harpy, the Philippine monkey-eating eagle is confined to the larger Philippine Islands and probably only survives

Bateleur, from the French for "circus performer." This African eagle performs somersaults in the air.

Earthworms pairing.

on Samar and Mindanao. Perhaps 50 pairs now exist, since this species is threatened by increasing destruction of habitat through human overpopulation, by the traffic in wild birds for zoos and by the prestige value of an eagle trophy. Efforts are now being made to conserve the remaining population.

The true, or booted eagles, with feathered tarsi, include the golden eagle, the African Verreaux's eagle, *Aquila verreauxi*, and the crowned eagle, *Stephanoaetus coronatus*. These eagles vary from very small species such as Ayres' hawk eagle, *Hieraetus dubius*, or Wallace's hawk eagle, *Spizaetus nanus*, of Malaysia to very large and powerful eagles such as the African crowned and martial eagle, *Polemaetus bellicosus*, and Siberian golden eagle, sufficiently large and powerful to kill wolves. Wild individuals feed upon large mammals (e.g., crowned eagles), game-birds (martial and golden eagles), smaller mammals (tawny and spotted eagles, *Aquila rapax*, *A. clanga* and *A. pomarina*), frogs (spotted eagles) and birds (Ayres' hawk eagle). They range the world's habitats from the Arctic to the equator, but no farther south than temperate Tasmania. South American *Spizaetus* species live in tropical forest. Half the species live in open country and half in forests or heavy woodland, with some intermediates. In Africa, the martial eagle inhabits savannas, while the crowned eagle lives in forests. Forest species are short-winged and long-tailed, resembling goshawks in silhouette.

The typical eagles, genus *Aquila*, are large or very large, varying from the rather small Wahlberg's eagle, *A. wahlbergi*, to huge golden, Verreaux's and wedge-tailed eagles, spanning 7-8 ft (2-2½ m) in large females, and able to kill mammals as large as a deer calf. The tawny eagle, *Aquila rapax*, widespread in open country in Asia and Africa, is probably the most common of the world's large eagles. It feeds on small mammals and carrion, and is strongly piratical, pursuing other species to obtain prey. All *Aquila* species are brown, immature birds and adults alike, except for Verreaux's eagle, which is a magnificent coal-black species with a white patch on the back. It feeds almost exclusively on rock hyrax (*Procavia* and *Heterohyrax*), the conies of the Bible.

Most species of the genus *Spizaetus* are also little known; the nests of five species have not been found. However, the large and powerful mountain hawk eagle, *S. nipalensis*, and the changeable hawk eagle, *S. cirrhatus*, are quite well known. The latter has several races and color phases, some crested. The melanistic crested types resemble the African long-crested eagle, *Lophaetus occipitalis*, which is black, with a long crest. The mountain hawk eagle is the Asian counterpart of the African crowned eagle and the South American Isidor's eagle, *Oreaetus isidori*. All are powerful birds capable of killing quite large mammals. Crowned eagles can kill mammals up to 35 lb (15½ kg) in weight; and at the nest they are extremely dangerous.

The martial eagle is the largest, but not the most powerful of all booted eagles. A female

may weigh 14 lb (6.3 kg), and span almost 8 ft (2½ m), as large as, or larger than the largest Siberian golden eagles. Martial eagles are rather shy, living on open plains or savanna and feeding mainly on gamebirds. They seldom kill large mammals and are not dangerous. A golden eagle, *Aquila chrysaetus*, needs about 9 oz (250 gm) of flesh per day, about 7% of the eagle's body weight. A pair and their offspring require about 550 lb (250 kg) of flesh a year, but most kill about 15% more, or about 660 lb (300 kg) altogether, to keep alive. Much of the winter and spring food is carrion, as then there are abundant dead deer and sheep. A pair might possibly kill two or three lambs per year at most. In South Africa, Verreaux's eagle can find so many hyrax in a territory that it has no need to kill sheep. In Australia, the main food of the wedge-tailed eagle is rabbits, the country's worst pest and any lambs taken may well be dead before they are picked up.

All eagles lay small clutches, one or two, the most prolific being the sea and fish eagles, which often lay three eggs. Many, including most snake eagles, lay only one. Eggs are usually incubated by the female, less often by the male, but when two are laid, the male is more likely to incubate. Incubation periods are long, from 43-49 days. This even applies to some small eagles such as Wahlberg's or Ayres', smaller than buzzards, that have much shorter incubation periods.

The eggs are laid, and hatch at intervals of several days. In many species, notably of the true or booted eagles, the eldest hatched young is much larger than those hatched later, which it invariably kills.

EAGLE RAYS, a family of raylike cartilaginous fishes with a whiplike tail usually with a venomous spine at its base. The expanded pectoral fins are used to propel the fish through the water with considerable grace, the tail being held stiffly behind. The eagle ray, *Myliobatis aquila*, of the Mediterranean and eastern Atlantic can be distinguished from similar sting rays of these coasts by its prominent head, which is raised well above the level of the pectoral wings. The back is smooth and brown, the undersides a rather dirty white and the tail black. It grows to about 4 ft (1.2 m) in length and feeds chiefly on mollusks, which are crushed with the powerful pavement of teeth in the jaws. As in all members of this family, the young are born alive, tail first, the poison spine being soft and sheathed in tissue until after birth. In certain areas, such as San Francisco Bay, eagle rays are a considerable pest of commercial clam and oyster beds, which must be protected with fences of stakes.

EARTHWORMS, so called because they burrow in the soil, are widely distributed and often occur in large numbers, for they feed mainly on decaying organic matter within the soil itself, which is usually relatively plentiful. One of the hazards of a terrestrial way of life is the risk of desiccation. Earthworms overcome this largely by burrowing in soil. The cuticle surrounding the worm is very thin and permeable unlike that of, for example, an insect. Consequently, it provides very little check on water loss, although it affords an efficient means of entry of oxygen for respiratory purposes and release of carbon dioxide. Earthworms have some behavioral adaptations, however, which serve to minimize water loss. Usually they confine themselves to damp soil and come out of their burrows only at night, when the drying influence of the sun is past and when the relative humidity is higher than during the day. They rarely emerge completely from their burrows. In hot or dry weather they move deeper into the soil, thereby tending to keep in moist conditions.

An earthworm has no eyes. There are, however, microscopic sense organs sensitive to light, distributed mainly near the ends of the worm, the areas most likely to receive light. There are also minute receptors sensitive to chemicals, touch and vibrations. There are probably taste receptors in the mouth, for an earthworm seems capable of some discrimination and choice over the leaves it pulls into the burrow for food.

The earthworm feeds either on such leaves pulled into the burrow with the aid of its suctorial pharynx, or by digesting the organic matter present among the particles of soil that it swallows when burrowing in earth otherwise too firm to penetrate. Undigested matter is extruded from the anus onto the surface of the soil as the familiar worm casts.

An earthworm moves by waves of muscular contraction and relaxation that pass along the length of the body, so that a particular region is alternately thin and extended or shortened and thickened. A good grip on the walls of the burrow is aided by the spiny outgrowths of the body wall called chaetae. The nervous system consists of ganglia in the head and other ganglia at intervals on a long nerve cord, lying below the gut, which gives off a series of paired nerves, one of the functions of which is to coordinate movement.

Oxygen for respiration is absorbed through the thin, moist cuticle, and passes readily into the blood, for the skin is well supplied with blood vessels. Circulation of the blood is maintained largely by five pairs of contractile vessels called pseudohearts, situated in the front part of the body.

Earthworms are hermaphrodite, each worm producing both sperm and eggs. During pairing, two worms come together, head to tail, and each exchanges sperm with the other. The sperm are stored by each recipient in pouches called spermathecae until after the worms separate. A slimy tube formed by a glandular region, the clitellum, later slips off each worm, collecting eggs and the deposited sperm as it goes, and is left in the soil as a sealed cocoon. In the cocoon the eggs are fertilized, the young worms develop, and eventually escape.

EARWIGS, slender insects, commonly ½-¾ in. (1¼-2 cm) long, with distinctive pincerlike structures at the end of the body (function unknown).

The name "earwig" is probably derived from the fact that it occasionally gets into the ear. If one pokes at an earwig with a finger it may curve its tail over its back and use the pincers (forceps) in a threatening manner. In this attitude it looks rather like a miniature scorpion, but it has no poison and the forceps are not strong enough to hurt. The ear is just another crevice to the earwig. The hind wing

is a semicircular transparent member marked with veins and folds and slightly resembling an ear. The arrangement of veins and folds arises from the fact that at rest the wings are folded beneath the short, leathery fore wings, and this necessitates transverse folds as well as longitudinal pleating.

Earwigs have jaws, which are used for biting and chewing, and they seem to eat almost anything. They are sometimes a minor horticultural pest because some flowers, like dahlias, provide ideal crevices in which to hide. When they are hungry they come out and gnaw holes in the petals, which spoils the flowers from the commercial point of view.

Female earwig.

The eggs of earwigs are laid in batches, and the young that hatch look like miniature versions of the adult except that they have no wings and the forceps are straight. Female earwigs are unusual among insects in displaying some sort of parental care over the eggs and young larvae. Usually, the female rests over her brood, and if the eggs become dispersed she will collect them together again.

ECHIDNAS, also known as spiny anteaters, are four-legged terrestrial animals, rather like hedgehogs in appearance. They belong to a special subclass of the Mammalia, the Prototheria or egg-laying mammals. A large echidna is about 18 in. (46 cm) long and 8 in. (20 cm) wide, rounded on the back, flat ventrally, and weighs over 10 lb (4½ kg)—the heaviest recorded weight is 14.3 lb (6.5 kg). The back is covered with hair interspersed with long sharp spines; the underparts, however, are covered with hair only. There is no neck, and two hairy

holes, one on each side of the head serve as ears; a true cartilaginous pinna is present in all specimens, but it is difficult to detect since it is buried in the musculature of the head. In some specimens, however, the pinna sticks out beyond the surface of the head to form an external ear. The eyes, found at the sides of the head at the base of the snout, are small and beady; the retina is made up of rods only, which probably enables the echidna to see well in shady and even in dark places. There is a short stubby tail devoid of hairs and spines, but there is no scrotum, the testes being internal, as they are in reptiles.

The legs are short and stout, the enormously strong fore feet bearing digits furnished with long spatulate claws. The hind feet also bear five claws, the second of which is always elongated and is used as a grooming claw. The femur is parallel to the ground and widely everted, giving the hindquarters a reptilian appearance. The grotesqueness of its appearance is enhanced by the outward eversion of the feet so that the strongly curved grooming claw actually points backward. The ankle in all males and some females bears a short spur of unknown function. The stoutness and strength of the musculature of the fore limbs is an adaptation for digging in hard earth or breaking up forest litter to expose the ants and termites that comprise their main food, hence the name spiny anteater.

Apart from their fore limbs echidnas are specialized in other ways for living on ants. The snout is elongated into a beak about 3 in. (7.5 cm) long, which houses a long whiplike tongue that can be thrust out 6-7 in. (15-18 cm) beyond the tiny mouth, which is found at the end of the snout. The tongue is smeared with a secretion of the sublingual gland that has the stickiness and consistency of treacle, so that the ants and termites exposed by digging stick to the tongue, which is quickly retracted, with its catch, into the mouth. Once there, the ants are scraped off, when the tongue is thrust out again, against a series of transversely arranged spines on the very long palate. There are no teeth on the jaws, and the ants and termites are pulped by the rubbing action of a set of spines on the base of the tongue against the spines on the roof of the mouth.

There is no separate anus for the passage of the feces, the urine, reproductive products and the feces all passing through a chamber called a cloaca to the exterior.

Echidnas occur only in Australia and New Guinea. In Australia they live in an extraordinary variety of habitats, ranging from the hottest and driest of deserts through humid rain forests to ridges and valleys at an altitude of 5,000-6,000 ft (1,500-1,800 m) in the Australian Alps, where the mean air temperatures for the three coldest months rarely rise above freezing.

The pouch or incubatorium of the echidna appears on the ventral surface in the females at the beginning of the breeding season, which lasts from early July to late September. There is equivocal evidence that the period of gestation in the uterus is 27 days. After this, the egg is deposited in the pouch (no one knows how) and is incubated there. The pouch egg has diameters of about 0.75 × 0.5 in. (16 × 13 mm). At hatching, a little animal looking most remarkably like a newborn marsupial, breaks out of the egg by means of an egg tooth and attaches itself by clinging with its relatively enormous forearms to one of two milk patches or areolae found on the dorsal surface of the pouch. From the areola it sucks milk secreted by the paired mammary glands, which have the many-chambered (alveolar) structure found in the mammary glands of other mammals.

The island of New Guinea harbors another kind of echidna quite different from *Tachyglossus*. This is *Zaglossus bruijni*. It is a very large echidna growing up to 39 in. (1 m) in length, and a weight of about 21 lb (9.5 kg) has been recorded. The snout is proportionately much longer than that of *Tachyglossus*, and it houses an extremely long tongue.

EELPOUTS, a family of blennylike marine fishes related to the wolf eels. They are found in the cold northern waters at all depths ranging from the low-tide mark to more than 5,000 ft (1,650 m). Their common name is derived from their eellike shape coupled with the word "pout," possibly a derivative of the Old Dutch *putt*, or toad, a reference to their toadlike head. The dorsal and anal fins and the tail are continuous to form one fin around the body. The pelvic fins are greatly reduced and lie in front of the bases of the pectoral fins. Many species of eelpouts give birth to live young.

The eelpout or viviparous blenny, *Zoarces viviparus*, of European shores is frequently found hiding under stones or weeds on the shore when the tide has receded. Like the butterfish or gunnel, this species has only a single ovary instead of the paired ovaries found in most fishes. Fertilization is internal, and the eggs are attached by small processes to the ovary walls. They hatch in about 20 days but remain for a further three months inside the mother, and the fry are $1\frac{1}{2}$ in. (4 cm) long when they are finally released. During that time they receive nourishment from secretions from the walls of the ovary. A female of 8 in. (20 cm) may bear 20-40 young, while a large fish may bear up to 300. The eelpout grows to about 16 in. (40 cm) in length.

The largest of the eelpouts is an American Atlantic species, *Macrozoarces americanus*, which reaches $3\frac{1}{2}$ ft (105 cm) in length. This species is not viviparous but lays large eggs, $\frac{1}{4}$ in. (6 mm) in diameter, which are guarded by one or both parents.

EELS, elongated fishes belonging to two suborders of the large order Anguilliformes, namely, the Saccopharyngoidei or Gulper eels and the Anguilloidei, which contains all other families of eels. Eels lack pelvic fins and were formerly placed in a large group, the Apodes (meaning "no feet"), which, however, contained a number of non-eel-like fishes, included because of the absence of pelvic fins. Characteristic of the eels is the long body and the long dorsal and anal fins, which are joined to the tail fin to form a single long fin. Pectoral fins are usually present, but the bones supporting them have lost the connection with the skull usually found in the bony fishes. As a result, the pectorals are often distant from the head, and the gill apparatus and branchial region of the head are elongated. The long and narrow branchial or gill chamber is used as a pump to force water over the gills and out of the sometimes small gill opening. Some species have smooth, naked bodies, whereas in others there are small but deep-set and irregularly scattered scales. There are very many vertebrae, normally 100-200 but sometimes as many as 500-600, and the sinuous body is extremely flexible. This eellike form, which is found in certain other bony fishes not related to the eels, is nearly always associated with a bottom-living, burrowing or crevice-dwelling mode of life. In most of the burrowing eels the fins are reduced or lost, and the tail may become hardened to become a digging tool. Eels are often rapacious feeders, and the jaws and teeth are well developed. Common to all the eels is a thin, leaflike larval form, the leptocephalus stage, from which the eel metamorphoses into the adult after a few months or even after two or three years. The muscle segments or myomeres of the leptocephalus not only remain constant in number when the adult form is attained, but also equal the number of vertebrae. The muscle segments are easily counted in the leptocephalus, while the vertebrae can be counted by dissection or X ray in the adult. Because the numbers of vertebrae are often characteristic of the species, this provides a most useful clue to identifying the leptocephalus larvae of different species.

Colorful moray eels resting in a drainpipe.

The eels are clearly a highly successful group. Modern classifications recognize 23 families of eels together with three families of gulper eels. There are over a hundred genera and several hundred species. The best-known family is the Anguillidae, which contains the freshwater eels, but the three largest families are the Congridae (conger eels), the Ophichthidae (worm eels) and the Muraenidae (moray eels), the two latter essentially tropical in distribution and characteristic of coral reefs. Eels are essentially warm-water species with a distribution that coincides with that of the tropical, shallow-water corals, i.e., bounded by the 68°F (20°C) isotherm. Some species have, however, penetrated into subtropical and temperate waters but return to warmer waters to breed. Probably all eels require a temperature of at least 65°F (18°C) and a salinity of 35 parts per 1000 in order to spawn, even though the adults may live in colder and less saline waters.

EELWORMS, colorless, transparent microscopic worms found wherever there is moisture and organic matter. They inhabit soils in all climates, often to a depth of 5-6 ft ($1\frac{1}{2}$-2 m) in sandy soils. They also live on seashores, in fresh water, sewage beds, vinegar vats, cardboard beer mats, mosses in the Arctic, lichens on walls and trees, in mushroom beds and the tunnels of bark-boring beetles, while some are parasites in insects and others are parasites in plants.

They are usually worm-shaped with a thick, superficially ringed cuticle. Most eelworms are only $\frac{1}{100}$-$\frac{1}{25}$ in. (0.2-1.0 mm) long but a few are $\frac{1}{5}$-$\frac{2}{5}$ in. (5-10 mm) or longer. Some have elaborate cuticular processes on the head but never true appendages on the body. The body wall has four longitudinal muscle bands. There is a nerve ring with associated ganglia around the hind end of the muscular pharynx. The excretory system consists of longitudinal canals leading to a ventral excretory pore in the front third of the body. Sex organs in females consist of single or paired ovaries, spermatheca, oviduct and uterus leading to a vagina and ventral vulva; males have one or two testes and a pair of copulatory spicules in the cloaca.

For movement they need water, which can be a film as thin as $\frac{1}{500}$ in. (0.005 mm), such as that covering the particles of a moist soil. Some survive desiccation for long periods: coiled, quiescent larvae of the stem and bulb eelworm, *Ditylenchus dipsaci*, can remain viable for 20 years in a dry cottonlike mass. *Anguina tritici* forms galls or "cockles" in ears of wheat and may survive in the dry galls for 30 years. The encysted eggs of the cyst-forming eelworms *Heterodera* also survive for many years, whether in moist soil or dry in tubes. Although some eelworms die in flooded soil through lack of oxygen, some survive immersion in seawater.

Eelworms living in the soil face an unusual hazard—that of predatory fungi. The eelworms are active forms that thread their way through the soil particles, and they fall prey to over 50 species of fungi, whose hyphal threads penetrate the eelworms' bodies. There are several ways by which the phyphae penetrate the eelworms. Some fungi form sticky cysts that adhere to the eelworm, then germinate and enter its body. Other fungi form sticky threads and networks, like a spider's web, and the eelworms become trapped on them. Some fungi form lassolike traps, consisting of three cells forming a ring on a side branch of a hypha. An eelworm may merely push its way into the ring and become wedged or the trap may be "sprung," the three cells suddenly expanding inward in a fraction of a second to secure the eelworm. It is interesting that these fungi do not need to feed on eelworms to flourish, and they only develop traps if eelworms are present in the soil.

EGG-EATING SNAKES. The true egg-eating snakes are adapted to living on eggs that are at least twice the size of the snake's gape. Modifications in the structure of these snakes

Echidna or spiny anteater of Australia.

First stage in the swallowing of an egg by an egg-eating snake.

allowing them to engulf, swallow, pierce and crush eggs then regurgitate the shells are to be found in the extreme reduction of teeth, flexibility of the jaws, development of special neck muscles and of long, downward and forward directed spines on some of the vertebrae. These sharp spines, which are tipped with dense bone, penetrate the wall of the gut and form a mechanism for cracking eggs.

Five species occur in Africa south of the Sahara and in Arabia and belong to the genus *Dasypeltis*. One species, *Elachistodon westermanni*, found in northeast India, is exceedingly rare; because it has grooved back fangs and a sensory nasal pit, it may not feed exclusively on eggs. All true egg-eaters lay eggs; there may be a dozen in a clutch. An adult 2 ft (61 cm) long, *Dasypeltis*, can eat a chicken's egg four times the diameter of its own head. Forcing the egg against a loop of its body, the snake may take 20 minutes to engulf it. Once worked into the throat, the egg is forced back, then as the snake arches its neck the shell is pierced by the spines on the backbone. As the snake moves its neck down again, bosslike processes on the vertebrae in front of the spiny ones flatten the egg, and by means of a valve in the esophagus the contents are squeezed toward the stomach, and the shell, compressed into a neat package, is regurgitated.

EGRET, name given to certain herons, usually those with an all-white plumage. The size and body shape of egrets varies considerably. The great white egret is the largest. It has very long legs and measures about 3 ft (1 m) in length including the long sinuous neck and daggerlike bill. Among the smaller species is the little egret, *Egretta garzetta*, which is only half the size of *E. alba* though with similar proportions. The cattle egret differs in having a short neck, giving it a somewhat hunched appearance. The long legs of the *Egretta* species are an adaptation to wading in shallow water; the long neck must have evolved to meet the need to reach the ground when feeding. The cattle egret, which normally feeds on dry land, has relatively short legs and therefore a short neck. Egrets are widely distributed in the tropics and warmer temperate regions of the world. The great white egret has a particularly extensive range, being found in the warmer parts of five continents. The distribution of the cattle egret is particularly interesting and gives us a well-documented example of an animal extending its range naturally. Late in the 19th or early in the 20th century, cattle egrets were found to have colonized part of the northern coast of South America. It is presumed that a party of these birds had crossed the Atlantic from Africa. They increased rapidly in their new home and began to spread. The cattle egret is now abundant and widespread in southeastern North America and in the north of South America and is still spreading. At the other extreme of its range the cattle egret is moving eastward into Australia.

Most egrets nest colonially, other in pure colonies or in mixed colonies with other water birds. In such a mixed colony in East Africa, over 40,000 nests of storks, cormorants, egrets and other herons were counted. These included some 10,000 nests of cattle egrets, over 1,500 of little egrets, and smaller numbers of great white egrets' and yellow-billed egrets' (*Egretta intermedius*) nests. Most egrets build a simple platform of sticks or reeds in trees, or in dense aquatic vegetation. The eggs are usually pale blue or white, and both sexes incubate. During the breeding season, male and female egrets become adorned with long white plumes, which grow from the nape, back or breast. The plumes are frequently erected and displayed during breeding activities and may function in species recognition. The beautiful white feathers were much sought after a few decades ago, when it was fashionable for ladies to wear ornate headgear.

The diet of egrets varies considerably. A number of them catch fishes and frogs, but some of the smaller species are largely insectivorous. The cattle egret is named from its habit of gathering in flocks around domestic stock. The birds walk close to the grazing animals, even perching on their backs, and catch the grasshoppers and other insects they disturb. The habit, which is simply an extension of their association with wild buffalo and other big-game animals, has brought them in close contact with man so that they are often extremely tame.

ELAND, the largest of the antelopes, differs from the kudu and its relatives in lacking preorbital glands, in the presence of horns in the female, a tufted tail, a pendulous dewlap, a tuft on the forehead of bulls, and in the entirely different spiral form of the horns.

A bull eland stands 5½-6 ft (165-180 cm) at the shoulder and weighs ¾-1 ton (700-1,000 kg). There is a short mane along the back of the neck. The dewlap is very long and extends all the way from the upper part of the throat to between the fore legs and is tufted for part of its length. The shoulders are slightly hunched, the neck is longer than that of an ox and the head is held higher than the withers. The general color is reddish-brown to buff, becoming a smoky blue-gray in old bulls. Most races have white stripes on the body. The nose

Smaller or lesser egrets, *Egretta intermedia*, with, on either side, openbill storks (*Anastomus lamelligerus*).

Common eland in East Africa. These are the largest antelopes, and both male and female have horns.

is dark, often with a white chevron between the eyes, and bulls have a dark thick mat of hair on the forehead.

The common eland, *Taurotragus oryx*, of East and South Africa has pointed ears. The horns are comparatively short, with two spiral twists near the base, which are so close that they overlap.

The giant eland, *Taurotragus derbianus*, is found from Senegal east to the Nile in Uganda and the Sudan. The giant eland has longer horns, which are as much as twice the length of the head, a large white spot on each cheek and broad, rounded ears with a dark bar on the inside. The neck is darker than the body, often bordered behind by a white stripe, and the dewlap is more extensive, beginning just behind the chin.

Lord Derby's eland is an extraordinarily beautiful animal, its dark rufous and black coloration contrasting well with the white stripes, nose stripe and cheek spots. It is almost extinct, being represented perhaps by a few dozen animals in two isolated areas.

Elands inhabit open forests and bush country, where they browse, breaking down high branches with their horns. They live in herds of 12-30, but sometimes gather into groups of 100 or more. Each herd has one or two bulls,

but many bulls are solitary. The herd is continuously on the move, restlessly moving at a fast walk, its members snatching food as they go and walking in single file. They are usually quite silent as they move, but occasionally a bull makes a low grunt to signal to the rest of the herd, and calves bleat to their mothers. During the dry season elands will dig up bulbs with their hoofs and eat melons, and at this time the big migrating masses are formed. When alarmed, they pause, then gallop away excitedly, often leaping over each other's backs. They have been known to clear obstacles at least 6 ft (2 m) high. They are placid and rarely defend themselves.

Elands, at least the females, become sexually mature at two years. The female's cycle lasts three weeks, during which she is in heat for three or four days. Gestation lasts nine months, and calves are born at different seasons in different parts of Africa.

ELECTRIC FISHES occur in several groups of quite unrelated fishes, which have independently evolved the ability to discharge an electric current. The two principal uses to which this is put are to incapacitate other creatures (for defense or feeding purposes) or to receive information about the environment in the manner of radar. At first sight, the

ability of an animal to generate electricity seems strange. In fact, it is merely an extension of the normal operation of muscles and nerves. Every time an impulse passes down a nerve to stimulate the contraction of a muscle a tiny electric current is involved. In the electric fishes, some of the muscles have lost their power to contract but have increased their electrical power. The size and number of nerve endings, where the electricity is normally released, are increased, while the electrical units are arranged to multiply the discharge from unit to unit so that a battery effect is achieved. The electrical discharge does not affect the fish itself, largely because of the great insulation around the nerves. The discharge of the organ is controlled by the brain.

Among the cartilaginous fishes, the best-known species are the electric rays (*Torpedo* spp.). These are Mediterranean and subtropical Atlantic fishes with a round, disc-shaped body and short tail, found over sandy or muddy bottoms. The largest is the black electric ray, *Torpedo nobiliana*, which may reach 5 ft (1.5 m) and weigh 100 lb (45 kg). The electric organs are in the winglike pectoral fins. They can produce a current of 200 volts, quite enough to stun small crustaceans and fishes on which they feed.

The best known of the bony fishes is the electric eel, *Electrophorus*, found in the Amazon basin. It is not a true eel, but is related to the carps and characins. This fish can grow to more than 6 ft (1.8 m). The body is elongated, but about $\frac{7}{8}$ of the total length is tail, the alimentary canal, heart, liver, kidneys and gonads being crowded into the first $\frac{1}{8}$. The "tail" part is largely occupied by the electric organs. These fishes are able to emit two kinds of discharge, a high voltage (over 500 volts) for stunning prey, and a much weaker regular pulse used as a direction finder and indicator for locating objects in the vicinity.

The electric catfish, *Malapterurus electricus*, found in the Nile system, the Zambesi and certain of the African lakes, has the Arabic name of *raad* which means "thunder." The electric organs are in the muscles over much of the body. These fishes grow to about 4 ft (120 cm) in length and can discharge 350 volts.

The electric organs found in the mormyrids or elephant-snout fishes are located in the caudal peduncle, the muscular base to the tail. In these fishes a continuous stream of electrical impulses is discharged from the organ, at a variable frequency (lowest when the fish is resting but rising to 80-100 impulses per minute if the fish is disturbed). The fish thus surrounds itself with an electromagnetic field, and any electrical conductor that enters the field will bring a response. In this way, fishes living in muddy waters can detect the presence of prey or of predators.

The knife fish, *Gymnarchus niloticus*, related to the elephant-snout fishes, is shaped rather like a compressed eel and also uses its electric organs for detection of prey and predators. It has rather poor sight and lives in muddy waters.

There is only one group of marine electric bony fishes, and this is the stargazers, bottom-living fishes in which the electric organs are located in deep pits behind the eyes. These organs generate currents of up to 50 volts, and this may be used both for defense and for stunning prey.

ELEPHANT, the largest land mammal, two species of which are known: the African and the Indian. The African elephant, *Loxodonta africana*, found only on that continent, has two subspecies, the bush elephant (subspecies *africana*) and the forest elephant (subspecies *cyclotis*), the former being larger, more abundant and better known. The bush elephant is the largest of the elephants, the female reaching an average mature height of 8 ft 4 in. (2.5 m) and the males, owing to a postpubertal growth spurt, 10 ft 2 in. (3.1 m). Forest elephants are about 2 ft (0.6 m) smaller, and the difference between the sexes is less pronounced. The body weight of an average mature female bush elephant is 5,900-7,700 lb (2,700-3,500 kg) and the male 10,000-11,700 lb (4,500-5,300 kg). There is a large seasonal variation as well as a variation in weight between populations according to the state of the habitat. The overall length of a large male (trunk to tail) is up to 27 ft (9 m).

In order to support this weight, the limbs are massive columns and are so constructed that the elephant cannot run or jump. The limb bones are heavy and have no marrow, and the soles of the feet cover a fatty cushion, which helps to distribute the load evenly. They usually have five and four toenails on fore and hind feet respectively, but they are reduced to three on all feet in some individuals. The ears are large, shaped like a map of Africa and measuring up to 3 ft × 5 ft (1 × 1½ m). They are important in thermoregulation and in aggressive behavioral displays. The ears of the forest elephant and the Indian elephant, *Elephas maximus*, are very much smaller, no doubt related to their more shady habitat. The skull is huge, being modified to support the tusks, which are rooted in large sheaths formed from the premaxillary bones. The brain case is massive, with walls that are thick but cellular in structure, to give strength with lightness.

The tusks first appear at two to three years of age. They are upper incisor teeth composed of

African elephant picking fruit, and demonstrating one of the many uses to which the trunk can be put.

African elephants playing in Lake Edward. Bathing sessions are healthful and offer an opportunity for elephantine fun.

dentine with a very small, 2-in. (5-cm) enamel cap that is quickly worn away. Their shape follows an equable spiral and they continue to grow in length at the rate of $3\frac{1}{2}$-$4\frac{1}{2}$ in. (9-11.5 cm) a year throughout life, but owing to breakage and wear they only reach about half their potential length. The average lengths of the tusks of the oldest males and females are about 8 ft (2.5 m) and 5 ft (1.6 m) respectively. Their rate of growth in weight increases progressively with age, and male tusks are much more massive than female tusks, reaching an average paired weight of 240 lb (109 kg) as compared with 39 lb (18 kg) in females. The world record single tusk weight (from East Africa) is 235 lb (107 kg) for males and 56 lb (25 kg) for females. In some populations, for example in Zambia, tuskless elephants are not uncommon.

The other teeth are also unusual. Because of their longevity and continued growth in size throughout life the elephants need a series of teeth, functionally covering their life span and increasing in size as the animal grows larger. This is achieved by having a series of six teeth in each side of each jaw (24 in all), which are formed and replace each other in succession throughout life. No more than one (or two) in each series are wholly (or partly) in wear at any one time, and they are progressively larger from the first to the sixth. The teeth themselves are unique, being constructed of a series of flat, vertical plates of dentine and enamel, held together by a matrix of cementum. The average number of these plates or laminae increases from 3 in the first tooth to 13 in the sixth.

The grinding area of the teeth in use in each jaw in a nine-month old calf is only 1.5 sq in. (9.4 cm²) and reaches a maximum of 50 sq in. (320 cm²) and 40 sq in. (260 cm²) in males and females respectively in their late forties. Subsequently, as no more teeth are produced, wear results in a reduction of the grinding area to 16 sq in. (100 cm²), and they are unable to feed efficiently and death due to "mechanical senescence" occurs. This sets a limit to the elephant life span of 60-70 years. The replacement and wear of the teeth can be used as criteria for estimating the age of elephants with surprising accuracy.

The senses of sight and hearing are only moderately developed, in contrast to the sense of smell, which is acute. Smell plays an important part in their social contacts, and the trunk is used to locate scents precisely. The calf is suckled by its mother for at least four years, but this is extended to six years or more in populations with low reproductive rates. The fat content of the milk increases with the age of the calf.

About half the mature bulls are solitary, and the remainder collect in bull herds usually containing 2-15 animals but occasionally over 100. Single bulls and bull herds form temporary associations with family units, and the family units may aggregate to form bigger herds, either an extended family of up to 20-30 fairly closely related animals, which frequently recombine, or chance aggregations of up to 100 or more. Close ties exist between members of family units, and examples of elephants supporting and assisting wounded companions are known. Large herds of as many as 1,000 elephants are found in certain situations, usually at the periphery of populations that have been displaced by human settlement or activity, or are otherwise in conflict with man. These often form spectacular tight-packed cohorts, leaving a trail of destruction in their wake.

The daily cycle is one of fairly continuous movement in search of food or water, and larger groups tend to be more mobile. In savanna regions, the hot hours of the middle of the day are often spent resting in the shade—and the destruction of shade by the elephants' own activity may lead to an increase in the mortality rate, especially of young calves, from heat stress. Elephants eat food equivalent to about 4% of their body weight daily, and cows with suckling calves about 6%. For a large bull this amounts to some 600 lb (270 kg). The preferred diet contains only about 30-50% grass, the greater amount being taken in the wet season.

The African elephant (note the large ears).

There is much less information on the biology of the Indian elephant, *Elephas maximus*. It differs from the African elephant in its shape (arched back and domed head), smaller tusks, often absent in the female, and much smaller ears. The end of the trunk has only a single process as compared with the two "fingers" of the African genus. In general it appears to be an animal of jungle or bush country, although it is found in grassland areas. Another point of interest is that, in contrast to the African elephant there is a progressive loss of pigment with age from the trunk and ears, which consequently develop pale patches. Albinos are also probably more frequent.

This species is now found in India, Assam, Burma, Siam, Malaya, Sumatra and Ceylon, with a few also in Borneo. They seem to frequent as wide a range of habitats as the African species.

ELEPHANT SEALS, largest of the earless seals. There are only two species: the northern, *Mirounga angustirostris*, and the southern, *Mirounga leonina*.

The northern elephant seal occurs on the islands off the coast of southern California and Mexico. Stray animals may be found much farther afield and have been reported from British Columbia and Alaska.

Breeding colonies of the southern elephant seal are to be found on the subantarctic islands. The largest breeding population is on South Georgia, where there are about 310,000 animals.

In size and general appearance the two elephant seals are similar. The adult male may be 16-20 ft (4.8-6 m) in nose to tail length and up to about 8,000 lb (3,628 kg) in weight. The adult female is smaller, being 10-12 ft (3-3.6 m) in length and weighing about 2,000 lb (907 kg). The pup at birth is about 4 ft (1.2 m) in length and 80 lb (36 kg) in weight.

The inflatable snout or proboscis of the adult male is its most characteristic feature. This is an enlargement of the nose and has the normal internasal septum and two nostrils. When fully developed it overhangs the mouth in front so

Indian elephant with its young bathing, a favorite and necessary routine for all elphants. It is distinguished from the African elephant by its much smaller ears.

African elephants at a water-hole. The long trunk, an elongated nose, enables the elephant to drink without kneeling.

that the nostrils open downward, and its elongated, cushionlike shape is marked by two transverse grooves. During the breeding season it can be erected by inflation, muscular action and blood pressure and may act as a resonating chamber to increase the volume of a big bull's roar.

The breeding season of the southern elephant seal starts at the beginning of September. The bulls come ashore first, followed by increasing numbers of females from about the middle of the month. By the end of the month there are enough animals present to form harems, each male presiding over a small number of females at first, later increasing his harem up to 30-40 females. Each harem has one dominant bull whose function is to mate with the females of that harem, and prevent other bulls from doing so. Younger mature bulls hang about the edges of the harem, sometimes managing to steal a female. A challenger to a reigning harem bull may threaten by roaring for some time. The reigning bull may then retreat without a fight or may stand his ground.

Most of the pups are born in October and are 4 ft (1.2 m) long and clad in black woolly hair. Their mothers feed them for about 23 days, and toward the end of this period each pup will be putting on about 20 lb (9 kg) a day, while the mother, remaining on land without feeding until the pup is weaned, may lose 700 lb (317 kg) during lactation. Elephant seal milk is of the very rich sea type, containing 40-50% fat (compared with 9% in a dog and 3.5% in

man), and enables the babies quickly to put on a thick layer of blubber as a protection against the cold. At about 35 days old, the pup sheds its black coat for a silvery gray one and is ready to enter the sea. The cows mate again about 18 days after the birth of the pup, and after it is weaned. They go off to sea to feed.

After the breeding season, the next gathering of adult animals is for molting, in December, January and February. The molt takes about 30-40 days, and again no food is taken during this time. At the molt large sheets of skin are shed with the old hairs embedded, pushed out by the developing new hair, instead of the more

Molting southern elephant seal on Bird Island, south Georgia, in December.

normal method, which consist of shedding single hairs.

Elephant seals have few enemies besides man, though killer whales and leopard seals may take the young of the southern elephant seal, and killer whales and sharks the young of the northern species.

ELEPHANT SHREW, a small mouse- or rat-sized animal confined to Africa.

The most distinctive feature is the long, pointed proboscis, adapted for nosing out ants, termites and other insects. Otherwise, they have a fairly normal mouselike build, adapted for running swiftly on the ground. The hind legs and feet are longer than the front, hence the alternative name of jumping shrews. This is, however, a misnomer, for they do not jump bipedally like jerboas but run with lightning dashes, using the powerful hind legs for thrust as in a hare. The eyes are large and bright, adapted to daytime vision. The tail is usually as long as the head and body and almost naked.

Most of the 15 or so species have uniform brown or gray fur on the upper side, usually closely matching the color of the local soil. Living in open grassland or semidesert and being diurnal, they need protection from aerial predators as they forage for food, usually within sprinting distance of shelter in the form of rocks or bushes. One group, however, the large forest elephant shrews, *Rhynchocyon*, are much more boldly patterned in black and rufous, or finely checkered to match the dappled background of the forest by day.

The reproduction of elephant shrews shows several peculiarities. The litter size is almost always limited to one or two, the young being very large and well developed at birth. Most mammals shed only a small number of eggs from the ovaries at each ovulation. Most elephant shrews do the same, but some species, very closely similar in every other respect, are remarkable in regularly releasing over 100 eggs at each ovulation in spite of the fact that the uterus is not adapted to allow more than two to develop. The significance of this remains a mystery.

Elephant shrews are found throughout east, central and southern Africa, but one isolated species, *Elephantulus rozeti*, is found in Morocco and Algeria.

ELEPHANT SNOUT FISHES, freshwater African fishes related to the bony tongues in some species, such as *Gnathonemus numenius* and *G. curvirostris*. The snout is elongated and turned downward, much like an elephant's trunk, with the small mouth at the tip. In other species only the soft lower lip is elongated (e.g., *Gnathonemus petersi*), and in yet others the snout is bluntly rounded with no elongation. In spite of this, all the members of the family Mormyridae have a quite unmistakable look, with smoothly scaled bodies, dorsal and anal fins set opposite to one another, the body brownish gray or slaty gray and a rather delicate forked tail on a slender base (caudal peduncle). There are about 100 species found in the lakes and rivers of Africa, mostly feeding on invertebrates at the bottom. Most are small, but some reach 5 ft (1.5 m) in length.

A number of species are found in the Nile, and these seem to have fascinated the ancient Egyptians, who depicted them in tomb draw-

ings, mummified them and produced amulets in the form of mormyrids.

Relative to body weight, the mormyrids have very large brains, the ratio of brain weight to body weight being about the same as in man. Mormyrids often live in murky waters, and their eyesight appears to be poor. In compensation, the muscles at the slender base of the tail are modified into electric organs, which can detect obstacles or predators. A small electric field is set up around the fish that acts in the same way as radar.

EMU, *Dromaius novae-hollandiae*, the largest bird inhabiting the Australian continent, is flightless, and its tiny wings are only $\frac{1}{10}$ of the length of the bird's body.

Emus are brown, although when the feathers are new after the molt they may appear nearly black, fading to pale brown with age. The bases of the feathers are white. Each feather has two identical shafts, with the barbs so widely spaced that they do not interlock to form the firm vane as do the feathers of most birds. Rather, they form a loose, hairlike body covering. The skin on the neck and head is often free of feathers and has a more or less bluish tinge. The sexes are similar in plumage except in the period prior to egg laying, when the female's head and neck are densely covered with black feathers, whereas the male's head and neck are largely bare.

Adult females weigh about 90 lb (41 kg) and males 80 lb (36 kg). The female of a pair is usually larger than the male. The legs are unfeathered and so long that a running bird can make a stride of 9 ft (2.7 m) with ease. Emus have three toes, compared with the two of the ostrich, *Struthio*, and the underside of each toe is flattened with a broad pad. The bill is broad and soft, adapted for browsing and grazing but with muscles too weak to hold any smooth heavy object.

The birds usually breed in May-August. The nest is a low platform of twigs or leaves, generally placed so that the sitting bird has a clear outlook, often downhill. The early eggs of the

clutch are covered and left, and the male does not begin sitting until between five and nine eggs have been laid. A hen will lay from 9-12 eggs, each weighing 1-1½ lb (0.5-0.7 kg) but in very good seasons the clutch may exceed 20, and in poor seasons be as low as four or five. Once the female has laid the clutch the male carries out the whole incubation process. Incubation takes about eight weeks from the time he begins to sit, and during this time he hardly eats and does not drink.

The tiny chicks leave the nest after two or three days. At first their plumage is cream with brown longitudinal stripes, and dark dots on the head.

Emus feed mainly on fruits, flowers, insects, seeds and green vegetation. Caterpillars are favored whenever they are available, and beetles and grasshoppers are taken in large quantities when they are abundant.

ENGRAVER BEETLES, or bark beetles, are closely related to both the weevils and the ambrosia beetles. The females burrow through the bark of trees, such as pine or elm, and then they excavate a large chamber or egg gallery. Along the edge of this chamber eggs are deposited, each of which hatches into a wood-eating larva. These larvae tunnel into the wood to produce a burrow which increases in diameter as the larvae grow in size. When fully grown the larvae pupate and hatch into adult beetles, which then have to burrow their way out of the timber. The adults then mate and infest a new tree.

Many of these beetles cause serious damage to timber, and one species transmits the fungus disease Dutch elm disease, which is fatal to those trees it infects.

EYED LIZARD, *Lacerta lepida*, also known as the Spanish or ocellated lizard. It lives in southern France and the Iberian peninsula and grows to 2 ft (60 cm), of which 16 in. (40 cm) is tail. It is brownish-green to reddish with black spots, which sometimes form rosettes with black centers. On the flanks are bluish oval markings, the so-called eyes or ocelli.

Elephant-shrews in a German zoo. Their alternative name of jumping shrews is a misnomer.

FAIRY SHRIMPS, crustaceans differing from other Branchiopoda in lacking a carapace and in bearing stalked eyes. The body is elongated and cylindrical. The anterior part, or thorax, usually consists of 11 segments bearing limbs, but in *Polyartemiella* there are 17 and in *Polyartemia* 19 such segments. Behind the thorax there are two genital segments, which in the female bear a brood pouch on the underside. The terminal part of the body, or abdomen, consists of five or six segments, together with the telson, which bears a pair of caudal rami. The normal swimming position for a fairy shrimp is with the back downward, and some, such as *Chirocephalus*, can adjust their limb movements so that the animal appears to hover in the water.

The male fairy shrimp has enlarged and often extremely complicated antennae, which serve as grasping organs when it transfers sperm to the brood pouch of the female. The male approaches the female from the side, seizes her in the region of the brood pouch and then bends his body around hers so that his own genital region comes into contact with the opening of the brood pouch.

Most fairy shrimps lay a single type of egg, with a tough outer coat and a capacity for resisting both freezing and drying. This is an adaptation to the usual habitat of fairy shrimps because most of them live in temporary ponds that either dry up or freeze solid according to the location. One species, *Branchinecta paludosa*, is widespread in arctic regions, where it lives in small pools that are frozen for six months or more of the year. This species also has two interesting relict populations, one in the High Tatra mountains of Czechoslovakia and Poland and the other in the Rocky Mountains.

FALLOW DEER, *Dama dama*, widely distributed in Europe and the countries bordering the Mediterranean, is the typical park deer, and at the end of the last century in England alone, over 71,000 fallow deer were being preserved in parks. They have also successfully been introduced to Australia, Tasmania, New Zealand and North and South America.

Standing about 36 in. (91 cm) high at the shoulder, the typical feature of the buck is the palmated antler. There are also more color variations in fallow deer than in any other wild mammal, and these include black, white, menil, cream, sandy, silver-gray and the normal fallow deer which is spotted in summer, but has little or no spotting in winter.

The rut usually takes place during October, at which time the bucks grunt huskily.

Another species of fallow deer is the Persian fallow deer *Dama mesopotamica* which has always had a rather limited distribution, and occurs only in Iran, where its total population is probably less than 50.

Only slightly larger than the European fallow deer and with spotted summer pelage, the antlers of the bucks are the main point of difference, the brow being very short, with a long tray point sprouting close to it. The upper points never palmate to the same extent as those of the European fallow deer.

FALSE KILLER WHALE, *Pseudorca crassidens*, a small whale closely related to the killer whale but with an all-black body and a small dorsal fin. First known from a half-fossilized skull found in 1846, the false killer whale has been known mainly from stranded specimens, until recent years when it has been kept in captivity. See killer whale.

FANWORMS, together with the feather-duster worms, are surely the most elegant of all the marine polychaete annelids. Pinnate or branched filaments radiate from the head to form an almost complete crown of orange, purple-green or combination of colors. The crown is developed from the prostomium, the anterior unsegmented region in front of the mouth, by a process of elaboration and subdivision. It forms a feeding organ and, incidentally, a gill. The remainder of the body is more or less cylindrical. All fanworms secrete close-fitting tubes in which they live. Typically, these consist of fragments of shell, sand or small stones or other debris cemented together with mucus secreted from a specialized building organ located just below the mouth at the base of the crown on the ventral side. *Myxicola* secretes an entirely gelatinous tube consisting of a mucus, which takes up water to form a thick protective jelly in which the worm lies and into which it can retract its crown.

FEATHER STARS, a group of echinoderms related to the sea lilies. In contrast to the sea lilies, which are attached to the substratum by a stalk, the feather stars are free living, stalkless and can swim with the aid of their long arms, although they are sedentary forms and most often are encountered sitting inactively on the sea bottom. The common genus *Antedon* is an example of this group.

FERRET, *Mustela eversmanni furo*, a domesticated form of the Asian polecat, a species usually smaller and lighter in color than the European polecat, *Mustela putorius*, but having similar behavioral and ecological traits. The ferret has now been spread throughout Europe,

interbreeding with the endemic wild polecat populations until the one can hardly be distinguished from the other. Because of a larger cranial capacity, the true polecats have rounder heads and "pop eyes" which the crossbred polecat-ferrets do not have.

FIDDLER CRABS. The male fiddler crab, *Uca*, has one of its pincers very much larger than the other. The general form of the body is rectangular, and the eyes are borne on stalks that can be folded sideways to lie flat against the front of the head. The small pincer is used in feeding. Sand is picked up and passed to the mouthparts, which are beset with spoon-shaped spines. After a number (usually 6-16) of small pincersful of sand have been passed through the mouthparts, a ball of sand is formed behind the mouth. This is then removed by the small pincer and placed carefully on the sand surface. This results in the sand surface around a feeding crab being dotted with small balls of sand. In some species they become arranged in a fairly regular pattern.

Fiddler crabs make burrows in intertidal sand

Bispira volutacornis, twin fanworm, with twin spiral fans exposed at mouth of tube.

and mud. When the tide comes in the burrows are usually, but not always, plugged up. Sometimes the plug takes the form of a small dome. Some species also build sand shelters over the entrance to the burrow.

The large pincer of the male is used in combination with various leg movements to produce a display that is characteristic for each species. In general, the movement of the pincer looks like a beckoning wave, and is generally considered to be a signal to any passing female that the male is desirous of mating. The males often have contrasting coloration, and this may change and become intensified during display. The display is also used to warn other males to keep away from the burrow and display ground.

FIERASFERS, also known as pearlfishes or cucumberfishes, elongated marine fishes, related to the cods that live inside sea cucumbers. In England they are often called fierasfers, derived from their former scientific name meaning "shining beasts." The most striking feature of these fishes is their habit of entering any small crevice tail first. Some species are very particular and *Carapus bermudensis* will only live in one species of sea cucumber, while others, such as *C. homei*, will live in any shell or sea cucumber. Although sea cucumbers are hollow a certain strategy is required in converting the animal into living quarters. When a young cucumberfish finds its host it searches for the anus and pushes its way in. As the fish becomes larger it tends to enter the sea cucumber tail first. In many other associations between two animals each partner gains something from the relationship and it is because of this advantage that the association has evolved. In the case of the cucumberfishes, the host seems to gain nothing and may even suffer unintentional damage to its organs. Some cucumberfishes are not above nibbling at their hosts while inside. *Carapus apus* from the Mediterranean is believed to spend its entire life inside the sea cucumber, presumably feeding on its host. This may not do as much damage as might be thought since the sea cucumbers have great powers of regeneration.

The eggs of the cucumberfish float at the surface and the young fish do not closely resemble their parents. While still young they search for a suitable home. The body of these fishes is naked and in some species the pelvic fins are lacking.

FIGHTING FISHES, species of fishes belonging to the genus *Betta* and members of a family of labyrinthfishes. The Siamese fighting fish, *Betta splendens*, is chiefly found in Thailand but occurs also throughout the Malayan Peninsula. Because of the pugnacity of the male, these fishes have been "domesticated" for a considerable time in Thailand and used for sport, wagers being laid on the outcome of a fight between two contestants. In the wild, the dorsal and anal fins are short and the color of the body is variable but dull. As in the case of the goldfish and many other fishes kept in captivity, however, special varieties have been bred that have long fins and vivid colors. The males are always more spectacular than the females.

During a fight between two males, the fins are spread as far as possible and the mouth and gill covers are opened wide.

Male fiddler crab has an enormous right claw used in courtship and in signaling occupation of a territory.

FINCHES, mainly arboreal, seed-eating, songbirds, with nine large primary feathers in the wing and twelve tail feathers, in which the female builds an open cup-shaped nest and is responsible for incubation. Usually, both the incubation and fledging periods last 11-14 days. Most species have sweet melodious songs.

Members of the subfamily Fringillinae feed their young on insects and hold large territories while breeding. There are three species, the chaffinch *Fringilla coelebs*, brambling *F. montifringilla* and blue chaffinch *F. teydea*, all of which are about 6 in. (16 cm) long and weigh around 1 oz (20-30 gm).

The Carduelinae form the largest branch of the finch family, with about 122 species. They are more specialized seed-eaters than the Fringillinae and feed their young mainly on seeds, sometimes supplemented with insects. They nest either solitarily or in loose colonies and feed away from their nests in flocks. Many feed directly from plants and are adept at clinging to stems or hanging from twigs. They show considerable variation in bill-shape, an adaptation for extracting the seeds from different types of seedheads. The largest are the *Mycerobas* grosbeaks of the Himalayas, which reach about 8 in. (20 cm) in length and 3½ oz (100 gm) in weight. The smallest is probably Lawrence's goldfinch *Spinus lawrencei* of eastern North America, which is about 4 in. (10 cm) in length and about ⅓ oz (8-11 gm) in weight. The cardulines vary greatly in color, but greens, reds, yellows and browns are prevalent, and the female is usually duller than the male. The eggs are usually whitish with pale brown spots and streaks.

Most species are found in temperate regions, with a few in the Arctic, deserts, tropics and subtropics. The Palearctic holds about 68 species in 21 genera; Africa about 36 species in nine genera; and the New World about

25 species in eight genera. But six of the New World species are also found in the Old World. The only regions where these finches do not occur are Madagascar, Antarctica and the south Pacific Islands; they have been introduced into Australia and New Zealand.

FINFOOTS, odd, grebelike birds related to the rails. The family is made up of three species, one in South America, another in Africa and a third in southern Asia.

The rather long but stout bill suggests a rail, the long neck and short legs placed well back on the body with flattened lobes bordering the toes are grebelike, while the long thin neck, low swimming position and longish graduated tail with stiff rectrices also suggest a cormorant or darter. The legs, placed well back for propulsion, give a horizontal body posture on land, but finfoots can run well and take to land when pursued. They swim with head-bobbing movements, presumably synchronized with simultaneous kicks of both legs.

They show a preference for still or slow-moving water at the edges of rivers, estuaries, or occasionally lakes, where thick vegetation comes to the water's edge or overhangs it. They can, however, cope with swift-moving water if necessary. They mostly keep to the cover of overhanging vegetation, and are adept at scrambling over low branches as well as swimming. When alarmed, they hurry into cover, at times fluttering along the surface with the help of the wings. The wings, although well developed, do not appear to be used for prolonged flights. Finfoots eat small aquatic animals mainly.

FIREFLIES, 1,100 species of nocturnal beetles famous for their ability to produce light. They are rather elongated, with nearly parallel sides and flexible wing cases. Although most species are tropical, about 60 occur in North America,

Lesser flamingos on Lake Nakuru, Kenya.

but only two, one of which is the glowworm, occur in Europe. In many species the male is fully winged and has huge compound eyes, whereas the female is commonly without wings, has small eyes and resembles a segmented larva. The adults feed very little, but the larvae are carnivores preying on snails and slugs.

The photogenic organs of the firefly are situated in certain segments of the abdomen, sometimes the thorax, and emit a yellow-green light or luminescence. Light is produced instantaneously when the substance luciferin in the photogenic organs is oxidized to oxyluciferin by atmospheric oxygen in the presence of water and the enzyme luciferase. This remarkable system of light production is extremely efficient, as about 95% of the energy released when luciferin is oxidized to oxyluciferin is in the form of light rays. In contrast, the sun produces about 35% light, and an electric lamp only gives out 10% of its energy as light. Some fireflies emit a continuous glow, whereas others show periodic flashes. The light mutually attracts the sexes for mating.

FISH LOUSE, not a louse, but a crustacean parasitic on fishes and remarkable for its disclike body and a pair of large adhesive suckers with which it clings to host fishes. The head bears a pair of sessile compound eyes, and the underside of the front part bears numerous triangular spinules. These spinules point backward and help the louse in adhering to the host. Members of the genus *Argulus* align themselves so that the head points in the same direction as that of the host and the spinules catch firmly in the fish's skin.

The fish louse has four pairs of swimming legs, so that it can swim actively to seek out its host. Once on the host, the louse clings by means of its suckers and spines and pierces the skin with its narrow mandibles, which are housed in a proboscis on the underside of the head. The physical damage inflicted by the mandibles is probably not very great, but the small wounds sometimes become infected by bacteria or fungi, and these may eventually kill the fish.

FLAMINGOS, a family of large, brilliantly colored, aquatic birds, which inhabit alkaline and saline lakes and lagoons of the Old World (except Australia), North and South America and some oceanic islands, including the Galápagos. They prefer to live in hot, dry regions, avoiding cool, moist, forested areas; but in the Andes they are found on freezing alkaline lakes at 14,000 ft (4,250 m).

All flamingos are large, 3-6 ft (1-2 m) in length, with long sinuous necks, long legs and webbed feet. The bill is highly specialized for filter feeding, sharply bent in the middle, with the lower mandible large and troughlike and the upper one small and lidlike. The plumage is pink, red and black, often brilliant. No other large gregarious birds are quite so colorful. The enormous flocks of lesser flamingos, *Phoeniconaias minor*, seen at some East African alkaline lakes, are probably the most remarkable of the world's bird spectacles.

Phoenicopterus, the greater flamingo, differs from the other two genera in its much larger size—large individuals standing nearly 6 ft (2 m) tall—and in the structure of the bill. In *Phoenicopterus* the upper mandible has a shallow internal keel, quite different from the deep triangular keel of *Phoeniconaias* and *Phoenicoparrus*. *Phoenicopterus* is adapted for feeding upon relatively large organisms on the bottom and in deep water, whereas the deep-keeled species are adapted for feeding on the surface or in shallow water.

Phoeniconaias contains only the lesser flamingo, *Ph. minor*, which is mainly African but does also occur in Asia. It is the smallest, least brilliantly colored, but by far the most numerous of all flamingos. There may be as many as

4-5 million lesser flamingos, of which about 3-3½ million inhabit East Africa.

The Andean flamingo, *Phoenicoparrus andinus*, and James's flamingo, *Ph. jamesi*, inhabit alkaline or saline lakes at the high altitudes in Peru and Bolivia. They breed only on these highland lakes. The Andean flamingo is relatively common, totaling perhaps 100,000, but James's flamingo is by far the rarest of all flamingos.

All flamingos feed by filtering small animals or microscopic plant life from the mud or water. The bill structure is fundamental to their method of feeding. In all species it is sharply bent in the middle, so that when the flamingo is walking and feeding in the water, the upper mandible is underneath and the lower uppermost. In the James's and Andean flamingos, the lower mandible is bulbous and full of cellular bone and may actually act as a float, helping the bird to feed steadily in choppy water. In all species the basic filter feeding method is similar. Water and mud are sucked into the bill, and special structures within the bill catch the small organisms on which the flamingos feed, while unwanted material is rejected.

Both sexes incubate, in all observed cases. The flamingo does not incubate, as in old travelers' tales, with the long legs hanging down beside the mud mound nest, but with them doubled under the bird and projecting behind like red drumsticks. The incubating flamingo continually bickers with its neighbors, picks up pieces of nest material, or may threaten a passing bird with a territorial display in which the feathers are raised so that the bird resembles a huge chrysanthemum. From time to time it will rise from the egg, move round the nest, then settle to incubate again. Incubation takes about 28 days in both the greater and the lesser flamingos, despite the difference in size.

The newly hatched chicks are clad in soft, silky gray down; occasionally albinos occur. The legs are swollen, soft and bright red. For the first few days the chick does not leave the nest mound; it may be unable to climb back onto it

Flamingo feeding chick on nest.

Profile of a flounder, from left side.

if disturbed. The legs harden and become blackish in four or five days, and the chick then becomes more active.

FLEAS, small blood-sucking wingless insects with streamlined, laterally flattened bodies, which are hairy and shiny and varying in color from yellowish-brown to almost black. Fleas that live on animals having dense fine fur are sleeker and more streamlined than those living on coarse-coated hosts. Other fleas that spend most of their adult life in the nest of the host (nest fleas) may lack eyes or have reduced powers of jumping. In "body fleas" the eyes are usually well developed, unless these fleas are parasites of wholly nocturnal hosts, such as bats, or of subterranean hosts. The short antennae lie in deep grooves, one on each side of the head, and are erectile in the male only (their function, in nearly all species, includes grasping the female during copulation). The mouthparts consist of two pairs of palps, while three stiletto-shaped parts together form the piercing-sucking tube. Each of the three thoracic segments bears a pair of legs that are modified for clinging and, especially the hind legs, for jumping.

FLOUNDER, *Platichthys flesus*, one of the best known of European flatfishes. It can be distinguished from other inshore flatfishes by the opaque, mother-of-pearl whiteness of the underside. The upperside (right) is brownish-green with some faint orange marks that are similar to those found in the plaice, but soon disappear once the fish is out of water. The body is lozenge-shaped, there is a strong spine in front of the anal fin and the scales are small and embedded except along the bases of the dorsal and anal fins, behind the eyes and behind the gill cover, where they are firmly attached and rough.

Most flatfishes live their entire lives in the sea, but the flounder migrates up rivers to feed. Anglers are sometimes surprised to catch

Flea, caught on man but probably of the species characteristically living on a cat or a dog.

flounders 40 miles or so (65 km) from the coast. They spend most of the summer in rivers feeding, and then in late autumn they make their way down the rivers, without feeding, to spawn in fairly deep water off the coast. Unlike the salmon, the flounder does not necessarily go back to the same river when it returns to feed in the spring. When in the sea, moderate migrations of several miles take place, and one marked individual was found to have traveled 70 miles (112 km) in 18 days.

FLOUR BEETLES, some of the many species of beetles occurring as pests in flour and associated products. The true flour beetles are exclusively members of the family Tenebrionidae. Those of the genus *Tribolium* are reddish-brown or blackish, rather flat insects, ranging from $\frac{1}{10}-\frac{1}{5}$ in. (3-6 mm) and live wherever flour and cereal products are manufactured or stored. The larvae are yellowish mobile grubs that burrow into flour, damaged grains, cereals, dried fruits, peanuts or spices. The confused flour beetle, *Tribolium confusum*, is perhaps the most widespread species, being tolerant of wide temperature extremes. The horned flour beetle, *Gnathocerus*, closely resembles *Tribolium*, but the males have large upward-curving mandibles. The confused flour beetle, so called because of its similarity to a near relative, *T. castaneum*, has been the subject of many experiments on the growth and limitations of animal populations. Its particular value is that it lives in a homogeneous, simple environment, namely flour. When a small number of flour beetles are placed in a box of flour, there is an initial rapid increase followed by a trailing off to a steady maximum population. This is known as a sigmoid growth curve, and the rate of increase and maximum numbers can be altered by varying temperature, humidity and so on.

FLOUR MITES, one of the many tiny arachnids normally found in collections of debris, such as dry leaves, stubble and animals' nests, which are attracted to man's food stores. The so-called flour mite, *Acarus siro*, shows a preference for the germ of wheat but can attack only damaged grain. Mechanical processes produce a good deal of this, and so the protection of bulk stored grain, flour and similar substances presents a very real problem. Associated with *Acarus siro* are found *Tyrophagus* spp. and glycyphagid mites. All these like a high relative humidity and, in fact, thrive in culture at relative humidities in the region of 80% at temperatures of 64-68 °F (18-20 °C). Consequently, the most effective preventive of infestation is storage under dry conditions. Flour and grain stored at 13% or less moisture content remains free of mites for a long time, even for periods of several years. The use of plastic sacks helps limit infection. Mites are not equally in need of moisture, glycyphagids being more so than the others, with *Tyrophagus* needing the least. Stringent precautions to keep stores clean are also necessary, as these mites will feed in organic dust and on the molds that grow thereon. The mites occur in the growing areas of grain and may often be on the freshly harvested material; they can also be carried on clothing and on sacks, so that their introduction into stores is all too easy.

FLUTEMOUTHS, small tropical marine fishes with elongated snouts, related to the trumpetfishes and seahorses. They are placed in a small family containing only a single genus, *Fistularia*. The three or four species are found near the shore in tropical and subtropical parts of all oceans. They can be easily distinguished from the related trumpetfishes by the very long filament stemming from the center of the tail, the filament often being as long as the fish itself (up to 6 ft or 1.8 m) in the case of the red flutemouth, *F. villosa*.

The flutemouths are long, cylindrical fishes with the dorsal and anal fins opposite each other and far back on the body. The most striking feature is the elongated snout, which is supported by the same series of bones as in other fishes, but all are greatly elongated or distorted to produce a tubelike mouth with which the fish sucks in its food. The bones of the snout can be separated slightly while the jaws are closed. This has the effect of enlarging the cavity of the mouth, so that when the jaws

are opened small invertebrates are sucked in. To pass the food along to the throat, waves of contractions can be set up along the snout.

FLYCATCHERS, term applied to certain perching birds of similar ecology but of two quite separate groups—one in the Old World and one in the New. Some members of each group capture insects by making short flights from a perch. The two groups are the Old World flycatchers, now placed in the subfamily Muscicapinae of the family Muscicapidae, and the New World flycatchers of the family Tyrannidae. The Old World flycatchers previously constituted the whole of the family Muscicapidae, but recent reorganization and consideration of intermediate forms has united other forms, such as the thrushes and Old World warblers, in the same family. Certain of the species of Tyrannidae, a general name for which is "tyrant flycatchers" may be known under the name of "tyrant," "kingbird," "phoebe" or "pewee"—the last two arising from the bird's call.

The best known of the Muscicapinae are the European species, such as the spotted flycatcher, *Muscicapa striata*, an ashy brown bird with a creamy breast, and the pied flycatcher, *Ficedula hypoleuca*, basically black above and white beneath, both birds being about 5 in. (13 cm) long. There are over 300 other species, however, spread over the whole of the eastern hemisphere, except for the extreme north of Asia, and reaching New Zealand, the Marquesas and Hawaii in the Pacific. The species are very variable, some are dull-colored, some very bright. A few are crested, others have face wattles. The tail may be extremely long, as in the paradise flycatchers of the genus *Terpsiphone* found around the Indian Ocean. In this species the males' central tail feathers may be elongated to give a total length of 21 in. (53 cm).

The range of form in the Muscicapinae is well illustrated by the 50-odd species of the subfamily that live in New Guinea. These species are extremely diverse, within the limits defined by their basic insect-eating habit. They vary in bill shape, behavior and tarsal length, so that some of them are more like warblers, chats or shrikes than flycatchers. Other species from elsewhere may be as well built as a European blackbird or an American robin, or almost as slight as a kinglet, genus *Regulus*.

The tyrant flycatchers—over 360 species, confined to North and South America—also show a considerable range of form. They vary from 3-16 in. (7½-40 cm) including tail, in length, and from the gray of the eastern phoebe, *Sayornis phoebe*, of North America, to the black, white, green, orange and scarlet of the many-colored tyrant, *Tachuris rubrigastra*, of South America. They also vary considerably in feeding habits, many being largely insectivorous, others taking small vertebrates as well as invertebrates, and others again eating fruits of various kinds.

The Tyrannidae cover a wide variety of habitats, from tropical rain forest to desert and from coniferous forest to pampas. Most of the species, however, are neotropical. A wide variety of nest structure is seen in the family— open or domed, in bushes or trees, or in holes, or on the ground.

These two groups of flycatchers illustrate the

Pied flycatcher at entrance to nest.

advantages of being the primary occupants of a particular ecological niche complex. Each group has evolved many different forms to take advantage of the varied opportunities offered to flying insectivores in wooded terrain.

FLYING DRAGONS, name commonly used for flying lizards. They are among the most bizarre and gaudy members of the family Agamidae. "Flying" is a misnomer, for unlike birds the "wings" of these lizards are not supported by the fore limbs nor can they beat in flight. All four limbs are free for landing and for climbing on tree trunks, but along the sides of the flattened body and between fore and hind legs are wings consisting of a thin membrane or skin stretched across greatly elongated and movable ribs. These ribs when extended provide the lizard with a taut patagium that can be opened and closed like a fan. When at rest on a tree the wings are closed; they open only when the lizard is displaying or when it is ready to launch itself from a trunk. As the lizard prepares to glide it turns, faces downward, dives steeply, straightens out at an angle of about 22° and then, as it is about to alight on another tree, it banks so as to land in an upward position; the lizard then folds its wings. Although a few kinds of lizard can descend from the ground by parachuting, only the flying dragon has the ability to glide and control the angle of descent.

FLYING PHALANGERS, or marsupial gliders, marsupials that glide from tree to tree and from tree to ground using a flap of skin that connects fore and hind limbs. The marsupial gliders range in size from the mouselike pygmy glider, *Acrobates pygmaeus*, to the lightly built but much larger greater glider, *Schoinobates volans*, which may reach a length of over 3 ft (1 m). Gliders have soft fur, long tails, rounded heads and large eyes—conditions so exactly repeated in the true gliding squirrels that some of the first marsupial gliders described were included in the genera *Sciurus* and *Petaurista* along with the true squirrels of North America. The flying phalangers have a generally eastern Australian, largely coastal distribution. The pygmy glider extends from Cape York, the extreme northern tip of the Australian continent, to the southern tip and westward to Spencer Gulf in South Australia. A single specimen recorded from New Guinea is thought to have been an introduction. The greater glider has approximately the same distribution but is not found as far north or west and does not reach New Guinea. The squirrel glider is of Eastern Australian distribution and overlaps in range there with the sugar glider, which has a more extensive range extending to Tasmania (perhaps introduced), southern South Australia, Arnhem Land and other parts of the Northern Territory and neighboring islands and New Guinea.

All, with the possible exception of the pygmy glider, about which there is little information, produce young in winter, which emerge from the pouch in spring and summer (August to February). The pygmy glider produces about four young, the squirrel and sugar gliders two and the greater glider one. Sugar gliders may produce two families in one breeding season. Their gestation period is 16 days and pouch life about three months.

FLYING SQUIRRELS, like all other "flying" mammals except bats, do not really fly, but they glide. They are true squirrels in that they belong to the same family of rodents, the Sciuridae, as the nonflying tree squirrels and the ground squirrels. There are about 30 species of flying squirrels, the great majority of them in the tropical forests of southeastern Asia, but with one species in temperate Eurasia and two in North America. Flying squirrels range in size from that of a small mouse, about 3 in. (8 cm) without tail, e.g., the pygmy flying squirrel, *Petaurillus hosei*, of Borneo, to that of a cat, e.g., the giant flying squirrel, *Petaurista petaurista*, which is found throughout southeastern Asia.

The gliding is achieved by a membrane on each side of the body, stretching from the wrist to the ankle and in some species also between the hind legs and the tail. The membrane is furred on both sides and is supported in front by a rod of cartilage attached to the wrist. Although the membrane looks like only two layers of skin, it does in fact contain a thin layer of muscle by which its curvature can be altered to control the aerodynamic properties. The tail is generally about equal in length to the head and body, and in most species the overall surface area is increased by the "distichous" arrangement of hairs on the tail, the hairs spreading sideways like the vanes of a feather. In most other details of structure flying squirrels closely resemble tree squirrels, but, in keeping with their nocturnal habits, the eyes tend to be proportionally larger.

Flying squirrels usually nest in holes in trees and emerge only after dark, in contrast to the largely diurnal behavior of most other squirrels. They tend to live entirely in the forest canopy and are therefore difficult to observe, and very little is known about many species. They glide from tree to tree, the larger species achieving glides of several hundred yards with very little loss of height.

Among the flying squirrels of southeastern Asia, one of the best known is the largest, *Petaurista petaurista*, which occurs, especially in montane forest, from the Himalayas and southern China through the Malaysia peninsula to Java and Borneo. It is a fairly uniform brown, and the tail is very long and bushy, not flattened as in some other species.

In the coniferous forests of Siberia is found a small gray flying squirrel, *Pteromys volans*, and two rather similar species occur in North America, a northern one, *Glaucomys sabrinus*, mainly in coniferous forest, and a southern one, *Glaucomys volans*, in deciduous forest throughout the eastern U.S.A.

FORAMINIFERANS, single-celled, largely marine Protozoa with calcareous shells that they secrete themselves. The shell is at first single chambered, but, as the animal grows, new chambers may be added, and the final shell takes a variety of forms ranging from a long straight shell to the more common flat spiral. The shells are perforated with tiny holes, or foramina, which give the group its name, and through these the cytoplasm is passed in the form of long thin filaments or pseudopodia, which join with each other to form a feeding net. Particles of food become entangled in the net and are drawn into the body of the animal. Foraminiferans reproduce sexually and there is a true alternation of generations between the sexual phase and an asexual phase. Gametes

The sugar glider *Petaurus australis;* note the fold of skin between the legs, which is stretched out during gliding flight.

are formed, and these are released to fuse in pairs and form zygotes. The first chamber of the shell forms around the zygote, and the animal increases in size simply by growing and adding new chambers to its shell. The asexual form possesses a number of nuclei, and eventually fragments of the parent, each containing a single nucleus, are shed and begin life on their own by secreting the first chamber of a shell.

The most intensively studied foraminiferan is *Elphidium crispum*, formerly called *Polystomella*. It is found on seaweed in shallower offshore waters. It is a large protozoan that measures about 1 mm in diameter and takes the form of a flat spiral in which the central chamber is the smallest, the remainder becoming progressively larger and tending to overlap the previous ones. Most foraminiferans live on the ocean floor, and their skeletons accumulate to form a thick calcareous layer known as foraminiferan ooze. It has been calculated that a third of the ocean floor is covered with this ooze. Fossil forms are common in limestones.

FOUR-EYED FISHES, freshwater toothcarps of the family Anablepidae found in Central America and characterized by appearing to have four eyes. The eyes are in fact divided horizontally into two distinct parts, the upper half for vision in air and the lower for vision under water. A different kind of lens is required in air than in water, and although these fishes have a single lens it is through the thickest part of it that underwater objects are viewed. The four-eyed fish *Anableps* spends its time cruising along at the surface with the upper part of the eyes exposed. This enables the fish to search for food and at the same time to keep an eye out, so to speak, for predators. In most of the land-living animals there is a tear duct that keeps the eye moist, but such a duct has never been evolved in fishes and the four-eyed fish must constantly duck its head under water to prevent eyes from drying out.

FOUR-HORNED ANTELOPE, *Tetracerus quadricornis*, a small antelope of the grasslands and open jungle of India, closely related to the nilgai. The male has two pairs of horns, the posterior pair 3-4 in. (8-10 cm) long, in the usual place, the front pair $\frac{1}{2}$-1 in. (1-2.5 cm) long, above or slightly in front of the eyes. The horns are straight, smooth and keeled in front. Females are, however, hornless. The animal stands 25 in. (65 cm) high and weighs 37-46 lb (17-21 kg). It is dull red-brown, white below, becoming lighter and yellower with age. There is a dark stripe down the front of each limb and a naked, black slitlike gland in front of each eye.

Also known as chousingha, the four-horned antelope lives a solitary life, sheltering in the tall grass and setting up a territory near water. It is found especially in hilly country from the Himalayan foothills to Cape Comorin, but not on the Malabar coast. In spring, females with their new fawns gather in small groups and the males make a low whistling call that seems to attract the females into the males' territories. Mating takes place in late spring, during the rains. Gestation lasts eight to eight-and-a-half months, and the young, one to three in number, are born in January and February.

FOWL, DOMESTIC, or chicken, considered to be descended from the red junglefowl *Gallus gallus* though some authorities think it may also be descended from one or more of the four species of junglefowl inhabiting southeastern Asia. The history of its domestication is lost in antiquity but chickens were probably kept in the Indus Valley in India as early as 3200 BC and they are known to have occurred in China by 1400 BC.

By the 5th century BC chickens were kept by most civilized countries both in the East and as far west as Greece and Italy. In Greece and Rome the fowl had a religious significance and its spread throughout western Europe by the 1st century BC was almost certainly the result of the expansion of the Roman Empire.

It seems likely that the use of the domestic fowl as a sacrificial or religious animal together with the popularity of the sport of cock fighting were responsible for its original spread.

In western countries there are more than 70 different breeds of chicken, a few being kept for ornamental purposes but the majority for the production of meat or eggs.

Chickens and domestic ducks and geese carry on laying if the eggs are removed from the nest regularly and, by means of selective breeding, hens of modern egg-producing strains produce more than 300 eggs in one year. Similar work on the conversion of plants into flesh has resulted in the well-known broiler chicken in which birds only about three months old are ready for the table.

No bird has had greater impact on man's economy than the chicken which is also the most widespread and possibly the most numerous bird in the world.

FOX, a general term referring to several species in the family Canidae, distinguished from the dogs by their smaller size, greater reliance on cunning and stealth in hunting and solitary habits. In addition, the foxes use sound and smell more than vision in locating prey and in communicating with others of the same species. Distributed over the globe, these small carnivores owe their success to their ability to adapt to changing environmental conditions, i.e., they take advantage of whatever food sources (both animal and vegetable) and shelter are available.

Although they prefer small mammals and birds, red foxes supplement their diet with plant material, and some have been known to survive while eating more than 50% fruits and vegetables.

Parturition in the red fox takes longer than in the domestic dog and usually occurs during a rest period. Since the four to ten cubs are helpless at birth and only weigh about 4 oz (100 gm), the female does not leave the den for the first few days. The male, however, often assumes responsibility for feeding both himself and his mate. The vixen may also take advantage of food caches prepared during pregnancy. At eight or nine days of age, the pups' eyes open, but the young remain relatively inactive until shortly before they emerge from the lair at five weeks. During this period, the cubs fight one another and thus establish a rank or peck order. Thereafter, they become intensely social and play in the vicinity of the den for many hours each day. There is a special greeting ceremony whenever the parents return to the lair, the cubs running to the vixen or dog fox,

Bat-eared fox.

wagging the thick brush, rolling over on the ground and pushing or licking the adult's snout. Finally, the litter begins to join the parents on the evening hunt and thus gains experience in locating and killing prey. Between three and four months of age, the cubs become less gregarious and wander farther from the den until eventually the family breaks up in the autumn. From then until the following spring, both young and adult foxes lead solitary lives and show little tolerance toward neighboring foxes.

The red fox, *Vulpes vulpes*, is distributed throughout the northern hemisphere in Europe, Asia, Africa, and North America, although the New World form is considered by some to be a separate species, *Vulpes fulva*. Living in wooded areas and plains country, this species hunts mainly at twilight and dawn, when its chief prey, rabbits, voles, and rats, are most active. Although they rely largely on sound and smell to find food in the dark, red foxes do employ vision when they switch to daytime hunting. During the summer, vixens who must feed their fast-growing cubs are often abroad in the day.

Another North American species, which is also found in Central America and northern South America, is the gray fox, *Urocyon* spp. Preferring woody or bush country, gray foxes are unusual in that they frequently climb trees, both for food and shelter. Rest sites may be in tree trunks or caves, but the gray fox is not choosy about the location of its den. Most burrows, however, are near a water source and under dense cover.

An inhabitant of the cold arctic regions of the Old and New World, the Arctic fox, *Alopex lagopus*, weighs 6-20 lb (3-9 kg) and has small rounded ears and a long bushy tail. It is the only canid that displays seasonal dichromatism, being a smoky gray during the summer and pure white in the winter.

The fennec, *Fennecus zerda*, is an inhabitant of the harsh desert environment of North Africa and the Middle East. Weighing a mere 3-4 lb (1.5 kg), the delicate fennec is noted for its pale buff-colored fur and unusually large pointed ears, possibly an aid in its search for lizards and insects. Other preferred foods are rodents, birds, eggs, and vegetable matter.

The bat-eared fox, *Otocyon megalotis*, a species dwelling on the arid plains and savannas of Africa, is an unusual canid. It is a small slen-

155

der animal weighing 7-10 lb (3-4½ kg). It has grizzled, gray-brown fur with black stockings and a black stripe running down the length of the tail. As in the raccoon dog, *Nyctereutes procyonoides*, there is a black mask around the eyes, which appears to be important in social behavior.

In the wild, bat-eared foxes run in pairs or small groups, although they are sometimes seen roaming alone over the dry African plains. They feed on a variety of small mammals, birds, lizards, eggs and insects (e.g., termites and their larvae). Fruits and other types of plant matter may occasionally be eaten.

FRANCOLINS, birds very closely related to the partridges which they resemble in general color and shape, although some are a good deal larger, about 1 ft (30 cm) long. They have large coarse bills and legs with one or several spurs. Most of them are dull brownish in color. There are five Asian species, of which the best known is the black or common francolin, *Francolinus francolinus*, from Asia Minor, Cyprus and the Near East through India to Assam. It inhabits grasslands or open country densely overgrown with bushes or scrub. There are various races, which differ from one another chiefly in size.

The gray partridge, *F. pondicerianus*, is another well-known bird from southern Iran, east to India and south to Ceylon. It is sedentary by nature and lives in open arid or dry country with scrub or grass, as well as in semidesert. It has been introduced into Arabia, the Seychelles and other islands.

The painted partridge, sometimes called the painted francolin, *F. pictus*, inhabits north-central India south to Ceylon, while the swamp partridge, *F. gularis*, is confined to eastern India, Nepal and East Pakistan. The Chinese francolin, *F. pintadeanus*, occurs in Manipur eastward through Indochina to southeastern China.

FRIGATEBIRDS, strong-flying, nonswimming seabirds. They are a homogeneous and distinct group of five species placed in a single genus *Fregata*. They are the most aerial of all the Pelecaniformes, and, unlike most seabirds, their feathers are not water-repellent so that they are unable to take off if forced to land in the sea. Associated with this loss of ability to swim is a shortening of the legs and a reduction of the webs that join all four toes together. The species are sedentary, and it is unusual to see them far from land. However, they have extraordinary powers of flight so if blown outside their normal range they can travel many thousands of miles. Unlike most seabirds they do not mind flying over land and regularly fly between the Atlantic and Pacific across the Panama Isthmus.

The species are similar and immediately recognizable as frigatebirds by their length, about 3 ft (1 m), very long thin wings which give a wingspan of some 7 ft (2.1 m), deeply forked tail and long straight bill with a hooked tip. In proportion to weight they have the largest wing area of any bird. Adult frigates are either black (most males) or black with white underparts (females of four species); immature birds have the head white, or white tinged with buff, and spend several years in this plumage before becoming adult.

The genus is restricted to tropical and sub-

tropical seas, and in most areas there is only a single species. Where there are two, as on Aldabra, Trinidad and Galápagos, they tend to be of different sizes. Frigatebirds normally nest in trees or bushes, even a fallen branch is sufficient, but they can nest on the ground. First the male selects a suitable nest site and then advertises for a mate by showing off a vivid crimson, balloonlike gular (throat) sac, which is only inflated during courtship. During the display the wings are spread, the whole body quivers, bill and wing quills rattle and the bird utters a falsetto warble. When it is joined by a suitably impressed female, they construct a flimsy nest of twigs picked from bushes or stolen from other nests. Later the single large white egg is laid here. The male then deflates his pouch and settles down for the first ten-day-long incubation period. Many eggs are lost to other frigates or fall from the nest, but if all goes well, an ugly naked chick hatches after six to seven weeks. This grows white down, and feathers gradually appear, but, even if food is plentiful, the chick does not fledge for five or six months. Even then it is still fed while it learns to find food for itself. This "weaning" is a critical period, and many young die at this stage. The time taken to raise young is so long that adults may be unable to breed every year. Although frigates eat the young and eggs of seabirds, their basic diet consists of fishes taken from the surface of the water and flying fishes caught just above the waves. They also harry boobies and tropic birds in order to make them regurgitate their last meal, which the frigates then eat. This habit has given them the sailor's name of man-o'-war birds.

FRILLED LIZARD, *Chlamydosaurus kingii*, an agamid lizard whose name is derived from the large ruff or frill of skin around its neck.

Although this frill normally lies in folds back along the body, when the lizard is alarmed or angered it opens its mouth wide and the frill is erected, supported by long extensions of the hyoid or tongue bone, which act like the ribs of an umbrella. The frilled lizard grows to nearly 3 ft (0.9 m) in length, with rather spindly legs and a long tail. Its color varies from gray or russet to almost black. Below it may be white to bright rusty red with a shiny black chest and throat. It is found throughout the northern half of Australia except in desert areas, and also occurs in the drier open forests along the southern coast of New Guinea.

The frilled lizard is partly arboreal, using a tree trunk as a vantage point from which it forages for the grasshoppers and other insects on which it largely feeds. The female lays about a dozen eggs in a shallow nesting chamber which she digs in soft earth.

FRILLED SHARK, *Chlamydoselachus anguineus*, a primitive shark with six gill slits (there being five in most other sharks), of which the first slit is continued right across the throat. Characteristic of this species are the partitions between the gill slits, each of which is enlarged to cover the slit behind, thus forming a frilled collar. The body is slender and elongated, with the tail continuing the line of the body (not bent upward as in most other sharks). The mouth is large and terminal, with well-developed teeth, giving the head a reptilian appearance. This species reaches a maximum recorded length of 6½ ft (2 m). It is a rather rare species which was first caught off Japan in 1884 but has since been found in the Atlantic (Portugal to Norway) and off the coast of California. It is essentially a deep-water species which feeds on octopuses and squids. It is an ovoviviparous form, the young hatching within the

A male frigate bird during the courtship season develops an inflatable sac used in display.

female and later being born, but little is known of its biology.

FRITFLY, *Oscinella frit,* a small black fly, 1-1.5 mm long, the larvae of which are serious pests of oats and other cereal crops in Europe and the New World. The eggs are laid at the bases of grasses and young oat plants in May or June and hatch into legless larvae that burrow into the young shoots and kill them. This either kills the plant or promotes the formation of many thin, weak seed-bearing shoots (tillers). This attack produces a late-ripening, poor yield of grain.

When the larvae are fully grown, pupation occurs in the soil at the base of the infected plant. A second generation of adults hatch from these puparia and lay their eggs in the flower heads of the oats, the larvae feeding on the developing grain and then pupating inside the grain husk. A third generation hatch from these puparia in the autumn and lay their eggs on the shoots of grasses. These eggs hatch to produce larvae, which feed throughout the winter, pupate in the spring and so give rise to the adults, which can then attack the young oat plants in the spring. This is an important pest species.

FROGS, a name that should refer only to members of the family Ranidae, the true frogs. Since "frog" and "toad" are the only common names available for the many species of tailless amphibians, the term "frog" is also used for many species not closely related to the Ranidae. In America, the green frog, *R. clamitans,* is found in the eastern half of the U.S.A. It is bright green on the head and shoulders and

North American wood frog.

olive brown on the back. It lives in any kind of pond or swamp. The leopard frog, *R. pipiens,* is found throughout North America except the Pacific coast states, and there are many subspecies. It is 3-5 in. (7.5-12.5 cm) long and has a striking pattern of dark olive spots edged with yellow on a bronze-green background It is often found far from water in fields or orchards. The hind limbs of many species of frog are considered a delicacy in many parts of the world. In Europe the edible frog is usually eaten al-

though, because it does not emerge from hibernation until late in spring, the common frog, which emerges earlier, is used in the early part of the year. In North America the large bullfrog, *R. catesbeiana,* is most commonly used, but in areas where it does not occur other large frogs such as the leopard frog or the pig frog, *R. grylio,* are caught.

The brown and green coloration of most frogs helps to render them inconspicuous, but in some the camouflage effect is more marked. The striped rana, *R. fasciata,* of South Africa, for example, lives in marshes and has gold and brown stripes making it very difficult to see in the grass. Its toes are only slightly webbed, and it is well adapted to climbing or diving into the dense mats of vegetation.

The wood frog, *R. sylvatica,* of northeastern U.S.A. and throughout Canada, lives among the dead leaves in oak woods where it is very difficult to find as its back is a plain light brown or buff, the color of dead leaves.

Female edible frog among water crowfoot.

Gaudy *Rana malabaricus* of West Africa.

FRUIT FLIES, true two-winged flies, the larvae of which feed on fruit. They belong to two families, the larger flies to the family Trypetidae and the smaller to the Drosophilidae. The large fruit flies are about $\frac{2}{5}$ in. (1 cm) long, often with patterned wings. In the female the end of the abdomen is produced to form a horny pen-shaped ovipositor, used to pierce the young fleshy fruit and to lay eggs. The larvae feed on the developing fruit, often reaching the pupation stage when the fruit is nearly ripe. They

then crawl out of the fruit, which has usually fallen, to pupate in the ground. Species of *Dacus* and *Ceratitis* are widespread and serious pests of stone fruits in particular. The Mediterranean fruit fly, *Ceratitis capitata,* will attack almost all commercial fruits and now occurs in all fruit-growing countries with a Mediterranean climate.

The drosophilid flies are much smaller, have a rather swollen appearance and red eyes. They are strongly attracted to fruity and fermentation odors and are commonly seen hovering rather slowly around overripe fruit. They can be a domestic nuisance when fruit is stored or alcohol brewed. The genus *Drosophila* is the most important. Its eggs are laid in the fermenting vegetable material upon which the larva feeds, and the life cycle is completed in two weeks. *Drosophila* species have been used extensively for the study of genetics and heredity because of their short life cycle, the ease with which they are bred and the large chromosomes of the larvae.

FULMARS, oceanic birds with some external resemblance to gulls to which, however, they are not related. The name "fulmar" is derived from "foul mew" (or "gull") because of their habit of ejecting stomach oil at intruders, and in life and after death their feathers retain a musty odor. Their effortless gliding on stiff, straight wings contrasts with the slower flapping flight of gulls, but on land, to which they are only attracted for breeding, they progress in an ungainly fashion, walking on the tarsus as well as the foot. The nostrils, contained within a tube on the upper mandible, characterize the order Procellariiformes (Tubinares). All are stocky in stature but capable of performing well in flight. They have distensible throat pouches, and all but one, the silver-gray petrel or Antarctic fulmar, *Fulmarus glacialoides,* possess lamellae on the inner edges of the mandibles. Probably all were plankton feeders originally, but some have now assumed the role of scavenger, feeding on the offal from man's fishing and whaling activities and from ships' galley waste. They nest in the open with little fear of predation, because they can defend themselves by spitting quantities of objectionable oil from the bill. *Fulmarus* is represented in both hemispheres; the other four genera are southern forms. The Antarctic fulmar is generally accepted as the ancestral type for the genus, and of the three northern forms the Pacific fulmar, *F. glacialis rogersii,* resembles it most closely. *F. g. glacialis* and *F. g. auduboni,* renamed by Dr. F. Salomonson in 1965, are found in the North Atlantic, and are presumed to have arrived there from the Pacific, via the Arctic seas.

The plumages of these different fulmars have a close resemblance to one another, the Antarctic being palest, and male, female and juvenile plumages are indistinguishable. Color polymorphism occurs in the fulmar—i.e., there are light and dark phase birds in the same populations, the proportions varying from one part of the species range to another.

The incubation period is a long one, of 49 days. The female rarely stays on the egg for more than a day, and generally for less than 24 hours after it is laid. The male then incubates it on average for the next seven days, but sometimes

Fulmars, seabirds that have spread spectacularly in recent years.

for as much as 11 days. The chick, in its thick powder-blue down is especially vulnerable to predators for the first two weeks of life, and most of the 14.5% of chick mortality occurs then. It is closely brooded by one or other of the parents in these early weeks.

FUR SEALS belong to the subfamily Arctocephalinae of the family Otariidae or earedseals, and have the two most obvious characters

Bull fur seal; the heavy neck identifies the male.

of this group—the small external ears and the ability to bring their hind flippers forward underneath the body. Characters particular to fur seals that distinguish them from sealions, the other members of the Otariidae, are their more pointed noses, the shape of their hind flippers and the quality of their fur. The hind flippers have all the digits of approximately the same length, whereas those of sealions have the outer digits longer than the inner three.

As in most mammals the coat is composed of two sorts of hair, the longer and stronger guard hairs, and the shorter finer fur hairs or underfur. The hairs grow in groups, each with a single, flattened, strong guard hair with a variable number of fur hairs underneath it.

The northern or Pribilof fur seal, *Callorhinus ursinus*, is the best known of the fur seals. They breed in the far north, on the Pribilof and Commander Islands on either side of the Bering Sea, and on Robben Island off Sakhalin, north of Japan, but although their breeding areas are restricted, they have a very wide migration route. Some animals cross the Pacific but most of the Commander and Robben Island herds move down the Japanese coast in winter, while the Pribilof animals go down the Canadian and American coast.

The adult male northern fur seal is about 7 ft (2.1 m) in length, with a peculiarly short pointed nose and long whiskers that give it a very haughty appearance. The color is rich dark

brown with a grayish tinge on the shoulders due to the presence of white hairs. The females are 5 ft (1.5 m) in length, grayer dorsally and more chestnut ventrally, and at 130 lb (59 kg) are more delicate and much slimmer than the male, weighing about 700 lb (317 kg).

When the adult males return to the rookery at the beginning of June, usually to the same place year after year, they fight and roar and establish their territories so that they are well in command by the time the females arrive in mid-June. Younger and immature bulls, known as the bachelors, hang about the edges of the territories while the harem bull keeps an eye on his 50 or so cows. He does not return to the sea for feeding for about two months. Bulls do not become harem masters until they are about 12, but they may continue until they are 20 years old. The pups are born, 2 ft (0.6 m) in length, 12 lb (5.4 kg) in weight and black in color, between about June 20 and July 20. Each is suckled for three months, but only for the first week does the mother stay close by it. After this she goes to sea for a day at a time, feeding, and the pups gather together in groups or "pods" in quiet areas of the rookery. Although able to swim at birth they do not normally enter the water until they are about a month old, when they spend much time playing in pools. All the other fur seals of the world belong to the single genus *Arctocephalus*, and all except one are restricted to the southern hemisphere.

G

GANNETS (*Morus bassanus*), large seabirds of temperate regions, weighing nearly 7 lb (3 kg), which spend much of their time on the wing and feed by plunge-diving for fishes from the air, a habit for which they are well adapted.

Gannets have dense white plumage with a buffish-yellow tinge on the head and upper neck, a bluish bill with the bare facial skin and feet black, as is the thin gular stripe of skin from the base of the bill down the center of the chin. The sexes are alike in plumage type (and also in voice), but males are slightly larger and heavier than females.

Gannets are streamlined for flight, with long, pointed and angled wings of some 6 ft (170 cm) span and a cigarlike shape due to the tapering bill in front and the long, wedge-shaped tail behind. They have binocular vision. The upper breast and neck carry air sacs below the skin that "cushion" the impact when the birds strike the water at great speed on diving. The skull is correspondingly strong and reinforced. The bill is stout, pointed and conical and lacks external nostrils; there is no terminal hook as found in many other pelicaniform birds, such as cormorants and pelicans, though there are toothlike serrations along the cutting edges that facilitate the seizure of fishes. The jaw and throat are widely distendable, and the tongue is minute, allowing the birds to swallow large prey easily. The neck is moderately long, the body stoutly elongated, the legs shortish but the feet large with long toes. Gannets have webs between all four toes. On the nail of the third toe a series of notches forms a functional comb used in scratching.

Gannets are birds of temperate waters. When at sea, gannets often congregate over shoals of fishes in large numbers and make spectacular mass plunging dives from 50 ft (15 m) or more above the surface, falling like projectiles and raising great spurts of spray as they enter the water vertically. They are even more social when breeding and form dense colonies, often of great size, on continental headlands, cliffs and islands at traditional sites that provide a measure of safety from predators by their inaccessibility or remoteness. At many gannetries, nests are spaced out over the ground at an even 3 ft (1 m).

The nest is a drumlike structure of flotsam and jetsam, and often including seaweed and grass solidified by excrement. A single white egg is laid, which is incubated by both sexes, not in the conventional way but by cupping and overlapping the feet around it. Incubation lasts six weeks, and the nestling is born blind and helpless with a scanty amount of down, which, however, soon develops into a thick, white fluffy coat.

GARDEN SPIDER, *Araneus diadematus*, also known as the diadem or cross spider because of a white cross on its body, which caused it to be regarded as holy in Germany during the Middle Ages. It builds a round web. See orb-weaving spiders.

GARFISHES, or needlefishes as they are often called in the United States, elongated, long-jawed fishes related to both the flying-fishes and the sauries. The body is long and slender and slightly compressed. The most striking features are the jaws, which are as long or longer than the head and bear needlelike teeth. The dorsal and anal fins are set far back on the body, and the lower lobe of the tail is larger than the upper.

Most of the garfishes are marine but at least two species enter fresh water. *Xenetodon cancila* is found in the rivers of southeast Asia and India. The young fishes are frequently imported for aquarists, but it must be remembered that the adults reach 12 in. (30 cm) in length. They require live food, especially fishes, and should be kept in a tank of their own. They are graceful, fast-swimming fishes that are capable of leaping almost vertically out of the water. The tank should be kept well covered. *Potamorhaphis guianensis* is another freshwater species and is found in the Amazon. Most garfishes grow to about 2 ft (61 cm) in length, but a few reach 4 ft (1.2 m).

GARPIKES, primitive fishes from the fresh waters of the southeastern states of North America. There are several species belonging to the genus *Lepisosteus*. All have long, thin bodies covered with thick, shiny scales that are diamond-shaped and, unlike the overlapping scales of other fishes, fit together like a mosaic.

Garpike, a freshwater fish of the United States, has scales which fit together as in a mosaic.

The scales can be fairly easily removed and have been used in jewelry. The dorsal and anal fins are short-based and far back on the long body. This positioning of the fins is typical of fishes that need to accelerate rapidly toward their prey, all the thrust-receiving surfaces being at the rear of the fish. The gars are usually rather lazy fishes living in weedy water, and it is only when lunging forward toward their prey that they move swiftly. The jaws in some species, for example in the alligator gars, are elongated and have rows of long, sharp teeth. In spite of the heavy covering of scales the gars are surprisingly flexible fishes.

GAUR, *Bos* (*Bibos*) *gaurus,* the largest species of wild ox. Bulls average 5 ft 8 in. (175 cm) at the shoulder, but are commonly up to 6 ft 4 in. (195 cm); a huge male shot in Burma was 7 ft (210 cm). Cows are smaller than bulls. A big bull may weigh 2,000 lb (900 kg). Gaurs are largely black or dark brownish-black, but the legs are white from the knees and hocks to the hoofs, and the forehead and arched ridge between the horns are gray-white. The face is sinuous, with a concave forehead and convex "Roman" nose. Bulls are extraordinarily muscular, with a strongly developed dorsal ridge, a dewlap between the forelegs and another just behind the chin. The horns go out sideways and are curved upward in a semi-circle. In big bulls they are yellowish with black tips, and strongly corrugated for about a third of their length.

Young gaurs are brownish in color, darkening at maturity. Their horns are orange, acquiring a greenish tint and corrugations with age. In bulls the horn tips turn upward, but in cows they point inward and occasionally cross.

They live in hilly forests in India, Burma, Cambodia, Laos, Vietnam, Thailand and Malaya.

GAVIAL, *Gavialis gangeticus,* a crocodile differing from all other living crocodiles in its extremely long snout, which resembles a beak. It may be up to six times longer than broad. Adult gavials may be up to 20 ft (6 m) long. There are 29 teeth on each side of the upper jaw and 26 on each side of the lower jaw, all of approximately the same size, even the fourth lower jaw tooth being not noticeably bigger, as is the case in alligators and the true crocodiles. This fourth tooth also does not fit into any gap or hollow in the upper jaw. When the gavial's mouth is closed, the teeth are locked between each other and point outward at an angle. The end of the snout broadens into an octagonal where, during the breeding season, the male develops a shell-like hump. This is the only known secondary sexual characteristic in crocodiles. The rear portion of the skull shows several differences from that of true crocodiles. The gavial's two upper temporal openings are round and the same size as the eye socket. Because of these and other characteristics some scientists are inclined to assume that the gavial is related to the now extinct sea crocodiles of the family Teleosauridae from the Jurassic period.

The food consists mainly of fish, but waterfowl and small mammals are also eaten.

European garden spider at the hub of its orb.

Gavial, a crocodilian of southern Asia with a long, narrow snout associated with fish-eating.

GAZELLE, the stereotype of the graceful antelope: slenderly built, long-necked with short, wagging tail. Gazelles are fawn above, white below, and characteristically have an alternating pattern of dark and light face stripes; the middle of the face is dark red-brown, on either side of this is a whitish stripe that reaches from the upper edge of the eye to the muzzle, and below this on each side is a blackish stripe. The rest of the face is fawn.

This pattern is known as the gazelline face pattern, and is seen sporadically among antelopes and in the chamois, too. It may be altered, so that the black lower stripes disappear and the white spreads partly around the eye, giving the eye ring of such forms as reedbuck. On the flanks, the white of the underside is always sharply marked off from the fawn of the upperparts. Sometimes there is a dark flank stripe between the two colors. Similarly,

Gecko after using the familiar lizard trick of throwing off part of the tail.

the white of the rump may be divided from the body color by a dark pygal band.

Gazelles have distinctive horns, tending to be lyre-shaped from in front and S-shaped from the side. They are tightly ringed almost to the tip.

All the 15 species of gazelle seem to have a similar type of behavior pattern, differing, however, in details. Basically, there are three types of socialization: territorial males, female herds, and all-male herds.

In the dry season (from late June), Thomson's gazelle move west to the dry-season refuge areas near Lake Victoria; in October they move back again to the Serengeti. They migrate in scattered herds, not massed like the gnu and zebra who accompany them. The most striking migrations are made by goitered gazelles, which, in autumn, travel as much as 300 miles (480 km) south from northern Kazakhstan into Uzbekistan and Turkmenistan, to escape the heavy snowfall. Territories are set up anew on the males' return.

Gazelles move by pacing, with both legs on the same side moving forward together; they also trot, with opposite legs moving; and "stot" or "pronk," suddenly springing aloft with all four legs straight and close together, the body being arched and the head held low. Stotting communicates excitement from one individual to another, and is contagious; its exact function is unknown, but may act to spread alarm. It is most developed in the springbok, which has been known to leap nearly 12 ft ($3\frac{1}{2}$ m) into the air.

Gestation lasts about 170 days in the Dorcas and Loder's gazelle, 190 days in Thomson's, and 200 days in Grant's. Births seem to occur at all seasons, both in the wild and in captivity. When the calf is born, the mother leaves it hidden and avoids it until it is time to suckle it—an obvious predator-distraction mechanism. Suckling takes place three or four times a day. The mother licks the calf while it sucks, drinks its urine and eats its feces. The calf begins to graze in four weeks, but continues to suckle for four to six months. Unlike many other antelope, gazelles do not foster: females will refuse to suckle a strange offspring.

When frightened, herding gazelles flee for 200-300 yd (180-275 m), then stop and look back. Territorial bucks, however, flee only to the edge of their territories except in an extremity. However, gerenuk at first stand motionless when alarmed, looking over the bushes; when fleeing they do not stop to look back.

GECKOS, a mainly primitive group of lizards, famous for their nocturnal habits and, more especially, for their climbing abilities due largely to the structure of their feet. A few retain the conventional claw and toe, but many have retractile claws and "snowshoe" pads for dashing across sand, but most geckos have developed "friction" pads that enable them not only to climb a vertical wall or a pane of glass but also to continue scampering across a ceiling upside down.

Geckos live in forests, swamps and deserts, on

Grant's gazelle, found from Somalia to Tanzania.

mountains, and on islands, as long as the nights do not get too cold. They range from 2 in. (5 cm) total length, to over 1 ft (30 cm), the majority being 3-6 in. (7.5-15 cm) long. Roughly half the length is tail.

Their bodies are covered by a soft skin with minute scales, among which, in many species, are larger scales. A few species have fishlike imbricating scales on all or some parts of the body. In most species some of the scales on the underside of each toe are specialized as broad pads, at the base, the tip or throughout the toe. Each pad resembles a miniature, densely packed pincushion; every microscopic bristle is split into delicate branches, each terminating in a disclike thickening, the free face of which is slightly concave. Recent studies show that the multitudinous tiny terminal discs adhere by suction, but on surfaces where this is not possible the bristles act like so many miniature hooks. The peculiar wriggling gait of the gecko is caused because, in order to lift its foot from the wall surface, it must curl each toe upward from the front to disengage the fringes without damaging them. This must be done each time the toes are picked up as it runs.

They are the only lizards that regularly vocalize. In many, usually the nocturnal species, the males utter special calls, something like "tsak-tsak" or "tik-tik," repeated a few times in succession. These masculine calls are different from the squeaking of either sex when seized. The social function of the male calls is not really known. The tokay, *Gecko gecko*, of southeastern Asia has a vocabulary of three different calls, all frighteningly loud.

GEESE, 15 species of water birds intermediate between the ducks and the swans, being longer in the legs and neck than most ducks. But they are not so aquatic as either ducks or swans, feeding largely by grazing on land.

As geese are quite large and their flesh is highly palatable, they have been important as food to the peoples of all countries where they have occurred regularly in appreciable numbers. Two species, the greylag goose, *Anser anser*, and the swan goose, *Anser cygnoides*, have been domesticated for meat production with considerable success.

There are two natural groups of true geese: the "gray" geese of the genus *Anser*, and the "black" geese, genus *Branta*. They are all confined to the northern hemisphere, typically breeding in arctic or subarctic regions. They are gregarious, particularly outside the breeding season, and migrate in large flocks. In flight they usually adopt a formation of wavy lines or, particularly, a V shape, each bird flying slightly to one side of its neighbors. These flight formations are known as skeins, from their similarity to strands of wool, and during peak migration they may cover much of the sky for several days. The flight formation is aided by the distinct tendency of family parties of geese to stay as units, even during the winter. The strength of this social bond may be seen clearly in parties of geese on their feeding grounds, where the members of a family stay together, and groups of families also tend to form larger units. The general term for a flock or party of geese on the ground is a "gaggle."

Canada geese with goslings, a species introduced into Europe and New Zealand.

The gray geese are typified by the greylag, which is widespread in the palearctic region and divided into eastern and western races. This is the largest of the gray geese, up to 35 in. (89 cm) long, with a large, heavy head and a stout bill, which is orange in the western race and pink in the eastern. Unlike the other gray geese it has no dark markings on the bill. The feet and legs are pink-flesh colored, and the fore wing is pale gray. Otherwise, the plumage is as in other gray geese: basically gray-brown, with light tips to the body feathers, which lie in such a way as to give a somewhat barred appearance on the back. The tail is gray with a white tip, and the upper and lower tail coverts are also white. The head and neck are no darker than the rest of the body, as they are in other gray geese, and the breast is generally unmarked. In some individuals, particularly older birds, the lower breast and belly are spotted with black, but not barred as in the white-fronted geese.

The greylag breeds in lowland marshes, moors and often on islands when they are available. The nests may be close together in small colonies, usually on the ground, hidden among heather or similar vegetation, with a lining of heather, mosses, and grasses, mixed with feathers. Sometimes the nests are built of dead reeds and similar materials among floating willow roots. Generally, 4-6 creamy-white eggs are laid, the clutch size being highly variable, the bigger clutches probably resulting from interbreeding with domesticated strains that tend to lay more eggs. These are incubated by the female alone for 27-28 days, while the male stays in the vicinity of the nest. The young are able to leave the nest shortly after hatching and are accompanied by both parents. They feed on a variety of food, taking more invertebrates when young.

The adult greylags feed on many kinds of plant materials, including grasses and berries, and they take very readily to agricultural land, visiting wheat fields at night and also feeding on other crops such as peas and turnips. In their winter quarters they tend to roost on inland lakes and fly only a few miles each day to forage.

The breeding range of the western greylag extends from Iceland and Scandinavia in the north to Yugoslavia and Macedonia in the south. The winter range is from Britain and Holland southward to North Africa. The population that winters in Britain has its origin in Iceland.

Another important palearctic gray goose is the bean goose, *Anser fabalis*, almost as large as the greylag, being about the same length but of slighter build. It is a more northerly species, breeding in the tundra and coniferous forest regions from East Greenland in the west right across the palearctic to eastern Asia.

The white-fronted goose, *Anser albifrons*, also breeds in the Arctic and flies south in enormous flocks for the winter. It is smaller and darker than the greylag, has broad black irregular bars across the belly, a white frontal patch on the forehead above a pink bill and orange legs.

A similar but smaller species, not exceeding 26 in. (67.6 cm) in length, is the lesser white front, *Anser erythropus*. This is also a palearctic species, breeding somewhat to the south of the white front and distinguished from the latter by a more extensive frontal patch and a yellow ring around the eye, as well as by those features which commonly accompany smaller size, such as faster wing beat and more highly pitched voice.

The swan goose breeds in Siberia, Mongolia and Manchuria, and has been domesticated in

China for at least 3,000 years. Apart from the greater bulk and a knob at the base of the bill of the domestic form, the two are similar, having the typical gray-goose plumage but with a neck cream-colored at the front and sides and chestnut behind. The bill of the wild form is longer than in any other goose. The swan goose is better able to stand heat than the greylag and is therefore the more common domesticated form in hot countries.

There are five species of black geese, of which one, the Canada goose, *Branta canadensis*, has been divided into 12 races, one probably extinct. All Canada geese have brown body plumage, a white rump, black head and neck, and a white cheek patch. The largest race, the giant Canada goose, *B. c. maxima*, is about twice the bulk of the smallest, the cackling goose, *B. c. minima*. Many of the races, however, interbreed where their distributions overlap, and there can be little doubt that they are all one species. The Canada goose breeds across the whole of Canada, except for the most northern arctic islands, part of the northern United States and in the Aleutian Islands. Like all other geese, except for the Hawaiian goose, *Branta sandvicensis*, they migrate south for the winter. The larger forms occupy the general ecological position in North America that the greylag goose takes in Europe.

The brent goose is the smallest and darkest of the black geese, measuring no more than 24 in. (61 cm) in length. The plumage is largely black or black and brown, paler beneath and with an incomplete white collar. The dark-bellied brent goose, *Branta bernicla bernicla*, breeds in arctic Europe and Asia; the light-bellied brant, *B. b. hrota*, in northeastern Canada, northern Greenland, Spitzbergen and Franz Josef Land; and the Pacific brent or black brent, *B. b. orientalis*, in northwest Canada, Alaska and Siberia.

The barnacle goose, *Branta leucopsis*, has a handsome gray, white and black plumage; basically black on the neck and upper breast, gray on the back, pale gray beneath and with a white rump and face. It breeds in the tundra regions of the western palearctic, including east Greenland and Spitzbergen, and usually makes its nest in inaccessible places such as cliff faces.

The most striking of the black geese is the red-breasted goose, *Branta ruficollis*, which has a bold black, white and chestnut-red plumage. The upperparts are black, the rump and flanks white, and the neck and breast red, picked out with white. It is even smaller than Ross's goose and breeds in the tundra of northern Siberia.

Finally, the né-né or Hawaiian goose is an example of a species saved from extinction by active conservation. The plumage is barred in brown, black and white, and it has a black head with rufous neck and cheeks. The neck feathers are conspicuously pleated or furrowed. This species, restricted to the Hawaiian Islands, is nonmigratory, and shows significant modifications adapting it for a terrestrial rather than an aquatic existence. Its stance, for example, is rather upright, and the webs on the feet are reduced. As a result it was at a disadvantage in competition with introduced predators, and by the 1950s there were fewer than 50 birds left. Conservation efforts in

Hawaii and in Britain have, however, built up the world population of né-nés to around 900 birds in 1970 (see Hawaiian goose).

The magpie goose, in a subfamily to itself, seems to represent a link between the other Anatidae and the screamers, Anhimidae, with which it has a number of structural features in common. It is a large black and white bird inhabiting southern New Guinea and northern Australia, and is unusual among geese in regularly perching in trees.

GELADA, *Theropithecus gelada,* a large baboonlike monkey of the Ethiopian highlands. Although it is as large as most baboons, has a long face and a similar type of sexual dimorphism, the gelada is not a true baboon, being as closely related to the mangabeys and macaques. The facial elongation, so marked in baboons, is effected in a different way in the gelada: instead of being lengthened into a long narrow muzzle, with the nostrils at the tip, the gelada's face is deep, with massive jaws and a snubby nose that is not at the tip of the thick, square snout. The face is grooved on either side of the snout. The fur is yellow to brown, with pink naked patches on the chest that meet in the middle. In the female there is a "necklace" of warty eminences around the chest patch, and at the height of oestrus these become very prominent and the chest patch turns bright red, a condition likened to a "bleeding heart." This cyclic indicator replaces the perineal swelling found in baboons, mangabeys and many macaques, and undoubtedly owes its origin to the prolonged sitting periods, during which the chest is more conspicuous than the rear end.

The gelada, baboonlike monkey of Ethiopia.

The gelada feeds mainly on seeds, blades of grass, and shoots, which it gathers by sitting in one place and picking them up, storing them in the hand until a fistful has been accumulated, when all the material is transferred to the mouth.

GENETS, related to the civets and mongooses and looking like a cross between a tabby cat and a mongoose. There are six species, the

best known being the feline or small spotted genet, *Genetta genetta*, which ranges over most of Africa apart from desert and semi-desert areas. It is also found in Spain and southern France. It is catlike but more slender, elegant in build and graceful in movement. It is up to 40 in. (1 m) in length of which nearly half is tail. Its fur is soft and spotted with brown to black on a light ground color. The head is slender and tapers to a pointed muzzle adorned with long whiskers, and the ears are large. The long tail has alternate black and white rings, while along the spine there is a crest of long black hairs that is raised in moments of excitement. The legs are short with small paws and retractile claws.

The blotched or tigrine genet, *G. tigrina*, is numerous throughout Africa. It is similar to the feline genet, but it has larger spots on a more yellowish ground color and the dorsal crest is missing. The rusty spotted genet, *G. rubiginosa*, is found south of Tanzania and is like the blotched genet except for its more reddish spots. Some authorities consider it to be merely a color phase of the blotched genet. The Abyssinian genet, *G. abyssinica*, of the highlands of Ethiopia is the smallest of the genets, not more than 20 in. (51 cm) in total length. It has ash-colored fur broken by longitudinal black stripes. It is very rarely seen. The Victorian genet, *G. victoriae*, and the water genet, *Osbornictis piscivora*, are known only from skins brought back by pygmy hunters in the Ituri Forest region of the Congo. The former is like the feline genet but with richer markings, while the water genet has a rich chestnut fur with white markings on the face and a black bushy tail.

Feline genets live in areas of bush and low scrub providing plenty of concealment and good food supplies. They are usually solitary or at most are seen in pairs, hunting by night and sleeping by day among the branches of bushes or trees. They are excellent climbers and skilled hunters, stalking their prey by gliding swiftly over the ground with the body held low and the tail held straight out behind, finally seizing their prey with a swift, sharp pounce.

Genets feed on small rodents, birds and insects, particularly night-flying moths and beetles. A certain amount of grass is also regularly eaten.

GERBILS, a group of mouse- and rat-sized rodents, numbering about 50 species, that are among the dominant small mammals of the great zone of desert, steppe and savanna, stretching from the Gobi Desert through central Asia to the Sahara and south throughout most of the drier parts of Africa.

Some species of gerbils are also known as jirds, and the name "sand rat" is often used, but they should not be confused with jerboas, which belong to quite a different family and are much less closely related to the true mice. There is no single external feature by which to distinguish a gerbil from a "murid" mouse or rat, but in general the fur is sandy-brown above and pure white below, the feet are white and the tail tends to be rather more thickly haired, although it is rarely so tufted as in a jerboa. The hind feet tend to be rather longer, giving a powerful thrust in running, but gerbils

Gerbil, a rodent typical of dry regions.

do not hop on the hind feet alone as do the jerboas. Another useful way of distinguishing most gerbils from most murid rats is by the presence of a longitudinal groove on each upper incisor tooth, but the one really constant characteristic of gerbils cannot be seen externally since it concerns the molar teeth. These have the tubercles on the grinding surface arranged in two longitudinal rows rather than three as in members of the family Muridae. Many gerbils, and especially those living under the more extreme desert conditions, have the bony capsule of the ear, the bulla, enormously enlarged. This is a feature found in many desert mammals, for which acute hearing over long distances and in dry air is of particular importance for survival.

Most gerbils are nocturnal, lying up in burrows during the day and emerging at dusk to feed

mainly on seeds, bulbs and succulent plants, although some species are also partial to insects. The breeding behavior is rather similar to that of the true mice and rats in that the gestation period is short, usually three to four weeks, and several litters, usually of three to six young, can follow each other in rapid succession.

GERENUK, *Lithocranius walleri,* a long-necked gazelle of Somalia and parts of East Africa. It stands 41 in. (103 cm) at the shoulder and has a fox-red coat. The males have short curving horns, up to 17 in. (43 cm) long. Gerenuks browse the dry thorn bushes, standing vertically on their hind legs and reaching up with their long giraffelike neck. They move about in ones or twos or in small groups of females led by a single male. When disturbed, a gerenuk runs away with its neck held horizontally, its body

somewhat crouched. Gerenuks seem able, when necessary, to go without water.

GHOST CRABS, crabs of the genus *Ocypode,* also known as racing crabs, and sometimes as sand crabs. The three names together give a guide to one of their most characteristic habits: they run at great speed over sandy beaches in the tropics; their color is very similar to that of the sand; and when they run in the bright tropical sun their shadows are often more conspicuous than their bodies. When they stop and lower their bodies to the sand the shadows disappear, and so, apparently, do the crabs.

As well as running over the sand, the ghost crabs also burrow, sometimes reaching a depth of 3-4 ft (90-120 cm). At low tide they emerge and scavenge, sometimes catching small fish at the edge of the sea. The eye stalks of ghost crabs are long and mobile, and in some species they are prolonged beyond the eye to form a sort of horn. One of the pincers is always larger than the other, and in many species bears on its inner side a ridge of granules that can be used to produce stridulations by friction against a smooth ridge on the third joint of the limb bearing the pincer. The noises produced in this way have been described as twittering squeaks or croaks. They are produced in order to warn other crabs that a burrow is occupied.

GIANT ANTEATER, *Myrmecophaga tridactyla,* the largest of the three living species of anteaters. Giant anteaters are unique in appearance and easily recognized. They are basically gray in color with a pronounced white-edged black stripe that runs diagonally from the throat across the shoulders to the back. The hair is coarse and particularly long on the bushy 2-3 ft (0.65-0.90 m) tail that resembles an enormous plume. Some of the hairs that radiate from the spine of the tail may be 16 in. (40.6 cm) in length. The tubelike head, with its small rounded ears and long tapering snout, ends in a tiny mouth through which the giant anteater can extend its pencil-slim tongue almost 2 ft (61 cm). The forefeet are equipped with stout claws, and because of these the giant anteater walks on its knuckles and the edges of its palms. Adults normally weigh from 38-51 or more lb (18-23 kg) and may measure 5.5-7 ft (1.6-2.1 m) from tip of snout to tip of tail. Males are usually larger than females.

Terrestrial in habit, giant anteaters are found in Central and South America's swampy areas and humid forests, but they mainly frequent open grasslands (pampas) or savannas from southern British Honduras to northern Argentina.

These anteaters give birth to a single offspring after a gestation period of about 190 days, and the newborn may weigh 3.5 lb (1.5 kg). The mother carries her youngster on her back, and the two remain together until the female is ready to breed again.

They feed principally on termites, ripping open the hard-walled nests with their claws, poking their snouts into the crumbled mounds, and picking up the insects with their sensitive probing tongues that are very sticky, being covered with viscous saliva when the animals feed.

GIANT SALAMANDER, *Megalobatrachus japonicus,* of Japan, is the largest living amphibian,

Ghost crab, best known from its shadow.

Giant salamander, largest living amphibian, which may be 5 ft (1.5 m) long.

Gibbon in southeast-Asian jungle.

reaching a length of 5 ft (1.5 m). A smaller relative, *M. davidianus*, lives in China, and a more distant relative, the hellbender, *Cryptobranchus alleganiensis*, in the United States. The first giant salamander was brought to Europe in 1829 by de Siebold and lived over 50 years.

The head and body are flattened, and skin folds are present along the sides of the body. The tail is laterally flattened, and the paired limbs are small in proportion to the body. When young the head region bears three pairs of gills, but a partial metamorphosis takes place and the external gills are absorbed. All gill slits close, whereas in *Cryptobranchus* one pair of gill slits remains open throughout life. The eyes of the giant salamander lack eyelids, and larval teeth are retained. This retention of some larval characters is an example of neoteny. The giant salamander lives in cool, swift streams. It is carnivorous, but instead of pursuing its prey it waits till the prey is within reach and seizes it with a swift lateral movement of the head. Prey consists of fish, smaller salamanders, crayfish and other vertebrates. The senses of smell and touch are probably more important than sight in food capture.

Being aquatic, and lacking external gills, the giant salamander has to surface at intervals to breathe. Lungs are quite well developed, and internal gills are lacking. The skin probably assists gas exchange.

Breeding takes place in late summer, and the eggs, 6 by 4 mm, form a string as they emerge to become externally fertilized. The larvae have three pairs of external gills.

As it is carnivorous, the giant salamander can be captured by fishing, using fish, frogs or large worms as bait. The bait has to be brought near to the animal for a bite to take place. The point of a baited hook is forced into the end of a wooden rod, and then, using the rod, the baited hook is directed to the spot where the salamander may be lurking.

GIBBONS, the smallest of the apes, have very long arms, nearly twice as long as the trunk and more than twice as long if the hands are included; they also have very long legs, 30% longer than the trunk, but shorter than the arms, only $\frac{2}{3}$ to $\frac{3}{4}$ the arm length. Nonetheless, the legs of gibbons are the longest, compared to the body size, found among the apes, and are as long, relatively, as in man. The hands, also, are very long and slender; the fingers are long and somewhat curved into a hook; the thumb, while short, is deeply cleft from the palm so that more of the gibbon's thumb is free than man's or apes'. Moreover, the joint between the thumb and the carpus (wrist) is a ball and socket, not a hinge joint as in most other primates, so that its mobility is very much increased. The foot is long and slender as well.

Gibbons are almost entirely arboreal, and move through the trees by their arms alone, grasping one branch and reaching over to the next with great rapidity, and often jumping between branches. It is a joy to watch a gibbon's easy, unhurried motions as it swings rapidly along a branch. Other apes brachiate only occasionally and then mostly when young; but a gibbon does so throughout its life cycle.

As well as for their ease of movement, gibbons

are known for their remarkable voices. The most characteristic of a gibbon's vocalizations is the great call, made in some species by both sexes, in others by the female only. It is made as a location signal; on sighting another troop; and generally in the early morning, as the sun rises, or at any marked change in the weather. In the best-studied species, the white-handed gibbon, the call is a series of musical whoops, rising in pitch and increasing in tempo and volume, finally dying away again. In this species, only the female calls. In the siamang gibbon, the call is made by both sexes: there is an inflatable sac in the throat, giving the call a harsh, resonant quality that is quite deafening from nearby; it is a two-tone call, a deep boom as the sac is filled, and an unpleasant shriek as the air is let out again, and this is repeated over and over for about half a minute or even more, increasing in tempo but not in volume or pitch.

When these gibbons come to the ground, which is very rarely since their preferred habitat is at the very tops of the trees, they walk upright with their arms bent and held to the sides, like balancing poles. Sometimes gibbons can be seen running along a branch in this fashion.

Most gibbons feed on fruit, with some leaves, buds, flowers, eggs and insects as well. Food is plucked by hand and then put in the mouth. The long canines, found in both sexes (unlike other apes, or Old World monkeys, where only the male has large canines) are used to pierce and peel tough fruits, as well as for fighting. There seem to be few predators of gibbons, but eagles possibly take young ones.

The hoolock or white-browed gibbon, *H.*

hoolock, is found in Assam, northern Burma and Yunnan. A few weeks after birth, both sexes turn black with white brow bands; then, at sexual maturity, the female becomes light brown, with a complete face ring. Thus the sexes are different colors when adult.

GILA MONSTER, *Heloderma suspectum*, one of the only two known venomous lizards, the other being the Mexican beaded lizard. They are restricted to the New World. The Gila (pronounced hee-la) is a relatively stocky reptile, with a thick short tail, head and body. This bulging body, giving a large surface area in relation to volume, serves to inhibit moisture loss and overheating in the event of prolonged exposure to desert heat and dryness.

This species, first mentioned by Baird in 1859, was called *Heloderma suspectum* because it was suspected of being a species of venomous lizard different from the already well-known Mexican beaded lizard. Subsequently, the Gila monster has been divided into two subspecies: the reticulate Gila monster, mottled with pink and black, ranging throughout Sonora, Mexico, into southern Arizona and into extreme southwestern New Mexico, and the banded Gila monster with a more clearly defined banded pattern, living in central western Arizona, southern Nevada and over the border into southwestern Utah. The brilliant banded pattern of the hatchlings gradually becomes mottled and diffused with maturity.

The primitive venom-conducting apparatus consists of a number of grooved teeth in the front portion of the lower jaw. When a Gila chews its victim, venom flows through ducts from glands in the lower jaw, into the mouth

between the teeth and lips, then by capillary action it follows the tooth grooves into the victim's wound.

GIRAFFE, *Giraffa camelopardalis*, one of the most striking animals in existence with its extraordinarily long neck, long legs, sloping back and tufted tail. The giraffe bulls may reach a height of 18 ft (5.5 m) and the cows 16 ft (5 m)—about three times the height of a full-grown man. The giraffe has one living relative, the okapi, and together these two species are classified in the giraffe family (Giraffidae) of the even-toed ungulates. The giraffe is also related, but less closely so, to deer, partly since the antlers of the deer when they are "in velvet" resemble the skin-covered horns of the giraffe. However, the antlers of the deer are shed each winter, whereas the horns of the giraffe are permanent.

Giraffes may have 2-5 horns, but usually have a pair on the forehead and a third boss in front.

Giraffe also resemble some deer, for instance the caribou, in having horns present in both sexes. Those of the males are longer, about 10 in. (25 cm) and thicker than those of the female, but even trophy giraffe horns are smaller than those of most other large-hoofed animals. Presumably larger horns would be too cumbersome and heavy for an animal with such a long neck. Deer and giraffes have the same number of teeth (32), which are adapted for browsing, but unlike giraffes, deer eat grass as well as leaves. The wide, lobed lower canine teeth of the giraffe are situated beside the incisors. They help to increase the width of the front teeth, which comb the leaves from trees and bushes rather than clipping them off. Members of the giraffe family are the only hoofed animals that possess a lobed canine, so this tooth is important to paleontologists in identifying fossil giraffids. Many fossil giraffes lived in Europe and Asia millions of years ago, but today both giraffes and okapis are found only in Africa. Even here the distribution is more restricted than it was 1,000 years ago. Then giraffes roamed south of the Orange River where drawings of them by Bushmen still exist, and throughout most of the Sahara, which was much less dry then than it is now. More recently, giraffes have been poached or shot out in some areas or driven back out of

The Gila monster, although venomous, feeds mainly on small defenseless animals and eggs.

A group of giraffes at a waterhole.

newly developed farmlands. However, they are still found throughout many parts of Africa where there are trees and bush to provide them with food. They do not live in the densely forested regions that okapi inhabit.

Giraffes spend much of their time eating, wandering from plant to plant pulling a few twigs from this bush, a few leaves from that tree. They eat many different kinds of leaves, but they prefer some to others. The preferred trees often show marked cropping downward from a height of 18 ft (5.5 m), the height to which a giraffe can reach, so that a zoologist can tell at a glance which species of trees have been heavily browsed. Giraffes often browse on acacia thorn bushes, reaching around the thorns with their prehensile tongues and lips to extract the leaves. The thickness of the lips and their abundance of stiff hairs prevent them from being torn on the thorns.

When water is available, giraffes may drink daily, bending their long front legs awkwardly or straddling them apart so that they can reach the water. However, giraffes can also survive in arid regions where there is no free water. Apparently they can satisfy their needs from the leaves they consume.

Like many ruminants, they often spend the

A female glowworm "calls" to her prospective mate with light signals.

hottest part of an African day chewing their cud, either while standing or while lying with their heads erect. They continue to chew their cud and also browse after nightfall, since they apparently need little sleep. Observations in zoos show that giraffes sleep deeply for less than an hour a night. To do this they lay their heads back alongside their bodies for a few minutes at a time.

Young are born at any time of the year, after a gestation period of about $14\frac{1}{2}$ months. Only one instance of twinning is known, and these were born dead prematurely. That a single young only is usually produced is not surprising, since the newborn animal is about 6 ft (2 m) tall and 130 lb (59 kg) in weight. It can stand and suck within an hour of its birth, and walk or run soon afterward.

GLASSFISH, *Chanda ranga*, a fish of fresh and brackish waters of India, Burma and Thailand, formerly known as *Ambassis lala*. The glassfish is a small stocky fish growing to 3 in. (7 cm) with two dorsal fins, the first being spiny. Its most remarkable feature is the transparent, glasslike body, which is so clear that not only can the vertebrae and bones supporting the fins be seen but also objects behind the fish can be viewed through the body. The abdominal cavity, however, is obscured by silvery tissue, possibly preventing the growth of algae inside the gut. The head is opaque. The transparency of this fish is probably an adaptation rendering the fish less conspicuous in water; transparency is found in other groups, such as certain catfishes.

GLASS SNAKES, sometimes called glass lizards. They are slender and elongate, legless lizards, snakelike in appearance but with movable eyelids and external ear openings, which are sometimes concealed under a fold of skin. The tail is very long, but in at least two species it is not fragile and it contains no fracture plane, unlike other related lizards in which the tail is readily thrown off when the animal is suddenly disturbed or held. In some species, minute, scalelike vestigial hind limbs are still discernible in the cloacal region, and both pectoral and pelvic girdles are present. The body is covered by squarish-rhomboidal

scales, arranged in straight longitudinal and transverse series. The jaws are armed with strong teeth. The scales are provided with separate, dermal bony plates or osteoderms. The tongue, indented in front, can be protruded and then retracted into a fleshy sheath. Most glass snakes are brown, some have a striped pattern. In *Ophisaurus harti*, of southern China, the sides are ornated with blue spots. The length varies between 15 in. (37 cm) in *O. gracilis*, of India and Burma, and 4 ft (125 cm) in *O. apodus* of southern Europe and Asia Minor. In *O. ventralis*, of the eastern United States, the tail is two-thirds of the total length. The typical habitat of glass snakes is wet or dry grasslands, such as wet meadows, coastal areas and rocky slopes. The lizards are oviparous, the female laying a clutch of 4-17 eggs, and in two species, *O. harti* and *O. ventralis*, the female stays coiled around the eggs until they hatch. Their food is insects and other small arthropods, snails, small lizards and mice, which they hunt by day, more especially at dawn and sunset.

GLOWWORMS, beetles belonging to the same family as the fireflies. The common European glowworm, *Lampyris noctiluca*, is typical in that the male is fully winged and has large compound eyes, whereas the female is wingless and has very small compound eyes and so more closely resembles the larval stage. In many glowworms both sexes emit a cold yellowish-green light, but in contrast to many of the fireflies, the females usually produce the greater brilliance. The production of light by adult glowworms almost certainly enables the sexes to locate each other, but it is difficult to account for the existence of luminescence in the egg and larval stages of some species.

Although adult glowworms seem to feed very little, the larvae eat slow-moving field snails and slugs. They have a pair of sickle-shaped mandibles with which they seize their prey and at the same time inject into its tissues a digestive fluid through a fine channel in each mandible. The digestive fluid is produced in glands associated with the gut and discharges from ducts opening near the bases of the mandibles. The mandibles and glands, therefore, operate

White-tailed gnu, a rare wildebeest.

in the manner of hypodermic syringes. The glowworm larva then sucks up the semifluid mass by a pumping action of part of the fore gut known as the cibarium.

GNU, or wildebeest, an ungainly-looking antelope, related to the hartebeest and impala. Gnu have short, thick necks, and the head, which is held low, is narrow, but not as elongated as a hartebeest's, and has a convex profile. The horns are smooth with expanded bases; the general direction is downward and outward, with the tips turning up. This shape is strongly

Goat and sheep at well in Umari, Jordan.

reminiscent of the Cape buffalo, and formerly a close relationship between the two was postulated.

The two species of gnu are very different, and have sometimes been placed in different genera. The white-tailed gnu or black wildebeest, *Connochaetes gnou*, is more slenderly built, about 4 ft (120 cm) high, short-bodied and long-legged. It is black in color with long fringes on the nape and throat and in the middle of the face; the tail is long, horselike and white; and the horns turn forward rather than

outward. The brindled gnu or blue wildebeest, *C. taurinus*, is shorter-legged, longer-bodied, blue-gray in color with brownish-black stripes on the neck; there is a short-haired mat of black color in the middle of the face, a beard extended backward into a throat and chest fringe; a loose floppy mane on the neck; and the tail is long-haired as in the other species, but black. The brindled gnu is larger, about 4 ft 10 in. (144 cm) high, and the horns are shorter, less down-curved, and point out instead of forward.

GOAT, *Capra aegagrus*, closely related to sheep but distinguished by the differently shaped horns, which are the same shape in males and females, by the absence of face and foot glands, by the presence of a beard, and by the possession, in the male, of a gland beneath the tail that gives a pungent and characteristic odor.

True goats are distinguished from other members of the genus *Capra*, the ibex and markhor, by their color pattern and horns. The color of wild goats is brown-gray in winter, yellow or red-fawn in summer, with a black line down the middle of the back, a cross stripe on the shoulders, and dark brown or black face, chin, beard, throat, tail, front of the legs (except for the knees) and a stripe along the lower part of the flanks. Their horns are scimitar-shaped, with a raised keel along the front edge.

Wild goats, often known as bezoar goats because they are one source of the "bezoar stone," a medieval panacea, are found in Turkey, the Caucasus, Oman, Iran and Pakistan, extending into Iraq, Lebanon and Turkmenistan. Goats thought to be truly wild also occur on Crete and on the Aegean island of Eremomilos. They stand 26-37 in. (68-95 cm) high, with horns about 45 in. (1.15 m) long in males but only 12 in. (30 cm) in females.

Males weigh as much as 40 lb (90 kg), females only 14 lb (30 kg). They live in dry, rocky areas with little water; unlike ibex, they are not necessarily mountainous animals, some living close to sea level. However, they do ascend to high altitudes: on Mt. Ararat, wild goats may be found at 13,750 ft (4,200 m). They are agile and active, feeding in the morning and evening but resting by day. The sexes live in separate herds of 10-30, even as many as 90 together, herd size increasing later in the year. The rut takes place in two weeks during the latter half of November; the males fight for harems. Young are born from May to June, at least in Iran. One to three kids are dropped.

It is uncertain where goats were first domesticated, or when, but their remains have been found in the Belt Cave, Iran, dating from about 9000 B.C.; according to Charles A. Reed, these were domesticated goats. It is possible that goats have been domesticated more than once. The existence of corkscrew-horned domestic goats in the Northwest Frontier region of Pakistan leads one to suppose that some goats at least may be descended from the markhor. The most diverse types of domestic goats are found today in India and southeast Asia; goats bred farther north and west tend to be better maintained and to conform to type more.

GOBIES, small fishes found in coastal waters in almost every part of the globe except the polar regions. Although the majority are marine, a few species live permanently in fresh water and others are able to pass from one environment to the other. The name *Gobius* was first used by Pliny the Elder nearly 2,000 years ago. Gobies are stocky little fishes with slightly depressed bodies, having two dorsal fins (the first rather flaglike) and pelvic fins, which are united partially or fully to form a sucking disc. As a group, the gobies present great problems in classification and identification, and there are undoubtedly many species awaiting discovery. Although there is a basic

Female rock goby *Gobius paganellus* of Europe.

similarity between all gobies, so that a goby can be very easily recognized as such, there is considerable diversity in their habits and in the ecological niches that they have exploited.

Gobies have the distinction of numbering among their species the smallest known vertebrate in the world. This is *Pandaka pygmaea* from the fresh waters of the Philippines. It grows to just under ½ in. (1.2 cm) when fully mature. A close rival for the title is *Mystichthys luzonensis*, also from the Philippines, the adults of which are only slightly larger.

GODWITS, four species of large wading birds of the genus *Limosa*, two palearctic and two nearctic. They are among the largest members of the family Scolopacidae, with long legs and bills up to a least 4 in. (10 cm) in length, that of the bar-tailed godwit, *L. lapponica*, being slightly upturned. Some populations of all species undertake long migrations, the black-tailed godwit, *L. limosa*, from western Europe, where it breeds, to west Africa, the bar-tailed godwit from Siberia to western Europe, the Hudsonian godwit, *L. haemastica*, from northwestern Canada to South America and the marbled godwit, *L. fedoa*, from North to Central America. One of each geographical pair of species nests in the tundra region (the bar-tailed and the Hudsonian godwits), the other in wet meadows in the temperate zone (the black-tailed and the marbled godwits).

GOLDCREST, *Regulus regulus*, very small warblerlike bird of European woodland, found particularly in conifers, and also occurring very discontinuously eastward as far as Japan. Two closely related species known as firecrests are also found in the palearctic, the European firecrest, *R. ignicapillus*, and the Formosan firecrest, *R. goodfellowi*. There are also two nearctic species in the group, the golden-crowned kinglet, *R. sapatra*, which resembles the palearctic firecrests and the ruby-crowned kinglet, *R. calendula*.

The goldcrest is a tiny bird, only 3½ in. (9 cm) long, quite common in certain areas in coniferous or mixed woodland but often passing unnoticed because of its small size. It is a plump little bird, olive green above, pale buff below, with a yellow, black-bordered crown. It hunts very actively for insects in foliage. In winter it is often found in company with mixed flocks of tits and tree creepers. It may often be detected by its shrill, high-pitched "zee-zee-zee-zee" call. The nest is a delicate ball of moss, pine needles and cobwebs, suspended under a branch, often high above the ground at the tip of a conifer bough. Seven to ten eggs are normally laid.

The firecrest is very similar, but its crown is more orange and it has a noticeable pale eyebrow stripe above a dark stripe through the eye.

The kinglets of North America breed in the spruce belt across Canada, the golden-crowned species a little farther south also. The crown of the latter is very conspicuous, yellow in the female, orange in the male. The ruby-crowned kinglet's crown is much less obvious and is scarlet, but it can be distinguished by a white eye ring.

GOLDEN EAGLE, *Aquila chrysaetos*, large predatory bird of mountainous areas in the northern hemisphere, is 30-35 in. (76-89 cm) long with a wingspan of about 6 ft (1.8 m), and its size together with its dark brown color and graceful flight make its identification straightforward over most of its range.

The sexes are alike, although the females are rather larger than the males and there is considerable individual plumage variation. The typical adult bird has dark-brown upperparts

Goldcrest, tiny woodland bird with golden crown.

Captive golden eagle, not everyone's choice as a pet.

The golden orfe, the counterpart to the goldfish.

and similar brown underparts, often with a paler breast, giving a general golden-brown appearance. The head is in marked contrast to the body, being much paler, with the lanceolate feathers of the crown, nape and sides of the neck having bright tawny buff edges and tips, which become abraded to a pale golden buff. The tail is blackish brown, banded with dark gray and dark brown, and each feather is tipped with black. The wings are brownish black. The long feathers of the legs are yellowish brown, with the exposed legs and feet being rich yellow and the claws black. The iris is yellowish brown or hazel; the bill black, paler at the base and the cere rich yellow.

The immature plumage is generally darker than that of the adult. It is lost by series of molts typically extending over three or four years, although it may take up to six.

The golden eagle is a holarctic species, widely spread throughout a large part of the northern hemisphere. Farther south it is replaced by other large species of eagle. In the Old World it occurs as far north as 70° in Norway, and it breeds in north Sweden, Finland and north Russia to the Urals. It is present in the mountain ranges of central Europe, but is absent from low-lying countries such as Denmark, Holland and Belgium. To the south it reaches as far as North Africa, Arabia and the Himalayas. In the New World its range extends from Alaska in the north to California in the south. It is sedentary over most of its range, but there is a movement south of the more northerly birds during the winter.

The breeding season begins in late March, the first sign often being an "advertisement" flight by the male soaring over the territory. Mutual display flighting follows, with the male and female soaring together in spirals, mewing to each other. Tremendous elevations are reached in these flights, which are interrupted by successions of headlong dives with half-closed wings, ending in an upward sweep at the end of the plunge.

The nest can be on a ledge or in a tree. Incubation is by both sexes and lasts for six weeks. The typical clutch is two, the second egg not being laid until three or four days after the first. As incubation starts with the first, it hatches before the other and under conditions of stress, such as food shortage, the older chick kills the younger. The eggs are large 3 × 2.3 in. (7.7 × 5.9 cm), blunt, oval, dull white, spotted and blotched red-brown. They are single brooded.

The nestlings at first are covered in thin white or pale gray down, which is gradually replaced by a second, thicker down. They remain in the nest for about ten weeks, but linger in the close vicinity for a further two or three weeks.

GOLDEN ORFE, *Leuciscus idus,* a golden variety of the orfe, a cultivated form that has been bred in much the same way as the goldfish. The orfe is a European fish which reaches 2 ft (60 cm) in length.

GOLDFISH, *Carassius auratus,* a carplike fish native to rivers and streams of China. In the wild state, the goldfish is a dull brown in color. In many carplike species, a form of albinism is known in which all the color pigments are missing except the reds. These erythritic varieties occur as a result of a chance mutation.

Some 2,000 years or more ago the Chinese appear to have kept such mutants or sports and found that they bred true. They were easy to keep in tanks and were popular pets. Different varieties were bred to produce forms with telescopic eyes, long or split fins and knobbly heads. The various forms of goldfishes are frequently found in Chinese art, where they are used as motifs for pottery and china decoration, paintings and jade and ivory carvings. When China was visited by merchants from the West in the 16th and 17th centuries, some goldfishes were brought back to Europe and became extremely popular. Among others, Samuel Pepys mentions goldfishes that he had seen, and when the Duchess of Portland wanted to impress a distinguished visitor from Sweden, she presented him with 100 live goldfish.

The standard color is now red-gold, but varieties have been bred from pure white through yellow to brown, orange, red and black, and individuals with mottlings of these colors are known. Much of the breeding of varieties of goldfish has been done by the Japanese. The veiltail goldfish has a characteristic three-lobed tail, the blackmoor is velvety black and has not only the veiled tail but bulbous eyes. In the celestial, the eyes are telescopic and point upward while the dorsal fin is missing. The lionhead goldfishes also lack the dorsal fin, and the head is knobbly. Producing and stabilizing such varieties has been the result of much patient work, and fanciers pay considerable sums for perfect specimens. The goldfish is now probably the most widely cultivated of all fish species, both in aquaria and in ornamental ponds.

GOPHER, name applied especially to the pocket gophers, a group of burrowing rodents confined to the more arid areas of North America.

Pocket gophers range from 4-10 in. (10-25 cm) in length and are mostly a rather uniform sandy brown, although the color may vary locally in one species according to the color of the soil. The name "pocket gopher" refers to the presence of large cheek pouches, opening by a slit on the outside of each cheek. These pouches are lined with fur and can be turned inside out to disgorge their contents. In most other rodents that have cheek pouches, e.g., in hamsters, they open inside the mouth and are not fur-lined. Pocket gophers resemble subterranean rodents in other parts of the world in having small eyes, and short ears, tail and legs. However, the tail is always quite noticeable, and the eyes and external ears are never reduced to the degree found in, for example, the African mole rats. The fore feet are developed as powerful digging tools, especially in the eastern gophers of the genus *Geomys*, where the claws are greatly enlarged. In digging, however, some assistance is given by the incisor teeth, which can be used for cutting roots and dislodging stones from the burrow.

Gophers are solitary animals, each living in its own elaborate network of tunnels and rarely emerging on the surface.

Breeding takes place in spring with sometimes a second litter later in the year. The litter often numbers 5, the young being small and naked at birth. They are weaned at six weeks and are then forced to leave the parental tunnels.

Gophers are vegetarians, feeding on bulbs, tubers and roots collected in their underground tunnels and also on green plants that are pulled down into the tunnels. Such food is stuffed into the cheek pouches and carried industriously to special storage chambers, where vast quantities may be assembled.

GORAL, with their relatives, the serow, are the east Asian representatives of the chamois group of "goat antelopes." Both have coarse hair with much woolly underfur and at least a slight mane on the neck. Serow have face glands; goral do not. Goral are 26-28 in. (65-70 cm) high and weigh 58-63 lb (25-30 kg);

A typical goldfish, a household pet for the last 2,000 years.

serow vary in size, those in Taiwan and Japan being similar to goral, but mainland forms are larger, 39-42 in. (100-110 cm) high, and weighing 200 lb (90 kg). The horns in both are 9-10 in. (22-25 cm) long, slightly curved back and ridged.

The gray goral, *Nemorhaedus goral*, is gray or gray-brown with white throat and chest patches. It is found in mountains at 8,000-12,000 ft (2,460-3,700 m) from Burma through Tibet to the Sikhote Alin Mts., near Vladivostok. The brown goral, *N. baileyi*, is entirely brown; it is found in the dry country of southeastern Tibet. The red goral, *N. cranbrooki*, is bright foxy-red and occurs in a small area of northern Burma and Assam.

The big maned serow, *N. sumatraensis*, is found from Sumatra north to the Yangtze in China; it is gray-black to red, often with a white mane and whitish or tan legs. The Formosan serow, *N. swinhoii*, restricted to Taiwan, is goral-sized, black-brown with the chin, throat and lower part of limbs red-brown, and has a bushy tail. The Japanese serow, *N. crispus*, has a bushy tail also, but is quite black with some white on the

head, and differs in skull characters. Serow are found on less steep slopes than goral and in altogether moister, more forested country; they are solitary, unlike goral.

GORILLA, *Gorilla gorilla*, the largest of the primates, and one of the closest to man.

A male gorilla stands about 68 in. (170 cm) high when bipedal, and occasional giants exceed 6 ft (180 cm); the weight of a wild male is about 350-450 lb (160-205 kg), but in captivity gorillas become immensely obese, and a top weight of 750 lb (340 kg) has recently been recorded. Females are not usually more than 5 ft (150 cm) high or 200 lb (90 kg) in weight. The gorilla differs from the closely related chimpanzee in its very large size, some different proportions (longer arms, shorter and broader hands and feet), and different color pattern. The build is much heavier, and many of the proportional differences are connected with this. In particular, the much larger teeth (especially the molars) needed to support such a huge bulk must in turn be worked by much bigger jaw muscles, especially the temporal muscles, which in male gorillas nearly always meet in the midline of the skull, throwing up a tall bony crest, the sagittal crest, between them. A small sagittal crest may occur in female gorillas or in chimpanzees, but a big one, meeting a big shelf of bone (the nuchal crest) at the back of the skull, is characteristic of male gorillas only, and considerably alters the external shape of the head.

The gorilla has small ears, unlike the chimpanzee, and much broader, more expanded nasal wings, which extend down onto the upper lip (a few chimpanzees have them nearly as broad). The gorilla's skin is jet black almost from birth; the hair is black to brown-gray, and adult males develop a broad silvery white

A gorilla's size and long arms distinguish it from the chimpanzee.

"saddle" on the back. Hair is short on the saddle, rather long elsewhere. Both sexes turn gray with increasing age.

Gorillas have a discontinuous distribution. One population is found in west central Africa.

The western or lowland gorilla, *G. g. gorilla*, lives in the western area of distribution. This race is distinguished by its brown-gray general colorations; in males the whitish saddle tends to extend onto the rump and thighs. The eastern lowland gorilla, *G. g. graueri*, is found in the eastern Congo lowlands and in the Itombwe mountains. This race is black in color; the male's white saddle is restricted to the short hairs of the back, and is well marked off from the surrounding black color; the jaws and teeth are larger, the face longer, and the body and chest are broader and stockier. Finally, the mountain gorilla, *G.g. beringei*, inhabits the Virunga volcanoes and Mt. Kahuzi (west of Lake Kivu), from about 9,000 to 12,000 ft (2,743.2-3,657.6 m); it is similar to the eastern lowland race, but with longer hair, especially on the arms; the jaws and teeth are still larger, but the face is shorter and broader; the arms are distinctly shorter.

Gorillas, contrary to popular opinion, are quadrupedal; they walk on the flat soles of their feet (like chimpanzees but unlike orangutans) and on their knuckles (actually the middle phalanges of the fingers) of their hands. This type of locomotion is called "knuckle walking." They spend most of their time on the ground, but young gorillas often climb trees, where they swing about by "brachiation" (arm swinging).

Gorillas live in groups, with one male, which vary in size according to habitat.

The home ranges of gorilla troops overlap, and there seems not to be much aggression between them. When gorilla troops meet normally, they may mingle, or else ignore one another. The adult male gorilla rises on his hind legs, hoots and beats his chest; this acts as a spacing mechanism between troops, a cohesive signal within the troop, and a displacement activity when two troops meet and the leading males wish to avoid a fight. The chest beating used to be construed by big-game hunters as signal for attack—which it definitely is not. It is often accompanied by other activities, such as tearing at the vegetation and a sideways run, giving a wild sideways sweep with the arm. Young gorillas beat their chests from about a year of age.

The gorilla troop wanders over about 10-15 sq miles (17-25.5 sq km) randomly without any distinct pattern. Wherever they happen to be when dusk falls, they make their nests; these are either in the trees or on the ground. Nests consist of a platform of interwoven branches or vegetation, and a rim around; ground nests often consist only of a rim, and make use of natural objects, such as fallen tree trunks, in their construction. A new set of nests is made every night. Gorillas under the age of about three share their mothers' nests. There is no specific nesting pattern, but from the adult male's nest can be seen most or all of the others.

Gorillas eat vast quantities of vegetable matter; they have never been seen to eat meat, insects or eggs in the wild, although they will accept them in captivity.

There is no special birth season. Gestation lasts about 260 days. Newborn infants weigh 4-5 lb (1.8-2.3 kg); they are grayish-pink in color with hair sparsely covering the back and thick on the head. The skin turns black within a few days. The young gorilla's eyes begin to focus in the first or second week; it crawls at nine weeks, and walks upright at 35-40 weeks. There is usually only one infant, but one case of fraternal and one of identical twins has been recorded. The infants play alone at first, then among themselves; juveniles (above three years of age) try to take infants from their mothers to play with them. Infants also play with the adult males, which are extremely tolerant of them.

Females are sexually mature at about six years, males at seven to eight; but a male may not be full-sized until he is 14 years old.

Gorillas are not savage and will not attack unprovoked; this is not to say that wounded or threatened males are not formidable adversaries. The troop male will always prefer to lead his troop away from danger rather than fight. If cornered, however, he will bite and scratch and cause terrible wounds; on occasion a female will too, but if the troop male has been killed, the females and young will submit to being clubbed or speared to death without much attempt at resistance.

GOSSAMER SPIDERS, cause of gossamer sheets on the fields and flecks of gossamer floating in the air, a phenomenon chiefly of autumn. Gossamer is caused by the restless wandering of innumerable small spiders, chiefly money spiders (Linyphiidae), each trailing a thread. These spiders also climb up grass or railings and are wafted into the air on silk threads. Sometimes they travel at great heights, even 14,000 ft (4,200 m) has been recorded, and sometimes for hundreds of miles.

GOURAMIS, tropical freshwater fishes found from India to Malaya, and members of the family of labyrinthfishes. They have moderately deep and compressed bodies, long dorsal and anal fins, and one ray of the pelvic fin is elongated and filamentous. These fishes are highly popular with aquarists, as much for their colors as for certain of their curious habits. The name "gourami" should strictly be applied only to *Osphronemus goramy*, a food fish that reaches 2 ft (60 cm) in length and has now been introduced from the East Indies to India, Thailand, the Philippines and China. The name "gourami" has now been applied, however, to a number of fairly similar fishes that have in common the filamentous pelvic ray.

The dwarf gourami, *Colisa lalia*, is a small fish of 2 in. (5 cm) from India. It is a pretty fish with two rows of blue-green spots on the flanks on a red background. One of the best-known species is the kissing gourami, *Helostoma temmincki*, from the Malay peninsula and Thailand, a species that reaches 10 in. (25 cm) in length and is a food fish in Malaya. In this species, the broad-lipped mouths are applied to those of another individual so that the two fishes appear to be kissing. This apparent gesture of affection is more likely to be a threat in the exercise of territorial rights or mate selection. A single specimen can be induced to kiss a mirror. The lake gourami, *Trichogaster leeri*, known also as the pearl or mosaic gourami, is found in Malaya, Thailand and Borneo. The general color is bluish with a fine lacework of white spots over the body.

The pearl or mosaic gourami of southeast Asia.

Black-necked grackle of North America.

The bases of the fins are red or yellowish. In the croaking gourami, *Trichopsis vittatus*, the males make a croaking noise when they come to the surface at night for air, for, like other species, there is an accessory air-breathing organ in the gill chamber.

GRACKLES, medium to medium-large perching birds of the New World, the species ranging from 8-18 in. (20-46 cm) in length, with all black plumage except for the red-bellied grackle, *Hypopyrrhus pyrohypogastor*. Most show a distinct iridescence of purple, green or bronze and have long tails and rather heavy crowlike bills.

Grackles are essentially birds of lightly wooded country, favoring parkland, suburban and agricultural areas. They were probably originally forest-edge dwellers. Between them the grackles cover most of the New World, the rusty blackbird, *Euphagus carolinus*, breeding in Alaska and Labrador in the northern hemisphere and the austral blackbird, *Notiospar curaeus*, reaching southern Chile and Tierra del Fuego in the south.

They feed in the open and breed in trees and well-grown scrub. Nesting is essentially colonial, but isolated pairs are found, especially in the blackbirds of the genus *Euphagus*. Like many colonial breeders mating tends to be promiscuous, polygamy certainly being widespread. Typical grackles of the genera *Cassidix* and *Quiscalus* are highly omnivorous, feeding on insects and small vertebrates, eggs of other birds, dead fishes, waste grain and other human refuse. They have also been seen to kill birds up to the size of snipe in hard weather when other food is scarce.

The nest consists of twigs and grasses held together with a certain amount of mud and lined with finer material, such as horsehair and rootlets. The site is often near water, even occasionally in reed-beds. The number of eggs laid varies between species: two or three in *Cassidix*, three or four in *Quiscalus* and four to six in *Euphagus*, for example. The ground color is pale blue or greenish with a certain amount of brown flecking. Incubation is by the female only.

GRASSHOPPERS, jumping insects of moderate to large size distantly related to the crickets and bush crickets. The antennae are relatively short, usually not longer than the body and composed of fewer than 30 segments. Behind the head is a saddle-shaped structure, the pronotum, protecting the front part of the thorax. There are usually two pairs of fully developed wings, but these may be reduced in size or completely absent. The hind wings are membranous and fold up like a fan when at rest beneath the tougher fore wings. The hind legs are greatly enlarged for jumping, and the attachments of the jumping muscles to the outer wall of the hind femora form a characteristic herringbone pattern. The females have a short, four-valved ovipositor at the tip of the abdomen.

There are about 6,000 species, and the distribution is worldwide; most are tropical, but many occur in temperate regions and a few of these extend their ranges into the Arctic Circle.

The eggs are usually laid in the ground in batches of four to several hundred, enclosed in a protective pod. The young grasshoppers or "hoppers" are similar to the adults in appearance, though lacking wings, and reach maturity after molting four to eight times during a period varying from a few weeks to several months.

Many grasshoppers can produce sounds by rubbing the hind femora against the folded fore wings, and, like bird songs, these are often characteristic of each species. The males of some species have a special courtship song, which can be very complex and may be accompanied by rhythmic movements of the body and antennae. It is usually only the males that have a fully developed sound-producing apparatus, but the females of some species produce sounds not unlike those of the males, and both sexes have a hearing organ on each side of the abdomen.

Grasshoppers are active during the day and feed entirely on grasses and other plants. A few are able periodically to multiply into vast migrating swarms; these species are known as locusts.

East African grasshopper in flight.

GRASS SNAKE, *Natrix natrix,* a well-known harmless snake. It is active and slender with a long tapering tail and grows to 3 ft (1 m) normally, although 6-ft (2-m) specimens occur. Females are larger than the males. The color pattern varies considerably with the locality, age and individual. There is often a yellow, orange, white or even pink crescent on the neck with dark patches in front and behind this. There are several rows of dark markings down the body on a ground color which can be green, gray, brown, bluish and even black. Partial albinos are found, but they are not as common as black specimens.

This snake has an extensive distribution from north Africa, Britain and Scandinavia in the west, to the Caspian Sea and Lake Baikal in the USSR.

After mating in April and June, about 8-40 eggs (according to the size of the female) are laid in June and July in manure or decomposing vegetation. The higher temperatures in such sites assist in incubating the eggs. In 6-10 weeks the hatchlings, which are 6-7½ in. (16-19 cm) long, escape from the parchmentlike eggshell by rupturing it in several places with the egg tooth.

The grass snake is basically terrestrial, but, its diet being fish and amphibians, it is often associated with water. It relies on speed for catching its prey and for escape. Its defensive behavior includes hissing, striking and playing dead, as well as emitting the foul-smelling contents of the cloacal gland.

GRAYLING, *Thymallus thymallus,* a beautiful salmonlike fish found in the arctic and temperate regions of the Old World; a related species *T. arcticus* being found in the New World. The scientific name records the slight smell of thyme exuded from the flesh of these fishes. The graylings resemble a trout in general body form, but the dorsal fin is very large. The color varies somewhat, the back being a greenish, bluish or ashy gray (hence its common name) with the flanks silver or brassy-yellow and irregularly scattered with black spots and dark yellow longitudinal streaks. In the breeding season the body has a green-gold shimmer and the dorsal and anal fins and the tail become deep purple. The juveniles shoal, but the adults become solitary. The grayling is essentially a river fish and avoids lakes and large ponds. It feeds on insects, worms and snails and grows to about 20 in. (50 cm). It is a shy fish with a soft mouth and is not easy to catch, but provides good sport.

GRAY SHARKS, a large family containing over 60 species of fairly typical sharks with five gill slits, two dorsal and one anal fin, the upper lobe of the tail larger than the lower and the mouth underneath the head. Members of this family are essentially tropical fishes, but some stray into temperate waters. The largest group of species are those belonging to the genus *Carcharhinus,* which includes the cub or bull shark, *C. leucas,* the white-tipped oceanic shark, *C. longimanus,* the black-tipped reef shark, *C. melanopterus,* the brown shark, *C. milberti,* and the dusky shark, *C. obscurus.* The identification of these gray sharks is not easy, and the presence of white or black tips to the fins may be misleading since there is a species of each in the Atlantic and Indo-Pacific regions. Members of this genus are also found in fresh waters. One of the best known is the Ganges shark, *C. gangeticus,* common in the Ganges and in the Hooghly River at Calcutta, which is known to be aggressive. Another well-known species is the Lake Nicaragua shark, formerly thought to represent a distinct species but now known to be merely a freshwater form of the bull shark. It grows to about 8 ft (2.4 m) and has been reported to be dangerous to man. Elsewhere, the bull shark has been found to enter fresh waters. It is this species that is known as the Zambesi shark, and it probably accounts for many other records of sharks in fresh waters throughout the tropical regions of the world.

The great blue shark, *Prionace glauca,* is a large species that reaches a length of 12 ft (3.6 m) and is colored dark blue on the back fading to white on the belly. Although this shark is found mainly in deeper tropical waters, it wanders into temperate regions, possibly to breed, and is there found near the surface in summer. It is found in the Atlantic, the Mediterranean and in the Indo-Pacific region. Its distribution appears to be closely correlated with water temperatures. The European angling record was for a fish of 218 lb (98.7 kg).

The gray sharks are all live bearers, the young hatching within the uterus of the female.

GRAY WHALE, *Eschrichtius gibbosus,* a whale showing some characters of both the right whales and the rorquals. It grows to some 45 ft (15 m), and as in other whalebone whales the females are slightly larger than the males. It has about 150 thick yellowish plates of baleen on each side, the largest of which are only about 15 in. (37 cm) long. There are two or three short grooves in the throat, not so many as in rorquals. The body is gray, but there is marked variation in shade and most animals show lighter markings. It carries heavy parasitic skin infestations of barnacles as well as whale lice.

As its old name *Rhachianectes* (swimmer on rocky shores) indicates, the gray whale is a shallow-water, inshore species, and as such it has been easy prey to inshore fishermen. It is now found only in the North Pacific, though from fossil records it once lived in the North Atlantic. In the Pacific there are two groups, the Korean, which spends the summer in the Okhotsk Sea and then moves south in the autumn, and the Californian, which spends the summer off Alaska as far north as Point Barrow and then moves south in the autumn inshore into Californian and Mexican waters to breed. The two populations appear not to mix. The migrations follow a remarkably accurate timetable each year, and the arrival of the whales in Californian waters has become an important tourist attraction. They are protected, and numbers are increasing.

GREAT WHITE SHARK, *Carcharodon carcharias,* also known as the man-eater or white pointer, the largest of all carnivorous fishes and probably the most dangerous of all man-eating sharks, certainly the most dreaded. It is found in all warm seas and grows to over 20 ft (6 m) in length; the largest specimen on record was one from Port Fairey (Australia) that was 36½ ft (10.9 m) in length, and, although the weight was not recorded, specimens half that length can weigh over 7,000 lb (3,000 kg).

Grayling, a European fish highly sensitive to the slightest environmental changes.

The jaws are enormous and are lined with triangular teeth, the edges of which are serrated. The great white shark has an unpleasant dossier on its feeding habits. The 16th-century naturalist, Guillaume Rondelet, recorded that at Marseilles and Nice specimens had been found with an entire soldier in the stomach, complete with armor. Another specimen was recorded with a complete horse in the stomach.

A grasshopper; note the large compound eye and the mouthparts with their jointed palps.

These sharks are chiefly found in open waters and are usually caught near the surface, although there is a record of a specimen caught off Cuba at a depth of 4,200 ft (1,260 m).

GREBES, highly specialized aquatic birds, gray, blackish or brown above and paler below on the throat, foreneck and underparts. The feathering is very dense and waterproof, particularly on the satiny underparts. The nuptial plumage of grebes is usually quite distinct from the eclipse plumage, both sexes developing special plumes, colors or markings, particularly on the head, which quite transfigure them and which are primarily associated with the courtship ceremonies for which grebes are famous. The great crested grebe, in addition to the elongated double crest on its crown, grows a long, beautiful, chestnut and black tippet (a frill or ruff) on each side of the head framing its white face; the foreneck and underparts remain white, but the flanks become chestnut. The Slavonian (or, better, horned) grebe, *Podiceps auritus*, has a black head and tippets with a small central crest on the crown and a golden-chestnut, wedge-shaped "horn"

projecting from above and behind each eye and meeting on the nape. The neck, upper breast and flanks are chestnut and the belly white. The black-necked (or, better, eared) grebe, *P. nigricollis*, has a black head, neck and upper breast, with highly erectile, "peaked" crown feathers and a long "fan" of golden plumes extending from behind the eye; its flanks are chestnut and belly white.

Grebes are small or medium to quite large birds, measuring 9-30 in. (23-76 cm) in length. Among the largest species are the western, great and great crested grebes; among the smallest, the dabchicks and the least grebe. Grebes are highly adapted for diving from the surface and for underwater swimming. Their legs are strong and situated at the rear of a rotund or somewhat elongated body. The large feet are partly webbed with well-developed, paddlelike lobes on each of the three main toes (the fourth being rudimentary). These are used to propel the bird at speed underwater with a powerful figure-of-eight movement.

In the breeding-season, grebes typically inhabit

both natural and artificial, standing fresh water (such as lakes, reservoirs, gravelpits, ponds and marshes). Some species, for example the little and lesser golden grebes, also frequent slow-moving streams. When not breeding, many grebe populations move away, especially from smaller waters, to larger lakes or coastal areas, both to complete their wing molt in safety and to avoid drought or ice. Populations of those species that summer in high latitudes, such as the western, Slavonian and pied-billed grebes, perform true seasonal migrations. When traveling, grebes usually fly by night and move on foot, swimming and feeding, by day.

Grebes' nests are usually built in the water, either tethered to aquatic vegetation (sometimes floating) or anchored to the bottom. They are simple structures of piled-up weeds, reeds and the like. The eggs are pale and unmarked but become stained brownish during incubation. This lasts from three to just over four weeks, according to species. Both sexes take turns in incubating the eggs, which are covered by material when the sitting bird is

Great crested grebe, with chick riding on its back.

disturbed from the nest, this habit serving both to conceal the clutch and reduce chilling.

GROUND BEETLES, fast-running predatory beetles. The smaller members feed on mites and springtails, but the larger species feed on larger prey, such as moth caterpillars, and they will also attack other beetles. A few species also eat plant material, and *Pseudophonus rufipes* has been recorded as a pest on strawberry beds.

The family is cosmopolitan, with over 20,000 species, and may be found in almost any terrestrial habitat from high mountains to the

Violet ground beetle *Carabus violaceus*.

seashore. Some maritime species hide in crevices when the shore is covered by the tide. The majority of ground beetles are dull in color, but some have brilliant metallic colors. *Calosoma sycophanta* is among the most splendid, with green and gold wing cases and a purple head.

The bombardier beetles, *Brachinus* species, which are included in this group, produce minute explosions in their anal regions. This remarkable effect is produced by the discharge of a fluid that immediately vaporizes with a distinct cracking sound. If these beetles are handled, the discharged fluid can burn and stain the skin, producing marks that last for several days.

The larvae of ground beetles have long, well-armored bodies and powerful jaws; like their parents they are active predators.

GROUND SQUIRRELS. Some of the better-known of these ground-living members of the squirrel family have special names such as chipmunks, prairie dogs, marmots and woodchucks, but in addition to these there are many less well-known species variously known as ground squirrels, sousliks or gophers, found in open country throughout much of North America, Eurasia and Africa.

As a whole, ground squirrels do not differ very much in structure from their arboreal relatives, but they tend to have shorter ears, shorter legs (adapted to running in constricted burrows) and shorter, less bushy tails. In keeping with the background of open plains they tend to be light in color, and many species have a longitudinal white line on the flank, serving to disrupt the outline of the animal when it freezes motionless on the approach of danger. Most species make extensive burrows, and they are accordingly equipped with powerful feet and strong claws in contrast to the more delicate and flexible digits of the tree squirrels. Ground squirrels feed on all kinds of vegetable matter, but especially seeds, and most species have internal cheek pouches in which they carry seeds for storage in the burrows.

GROUPERS, or epinephelids, large perchlike marine fishes of temperate and tropical seas. The family Serranidae contains some of the most generalized perchlike fishes and it was from forms of this degree of primitiveness that all the more specialized perchlike fishes evolved. Groupers are usually large, bulky fishes with slightly compressed bodies and enormous mouths. Many have remarkable color patterns, often made up of regular spots or mottlings but usually involving rather somber colors. Color variations between adults and juveniles, or between adults of the same species from different regions, are sometimes so great that identification is difficult. Some species, such as the Nassau grouper, *Epinephelus striatus*, can change color with chameleon-like rapidity. The estuary rock cod, *E. tauvina*, is found throughout the Indo-Pacific region. So variable is its coloration that what were formerly described as 24 distinct species are now all recognized as one. Like most groupers, this species lives among coral reefs, but it also enters estuaries. It is a large species, reaching a length of 7 ft (2.1 m) and a weight of 500 lb (220 kg). It has a reputation for being dangerous. An Australian grouper, *E. lanceolatus*, attains 10 ft (3.5 m) and has been known to stalk divers. Pearl fishermen in the Torres Straits are occasionally killed by these fishes. The dusky perch, *E. gigas*, is found in the Mediterranean but sometimes comes as far north as the English Channel. It is eagerly sought by underwater spear-fishermen. All groupers are considered valuable food fishes.

GRYSBOK, *Raphicerus melanotis*, a medium-sized antelope related to the steinbok. There is a second species, Sharpe's grysbok, *R. sharpei*. Both are similar to the steinbok, up to 22 in. (55 cm) at the shoulder, with large eyes and ears, the males with straight erect horns, 5 in. (12.5 cm) long. The coat is reddish stippled with white. The grysbok ranges from South Africa to the Zambezi, Sharpe's grysbok from Tanzania to the Transvaal. They are mainly reported from reserves.

GUAN, common name for the 13 species of birds of the genus *Penelope* in the family of

curassows, which range from Mexico through Central America to northern Argentina.
They are larger than chachalacas, about the size of a turkey and are similar in color, being brownish to dark olive green. They have short crest feathers and the chin and the throat are naked. The habits of guans are very similar to those of other curassows.
In the Marail guan, *Penelope marail*, which occurs from southern Venezuela and the Guianas to the north bank of the Amazon, the upperparts are dark glossy olive green with whitish spots on the hind neck and mantle. Its underparts lack the gloss in the feathers and the lower throat and breast feathers are edged with white giving a speckled appearance. The wings and tail are uniform brown and the legs are red. It is a well-known bird in the forests of the Guianas where it is the most numerous of the curassows.

GUANACO, *Lama guanicoe*, a wild form of the llama, living in South America, slightly smaller and more delicately built than the domestic type, *Lama glama*, with a single-colored reddish-brown coat, blackish-gray muzzle and ears, and belly and insides of the legs white. It stands about 3½ ft (1.1 m) at the shoulder. It is the only one of the four types of llama that thrives not only at high altitudes but also on the plains, rarely going above 9,000 ft (3,000 m). It ranges from southern Peru to Tierra del Fuego in the south, and at the present time it is found particularly in Patagonia.
Guanacos usually live in family groups, forming small herds of up to 20 led by a stallion. These cuadrillas sometimes join up into larger herds, but do not mix. The stallion gives warning of danger by characteristic penetrating "neighing" and guards the rear when the herd is in flight, in close formation. If the stallion is killed, the herd rushes off in great disorder. Deposed stallions often live alone.

The chief enemies of guanacos are the pumas and, for the very young, the condors, but above all man, who hunts the very young animals for their soft coats. Of all four kinds the wool of least value is that of the older animals of all four llama types.
The mating season is November to February. The mare reaches sexual maturity at one year old and is in heat for only a few days. She produces a calf after 11 months.

GUENONS, common monkeys of the African forest, and their savanna relatives. About 20 species have been recognized. They are all small with long tails, small faces and round skulls, and small teeth. Many of the Old World monkeys have five cusps on the last lower molar, but guenons never do. Their fur is always speckled, with alternating light and dark bands, and generally brightly colored with white spots or bands on the face or thigh.
The 16 species of *Cercopithecus* can be arranged into a number of species groups, of which one contains over half the species. They can be briefly characterized as follows:
1. *Aethiops* group. Only one species: a. *C. aethiops*, green, grivet or vervet monkey. Large, greenish with a black face surrounded by white brow band and cheek whiskers. Lives in savanna and woodland savanna of west, east and South Africa.
2. *Mitis* group. Big, sedate monkeys with a smooth mat of hair on the cheeks, black underside, black hands and feet. b. *C. mitis*, blue, Sykes or pluto monkey. Blue-black, commonly with a white brow band; east African races have a white throat. Lives in east and central Africa, in mature forest (including isolated forest "islands"), where it is most at home in the upper layers of the trees. c. *C. nictitans*, spot-nosed monkey (French, *le pain à cacheter*). Gray-black with a long oval spot down the nose. West African high forest.

The wreckfish *Polyprion americanum*, a grouper.

The golden-mantled ground squirrel *Citellus lateralis*.

3. *Cephus-mona* group. Small monkeys with light undersides, a black stripe from eye to ear, and flesh-colored mouth. The rest of the face is blue-gray. d. *C. cephus*, mustached monkey. Gray-red with a white "mustache"; one race has red ear tufts, another has a red tail. Nigeria to Gabon. e. *C. petaurista*, putty-nosed monkey. Gray-black with a white heart-shaped nose spot. West Africa, preferring younger forests and lower levels of the trees than blues and spot-nosed (mustached is similar to putty-nosed). f. *C. ascanius*, red-tailed monkey. Dark gray with a white nose and red tail. East and central Africa, especially in secondary forest. g. *C. erythrogaster*, red-bellied monkey. Reddish agouti, often with a red belly and a red or whitish spot on the nose. A variable species, restricted to the lower Niger region. h. *C. mona*, mona monkey. Reddish with a yellow crown and a white thigh stripe. Ghana to Cameroun, in all types of forest but favoring secondary growth and mangroves, usually at low levels.

Guenon *Cercopithecus cephus*, of Nigeria.

i. *C. campbelli*, Campbell's mona. Yellow crown and fore parts, becoming gray on hind parts. Sierra Leone to Ghana. As the last, but not in mangroves. j. *C. pogonias*, crested mona. Reddish yellow with a crest on the crown and a black line from forehead to nape. Nigeria to the river Congo, usually at high levels in mature primary forest. k. *C. wolfi*, Wolf's mona. Blackish with sharply marked white or yellow underside. Congo and Uganda, at low levels.

4. *Diana* group. Small monkeys with a white beard, cheeks and brow band around a black face. Both species are rather rare, and restricted to high levels in primary forest. l. *C. diana*, Diana monkey. Deep purple color with red "saddle" on hind parts; a long white beard curving backward. West Africa. m. *C. dryas*, dryas monkey. Yellowish without saddle or such a long beard. Congo.

5. *Lhoesti* group. Large monkeys with long, soft coats and big jaws and teeth. Found typically in montane forest but extending into nearby lowlands, and largely terrestrial. n. *C. lhoesti*, L'Hoest's monkey. Gray-black with a bushy white throat ruff. Discontinuous distribution: eastern Congo, and Mt. Cameroun and Fernando Po. o. *C. hamlyni*, Hamlyn's monkey. Gray with a long cape of hair covering the ears, and a white streak down the nose onto the upper lip. Eastern Congo.

6. *Neglectus* group. Only one species: p. *C. neglectus*, De Brazza's monkey. Gray with a white thigh stripe and orange rump; white nose, mouth, chin and beard; an orange-red diadem on the brow, outlined behind by white and by black behind this. A stocky, thickset monkey with a short tail, found sporadically through Cameroun and the Congo, favoring riverain forest.

The largest species of the genus grow up to 26 in. (65 cm) in head and body length, the smallest to only about 16 in. (40 cm). The biggest species, L'Hoest's and De Brazza's, may weigh 20 lb (9 kg).

The other genera of guenon have a single species each. The patas monkey, *Erythrocebus patas*, lives in open-country, long-grass regions of west Africa, as far east as the Rift Valley in Kenya. It is tall and rangy, built like a greyhound. Males weigh as much as 30 lb (13½ kg) and are bright red with a white underside and a blue scrotum (like some other species). The female, more drably colored, is only half the size.

Allen's swamp monkey, *Allenopithecus nigroviridis*, is 20 in. (50 cm) long, stockily built, dark greenish in color with broad dark cheek whiskers. The female has cyclic sexual swellings, and the male appears to have a permanent swelling under the tail. It is found only in the swamp forests of the Lower Congo.

The talapoin, *Miopithecus talapoin*, is much the smallest species of guenon, rarely above 12 in. (30 cm) long (head and body length) and weighing 2-3 lb (1-1.3 kg). It is light green with fan-shaped cheek whiskers. It lives also in swamp forests, mangroves and thick secondary growth, from Cameroun to the Congo River.

GUILLEMOTS, two closely related diving seabirds. The common guillemot or common murre, *Uria aalge*, and Brünnich's guillemot, or thick-billed murre, *U. lomvia*, are very similar in appearance, both being about 16 in. (24 cm) long and weighing about 2 lb (900 gm). They are black above with black heads, and

Helmeted guinea fowl.

A female guenon monkey with her growing ▷ offspring.

white beneath, with black bills and black and yellow feet. In the common guillemot the bill is long and pointed, in Brünnich's guillemot it is shorter, thicker and has a white line extending along the basal half of the upper mandible. Both guillemots are well adapted to swimming underwater. The body is highly streamlined, the large webbed feet are placed far back and the wings are short and narrow, forming efficient paddles. Guillemots in fact "fly" underwater, beating half-open wings and using the webbed feet for steering and as auxiliary propellers. As a result of these adaptations, guillemots are rather clumsy on land and have to stand upright. When young they are more active, having more mobile limbs, but as they reach maturity the mobility of the legs is decreased as the muscle blocks and their specialized attachments increase in size.

The guillemots, which feed on fish, are restricted to the northern hemisphere, and they are circumpolar in distribution although only the common guillemot nests on European coasts. They are highly social, particularly during the breeding season, and breed, sometimes in enormous numbers, in colonies, also known as loomeries or bazaars. The nests are usually on ledges of steep cliffs, sometimes hundreds of feet above the water.

Guillemots make little or no nest, the one egg being placed on the bare rock, sometimes with a few stones that the bird has accumulated, but the irregularities of the rock surface are usually enough to prevent the egg from rolling away. The egg is pear-shaped, but this does not seem to be an adaptation to prevent the egg rolling away as has often been thought. In fact, many guillemot eggs are lost through being knocked off by the adults.

GUINEA FOWL, a family of game birds about the size of domestic fowl and characterized by mainly featherless or ornamented heads and

usually spotted plumage. Under natural conditions they occur in Africa and the associated islands and also in part of Arabia, where they may have been introduced in the distant past. The plumage is thick and smooth; the rounded wings and a tail partly concealed by long coverts are not conspicuous. The outline of the body is generally rounded and topped by a bare head, which, from lack of feathers, seems proportionally small.

The plumage pattern frequently consists of slanting rows of small, black-edged, white spots on a gray ground, in some plumages tinted blue. On the vulturine guinea fowl, *Acryllium vulturinum*, the rows of spots coalesce to form streaks, and the neck feathers are white striped. This species also has vivid blue and violet color on parts of the plumage. The patterned plumage is lost in the west African forest species, the black guinea fowl, *Agelestes niger*, having black feathers, and the white-breasted guinea fowl, *A. meleagrides*, being white on the breast and mantle and otherwise black.

The head ornamentation is diverse. *Agelestes* species have bare red heads. The vulturine guinea fowl has a gray-blue head and neck, ornamented with a narrow strip of fluffy brown feathering around the back of the cranium like a monk's tonsure, and a crimson iris to the eye. The crested guinea fowl, *Guttera edouardi*, is like a parody of a fashion plate with a bare gray head arising from a collar of black feathers and topped with a cluster of curly black feathers on the crown, while a related species has a head tuft of straight feathers.

GUINEA PIG, *Cavia porcellus*, one of the most familiar of all rodents and probably the first to have been domesticated. The plump body, absence of tail and extremely short legs make the guinea pig very distinctive even among the multitude of rodent species.

Guinea pigs were domesticated by the Incas in Peru before the European conquest, but the reason at that time was purely culinary. The guinea pig lends itself well to captivity: it needs little space, it can tolerate the company of its fellows better than most other rodents and it breeds quickly, although in the last respect it falls far short of the capability of rats and mice. The gestation period is long, about nine weeks as compared with three in a house mouse, and the litter size is small, rarely over three. However, this is balanced to some extent by the female's ability to mate and conceive again immediately after the birth of a litter, and by the advanced stage of development at birth. Newborn guinea pigs can run within a few hours of birth, and the females are capable of breeding at an age of a little over a month.

A great diversity of breeds of guinea pigs has been developed, as in any animal that is such a popular pet. The three main groups are the English, which is short-haired, the Abyssinian (another peculiar misnomer), which has a rough coat with rosettes of hair, and the Peruvian, which is long-haired. Most of the animals that are now kept for laboratory work are pink-eyed albinos of the English variety. Guinea pigs feed on most vegetable food and can be kept on a mixture of roots and greens. They will crop grass industriously if kept on a lawn, and when kept caged it is necessary to provide them with hay, whatever else they are given to eat.

GUITARFISHES, sharklike fishes whose true affinities lie with the rays, the group that contains the skates and other flattened cartilaginous fishes. Unlike most rays, however, the guitarfishes have rather small pectoral fins that are not used to propel the fish, the motive power deriving from sinuous movements of the body, as in sharks. The gill slits are on the underside of the head, and the leading edge of the pectoral fins joins smoothly onto the head, as in the rays. The guitarfishes are found in warm and temperate coastal waters throughout the world, and about 40 species are known. The guitarfish *Rhinobatos rhinobatos*, of the eastern Atlantic and Mediterranean, is typical of this group. The snout is pointed, and the pectoral fins are small, tapering from the head. There are two dorsal fins, and the teeth are blunt and formed into a pavement. It is a bottom feeder that browses chiefly on crustaceans. Like all members of this family, the young are hatched inside the female and are born later.

Most of the guitarfishes are fairly small, only occasionally reaching 6 ft (1.8 m) in length, but *Rhynchobatis djiddensis* of the Indo-Pacific region has been reported to reach 10 ft (3 m) and to weigh up to 500 lb (227 kg).

GULLS, medium to large birds with moderately long wings and tail, the adult plumage being basically white with darker wings and back. The outermost wing feathers either are black or have black tips, as black feathers are stronger than white and these outer feathers become very worn. All gulls can fly and swim strongly.

There are many more species in the northern hemisphere and in the temperate regions than in the southern hemisphere or in the tropics. Some species have very restricted ranges and small populations; the lava and swallow-tailed gulls are both restricted to the Galápagos Islands and have total populations of about 400 and 10,000 pairs respectively. However, many temperate species are widespread. The common gull, *Larus canus*, has a breeding distribution about 1,000 miles (1,600 km) wide, which stretches from central Canada westward to Iceland.

The distribution of the herring and lesser black-backed gulls, the *L. argentatus-fuscus* group, provides a classical example of geographic speciation and of a ring species. Between them the subspecies of the two species encircle the globe in Europe, north Asia and northern North America. In Britain the ends of the chain overlap, and there are two species that are very different in plumage and habits and that breed side by side without interbreeding. Therefore they can be considered separate species. Even though there are several well-marked geographical forms of both species in other areas, it is difficult to draw a line separating the two species because there is a perfect series of forms linking the two extremes. Although climatic changes have altered the distribution of some forms, it is possible to piece together the likely development for this ring species. In geologically very recent times there was probably a single species of this type of gull in eastern Siberia or the Bering Sea. This spread eastward and westward, and as birds colonized new areas, they altered slightly. Gradually these new populations stabilized as separate races, and in due course the two ends of the chain developed into what we now know as the herring gull (with a pale gray back, flesh-colored legs, yellow eye ring) on the Atlantic seaboard of Canada and the lesser black-backed gull (yellow legs, dark gray back, red eye ring) in western Europe and Britain. By the time the herring gull colonized Europe, the two forms had become so distinct that they coexisted as separate species without interbreeding.

The nest is normally an untidy mass of vegetation, usually on the ground, less commonly on a cliff or tussock of vegetation, and rarely in a tree or on a building. The normal clutch of three eggs is incubated for about 28 days. On hatching, the downy young are quite mobile but must be protected and fed by the parents. Fledging normally takes six to eight weeks.

Behavior of animals can be modified by natural selection, so that the best-adapted individuals will leave the most offspring. Chicks of many species of gulls are fed near the nest by the parents regurgitating food onto the ground. As the young may be hiding some distance away, both the adults and young have special feeding calls so that they can contact each other. The kittiwake, *Rissa tridactyla*, nests on very small ledges on precipitous cliffs, and the young take food from the back of the parent's throat since, if food was regurgitated, both it and the young would be in danger of falling from the nest. As the chicks cannot move from the nest, no feeding calls have been developed, and, unlike most gulls, adults do not learn to recognize their own chicks.

GUPPY, *Lebistes reticulatus*, also known as the millions fish, a small live-bearing toothcarp from the fresh waters of the southern West Indies and parts of northwest South America. Its common name records its discoverer, the Rev. Robert Guppy, who found this little fish on Trinidad in 1866. Guppies grow to about 2½ in. (6 cm). The female is a rather dull olive, but the males, which are smaller and slimmer, are brightly colored with all kinds of variable spots and patterns (in orange, blues, green-blues, white, etc.). In the male, as in all the live-bearing toothcarps, the anal fin is modified into a copulatory organ or gonopodium, with hooks and spines, for the transfer of sperm to the female. Guppies are surface-feeding forms, and this has led to their introduction into tropical countries in order to control mosquito larvae. It is possible, however, that they also eat the eggs and small larvae of other fishes. This has certainly been found to be the case in the related species, *Gambusia affinis* (often called "guppy" but better known as the mosquitofish).

The variants, especially in the shape of the tail and the colors, that appear in aquarium populations of the guppy have been bred into distinct and true-breeding varieties, and special clubs have been formed for the culture of this species.

GURNARDS, a family of marine bottom-living fishes with strongly armored heads bearing spines. They are known as sea robins in the United States. There are two dorsal fins, the second opposite to the fairly long anal fin. The pectoral fins are large, and the first few rays are separate and can be used to support the body when the fish moves slowly along the bottom; they may also be used to probe the substrate. The body is heavily scaled, often with a row of spines at the base of the dorsal fins, and the colors are characteristically reds, yellows and oranges, with areas of blue or green on the fins. The gurnards are found in all tropical and temperate seas in shallow to moderately deep water. They grow to 2-3 ft (60-90 cm) in length, and many of them are able to produce noises.

The northern sea robin, *Prionotus carolinus*, of the Atlantic coasts of the United States, is found most commonly in shoal waters from Cape Cod to Cape Lookout. It is a typical bottom dweller, showing a preference for smooth sand and often becoming completely buried. It is capable of making noises similar to those of a wet finger drawn across an inflated balloon and is in fact the noisiest fish along the American Atlantic seaboard. The noise is produced from muscular vibrations of the large bilobed swim bladder, which occupies no less than half the body cavity.

The striped sea robin, *P. evolans*, is another noisy species found along the western Atlantic coast.

The gray gurnard, *Trigla gurnardus*, is a gray, but occasionally reddish, species that reaches 16 in. (40 cm) in length and is found from the Mediterranean northward to Iceland. The tub gurnard, *T. lucerna*, has much larger pectoral fins, which are orange and reach as far as the vent. Occasionally species common to more southerly waters wander into the North Sea. One of the most easily recognized is the piper, *T. lyra*, in which bones from the upper surface of the head project forward as two large serrated plates.

One of many colorful varieties of the millionfish, better known as the guppy, bred by aquarists.

HADDOCK, *Melanogrammus aeglefinus,* a codlike fish of the North Atlantic and economically one of the most important of all the species caught by the countries fishing the North Sea. It resembles the cod in having three dorsal fins and two anal fins but can be easily distinguished by the presence of a black blotch above the pectoral fins, the lack of other spots on the flanks and the black lateral line. There is a small barbel under the chin. The haddock grows to about 24 lb (11 kg) in weight but is generally much smaller than this.

It is found throughout the North Atlantic but is more common off European coasts. It spawns early in the year, the female producing half a million small pelagic eggs. After

hatching, the young live in the upper waters for the first year or so and then migrate toward the bottom in shallow waters, finally making for deeper waters. They are shoaling fishes that prefer sandy bottoms, where they feed on shellfish, sea urchins and small fishes. They are often marketed smoked. After the head has been removed, the fish is split down the back, gutted, steeped in strong brine for a short time, dried and then spread on sticks to be smoked for five or six hours over a fire of peat and sawdust. The black spot on the side of the haddock is said to represent the thumb mark of St. Peter, left when he held the fish and found the tribute.

HAGFISHES, marine eel-shaped fishes that constitute one of the two surviving groups of

jawless fishes. There are about 25 species, and all are found in colder seas at depths of 60-2,000 ft (20-650 m). Hagfishes have a small and fleshy fin around the tail and barbels around the mouth. The eyes are rudimentary in some species, merely small pigmented cups capable of distinguishing between light and dark but nothing more.

Occasionally in the North Atlantic a trawler will find among its living catch a dead fish, which, when opened up, will be almost hollow except for a hagfish. Hagfishes feed on dead and dying fishes, into which they burrow with the aid of their rasping tongue covered with horny teeth. They usually live burrowed in muddy areas and are very sluggish, rarely moving unless they have to. One striking characteristic of the hagfishes is their ability to produce copious secretions of slime.

The backbone is composed of cartilage (as indeed is the entire skeleton), and the fish is able to tie itself into a knot and then to flow through the knot and thus clean off the slime.

Hagfishes are usually putty-colored and reach 30 in. (80 cm) in length. They find their food by smell, and there is a single nostril above the mouth. Although both sex organs (ovaries and testes) are present in the same individual, only one of these develops. So far, no larval form comparable to the ammocoete larvae of the lampreys has been found.

HAKE, *Merluccius merluccius,* an elongated deep-water fish of the cod family found in the eastern North Atlantic. It differs from the other codlike fishes in that the second dorsal fin and the anal fins are single and not split into two. It somewhat resembles the ling, but there is no barbel under the chin, the scales are larger and the tail is truncated and not rounded posteriorly. The back is grayish, the belly white and the lateral line scales are black. The hake is found at depths down to 2,400 ft (700 m) and occurs in the Mediterranean and northward to Trondheim and Iceland. It is a migratory species that appears in the North Sea in summer. Hake are voracious fishes with sharp teeth in the mouth, and they feed chiefly on other fishes. Large, commercially caught fishes can weigh as much as 40 lb (18 kg), and the European record for a rod-caught fish is a little over 25 lb (11 kg). A large female may produce over 1 million eggs, which are small and pelagic.

HALF-BEAKS, members of the group that includes the flying fishes and the needlefishes, they are small, rarely attaining more than 12 in. (30 cm), and are found in both marine and

The underside of the head of a bonnet shark or shovelhead, showing the mouth, the nostrils and one eye.

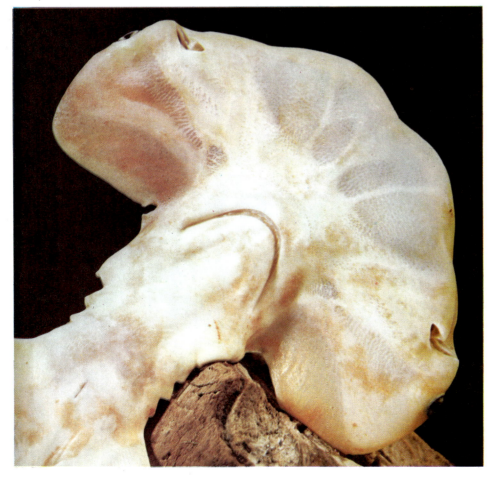

fresh waters in most of the tropical and temperate regions of the world. They derive their common name from their jaws. The lower jaw is elongated and beaklike, but the upper jaw is much shorter; teeth occur in the lower jaw only where it is in contact with the upper. In the needlefishes the jaws are equal in length, while in the flying fishes the jaws are normal. The half-beaks are surface-living forms, the elongated lower jaw apparently being an adaptation to catching food at the surface. A few species, such as *Arrhamphus sclerolepis* of Queensland, have the jaws almost equal in length. Normally, the lower jaw is slightly wider than the upper, but in the Spatulate "garfish" *Loligorhamphus normani*, from the Indo-Pacific region, the lower jaw is very much wider than the upper.

Half-beaks are clearly closely related to the flying fishes, and the juveniles are often very similar indeed. Like the flying fishes, the half-beaks have the lower lobe of the caudal fin enlarged, and they not only skitter along the surface but also sometimes leap into the air.

HALIBUT, *Hippoglossus hippoglossus,* the largest of the North Atlantic flatfishes and one of the largest fish in this order. It is reported to reach 9 ft (2.7 m) and to weigh up to 600 lb (270 kg), and specimens of 7 ft (2.1 m) and 300 lb (135 kg) are by no means rare. The body is fairly elongated, the jaws are symmetrical and the scales are smooth. It is dark olive above and pearly white below. The halibut is carnivorous, feeding on other fishes (chiefly codlike fishes) as well as squids, crabs and octopuses.

The Pacific halibut, *H. stenolepis,* is a slightly smaller fish. The females, which are larger than the males, have been reported to reach a weight of 470 lb (210 kg) in 35 years. These fishes spawn in winter, producing over 2 million eggs, which float in midwater, the fry swimming to the surface when they hatch and drifting toward the shores before settling on the bottom.

The long rough dab, *Hippoglossoides platessoides,* a relative of the halibut that is found on both sides of the North Atlantic, is a much smaller fish, usually growing to about 12 in. (30 cm). It can be distinguished from a small halibut by its rough scales and almost straight lateral line (the front part curves up toward the dorsal fin in the halibut).

HAMMERHEAD SHARK, most easily recognized of all sharklike fishes, possessing, as it does, a curious lateral extension of the

Dew-bespangled hammock web of a linyphiid spider at sunrise.

The golden hamster is a popular pet for children - but active at night!

head, whence its common name is derived. It was known to early writers as the balancefish because of the resemblance of the head to balance scales. Hammerheads are found in all warm seas throughout the world and show a tendency to move into temperate waters during the summer. In general form they closely resemble the gray sharks except for the two platelike out-growths of the head with the eyes set at their tips. Nine species are known, all placed in the genus *Sphyrna* but separated, chiefly on the basis of the shape of the head, into three subgenera. In the common hammerhead, *S. zygaena,* the flattened head is almost rectangular. It is virtually worldwide in its distribution, occasionally wandering as far north as Great Britain. It reaches 13 ft (3.9 m) in length and feeds on fishes, but like many other sharks is a scavenger and has also been known to attack men. The great hammerhead, *S. mokarran,* also having a worldwide distribution, is the largest of the nine species, reaching 15 ft (4.5 m). In the bonnet shark or shovelhead, *S. tiburo,* the head is rounded in front, giving the appearance of a shovel. This is a common species of the western Atlantic, which grows to about 6 ft (1.8 m). The hammerheads are probably all live bearers, the young being hatched within the uterus of the female.

HAMMOCK-WEB SPIDERS, spiders of several families whose sheet webs have been evolved independently by each family. In those that evolved from burrow excavators, like the Agelenidae, the spider runs on the upper surface. The delicate hammock users of the Liny-

phiidae sprang from ancestors that hung upside down in scaffolding snares, like those of the Theridiidae, and the spiders run upside down below the platform. Scaffolding threads above this platform help to cause flying insects to fall on the platform, where they are bitten from beneath through the fabric.

The platform is not always hammock-shaped as in *Linyphia montana.* Sometimes it is domeshaped, as in the American and European *L. marginata,* and sometimes flat as in *Floronia bucculenta.*

The Linyphiidae are the dominant family in temperate and arctic regions. They include the money spiders, which cause gossamer, and all are small.

The hammocks are so fine that few are noticed in shrubs or on the ground until dew or frost displays their fragile beauty.

HAMSTERS, short-tailed rodents, similar in many ways to gerbils and voles. The most familiar to many people is the golden hamster. It makes an attractive, easily-cared-for pet whose only drawback is that it is most active at night. It is, however, only one of about 14 species of hamster. The largest of these is the common or black-bellied hamster, which is unusual in having, as its name implies, black fur on its underparts with lighter, brown fur above, and white patches on its sides. It is about the size of a guinea pig and is found in Europe and Asia over a wide area between Belgium and Lake Baikal. The range of the golden hamster is from Rumania and Bulgaria through the Caucasus and Asia Minor to Iran.

Its coat is a light reddish-brown above and white underneath. The tail is short, and the skin covering the body loose, which gives the hamster an extraordinary waddling look when it moves about.

There are seven species of gray hamster, ranging from Greece and Bulgaria to the Altai Mountains on the borders of Outer Mongolia. They have long tails, up to 4 in. (10 cm), compared with their body length which is 3-8 in. (8-20 cm), and generally mouse-colored fur, although this may be reddish, with white underparts. The dwarf hamster of Siberia, Manchuria and northern China is the smallest of the species, measuring only 2-4 in. (5-10 cm) head and body, with a short tail. Its color is grayish or buff above and white on the underparts.

Hamsters are mainly vegetarian, their diet consisting chiefly of cereals; in addition they will eat many different kinds of fruits and roots, green leaves and other plant material. The common hamster eats frogs and insect larvae as well. The large cheek pouches that are a feature of these rodents are used for carrying food back to the nest. A surprisingly large amount of food can be fitted into the pouches, which may be filled so full as to give the hamster a grotesque appearance. Food is stored in their burrows by hamsters against the winter. The normal habitat of hamsters is dry, for example in steppe country, among sand dunes or on the edges of deserts. The common hamster, however, often lives in cultivated country, among crops or in plowed fields. It can also be found along riverbanks and will even swim, blowing out its cheek pouches to give itself extra buoyancy. Pet golden hamsters, on the other hand, may well die if they are exposed to damp conditions, whereas guinea pigs or white mice in the same situation remain quite unaffected.

Common and golden hamsters are nocturnal, but as with most nocturnal animals, they can sometimes be seen about during daylight hours. Also, although they hibernate, their sleep is not continuous, for they must wake up from time to time to eat, because they do not build up a layer of fat in their bodies prior to hibernation. During hibernation, the golden hamster breathes only twice a minute, and its pulse rate drops from 400 to 4 beats a minute. In captivity, golden hamsters have to be separated once mating has taken place, for the female is likely to kill the male. Gestation of the golden hamster, the period of which is known only for individuals in captivity, is 17 days, while that of the common hamster is 19-20 days. The young are born naked and helpless. There are 6-12, and occasionally as many as 20 to a litter. They leave their mother as soon as weaning is completed, at three to four weeks old, and are then independent. There may be two or three litters in a year. Golden hamsters may live for two years, but not many females of over nine months of age breed. Thus the average for a female is four litters altogether. Although there must be many thousands of golden hamsters alive in captivity today, they all stem from one family. This remarkable story began in 1930, prior to which the golden hamster was known only from museum specimens. But in that year a female and 12 young were dug up near Aleppo in Syria. They were

The black-tailed jack rabbit.

taken to the Hebrew University in Jerusalem and became the ancestors of all present-day captive golden hamsters. Besides those kept as pets—and they are very popular, being easy to keep, clean and attractive—a great many are used as laboratory animals, especially for research into skin grafting and hibernation.

HARES, including true hares and seven species of jack rabbits, all belonging to the genus *Lepus.*

True hares are medium-sized animals, adapted for swift running, characterized by their long ears, short tails, long hind legs and feet, and slender bodies. They are distinguished from rabbits by skull characteristics. Young hares, or leverets, are born at an advanced stage of development, fully furred, with eyes open and capable of independent movement. No elaborate nest is prepared for them. The gestation period lasts 40-50 days compared with about 30 days for rabbits, which are born naked and blind. Hares are herbivorous, inhabiting mainly grasslands, although some species occur in timbered habitats. They have the widest distribution of all lagomorphs, being represented

The snowshoe rabbit or varying hare *Lepus americanus,* one of the North American hares, halfway between its white winter coat and the agouti coat of spring and summer.

Young hen harrier in aggressive display.

by many species in Eurasia, Africa, North America and, by introduction, South America, Australia, and New Zealand.

The Arctic hare, *Lepus arcticus*, weighing up to 12 lb (5½ kg), is the largest of the family. It is found north of the tree line in Canada, Alaska and Greenland. In the northern part of its range its fur is white all the year round with only the tips of the ears black. In the southern parts of its range the hair on the upperparts of the body is gray in summer. It inhabits open upland, tundra, rocky slopes and lowland areas, and shelters under hedges, rocks and depressions in the snow, in which it sometimes digs tunnels. Arctic willow, on which it feeds, strongly affects its distribution. It has a characteristic habit of walking on the hind feet. Heavy black claws and extraordinarily long, obliquely projecting incisors are adaptations to feeding in snow-covered habitats. The breeding season is short, and the 1-8 young are born in June and July.

The European hare, *L. capensis*—formerly known as *L. europaeus*—weighs up to 9 lb (4 kg). It lives in open grassland, and its natural range is Eurasia and Africa but it has been

Harlequinfish about to spawn under a leaf.

introduced to South America, Australia, New Zealand and parts of North America. It is solitary except in the breeding season when small groups form. Three to four litters are born between spring and autumn. Superfetation, that is, the conception of a second litter before the first is born, may occur, the pregnant females allowing copulation 3-4 days before birth, this resulting in a normal pregnancy. The minimum individual range of movement is about 1 mile (1.7 km). Homing to the point of original capture from distances up to 290 miles (460 km) has been reported. Little is known of its territorial behavior.

The mountain hare, *Lepus timidus*, is stockier than the European hare, has much shorter ears and much white on the top of the tail. It weighs up to 10 lb (4½ kg). In summer and autumn the fur is dusky brown, sometimes with a bluish appearance, hence the alternative common name, blue hare. There are three annual molts, the animals changing to full white in the winter with only the tips of the ears black. The occurrence of a white pelage is controlled by temperature and intensity of light. The mountain hare extends over the whole northern part of Eurasia and is also found in the European Alps. Its chief habitat is wooded areas, and it feeds mainly on grass and on bark and twigs as well in winter. It breeds from February to July, usually with 3 litters per year with 3-5 young in each litter. Gestation is approximately 50 days. It is sedentary, and marked animals were rarely found outside a circle of 100 yd (90 m) diameter from the point of original capture.

HARLEQUINFISH, *Rasbora heteromorpha*, an attractive small carplike fish from the Malay Peninsula and Thailand. It grows to 1¾ in. (4.5 cm) and has a rather deeper body than in other species of *Rasbora*. The flanks are pinky-silver with violet tints and there is a distinctive dark triangle on the posterior half of the body. It is a shoaling fish popular with aquarists.

HARRIERS, medium-sized diurnal predatory birds, possessing long narrow wings, a long rounded tail and an owllike face. They occupy a wide variety of open habitats over which they hunt at low level, with a leisurely flapping and gliding flight, in search of small ground prey. There are ten species. Harriers have a slender appearance and weigh about 12 oz (350 gm). The head is relatively small, the owllike face resulting from a facial ruff of feathers. The bill is small, compressed and sharply curved, and the nostrils are large and covered with bristles. The feet are yellow and long, the toes are also long and carry sharply curved claws. The average body length is 18½ in. (47 cm). The largest species are the hen harrier, *C. cyaneus*, and the spotted harrier, *C. assimilis*, which reach 22 in. (56 cm); the smallest is the pied harrier, *C. melanoleucus*, of only 15 in. (38 cm). The females are larger than the males.

Their coloration is mainly brown, and, with the exception of the spotted harrier and female marsh harrier, *C. aeruginosus*, all have a conspicuous white rump. The adult males of the hen, pallid (*C. macrourus*), Montagu's (*C. pygargus*), and cinereous harrier (*C. cinereus*) have gray upperparts and are light underneath, although in immature plumage they are brown and resemble the adult females. The male pied harrier is unusual in being glossy black with white shoulders and rump, and the black harrier, *C. maurus*, has a generally black rather than brown appearance. The most distinctively colored species is the spotted harrier, which is slate gray, spotted white above, and chestnut, spotted white below.

The hen harrier (known as the marsh hawk in North America) is holarctic in distribution, and the cinereous harrier and long-winged harrier, *C. buffoni*, are confined to South America. The remaining seven species occur in the Old World: the pallid and Montagu's have a Eurasian distribution; the pied inhabits

Kongoni or Coke's hartebeest on a savanna near Nairobi, Kenya.

northern Asia; the black harrier and African marsh harrier, *C. ranivorus*, are found in South Africa; the spotted harrier in Australia; and the marsh harrier almost everywhere.

At the start of the breeding season the males are usually the first to arrive at the breeding territories, where they perform aerial display flights. Later they are joined by the females, who may accompany the males in these flights. The display flights include circling, soaring and spectacular dives and loops. With one exception, the nests are relatively small and built on the ground. The exception is the Australian spotted harrier, which builds a large flat nest of sticks in the main fork of a stunted tree about 40 ft (12 m) above the ground. The typical clutch is four eggs, but clutches vary from 3 to 8. The males, which rarely help with incubation, bring food to the sitting females. The food is exchanged in flight, the male dropping it from his claws to be caught by the female in hers. This aerial food exchange is a noted feature of the breeding season. The eggs hatch after about a month, and the young fly when just over five weeks old, although even before this they will have often left the nest to hide in the surrounding vegetation. When the young are two weeks old the females join the males in hunting.

When hunting, they fly a few feet above the ground, skimming the vegetation and suddenly dropping onto prey. This is most usually small mammals, but also includes amphibians, small reptiles, insects, nestling birds and small birds surprised as they dart from cover. They do not pursue birds in full flight. They roost and devour their prey on the ground.

HARTEBEEST, one of the least typical of African antelopes. The name "hartebeest" is given to two genera of the tribe Alcelaphini, the "true" hartebeest *Alcelaphus* and the "bastard" hartebeest *Damaliscus*. Both have upright horns with a sharp kink in the middle (like the impala; the related gnu has similar horns, but not upright), long faces, rudimen-

tary face glands and foot glands in the fore feet only. The "true" hartebeests are more specialized, with exaggerated long thin faces, and a high horn pedicle, which makes their appearance even more bizarre. The horns tend to be short and stout, and the back slopes sharply from the shoulders to the rump. The "bastard" hartebeests are more ordinary-looking with shorter faces, horn pedicles that are not raised, longer and more slender horns and horizontal backs. In both kinds of hartebeests, females have horns as large as those of males.

There are three species of "true" hartebeest. The most widespread is the common hartebeest, *A. buselaphus*, which formerly was distributed throughout north Africa and south of the Sahara as far as Tanzania. It is reddish or yellowish brown in color, varying in the different races. The races of common hartebeest are as follows: (1) Bubal hartebeest, *A. b. buselaphus*. The extinct race of north Africa. It was 44 in. (110 cm) high, pale yellow-fawn with horns forming a U when seen from in front. (2) Western hartebeest, *A. b. major*, Senegal, east to northern Cameroun, 48 in. (120 cm) high, red-fawn with horns forming a U. (3) Lelwel or Jackson's hartebeest, *A. b. lel wel*, Central African Republic to southwestern Ethiopia, Rwanda and western Kenya. Deep foxy-red, the horns forming a V from in front. (4) Tora hartebeest, *A. b. tora*. East Sudan, northern and western Ethiopia. Tawny with a light rump; horns forming a high, inverted bracket. (5) Swayne's hartebeest, *A. b. swaynei*. Central, southern and eastern Ethiopia and Somalia. Similar to the tora but with no rump patch; dark face and limbs and white-speckled flanks. (6) Kongoni or Coke's hartebeest, *A. b. cokii*, southern Kenya and northeastern Tanzania. Horns bracket-shaped, but with the mid-portion long and horizontal instead of upwardly directed as in the last two. The horns are also rather stout. Tawny color, paler on the rump.

The second species is Lichtenstein's harte-

beest, *A. lichtensteini*, which meets Coke's southwest of Lake Kitangiri in Tanzania without any interbreeding taking place and extends south into Katanga, Zambia, Rhodesia and Mozambique north of the Save River. It has horns that are on a short, broad pedicle, and are very stout and flat at the base, then bend sharply in and as sharply out again. It is tawny in color, with the chin and limbs blackish. The hair is reserved in the middle of the face.

The third species is the Cape, red or rooi hartebeest, *A. caama*, from the Cape north to Botswana, western Rhodesia and southern Angola. The horns are rather like those of the tora, but on a higher pedicle; the color is red, with dark shoulders, thighs, fore legs and face, and a white buttock patch. Now almost extinct in South Africa, it still exists in fairly large numbers in the Kalahari Desert.

Hartebeest feed by nibbling until the mouth is full, then raising the head to chew, scanning the horizon. When alarmed, they stand surveying the scene on a mound or termite hill, then flee—but not usually beyond the edge of the territory. They flee in a zigzag pattern (usually in rank order), the adults having a stilty gallop, the young running straight.

The male marks his territory by standing on a high point such as a termite mound, in order to be seen. When another male enters the territory, the resident male chases him at a fast gallop, continually overtaking him, turning and threatening with his horns, until the intruder has been seen off the territory.

When the young are born, they lie still in the grass; the mother watches her calf without coming too near except for suckling.

Hartebeest wallow regularly, pushing the head and horns into the mud and rubbing them against the flanks. They have no face glands, and hence do not mark grasses or twigs; but their foot glands leave a scent trail. Real fights occasionally occur among the males.

"Bastard" hartebeest also come in three species. The best known are the bontebok and blesbok.

HARVESTMEN differ from spiders, to which they are related, in having no "waist," the prosoma (cephalothorax) and opisthoma (abdomen) are broadly joined to form an elliptical body. The prosoma of six segments is usually covered by a single shield or scute, whereas the opisthoma has visible segments. The prosoma bears six pairs of appendages, a pair of chelicerae, a pair of pedipalpi and four pairs of long walking legs. The chelicerae are three-jointed, the last two joints forming pincers, quite different from the piercing chelicerae of spiders. The pedipalpi are leglike with six segments. The body is $\frac{1}{25}$-1 in. (1-20 mm) long, whereas the legs, each with seven segments, may be from one to eight times the body length. The pedipalpi and the first pair of legs each bear a forwardly projecting process at its base. Most harvestmen have in the center of the prosoma a tubercle of varying shape and size with an eye on each side. They have a breathing pore on either side of the abdomen, the openings of the tracheal system, and there are accessory pores on the legs of some species, which possibly aid muscular respiration in the extremities of the long legs.

Harvestmen occur over most land surfaces, although they are absent from Antarctica. Certain species have a very wide distribution; for example, *Mitopus morio* has been recorded throughout Europe including Spitzbergen and Iceland, from Siberia, China, Iran, Mongolia, north Africa and North America. Nevertheless, the bulk of species inhabit tropical regions, to which many families are wholly restricted. Over 2,300 species of harvestmen have been described.

HARVEST MITES, *Trombicula* (*Neotrombicula*) *autumnalis,* free-living in the nymphal and adult stages but frequently parasitic as larvae. The larvae parasitize mammals, ranging from man to mice, in a number of parts of Europe. In man, they cause a form of dermatitis. The six-legged larvae pierce the skin with styletlike mouthparts and inject saliva, which dissolves a feeding canal down to the deeper layers of the skin, up which tissue fluid and cell debris is sucked. On small mammals, the ear is a favored site, on man the wrists, armpit, groin and ankles. The larvae appear in July, reach a peak in numbers by September, and a few may be found on hosts during the winter. After feeding for about three days the larvae drop off, become quiescent and undergo a molt inside the larval skin. From this, the nymphanochrysalis, emerges the one free nymphal stage with eight legs and nonfunctional genital apparatus. This stage, like the adult, lives in the soil, most probably in the eggs of small arthropods. They have been raised on mosquito and collembolan eggs in the laboratory. The adult emerges from an imagochrysalis, which is formed inside the nymphal skin. There is one generation per year; distribution is very discontinuous for no obvious reason. In Britain, it is more frequent in chalk areas than on clay; the reverse is said to be true over Europe generally. The mites are said to favor the areas around old-established rabbit warrens. Although not carriers of disease in Europe, trombiculid larvae of related species act as vectors of scrub typhus in the Far East and in Australasia.

HARVEST MOUSE, an attractive mouse, lacking disagreeable odor, 5 in. (12.5 cm) in length, including its long, scaly prehensile tail, and weighing $\frac{1}{4}$ oz (7.5 gm) or less. Its thick, soft coat is yellowish red with white underparts. The face has a blunt nose, black bright eyes and large rounded ears, and the voice is a low chirp.

The range of the harvest mouse includes all of Europe except for the Mediterranean region, Siberia and China to Formosa. Over parts of its range the population has decreased, because, it is thought, of earlier harvesting and the use of reaping machines and other modern agricultural methods.

Formerly, the Old World harvest mouse was considered to be mainly diurnal, but it is now known to have 3-hourly cycles of rest and activity throughout each 24 hours. Every third hour, night and day, it feeds for half an hour and it spends $1\frac{1}{2}$-2 hours of each 3-hour period sleeping in its nest of shredded grass blades, massed on the ground or in a shallow burrow. During its waking period, a harvest mouse will climb about on the stalks of stout herbaceous or cereal plants. Its usual habitat is among the rank herbage of ditches, in pastures

or fields of cereal crops, and it is sometimes to be found in salt marshes, reed beds and dikes. Its staple diet is seeds, especially those of grasses and grain, supplemented in summer by a certain number of insects. It does not hibernate but winters in burrows in the ground or in hay or grain ricks.

Breeding is from April to September, and there may be several litters in a season. 5-9 young are born after a gestation of 21 days. The nest, built by the female, is circular, 3 in. (7.5 cm) in diameter, woven from grass or wheat blades and slung between two or three stalks. There is no entrance to it, the female merely pushing her way through the woven blades to get in or out. She does not allow the male to enter the nest. The harvest mouse is remarkable for its agile climbing, in which the tail plays a large part. It has a good grip with its hind feet, the outer of the five toes on each of them being opposable to the rest, but when the mouse is moving, its tail is constantly taking a partial grip on stems, ready to take firm hold if necessary.

HATCHET FISHES, small freshwater South American fishes that are capable of flight. Their common name is a reference to their slim bodies and very deep chests, which rise up toward the tail to resemble a hatchet in shape. The anal fin is long, and the pectoral fins are large and sickle-shaped. These fishes are quite small, rarely exceeding 4 in. (10 cm) in length, and are frequently kept by aquarists who are often unaware that the hatchet fishes are in fact flying fishes. The chest region contains the greatly enlarged bony supports for the pectoral fins, to which very large muscles are attached. These extra large muscles enable the pectoral fins to be flapped during flight, and it is this flapping of the fins that produces the buzzing noise made by these fishes when they are in the air. The fishes make a short dash before they take off, but they cannot change direction once in flight. The marbled hatchet fish, *Carnegiella strigata,*

and the related black-winged hatchet fish, *C. marthae,* are two of the smallest species, growing to less than 2 in. (4.5 cm) in length. The silver hatchet fish. *Gasteropelecus levis,* and the common hatchet fish, *G. sternicla,* reach about 3 in. (7.5 cm) and are slightly less graceful in build than the two *Carnegiella.*

HAWAIIAN GOOSE, *Branta sandvicensis,* one of the world's rarest birds, restricted, in its natural state, to the Hawaiian archipelago, where it is known as the né-né. The plumage of the né-né is rather somber, with broad transverse barrings of brown or black and pale buff, the dark bars being less obvious beneath. The sides of the face and the neck are a richer buff, with conspicuous folds or pleats in the feathers arranged in a pattern diagonally backward and downward. The head and back of neck are black, as are the feet, tail and bill.

Since the 18th century the history of the né-né has been one of catastrophic reduction in numbers as a result of man's thoughtlessness, followed by a near-miraculous recovery due to last minute efforts at conservation. The né-né, an offshoot of the migratory Canada goose stock, is a sedentary island form, adapted to the sparsely vegetated volcanic areas of the Hawaiian islands, where it is relatively little permanent water. It shows a number of features adapting it for a terrestrial, rather than an aquatic existence, such as the reduction in the webs, the strong legs and gait, and the rather upright stance. Also, like many other island animals, it is fairly tame. Thus, with the introduction of firearms to the region it suffered considerably, and its numbers were further reduced by its habitat being lost to agriculture and its eggs and young being taken by introduced predators such as rats, pigs, and cats. From a population strength of 25,000 or more it almost became extinct in the 1950s, when only 35 individuals remained alive in the wild and the total world population was only about 50 birds.

Fortunately the né-né breeds readily in captiv-

Female harvest mouse on her nest of woven grass slung between wheat stems.

ity, and from 1949 onward captive stocks in Hawaii and at the Wildfowl Trust in England have produced up to three, sometimes four clutches of eggs per female per year with 3-6 eggs in each clutch. English-bred birds were released on the island of Maui in 1962, and successive releases have built up the Hawaiian population so that it now numbers hundreds instead of dozens.

HAWK, the name properly applied to small or medium-sized, fierce, broad-winged, long-tailed diurnal predatory birds of the genus *Accipiter* and near allies. It was originally applied to the European goshawk, *Accipiter gentilis* (Dutch *Havik*), and the sparrow hawk, *Accipiter nisus*. The hawks were formerly separated into two genera, *Astur* and *Accipiter*, but all are now included in the largest falconiform genus, *Accipiter*, with 44 species. *Accipiter* is worldwide, found in many islands and from the equatorial forests to the Arctic. Near relatives include the African chanting and gabar goshawks (*Melierax* with three species) and the long-tailed hawks, *Urotriorchis macrourus*; the Australasian red goshawk, *Erythrotriorchis radiatus*, and Doria's goshawk, *Megatriorchis doriae*.

All are swift, active species of woodland or forest, rarely of more open country. They hunt by approaching prey behind cover and making a swift dash to kill, only rarely flying down birds in the open. Most hawks feed on birds, but some prey on lizards, frogs and mammals. Chanting goshawks feed on lizards, spend much time on the ground and have unusually long legs. The largest *Accipiter* is the European goshawk, the smallest the African little sparrow hawk, *A. minullus*. In all species the female is much larger and more powerful than the male, sometimes being twice as big.

Many other species are called hawks, especially in North America, where the term is also applied to buzzards (*Buteo* and near relatives), falcons (*Falco*) and harriers (*Circus*). This usage is especially confusing in, for instance, "sparrow hawk" for the American kestrel, *Falco sparverius*, and "duck hawk" for peregrine falcon, *F. peregrinus*. The usage has spread to many South American species, e.g., the Mexican black hawk, *Buteogallus anthracimus*. It has become a convenient term for any medium-sized diurnal raptor not obviously an eagle, buzzard, falcon or harrier—for instance, such specialized birds as the African harrier hawk, *Polyboroides typus*. A further specialized usage is in "hawking." Falconers may take out falcons (*Falco* spp.), but they do not go falconing, and may refer to their falcons as "hawks."

The true hawks of the genus *Accipiter* and allies make a solid stick nest in trees, lined with green sprays. In large species the same nest may be used again; but in small species, it is usually built afresh each year. In northern species up to five or six eggs may be laid, and large broods reared; tropical species lay smaller clutches of two or three. The female typically incubates alone and molts while doing so. She is fed by the male, but by the time the voracious brood is half-feathered, the female has grown new feathers herself and will hunt during the latter part of the fledging period. Incubation and fledging periods are usually short, 28-35 days. The young remain near the nest until their feathers have hardened, then disperse. If they remain too long they may be eaten by the fierce female, who will eat her mate if he is caged with her. Some females, e.g., goshawks, are very aggressive to man at the nest, but are normally very retiring and secretive, even where they are never persecuted.

Most northern species are partially or totally migratory, and even some tropical species make regular transequatorial movements, although they are usually sedentary. Because of their secretive behavior they are difficult birds to observe, but several are intimately known.

Roadside hawk *Buteo magnirostris* of South America is another buzzard always called a hawk.

HEDGEHOG, *Erinaceus europaeus*, a small, spine-covered insectivore of the family Erinaceidae, a group which is divided into the hairy hedgehogs or gymnures of Asia and the true prickly hedgehogs of Asia, Africa and Europe. It is one of the oldest families of mammals with living representatives.

The hedgehog is 5-12 in. (13-27 cm) long, and weighs 1-3 lb (400-1200 gm). It has a short tail, small ears and fairly prominent eyes. All feet have five toes with strong claws, and the hind feet are laid flat on the ground when walking (i.e., plantigrade). The skeleton lacks any great specializations, and there is little that is anatomically remarkable about the animal apart from its spines, which cover its dorsal surface.

Many mammals have evolved spines, but those of the hedgehog are unique. Each is a modified hair, about $\frac{3}{4}$-$1\frac{1}{4}$ in. (2-3 cm) long, 1-2 mm in diameter and internally composed of partitions and air-filled chambers, giving strength, but lightness. Typically, the spine is a cream color with a broad subterminal dark-brown band. The tip is sharp, but the body end of the spine forms a narrow neck, bent at an angle of about 60°, and terminating in a large round knob buried in the skin. The hedgehog has up to 6,000 spines depending on its age; they are shed singly at irregular intervals, there being no regular molt as in other insectivores. The fur of the face and underside is coarse and usually pale brown.

The hedgehog's skin musculature is also unique, enabling the animal to roll up very tightly and become entirely enclosed by the spiny part of its skin. Skin muscles also serve to erect the spines to form a prickly mass.

The distribution of the common hedgehog covers the whole of the palearctic, from Britain eastward to China.

The European hedgehog, although a member of the insectivora, feeds largely on earthworms and slugs.

Hedgehogs are usually fertile from April until September. In springtime, the male accessory reproductive glands undergo a temporary but enormous growth till they occupy a large part of the abdomen, a state of flamboyant development unrivaled by any other mammal. Mating takes place after a protracted shuffling "courtship" maneuver during which the male continuously walks around the female. The gestation period is 30-35 days, and the young are usually born in June or July. Some females have very early litters (April or May), and some may bear a second litter as late as October. The average litter size (in Britain) is 4 to 5, but larger families are produced on the Continent.

The young are born blind and pink with no spines showing. A first set of soft white spines breaks through the skin within hours of birth, and these are supplemented by two further sets of normal brown spines that develop as the infants grow. The mother begins to lead the family on foraging excursions after about a month, and at the age of about six weeks the young disperse to lead solitary lives.

All hedgehogs are nocturnal and have a keen sense of smell and acute hearing. Their eyesight is not so good, though they can be taught simple tests of visual discrimination. They are surprisingly agile and can run, climb and swim. However, most of their activity consists of meticulously searching the ground for food, mainly invertebrates and especially worms, beetles and slugs. They will eat almost anything edible they come across and are persecuted on suspicion of taking gamebird eggs and chicks. The hedgehog hibernates in the colder parts of its range for four to five months in a nest, usually sited under logs or thick undergrowth. Hibernation is not merely a deep sleep, but involves the most profound changes in the animal's physiology. Its heart rate falls from 180 beats per minute to 20 or fewer, breathing almost ceases and the body temperature is greatly reduced.

HELIOZOANS, single-celled animals, also called sun animalcules because of their distinctive appearance. Their characteristic feature is the presence of a number of radiating cytoplasmic projections called axopodia that emerge from the spherical single-celled body. It is this sunlike appearance that gives the group its name. Some forms possess shells, which are made out of grains of sand or similar small particles, and others possess skeletons made of silicon. Neither the shells nor the skeletons are as elaborate as those seen in other orders of the Sarcodina. The body itself consists of two layers of cytoplasm, the ectoplasm and the endoplasm, as in other orders of the Sarcodina. The ectoplasm is very vacuolated whereas the endoplasm is dense and contains a number of inclusions among which are the nucleus or nuclei. There is no capsule between the ectoplasm and the endoplasm, and this distinguishes these animals from the radiolarians which characteristically possess such a capsule. Heliozoans live in the sea or in fresh water and are found typically in the plankton, although a few forms are stalked and attached to the substratum.

HELLBENDER, *Cryptobranchus alleganiensis,* the largest salamander in North America, is

The hermit crab *Pagurus bernhardus* photographed in an artificial glass shell. The tip of this shell also serves as the home for a polychaete worm *Nereis fucata.*

18 in. (45 cm) long, seldom seen, nocturnal and lives strictly under the water in the eastern United States. By day, it hides in cracks and crannies found around submerged rocks and logs. Its food consists of any smaller aquatic animals. Although possessing lungs and gills, it breathes mainly through the skin. Its fat, wrinkled, slimy, olive-colored body is more repulsive than pretty. It reproduces by external fertilization of eggs placed in a scooped-out hole at the bottom of the pond. The male stands guard until the eggs hatch.

The hermit crab *Pagurus bernhardus* occupying the empty shell of the whelk *Buccinum.*

HERMIT CRABS, basically small modified lobsters, that, instead of having the abdomen straight and hard generally have a soft abdomen twisted to the right and inserted into an empty shell of a sea snail. The appendages on the right side of the abdomen do not develop, except for the last one, which forms part of a modified tail fan. This sticks out sideways from the end of the soft swollen abdomen, and each branch bears special antiskid surfaces, covered with minute spines that can be pressed against the inside of the shell. It is virtually impossible to pull a hermit out of its shell without tearing the thorax away from the abdomen because it holds the spines firmly in place.

A hermit crab has well-developed pincers and two pairs of functional walking legs. The other two pairs of walking legs have become reduced and modified for use as struts against the inner wall of the snail shell. They are also used to aid in carrying the shell. When danger threatens, the hermit can withdraw into the shell, leaving only the tips of its pincers showing.

The large, swollen abdomen hidden inside the shell has allowed the rearrangement of some of the internal organs. It contains the large digestive gland, the gonads and a large part of the urinary bladder, which other decapods have in the thorax.

Hermit crabs are generally detritus feeders and scavengers. The common *Pagurus bernhardus* of European shores sometimes uses its pincers to break open barnacles and tube worms, but most of its food is collected by scraping and brushing surfaces with its outermost mouthparts (3rd maxillipeds) and passing any edible material up to the other mouthparts.

HERONS, long-billed and long-legged, typically wading birds. Most of the 64 species of the subfamily Ardeinae inhabit tropical or subtropical regions, but some are found in the north temperate zone and, as some occur on every continent except Antarctica, they may truly be termed cosmopolitan. While some species have evolved as marsh-dwelling waders, others make more use of trees both to support their nests and as perches. At nesting time they are mainly gregarious, several species often living together in mixed colonies. In some parts of their range they are migratory, whereas in other areas they may wander only so far as the quest for food drives them.

Among Old World species, the purple heron, *Ardea purpurea,* the great white heron, *Egretta alba,* the little egret, *Egretta garzetta,* the cattle egret, *Ardeola ibis,* and the squacco heron, *Ardeola ralloides,* breed in Europe and are rare or very rare vagrants to Britain. Of the European herons only the gray heron, *Ardea cinerea,* nests in Britain, where its insular popula-

The pond heron or paddy-bird *Ardeola grayii* of India.

tion disperses but does not migrate.

New World species include the American egret, *Casmerodius albus egretta*, the snowy egret, *Leucophoyx thula*, the Louisiana heron, *Hydranassa tricolor*, the little blue heron, *Florida caerulea*, and the green heron, *Butorides virescens*.

Herons are slender, with long necks and bills and short tails; their wings are long and broad. The green heron of North America may measure only 16 in. (40 cm) whereas the largest members of the subfamily measure up to 50 in. (142 cm). The gray heron reaches a height of 3 ft (91 cm), has a wingspan of over 5 ft (152 cm) and varies in weight from 3-4½ lb (1.3-2 kg). In flight the feet extend beyond the tail and the neck is withdrawn. A kink in the neck is caused by the sixth vertebra articulating differently from the others, an adaptation that permits the gullet to slide before or behind the vertebrae to facilitate swallowing large prey. It also acts as a fulcrum in catching prey.

Altogether more slender than the gray heron, the purple heron has adaptations to aid its foraging in swamps. Its toes, much longer than those of its relatives, span the aquatic vegetation but are apparently less suitable for perching in trees. When flying, purple herons are easily distinguished from gray herons by their extremely slender necks, and even their longer toes can often be seen. Tree nesting is exceptional in western Europe, where nests are usually on or close to the ground, built of reed stems, or on a low bush among reeds. Curiously enough, some birds of the eastern race, *A. p. manilensis*, not infrequently nest in trees.

Adapted for wading, herons feed mainly on fishes and other animal life caught near water, although they also catch a surprising number of moles. The disruptive pattern of the gray heron is an ideal adaptation for a wait-and-watch method of hunting. The green heron has no such adaptation and catches its prey by diving. The gray heron normally stands motionless at the water's edge, waiting for prey to swim within grasping distance. It then uses its bill as a vise, not a spear. Occasionally it varies its method of hunting, as in deeper water where it may swim or dive after fishes.

About the end of February the adult birds gather in the vicinity of the nesting trees, but do not immediately take to the treetops. Adjacent to the heronry is an area that can be called the "standing ground," and there synchronization of breeding condition is achieved. At first only a few birds arrive, the number growing until the full complement is present. They stand all facing one direction, just waiting. Gradually, they begin to occupy the treetops—the males, now in full sexual condition, going first and each taking possession of an existing

nest or a stance where one will subsequently be built.

The gray heron's nest is a near-permanent structure, its dimensions increasing yearly as more material is added, until eventually it becomes so heavy that it topples in a winter gale. A first-year nest may not exceed 18 in. (46 cm) in diameter, although one that has been occupied several seasons may measure 3 ft (91 cm). Treetop nests are built of sticks, but in some parts of their range gray herons nest in marshes where reeds, docks and similar vegetation are used.

At the nest a male advertises for a mate by formalized displays, vocal and visual. Gradually the female responds and, after some rebuff, is accepted, and the nest is prepared for the reception of the clutch, which, in Europe, averages four eggs. Incubation is shared and extends to the 26th day. The gray heron, which breeds for the first time in its second or third year, is single-brooded but may lay a replacement clutch.

The young spend eight weeks in the nest, followed by a further spell in the vicinity of the heronry before becoming independent.

HERRINGS, or clupeid fishes, a large family of highly important food fishes of worldwide distribution, the best-known member being the herring of North Atlantic and Pacific waters. About 200 species are known, many of them small and confined to tropical waters but often of great importance to fisheries. The herringlike fishes have single soft-rayed dorsal and anal fins, the former usually set near the midpoint of the body and over the pelvic fins. Typically, the belly has a serrated "keel" made up of a series of sharp scutes running from the throat to the anus.

Hippo nearly submerged, showing how eyes and nostrils are last to disappear under water.

The herring, *Clupea harengus*, is the best-known member of this family. It is found throughout the North Atlantic, reaching southward to Cape Hatteras in the west and the Bay of Biscay in the east. There is a distinct population in the White Sea, but this is more closely related to the form found in the northern parts of the Pacific. The Atlantic and Pacific forms are recognized as distinct subspecies. The herring has a cylindrical body with the belly rather smooth and the scutes barely forming a keel. Rather rarely it may reach a length of 17 in. (43 cm), but usually the adults are about 12 in. (30 cm) in length. The herring is a shoaling species that congregates in enormous numbers for feeding or spawning at certain times of the year. It was once assumed that the successive appearance of the herring shoals down European coasts resulted from a gigantic army of these fishes in the Arctic that spread southward in spring. It is now known, however, that although the herrings undertake some migrations, the shoals in each area are local phenomena and it is the southward spread of the shoaling behavior and not the southward spread of the fishes that gives the impression of a vast migration.

Off European coasts there is no time of the year when spawning is not taking place. The eggs may be laid some distance from the shore and down to 600 ft (200 m), or close to the shore in bays, brackish water or in the nearly fresh water of the northern Baltic. The coastal herrings spawn mostly in brackish water during spring and early summer, whereas the sea herrings spawn in the open sea chiefly in late summer and autumn but continuing into winter. Off the North American coasts the more northerly fishes spawn in spring while those to the south spawn in summer and autumn. The herring is unique among the commercially important bony fishes of European seas in that the eggs are demersal and not floating. They are laid in enormous numbers at the bottom on stones, shells or weeds, either in irregular layers or clusters, and are heavily preyed upon by such fishes as cod, whiting and mackerel. A large female may deposit up to 30,000 eggs, which is not a great number compared with the several millions laid by certain of the codlike fishes but is made up for by the great numbers of fishes in any one shoal. The larvae hatch at the bottom and later migrate to the surface layers where they feed on members of the plankton (diatoms and other single-celled organisms at first, but later small crustaceans).

HIPPOPOTAMUS, *Hippopotamus amphibius*, one of the largest living terrestrial mammals, distantly related to the pigs. The general impression is of a large barrellike body and massive head, set on short legs, each with four toes with hooflike nails. In spite of its unwieldy appearance, the hippo is remarkably fast and agile and can overtake a running man. It is adapted to a life spent largely in water, with the sense organs—nostrils, eyes and ears—set on top of the head. All three senses are good. The nostrils can be tightly closed when it submerges, which is usually for not longer than 5 minutes at a time. In color it is a slaty copper brown, shading from very dark above to pink below the body, face and neck. Albinos are

known and are colored a brilliant pink due to the superficial blood vessels. It is hairless, except for sparse bristles on the tail, ears and muzzle. The voice is a loud repeated grunting "moo" which carries a long way over water.

The largest recorded, a Kenya male common hippopotamus, weighed 5,872 lb (2,664 kg) and was over 15 ft (4½ m) long, but in general they are much smaller. In a very large sample from western Uganda fully adult males averaged about 11½ ft (3½ m) in length from nose to the tip of the tail, the females being about 6 in. (15 cm) shorter. The average shoulder height is about 4½ ft (1½ m). Average mature weights

Although hippos spend most of their time in water, they come onto land to graze on tall grass, moving along regular pathways.

of males and females were about 3,500 lb (1,600 kg) and 3,100 lb (1,400 kg) respectively, the heaviest recorded being 4,552 lb (2,065 kg). The average weight over all age groups was about a ton (1,000 kg).

The continuously growing lower canine teeth are enormously enlarged and are exclusively used for fighting, being kept razor sharp by wear against the upper teeth. In the male the lower canines reach a combined weight of 5½ lb (2½ kg), but they grow to only half this size in the female. Their curved length is generally about 25 in. (60 cm), but less than half protrudes from the gum. The large incisors are used for digging for salt and other minerals.

The common hippopotamus was formerly found throughout Africa, including north Africa and the Sahara, along rivers and in lakes and swamps. It is still widespread south of the Sahara, but is very vulnerable to hunting and is becoming increasingly scarce in most areas outside national parks. It is restricted to water during the day, but feeds on land at night, intensively grazing a strip extending up to 6 miles (10 km) inland. The width of this grazing zone varies seasonally, increasing in the dry season as grazing quality deteriorates. The hippo trails leading inland are punctuated by dung heaps, which probably serve as chemical markers in the dark.

The gestation period is about eight months, the calf weighs an average of 110 lb (50 kg) at birth and suckles for about a year. Births usually occur on land, but have been observed in and under water. Twins are known.

Although the hippo ranges far inland, it is seldom seen far offshore. It can swim and is able, by trimming its buoyancy, to walk along the bottom. Hippo schools usually consist of

about ten animals, but there may be over 100. They contain adult females, calves and juveniles of both sexes and a few adult males. Solitary animals, usually adult males, are found in the vicinity of the school. In deeper water calves often lie across their mother's back. Territorial fights are common and always start at the water's edge or in the water. They start with an aggressive "yawning" display, but if the challenger is answered by the submissive posture (a lowering of the head), fighting is averted.

One point of particular interest about hippos is the fact that they are said to "sweat blood." In fact, the hippopotamus has no sweat glands and the "blood" is a pink sticky secretion from glands under the skin. This dries to form a protective lacquer over the skin when the animal is on land.

The average individual nightly consumption of grass is about 150 lb (70 kg), and at very high densities about 5-6 tons of grass is consumed by the population per sq mile (2.7 sq km) each night. The fishery based on Lake George in western Uganda is one of the most productive in the world, mainly because of the very large hippopotamus populations along its shore fertilizing the water with their dung.

The pygmy hippopotamus, *Choeropsis liberiensis*, is now only known from Liberia and Sierra Leone, where it frequents forest streams singly or in pairs. It closely resembles a young common hippopotamus, which, compared with the adult, has a shorter body, relatively longer legs and a small head with the eyes and nostrils not much raised. The pygmy adult is about 5-6 ft long (1.5-1.8 m), 2 ft 6 in. (76 cm) high and weighs about 400-500 lb (180-230 kg). Very little is known of its ecology, but it breeds freely in captivity, the calf weighing from 6-10 lb (2.7-4.5 kg) at birth. The gestation period is about seven months.

HOGFISH, *Lachnolaimus maximus*, a shore fish of the western Atlantic and a member of the wrasse family. It is known as the capitan or pex perro in the Spanish-speaking parts of the Caribbean. This species has a fairly deep and compressed body and can be easily recognized by the elongation of the first three dorsal rays. In general color the hogfish is reddish with a dark spot far back at the base of the dorsal fin. Its common name is said to stem from the profile of the head, but the resemblance to a hog is not very convincing. These fishes are most abundant in open areas near reefs, especially where gorgonians are present, but are also found less commonly over mud or broad stretches of sand. Hogfishes feed chiefly on mollusks. It is a highly esteemed food fish, although on rare occasions it has been implicated in ciguatera poisoning.

HONEYCREEPERS, a family of small, brightly colored birds of the New World. There are thought to be 39 species.

Honeycreepers are about the size of a tit, and in many species the males are brilliantly colored. They have a thin and somewhat decurved bill. Although they are confined to the New World, there their distribution is wide, extending from Mexico through Central and South America (including the West Indies) to Argentina. One of the best-known species is the bananaquit, *Coereba flaveola*.

Another brilliantly colored species is the green honeycreeper, *Chlorophanes spiza*, which is somewhat larger than the previous two. The upperpart of the male's head is velvety black, and the red eyes are prominent, the remaining parts are glossy green, somewhat darker on the back. The female's head is green where that of the male is black. These three species are often kept in captivity and are known among aviculturists as "sugarbirds."

The black-faced dacnis, *Dacnis lineata*, and the blue dacnis, *Dacnis cayana*, are again brightly colored, the male of the former being turquoise blue, except for the forehead sides of the face, upper back, wings and tail, which are black. It has conspicuous yellow eyes. The male blue dacnis is turquoise blue except for the base of the forehead, the throat, the upper back, wings and tail, which are black. In both species the female is green, but the female of the blue dacnis has a grayish-blue patch on the upper part of the head.

HONEY EATERS, a large family of small to medium-sized, mainly nectar-feeding birds. There are 167 species. Most of the species are found in Australasia and the Pacific islands, but one Australasian species reaches Bali, one species occurs in the Bonin Islands near Japan, and the two sugarbirds, *Promerops* species, are confined to the southern tip of Africa and isolated from the rest of the family. Basically, honey eaters are fairly slenderly built birds, with strong feet enabling them to cling to twigs and plants in a variety of postures, a rather elongated and often slender bill, and a specialized brush tongue designed for extracting nectar from flowers. This tongue is long and narrow and can be extended beyond the tip of the bill. It is curled upward and inward along the edges to form a double groove. The end is split into four main segments in the honey eaters, other nectar-feeding birds in different families varying in this respect. Each of the terminal segments is split again into a number of fine filaments, the whole forming a brushlike tip to the tongue.

In size, the honey eaters range from tiny species about as big as small warblers or wrens to large ones approaching jays or magpies in size. Their habitats are equally diverse, ranging from moist tropical forest to dry heathland and semidesert, although usually associated with flowering trees or shrubs, however small. In addition to nectar, many species take insects and various small fruits. In some instances where honey eaters of various species have been seen apparently feeding from flowers, it has been discovered that they were, in fact, collecting the insects that were feeding in the flowers.

The smallest species is the pygmy honey eater, *Oedistoma pygmaeum*, a tiny olive-colored bird less than 3 in. (7 cm) long with a thin, decurved bill and very similar in appearance and behavior to sunbirds, family Nectariniidae. *Myzomela* is a large genus of similarly small species, 3-4 in. (7-10 cm) long with thin, curved bills for probing flowers. They are boldly patterned in contrasting areas of scarlet, black and gray, and are the only honey eaters to show red on the plumage.

Female cape sugarbird feeding her chicks.

Most species are gregarious, living in groups often associating with other species of the family at the tops of trees in light forests, forest edges and gardens. They feed not only on nectar but also on fruits and insects.

Among the most brightly colored are the red-legged honeycreeper, *Cyanerpes cyaneus*, and the purple honeycreeper, *Cyanerpes caeruleus*. The former has a wide distribution, as it occurs from Mexico south to Bolivia whereas the latter has a more restricted range from Colombia and Venezuela south to northern Bolivia. The male red-legged honeycreeper has purplish blue underparts, the crown of the head being turquoise blue and the back, wings and tail black. It has bright red legs. The male purple honeycreeper is purplish blue except for the throat, wings and tail, which are black. It has bright yellow legs. The females are totally different, as they lack the purple-blue coloration, being green on their upperparts and streaked green and white on the underparts. The female purple honeycreeper can be easily distinguished from the female red-legged honeycreeper, as its chin and throat are fawn and it has a narrow blue streak on the side of the face.

The purple honeycreeper *Cyanerpes caeruleus*.

Honey eater of the genus *Myzomela*.

A hoopoe with its crest lowered.

Species of the genera *Meliphaga*, *Lichmera* and *Melithreptus* are more typical of the honey eaters in general. They are finch- to thrush-sized birds, mainly brown, olive, green or yellow, with prominent yellow or white tufts on each side of the head in *Meliphaga*, and bold black and white headpatterns with colored skin around the eye in *Melithreptus*. They are agile and mainly arboreal. Some *Melithreptus* species, although showing no obvious outward adaptation, have a tendency to climb up tree trunks and branches in tree-creeper fashion, searching the crevices for insects. A Hawaiian species, the o-o-aa, *Moho braccatus*, a large, black honey eater with conspicuous yellow tufts at the sides of the breast, has become fur-

ther modified in this respect, and uses its long stiffened tail feathers to prop it against the tree trunks while it hunts for insects woodpecker-fashion.

The largest honey eaters are the Australian wattlebirds; the yellow wattlebird, *Anthochaera paradoxa*, of Tasmania being about 18 in. (46 cm) long. The wattles in this species are yellow pendent vermiform structures about 2 in. (5 cm) long, one on either side of the head behind the ear coverts. The little wattlebird, *A. chrysoptera*, lacks wattles in spite of its name.

The nests of honey eaters are usually open and cup-shaped, varying from thicker structures placed in tree forks to fine, thin-walled pouches

that may be partly slung between horizontal twigs, suspended from them by their edges. Fine strips of bark are often used for building such nests. Hair is used as a lining in many, and the birds will go to exceptional lengths to obtain this. There are a number of records of hair being plucked from live animals and even from human beings. Nests are normally in trees or bushes, the height varying with the local vegetation, down to almost ground level. They may be grouped in loose colonies in some species. Clutches are of one to four eggs, and the eggs are pale pink or buff with sparse reddish-brown markings.

HONEY GUIDES, related to the barbets and woodpeckers, are small, dull brown or grayish birds with short stout bills. Although undistinguished in appearance, they are remarkable in their feeding and reproductive behavior. Most of the 12 species of honey guides live in Africa, where they are widely distributed in the forests and well-wooded savanna areas. The remaining two species are found in southern Asia.

Five of the honey-guide species are known definitely to be brood parasites, and the rest will probably be found to be so when details of their breeding habits are known. The black-throated honey guide, *Indicator indicator*, of Africa lays its eggs in the nests of barbets, bee eaters, woodpeckers, starlings and other hole-nesting birds. Most other honey guides also select hosts that nest in holes. Cassin's honey guide, *Prodotiscus insignis*, is exceptional in parasitizing such birds as white-eyes and warblers, which nest in bushes. Honey guide eggs are white, like those of many of their hosts that lay in dark nest chambers. When the female honey guide enters the nest to lay, she destroys all the host's eggs. Nestlings produced from eggs of the host laid subsequently are killed by the baby honey guide, which has hooked tips to its upper and lower mandibles. This adaptation is peculiar to the family Indicatoridae and is seen only in the nestlings. The hooks disappear before the young honey guide leaves the nest.

Honey guides are best known for their ability in finding wild bees' nests and leading a man or honey badger, *Mellivora*, to the place. So far, only two species are known certainly to show this behavior: the black-throated honey guide and the scaly-throated honey guide, *Indicator variegatus*. Upon finding an active bees' nest, which they are unable to break into themselves, both these species will guide a man or honey badger to the spot by calling insistently. The call can easily be recognized and followed through the bush. The bird keeps ahead by making short flights that eventually lead to the wild honey. After the combs have been broken open, and the honey removed by the animal in attendance, the bird consumes bee grubs and pieces of wax, which it can apparently digest. Recently, it was discovered that honey guides, unlike most other birds, have large olfactory lobes and can almost certainly smell beeswax. There are anecdotal records of honey guides being attracted to the altar in a mission—but only when beeswax candles were being used. Wax eating is not confined to the two African honey guides mentioned above. Pieces of comb, bee larvae and adult bees have been found in the stomachs of members of a Malaysian species, *Indicator archipelagicus*, which is not known to guide other animals, and presumably tackles the nests unaided. Even the two "guiding" species can probably open many of the bee nests they discover. It has been suggested that their remarkably thick skin is an adaptation for withstanding the stings of the bees during such raids.

HOOKWORMS, intestinal parasites of man and his domestic animals, which belong to the roundworms. In past times they have been the most widespread and harmful worm parasite of man, infecting and debilitating many millions of people over wide areas of the world wherever social and climatic conditions were suitable for transmission of infection. Modern methods of treatment and improved standards of hygiene have greatly reduced the extent of infection, although these parasites remain a significant source of human disease.

The life cycle of the hookworm involves free-living larval stages, which lead an independent existence in the soil, and a direct mode of infection of the final host, no intermediate hosts being involved. Because of these factors, transmission of the parasite is possible only in the warm, humid regions of the world, and hookworm infections are to be found in all countries lying between latitudes 30° N and 30° S. Two species of hookworm infect man: *Ancylostoma duodenale* and *Necator americanus*.

Although similar in many respects, the two differ in their biology, distribution and effect upon the human host. *Ancylostoma* is primarily an Old World parasite and is probably the more harmful of the two. *Necator*, the New World hookworm, is found most often in tropical regions and was possibly taken to America by infected slaves. The worms of both species are relatively small, being about ½ in. (10-12 mm) in length. The body typically shows a flexure at the front end, producing the "hook" appearance from which the common name is derived.

The adult females produce large numbers of eggs each day during their reproductive lives. It has been estimated that *Ancylostoma* lays 10,000-20,000 eggs per day and *Necator* 5,000-10,000. The eggs pass out of the intestine with the feces and, in warm, humid conditions, develop rapidly to the first larval stage. When a person walks barefoot over infected ground, the larvae penetrate rapidly through the skin of the feet by secreting digestive enzymes and dissolving the layers of the skin. Once through the skin they enter blood vessels and are carried passively to the heart and then to the lungs. Here they break out of the blood capillaries, move through the air spaces of the lungs, are carried up the trachea and swallowed. They pass unharmed through the stomach and are finally able to establish themselves in the wall of the intestine.

HOOPOE, *Upupa epops*, a pink-buff and black-and-white striped bird in its own family, Upupidae, widely distributed in warmer parts of the Old World. Throughout their range, the four races of the hoopoe, which differ only in slight variations of size and shade, cannot be mistaken for any other bird. Because they are handsome and striking and frequent gardens, lawns and parks near human habitation and are not particularly shy, they are well known and liked. They feature widely in folklore and mythology, receive mention in the Old Testament and formed an Egyptian hieroglyph. Hoopoes are 10-12 in. (25-30 cm) long, with a body about the size and proportions of a dove. They have a slender decurved bill set on a small head, but the most diagnostic feature is a large fan-shaped crest, rich buff with a black tip, which can be sleeked down on the crown or raised fore and aft. The crest, which is generally depressed, is erected when the bird is at all excited and alone suffices to identify the hoopoe, but the rest of the plumage renders it equally unmistakable. Head, neck, back and breast are pinkish-buff, richer above and paler below, and the wings and tail are broadly barred black and white. This feature is particularly striking when the bird flies, for the flight on rounded wings is slow and undulating, with the irregular flicking action of a butterfly, and this seems to enhance the pied pattern of the wings and tail.

Hoopoes are not sociable, except sometimes during migration, and the sexes are similar. For their small size, their gait is stately, and they walk over lawns, gardens and arable land probing in search of the insects, larvae, spiders and worms that comprise their diet. Occasionally, small vertebrates, such as young lizards, are taken too. Although they spend much time on the ground, hoopoes roost in trees, and often nest in tree holes.

The breeding range of hoopoes embraces all of Europe except Scandinavia, Finland and Britain, although they have nested sporadically in most of these countries and have occurred as drifted migrants as far off course as Iceland. From Europe they range across Asia to the Pacific, but are summer visitors only to these latitudes, wintering in Africa and southern Asia. In addition, there are breeding populations of hoopoes throughout Africa, occurring everywhere except desert, highland and forest, and in India.

The nest is always a natural cavity—in a tree, wall, the ground or a termite hill—and is seldom lined. Four to six or more immaculate pale blue or brown eggs are laid. The nest becomes foul, as it is not cleaned in any way.

HORNBILLS, an Old World family of birds with large, brightly-colored bills. They are divided into two subfamilies: the Bucoracinae, which includes the two mainly terrestrial African ground hornbills (*Bucorvus*), and the Bucerotinae, of somewhat more than 40 mainly arboreal species.

Hornbills are characterized by a relatively large bill, which is usually surmounted by a large decorative casque from which their name is derived. Both bill and casque are often brightly colored, and the bill itself may be ridged and grooved. The casque is sometimes enormous and reaches its extreme development in males

Male of Van der Decken's hornbill, of Africa. The female's bill is all black.

A horned frog *Ceratophrys dorsata*, of South America. It looks more like a toad than like a frog.

of the black-casqued hornbill, *Ceratogymna atrata*, of west Africa, the great hornbill, *Buceros bicornis*, of India and much of southeast Asia and the rhinoceros hornbill, *B. rhinoceros*, of Malaya, Sumatra, Java and Borneo. The general effect is of being unwieldy and top-heavy, although the casque is very light, being composed of a thin outer covering of horn that is filled with a spongelike cellular tissue. An exception to this rule is the helmeted hornbill, *Rhinoplax vigil*, of Malaya, Sumatra and Borneo. It has a solid casque, with the consistency of ivory, red on the outside but golden inside. After being specially processed, this is known as hornbill ivory, or *ho-ting* to the Chinese.

Young of the common horse mackerel in the Plymouth Aquarium.

Many hornbills have brightly colored patches of bare skin on the throat and around the eyes. The most common colors are blue, red and yellow, and the colors may differ between the sexes. In addition, the black-casqued hornbill and the yellow-casqued hornbill, *Ceratogymna elata*, both of west Africa, have bright, cobalt-blue neck wattles. Another rather bizarre peculiarity of hornbills is their eyelashes, which are long, thick, black and curly and quite as attractive as the false variety worn by some women. Most species also have a distinct and rather hairy crest. The plumage of hornbills tends to be boldly patterned in black or brown and white. The sexes are usually similar in general appearance, although the casque is often bigger and more brightly colored in the male.

Hornbills vary greatly in size, ranging from the 15-in. (38-cm) long red-billed dwarf hornbill, *Tockus camurus*, of west African forests to the turkey-sized ground hornbills of African savannas and the 4-ft (1.3-m) great hornbill of Asian forests. Most species have rather thick tarsi, while the toes are broad-soled, the three that point forward being partially united to form a pad. The two ground hornbills have much longer and thicker tarsi, clearly an adaptation to their terrestrial habits. The flight of the larger species is rather slow and labored, consisting of a series of wing beats followed by a glide. In the smaller species flight is light and swooping, the tail appearing disproportionately long and cumbersome. The larger species are remarkable for the great rushing noise their wings make. This noise is said to be caused, or at least accentuated, by the lack of feathers covering the bases of the flight feathers, which allows air to rush between them.

HORNED FROGS, very distinct and bizarre frogs of the genus *Ceratophrys*, found in most of South America east of the Andes, from Colombia to Argentina. Once seen, they are easily remembered and identified. They resemble toads and are very stocky, almost round when sitting with their legs tucked in. The legs, particularly the fore legs, are short and powerful. The most characteristic feature, however, is the large head with a blunt snout and very wide jaws. Although the eye is moderate in relation to the overall size, the distance between the eye and the jaw may equal two and a half to three times the diameter of the eye. It is this and the horizontal pupil that gives these frogs their peculiar facial expression and their unique appearance. The "horns," for which the horned frogs are named, are not present in all members of the genus nor are they found only within it.

Different species of horned frogs vary greatly in adult size. The smallest are little larger than a bottle cap; the largest would nicely fill a large soup bowl. They have a toadlike warty skin. Many species have a bold pattern of blotches or bars, and some are vividly colored. They feed readily on frogs and mice, which they easily subdue and stuff into their capacious mouths with their stout fore limbs, but they also consume insects.

Reproduction takes place in the typical frog manner, with the male grasping the female from behind with his fore legs. Several hundred eggs in a clear jellylike mass are deposited.

Shetland ponies, the smallest of the ponies.

HORNED TOAD, the confusing common name of the iguanid lizard, genus *Phrynosoma*, which is preferably known as the horned lizard. Horned "toads," which are very widespread in North America and Mexico, have short, flattened bodies armed with spines. Typical species are small, about 3-4 in. (7.5-10 cm), with a very short tail, and have long hard spines on the head. The general appearance is somewhat toadlike. They have the odd habit of squirting thin jets of blood from the eyes, which is believed to constitute a passive defense mechanism. This behavior is, however, sporadic, and some people who know these reptiles well have never seen it happen.

HORNETS, wasps that construct a large, globular nest of papery material. Originally the name was given to the European *Vespa*

crabro, which has been introduced to America, but it has been applied to many large wasps, such as the American *Dolichovespula*. Unlike many of the commoner wasps, which nest underground, the hornets build their home in trees or in human dwellings attached to rafters or the underside of the eaves. The nest is constructed of a series of horizontal combs, each comb lying parallel to its neighbors above and below it, and connected by a central spindle made of toughened paper: the whole structure is enclosed in a paper shell. Each comb consists of a cluster of cells in which the young are reared, and the entrance to each cell is closed by a wafer-thin paper seal. When the adult stage is reached, it bites through this delicate envelope and emerges. The papery material used in the construction of the nest is manu-

factured from woody plant material chewed by the hornets and mixed with saliva until it is transformed into a soft pulp. The outer paper shell that envelops the nest serves to protect the combs and their occupants from the weather.

In some countries, notably the warmer parts of the Americas, hornets are treated with a considerable amount of fear and respect on account of their painful stings and their habit of congregating around human dwellings. The popular remedy for removing hornets by burning down their nests is not always successful.

HORSE, *Equus przewalskii,* wild horses that occurred, in prehistoric times, over most of the Eurasiatic continent and, as can be judged from the huge piles of bones near the dwellings of Stone Age cavemen, in immense numbers. They survived until historic times in eastern Europe and in Siberia and Mongolia but are now on the verge of extinction. Wild horses have been exterminated by man because they damaged the crops and were competitors with the domestic stock for food and water.

Three subspecies are recognized: the steppe tarpan, *E. p. gmelini,* the forest tarpan, *E.p. silvaticus,* and the eastern wild horse or Przewalski horse, *E.p. przewalskii.* The domestic horse originates from these three wild horses, and its scientific name is *E. przewalskii caballus.* It is, however, usually referred to as *E. caballus.*

The steppe tarpan lived in the open steppe country of southeastern Europe. They were gray animals with black manes and tails and with a dark line along their backs. They were finally exterminated in the 19th century.

The forest tarpan was an animal of the wooded areas of western, central and eastern Europe, and only in the east did it survive the Middle Ages. Some small populations still existed, at the beginning of the last century, in the forests of Bialowieza and eastern Prussia. Forest tarpans were taller than the steppe tarpan; they had a shoulder height of about 48 in. (1.20 m).

The Przewalski horse inhabited the steppes and semideserts of southern Siberia, Mongolia and western China. In the wild they also are now probably extinct, although occasional resightings have been reported during the last few years from the Gobi desert on the Mongolian-Chinese frontier. It is, however, doubtful that these animals are pure-blooded wild horses, as these are known to abduct domestic horse mares and to interbreed with them. Przewalski horses vary in color from yellow-brown to reddish-brown and gray. Their legs are dark brown or black, the muzzle is whitish, and a brown line down the back is prominent in their summer coats. Mane and tail are black or dark brown; the mane is erect and bordered by lighter hair. The shoulder height is from 48 to 58 in. (1.20 to 1.46 m), about the size of a pony.

HORSE MACKERELS, also known as scads, jacks, cavallas and pompanos, are not true mackerels (Scombridae), and are distinguished from them by two small spines before the anal fin and the absence of small finlets behind the dorsal and anal fins. The horse mackerels are usually fast-swimming fishes, well streamlined, but in some species deep-bodied and compressed. The body has a line of little keeled scutes along part or the entire length of the

Mare with foal.

The horseshoe crab *Limulus polyphemus* crawling toward the sea on the sandy shore at low tide. Note the joint between the head and the rest of the body.

flanks. They are found in temperate and tropical waters throughout the world.

The common horse mackerel, *Trachurus trachurus*, of the Mediterranean and eastern North Atlantic grows to about 14 in. (36 cm), and is found as far north as Trondheim in Norway. Similar forms are found off the coasts of South Africa, China, Australia and western America. It has a short first dorsal fin, with the first spine directed forward, and a long second dorsal fin with soft rays. The back is gray-blue or green, the flanks silvery, and there is a dark spot behind the gill opening. It feeds on fishes and invertebrates. The young take shelter in the bell of the sombrero jellyfish, *Cotylrhiza*, probably for protection.

HORSESHOE CRABS, also called king crabs in North America, are relatively large marine arthropods, up to 2 ft (60 cm) in length, closely related to the extinct water scorpions (Eurypterida). Despite their common name they are not crustaceans, and, of the various groups alive today, their closest affinities lie with the arachnids. These animals are easily recognized by their horseshoe shape because the upper surface of the body is covered by a semicircular scoop-shaped plate, or carapace, fringed on the abdominal region with a series of short, stout movable spines, and terminating in a long tail spine that articulates with the posterior end of the abdomen. The mouth is situated on the underside of the cephalothorax,

almost in the center, and is surrounded by the insertions of six pairs of appendages, namely a pair of chelicerae and five pairs of walking legs. The abdomen also carries on its underside six pairs of appendages in the form of flattened plates, the first pair being fused together as a genital operculum carrying the reproductive pores, and the remaining five pairs bearing leaflike gills on their posterior faces. They possess a pair of compound eyes sited laterally on the carapace, in addition to a median pair of simple eyes.

Horseshoe crabs are carnivorous, preying on worms and mollusks. During their search for food they frequently burrow into soft sand and mud by pushing the scoop-shaped carapace into the substratum. The cephalothorax and the abdomen are hinged together so that the body can jackknife, aided by the tail spine. By flexing and straightening in this way, additional thrust for burrowing can be obtained, and the animal often buries itself completely in the sand or mud. The usual method of locomotion is by walking, although small horseshoe crabs can swim on their backs using the platelike abdominal appendages as swimmerets.

These curious creatures are truly living fossils, for they have remained virtually unchanged for over 300 million years. The four species living today belong to three different genera, namely *Limulus polyphemus* of the Atlantic seaboard of

North America and Mexico, *Tachypleus gigas* and *T. tridentatus*, and *Carcinoscorpius rotundicauda* of southeast Asia. All of these forms live in shallow coastal waters and come into the intertidal zone to mate and lay their eggs, which are buried in a depression in the sand. The larvae hatching from these bear a close resemblance to the extinct trilobites.

HOUNDS, also known as smooth hounds or smooth dogfishes, a family of small sharks externally resembling the cat sharks but being distinguished by the arrangement of the teeth, which are in the form of a pavement as in the rays. Most sharks have triangular or pointed teeth used for cutting or grasping their prey, but the flat pavement of teeth in the hounds is used for grinding and crushing food found on the bottom (mostly mollusks and crustaceans). They have a worldwide distribution, but the best-known species of European coasts is the smooth hound, *Mustelus mustelus*. It has a supple, streamlined body with two dorsal fins and one anal fin and grows to 6 ft (1.8 m) in length. It is extremely abundant, as is its counterpart in the western Atlantic, the smooth dogfish, *M. canis*. Both are inshore species that frequent shallow waters and mostly browse on the bottom for food.

The hounds are viviparous fishes in which the young are not only hatched within the uterus but are nourished through a placentalike connection between the embryo and the uterine wall of the mother. There are about 30 species known in this family, none of which grow to more than 6 ft (1.8 m) in length. They are a considerable pest to commercial fishermen because of their predation on lobsters, crabs and other fishes.

HOUSEFLY, *Musca domestica,* one of many species of flies, a few others of which are minor pests, in one way or another, although most are harmless, obscure flies that usually pass unnoticed. The housefly is unique, simply because both adult and larva happen to like the sort of surroundings in which man lives, and as a result the housefly has become attached to man wherever he goes. It almost certainly originated in the tropics, but now it occurs throughout the world. The housefly is rather small, ashy brown and gray in color, with four thin, blackish stripes on the thorax. Indoors it is most likely to be confused with the lesser housefly, *Fannia canicularis,* the males of which fly around and around in the middle of the room. The true housefly can be distinguished from most other members of the family Muscidae by the conspicuous bend in the vein at the tip of the wing.

The housefly has a spongy proboscis, with large lobes, or labella, richly supplied with branching grooves called pseudotracheae, and can feed only by mopping up liquids or semiliquid food. Solid food must be softened by expelling saliva over it, as well as regurgitating liquid from the crop. These fluids contain digestive enzymes, which predigest the solid food until it can be sucked back into the crop. It is this alternate vomiting and sucking that makes the housefly so dangerous to health, because any disease organisms that the fly has picked up with its food are likely to be voided again onto its next meal.

HOUSE MOUSE, *Mus musculus,* probably the most familiar of all rodents. Not only has it followed man throughout the world as a pest, but it is also the ancestor of the domesticated mouse, kept alike as pets by children and as experimental animals by almost every laboratory engaged in teaching or research in genetics, physiology, biochemistry or pharmacology.

Most wild house mice that infest buildings can be distinguished from other species of mice by having the fur smoky grayish-brown above and below. This does not apply, however, to many of the ancestral races of the species that continue to live predominantly outdoors. These races, e.g., in eastern Europe and in the Mediterranean region, have the underside white or pale gray as in most other wild mice.

The house mouse probably began its association with man soon after he first began to cultivate cereals more than 8,000 years ago. It is therefore difficult to know what was the original distribution of the species, but it seems likely that it included the grassland areas of the Near and Middle East, and probably also the Mediterranean region and the steppes of eastern Europe and southern Russia, perhaps as far as western China. In most of this area small, pale-bellied, short-tailed outdoor forms are still found, but often in company with the larger, darker, longer-tailed forms that seem to be more closely adapted to an indoor or farmyard existence. Throughout most of the rest of the world it is the larger and darker forms that have been spread by man. Provided that ample food is available, house mice are adaptable animals. They occur from the Arctic to the equator and have even been found living and breeding in frozen meat stores.

House mice when they are living indoors or in hay-ricks with an abundance of food and protection from cold, continue to breed throughout the year. Under these conditions five or six litters per year of five or six young is quite normal, and the potential rate of increase is therefore enormous, but natural predators keep the numbers within bounds.

HOVER FLIES, characteristically seen hovering, especially around flowers where they are a familiar sight, their wings beating so quickly that they are more or less lost to sight. Some are more common in woods, where they remain poised at different heights above the ground apparently motionless, but occasionally darting away in any direction to a new position, and back again. From the hovering position they can fly forward, backward, sideways, up or down with equal ease. Many of them are colored with black and yellow bands rather like bees or wasps so that they are not immediately recognized as flies.

Flowers are essential for hover flies. The constant activity requires a high intake of carbohydrate, and they feed freely on the nectar. At the same time they pollinate the flowers and are next in importance to bees for this.

The larvae of hover flies have very varied habits. One type, the rat-tailed maggot, lives in water or mud, feeding on decaying organic matter. This has a long, telescopically extensible breathing tube so that it can breathe atmospheric air while remaining submerged at some depth. The adult of one of these is the drone fly, so called because it resembles the drone of the honeybee. Other hover-fly larvae live in ants' and wasps' nests where they feed on excreta, while a large group are actively carnivorous, feeding largely on aphids. These aphids are easy prey for a blind hover-fly maggot as they move so little and do not notice if a neighbor is devoured. Only the soft, easily digested body contents are sucked out, and up to 60 may be eaten daily.

HOWLER MONKEYS, South American monkeys of which there are five species. In the four northern ones the sexes are similar in color: in the red howler, *Alouatta seniculus* (South America, north of the Amazon and Madeira rivers), the color is red; in the brown howler, *A. fusca* (Brazil coast from Bahia to Santa Catarina), brownish; in the red-handed

House mouse with young in the nest.

A hover fly, master of helicopter flight, at rest on a flower, which is essential to its life.

howler, *A. belzebul* (south of the Amazon in the interior of Brazil), black or brownish with red hands, feet and end of tail; and in the mantled howler, *A. villosa* (Central America and western Colombia), black and long-haired, with usually a long yellowish flank-fringe. In

Male mantled howler monkey resting on a branch. Howlers are the largest of the New World monkeys. Their calls can be heard for several miles.

the southernmost species, the black howler, *A. caraya*, from the interior of southern Brazil, Paraguay and northern Argentina, the male is black with light brown tones, and the female yellowish. These five species differ in the form of the hyoid, that of the mantled howler being smallest and least specialized, and that of the red howler by far the largest, most inflated and most evolved, measuring 38-55 mm in width and 55-80 mm in depth in the male; in the female the volume is only $\frac{1}{4}$ that of the male's. The voice of the red howler carries for three miles.

Howlers live in groups of 15-20, at least in the mantled species, with a proportion of $3\frac{1}{2}$ females to one male, but nearly always more than a single male in the group. They are entirely arboreal. They move slowly, both quadrupedally and by arm-swinging (brachiation), occasionally leaping but usually making sure to have a firm handhold on one branch before leaving go with the feet and tail from the other. When crossing gaps in the foliage, large howlers will allow themselves to be used as a bridge by juveniles, who walk across them as they cling with hands and tail. When walking through the trees, the adult male goes in front and the whole troop follows in the same path, conga-style. They feed in the high canopy and are even more highly selective of their food than spider monkeys, up to one-third of the food plucked being dropped half-eaten. Food is held between the fingers and palm, or between the second and third fingers.

Howling precedes early morning movement; it is initiated by the leading male and taken up by other troop members. As the animal howls

the jaws are opened and thrust forward with lips protruded, saliva flowing copiously. When two troops meet at the territorial boundary a vocal "battle" ensues, both troops howling and breaking off branches and shaking branches violently, but very rarely is there any fighting. Although territorial boundaries are stable over short periods, in the long term the larger troops, whose leading males are stronger and can keep a larger number of animals together, manage to shift boundaries in their favor so as to include not merely a larger area but more particularly an area with more abundant food for the particular season of the year.

HUMMINGBIRDS, a family of 319 species of tiny, nectar-drinking birds of the New World, which take their name from the noise made by their incredibly rapid wing beats. They are mainly aerial, feeding on the wing, and are related to the swifts. Like the swifts they have very small legs and feet, and the bony structure of the wings is also very much reduced, the greater part of the visible wing consisting of a large area of elongated primary flight feathers. The hummingbirds move about by flying, the short legs and little feet merely being used for perching. The head is large in relation to the rest of the body, but this is not obvious in most species because of the relatively large size of the wings and tail.

Hummingbirds are, in general, small, fast and extremely active. The largest species, the giant hummingbird, *Patagona gigas*, of the Andes, is a little over 8 in. (20 cm) long, of which half is tail. The smallest, the bee hummingbird, *Mellisuga helenae*, of Cuba, is virtually the size of an insect. The body is about 1 in. (2.5 cm) long, the bill and tail adding another inch to this. In a majority of species the body size is about 2 in. (5 cm) or less.

The hummingbird has a problem associated with its mode of feeding. When taking nectar from blossoms there is no convenient perch nearby so it must hover while feeding, then move backward to withdraw its bill from the flower. As a result, the flight is specialized. The wing beat is highly efficient, and, proportionally, the hummingbird uses fewer beats than other birds. Some of the larger hummingbirds have a rate of 20-25 beats per second, comparable with that of the considerably larger and slower tits. In small hummingbirds the rate rises to about 70 beats per second, but in the giant hummingbirds it is the surprisingly slow 8-10 beats per second.

The bill is long and fine, an exceptionally short one being only about half as long again as the head, while the longest, that of the sword-billed hummingbird, *Ensifera ensifera*, is straight and as long as head, body and tail combined. Such a bill is designed for probing long tubular flowers. The tongue is slender and elongated. It can be extended well beyond the tip of the bill, and the edges are rolled in to form a double tube up which the nectar is sucked.

Many hummingbirds are forest dwellers, but they may also occur in a wide range of habitats, extending into open country where flowering plants are present and also to high altitudes in mountain regions. In some areas they are nomadic or subject to seasonal movements in order to take advantage of the flowering seasons of different plants. Several species are migratory, and three move into North America to nest in the southern parts of Canada or Alaska.

Hummingbirds usually bathe on the wing, in fine spray or rain, or by brushing against wet foliage. They do not tuck the bill into the feathers of the back when they sleep, but squat on a twig with body feathers fluffed, the head tilted back a little and the bill pointing up at an angle.

The displays of the hummingbirds are relatively unspectacular in terms of movement, usually being very rapid swoops terminating in a hover in front of the female, but the brilliance of the display lies in the vivid iridescent colors of the male's plumage and the often elaborate plumage decorations such as crests, ruffs, beards and tail streamers. Most of the more vividly iridescent plumage is on the throat and crown—brilliant shapes of red, yellow, pink, purple, blue or green. In many cases this is really conspicuous only from one angle, usually from the front, so that it shows to best advantage when the male hovers before the female. The head color may be enhanced by decoration, such as the long tapering green or violet crests of the plover-crests, *Stephanoxis*, the green and red paired horns of *Heliactin cornuta* or the black and white paired crest and pointed beard of *Oxypogon guerinii*. In addition to a bright crest, the coquettes, *Lophornis*, have erectile fan-shaped ruffs on the sides of the neck, boldly colored and with contrasting bars or spots at the feather tips. When the male hovers before the female, the other visible parts of the plumage are the undersides of wings and tail. The body plumage

of most hummingbirds is glossy green or blue, and the underwings may have a contrasting chestnut-red tint. The tails of some show an overall bronze or purple iridescence on the undersides of tail feathers, visible in display but not obvious from above; while on others the feathers may be tipped, streaked or blotched with white. In addition, the tails may have a variety of shapes. Broad fans are frequent, and forked tails vary from blunt forks to the long scissor shapes of the trainbearers, *Lesbia*, and the slender tail streamers of the streamer tail, *Trochilus polytmus*. Others show different degrees of tapering and elongation of the central tail feathers, as in the hermits, *Phaethornis*. The racket tails, *Ocreatus*, have the two outer feathers elongated to long wires terminating in large rounded vanes, and show the additional decoration of white down patches like powder puffs around the legs. The latter feature is not unique, being found in some other species.

The nests of hummingbirds are built of fibers, plant down and similar fine material and moss and lichen, bound together with spiders' webs. Some nests are smooth, neat cups on twigs, others are domed, and some are pendent, or built onto hanging plants. Nests are often decorated externally with lichens. The female undertakes the building of the nest and the incubation and care of the young. She lays two white eggs, which are very large in proportion to her body size and very bluntly elliptical. Incubation takes about a fortnight, and

The black-throated trainbearer, *Lesbia victoriae*.

the young are born naked. The female feeds them by inserting her bill well into the chicks' gullets and regurgitating food. The young take three to four weeks to fledge, the period varying, apparently in response to the food supply available.

HUMPBACK WHALE, *Megaptera novaeangliae*, a whale with long flippers, hence *Megaptera*—large wings—with serrated hind edges, which are about a third of the total body length of some 50 ft (16 m). Nodules are present on the head, and, in addition, the body is usually covered with large barnacles. As the body is black dorsally and white ventrally the barnacles are very obvious on the back, giving a rough black and white spotted appearance. The humpback is a bulky animal and appears to hunch its back on diving, hence its name. Humpbacks are widely distributed throughout the oceans and follow the rorqual migratory pattern. This follows coasts in certain parts of the world, for example, in New Zealand, where the schools move close inshore, frequenting bays and inlets. In spite of this the humpback rarely becomes stranded.

An indication of its bulk compared with length is given by its oil yield. The amount of oil obtained from one blue whale is obtained from two fin whales, two and a half humpbacks and six sei whales. But the sei whale exceeds the humpback by 10 ft (3.3 m) in length, and the fin whale by 30 ft (10 m).

In spite of the great bulk, humpbacks may be seen sporting in the sea, often leaping and rolling to fall back in the water with an enormous splash.

The gestation period of about one year is rather longer than that of the rorquals, and in this the humpback compares with the Californian gray whale. However, like them it breeds in alternate years.

HUNTING WASPS, usually large and, typically, with relatively long legs that hang down below the body in flight. They are often common in hot desert regions. As their common name implies, they are active predators and show a particular predilection for spiders, which they paralyze with a sting, prior to storing them in their underground nests for their larvae to feed on. This feeding preference is reflected in an alternative name, the spider wasps. Some of the larger species provision their nests with the largest of the so-called tarantula spiders, agilely paralyzing the spider before it can retaliate with its poison fangs.

HUTIA, fairly large terrestrial rodents of the genera *Capromys* and *Geocapromys*, of primitive form, restricted to the West Indies with one species in Venezuela. Other species live in Jamaica, Little Swan Island, Isle of Pines, Cuba and the Bahamas. They look very like the coypu, and their continued survival is threatened by dogs, cats and introduced mongooses as well as by being hunted for food by man.

HYDRA, a solitary polyp, with a long cylindrical body, closed at one end and with a mouth borne on a projection or hypostome at the other. Surrounding the mouth are tentacles, generally five or six in number, of differing lengths. As in all members of the Cnidaria the body wall of the polyp is composed of two cell layers, separated by a thin, noncellular layer. This body wall encloses a single body cavity and the body is a simple tube.

The tentacles bear "batteries" of nematocysts, or stinging cells, which are used to capture small Crustacea. The nematocysts are of four kinds with different functions. The penetrants or stenoteles are probably the best-known variety of nematocyst with a long, barbed thread arising from a thick basal region, also bearing barbs; they are used in the paralysis of prey. The volvents or desmonemes have a long, coiled thread and are used for holding the prey, while the holotrichous izorhizas with long barbed threads are used for defensive purposes. The fourth type or atrichous izorhiza, which was believed before the advent of electron microscopy to have a bare thread, is used to anchor the polyp to the substrate, together with mucus secreted by the basal region of the polyp. On the capture of prey, the tentacles curve towards the mouth, which opens wide, and this opening of the mouth has been called the "feeding reflex."

Hydras are noted for their power of regeneration. Almost any piece cut from the polyp will reorganize and form a complete, although small, individual. Similarly tentacles, if removed, will be replaced by the polyp. As noted by van Leeuwenhoek, hydras when well fed will reproduce asexually by budding. A small protuberance arises somewhere along the long axis of the polyp, the site depending on the species. This bud grows and develops mouth and tentacles at the end furthest from the parent polyp. When fully developed the new animal is constricted off the parent. Hydras also reproduce sexually.

HYENAS, a much reviled group of animals owing to their habit of eating carrion and to their rather unlovely appearance. There are three species: the spotted or laughing hyena, *Crocuta crocuta*, of Africa south of the Sahara, the brown hyena, *Hyaena brunnea*, of southern Africa, and the striped hyena, *H. hyaena*, which ranges from northern and northeast Africa through Asia Minor to India. A feature of all the hyenas is the shoulders being noticeably higher than the hindquarters, giving them a somewhat ungainly, hunchbacked appearance. They are also remarkable for their gait, which is usually associated with camels. This is known as pacing, and it involves moving both limbs of one side of the body together, instead of the normal quadrupedal action. All three

The striped hyena of Africa and Asia is the smallest of the hyenas.

species have massive heads with large ears and powerful jaws equipped with large teeth that are capable of shearing through a zebra's thigh bone. Their tails are short, and each foot bears four toes.

The spotted hyena is the largest and most aggressive of the three species. The male may be 5 ft (1.5 m) long in head and body with a 13 in. (33 cm) long tail and 3 ft (91 cm) at the shoulders. It can weigh up to 180 lb (81 kg). The female is slightly smaller. The fur is scanty, varying from gray to tawny or yellowish-buff broken by numerous brown spots. There is only a slight mane. Spotted hyenas are nocturnal, spending most of the day in holes in the ground. They live in clans of up to 100 in defined territories marked by their urine and droppings. Members of other clans are driven off. They often hunt in packs and can run up to

40 mph (65 kph). The usual call of the spotted hyena is a mournful howl, made with the head held near the ground, beginning low but becoming louder as the pitch rises. When excited, it utters what can only be described as a demented cackle, and it is this that has earned it the name of the "laughing hyena." It also appears to have the ability to project its voice, thus making it hard to locate the animal from its call. Spotted hyenas eat carrion, and, although their sight is poor, their sense of smell is remarkably acute and it is believed that they can detect a carcass over a range of several miles. They also kill sheep, goats, calves, young antelopes and even smaller prey. They may even eat locusts. They have been observed at night in a pack of up to 20 following a herd of wildebeest and harassing them until they have managed to slow down one of them, after which all the hyenas concentrated on it. This refutes the idea that hyenas are cowards and live only by feeding on the remains of the lion's kill. Immensely strong in the jaws and shoulders, they are said to be able to carry away a human body or the carcass of an ass.

Mating occurs in January, and after a gestation of 110 days one or two young are born. At the time of birth, the young are blind but covered with a thick brown woolly coat. At this stage there are no stripes or other markings, but after about a month the fur starts to lighten in color in some areas, leaving the characteristic brown spots. The spotted hyena has no natural enemies apart from man, and in the wild the life span may be as much as 25 years.

The striped hyena is the smallest species, being not more than 4 ft (1.2 m) long in head and body with an 18-in. (46-cm) long tail. It stands 30 in. (76 cm) at the shoulder and a full-grown male may weigh up to 85 lb (38.2 kg). The coat is gray to yellowish-brown, broken by dark, almost black stripes. Along the line of the spine there is a crest or mane of longer hairs. The variation in the thickness of the coat bears witness to the variation in climate that this species encounters throughout its range. In the southerly parts of its range the coat is thin and sparse, while the more northerly individuals may have an almost woolly coat. The diet is similar to that of the spotted hyena, but the breeding differs in that the gestation period is only 90 days and there are two to four young in a litter, sometimes as many as six. The babies are born in a hollow in the ground, blind and with their eyes closed. The coat is woolly and unmarked. The striped hyena is remarkable for its habit of shamming dead when cornered by dogs. It will lie perfectly still until the vigilance of the dogs wavers. As soon as an opportunity presents itself, the hyena will leap to its feet and make off with all speed.

The brown hyena is halfway in size between the spotted and striped hyenas. Its coat is dark brown with indistinct stripes but with dark rings around the lower part of the legs. Like the striped hyena it has a long-haired erectile mane. Its feeding and breeding habits are like those of the striped hyena, but it lives near the shore and feeds on carrion and marine refuse left by the receding tide, eating anything from dead crabs to the carcasses of stranded whales. For this reason it is also known as the strand-wolf.

The broad-tailed hummingbird of southern North America.

I

IBEX, the name given to seven species of wild goat living in high mountains. They mostly differ from the true wild goat in the smaller amount of black in their coloration, in their flattened foreheads, and in their broad-fronted horns, although this last difference does not apply to the Spanish ibex.

The Alpine ibex, *Capra ibex*, is 32-34 in. (80-85 cm) high, with backcurved horns averaging 26 in. (67.5 cm) long. The horns have a broad front surface with even-spaced knots on it; the outer angle is somewhat beveled off, so that the knots fade toward the outer side. Alpine ibex are dark brown, darker on the underparts and face. Like many ungulates they tend to develop a hard stomach concretion of indigestible matter called the "bezoar stone" and in former days this was considered to be an antidote to poison, among other things, and because of it ibex were nearly exterminated. At the beginning of the century the Gran Paradiso National Park in northern Italy was founded for their preservation; from there ibex have been reintroduced all over the Alps.

The Siberian ibex, *Capra sibirica*, is larger, 40 in. (101 cm) high, with horns averaging 53

breast and front of the legs. It is smaller, with horns not above 3 ft (91 cm) long. The southern race, *C. s. sakeen*, from the Tienshan, Pamir and Kashmir ranges, is dark brown in winter with a whitish "saddle" in the male, and long horns. Both races are darker in summer. In the Pamir, ibex ascend to 16,000 ft (5,000 m); elsewhere, they go less high.

The Caucasian ibex, known locally as tur, are about the same size, but more heavily built. There are two species there, whose interrelationships have recently become clear thanks to the work of Vereschagin, Zalkin and Heptner. The Kuban tur, *C. caucasica*, is found in the western Caucasus; its horns resemble the Alpine ibex (as does its color pattern), but the knots are lower, less accentuated, and the horns are shorter and thicker. The Daghestan tur, *C. cylindricornis*, is found in the eastern Caucasus. Its horns are thick and rounded, with no trace of knots. They turn out and up, then back and down, finally the tips turn in and up—quite unlike any other ibex, but more like the blue sheep or bharal. In the central Caucasus, the two species interbreed, and as much as 10% of the central ibex population

consists of hybrids. However, the hybrids do not themselves appear to breed—they are very uniform in appearance, with an intermediate type of horn—so that there is no actual gene exchange between the two species. The two species of tur are fairly similar in other respects, but the Daghestan tur differs slightly more from other ibex, and it seems likely that it has diverged fairly recently.

The Nubian ibex, *C. nubiana*, is small, only 33 in. (84 cm) high, with horns 43-47 in. (109-119 cm) long; it is light fawn, with the dark markings well pronounced, especially on the front of the legs. It is found in the mountains on both sides of the Red Sea: northern Eritrea, north to Cairo, Sinai and Arabia.

The walia or Abyssinian ibex, *C. walie*, is 38 in. (96 cm) high with horns 40-45 in. (101-114 cm) long. Like the Nubian and Siberian ibex, the front surface of the horns is broad and square with prominent knots. It has a marked bony boss on the forehead and is reddish-brown in color with well-pronounced leg-markings. It

The alpine ibex in Switzerland, a wild goat living in high mountains.

Siberian ibex.

in. (133 cm) in the large southern race. Big males may weigh 200 lb (90 kg). The horns curve more than those of the Alpine ibex, forming at least a semicircle, and the outer angle is not beveled off, so that the front surface is square with the knots very prominent. The northern race *C. s. sibirica*, from the Altai system, is light yellow-white in winter, usually darker on the shoulders, lumbar region, throat,

occurs only in the Simien mountains of Ethiopia, at 13° N, 38° E. In 1965 only 150-200 were thought to have survived.

The last and most distinctive species is the Spanish ibex, *C. pyrenaica*. This is a small species, 28-30 in. (70-75 cm) high, weighing about 100 lb (45 kg). The male has horns 48 in. (120 cm) long; not dissimilar to the Daghestan tur but closer together and more upright, compressed, with a sharp anterior keel. The color pattern is sharp, like a true wild goat, with clearly marked black flankband and black legs and face and a white belly. Spanish ibex are extremely rare and the only places where they are at all common are on some of the high ranges of southern Spain, such as the Sierra de Gredos and Sierra Nevada. The Pyreneean race has been reduced to a herd of about 20 in the Mt. Perdido area. Ibex from other localities have recently been released there and although they are of different races, at least the species will survive in the Pyrenees. In spring, ibex live below the snow-line in separate-sexed herds; the old males live near the female/kid herds, however. In late summer, the males retreat to high altitudes, where they live alone or in groups of three or four. In October to January (which is the rutting season; the exact time depending on the locality) the males rejoin the females, collecting harems of 5-15 females around them. Newborn ibex weigh 8-9 lb (3½-4 kg); one or two are dropped per female, from May to June. They continue suckling until autumn, but begin to graze at one month.

IBISES, storklike birds of moderate size characterized by long, thin and markedly decurved bills, resembling that of the curlew and similarly used. The smaller species are also comparable with the curlew in general dimensions; others are considerably larger. The neck is long, as are the legs; the feet are slightly webbed, between the three forward-directed toes. There is a considerable range of plumage coloration among the species.

Ibises tend to be gregarious at all times and breed in colonies. The nesting site may vary even for a single species, but is commonly in trees, bushes, or reedbeds and sometimes on cliffs, or on stony islands. There are usually three or four eggs. In the different species the eggs range in color from off-white to blue, and they may be marked or plain. The sexes share parental duties.

The worldwide species is the glossy ibis, *Plegadis falcinellus*, found in all six continents although not in the colder temperate parts. The distribution of its breeding colonies is, however, strangely sporadic, and there seems to be a tendency to irregular population movements as well as annual migrations. In Europe it is native only to some southern countries and is now decreasing owing to reclamation of marshes. As in several other ibises, but here to a notable extent, the feathers show a metallic gloss of bronze and green. The prevalent color of the plumage is otherwise purplish-brown, but at a distance, or in flight, or in a poor light, the general effect is black. The sexes are similar, but the glossiness is reduced in winter; im-

The white ibis of tropical America.

mature birds are also duller. Unlike some others, the glossy ibis is rather silent.

The hermit ibis, *Geronticus eremita*, also called waldrapp, is now found only in the Middle East and north Africa, but until the 17th century it bred in Switzerland and other parts of central Europe. It is peculiar in living in dry country and nesting on cliffs. It has a very disagreeable and persistent smell that has been likened to that of putrid carrion, and its flesh tastes equally bad. The food consists mainly of insects, especially beetles. The plumage is mostly bronze-green and purple, with a metallic sheen, but the face and crown are bare of feathers, the skin being dull crimson like the bill. There is a related species in the mountains of southern Africa.

The best-known species is undoubtedly the sacred ibis, *Threskiornis aethiopica*. It is now extinct in Egypt, but it was revered there in ancient days, and many mummified bodies and mural pictures have been found, some showing eggs and young. Farther south in Africa, it is still a common bird, feeding in flocks on the river banks or on open ground. Its contrasting pattern of black and white makes it conspicuous and easily recognizable. The plumage is all white, except for dark wing tips and dark plumes on the lower back. The head and neck are bare of feathers, with black skin, and the dark red legs may appear black from a distance. The birds fly, often in V-formation, with necks outstretched as do other ibises. Nesting is in trees or on the ground.

Of the now very rare Japanese ibis, *Nipponia nippon*, a white bird with a red face, only one small colony is known to exist. The giant ibis, *Thaumatibis gigantea*, of southeast Asia is notable for its size.

The most beautiful species is the scarlet ibis, *Eudocimus ruber*, of the Caribbean area, the adults of which are bright scarlet. The related white ibis, *E. albus*, has a range extending northward to the southern United States; the young are dark, but the adults are white except for dark wing tips, dull red bare skin on the face, and pink legs.

ICE FISHES,

or bloodless fishes, Antarctic fishes that appear to lack blood. Reports of these bloodless fishes by men back from whaling expeditions were not believed until about 1930, when scientists examined specimens on the spot. They found that the fishes were not, in fact, without blood but that the blood contained no red cells. This was most noticeable in the gills, which were a pale cream instead of the usual red. All are slim-bodied fishes, less than 12 in. (30 cm) long, and have large heads and mouths. The shape of the head has given rise to another common name, the crocodile fishes.

The red color of blood is caused by the presence of hemoglobin, a chemical agent capable of uniting with oxygen, transporting it around the body and releasing it where it is required for muscle action, etc. It is difficult at first to see how the ice fishes can survive without any oxygen carrier. Their blood is almost colorless and contains only a few white corpuscles. Experiments have shown that the only oxygen in the blood is the very small amount that is dissolved in the blood plasma, the watery fluid that normally causes the red blood cells to circulate through the body.

An adult ichneumon *Rhyssa persuasoria* with her long ovipositor with which she is able to bore into wood in search of suitable hosts for her young.

ICHNEUMON FLIES,

parasitic wasps, largely parasites of caterpillars of moths, but they also attack aphids and beetles. Ichneumons have a wasp-waisted appearance owing to the very slender first and second segments of the abdomen. They are generally red, black or yellow in color and are active fliers. They have long, many-segmented antennae, which are not elbowed. The wings are divided by veins into a number of small distinct cells, often having a hardened patch called a stigma near the apex on the leading edge of the fore wing. The ovipositor protrudes from the ventral part of the abdomen some way before the tip.

Adult ichneumons are free living, and as liquid feeders they are often to be found in the summer visiting flower heads, especially those of the Umbelliferae. When laying eggs, the female unsheaths the ovipositor and thrusts it into the body of the host, thus depositing the egg hypodermically. The young ichneumon larva grows rapidly inside the host, feeding first on its blood and later on other body tissues. Eventually the host is entirely consumed from inside, and the ichneumon fly larva breaks through the now shrunken skin of the caterpillar to make its own cocoon and to pupate. Many ichneumons attack wood- and stem-boring larvae of moths, beetles and other Hymenoptera. One such ichneumon fly is *Rhyssa persuasoria*, a parasite of wood wasp larvae, which can attain a total length of 4 in. (10 cm). Adults may be seen in conifer forests in spring and early summer, feeding on nectar at flower heads. The males emerge from the timber first, flying up and down infested trees searching for signs of emerging females. After mating, the female seeks trees containing larvae or pupae of the wood wasp, which lie in small chambers within the wood at the end of long, frass-filled tunnels. The wasp probably finds the infested tree by responding to odors emanating from the wood, which is permeated by a symbiotic fungus introduced by the wood wasp when she lays her eggs. Having located a tree, *Rhyssa* uses her antennae to find indications of the larvae beneath the surface, and having found a suitable place, she drills her fine, hypodermiclike ovipositor through the bark and into the wood. When the ichneumon larva hatches, it feeds on the immobilized host until it is consumed, and then spins a cocoon within which it passes the winter months before becoming adult in the spring and chewing its way out of the wood.

Many ichneumons are hyperparasitic, that is, they attack only caterpillars that have already been parasitized by another parasitic species; and their larvae feed on the contained parasite larva.

Larvae of an endoparasitic ichneumon emerging from a caterpillar.

IGUANAS, the largest and most elaborately marked group of lizards in the New World. The family is almost totally restricted to that area, except for two genera in Madagascar and one genus in Polynesia. Iguanids range from species 3 in. (7.5 cm) long to those 6 ft (2 m) in length. Their diets may be insectivorous, carnivorous, herbivorous or omnivorous. Many forms are territorial. All are oviparous, except for the swift lizards, *Sceloporus*, and horned lizards, *Phrynosoma*, some of which are ovoviparous. The eggs are soft-shelled and buried underground.

The common iguana, *Iguana iguana*, ranges from the lowlands of central Mexico south into southern South America. It lives in the vicinity of ponds or rivers at altitudes from sea level to the mountains. Iguanas bask on branches of trees during the day, usually over water so that if danger appears the reptile can drop (sometimes for a considerable distance) into the river or pool, dive, and remain submerged on the bottom for many minutes. These reptiles are fast runners, good climbers, swimmers and divers.

A number of species of large ground iguanas, *Ctenosaura*, occur throughout Mexico and range farther north than the common iguana. They are also found south into Central America. Although they do ascend trees, they prefer to hunt for food on the ground as they are mainly carnivorous. One form, the black-and-white ground iguana, *Ctenosaura similis*, has large muscular jaws enabling it to prey on small rodents, other lizards and an occasional bird. These iguanas are also used as food by the local people.

In the Caribbean area, island forms of the ground iguanas of the genus *Cyclura* have remained isolated from one another for long enough to evolve into several distinct forms. All of these primitive forms have a similar body build, are mainly ground dwellers, and

Young impala rams.

The common iguana is widespread throughout tropical America.

feed both on vegetation and small animal life. However, on the Lesser Antilles and the Virgin Islands near northern South America, the ground iguanas are extinct and are known only from fossils found in caves. These islands are occupied by the common iguana, and it is possible that some overlap in competition resulted in the eventual extinction of the cyclurids in these areas.

One of the largest of these insular iguanas is the rhinoceros iguana of Haiti and neighboring areas, which attains perhaps the greatest weight (although not the greatest length) of the iguanids. This reptile has heavier jaws than other members of this group. This, plus a grotesque pattern of enlarged scales on the head, including three prominent pointed scales on the snout, accounts for its popular name.

One island form of iguana in the Pacific, the marine iguana, *Amblyrhynchus cristatus*, has evolved on a number of the Galápagos Islands. A lack of natural enemies has resulted in a reptile that is apparently unafraid of man, but an inhospitable environment has produced some alterations in the basic iguana pattern resulting in a unique reptile. It is vegetarian, but feeds mainly on various seaweeds growing at the bases of cliffs and in vast underwater "fields."

IMPALA, *Aepyceros melampus*, one of the most abundant and most graceful of African antelopes. Impala are around 39 in. (1 m) high, red-brown in color with a well-marked fawn band on the flanks and a white underside. There are no lateral hoofs. Only males have horns; these are 20-30 in. (50-75 cm) long, turning first up, then out and back, then up again, with long tips. The horns have well-marked, spaced-out ridges on them.

The common impala is found from the northern Cape Province to Natal, Zambia, western Kenya, and Uganda.

In the dry season, impala live in big herds of hundreds of individuals; in the wetter months these big herds break up into small one-male units, with 15-25 females to every male. The surplus males live in bachelor bands, up to 50 or 60 strong.

INDRI, *Indri indri*, a large, gray and black lemur, which, with the sifaka and the avahi, makes up the family Indriidae, characterized by a vertical clinging and leaping habit and a specialized diet of leaves, buds, flowers, fruit and bark. It is the largest of the Madagascar lemurs, with a head and body length of 28 in. (70 cm), and is peculiar in having a very short tail, 1 in. (3 cm) long. Its long back legs, used for its enormous vertical leaps, give the impression that it is much bigger.

Indris live in small social groups, probably family units, of 2-4 adults. These seem to be territorial, and the indri is most famous for its eerie howls, which carry over great distances and are answered by howls from neighboring groups. A group of indri can be induced to produce this howl by playing a tape-recording of their own calls. It also utters short, intermittent grunts as a mild alarm call. There is a single offspring at birth, which is carried on the mother's fur. The indri is found only in the northern part of the east coast rain forest.

The avahi, *Avahi laniger*, is a grayish or brownish, soft-coated, slightly smaller lemur with a head and body length of 12 in. (30 cm) and a tail of 16 in. (40 cm) long. It differs from the indri and sifaka in being nocturnal.

The avahi also forms small groups, probably family units, of 2-4 individuals, although it is most often encountered singly at night. As with all nocturnal lemurs, its eyes reflect a bright-red glow, and it can easily be spotted with a flashlight in the dark. High-pitched, whistling calls are produced at night, and these may serve as territorial advertisement between social groups. There is one baby produced toward the end of the dry season, and the infant is carried on the mother's fur. The avahi occurs all over Madagascar, and, as is usual with Madagascar lemurs, there is a rufous rain-forest form, *Avahi laniger laniger*, on the east coast and a grayish drier forest form in the northwest, west and south of the island.

J

JACAMARS, a family of tropical American birds resembling bee eaters. There are 14 species in five genera, ranging from southern Mexico to southeastern Brazil. They are small to medium-sized, bright-plumaged hunters of airborne insects, which the birds await sitting patiently at favorite vantage points in savanna woodland, farmland and at the edges of forest, often near water. Most species have an iridescent dark green head and upperparts, rufous or green underparts, and a contrasting white patch on throat or breast. Sexes are similar, except that any throat patch in females is buff. The bill is straight, sharp and pointed, and one to three times the length of the head. The legs are very short, with the toe arrangement (two forward, two backward) that characterizes the order, and the tail is moderately long and graduated. Jacamars are quiet birds, not gregarious and nonmigratory.

The 10-in. (25-cm) long rufous-tailed jacamar, *Galbula ruficauda,* is a typical member. A pair will perch low down at the edge of gallery forest, or in the shade of a riverside shrub, and dart out after a passing insect, which, despite the slender 2½-in. (7-cm) bill of this species, is deftly snapped up and brought back to the perch where it is beaten until inactive. Their diet is almost exactly the same as that of bee eaters in Africa, that is, chiefly hard-bodied insects with wasps and bees comprising about 80% of their intake.

All jacamars excavate their nest cavities at the end of a tunnel 1-3 ft (1 m) long in earth banks and shelving ground. In the oval terminal chamber is laid the clutch of two or more round white eggs.

JACANAS, a small family of mainly tropical aberrant wading birds. There are seven species. The American jacana, *Jacana spinosa,* has several subspecies from Mexico through Central and most of South America. In Africa there are the lily trotter, *Actophilornis africana,* the lesser lily trotter, *Microparra capensis,* with *Actophilornis albinucha* in Madagascar. From India through southeast Asia to the Philippines is found the pheasant-tailed jacana, *Hydrophasianus chirurgus,* while the bronze-winged jacana, *Metopidius indicus,* has a similar but more restricted range. Lastly, the lotus bird, *Irediparra gallinacea,* inhabits Indonesia and Australia. The jacanas form a fairly uniform group of ploverlike or crakelike birds inhabiting the fringes of lakes and only differing among themselves in relatively trivial points, such as the conformation of the bill and presence or absence of a horny frontal shield. In length they vary from about 6½-12 in. (17-30 cm), except

for the pheasant-tailed jacana, the name of which suggests its principle feature, an 8-in. (20-cm) tail, which makes the bird 21 in. (53 cm) long overall. The wings are rather long but rounded, the tail short in most species, the bill about the same length as the head or rather shorter and straight, and the legs long and thin. In addition, the toes are very long and the nails long and straight, especially the hind toenail, so that the jacana's weight is distributed widely by the foot, enabling it to walk nimbly over lily leaves.

Lily trotters can usually be seen picking their way along the margins of sluggish rivers, water meadows, or well out in the open water of lakes with good water-lily beds, feeding on a variety of invertebrates and seeds.

Breeding takes place during the rainy season, and the sexes participate equally in building the nest, a simple platform of leaves of water plants concealed in sedges at the edge of the water or built on floating vegetation, and in incubation. The four eggs are pointed and highly polished and are brown, heavily penciled with irregular dark lines.

JACKALS, carnivores very similar to dogs and wolves. There are four species distributed throughout Africa and southern Asia, which, although similar in habits, differ markedly in appearance. The golden jackal, *Canis aureus,* ranging through southern Asia and north Africa, looks like a small wolf or coyote with a tawny coat and black markings on the back and tail. The much smaller black-backed jackal,

Juvenile American jacana already has the adult's long toes.

Despite their reputation for meanness, jackals are handsome dogs; the side-striped jackal is the most handsome.

Canis mesomelas, inhabiting the plains of central and South Africa, has a black saddle extending along the back and a black stripe along the length of the tail; its sides and legs are bright rufous, and it has large pointed ears. The side-striped jackal, *Canis adustus*, grayer in color and with a thin black and white stripe extending along the sides of the body, a black stripe along the tail, and a white tail tip, inhabits the African tropical forest or woodland areas with dense vegetation. Finally, in a very limited area in the highlands of Ethiopia, is the little-known Simenian jackal, *Canis simensis*, bright rufous in coloring and with a cream belly and black-tipped tail.

Like most dogs and wolves, the gestation of jackals is about two months, but the timing of breeding is flexible and depends upon the area of Africa that each species inhabits. Golden jackals, living north of the equator, give birth in May or June; the side-striped and black-backed jackals may also give birth from May to June in central Africa, but they bear their young in September and October if located in southern Africa.

All of the jackals are extremely opportunistic, obtaining their food from whatever sources are available. Although often considered to be mainly scavengers, they will also hunt and kill birds, hares, mice, lizards, turtles and various insects. Moreover, several individuals may band together in a small pack (especially the side-striped jackal) to prey upon larger game like sheep, goats and small antelopes.

JACK RABBITS, true hares of the genus *Lepus*. All seven species are found in western North America and Central America.

The antelope jack rabbit, *Lepus alleni*, has enormously large ears, $8\frac{1}{4}$ in. (21 cm), with white edging but no black spot on the tip. The distribution is principally Mexican, but it also occurs in southern Arizona.

The breeding season from January to September is correlated with the average monthly rainfall. Three to four litters, averaging two young each, are produced annually. The precocial young are dropped in shallow, well-concealed nests, but scatter soon after birth. It is active at night, the day being spent in the shade.

When escaping, a long high "observation" jump is performed every four to five leaps. It has been reported to clear a $5\frac{1}{2}$-ft (1.8-m) obstacle, and reach a speed of over 35 mph (57 kph).

The black-tailed jack rabbit, *Lepus californicus*, has ears 6 in. (15 cm) long and black tipped. Open plains with short, poor mixed grasses are the favored habitat, and it flourishes in drought-stricken overgrazed areas.

It breeds from January to April, and the gestation period is 43 days. Four to five litters are dropped annually; averaging three young each. Often no nest is prepared, and the scattered leverets are suckled for 17-20 days.

The black-tailed jack rabbit is solitary except during the breeding season. It is sedentary, the home range of adults being about 42 acres

(17 ha), and of juveniles about 35 acres (14 ha). Density is 0.3-0.5 per acre (0.5-1.0 per ha). One of the most numerous species, it reached plague proportions in California at the turn of the century in response to changes in agricultural practice. Bounty was still offered in Kansas and other states until recently. It has been found that 62 black-tailed jacks consume as much as one cow.

The black-tailed jack rabbit has been shown to be a reservoir of diseases pathogenic to man and domestic animals, such as tularemia, bubonic plague, brucellosis, Q fever, and Rocky Mountain spotted fever.

JAGUAR, *Panthera onca*, sometimes confused with the leopard because both have spots in rosettes, but those of the jaguar have a black spot at the center. The yellowish-buff ground color is similar in both, but in the jaguar the rosettes tend to be concentrated along the back. The belly and chest are pale and have irregularly placed black spots, and the lower half of the tail is ringed. There is a black mark near the mouth on the lower jaw, and the backs of the ears are black. The variations in color range from an almost white ground color to black forms, in which the rosettes can be seen only as a variation in texture. Jaguars vary in size from $5\frac{1}{2}$-9 ft (1.5-2.7 m) overall, the tail being one third of the total length. At the shoulder, they are $2\frac{1}{4}$-$2\frac{1}{2}$ ft (68-76 cm) high. The weight is 125-250 lb (56.7-113.4 kg), which makes it a heavier animal than the leopard or the puma. It is a good climber.

The jaguar inhabits an area that is bounded to the north by the southwestern states of the U.S.A., and to the south by Argentina. It lives in thick cover, in forests or swamps, but it is also found in desert and savanna areas in the north and south of its range.

The jaguar is able to breed at about two years old. The individuals are solitary, only coming together at the mating time, which is probably in January in the northern part of the range, but at no fixed time elsewhere. After mating, the female has no more to do with the male and, after a gestation period of 95-105 days, the cubs are born. The usual size of a litter is two, but there have been as many as four. The young are fully furred at birth and blind. They first emerge from the den at two weeks of age.

The jaguar is one of the roaring cats, but it appears to make very little use of this ability, and in this it is similar to the leopard. The most common sound is a grunt, usually made when it is hunting. In addition to this, the male makes a mewing noise in the mating season, and a cornered animal will snarl and growl at its enemy. Like many of the other cats, the jaguar is an accomplished swimmer and has no fear of water, although it does most of its hunting on the ground or in the trees. It feeds mainly on peccaries, but the range of food taken is very wide and it will hunt deer, capybara, tapir, agouti, sloth, monkey, and a wide variety of birds. It also eats fish, which it catches by scooping them from the water, turtles and small alligators and crocodiles. Like so many of the cats it cannot resist the chance of an easy meal, and will raid domestic stock whenever the opportunity arises. The jaguar demonstrates remarkable strength, and has been recorded as pulling a dead horse through 70 yd (64 m) of

Jaguar kittens, with characteristic rosettes in their coat.

brush and then towing it across a river.

The method of killing prey is very similar to that used by the leopard, but the jaguar, if the first attempt does not succeed, will often leave the intended victim and go to seek food elsewhere. Once a kill has been made and has been dragged to a place of the animal's liking, it will usually be buried, but should the ground be too hard or rocky, no further attempt will be made to conceal the carcass. The jaguar seems unaffected by the smell of high meat, and it will return several times to a kill should it be large enough.

JAGUARUNDI, *Felis yagouaroundi,* jaguarondi or yaguarondi in many ways approaches the general forms of weasels and otters. The name is misleading, as the jaguarundi has very little in common with the jaguar. The body is much elongated, and the head is long and has a low silhouette and the ears are small. The average length of the jaguarundi is about $3\frac{1}{2}$ ft (137 cm), of which a third is tail, while the height at the shoulder is only 11 in. (28 cm). A well-grown individual will weigh about 20 lb (9 kg).

There are two color phases, and for many years they were thought to be distinct species. They are found with either a black to brownish-gray coat, or they are tawny to chestnut. Neither is spotted. The brown form is known as the eyra. The gray form becomes somewhat darker in the winter months.

The jaguarundi is found in the extreme southwest of the U.S.A., but is far less common than it used to be, because of the thinning out of the cover along the delta of the Rio Grande. The southern limit of its range is Paraguay. It appears to be an animal of the swamp and forest and is known to be an excellent swimmer, from which it derives one of its many local names, "otter cat." In North America its favorite habitat is the chaparral scrub near water. This animal is never found very far away from water. In its normal habitat, it moves along clear and well-defined trails, which makes it easy to trap.

They are solitary animals for most of the year, meeting only in the mating season, which occurs either in November and December, or May and June. The young are born after a gestation period of about 70 days, and the litter usually consists of 2 kittens, but there can be as many as 4. The den is similar to that of the ocelot, in a hollow log or among rocks. The young are unmarked at birth, and the one litter may well contain both the gray form and the brown.

Jaguarundis prey on a variety of small mammals and on game birds, which they appear to prefer. They often cause havoc by breaking into the pens of domestic fowl.

JAWFISHES, small, rather elongated marine fishes related to the weeverfishes. They derive their name from their very large mouths. Some species have a backward extension of the upper jaw like that found in some of the tropical anchovies. There is a single, long dorsal fin, the first part supported by spiny rays. The most remarkable feature of this otherwise rather undistinguished group is their habit of constructing burrows. The burrows are very elaborate, with a chamber at the end several times larger than the fish itself and lined with pieces of rock and coral. One or two of the species are mouth brooders. The yellow-head jawfish,

Female jaguar.

Opisthognathus aurifrons, from the Virgin Islands, which grows to about 4 in. (10 cm), has a beautiful blue body, which becomes darker toward the tail and merges into a striking yellow on the head. It builds its burrows in sand and spends much of its time hovering obliquely over the mouth of the burrow, eating small animals and larvae that float past.

Jawfishes also live in shallow waters in the Indo-Pacific region.

JAYS, a diverse group of birds in the crow family, many of which are brightly colored—particularly in shades of blue—and have screeching, raucous voices.

The jays are most numerous (in the number of species as well as individuals) in America, particularly in the tropics. The original bearer of the name "jay," however, is the European jay, *Garrulus glandarius,* a shy species found in most of Europe and east through Asia to Japan. It is a strikingly beautiful bird some 13 in. (33 cm) long, with a pinkish-brown body, black, white and blue wings and a black tail. The rump is white, and the erectile crown feathers are streaked in black and white. Like many other jays it is an omnivorous feeder and will take eggs and chicks of other birds. It is basically a bird of woodland and feeds to a large extent on acorns, of which it makes large stores in the autumn for use in the following winter and spring. The nest is a cup of twigs in which five to seven eggs are incubated for 16-17 days. There are two species closely related to the European jay: *G. lanceolatus* of the Himalayas and *G. lidthi* of islands in the China Sea.

The other genus of Old World jay is *Perisoreus,* and this genus spreads also into North America. The Siberian jay, *P. infaustus,* reaches the

Pacific coast in the east and Scandinavia in the west. The Szechwan gray jay, *P. internigrans*, lives in eastern Tibet and western China, particularly in coniferous forests. In similar areas of North America is found the Canada jay, *P. canadensis*. These species have duller plumage than most other jays, being typically brown or gray.

There are nearly 30 other species of jays in America, 20 of them in the tropical areas. The most common species is the blue jay, *Cyanocitta cristata*, widespread in North America, mostly blue, with black and white on wings and tail. The Florida or scrub jay, *Aphelocoma caerulescens*, is restricted to the scrub-oak areas of the Florida peninsula. Another ecologically restricted species is the piñon jay, *Gymnorhinus cyanocephala*, of the mountains of the western United States.

Some of the neotropical species are most strikingly colored. The turquoise jay, *Cyanolyca turcosa*, for example, which is a bird of the Andes, is almost entirely turquoise blue. The green jay, *Cyanocorax yncas*, found from Texas southward to Peru, is green above, yellow below, with a blue crest and yellow outer tail feathers.

JELLYFISH, marine animal often seen stranded in rock pools at low tide or swimming under the surface of the sea with a pulsating bell and long trailing tentacles.

Externally, the jellyfish shows four-rayed symmetry. The bell varies in shape from a wide saucer to a cube and may be grooved on the upper or exumbrella surface. The gelatinous consistency results from the extensive formation of a jellylike mesoglea separating the outer ectoderm from the inner endoderm. Round the edge of the bell are a varying number of tentacles, and the margin is not smooth but scalloped into lappets. From the center of the subumbrella or undersurface of the bell hangs an extension

or manubrium bearing a four-cornered mouth. The mouth may be drawn out at the corners into short lobes or long, frilly extensions or oral arms, which become fused together in one order, the Rhizostomae, giving rise to "suctorial mouths." The ectoderm of the bell and oral arms is covered with nematocysts or stinging cells, enabling the capture of food organisms over a large surface area. Prey captured on the exumbrella surface is carried to the edge of the bell by cilia, removed by the oral arms and conveyed to the mouth.

One of the most striking features of the typical jellyfish is the elaborate canal system that has been developed. Radial, circular, and branched canals insure circulation of food materials and oxygen throughout the whole region of the bell of the jellyfish.

JERBOAS, rodents living in the deserts and steppes from central Asia to the Sahara. They progress by jumping on their very long hind legs and feet in the manner of a kangaroo, a habit that they share with the kangaroo rats of North America. The jerboas should not be confused with the gerbils, which, although adapted to similar habitats, look more like typical rodents and are more closely related to the hamsters and voles.

Jerboas range in size (head and body) from 1½ in. (4 cm) excluding tail in some of the dwarf jerboas of central Asia (among the smallest of all rodents) to 6 in. (15 cm) in, for example, the north African jerboa, *Jaculus orientalis*. The tail is always very long, sometimes over twice the length of the head and body, and in most species has a tuft of long white hairs at the tip. It is an important organ of balance when the animal is jumping. In one group, however, the fat-tailed jerboas, *Pygeretmus*, of Turkestan, the tail is shorter than the head and body and contains large deposits of fat, a condition parallel to that in the fat-tailed gerbil of north

Head-on view of a John Dory.

Africa. The front and hind legs are very disproportionate in size. The hind legs are enormously elongated, and in the larger species they enable the jerboa to make prodigious leaps of up to 8 ft (2.4 m). The hind feet usually have only three toes, and these are supported by a single strong foot bone, which results from the fusion of the three central metatarsals, a condition similar to that found in the cannon bones of cows and sheep.

Like most desert animals, jerboas have either large external ears or enormously inflated auditory chambers in the skull. An extreme example is the long-eared jerboa, *Euchoreutes naso*, of Mongolia, which is one of the most bizarre species of mammals, with ears as big as a hare's, a rather piglike snout, whiskers that reach to the base of the tail and an exceedingly long thin tail with a bold black band just before the tip. Equally bizarre are the dwarf jerboas, *Salpingotus*, with very short external ears but enormous tympanic chambers that dwarf the rest of the skull and make the whole head as large as the rest of the body.

Most jerboas are sandy colored, closely matching the color of the environment in which they live, and they lack bold markings except for the contrasting black and white tail tip. The fur is extremely soft and silky, and the ear opening is protected from sand by long hairs. The hair of the feet is also closely adapted to life on sand, with dense fringes of hairs on the toes. Jerboas are predominantly nocturnal animals, spending the day in burrows that may be quite elaborate, with several entrances and with one or more nest chambers.

Jerboas are predominantly vegetarian, feeding on seeds of grass, salt bush and other desert plants, and especially on bulbs and tubers.

Like most highly specialized rodents, jerboas are less prolific and are longer-lived than most of the mouselike rodents. Nevertheless there may be two or three litters per year, the timing depending very much on the climate and the annual cycle of rainfall. The litter size is usually

The hen of the European jay is like the male in plumage.

about three or four and the gestation period about four or five weeks.

JOHN DORIES, rather grotesque fishes of temperate oceans related to the boarfish, also called the St. Peter fish because legend has it that it was in this fish that the apostle found the tribute money, the dark blotch on the flank being St. Peter's thumb print. Yet another name given to the fish, by fishermen in northern Germany, is "King of the herrings," since it is reputed to shepherd the herring shoals. In reality, the John Dory is a fish eater, feeding on herrings, pilchards and sand eels.

The John Dory, *Zeus faber*, is an almost oval, compressed fish with the rays of the anterior spiny dorsal fin greatly elongated into filaments. The pelvic fins are also long. The jaws are protrusile and can be thrust out a surprisingly long way. This fish is found in moderate depths down to 600 ft (200 m), and it is widely distributed in the Atlantic, occurring as far north as Scandinavia and as far south as South Africa. It reaches 3 ft (100 cm) in length, and although its flesh is delicious its grotesque appearance discourages would-be purchasers. Other species of dories are fishes of deeper water, which are infrequently caught.

JUNGLE FOWL, the ancestors of modern domestic fowl, which have been domesticated for more than 4,000 years, and, as one would expect, their general appearance closely resembles that of chickens, especially leghorns. Jungle fowl are found all over the warmer parts of Asia and Malaysia but are absent from Borneo. They inhabit a wide variety of country from low-altitude forest, dry scrub and bamboo groves to small woods and rough ground near villages. They are always wild and extremely wary so that they continue to survive despite persecution by man. During the breeding season they are often found in family parties consisting of one cock and several hens and they congregate in larger flocks during the winter.

The male's courtship is similar to that of the domestic fowl and includes circling the hen accompanied by the rasping sound produced by the movement of the lowered primary flight feathers. The crow of the male jungle fowl is similar to a domestic cock's but only that of the red jungle fowl resembles it closely. The hen jungle fowl makes her nest on the ground under the shelter of vegetation, the number of eggs in the clutch depending upon the species. Wild jungle fowl living near villages sometimes cross with domestic chickens.

All jungle fowl feed chiefly upon seeds, grain, shoots and buds, as well as insects.

There are four distinct species of jungle fowl. The best known and most widespread is the red jungle fowl, *Gallus gallus*. The male resembles a brown leghorn cockerel with his fiery red and golden-brown plumage. Red jungle fowl exist in the genuine wild state from northwest India through Assam and Burma, Thailand and Malaya, to Indochina in the east and also to south China, Sumatra and Java. Males of the red jungle fowl, if of pure stock, undergo a molt into dull plumage in the summer. They have a short shrill crow which ends abruptly. The Ceylon jungle fowl, *G. lafayettei*, also called La Fayette's jungle fowl, is confined to Ceylon and occurs wherever there is sufficient cover

A burrowing hairy-footed jerboa *Jaculus jaculus*. Note extremely long hindlegs.

from sea-level to 6,000 ft (1,800 m). It is similar in habits to the red jungle fowl but never gathers in large flocks and keeps away from cultivation and human habitation.

Sonnerat's jungle fowl, *G. sonnerati*, sometimes called the gray jungle fowl, is confined to western and southern India where it lives in bamboo groves and forests on mountain slopes up to 5,000 ft (1,520 m). It is usually found in pairs or family parties but appears to be monogamous and does not normally congregate in flocks. The

hackles or neck feathers are much in demand for dressing flies for salmon and trout fishing and as a result of trade in its plumage Sonnerat's jungle fowl has declined drastically in numbers. The Indian government has recently banned the export of the plumage.

The green jungle fowl, *G. varius*, is found in Java and other East Indian islands and, although small numbers inhabit inland forests, it is primarily a bird of the seashore and coastal valleys.

Cock and hen jungle fowl drinking.

K

KAKAPO, *Strigops habroptilus,* a strange nocturnal New Zealand parrot, which has presumably been isolated from other parrots for a considerable time, as it has developed unique characteristics. The kakapo is about 2 ft (60 cm) long, chiefly greenish-yellow in color with darker barring. It is almost flightless, the rounded wings being used only for gliding. The bill is unlike that of any other parrot in having strong ridges on the lower part. The kakapo was called the "owl parrot" by the European colonists because of its soft plumage, "facial disc" of feathers and nocturnal habits.

When Europeans first arrived in New Zealand the kakapo was widespread, living in the southern beech, *Nothofagus,* forests throughout the country, though bones found in the North Island and on the Chatham Island group indicate a previously wider distribution. The kakapo is now drastically reduced in numbers, being found in only a few remote refuges in the fiordland region and on Stewart Island in the south.

The kakapo runs along the paths through the forest and grassland, but occasionally climbs trees and glides for some distance. The paths are maintained by constantly trimming the surrounding vegetation, and they lead from daytime forest resting places in holes among tree roots and rocks to feeding areas that are often in montane grassland above the forest or below in the valleys. Along these paths are numerous dust-bathing hollows.

Kakapos breed in natural holes or construct tunnels of their own: sometimes quite deep, one of 9 ft (2.7 m) having been recorded. The female incubates two to four white eggs in a bare nest rather late in the year, eggs having been found from January to May.

Their food consists mainly of the leaves of tussock plants, berries and nectar. Fibrous material is not swallowed, but is thoroughly chewed to extract the juices and then rejected as a ball. These balls, found attached to the plants or in the resting places and dusting hollows, provide a clue to the presence of the now rare bird.

The call is a weird bitternlike booming.

KANGAROO, a marsupial animal with large hind feet and strong hind limbs and tail, which adopts a bipedal method of locomotion when moving quickly. The female bears a pouch, containing the teats and mammary glands, in which the young is raised. In its widest sense the name "kangaroo" is applied to 50 kinds of animals grouped in the family Macropodidae (big-footed marsupials). These range in size from the tiny musky rat kangaroo, *Hypsiprymnodon,* weighing a little over 1 lb (500 gm) to the largest gray and red kangaroos, which approach 200 lb (90 kg) in weight and reach a height of 6 and occasionally 7 ft (2 m).

Living kangaroos may be divided into the rat kangaroos, commonly called kangaroo rats, and the true kangaroos. It is customary to call the largest of them kangaroos, while the remaining, smaller members are called wallabies, pademelons, etc., but there is no real criterion by which kangaroos and wallabies may be distinguished.

Kangaroos are heavily built in the hindquarters and lightly built in the forequarters: the fore limbs are thin, mobile and frequently used in bringing food to the mouth. Strong muscular

Female red kangaroo *Macropus rufa*.

development of the upper fore limb occurs only in the males of some species. The tail is heavily built in the largest kangaroos and serves to balance the forepart of the body, being held clear of the ground during bipedal locomotion. It is used as a prop during bipedal stance and may be the only part touching the ground during fighting, when a kangaroo can kick an opponent using both hind feet together. During quadrupedal locomotion, the tail "walks" along, behind the kangaroo, and leaves a characteristic track on soft ground. The foot is very long and bears four toes. The second and third digits are very small and bound together by skin except in their terminal regions. This condition is known as syndactyly, and the fused toes as the syndactylous digits.

The teeth of kangaroos are characteristic. There are three incisors on each side of the upper jaw separated by a long toothless gap (diastema) from the single "permanent" upper premolar and the succeeding molars.

The gestation period, or interval between insemination and birth of young, varies between 29 and 38 days in the various large kangaroos. Kangaroos generally produce a single young at one time, although about one in each thousand red kangaroos have twin young in the pouch. The nonlactating female kangaroo produces a single ovum each four to six weeks, depending on the species. The female kangaroo about to give birth cleans the inside of her pouch by licking it, and. then assumes a resting position with her back supported, hind legs extended forward and tail passed forward between them. Birth is preceded by the appearance of a little straw-colored fluid and the embryonic allantoic sac containing waste products accumulated during pregnancy. The young is born enclosed in a fluid-filled amnion, which bursts, allowing the young to grasp the mother's fur, up which it climbs to enter the pouch between one and five minutes later. The mother pays no attention to the young during its climb to the pouch but remains in the birth position licking the blood and embryonic membranes expelled after birth from her fur. Once in the pouch the young attaches to one of the four teats, from which it draws nourishment throughout pouch life.

KEA, *Nestor notabilis,* a New Zealand mountain parrot, placed with its close relative the kaka or New Zealand forest parrot, *Nestor meridionalis,* in a suborder that is considered primitive and not closely related to other parrots. The bill of both species is rather long and a little curved. The tongue has a hairlike fringe at its tip, and the tail feathers have projecting shafts. The kea is not as gaudy as some parrots, being largely olive-green with dark edges to the feathers. The wing and tail quills are bluish-green, and it is only in flight that the scarlet underwing comes into view. The male is distinguished by a longer and stronger bill, while that of immature birds has a yellow base.

Keas are now found only in the South Island of New Zealand, though there are subfossil remains in both the North Island and the Chatham Island group, 500 miles (800 km) east of New Zealand. This wider distribution possibly occurred during the Ice Ages.

The kea, a New Zealand mountain parrot, has a long, little curved beak which distinguishes it from other parrots.

The nest is usually placed in a rock crevice or in a hollow log in the forest, and is often near a rock vantage point. The two to four white eggs are usually laid between July and January, although they have been found throughout the year. They are incubated mainly by the female for three to four weeks. The young take about 14 weeks to fledge. Males may be polygamous, but adult females do not necessarily lay every year, and on average each female raises one young bird every second year. A life expectancy of at least six years is thus needed to maintain the population.

The kea is responsible for much controversy in New Zealand. While there is little doubt that there is some factual basis for the kea's reputation as a sheep killer, it seems that the problem has been grossly exaggerated.

KESTRELS, a distinctive group of falcons, contrasting with most other falcons by persistently hovering for ground prey, rather than taking birds or insects in flight. There are about ten species of true kestrel, together with four others regarded as aberrant kestrels and usually included with them. The average size is $11\frac{1}{2}$ in. (29 cm), the largest being the fox kestrel, *Falco alopex,* of Africa reaching 15 in. (38 cm), and the smallest the Seychelles kestrel, *F. araea,* of 8-9 in. (20-23 cm). Females are usually larger than males. The coloration of the upperparts is typically reddish-brown, spotted black; underneath they are more buff, streaked and barred black. Males are often distinguished by the inclusion of gray on the upperparts.

They are worldwide in their distribution. The common kestrel, *F. tinnunculus,* has many races and occurs throughout Europe, Asia and Africa, being replaced in the New World by the similar American kestrel *F. sparverius,* commonly known as the sparrowhawk in America. Throughout their range kestrels occupy rather open habitat, but this can be desert, savanna, cultivated, partially wooded

or mountainous. They also now occur commonly in cities and suburbs.

Kestrels do not build nests, but lay their eggs on ledges, in holes and in the old nests of other species. The lesser kestrel and red-footed falcon are unusual in being colonial nesters. Incubation is usually by the female alone, the male hunting and bringing food to the vicinity of the nest. The principal food is small mammals, but in certain areas or at certain times of the year their diet may be largely insectivorous; small reptiles, frogs, worms and birds are also taken.

KILLER WHALE, *Orcinus orca,* or grampus, an example of a true dolphin without a beak. The males reach truly whale proportions of 30 ft (10 m), and it has always been said that the females are only about half this size, rarely reaching 20 ft (7 m). So it is always quoted as one of the few species of Cetacea in which females are smaller than the male, the sperm whale being the only other species to show

Sparrowhawk.

this. From 1948-1957, however, Nishiwaki measured 600 killer whales caught in Japanese coastal waters and found the average adult lengths to be 21 ft (7.5 m) for males and 20 ft (7 m) for females, a difference which is negligible.

The head of a killer whale flows straight back onto the trunk, but the powerful streamlined appearance is offset by very large rounded fore flippers, a large dorsal fin and large tail flukes. In the males the flippers and dorsal fin continue to grow throughout life until they become very large. In a young male or a female, the flipper is about a ninth of the body length but has increased to a fifth in an old male. The dorsal fin grows to about 2-6 ft (0.3-0.9 m) in height, and for this reason, and because of the shape, the killer is sometimes called a "swordfish." Killer whales have a black back and white belly. There is a large white patch behind the eye and a large white invasion of the flank from the belly as well as a pale-colored saddle behind the dorsal fin.

The killer whale is a fast and voracious feeder. It eats dolphins and porpoises, seals and sea lions, penguins, fish and squid. Even the largest whalebone whales may be attacked and killed by a hunting pack of killer whales. They may hunt in small groups or in packs of up to 40 or more. A reported attack on sea lions illustrates their hunting method. Seals and penguins are swallowed whole where possible, but larger animals such as the larger whales are seized by the tongue, mouth and flukes first, and then the helpless animal is eaten in large pieces bitten off by the killers' powerful teeth and jaws. Records such as that of the killer whale with remnants of 27 porpoises and seals in its stomach are not uncommon.

Killer whales are found throughout the world but most frequently in Arctic and Antarctic waters.

The false killer whale, *Pseudorca crassidens*, although closely related to the killer whale, has external differences which make it closer in appearance to the pilot whale. There is little difference in size between the sexes, and each may reach up to 16 ft (5 m). The false killer is black almost all over, though there may be white scar marks. It is slimmer than the killer whale with a narrower flipper and smaller dorsal fin, while the snout is rather rounded over the lower jaw. It is a deep-water animal of worldwide distribution, except for the polar regions, and is found in large schools of up to several hundred individuals. It feeds on squid, cod and other fish that move in the oceanic currents.

KINGFISHERS, a family of birds related to the bee eaters, hornbills, motmots and others, and sharing with them a syndactyl foot, the three front toes being joined for part of their length. Europe has only one species, the common kingfisher, *Alcedo atthis*, which is also found in Africa and the Far East, eastward to the Solomon Islands.

The common kingfisher is a dumpy bird only $6\frac{1}{2}$ in. (16.5 cm) long. The azure feathers of its back can also look emerald green depending upon the angle of the light. The tail feathers are a darker cobalt, as are those of the head and wings, where the striae glow with rows of azure speckles. In contrast, the underparts are

a warm chestnut orange. It has a white throat or bib, white neck patches and orange cheek patches behind the eye. The $1\frac{1}{2}$-in. (3.8-cm) dagger-shaped bill is the only external indication of the sex. In an adult male it is wholly black, but the female usually has a partially or completely rose-colored lower mandible. The small feet and legs are sealing-wax red. Juveniles are duller than their parents, slightly smaller with shorter black bills and black feet.

Kingfishers live mainly on unpolluted rivers, lakes and streams, canals and fen drains. They also inhabit tidal estuaries, salt marshes, gutters and rocky seashores, especially in the winter when driven to the coast because fresh water has frozen over.

The chief prey is fish, which the kingfisher secures underwater by grasping the fish between its mandibles, not by stabbing as is the popular misconception. The bird will watch from a perch, usually overhanging the water, until it sights its prey, and then, having aimed, it tenses and dives headlong into the water straight as an arrow, beak open and the opaque nictitating membrane or third eyelid closed.

Early in the year the pair of kingfishers fly high in a courtship flight, after which they look for a suitable place to nest.

The eggs are a delicate translucent pink, about $\frac{3}{4}$ in. (1.9 cm) long, and are laid upon the bare soil. The normal clutch is seven eggs, and incubation starts once the clutch is complete. Again the work is shared, each bird spending $1\frac{1}{4}$-3 hours on the eggs. While on the eggs the parents throw up pellets of clean fishbones, which accumulate around the eggs, until it looks as though the eggs may have been laid upon a bed of fishbones, so giving rise to the old fallacy that the kingfisher makes a nest of fishbones.

KINKAJOU, or "honey bear," *Potos flavus*, South American member of the raccoon family, which is almost exclusively arboreal. Living in forests from Mexico to Brazil, this agile climber is a long, low-bodied animal the fore legs of which are shorter than the hind ones. The most outstanding feature is the prehensile tail, which serves as a fifth grip when moving cautiously through the treetops. Cat-sized, kinkajous measure $31\frac{1}{2}$-$44\frac{1}{2}$ in. (81-

The pygmy kingfisher *Ispidina pictus* of Africa, recognizable by its violet collar.

A pied kingfisher, of southern Africa, looks out from its nest burrow.

113 cm) overall, the tail being at least as long as the body, and weigh from 3-6 lb (1.4-2.7 kg). The ears are small, placed low on either side of the round head in line with the dark, sparkling eyes. General coat color varies from golden yellow to brown, and the fur's texture is soft and woolly. Kinkajous are sometimes confused with olingos (*Bassaricyon*) but can be distinguished as the olingo's tail is ringed and bushy whereas that of the kinkajou is short-haired and uniformly colored. Further, the kinkajou is stocky with a broad head while the slender olingo has a pointed snout. Finally, unlike this other South American procyonid, kinkajous are not gregarious. During their nocturnal forays, they travel singly or, during the breeding season, in pairs. They sleep in elevated tree crevices and, when disturbed during the day, hiss and spit loudly. When cornered they will inflict painful stab wounds with their grooved canines and scratch with their powerful claws.

Favorite food items include fruit, such as bananas and pulpy citrus fruits, but a variety of insects and even bird nestlings are also eaten. Kinkajou "troops" have been observed, on occasion, congregating in ripe fruit trees until the supply is depleted, but, at dawn, each one returns to its nest separately. A particular adaptation to the frugivorous diet is a very long tongue, used to lick sticky tropical fruits, which are held between the forepaws. It is possible that insects under bark are also retrieved in this way.

KIWI, flightless New Zealand bird standing about 18 in. (45 cm) high and lacking a tail and visible wings. So strange a bird is the kiwi (*Apteryx*) that many doubted its existence long after it was described in England by Dr. Shaw from a specimen obtained by Captain Barclay in 1813.

The feathers are gray or brown and hairlike in texture. Their eyes seem to be rather ineffective, at least during the day, but they have prominent ear openings and a keen sense of smell. The long slender bill is remarkably adapted for probing into soil, with the tip of the upper mandible covering that of the lower and bearing the nostrils on either side. This tip seems also to have a highly developed sense of touch, as Buller records that captive kiwis used it extensively for investigating their surroundings. Of the three species, the brown kiwi, *A. australis*, is found on New Zealand's three main islands: a separate subspecies on each. The little-spotted kiwi, *A. oweni*, and the great-spotted kiwi, *A. haasti*, are found only on the South Island. The habits of all three are so similar that one account will suffice for them all.

They are birds of the damp New Zealand forests, feeding at night on worms, insects and berries. They are rarely seen, not because of rarity, but rather because they are shy and can move nimbly through the forest at night. Those who can recognize the calls know that they are quite common in many forested areas and even in scrub and in grassland bordering on forest.

The male finds or excavates a hole beneath the roots of a tree, or in a bank, and builds the nest. The female's sole task is to produce the enormous eggs (usually two), each about

Koala, Australian teddy bear, lives in trees and eats only eucalyptus leaves.

a quarter of her weight. The male incubates the eggs for about 75 days, and though the chicks can find their own food soon after hatching they stay with their father for a considerable time.

KNIFEFISHES, three families of South American freshwater fishes related to the electric eel. The knifefishes have an eellike body but are in no way related to the eels. There is a very long anal fin, and these fishes swim gracefully by passing a series of undulations along the anal fin. The elongated and whiplike tail is in some species used as a probe when the fish is swimming backward. Often, the tail is bitten off by other fishes, but this does not seem to cause the owner much distress, and a new tail is grown. Like the electric eel, the knifefishes have electric organs along the flanks. Impulses are discharged intermittently at a rate of up to 1,000 per second, and this system is used for the location of food or other objects around the fish.

Members of the family Gymnotidae have elongated, compressed bodies with no dorsal, pelvic or tail fins. One well-known species is the banded knifefish, *Gymnotus carapo*, which is often imported for aquarists. The body is light brown with vertical dark bars that coalesce on the back. It reaches 2 ft (60 cm) in length and is peaceful with other species but is liable to attack members of its own.

The family Rhamphichthyidae contains one of the largest of knifefishes, *Rhamphichthys rostratus*, a food fish of the Amazon region that reaches 4 ft (1.2 m) in length.

In the family Apteronotidae there is a small tail fin and a rather curious filamentous dorsal fin that can be folded into a groove along the middle of the back. *Sternarchus albifrons* has a velvety black body with two off-white vertical bars on the tail and reaches 18 in. (46 cm) in length.

KOALA, *Phascolarctos cinereus*, a large arboreal, superficially bearlike marsupial averaging about 20 lb (9 kg) in weight. The present-day range of the koala is in eastern coastal

Australia from about latitude 20°S to the southeastern corner of the continent. It formerly occurred in the states of Western Australia and South Australia, where it was extinct until successful reintroductions were made into South Australia.

The mating period is from September until late January. The single young is born after a gestation period of about 35 days and makes its way, unaided by the mother, up the fur from the external opening of the birth canal to the pouch. In the pouch the young grows from a birth weight of about 5 grains (0.3 gm) to about 1 lb (450 gm) in approximately four months. It begins to leave the pouch when it is lightly furred and is four to five months old, returning periodically until about six months old. The young is carried on the back after it leaves the pouch but returns to drink from the teat at the mouth of the pouch for an extended period. Young koalas have been observed eating a soft substance that exudes from the mother's anus. The exuded substance differs in texture from feces and is said to consist of "peptonized gum leaves from the upper bowel." The young koala has a growth period lasting for a total of about four years and is sexually mature when three to four years old. The total life span may last 20 years.

The koala has been considered an endangered or vanishing mammal, but it is one of the few marsupials that are today definitely increasing in numbers. It has been saved by public interest and in places by an active management program.

The diet was formerly considered to be restricted to the leaves of a few species of *Eucalyptus* but is now known to include many species of that genus and also the foliage of other trees.

"Koala" is an Australian aboriginal word said to mean "no water" and to refer to lack of drinking. However, the koala is an able swimmer that enters water voluntarily and that, after a river crossing, licks and swallows the water from its fur. It also drinks from pools of water left after rain.

KOMODO DRAGON, *Varanus komodoensis,* of the small island of Komodo, to the east of Java, is usually regarded as the largest living lizard. It reaches at least 10 ft (3 m) long and weighs 360 lb (163 kg). However, Salvador's monitor, *V. salvadorii,* although of more slender build, is believed to reach 15 ft (4.6 m) long. See monitors.

KOOKABURRA, *Dacelo gigas,* also known as the laughing kookaburra, laughing jackass, or simply as jack, the largest Australian kingfisher, 43 cm long, named for its wild laughing cry. It is brown, paler on the under-parts and around the neck. There is a second, slightly smaller species, the blue-winged kookaburra, *D. leachi,* in northern Australia.

It is popular among the bushmen, though it loses its good name by robbing farmyards of chicks and ducklings. Besides the name laughing jackass, from its humanlike laughing voice, it is also called the "bushman's clock," as its weird laughing notes can be heard as it goes to its treetop roost at dusk dawn.

KOUPREY, *Bos (Bibos) sauveli,* species of wild ox closely related to the banteng, and less closely to the gaur. Kouprey bulls stand up to 6 ft 4 in. (190 cm) high and are blackish in color with white "stockings" on the legs; older ones become grayish on the sides. Cows are smaller, silvery- or mouse-gray to brownish, sometimes irregularly blotched with black. Bulls have a long dewlap between the fore legs which drags through the grass as they walk. In cows the dewlap is only 3-4 in. (8-10 cm) long, but this is still longer than that of the banteng or gaur.

Koupreys are confined to a small area in Cambodia on either side of the Mekong River in the vicinity of Kosker and Chep, and extending slightly into Vietnam around Ban Methuot and Laos, north of Lake Repou. They inhabit open forests, being especially common in parkland—in more open country than that frequented by banteng. They form small herds that in the rutting season, April, consist of both sexes, but by June the cows and bulls begin to form their own herds. The young are born in December and January, and are red in color; at the end of May the young turn gray.

KUDU, a large striped African antelope. The greater kudu, *Tragelaphus strepsiceros,* is found in Africa, from north of the Orange River to Somalia, southern Sudan and Lake Chad, is 4 ft 2 in.-4 ft 5 in. (127-132 cm) high, weighing 600 lb (270 kg). Horns average 40 in. (1 m) long, with an open spiral of at least two and a half turns. There is a long throat fringe and mane on the nuchal and dorsal midline. The kudu is grayish, with legs rich fawn below the knees and hocks, white spots above the hoofs, and four to ten white transverse body stripes.

The lesser kudu, *T. imberbis,* inhabits northern Tanzania to Sudan, northern Ethiopia and Somalia. It is 3 ft 4 in. (1 m) high, and its horns average 2 ft (61 cm) long, with a close spiral of two and a half turns or more. There is no throat fringe, but a white crescent on the throat instead.

Kudus are wary animals, living mostly in hilly or broken country with thorn bushes or tall grass. Lesser kudus form herds of up to six animals, which remain together all the year round, but greater kudus split up into small groups of one to four in the rains, and form bigger herds up to 14 head in the dry season. Bulls are more often solitary than cows, and are always extremely shy and difficult to approach. They hide by day and browse, rather than graze, at dawn and dusk. When disturbed, kudus pause in midstride, then race away, leaping over bushes as high as $8\frac{1}{2}$ ft ($2\frac{1}{2}$ m); a lesser kudu has been recorded as jumping 30 ft 8 in. (9.2 m) on the ground, while clearing a bush of 5 ft (1.5 m) high. Kudus run with their tails, which are white underneath, curled back over the rump, and after a short dash, they invariably stop to look back.

Australian blue-winged kookaburra, a land kingfisher with a boisterous laugh.

Bull greater kudu, one of the largest antelopes. The animal is alert, shy, and very difficult to approach.

L

LACEWING, the common name given to 1 in. (2½ cm) long insects.

The green lacewing is known to fishermen as the goldeye gauzewing, for the adult is a pale green color, and its four translucent wings are iridescent and it has large, gold-colored eyes. The common European species, *Chrysopa flava*, is abundant as an adult in late summer and is often seen when it flies indoors to hibernate in September and October. Lacewings emerge from hibernation in the late spring and lay their stalked eggs singly on leaves. The fine stalk that is attached to the leaf is about four times as long as the egg so that the latter appears as a pale knob at the end of the bristlelike stalk. As in all Neuroptera the larvae are active carnivores and feed principally on aphids. The larva can be found among clusters of aphids, covered, in some species, by the dead skins of those it has eaten, which form an effective protection against predation by birds. Since they are common insects and their larvae destroy large numbers of aphids, they are an important natural control of such pests.

LADYBUGS, small, tortoise-shaped, brightly colored insects. They are among the best-loved of all insects, although very few people are able to associate the tuberculated or spiny larvae with the attractive adult beetles. Most of the 5,000 species of ladybugs are between $\frac{1}{10}$-$\frac{3}{10}$ in. (2.5-7.5 mm) long. The larvae are often a grayish color with darker spots, but others may be quite dark in color with white or orange markings. Many species of ladybugs are fully winged and readily take to flight during the daytime. They have a worldwide distribution. The two-spot ladybug, *Adalia bipunctata*, is one of the most common species in Britain. It is either red with two black spots or black with four or six red spots. Also common is the seven-spot ladybug, *Coccinella 7-punctata* (red with seven black spots), and the small yellow and black spotted ladybug, *Thea 22-punctata*.

The eggs of ladybug beetles are yellow or orange and are laid in masses on leaves, with each of the rather elongated eggs placed at right angles to the surface. They hatch in a few days, and the larval stage is followed by the pupa, which is formed in an exposed position, often on a leaf surface. The pupa is attached only at its extreme hind end, and if disturbed it can lift the whole body upward from its point of attachment. The rate of development of ladybugs varies with the species and weather conditions, but in Britain

and the United States the life cycle of the two-spot ladybug is about one month during the summer period.

The female ladybug lays her eggs close to thriving aphid colonies, so that when the larvae emerge their prey is close at hand. Aphids of all ages spend much time with the mouthparts embedded in plant tissues sucking sap, and the young ladybird larva only has to seize the first aphid of a suitable size that it comes across. The aphids are utterly defenseless, and very few escape when attacked by a larval or adult ladybug. Each larval ladybug consumes 200-500 aphids, and the adults eat many more. The Coccinellidae have considerable economic value in controlling aphid populations but, as is apparent to anyone who grows roses, ladybugs by no means always prevent the build-up of large, damaging populations of these pests. The initial proportions of predator and prey are critical in this respect.

LÄMMERGEIER, *Gypaetus barbatus*, or bearded vulture member of the Old World vulture subfamily of the family Accipitridae. It is a carrion-feeding bird like most other vultures. It is sparsely distributed through the mountainous regions of Asia, Africa and southern Europe. It is very large, with a wing

span up to 9 ft (3 m), has loosely feathered plumage and a beard of black bristly feathers beneath the bill. The Lämmergeier breeds on rocky cliffs and is known to obtain part of its food by dropping large bones, and even tortoises, onto rocks from a height, thus enabling it to feed on the marrow, or other tissues then exposed. It has an unusually long tongue, which aids in its method of feeding.

LAMPREYS, primitive jawless fishes found in both fresh water and the sea. The body is eellike, and there is a round, sucking mouth lined with horny teeth with which the lampreys rasp away at their prey. In some parts the name for lamprey is "nine eyes," a reference to the seven external gill openings, the median nostril and the eye. In Britain there are three species of lamprey, the river lamprey, *Lampetra fluviatilis*, the brook or planer's lamprey, *L. planeri*, and the sea lamprey, *Petromyzon marinus*, which is found on both sides of the Atlantic. There are ten other species in Europe. Many species of lamprey are parasitic when adult, feeding on the flesh of living fishes. A few freshwater species in North America are nonparasitic and do not feed at all as adults, merely breeding and dying. Lampreys lack paired fins (pectorals and pelvics) but have either one or two dorsal

A South African ladybird *Chilomenes lunata* laying her eggs. Ladybirds and their larvae feed on aphids.

fins and a tail fin. The gills are contained in muscular pouches that open to the exterior by a series of seven small apertures but are connected internally to a canal which opens in the mouth. The skeleton is of cartilage.

The river lampreys move upstream to breed, but the sea lampreys make a major migration from the sea into fresh waters, using their sucker mouths to ascend rapids and waterfalls. Crude nests are made by both parents, the mouths being used to remove stones to form a hollow of about 2 ft (60 cm) in diameter. (*Petromyzon* means "stone sucker" and *Lampetra* means "rock licker.") Many thousands of eggs are laid by the female, but after they have been fertilized both parents die. From each egg hatches a blind, wormlike larva quite unlike the parents and known as an ammocoete or pride. So different is it from the adult that it was at one time given the scientific name of *Ammocoetes branchialis*. It lacks the sucking mouth of the adult, and there is a kind of hood present with structures adapted for filter feeding. For six, seven or more years the larva lives a burrowing life, blind yet shunning light. At the end of this period a most remarkable metamorphosis occurs, the larva changing into the adult. The eyes appear, the circular sucking mouth develops and part of the former ciliated filter-feeding mechanism becomes associated with the pituitary region of the brain.

The suctorial mouth of a sea lamprey.

LAMPSHELLS, marine mollusklike animals, so named from the resemblance between their shells and ancient Roman oil lamps. They differ so much from all other animals as to constitute a separate phylum, the Brachiopoda. Although they were included in the Mollusca for a long time because of the superficial similarity of their shells with those of bivalves, the two are not related. Their closest relatives are two rather obscure phyla, the Phoronida and the Ectoprocta.

Lampshells are represented today by about 260 species, none of which is very common.

Their fossil history is, however, extensive; at least 30,000 extinct species are known, and their remains are quite abundant in certain deposits. The phylum is divided into two classes: the hinged brachiopods or Articulata, in which the valves of the shell are located upon each other by means of a tooth and socket arrangement; and the unhinged brachiopods or Inarticulata, in which the valves lack any locating mechanism.

All living forms, and most fossils, are fairly small, up to 3 in. (8 cm) long, but a few extinct species reached over a foot (37 cm) in width. Lampshells are solitary and usually live attached to the substratum, although some can move around to some extent. The animal is enclosed within a bivalved shell and usually attached by means of a long flexible stalk, the pedicel.

LANCELET, a small, marine fishlike animal about 1½ in. (40 mm) in length, with one species on the Californian coast growing to 3 in. (80 mm). Lancelets usually burrow in rather coarse sand or shell gravel at a depth of 15-90 ft (5-30 m), but they may occur in shallower waters or even in intertidal sands in coastal regions or brackish lagoons that are protected from wave action. Their distribution is worldwide between latitudes of approximately 70°N and 50°S where the surface temperature of the sea does not fall much below 50°F (10°C). About 30 species have been described, the majority belonging to the principal genus, *Branchiostoma*.

The animal has no head and is pointed at both ends, a feature that has given rise to the alternative commonly used name, *Amphioxus*. It has a prominent postanal tail for swimming and a median fin extending down the back, around the tail, forward along the ventral surface,

The lancelet *Branchiostoma lanceolatus* has oral tentacles and gill pouches.

past the anus to the ventral aperture called the atriopore. There are no paired fins. Water enters the pharynx through the mouth and leaves via the gill slits, the atrium and the atriopore. The mouth is subterminal and protected by an oral hood fringed with fingerlike cirri that can be crossed, those of the one side over the other, to form a sieve excluding the larger particles from the pharynx. The lancelet is a suspension feeder. The current of water through the pharynx is generated by cilia on the gill bars. Fine particles are strained from it by the gill basket and, entrapped in

mucus, are carried up the gill bars and backward in a dorsal pharyngeal groove to the intestine by ciliary beat. The food string of mucus and particles passes through the gut, where digestion takes place, by the action of the ciliated intestinal epithelium, and thence to the exterior through the anus.

LANCET FISHES, oceanic fishes of moderately deep waters, related to the lantern fishes and the lizard fishes. They have long, thin bodies with large mouths armed with rows of fierce daggerlike teeth. Characteristic of the lancet fishes is the huge, saillike dorsal fin, behind which is a small adipose fin. They are large fishes, reaching a length of 6 ft (200 cm), and are very voracious, appearing to eat almost anything.

There are two species of lancet fishes. *Alepisaurus ferox* is found in the Atlantic, and the closely related *A. borealis* occurs in the Pacific. Off Madeira, *A. ferox* is not infrequently caught by fishermen who have set their long lines for tunas or other species. It was here that the first lancet fish was caught in the early part of the last century.

LAND CRABS, term applied indiscriminately to any large crustaceans which live much of their life on land. In parts of North America crayfish, those small freshwater relatives of the lobster, are often known as crabs. Several species, chiefly of the genus *Cambarus*, have deserted the water to take up life burrowing in marshy or swampy grounds. In parts of the United States these "land crabs" can be a serious pest of cotton, nipping off the young shoots as they emerge in the marshy irrigated ground.

Both the west African and West Indian land crabs are true crabs but are quite unrelated to one another, having independently taken up terrestrial life. The west African land crab is a member of the genus *Sesarma*, a large genus most members of which live in mangrove swamps. It has deserted the mangroves to run about on land and burrow in marshy ground where it can do great damage to newly planted crops. *Cardisoma*, the land crab of the West Indies, Florida and also some of the Pacific islands, is another crab which has turned from the sea to live on land. Large specimens may measure 2 ft (60 cm) across their outstretched legs and can excavate burrows several feet deep. In Puerto Rico land crabs have hampered

Temminck's horned lark, with nestling gaping to be fed.

Lesser short-toed lark *Calandrella rufescens*.

rice cultivation by burrowing down through the impervious stratum of clay, which serves to retain the irrigation water in the paddy fields, thus draining the impounded water and destroying the crop at one fell swoop. In southern Florida they are a major pest of tomatoes since they bite off the sprouting tomato plants as soon as they push their way above the surface of the ground. As many as 10,000 per acre (per 0.4 ha) have been recorded in southern Florida and even cement retaining walls of 30 in. (76 cm) present no barrier since these crabs are good climbers. A close relative is *Gecarcinus*, which is remarkable for the mass migrations undertaken from burrows on rocky hillsides down to the sea to breed.

The biggest and perhaps the best known of the land crabs is the robber crab of the Pacific islands, *Birgus latro*, also known as the coconut crab. It is not a true crab but a highly modified hermit crab, a member of the sub-order Anomura. At one time in its ancestry it must have inhabited some kind of mollusk shell, in the way that most hermit crabs do, for the abdomen is asymmetrical and the female possesses abdominal legs only on the left side. As in most hermit crabs one claw is much the

larger. Robber crabs may grow to 18 in. (45 cm) long and weigh as much as 6 lb (2.7 kg) but even the largest are expert climbers. By day they may wander over the ground until they come into the shadow of a palm trunk. They then run along the shadow towards the sun and climb the tree on the shady side. Their habits vary according to which islands in the Pacific they inhabit. On some islands they sleep in the trees at night, on others they may live in burrows during the day and search for food chiefly at night. On islands where coconuts are abundant they feed on the kernels of those that are broken open by other agencies.

LANGOUSTE, *Palinurus vulgaris*, or craw-fish, a spiny lobster with small claws and very long antennae, up to 10 in. (25 cm) long, reddish-brown blotched with purple, lives on rocky bottoms offshore in the warmer eastern Atlantic waters. It has long been famous as a French delicacy, and it has a remarkable larva, broad and flattened, paper-thin and transparent as glass, with long spidery legs and large black eyes on long stalks. In warm seas the larva may be 3 in. (7.5 cm) long but usually it is $\frac{1}{2}$ in. (12 mm). Because it is so unlike the adult it was regarded at first as a distinct animal.

A land crab of the family Grapsidae, of Nepal, a long way from the sea.

LAPWINGS, a genus (*Vanellus*) of plovers, usually occurring on grass plains or pastures. Some authors treat the lapwings as a subfamily, Vanellinae, containing many different genera, not as a single genus. The latter, simpler, classification is used here. Unlike other plovers, many species of lapwings possess either crest or wattles, and all have a bony lump beneath the skin at the carpal joint, this being developed further in some species into a wing spur. Lapwings are of medium size, about 12 in. (30 cm), and in most species both sexes are alike or very similar. They share with the plovers the strongly patterned, rather than uniformly cryptic plumage and the lack of specialization of bill structure; the bill is straight and of only moderate length in comparison with the bird's head width. Plumage patterns vary from species to species, but in many the fairly broad bands of adjacent dark and light colors break up the overall shape of the bird (disruptive coloration), so camouflaging it.

In Europe and Asia are found two species, in addition to the few spur-winged plovers: the sociable plover, *V. gregarius*, and the lapwing, *V. vanellus*. The lapwing breeds on arable land and wet meadows right across the palearctic, chiefly in the temperate and boreal zones. It has bred once also in Iceland, and flocks of vagrants have reached North America in some winters, but have not survived to colonize the continent. The breeding range in northern Europe has extended northward this century, coincident with a slight increase in mean summer temperature. Lapwings move westward from some areas of central western Europe as early as the end of May (after breeding), perhaps to avoid food shortage in high summer, when the soil animals that form their main food become inaccessible through drying and baking of the soils. Other movements take place in autumn, notably from Europe to the British Isles, and if the ground becomes frozen in winter, large flocks move either west to Ireland or south to Iberia, depending on whether the cold conditions are accompanied by east or north winds, respectively.

LARKS, small terrestrial songbirds, usually streaked brown above and more or less streaked below. The hind claw is often much elongated, an adaptation to a terrestrial way of life. They rarely hop, but usually walk or run, sometimes at speed, many species preferring to do this than to take flight. They are renowned the world over for their beautiful songs, usually delivered on the wing, and have provided endless inspiration for poets.

The lark family differs from all other perching birds in that the back of the tarsus is rounded and covered in scales, instead of being sharp and unsegmented. Together with certain primitive features of the syrinx, this has led to larks being generally regarded as the most primitive of true songbirds. They are almost entirely Old World in distribution, and are particularly well represented in Africa, all genera and no less than 80% of the species occurring there. Larks live in a great variety of open habitats, from cultivated meadows and steppe country, wasteland and semiarid sparsely vegetated areas, to sandy and rocky wastes, arctic tundra and absolute desert. They are widely represented in Europe, Asia and the subcontinent of India, but only one species has reached Australia and the Americas naturally, and, apart from introductions, the family is absent from New Zealand and most oceanic islands.

Larks are ground nesters. The nest is a loose cup of dead grass and occasionally hair or wool, placed in a depression in the ground, usually in the shelter of a tuft of grass or a rock. The usual clutch is three to five eggs, which are generally white or whitish, heavily streaked or speckled with gray or brown.

The song of many larks is remarkable for its melodious quality and the vehemence with which it is delivered, often from a prolonged song flight. The skylark delivers a loud and clear warbling, which may be sustained for as long as 5 minutes, while hovering in the air above its territory, sometimes at a great height.

LEAF BEETLES, a very large group of beetles comprising 30,000 species. Many are small, often no more than a few millimeters long, but they are strikingly patterned and colored. They feed on living leaf material and include some of the most destructive agricultural pests, such as the Colorado beetle, the potato beetle, the asparagus beetle, the beet tortoise beetle and the elm leaf beetle.

LEAF BIRDS, eight species of *Chloropsis*, related to bulbuls, living in southeast Asia from southern China to the Philippines. They are mainly bright but pale green, with a slightly curved bill, and they feed on fruit, berries, insects, nectar and pollen. The males have more colorful markings, with more black, orange and blue on the head. The largest and most brilliantly marked is the orange-bellied leaf bird, *C. hardwickii*, 8 in. (20 cm) long, which ranges from northern India to Malaya. The golden-fronted leaf bird, *C. aurifrons*, of India eastward into southeast Asia, is popular as a cagebird. They are good singers and accomplished mimics.

LEAF FISHES, a family of tropical freshwater fishes from South America, west Africa and southeast Asia the members of which, to a greater or lesser degree, resemble floating leaves. The most famous of all is the leaf fish *Monocirrhus polyacanthus* from the Amazon

Leaf-fish *Nandus nebulosus* of southeast Asia.

and Rio Negro basins of South America. The body is leaf-shaped and tapers toward the snout where a small barbel increases the camouflage by resembling the stalk of the leaf. The general body color is a mottled brown, sometimes lighter and sometimes darker, but always matching the dead leaves that subside in the water where the fish lives. It grows to about 4 in. (10 cm) and drifts with the current, head usually downward, until it approaches a small fish on which it can feed. The mouth is large, and the jaws can be extended to engulf fishes half its own size. It can show a remarkable turn of speed once its prey is within reach.

The African leaf fish, *Polycentrus abbreviata*, is found in tropical west Africa, and a related species, *P. schomburgki*, occurs in Trinidad and the Guianas. One species from India and Burma, *Badis badis*, shows an extraordinary ability to change color, often very rapidly, but does not resemble a leaf.

LEAF HOPPERS, insects closely related to the spittle bugs, tree hoppers and aphids. They are often brightly colored, sometimes green, and they feed on leaves by sucking the sap. The eggs are laid on the undersides of the leaves, and they hatch to give nymphs, which eventually develop into winged adults. These animals transmit virus diseases from one host plant to another and so are important agricultural pests, especially in the tropics.

LEAF INSECT, the name given to some large tropical insects which have the body flattened and the wings expanded in such a way that they look like leaves. Usually they are green, although some are brown, and the veins on the wings are arranged and picked out in another color so that they resemble the veins of a leaf. Most of these insects belong to the order Phasmida, but some are true Orthoptera, related to the katydids. It is easy to tell the difference between the two because phasmids usually have short antennae while the katydids have long antennae and have the back legs modified for jumping. The phasmid leaf insects also have the legs expanded into leafy lobes so their general appearance is quite unlike an insect. It is fairly certain that the leaflike appearance of these insects is a useful camouflage, concealing them from possible enemies.

Plasmids lay one egg at a time, and since the eggs are not stuck to leaves they often drop to the ground. Here they may remain for some months before the young hatch. The latter look very like small versions of the adult, but have no wings. They feed on leaves and grow slowly, the female having about six molts before becoming adult, but the male only four or five. The whole life history is rather drawn out and may take over a year.

LEATHERY TURTLE, *Dermochelys coriacea*, also known as the leatherback, the largest of the marine turtles. Its occasional appearance in northern waters, off the west coast of Scotland, has given rise to stories of the Soay Beast.

LEECHES. Some years ago the most familiar leech—at least to the wealthier classes—would have been the medicinal leech, *Hirudo medicinalis*, much used by physicians in blood letting. This practice was believed to be a relief for almost all imaginable ills, "vapours" and "humours." Their medicinal value may be questioned, but the ability of the leeches to gorge themselves with blood in a relatively short time, and their related ability to fast for long periods between meals, are certainly adaptations to their more normal existence in the wild. There, meals are available only at intervals and entirely in the lap of chance. The medicinal leech has three sharp teeth, which form a characteristic triradiate wound in the skin. When a leech bites, an anticlotting saliva is poured into the wound, and this prevents the blood from clotting and keeps it liquid within the leech's body.

The buccal cavity (or mouth cavity) leads into an enormous crop where the meal of blood can be stored. The crop is capable of great distention, to enable large meals to be taken as opportunity permits. Most of the gut in fact is storage crop, the part where absorption of the digested contents takes place being quite restricted. This is perhaps not surprising, for the meals are not varied in composition, and small amounts can be allowed to enter the "stomach" from the crop from time to time. So, once having fed, the leech can live for long periods on what is stored in the crop.

A fully grown medicinal leech can survive for a year on a single meal, though it probably feeds more often than this.

Leeches are easily recognized. Their soft bodies are annulated, usually without external projections and with a prominent, often circular sucker at the posterior end. Anteriorly there is another sucker around the mouth, and, although this may be quite prominent as in fish leeches, it frequently is not.

LELWEL, *Alcelaphus buselaphus lelwel*, or Jackson's hartebeest, one of the subspecies of the common hartebeest.

LEMMINGS, small rodents, about 3-6 in. (7.5-15 cm) in length, closely related to the voles. They are the characteristic rodents of the Arctic tundra, although a few species extend south into the coniferous forest zone in both America and Eurasia.

Lemmings are adapted to the severity of arctic conditions in many ways. The fur is long, soft and dense, and the extremities are short, the ears being scarcely visible in the long fur and the tail just long enough to be seen. The collared lemmings, *Dicrostonyx*, extend farther north than any other rodent, being found in such unlikely places as the north coast of Greenland at a latitude of over 80° North, and it is significant that these are the only true rodents that turn white in winter, although other animals that do this, e.g., the Arctic hare and

the stoat, extend much farther south. In summer collared lemmings are mostly a rather obscure grayish brown, but other species are more colorful. The Norway lemming, *Lemmus lemmus*, for example, is boldly patterned with a black head and a large black patch on the back on a yellowish-brown background, while the smaller wood lemming, *Myopus schisticolor*, is slaty gray with a streak of rust-red on the back. All lemmings have strong claws, by which they dig tunnels among the roots of the tundra heaths and grasses. In summer these claws are normal in shape, but in winter the collared lemming develops two peculiar, enlarged, forked claws on the front feet, well adapted for digging in frozen snow.

Lemmings are known, above all, for their periodic mass emigrations. These are best known in the Norway lemming of Scandinavia, but the phenomenon occurs to some extent in all species. One reason why it appears more spectacular in Scandinavia is the mountainous nature of the country. When the mountain lemmings begin to disperse they can move only downhill, and they are then concentrated in the valleys until, in the major "lemming years," great rivers of lemmings seem to be flowing down the valleys toward the lowland

fields and forests. The reasons for these periodic eruptions are still poorly understood, but a great many of the myths that surround the subject can safely be exploded. They do not move persistently in one direction; they do not follow leaders; they do not plunge recklessly into the sea and commit suicide. Mass movements of lemmings are very aimless, although they may travel several miles each day. The animals are very aggressive, especially when they become crowded. When they come to water, they are very reluctant to swim, and it seems that they will only take to the water if it is calm and they can see the other side. They are competent swimmers but their small size makes them vulnerable, and, if the water is at all rough, they drown, so that it is not uncommon for large numbers of dead lemmings to be found washed up on the shores of fjords. Others, however, survive and may settle down to form temporary breeding colonies far from the normal range.

LEMURS, small relatives of monkeys found only on Madagascar and small nearby islands, with the mongoose lemur ranging to the Comoro islands to the north. They are mostly cat-sized, a few being mouse-sized, usually with a long tail and foxlike muzzle, and mainly nocturnal and arboreal. Their food may be insects to small mammals and fruit, and the incisors of the lower jaw, projecting forward with the canines lying along each side, are comblike and used for scooping out soft fruit as well as for combing the fur.

The family Lemuridae includes two subfamilies: the Cheirogaleinae or mouse lemurs and the Lemurinae. Mouse lemurs are among the smallest of primates, the smallest of them weighing only 2 oz (60 gm) and being 5 in. (13 cm) long in both body and tail. They are named solely for their small size since they in no way resemble mice.

Mouse lemurs are entirely arboreal, leaping around in fine branches and lianas at night with tremendous accuracy in primary and secondary forest and even in scrub. They are omnivorous, eating berries, flowers, buds, leaves, bark, insects and other animal food. The typical species includes two subspecies, one *Microcebus murinus rufus* living in the rain forest of the east coast and a grayish form with large ears, *M. murinus murinus*, in the drier forest areas. The brown subspecies is adapted for higher humidity and seems to be more insectivorous, while the gray is adapted to withstand dry conditions and has better developed teeth and musculature for eating plant material in the semiarid zones.

Both subspecies construct spherical leaf-nests wedged between fine branches, or live in hollow trees lined with a few leaves. The gray has a loosely organized social system. Males and females have home ranges around the nest, and the females nest in groups of up to 15. Males usually occur singly, but sometimes two are found to share the same nest and home range. Members of a group may groom one another, by holding the ears of the partner in the hands and raking the top of the head with the horizontal lower incisors ("toothcomb"). At night, several members of a female group may be seen feeding together in the same food-tree. The home ranges of males probably

The famous Norwegian lemming.

Steppe lemming in summer coat.

overlap with those of the females, at least during the mating period, and a male may very rarely be found in a female group nest.

Usually, mouse lemurs have two offspring at birth, occasionally one or three, after a gestation of about 60 days. They are born about one month after the start of the rainy season in mid-October to mid-November in a leaf-nest or hollow tree, and the female groups do not seem to split up. This means that the nest is turned into a kind of nursery, with up to a dozen females and their offspring. The mothers are in the nest with their offspring during the daytime "sleeping period" and leave them behind at night when they go off to forage. Suckling occurs frequently, and the mother ingests the urine and feces discharged by the offspring, thus keeping the nest clean. There are indications that the mothers in a female group recognize their offspring individually. Nesting in groups may help solve the problem of maintaining a constant temperature in such tiny animals, and it may also serve for mutual defense.

The related dwarf lemur is about the size of a large rat and weighs about 8 oz (250 gm). As with the mouse lemur, there is a brown rain-forest form along the east coast, the greater dwarf lemur, *Cheirogaleus major*, and a gray drier-forest form, the fat-tailed dwarf lemur, *C. medius*. The name of the second species is based on a feature common to all mouse and dwarf lemurs of storing fat in the tail. In the northwest, west and south of Madagascar, in the drier-forest regions, there is a marked dry season during which food is hard to find. In these areas, and apparently even in the rain forest, dwarf lemurs "hiber-

nate" in hollow trees for several months. Mouse lemurs seem not to show a clear-cut hibernation, but they probably use fat reserves in the tail to tide over lean periods.

The largest and best known of these is the ring-tailed lemur (*Lemur catta*), gray with white underparts and black markings on the hands, face and tail. The tail has alternating rings of black and white fur. It is cat-sized, 14 in. (35 cm) head and body length with a tail 16 in. (40 cm) long, and one of its many vocalizations is a catlike miaouw (its German name is Katzenmaki or cat-lemur). The eyes are large and forward-pointing, the muzzle doglike. There are large, canine teeth, slightly exposed when the animal is in a mildly threatening mood, this aggressive impression being reinforced by the glaring quality of the eyes, which have a yellow iris. This is the commonest species of lemur seen in zoos, and it also seems to be one of the most dangerous. Many a zoo attendant has been badly bitten by a ring-tailed lemur.

The ring-tailed lemur, by contrast with all other lemurs that rarely leave the trees, seems to be equally at home on the ground and in the trees, and can be seen eating, playing and resting at all levels in the forest. When walking on the ground, it holds its tail erect in an S-shape, and it is a common sight to see a group walking along the ground in the sunshine with their tails held aloft like flags.

This species is basically a herbivore, and animal food is rarely, if ever, taken. Its basic diet is provided by a wide range of fruits, flowers and leaves. It does not become active until sunrise and starts to retire for the night at sunset. It lives only in the arid forests of the south in

which foliage is very restricted and undergrowth usually sparse, where it can move around more easily on the ground.

Lemur catta forms some of the largest social groups known among Madagascar lemurs, assembling in troops of up to two dozen, which move around, eat and sleep together. Social grooming between group members is common and the group moves around, feeds and sleeps in a fairly distinct home range which is defended against neighboring groups. In the evening, the group moves off in close formation to one of a small number of traditional sleeping trees. The ring-tailed lemur has the richest vocal repertoire of all lemurs, including clicks, grunts, explosive snorts, squeaks, miaouws and howls. There is also a distinctive shriek uttered when an individual sees a hawk or other danger, and calling of this kind is most frequent in the evening, when the group is gathering for the night. The miaouw call is most frequent, and it apparently acts as a signal to keep the group members together as they move off.

In its terrestrial and social life, the ring-tailed lemur provides an interesting parallel to some African baboons which are terrestrial, form large troops and have a large vocal repertoire. They also use visual signals for communicating in clear areas, and possibly the striking black-and-white tail of *L. catta* plays the same role.

Mating takes place in April-May and births occur about $4\frac{1}{2}$ months later in September-October. There is usually a single baby, but twins are fairly frequent. At birth, the baby crawls directly onto the mother's fur and is thereafter carried by her or (later) by other adults until reaching independence. At first, the baby rides on the mother's belly, but after about two weeks it changes to riding on her back. The mother pays a lot of attention to her offspring, licking and suckling it frequently. As it grows older, the baby begins to venture away from its mother, both on the ground and in the trees, but at the slightest sign of danger it will flee back to cling on her fur. *L. catta* never constructs a nest, and it sleeps clinging to the higher branches of the trees of the forest.

LEOPARD, *Panthera pardus*, one of the smaller of the big cats similar to the jaguar, with which it appears to have had a common ancestor in the recent past. This had led to tales of the leopard being seen in South America, the home of the jaguar.

The leopard is about 7 ft (2.1 m) in overall length, exceptionally 8 ft, of which some 3 ft (1 m) is tail. The ground color is usually a dark shade of yellow, and this is marked by black rosettes on the flanks and back, the centers of the rosettes being a darker shade than the general ground color. The jaguar has a black center to the rosettes. The leopard is not as stockily built as the jaguar, but it is extremely agile. There is a considerable variation in the degree of spotting and in the ground color between the different races, and the melanistic, or black form, known as the panther, is quite common. In some extreme cases, the dark form has become so pronounced that the dark pigmentation extends to the gums, tongue and palate, and the eyes may even be blue. Albinism is almost unknown and has been

The fat-tailed dwarf lemur *Cheirogaleus medius*.

The fearsome snarl of a killer, much feared because of its ability to hide in ambush.

encountered only on one or two occasions.

The skeleton of the leopard shows it to be an animal that relies on its power to spring quickly, while the structure of the bones allows it to drop from a high branch and absorb the considerable shock of landing. The upper limit of the weight of a fully grown leopard would be unlikely to exceed 200 lb (90 kg), and the average weight for the species is about 100 lb (45 kg). For some time it was thought that the leopard and the panther were two separate species.

Leopards live in a variety of habitats, from jungle to grassland and from semidesert to snow-covered highlands. The leopard is found throughout Africa and through Asia Minor, the Caucasus, the Transcaucasus, Persia, India and the Indian highlands, China, Korea, Siberia and Manchuria. It is therefore the most widespread of all the big cats, and this wide distribution accounts for the great difference in the quality and texture of the coat, which varies from the very thick rich fur of the Siberian and Chinese races to the short, coarse fur of the tropical and subtropical races. The dark form, generally known as the black panther, appears to be more common in areas where there is a heavy rainfall. In Pleistocene times the leopard was even more widespread than it is now, and was common in Europe, extending south into Spain.

Individuals of the species are solitary except during the breeding season. The cubs are born after a gestation period of 90-95 days and are fully furred at birth. They are born blind, and the eyes open after about ten days. The number of young in a litter is normally two or three, but it may be as many as five. The family will remain together until the cubs are nearly mature at the age of about two years.

Leopards employ a number of stratagems to obtain food, and appear to be very quick to learn new "tricks." They have been observed to roll about on the ground when near a herd of deer to arouse their curiosity and so entice one of them to within striking distance. They are also noted for their fondness for dog meat, even to the point of entering bungalows to carry off the dogs inside. Their favorite prey in India is the spotted deer, while in Africa they are even known to take baboons, but only a lone baboon, as a whole troop would inflict serious damage, or even kill the intruder. Other prey include monkeys, dogs, pigs, deer, domestic stock, antelope, porcupines and various species of small game. The leopard is well known for its habit of dragging its kill up into a tree, out of the reach of jackals and hyenas. Often the kill weighs more than the leopard, a good indication of the animal's strength. The average consumption of the leopard in India has been estimated at one spotted deer every two weeks. Because of their fondness for pigs, deer and baboons, leopards

help to control these species, which, when locally numerous, can do great damage to crops. When a leopard turns man-eater it is far more troublesome than a man-eating tiger or lion. The record for a man-eating leopard, or any man-eating cat, is held by the leopard of Rudrapranag, which, over a period of nine years, killed 125 people before it was finally shot. Man-eating is luckily rare, and usually arises because the animal either is too old to obtain its normal food, or has been injured and has had to take the easier, slow-moving prey, like man. Alternatively, it may be a cub of a female that had become a man-eater before she had the litter. In the latter case, the young will be partly raised on a diet of human flesh.

The clouded leopard, *Felis nebulosa*, is the largest of all the Asian purring cats, and differs from the big cats in having hyoid bones that are continuous. A large specimen may measure as much as $6\frac{1}{2}$ ft (1.9 m) overall, of which about 3 ft (1 m) is tail. The height at the shoulder is about 21 in. (53.3 cm) and the weight some 44 lb (19.9 kg). The most noticeable feature of this animal, apart from the distinctive coloring, is the broad spatulate paw, which appears to be used as a kind of swat when capturing birds. The coat varies in color from a dark brown or gray to a yellow buff. The coloring shades to practically white on the belly and underparts, which are marked

The Polyphemus moth *Telea polyphemus*, of North America, is 5 in. (12.5 cm) across the spread wings.

The intricate antennae, sensitive to smell, enable many moths to home in on their potential mates over several kilometers.

The elephant hawkmoth *Deilephila elpenor* of Europe.

are composed of a double membrane: yet they do not swell into bags when the blood is forced between the layers, for these are held in close proximity by strands. Some hours elapse before the insect can fly, for the wings must harden, being at first soft and damp. They are strengthened by struts or nervures (an unsatisfactory name, because they are not nerves), the arrangement of which, being characteristic of different groups of the Lepidoptera, is useful in classification: so too are the external parts of the reproductive organs, especially those of the male.

In certain groups, patches of scales are modified in the male to produce the scents used to stimulate the female, while in some species female scents may be perceived by, and attract, males more than a mile away. The organs responsible for the sense of smell are situated in the antennae, which are a pair of many jointed rodlike structures, branched in some moths, but never in butterflies.

Lepidoptera are extremely widespread, wherever vegetation exists in the world, up to the region of permanent snow on mountains and in the Arctic, in forests, marshes and open country and on sandy shores, while some even feed upon the smoke-polluted trees growing in towns. Indeed a few moths such as the sycamore moth, *Apatele aceris*, and the brindled beauty, *Lycea hirtaria*, are actually commoner in London than elsewhere.

with dark spots and blotches. The back is marked by two broad dark bands, with narrow bands in between, running longitudinally. The clouded effect on the flanks is caused by large areas etched in brown or heavy gray and surrounded by areas of lighter colored fur. The tail is long, bushy and marked with rings. The head and face are marked with dark streaks and spots, which are particularly noticeable on the face and forehead. The backs of the ears are black, and each has a buff spot. The upper canines are remarkable for their length. The fur is short.

The clouded leopard is found in Assam, Bhutan, Sikkim, Burma, Thailand, the Malay peninsula, Borneo, Sumatra, Taiwan and parts of China. Nothing is known of the breeding habits of this animal in the wild, and there is very little information of its habits in captivity.

LEPIDOPTERA, the insect order that includes butterflies and moths. Their bodies and wings are covered with minute scales of chitin which overlap like slates upon a roof. These carry the pigments that produce the colors and patterns, often so striking a feature of these insects. The mouthparts of the adults take the form of a proboscis; a tube for sucking up liquid, such as the nectar from flowers. When not in use it is coiled under the head like a watch spring.

The life history includes the egg, the larva or caterpillar, the pupa or chrysalis and the usually winged adult or imago. The eggs vary greatly from one group of butterflies or moths to another but are fairly constant within each, so that their external structure is useful in classification. They may be spherical, bottle-shaped or flat, while their surface may be smooth or sculptured into complex patterns. Usually about 200 eggs are laid. The caterpillar can grow to a limited extent by stretching its chitinous cuticle, but at intervals this is shed and a new one, which has formed beneath it, is disclosed. It may differ considerably in shape from the old, and rapid growth takes place before it hardens. This process generally occurs four times. Finally, a much more striking change takes place. The last larval skin is shed and it is found that the chrysalis has formed beneath it and on its surface can be seen cases which will contain the main externally visible structures of the imago: the eye, legs, antennae and wings.

The chrysalis is an inert stage and it may take several different forms. In moths it is often enclosed in a silken case, or cocoon. This is sometimes buried in the earth, or it may be constructed of leaves sewn together with silk threads. In butterflies the chrysalis is generally exposed. It may be attached by the tail and held head upwards by a silken girth or it may hang head downward by the tail alone.

Sooner or later a great transformation occurs: the chrysalis bursts and the imago, which has formed within it, creeps forth. Yet the wings cannot attain their large size prior to that occurrence. They have to expand subsequently, when the butterfly or moth has emerged. That is done by pumping blood into them, for they

Some of the species are capable of flying for many hundreds of miles. Indeed, those adapted to warm countries may migrate regularly or occasionally to places where their larvae cannot survive: from North Africa to England, for example.

Some Lepidoptera are poisonous or have an unpleasant flavor. Since, therefore, they may escape predation because they are unfit for food, they adopt "warning coloration." That is to say, they make themselves conspicuous: their colors are bright and distinctive and they fly slowly. Thus they are easily recognized and avoided by insect-eating birds. An example of this kind is provided by the cinnabar moth,

Callimorpha jacobaeae. Its colors are brilliant red and blackish-green and the body contains poisonous alkaloids. Other species possess irritant hairs as larvae and are avoided by most birds on that account.

Much more often, the colors of the Lepidoptera are "cryptic," or concealing, when at rest and they may match their background to perfection. Great numbers of moths resemble bark or lichen and are to be found with the wings spread and pressed flat against a tree trunk. They may then be extremely difficult to detect. Others resemble a piece of wood or, as in some moorland species, a small stone and these latter spend most of the day upon the ground.

Those which obtain protection by copying rocks and soil may adjust their coloring appropriately. Thus the annulet moth, *Gnophos obscurata*, is reddish-brown on the red sandstone of Devon, white on the chalk and almost black on moors where peat is exposed.

There are two other types of protection. One is "flash coloration" in which the forewings have a concealing pattern while the hind pair, hidden when the insect is at rest, are brilliantly colored. Their sudden appearance when the moth flies and their disappearance when it alights is confusing and difficult to follow. Another method of protection, relatively common in the tropics is "mimicry." That is

Male purple emperor butterfly *Apatura iris.*

British butterflies: Small tortoiseshell *Aglais urticae,* wings open *(left)* and closed *(center),* and the ringlet *Aphantopus hyperanthus (right).*

to say, some species which do not possess poisonous qualities, an unpleasant flavor or a sting, copy those that are endowed in one of these ways. Some have even lost the scales on their wings and closely resemble bees and wasps; although harmless, they tend to escape predation by this means.

Some moth caterpillars bore into the wood of trees and shrubs while others are common enough to defoliate forests: a small green moth, *Tortrix viridana*, does great damage to oakwoods in this way. Perhaps the most economically important species is the cotton boll worm, or tomato fruit worm, *Meleothis armigera*, the larva of which causes great damage to wheat, cotton, tomatoes and other crops in the U.S.A. where it is responsible for immense financial loss.

White ermine moth *Spilosoma lubricipeda*.

A species may become a dangerous pest if introduced into a new area where the natural checks normally adjusting its numbers, including the parasites that prey upon it, may be absent. This occurred when the gypsy moth, *Lymantria dispar*, was imported into North America. It proved far more destructive to orchards and forests than in Europe, whence it came.

The small white butterfly, *Pieris rapae*, is responsible for great damage to cabbages and related species and has unfortunately become widespread in the more temperate regions of the world. Originally restricted to Europe and parts of Asia, it has become one of the commonest butterflies in North America, Australia and New Zealand. The related large white *Pieris brassicae* is similarly very destructive, but has not been introduced so extensively. The larvae of these two butterflies illustrate the remarks already made about cryptic and warning coloration. Those of the large white have an offensive smell and a conspicuous color pattern, and they remain exposed on the outside of the cabbage leaves. In the small white the larvae, which lack the unpleasant odor, are green, matching the cabbage leaves well and they often burrow into the plants.

The Lepidoptera tend to fluctuate greatly in

numbers from season to season and on a long-term basis also. In Britain many became restricted in their distribution about the middle of last century. Some altogether disappeared in their old haunts, but some spread widely, returning to them again, from about 1912 onward. A more serious threat has arisen today from the wholesale cutting of timber, particularly in deciduous woods, and from the use of insecticides and selective weed killers in the form of sprays. These have caused much destruction to plant and animal life.

Although a few Lepidoptera constitute pests, the majority do not, and they add beauty and great interest to the countryside. Active steps must be taken to extend the work of nature conservation if some of those now in danger, whether nationally or on a wide local scale, are to be preserved.

LICE habitually pass their whole lives on the body of their host and are thus only likely to transfer themselves from one host to another when the latter associate very closely. In this way they contrast with fleas and bedbugs, which occur on the host's body only while feeding, passing most of their time in the nest, lair, etc., of the host. The rather large, oval eggs of lice are firmly cemented onto the host's hair or fur and hatch into larvae that closely resemble the adult lice in general form—that is to say, lice undergo an incomplete metamorphosis in their development from a newly emerged larva to an adult insect. When the larva is fully formed within the eggshell, it sucks air in through the mouth, pumps it along its gut and out through the anus. Thus, air accumulates behind the larva, and the steadily increasing pressure causes the egg cap, or operculum, to be forced open. The larva then frees itself by muscular action. The larvae take frequent blood meals, and after molting three or four times assume the sexually active adult form.

LIMPET, a name usually applied to any apparently sedentary mollusk with a conical,

Linnet at nest in black currant bush.

or flattened, shell. It is applied to a wide variety of animals including many common seashore animals, such as the slipper limpet, *Crepidula fornicata*, and the common limpet, *Patella vulgata*.

LIMPKIN, *Aramus guarauna*, so called because of its gait, is a long-legged wading bird of marshes and wet woods found from southern Georgia and Florida through Central America and the West Indies as far south as Argentina. It is the only species of the family Aramidae, related to the crane family on the one hand and the rails on the other.

The limpkin is 23-28 in. (58-71 cm) long, olive brown with white streaks. The bill is long and curved, and the feet are strong for perching in trees. It feeds almost entirely on water snails, which it extracts from the shell before swallowing. The nest is made on or near the ground.

It swims well but flies weakly, which, together with its rather trusting nature and palatability, has resulted in a dangerous decrease in numbers, but the species is now responding to protection.

LING, *Molva molva*, an elongated marine fish of the cod family found in the eastern North Atlantic. The common name is believed to be a corruption of "long" in reference to its rather long and slender body in comparison with other codlike fishes. There are two dorsal fins and one anal fin, the scales are small and embedded and there is a barbel on the chin. The general body color is brownish, sometimes marbled, with the flanks a little lighter and the fins edged with white. The ling is found chiefly over rocks in fairly deep water from 120-300 ft (40-100 m). It occurs from the Bay of Biscay northward to Scandinavia. It grows to about 6 ft (1.8 m) in length and may weigh up to 70 lb (32 kg). Ling are frequently fished for by trawlers, but the mottled brown skin is usually removed before the fishes are placed on sale in shops. It is a most prolific fish, a large female producing up to 50 million eggs. The blue or lesser ling, *M. dypterygia*, is a smaller species with a similar distribution. The body color is bluish-brown, and it lacks the mottling found in the ling.

LINNET, *Acanthis cannabina*, a small finch of Europe and western Asia, generally associated with agricultural areas and particularly fond of waste land where it feeds on weed seeds. The breeding dress of the male is basically brown above, buff beneath, with crimson breast and crown and white edges to wings and tail. The female is duller. There is a tendency to breed socially, the nests being usually placed in gorse bushes or hedges. At one time the linnet was a popular cagebird, its song being considered second only to that of the nightingale.

The name has also been applied to other finches, the greenfinch, *Chloris chloris*, being called the green linnet and the North American siskin, *Spinus spinus*, the pine linnet.

LINSANG, *Prionodon*, probably the most slender and one of the most beautiful members of the family Viverridae, which includes the mongooses, civets and genets. There are two species, the banded linsang, *Prionodon linsang*, which ranges from Tenasserim, Burma, through Malaya to Java, Sumatra and Borneo, and the

A group of lionesses moving across the African savanna at the foot of Mount Kilimanjaro, evidently starting out on a hunting foray.

spotted linsang, *P. pardicolor*, which is found from Nepal through Assam to Vietnam.

The banded linsang is about $2\frac{1}{2}$ ft (76 cm) long, of which nearly half is tail. It weighs just over $1\frac{1}{2}$ lb (0.7 kg). Its body is long and slender, its legs short and its head slender with a sharp snout, containing distinctive saw teeth for which the genus is named. The claws are retractile, those on the fore paws being sheathed as in a cat, those on the hind feet being protected by lobes of skin. The eyes are large, and the big ears are delicate and sensitive. The fur is short and thick and has a velvety quality. Its ground color is whitish to brownish gray with six broad, irregular, brownish-black bands across the back. There is a stripe of similar color along each side of the neck, and the lower flanks and the outsides of the legs are marked with dark spots. The tail is marked with alternating light and dark rings. The slightly smaller spotted linsang has the same close, velvety coat, but the ground color is either pale brown or orange-buff with rows of black spots on the back and flanks. The tail is similarly ringed.

Linsangs are solitary and nocturnal, resting most of the day in hollows in trees. They are expert tree climbers, hunting lizards, small mammals and birds, frogs and insects, either among the trees or on the ground. At times they will also take fish. When hunting, the linsang slips through the grass or along the branches of trees with its body slung low between its short legs and its slender head stretched forward on its long neck. It can then often be mistaken for a poisonous snake.

LION, *Panthera leo*, the largest of the big cats, after the tiger. It is one of the "big five roaring cats" and therefore is closely related to the tiger, leopard, snow leopard and jaguar. The lion also has the distinction of being one of those species that originated in a climate considerably colder than the one in which it now lives. One does not have to go back very far in time to find traces of the lion in Europe.

Lions are solidly and somewhat stockily built. Living on open plains they do not need the slender sinuous form adapted for gliding noiselessly through heavy cover. They stand about 36 in. (0.9 m) at the shoulder, having an overall length of between nine and ten ft (2.7 m) at maturity and weighing up to 450 lb (204 kg). The female is smaller than the male, being some 6 in. (15 cm) shorter at the shoulder and weighing from 50-75 lb (22.7-34 kg) less. The male bears a mane, which gives an exaggerated impression of the size of the animal and has helped to foster the erroneous idea that the lion, not the tiger, is the largest. The lack of a mane in the female tends to make her look far smaller than she really is. In captive individuals the mane is usually thicker and more luxuriant than in the wild, because a captive animal does not have to travel through thorn scrub or fight other males for possession of territory or of females. Lions are a uniform sandy buff, shading to a paler cream on the belly. The ears are black and often give the animal away when it is otherwise hidden. The tail is long, lightly furred and ends in a tuft of black hair, which sometimes conceals a spur or bony spike, which gave rise to the old story that a lion could use this "built-in goad" to lash itself into a fury. The spur is, in fact, the deformed last few tail bones.

The fore legs are enormously powerful, probably matched only in strength by those of the tiger. A lion can break the neck of a zebra with one blow of a paw. The dew claw is the usual shape and in the same position as in other cats, but the lion has learned to use it as a sort of toothpick. If a feeding animal gets too large a chunk of meat in its mouth, the dew claw is used to hook it out again. Although the lion is not as specialized a leaper as the tiger or the lynx, it can still perform some remarkable jumps. It is on record that lions have jumped as much as 12 ft (3.7 m) vertically and 36 ft (10.8 m) horizontally. Lions can, and the young animals frequently do, climb trees, mainly using the pulling

power of the front legs to haul themselves up the trunk. At a rough estimate the pulling power of a lion would match the combined strength of ten men.

The gestation period is about 105 days, at the end of which time two or three cubs are born. At birth, each is about 1 ft (30 cm) long and weighs about 1 lb (0.45 kg). They are fully furred and have gray spots and rosettes. The tail is rather broad and has no tuft at first. There is some doubt as to whether the eyes are open or closed at birth, but it would seem that the cubs do not have full sight until they are about two weeks old. The teeth start to appear at about three weeks. As soon as the cubs are able to leave the den, they go back to their mother's pride, and by the time they are six months old, they join in the hunt under their mother's guidance. They become independent and are ready to fend for themselves when they are one and a half to two years old. They will reach maturity at between three and four years old. It is unlikely that lions live much beyond the age of ten in the wild, but in captivity they have survived to 20 years and beyond.

Lions live in groups called "prides" which can vary in size from as few as three or four individuals to as many as 30. The pride is not strictly a family group, although the nucleus is typically formed of the leader, or king, several lionesses and their young. This nucleus may be added to by the arrival of other individuals, and in some cases two prides may join, in which case it is possible to find two leaders within one group. The position of king is achieved and maintained by physical superiority, and when a king is deposed, he may well leave the pride and join forces with a young bachelor male, and the two will then hunt together.

The lion was far more widespread in former times than it is now, and it is only comparatively recently that it has become regarded as a primarily African animal. There are still a few surviving Asiatic lions, *Panthera leo persica*, in the Gir Forest Reserve on the Kathiawar peninsula of western India, where they are maintained at a constant population level of about 300 animals. Apart from these, the lion is now confined to the African continent.

The lion roars more than the other four roaring cats, but it does so to announce possession of a territory. It also makes other noises, all of which have a direct relation to a given set of circumstances. A beast that is merely moving from one place to another may grunt as he walks along, while the lioness with cubs calls to them with a soft mewing noise. An angry lion will growl in warning, and one that is in a fury and is about to charge will cough with rage.

Lions, like all carnivores, have a large territory in which they live and hunt, and the territory of individual prides may overlap. When moving normally they travel at about 3½ mph (5 kph), but they can also cover as much as 30 miles (48 km) in the course of one night. They have also been credited with fantastic turns of speed when charging, but it is unlikely that any lion

A lioness with kill in Africa.

can manage more than about 30 mph (48 kph) and that only over about 50 yd (46 m).

LION FISH, *Pterois volitans*, an extremely colorful and dangerous fish of the tropical Indo-Pacific and a member of the family of scorpion fishes. The body and fins are brilliantly striped in red, reddish-brown and white or cream, and the pectoral fins and spiny part of the dorsal fin are produced as separate rays, those of the pectoral appearing soft and feathery. When danger threatens, the lion fish will erect the fins, and it is then a magnificent sight. Its beautiful appearance, however, is deceptive because the spines are armed with poison glands and the fish has been known actually to attack by jabbing with its dorsal spines. The poison is strong and causes great pain to humans, although no fatalities have been recorded. The bright colors of this fish serve as a warning to other species. It should be noted, also, that the poison is still effective when the fish is dead, so that even preserved specimens must be handled with care.

In the United States this species is known as the turkey fish, a reference to its appearance when the colorful pectoral fins are erected.

LITTLE AUK, *Plautus alle*, a small, diving auk of the North Atlantic also known as the dovekie. It has an unusually chubby form, 8 in. (20 cm) long, with no apparent neck and a very short bill. Like other auks it dives beneath the surface of the sea for its food, and is black above, white beneath. It breeds in colonies, some composed of millions of birds, in rock crevices on high, rocky coasts, or sometimes inland, along the arctic coasts of the North Atlantic. Outside the breeding season it disperses over the North Atlantic, and sometimes, during extended periods of severe weather, is "wrecked" on adjacent coastlines, sometimes being found far inland.

LIVER FLUKE, *Fasciola hepatica*, a parasitic flatworm that causes "liver rot" or fascioliasis especially in sheep and cattle. It can infect dogs, horses, deer, rabbits and man. The rabbit may act as a reservoir host for the disease. Besides making the liver unfit for human consumption, the infestation causes loss of condition in the host and reduces the amount and quality of meat and milk produced. In severe cases infection may cause death, either directly or by facilitating secondary bacterial infections as in the case of "black disease" caused by *Clostridium oedematiens*. This bacterium is introduced from the host's gut as the larval flukes bore from the gut wall and excavate long, twisting channels into the liver. The name of this disease probably derives from the fact that the hides of infected carcasses turn black. In years with warm, wet summers, which favor development both of the snail (the intermediate host of the fluke) and the larval stages of the parasite, infection may reach epidemic proportions.

Probably the largest single European outbreak of fascioliasis in man occurred in Lyons, France, in 1956-1957, when five hundred human cases were reported. The infection can be acquired by eating watercress contaminated with metacercarial cysts of the parasite. The symptoms are rashes, anemia, loss of weight, sweating and jaundice. Cirrhosis and secondary infection may also occur.

LLAMA, *Lama glama*, domestic form of a *Lama* species nearly 4 ft (1.20 m) at the shoulder, a little larger than the wild guanaco. It has a thicker fleece than its wild ancestor, the guanaco, and varies from white to black, with many combinations. From southern Peru through western Bolivia as far as Catamaca in Argentina and the tableland of Atacama in Chile it is the principal beast of burden of the Indians and may be found at altitudes of 7,600-13,000 ft (2,300-4,000 m), but it thrives also in less cold regions, with higher air humidity, than the vicuña and alpaca. Of all the domestic animals it is the most suitable for the steep mountain paths and the hard ground and can go without food and water longer than any other of its relatives. Only the stallions are made to carry loads, and they can do a day's march of 19 miles (30 km) with a maximum load of 110-176 lb (50-80 kg). If the weight is more than that they obstinately refuse to go on. A llama will also only allow itself to be loaded when it is one of a group. On the march the animals go in single file, like camels, and are driven by the Indians merely by means of certain calls and whistles, as the desert nomads do with their camels. The wool production is of no economic importance and only of interest to the Indians, but the leather is highly valued for its durability. The dung is also used as fuel. In earlier times it was the only fuel in the mines for extracting the lead from galena; it was even used for firing the steamers on Lake Titicaca. The Incas, whose empire, founded in the 13th century, covered the whole range of the llamas, were already giving great care and attention to breeding. The llama was the chief pack animal and wool supplier, and only a small proportion were slaughtered for food. Remains of old llama and alpaca reserves show the size of the herds at that time. The llama played an important part, particularly in the religious life of the time. In Cuzco alone, the capital city of the Incas, more than 200,000 llamas and about the same number of alpacas were sacrificed every year to the gods. The meat was then distributed to the crowd. Hundreds of thousands of llamas were also used in the mines as pack animals. The building of the irrigation systems, magnificent roads and temples of the Incas would have been impossible without the llama. However, only the government and the priests had the right to own large herds, which were carefully registered.

LOACHES, fishes related to the carps and living in fresh waters of Asia and Europe, with a few species found in Africa but confined to Ethiopia and Morocco. Loaches are elongated and compressed bottom-living fishes, many being nocturnal in habit. Most species have barbels around the mouth, and the dorsal and anal fins are usually short and placed in the rear half of the body. There is no adipose fin, and the pelvic fins lie at about the midpoint of the body. They differ from the carplike fishes (cyprinids) in having the swim bladder reduced and encased, either partially or totally in bone. Some species of loach use their intestine as an accessory breathing organ.

There are two species, the stone loach, *Noemachilus barbatulus*, and the spined loach, *Cobitis taenia*. The stone loach is common in most clear brooks, and it has a very wide distribution, being found all over Europe and eastward to China and Japan. It hides in cracks or under stones during the day, coming out to feed at night. Although it grows only to 5 in. (13 cm), it makes excellent eating.

The spined loach derives its common name from the presence of a small spine below the eye. Its range also extends across Asia to China, but in England it is much rarer than the stone loach. Reaching about 4 in. (10 cm) in length, it has a light pattern of horizontal rows of spots along the flanks. It is not considered to be edible.

A most interesting European loach is the weather fish, *Misgurnus fossilis*. It has earned its common name from its habit of becoming highly agitated when the atmospheric pressure rises, as for example before a thunderstorm. Under these conditions the fishes constantly come to the surface. This unusual behavior arises from the great sensitivity of the bone-encased swim bladder. This has a canal filled with a gelatinous substance, which passes through the swim bladder capsule and reaches the skin near the pectoral fins. In this way, pressure changes are transmitted to the swim bladder and thence to the Weberian apparatus and thus to the inner ear. Weather fishes were formerly kept by country people in tanks as weather guides. Compared with other loaches, this species is fairly large, reaching 20 in. (51 cm) in Europe and Asia.

LOBSTERS, large marine crustaceans with ten legs, the first pair bearing large claws. The true lobsters, of which there are several species, belong to the genus *Homarus*. The best known are the European lobster, *Homarus vulgaris*, and the lobster of the Atlantic coast of North America, *Homarus americanus*. The true lobsters are animals of shallow waters living among rocks in holes and crannies and venturing out from their retreats only to feed. Their favorite haunt is usually underneath a rock lying on sand where they can dig the sand and adapt the hole to their own convenience. The dark blue pigment of the living lobster is a complex compound between a protein and a carotenoid closely related to the pigment of carrots. When a lobster is boiled, the linkage between the carotenoid and the protein is

broken, and the color changes from blue to the red of the free carotenoid. The two large claws of the lobster, much larger in the male than in the female, are not alike either in structure or in function. One of them is always adapted as a crushing claw, the other as a fine picking or scraping claw. There is no uniformity, from one lobster to another, as to which is which. Lobsters feed on carrion, small crabs, worms, fishes, or indeed any animal matter dead or alive that the lobster can pick up. Shells are crushed in the large claw, and the finer claw serves to scrape the flesh off the bones or shell of the prey. It is clear that the claws must have another function too, since they are so much larger in the male than in the female, probably playing some part in courtship.

LOCUSTS, term restricted to the swarming gregarious stage of about 50 species of tropical grasshoppers. Grasshoppers in temperate latitudes feed chiefly on grasses, but many of their long-legged tropical relatives feed on all kinds of other plants, even on some like the Sodom apple, which are highly poisonous to other creatures. The name "grasshopper" is, therefore, something of a misnomer. Typically, grasshoppers live out their lives near to the place where they hatched from the egg. In an English meadow the eggs are laid in the damp earth in the autumn and remain underground during winter and spring. They hatch out in early summer, and the young hoppers (as the wingless juveniles are called) feed on the grasses and grow to adulthood. In the tropics the long winter pause in development need not occur, and many grasshoppers can produce three or even more generations a year if conditions are favorable. Year-round lush green vegetation in the damper parts of the tropics could provide a copious food supply for the hoppers, but under these conditions the earth is so full of the roots of plants that the female grasshoppers have difficulty in burying their eggs, and so breeding is restricted. In the more arid areas, where the rainfall is not sufficient to maintain a continuous ground cover of vegetation, there may be plenty of bare sand for egg laying, but here there is no year-round supply of food. Grasshoppers in

such areas tend to be seasonal, producing only one generation a year. In general, throughout the tropics we find that if there are plenty of suitable sites for grasshoppers to lay their eggs, there is little food available, while if there is plenty of food there are few egg-laying sites. We apply the name "locust" to those species of tropical grasshoppers (mostly large) that have found a way around this difficulty and become opportunists, taking advantage of good conditions for breeding and then flying in large swarms to areas where food may be available.

LONG-TAILED TITS, small perching birds comprising about eight species of the family Paridae in the subfamily Aegithalinae. There are six species in the Old World and two (the bush-tits) in the southern and western parts of the United States and in Mexico.

Rainbow lorikeet *Trichoglossus moluccanus*.

LORIKEETS, the name given to both the smaller species of lory and the hanging parrots or bat parrots, *Loriculus*.

The three lorikeets of the Australian genus *Glossopsitta* are 5½-8 in. (14-20 cm) long and have predominantly green plumage with short tails. All three have red and yellow plumage markings, and the purple-crowned lorikeet, *G. porphyrocephala*, has a dark blue patch on the crown. They feed mainly on nectar and pollen, sometimes on fruit.

The hanging parrots or bat parrots are most commonly called lorikeets, although they are only distantly related to the subfamily Loriinae. These are small to very small, green-plumaged parrots, which feed mainly on the nectar and pollen of flowering trees, some species also eating fruit. They are distributed from northern India and the Malay peninsula to New Guinea and the southern Moluccas. Their names relate to their habit of hanging upside down while asleep among the foliage at the ends of thin tree branches, apparently as an adaptation to reduce predation by hawks.

LORISES, nocturnal mammals related to lemurs with large, forward-oriented eyes and well-developed grasping hands. The large eyes are emphasized by black, pear-shaped markings

A loach *Botia hymenophysa*, very similar to the tiger botia or clown loach, both of southeast Asia.

Caribbean lobster.

Hand of loris shows reduction of index finger.

that almost give the impression of a pair of spectacles. There are two kinds: the slender loris, *Loris tardigradus*, and the slow loris, *Nycticebus coucang*. The slender loris owes its name to its long, thin limbs. In contrast to what is generally thought, its body is not slim and is little different from that of the slow loris. Both lorises are normally very slow-moving, though they can move fast when threatened. The slow-moving gait is dependent upon the deliberate, powerful grasp of the hands and feet. In association with this, the

index finger is reduced to a small stub in both species, and in the slender loris there is also reduction of the second toe. Neither species has a tail long enough to show through the fur. The slow loris has mainly dark brown fur, and the head and body measure 13 in. (33 cm). The slender loris is somewhat smaller, with a head and body of 9 in. (23 cm), and ranges in color from grayish-brown to rufous. There is a white stripe between the eyes.

The slender loris, confined to Ceylon and southern India, is completely arboreal, occurring in both wet and fairly dry tropical forest. The slow loris is also completely arboreal and is found in relatively wet forest regions of India (Assam) and the whole southeast Asian region from Burma to the Philippines. Both species are omnivorous, eating insects, fruit, leaves, birds' eggs and small lizards and mammals. The slender loris has pointed cusps on its cheek teeth (premolars and molars) and long, slender canines. It therefore relies fairly heavily on insects in its diet. In the slow loris, the canines are more powerful, and the cusps on the cheek teeth less sharp. It probably eats a fair amount of larger animal prey.

There is only one offspring at birth after a gestation period of about five months. The

baby climbs onto its mother's fur at birth and is afterward carried around by her until becoming independent. Both species are solitary or pair-living. There is no nest, and the adults sleep either in tree hollows or clasped to the fork of a branch.

LORY, a group of about 65 species of parrots all of which have a brushlike tip to the tongue. The smaller species, along with some unrelated small parrots, are called lorikeets.

The lories are probably the most brilliantly colored of all parrot groups. Bright red, yellow, green and blue figures prominently in the plumage of many species, although some are predominantly red, green, blue or black. They are 5-20 in. (12.5-50 cm) long, and have tails of long, short or medium length. The central tail feathers of three species of the genus *Charmosyna* form long wispy plumes. The distribution of the lories is centered on New Guinea and the central Moluccas, with smaller numbers of species in Australia, the Solomon Islands, and small islands from the southwest to the central Pacific Ocean. The peculiar large brown kaka, *Nestor meridionalis*, and kea, *N. notabilis*, of New Zealand are more closely related to the lories than to other parrots, but they are currently placed in a

Portrait of a slow loris in Malaya showing the large eyes of a nocturnal animal.

separate subfamily. They are nearly all noisy and gregarious, feeding on the nectar of flowering forest trees, although many of the large species also eat fruit, and a few also take seeds or insects. Their narrow bills, brush-tipped tongues and compact head feathering are all neat adaptations to a nectar and pollen diet, enabling them, respectively, to pierce and penetrate flowers, scoop up pollen, and keep their feathers clean of this sticky food.

So far as is known the breeding behavior of lories is similar to that of other parrots.

LOUVAR, *Luvarus imperialis,* a large pelagic orange and red fish related to the mackerels and tunas. It is common in the Mediterranean and the warmer waters of the Atlantic, and will occasionally stray as far north as the British Isles. The body is powerful, tapering evenly from the rather blunt head to the tail. The pectoral fins are sickle-shaped, and the pelvic fins are minute. The dorsal and anal fins are long and low and are set far back on the body, each consisting of about 12 strong spines connected by a membrane. The caudal fin is deeply forked as in most strong oceanic swimmers. In spite of all this, however, the fish is in fact rather fragile, since the skeleton is composed of cartilage and the body is liable to tear under any undue external strain. The small mouth is adapted for feeding on jelly-fishes. To obtain the maximum nourishment from these, the intestine is very long and the stomach is lined with numerous internal projections that serve to increase the absorbent surface area.

The louvar is easily identified by its color. The fins are scarlet, and the body is a metallic orange fading to yellow on the belly. Dead fishes, however, tend to turn an indifferent silver. The louvar grows to about 6 ft (180 cm) in length and may weigh 300 lb (136 kg).

LOVEBIRDS, 8-9 species of African parrots. The name is used sometimes for the budgerigar and other small parrots. The true lovebirds are small, about 4 in. (10 cm), short-tailed

Fisher's lovebird *Agapornis fischeri* of Africa and Madagascar.

parrots found from Ethiopia to the Congo and Madagascar. The gray-headed or Madagascar lovebird, *Agapornis cana,* is a small parrot found in woodland and scrub. The male is mainly green, with a white bill and a pale gray head and neck; in the female the head is pale yellow-green.

The black-winged or Abyssinian lovebird, *A. taranta,* is found only in Ethiopia and Somalia. It is one of the largest lovebirds, predominantly green, with black wings, and red markings on the head of the male. It is found in scrub and wood edges where it feeds on fruit and seeds. The red-headed or red-faced lovebird, *A. pullaria,* of west Africa is predominantly green, the front of the head being red in the male, orange in the female and yellow in immatures.

Young lumpsucker.

Lovebirds are so called because of the close pair bond and the frequency with which paired birds preen each other. In some species this preening is mutual, but in other species it is only the male that preens the female. The more gregarious lovebird species have highly ritualized fighting behavior, which prevents injury. In these, fighting consists of an elaborate bill fencing, the object being to bite the opponent's foot. Other parts of an aggressor are never bitten, although the opportunity to do so must arise almost every time they fight.

LUGWORMS, fairly large marine worms, the coiled casts of which are a very familiar sight on sandy shores. The worms are much used as bait by fishermen in all parts of the world. *Arenicola* and its allied genera are superficially rather similar. Most species never attain more than 2¾-3 in. (7-8 cm), though some are much larger than this, and most have a fat cylindrical body with a narrower tail, though *Arenicolides ecaudata* (as its name implies) does not have a tail.

There are three main regions of the body: the head, formed from the prostomium and the first two segments; the trunk, comprising the bulk of the body and consisting of a fixed number of segments; and the tail, which has no chaetae or parapodia and is much narrower than the trunk. The common European lugworm, *Arenicola marina,* has 19 trunk segments with gills on the last 13. Other species have the same general arrangement, but have different numbers of gills on the trunk. Each of the trunk segments is annulated, and these annulations in the most anterior segments are important in burrowing.

LUMPSUCKER, *Cyclopterus lumpus,* one of the mail-cheeked fishes, deriving its name from the warty lumps on its body and the presence of a ventral sucker. It is a bulky and ungainly-looking fish that is commonly found stranded in rock pools around British and European coasts, especially in spring when it comes into shallow waters to breed. It is also found on the western side of the Atlantic.

The lumpsucker grows to a maximum of 2 ft (60 cm) in length, but specimens over 18 in. (46 cm) are very rare. The body is rounded and rather flaccid with warty tubercles, the largest arranged in three distinct rows down the flanks. A median soft crest is present in front of the small dorsal fin, the latter set far back and opposite the anal fin. The pectoral fins are large and extend under the head. The pelvic fins are greatly reduced and contribute to the sucker, which is formed on the chest. During the breeding season, the time when these fish are most often encountered, the male and female differ somewhat in color. The male has a dark, almost black back and a red belly, while the female is brown on the back and has a yellow belly. In summer the stomachs of lumpsuckers seem to contain little else but water, and as yet there is no satisfactory explanation for this.

The lumpsucker has unusual breeding habits. The females produce a very large number of eggs (up to 100,000), and these are laid in a loose ball with gaps between the eggs so that water can percolate through the mass. The eggs are laid just above the low water mark, and they are guarded by the male, even though the eggs may be exposed at low tide. The males guard the egg mass and will continue to do so even during attacks by seabirds and in one case even while seagulls were pecking its liver. If rough weather scatters the eggs, the males search frantically for them. This behavior has earned the lumpsucker the name "henfish" in some parts of England.

Lumpsuckers are not eaten in England, but in some parts of the continent of Europe they are considered a delicacy when smoked.

LUNGFISHES, primitive bony fishes that were worldwide in their distribution in Devonian and Triassic times but are now restricted to South America, Africa and Australia. They are characterized by the presence of one lung or a pair, which are used for breathing air. The lungfish of Australia, the Burnett salmon,

Neoceratodus forsteri, is confined to a few rivers in Queensland and is both the rarest of the modern lungfishes and the one that has changed least with the passage of time, at least in outward appearance. Large scales are present on the body, the paired fins still retain their fleshy, lobelike bases, but the dorsal and anal fins form a continuous fin around the hind end of the body. The name "lungfish" for this species is not entirely apt since in well-oxygenated water this species uses its gills for breathing. There is a single lung lying above the esophagus and connected to it laterally by a tube. In its position and shape the lung resembles a swim bladder more than a lung, but the lining of the "lung" is composed of absorbent tissue. This rather docile species normally reaches a length of 3 ft (90 cm), but large specimens of almost twice that length have been recorded. It is principally found in the Burnett and Mary rivers of Queensland, where it is protected, but recently attempts have been made to establish it elsewhere. It differs from the other lungfishes in being less tolerant of foul conditions lacking oxygen and in being unable to aestivate in a cocoon of mud and thus survive droughts, as can the African lungfishes.

The South American lungfish, *Lepidosiren paradoxa*, and the African lungfishes, *Protopterus* spp., are more elongated fishes and share a number of common features. A median fin fold surrounds the hind end of the body, but the paired pectoral and pelvic fins are reduced to thin fleshy feelers and are filamentous in the African species. In the South American species the pelvic fins grow branch-

ing vascular filaments in the breeding season, but the function of these is not known. In both the African and the South American forms the lung is a paired structure lying below the esophagus (as in man), and these fishes rely to a considerable extent on air breathing. They are thus well equipped to survive in foul conditions, and both genera can aestivate in the dry season, the South American species making mud tunnels and the African species forming a hard cocoon in the mud.

These fishes also care for their eggs. Nests have been described for the African lungfishes *Protopterus aethiopicus* and *P. annectens*. In some cases these nests comprise holes in the matted roots of the papyrus about 2 ft (60 cm) in diameter and a little deeper, with either a surface entrance or a tunnel. Up to 5,000 eggs are laid in the nest, and the young may remain in it for as long as eight weeks. The young have external gills, but in *P. aethiopicus* air breathing starts when the juveniles are $1\frac{1}{2}$ in. (35 mm) long, and the external gills are gradually lost, the fourth and last external gill disappearing when the fishes are 7-8 in. (18-20 cm) long.

The African lungfishes are widespread and much more common, in certain areas providing a useful source of food. The most elongated of the African species is *P. dolloi*, a species from the Congo basin that reaches 33 in. (85 cm) in length. A rather similar west African species is *P. annectens*, which has a deeper body. The most widespread is *P. aethiopicus*, found from the eastern Sudan to Lake Tanganyika and reaching a length of 6 ft (1.8 m). Finally, *P. amphibius* is found in Lake Rudolf and the eastern part of Africa. The African lungfishes are found chiefly in marshy areas, but *P. aethiopicus* is essentially a lake dweller; nevertheless, it regularly rises to the surface to breathe air. In spite of their appearance of sluggishness, these fishes are surprisingly aggressive carnivores, and large specimens are treated with great respect by local fishermen.

LYNX, bobtailed members of the cat family, including the Canadian or northern lynx, European lynx and Spanish lynx, as well as the bobcat and caracal, which stand somewhat apart from the others.

The Canadian lynx, *Lynx canadensis*, gives the appearance of being larger than it really is, because of the heavy fur. The total length is 3-4 ft (0.9-1.2 m) and the height at the shoulder about 2 ft (0.6 m). A fully grown male weighs about 20 lb (9 kg) but may go as high as 40 lb (18 kg). The coat is a mixture of black, dark brown and a tawny yellow, with a silvery frosted appearance because the long guard hairs are tipped with a silvery white shade. The underparts are cinnamon, the cheek ruffs white marked with black stripes and the tip of the tail entirely black. There is also a rather rare dilute form that occurs in about one out of every thousand pelts, and this is known as the "blue" lynx. The Canadian lynx is found mainly in Canada and in particular in the region of the muskeg bogs of Alaska and the northern fir forests. It is also found in the U.S.A. in northern New England, close to the Great Lakes and in the Adirondack Mountains of New York State.

The European lynx, *L. lynx*, and the Spanish

lynx, *L. pardellus*, are smaller than the Canadian lynx but similar in coloring except that their fur is shorter and is usually spotted with black. These two species are becoming increasingly rare. The European lynx ranges through the wooded parts of Europe, except south of a line from the Pyrenees and Alps. It is also found in Asia extending eastward to the Pacific coast of Siberia and southward to the Himalayas. The Spanish lynx now survives only in Portugal, the Pyrenees and in the Spanish Coto Donaña, which is a reserve.

Lynxes live in forests, especially of pine, usually leading solitary lives, hunting by night using sight and smell. Indeed, their keen sight has become proverbial. Although they run very little they are tireless walkers, following a scent trail for miles in pursuit of prey. They are good climbers and swim well, and their large paws have a snowshoe effect to carry them easily over deep snow. The voice is a caterwauling like that of a domestic tomcat but louder, and, like a cat the lynx uses its claws and teeth in a fight. The natural food of the lynx includes small deer, badgers, foxes, hares, rabbits, squirrels, small rodents, fish, beetles and occasionally ground-living birds. It will eat carrion, but it prefers fresh prey. It has often been persecuted because of its alleged raids on livestock. The snowshoe rabbit, one of the North American hares, is the main prey of the Canadian lynx.

Mating takes place in March, the young being born after a gestation of 63 days in the European lynx and 60 in the Canadian lynx, in a den among rocks or in a hollow log. There are usually 2-4 kittens in a litter, born blind but well furred. The eyes open after ten days, and by the time they are two months old they are ready to leave the den and go out hunting with the mother.

LYREBIRDS, Australian birds, pheasantlike in appearance, which, together with the scrubbirds, form the suborder Menurae of special interest because of the spectacular nature of their song and display and their remarkable use of mimicry.

Though normally shy and difficult to observe, the superb lyrebird, *Menura novaehollandiae*, has become well known through a small colony in Sherbrooke Forest, a few miles from the city of Melbourne, where many individuals have become remarkably tame and can be seen in display and song in a very beautiful setting. The general color above is brown, the underparts being a lighter brown. The throat and tail coverts are rufous. Adult males can be distinguished from females and young males by the tail, which grows up to 30 in. (76 cm) in length and consists of two large outer feathers, which have brown crescent-shaped markings on the upper side, two black wirelike feathers and 12 filamentary feathers, which are dark on top and silver underneath, as are the outer tail feathers. This combination of feathers can be erected to give the shape of a lyre, the musical instrument after which the bird was named, though it is seldom carried in this position.

This species is still plentiful in an area between the Great Dividing Range and the sea, from southern Queensland through to the Dandenong Ranges in southern Victoria.

The European lynx.

The superb lyrebird, an Australian bird of pheasantlike appearance, in full courtship or sexual display.

Breeding takes place from May to October, but principally in the winter months of June and July when the single egg is laid. Where it breeds at high altitudes the large domed nest is often covered with snow. The female builds the nest, incubates the egg and feeds the young unaided. The egg is remarkably resistant to cold and in the early stages of incubation can be left for 24 hours or more without damage. Although the nest is large and may measure up to 24 in. (61 cm) in width, nesting females can be distinguished by their bent tail feathers, which are curled around the body when in the nest. Incubation takes about six weeks, and the chick spends a further six weeks in the nest, which may be on the ground, in an old tree stump, on a cliff ledge, or even at times in a tall tree. Nesting areas are established by the females in the thickly vegetated damp gullies of the forest, near a pool or running stream into which they place the droppings collected from the nestlings.

Males occupy territories on the slopes of the gullies, and each will defend an area of 2-10 acres (0.8-4 ha). They make small clearings from which all vegetation is removed and scratch earth up into the form of a low mound

some 6 in. (15 cm) high and 36 in. (90 cm) in diameter, on which they sing and display. Usually from four to six such mounds, at various vantage points in the territory, are in use at any one time, but several more may be built and discarded during the breeding season. These mounds are kept well tended, and before a male commences to sing he carefully scratches the surface of the mound until any accumulated litter has been removed. Scratched earth is attractive to the females, who will visit mounds and scratch around even when the male is not present. Much of the song and display takes place in the early morning when, in the misty glades of the silent forest, the silver of the quivering tail, extended like a fan right over the head, is a sight of breathtaking beauty. As the male bird pours out its continuous song, which incorporates the calls of many of the forest birds, and slowly gyrates on its mound, even the most objective observer is apt to be carried away, and it is easy to understand the many romantic misconceptions that are to be found in some of the lyrebird literature. Most of the male's song and display is directed to rival males, the territorial song with its accompanying stream

of mimicry, and the very loud "pilik pilik" calls that accompany full display are largely directed to this end.

LYRETAILS, species of *Aphyosemion*, small freshwater fishes from tropical west Africa belonging to the family of toothcarps. They are extremely popular with aquarists, not only for their bright colors but also for their interesting breeding habits. In the wild they live in small pools and ditches that are liable to evaporate during the dry season. To ensure the continuity of the species under such conditions the eggs are deposited in the mud and hatch only when the pool is refilled at the beginning of the rainy season. Those species that live in a more stable environment deposit their eggs on the leaves of water plants. Usually a species is either a bottom spawner (*A. arnoldi, A. sjoestedti*) or a plant spawner (*A. australe, A. multicolor*), but an alteration in the conditions of the pool may lead to a change in the habits of the fish. Some species (*A. calabaricus*) are liable to adopt either breeding habit. The courtship "dances" of these fishes are said to be most spectacular. The brilliant markings and the lyre-shaped tail are possessed only by the male. The female is rather drab.

M

MACAQUES, the common omnivorous monkeys of Asia, related to the baboons. They are smaller than baboons with shorter faces, smaller teeth, no facial fossae; and in the male the ischial callosities do not fuse across the midline. Some are largely terrestrial, others at least partially arboreal; the dominant color is brown, and instead of the bright and contrasted colors of the large African genus *Cercopithecus* there are flamboyant hair patterns in the form of facial fringes and crown tufts. The brow ridges are well developed, and the canines of the adult males are large and sharp. Macaques of one kind or another are the monkeys most familiar to the general public, both in zoos and in medical research institutes: the rhesus, crab-eating and pigtailed monkeys being the commonest.

Macaques are mainly tropical and subtropical, living in Indonesia, the Philippines, southeast Asia, India, Ceylon, and southern China; but also in the Szechwan mountains, near Peking, on Taiwan, Japan and north Africa—the last area is the western remnant of a formerly wide distribution of macaques across Eurasia, as demonstrated by fossils of macaques from the Pliocene and Pleistocene of China and Europe. Exactly how many species of *Macaca* there are is a matter for dispute, but at least a dozen.

Beginning with the medium- to long-tailed species, the recognizable species (some may be only subspecies) are as follows.

Rhesus monkey, *Macaca mulatta*, medium-sized 20-24 in. (50-60 cm) head and body length, and a tail 8-12 in. (20-30 cm) long. It has simple hair patterns, with the hair on the crown directed backward from the brows, is brown with brighter, orange-red hind parts, a pink face, and reddish perineal-genital skin, which in the female turns red at the height of estrous. Estrous females also tend to go blotchy red on the face. The rhesus monkey is found in India north of the Godavari River, west to Afghanistan, north to the Himalaya foothills, which it follows, reaching the banks of the Yangtze in China, with an isolated population near Peking. A related form, *M. cyclopis*, lives in Taiwan, and may be a subspecies of *M. mulatta*.

Crab-eating monkey, *M. fascicularis*, sometimes known as the irus macaque, Java monkey or cynomolgus, smaller than the rhesus, not more than 20 in. (50 cm) long, with a tail 1-1½ times the head and body length. It lacks the bright hind parts of the rhesus, but sometimes has a little peak of hair on the crown or gray cheek whiskers, these being features that vary geographically. It replaces the rhesus from Thailand and South Vietnam, south through Malaya, Sumatra, Java, Timor and neighboring small islands, Borneo and the Philippines. Jack Fooden has claimed that rhesus and crab-eating monkeys hybridize in Thailand, on the basis of three specimens; although these may be merely casual hybrids, the two species certainly represent each other in their respective areas and may indeed turn out to belong to the same species.

Bonnet and toque monkey, *M. sinica*, like the crab-eating monkey, replaces the rhesus geographically, in India south of the Godavari River and in Ceylon. It is small and long-tailed with a red face and circular cap of hairs radiating from a whorl on the crown. The two geographic forms of this species are the toque, *M. s. sinica*, of Ceylon, in which the whorl is set well forward on the crown, and the bonnet, *M. s. radiata*, of southern India, in which it is set back, leaving the forehead bare; each geographic type shows variation in color and tail length in different environmental zones.

Assam macaque, *M. assamensis*, larger than the rhesus, over 24 in. (61 cm) long, with a tail only half as long. It is yellowish brown and more robustly built than the rhesus and has a parting on the crown and a short fringe around the face. Although it superficially resembles the rhesus, there seems to be no close relationship. It is found in the Himalayan foothills, the Sunderbans and east to North Vietnam.

Pig-tailed monkey, *M. nemestrina*, probably quite closely related to the Assam macaque, has a shorter tail, which is held up in a loop. It is robust and long-legged, with a long snout; the crown hairs are dark brown and diverge from a parting, meeting the somewhat elongated gray cheek fringe. Big males weigh up to 14 lb (6 kg). It is found from Burma south to Sumatra, Borneo and Bangka, and also occurs on the Mentawei Islands and the Andaman Islands.

Lion-tailed monkey, *M. silenus*, black instead of the usual "macaque brown": it is in fact very close to the pig-tailed, and they probably form a single species. It is rather smaller than the pig-tailed, and has the same tail shape and hair arrangements, but the cheek whiskers are longer and stand out in a gray mane around the face, sharply contrasting with the black body. It is found only in the Western Ghats of south India.

Stump-tailed or bear macaque, *M. arctoides* (formerly, but incorrectly, called *M. speciosa*), has a very short, stumpy tail only 2½ in. (6 cm)

Siesta for rhesus monkeys, long-tailed species of the genus *Macaca*.

long, or less. It is short-legged and stoutly built, with a shaggy dark brown coat that is very thin underneath; the forehead is very short-haired, and adults are bald here; the face is pink, becoming red under stress and developing brown freckles in sunlight. It occurs from southern China into Thailand.

Tibetan macaque, *M. thibetana*, often confused with the stump-tailed but larger, with an even shorter tail, a brown face, and a shaggy fringe around the face that becomes a bushy beard in the male. It is found in the mountains of Szechwan and east to the hills of the Fukien coast.

Japanese macaque, *M. fuscata*, similar to the Tibetan macaque but smaller, with a pink face and a mournful expression. The fur underneath is thick and white. It is found on most of the Japanese islands, from Yakushima to the northern tip of Honshu, but not on Hokkaido.

Barbary ape, *M. sylvanus*, a distinctive species, stockily built with short thick legs and a thick body; the face is short and brown, the fur is fawn-brown, and there is an irregular black line from eye to ear. It is at present found in the Atlas Mountains of Morocco, Algeria and Tunisia, and in suitable areas north to the Mediterranean coast; it has been long famous as the monkey of the Rock of Gibraltar, where a stock is kept for touristic purposes, replenished from time to time with fresh blood from north Africa. It was plentiful in Europe in the Pleistocene.

Celebes macaques. These belong probably to several species; all have very short tails, long protruding snouts with raised maxillary ridges, and a predominantly dark coloration. However, there are a number of distinctive types; one, the Celebes black "ape," *M. nigra*, was at one time put into a separate genus, *Cynopithecus*, because of its long muzzle, jet-black color, long narrow crest on the crown, total lack of a tail, and peculiar pink kidney-shaped ischial callosities; but all these features are progressively diluted as one travels farther south on Celebes, and the southernmost from, the moor macaque, *M. maura*, is a much more typical-looking macaque. The true black ape is restricted to the tip of Celebes' long, narrow northern peninsula, and the five or six other species, which may turn out to be subspecies of a single species, are found over the rest of the island.

MACAWS, 15 species of large gaudily colored birds of the parrot family living in the tropical rain forests of Mexico and Central and South America. They include the largest members of the family, a number of them approaching 40 in. (100 cm) in length, a large proportion of which is made up of the long, graduated tail. As in other parrots, the bill is very large—in bulk about half that of the head—and is strong, stout and much curved. It is capable of some movement at its articulation with the skull, and this, with the strong, muscular tongue, gives macaws a considerable degree of manipulative ability, which is of use to them in climbing about the branches as well as in feeding. In climbing they are also aided by the structure of the feet, which are zygodactyl—

Rhesus monkey taking a chew at a twig.

Blue-and-yellow macaws carrying out reciprocal preening. One bird nibbles the feathers which the other bird cannot reach itself.

MACKERELS, small members of the family that includes the bonitos and tuna fishes, beautifully streamlined fish with pointed jaws and a body tapering to the slender base of the forked tail. The best known and economically the most important are members of the genus *Scomber*. The common mackerel, *S. scombrus*, has two dorsal fins well separated from each other and a series of small finlets behind both the dorsal and anal fins. The back is dark green-blue with dark wavy lines on the upper part of the flanks, and the undersurface is pearly white shot with rosy tints. It occasionally reaches 6 lb (2.7 kg) and is found on both sides of the North Atlantic, ranging from the Mediterranean to Ireland on the European side. The mackerel is a pelagic fish, forming enormous shoals at the surface near coasts in summer and feeding on small crustaceans and other planktonic animals (fish and fish larvae). In winter the shoals disband and move into deeper water, where they remain in a state approaching that of hibernation. Off the Atlantic coasts of North America huge fisheries exist for the mackerel, and shoals that are 20 miles long and ½ mile broad (32 km by 0.8 km) have been seen, representing a million barrels of fish. Spawning takes place in coastal waters, large females laying about half a million minute eggs each.

MAGPIE LARK, *Grallina cyanoleuca*, a pied ground-feeding songbird, about 1 ft (30 cm) in total length, found only in Australia. The magpie lark is a bird of open country, found particularly by water, occurring in open grassland and pastures, roadsides and the larger lawns of gardens and sports fields in suburban areas. It feeds on insects and other invertebrates, and snails are taken from freshwater margins. Although feeding mainly on the ground, walking rapidly with the body well raised on slender legs, the magpie lark nests in a tree and is, therefore, most usually found where some scattered trees are present. They generally occur in pairs, apparently with a permanent pair bond, occupying distinct territories. Immature birds may gather in large flocks, sometimes joined by pairs outside the breeding period. Within their territories pairs are conspicuous and noisy. They fly with a flapping of broad black and white wings in a manner reminiscent of the lapwing. They can also glide and soar.

The voice is striking. A frequent "pee-wit" call is uttered, with shriller alarm and mobbing calls. Pairs have an antiphonal call: a trisyllabic utterance given by one bird being immediately answered by a bisyllabic response, variously written as "pee-o-wit, te-he" or "dillipot, peewit."

The nest is a well-shaped bowl of mud, strengthened with grass strands mixed in during building; hence the species' alternative name of "mudlark." It is about 6 in. (15 cm) across and more than half as deep, with walls ¾ in. (2 cm) thick. It is placed well up in a tree, preferably a high one, and built directly onto a bare horizontal branch or a bare fork.

Three to four eggs, blotched purple and brown on a white ground, are laid. Both parents cooperate in incubating and rearing the young, which resemble the parents.

that is, they have two digits pointing forward and two back.

Macaws are among the most gaudy birds, many species displaying the brightest yellow, green, blue, and red, often with the most striking contrast, and with a bare white face. A number of species are also favored by aviculturalists, being kept for their plumage and also for their intelligence and the readiness with which they take to captivity. But they are even more impressive in their natural environment, flying strongly, at least for short distances, through and over the tops of the tallest trees. Their calls are very powerful, though raucous. They feed on a variety of fruits and seeds, and nest in holes in trees, laying from two to four eggs according to the species.

The largest is the scarlet macaw, *Ara macao*, which may be over 36 in. (92 cm) in length and is found from Mexico to Bolivia. Its basically scarlet plumage is enhanced by blue in the wings and tail and by yellow and green on the wings also. This and the closely related blue-and-yellow macaw, *Ara ararauna*, are perhaps the most commonly kept parrots in zoos. The latter species may be over 32 in. (82 cm) in length, and is found through South America from eastern Panama to Argentina. It is a rich sky blue above and bright golden yellow beneath, and is particularly amenable in captivity, having been kept in Europe since the 16th century.

The hyacinthine macaw, *Anodorhynchus hyacinthinus*, is separated from the other large species by a number of structural differences including the full feathering of the face. It may reach 35 in. (89 cm) in length, and is found from the mouths of the Amazon into Brazil. The plumage is unicolored yet impressive, for it is a deep but rich, bright and glossy blue. A small area of skin at the base of the lower jaw is devoid of feathers and colored bright yellow, as is a ring around the eye. By their size, coloration, strident calls and powerful flight, these birds create a very strong impression, and they are becoming more and more popular with aviculturists. In a number of collections they may be seen to their best advantage—in a fully-winged, free-flying state. The macaws may be regarded as one of the highly successful branches of the flourishing parrot group, their size and powerful build giving them a degree of independence from predators that has enabled them to develop advanced features of plumage and behavior.

MAGPIES, birds of the crow family, most of which have long tails and many of which are brightly colored or are strikingly patterned in black and white.

The common magpie, *Pica pica*, is widely distributed in the palearctic and in western North America in a wide variety of wooded habitats. It is black with metallic sheens of green, blue and purple, with white on the belly and flanks and a white flash on the shoulder region. Like other magpies it is omnivorous. Its nest is of sticks lined with fine roots and earth. There are usually between five and eight eggs, and these are incubated by the female only.

The yellow-billed blue magpie, *Urocissa flavirostris*, of the Himalayas has a particularly long tail—more than twice as long as the body. The azure-winged magpie, *Cyanopica cyanus*, which is brown-gray with blue wings and tail, has an unusually discontinuous distribution, being found only in the far west and the far east of the Eurasian land mass. It is not known if it once covered the whole of Eurasia.

In southeast Asia, which is particularly rich in magpie species, are found the racquet-tailed magpies, in which the tips of the central tail feathers are expanded to form "racquets."

In many places the opportunistic feeding habits of magpies with the consequent utilization of birds' eggs or chicks, or cultivated fruits, have resulted in their being regarded as pests. In these situations they become very shy, but they remain quite successful being able to utilize a wide variety of foods and often nesting in thorn bushes or other places difficult of access. A number of species are scavengers, feeding with other corvids on carcasses, and it has been shown that the common magpie takes a certain proportion of carrion in its diet all the year round. In many ways magpies may be regarded as another example of the corvid principle of success by adaptability.

MALARIA PARASITE, a protozoan, or single-celled animal, of microscopic size but with a complicated life cycle, responsible for the disease after which it is named. Only the members of one family, the Plasmodiidae, should really be called malaria parasites. Two hosts are involved in the life cycle, a vertebrate and a blood-sucking insect, such as the mosquito. The cycle in the vertebrate begins when the infective stages, or sporozoites, are infected into the host by the bite of the insect. The sporozoites make their way to various cells of the body, where they undergo one or more cycles of division, which give rise to many minute invasive stages, or merozoites. The merozoites may invade further cells, or they may infect red blood cells. In the red blood cells sexual stages, or gametocytes, are formed, and these are taken up by the insect host when it bites. In the malaria parasites proper, the Plasmodiidae, cycles of division in the blood occur, and the merozoites produced invade fresh red cells. In the insect the gametocytes mature into male and female micro- and macrogametes, which fuse to produce a zygote. The zygotes become motile and pass through the gut wall of the insect to form an oocyst within which a further stage of division occurs, resulting in the production of sporozoites, which make their way to the salivary glands.

In the case of the malaria parasites of man,

the first stage of multiplication takes place in the liver; the merozoites then invade further liver cells and also red blood cells where the second stage of multiplication occurs. The gametocytes occur in red blood cells, and the invertebrate hosts are mosquitoes. Man is not the only animal to be infected by malaria parasites, and the 85 or so recorded species are distributed through reptiles, birds and mammals, especially rodents and primates. Only the species living in mammals multiply in the liver.

White-backed magpie *Gymnorhina hypoleuca*, from south and central Australia and Tasmania.

Four species of malaria parasites occur in man: *Plasmodium falciparum*, which causes malignant tertian malaria; *P. vivax*, which causes benign tertian malaria; *P. ovale*, which causes ovale tertian malaria; and *P. malariae*, which causes quartan malaria. The diseases are named after the fevers that the parasites cause, tertian fevers occurring every three days or 48 hours and quartan fevers occurring every four days or 72 hours. The naming of these fevers is based on the Roman system of calling the first day one, whereas we would call the first day zero.

MALLARD, *Anas platyrhynchos*, the common wild duck, which is found in large numbers throughout most of North America, Europe and central Asia. Members of various groups winter as far south as the Gulf of Mexico, the Mediterranean and Borneo. The name "mallard" now refers to both sexes of this very common species, though previously it referred only to the male, the species generally being called the "wild duck," a name still applied in some areas.

The mallard is probably the most abundant of all the ducks. It is highly adaptable, being found wherever there is fresh or brackish

water from sea level to high mountains. It is abundant in agricultural areas. One of the reasons for the abundance of this species is the catholic nature of its food requirements. It will take anything edible, animal or plant, and is well able to profit from man's agricultural activities. In Greenland in the winter it even eats seaweed and large mollusks.

The mallard drake is a striking bird with a green head and yellow bill, a white neck ring, a purple-brown breast, gray back and belly with brown at the sides. The tail is white, tail coverts black and the four central tail feathers also black and distinctively curled upward. The female is attractively mottled buff and brown. Both sexes have gray wings with a violet-purple speculum bordered by black and white roughly in the center of the wing. They also have orange feet and legs. Like most of its near relatives, the mallard drake molts into an eclipse plumage for two months or so after the display season of the spring and early summer, and at this time it looks very similar to the female.

Mallards nests are usually built on the ground, well hidden in dense vegetation. However, they may also be built in the forks of trees or in old nests of other tree-nesting species, sometimes among rocks or on buildings. The nest, built by the female only, is lined with dead leaves and grasses and a plentiful supply of down plucked from the breast and belly. An average of 10-12 eggs is laid, and the clutch is incubated by the female only, who pulls the down over the eggs when she leaves the nest. Incubation takes about 28 days. The young are taken to water soon after hatching and obtain their own food under the protection of the female —sometimes accompanied by the male.

This species has given rise to most of the breeds

The Australian mound-builder lays its eggs in an incubator and uses its tongue like a thermometer.

Eastern green mamba walking along a branch.

of domesticated ducks in the world, the Muscovy duck being the only exception.

MALLEE FOWL, *Leipoa ocellata,* a large chickenlike ground bird, also known as the lowan, the best known of the megapodes or incubator birds. It is unusual among incubator birds in inhabiting semiarid regions where vegetable ground litter is sparse. It is found largely in the inland scrub country in Australia, which consists primarily of a number of species of dwarf eucalyptus and is known as mallee scrub. The mallee fowl, like other incubator birds, does not incubate its own eggs, which are incubated by natural means, principally by the heat generated by the fermentation of dead leaves and other plant materials. It is an extremely complicated process, our present understanding of which is almost entirely the result of the work of H. J. Frith.

In the habitat of the mallee fowl suitable material with which to build a mound of fermentable vegetation is scarce, added to which the temperature fluctuates widely and the air is very dry for most of the year, so that dead plant material tends to wither rather than ferment. The mallee fowl overcomes this by digging a hole in the ground that may be 15 ft (4.25 m) in diameter and 4 ft (1.2 m) deep. Over the winter it collects plant material from a radius of up to 150 ft (45 m) and deposits it in the hole. Then, after it has been damped by rain, it is covered with a 2-ft (0.6-m) thick layer of sandy soil which enables fermentation to take place, generating heat.

Fluctuations in the temperature of the egg chamber at the top of the mound are checked by the male bird, which tends the mound throughout the breeding season. He tests its temperature with the bill (the nature of the thermometer mechanism is unknown) and regulates the gain or loss of heat. In the spring there is rapid fermentation, and the male cools the eggs when necessary by digging into the top of the mound in the early morning to allow the heat to escape. In summer there is overheating from the sun also and the male adds more insulating soil, sometimes also scattering the covering soil very early in the morning and returning it when the mound has cooled and before the sun is too hot. In autumn there is not enough heat, and the bird scoops out the center of the mound allowing the sun to reach the eggs. He also turns the soil over in the sun to warm it and scatters it, returning the heated soil to the mound. By these various means the egg temperature is maintained very close to the required 92 °F (33 °C) for the whole of the incubation period from September to April. The male is busy building or maintaining the mound for 11 months of the year.

The eggs, which vary considerably in number from mound to mound, are laid over a period of several months and are incubated from the time of laying. They therefore hatch at intervals. The young are extremely precocious. They dig themselves out of the mound and fend for themselves entirely, never seeing the parents and being able to fly within 24 hours. This may be regarded as another example of the modifications that are possible when an animal is isolated from significant predators.

MAMBAS, large elapid snakes belonging to the genus *Dendroaspis.* They have a short maxillary bone on each side in the upper jaw that carries a long hollow fang with no small solid teeth following it, while there are a pair of elongate teeth at the front of the lower jaw. Mamba venom is a nerve poison, and bites from the black mamba, *D. polylepis,* are particularly dangerous. This is the largest of the mambas, reaching a length of 14 ft (4.3 m). It has a nervous disposition and secretes the most potent venom. When disturbed, it will rear up, spreading a narrow hood and gaping the jaws to show the blackish interior of the mouth. Any sudden movement will provoke a strike, which tends to be delivered on the head or body and will, therefore, prove fatal in a short time unless mamba serum is immediately available.

The black mamba is a relatively slender snake with a long coffin-shaped head, oblique smooth scales on the body and a long tail. Despite its name, this snake is pale gray, olive or dark gray-brown, never a true black. Just after sloughing its skin it may be olive green with a bluish "bloom" like that on the skin of a plum.

Jameson's mamba, *D. jamesonii,* a slightly smaller species, is olive green and blackish posteriorly in the eastern part of its range, which extends from west Africa to western Kenya.

The eastern green mamba, *D. angusticeps,* rarely exceeds 7 ft (2.1 m) in length. It is brilliant emerald green to yellow-green above, greenish white below, and extends along the east African coastal plain from Kenya south to Natal.

The western green mamba, *D. viridis,* is restricted to west Africa and is usually speckled green and yellow.

Mambas lay 9-14 elongate eggs at the beginning of the rains, a termitarium being a favorite nesting site. Newly hatched young are about 20 in. (50 cm) in length.

The two green mambas and Jameson's mamba

are arboreal and are usually found in forested regions. The black mamba inhabits dry savanna and, although frequently found in big trees along rivers, it is equally at home on the ground and is particularly common on rocky hills covered with thick bush.

It seems to feed largely on small mammals, especially rats, squirrels and young dassies, but birds and reptiles are occasionally taken. The other three species feed on birds and their eggs, together with chameleons and other arboreal lizards.

MANAKINS, New World birds about the size of a tit, with a short bill and short wings. In most of the 59 species the tail is also small. Usually the males differ greatly from the females.

Manakins are confined to Central and South America where they range from Nicaragua and Costa Rica south to northern Argentina. They live in dark forests but are also found in secondary growth. They feed mainly on berries, which are snapped off in flight, but some species also eat small insects.

One of the main characteristics of the manakins is their complicated courtship behavior, in which the male makes a small clearing on the forest floor by removing all leaves and twigs. In this small court a few vertical twigs are used as display perches on which the males jump and make strange sounds.

Among the most common species is the white-bearded manakin, *Manacus manacus*. It is black on its upperparts with a white ring on the hind neck. The underparts are white, and the feathers on the throat are long and fluffy. These are extended in courtship to form a widely spread "beard." The female is completely green but somewhat darker on its upperparts. In both sexes the legs are bright orange-yellow. When courting, the male jumps between its display perches, with its white "beard" extended, making an odd clicking sound that resembles the noise made by the snapping of two fingers. It is quite a loud sound and very strange when heard in the midst of the forest. The females visit the display ground to mate. The male is polygamous and the female promiscuous. The male plays no part in any of the nesting phases: it is the female alone that builds the nest, incubates the eggs and rears the nestlings. The nest of the white-bearded manakin is like that of most other manakins, being an open cup between the fork of two twigs. It lays only one or two eggs, which are white with brown blotches. The incubation period is 18-19 days, and the nestlings stay in the nest for about 13-15 days.

MANATEES, large and fully aquatic herbivorous mammals of the tropical and subtropical Atlantic coasts, estuaries and great rivers.

The genus *Trichechus* has three species. *T. senegalensis* inhabits the coast and certain large rivers of west Africa, while *T. inunguis* is confined to the fresh waters of the Amazon and perhaps the Orinoco. The more widespread *T. manatus* of the western Atlantic and Caribbean coasts, estuaries and rivers from Virginia to Cayenne is subdivided into two subspecies, *T. m. latirostris*, and the more northerly *T. m. manatus*, best known from Florida. All the distinctions are minor.

T. manatus grows to a length of around 12 ft

Longtailed manakins, *Chiroxipha linearis,* from South America.

(4 m), whereas *T. inunguis* is distinctly smaller. Manatees are heavily built and of torpedo shape with powerful rounded tails, flattened in the horizontal plane. The tail provides the main propulsion, the hind limbs being quite absent and the fore limbs being small. The teeth are a series of 20 to 30 crushing molars and are remarkable in that, like in the distantly related elephants, new teeth are formed posteriorly while the front teeth are lost successively by a progressive forward movement of the whole series. The manatee is exceptional, too,

among mammals in having only six, instead of the standard seven, vertebrae in the neck. The skin is tough, almost hairless, and of a gray-brown color. There is no substantial blubber layer, but the body does contain much fat. The head demonstrates several specializations in connection with aquatic life. The flattened, rather piglike bristly snout has a much enlarged and intensely muscular upper lip, and it is with the corners of this lip acting like mandibles that the animal plucks the vegetation on which it feeds. The nostrils are high on the snout and

The hideous face of the manatee — the seacow mistaken by Columbus for a mermaid.

equipped with powerful valves, which open only when, for a few seconds, they break the surface for air. The eyes are circular and the ear holes minute.

The mammary glands consist of a single pair and are pectoral in position, close to the flippers. It is doubtless because of this feature that stories of mermaids arose. Observations suggest that suckling takes place normally with the mother lying in a horizontal position. The manatee is totally herbivorous. It will feed upon green higher plants when obtainable whether they be marine or freshwater, growing on the bottom or floating at the surface, or even on banks of rivers up to a foot above water level. It is this catholicity which has led to the experimental use in Guyana of manatees to keep clear the irrigation and transport channels near the cane fields, channels that otherwise have to be cleared by hand. But so far manatees have never bred under such confined conditions, so that their use in this way would still further deplete the stock of an already increasingly rare animal.

MAN-EATER SHARKS, sometimes spoken of as man-killers, include several sharks notorious for alleged attacks on human beings. One species more especially that has been given the name of man-eater is the great white shark, *Carcharodon rondeleti*, found in all warm seas and reaching a length of 40 ft (12.2 m). There have been reports of human limbs found in their stomachs, even of entire corpses, and one report of long ago told of a man in armor being found in the stomach of a great white shark. Other sharks said to attack men or boats include the tiger shark, *Galeocerdo arcticus*, up to 20 ft (6.1 m), the blue shark, *Prionace glauca*, 25 ft (7.7 m) or more, the Ganges shark, *Carcharinus gangeticus*, the mako, *Isuropsis mako*, one of the mackerel sharks, and the hammerhead *Sphyrna zygaena*. Although the Ganges shark is only 7 ft (2 m) long, it seems that for centuries it attacked partially burned bodies thrown into the Ganges from the ghats, and when this was discontinued it began attacking people in the bathing ghats, especially in April and May when the river is very salty.

MANED WOLF, *Chrysocyon brachyurus*, one of the least-known but most striking-looking of all canids, it resembles a long-legged fox, which gives rise to its native South American name, the "fox-on-stilts." Its thin muzzle, narrow chest, long ears and very long legs are combined with a beautiful coloring of bright reddish-brown contrasting with a pure white throat patch and tail and black hackles, stockings and ears.

Maned wolves are found throughout the plains region of South America, in Brazil, Paraguay and Argentina. Their height is an adaptation for moving through and seeing over the long grass of the pampas, but surprisingly, they are not fast runners and cannot, in contrast to the wolf, hunt large prey. Instead, they feed on a variety of small animals like wild guinea pigs, birds, lizards and frogs, which they catch either by a swift pounce or by digging into burrows or banks. They often immobilize their prey by striking it with a fore paw, and when the victim is held in the mouth they shake the head vigorously. The diet of the maned wolf is supplemented by fruits such as figs, and even sugar cane. Excess food may be cached in small holes dug with the fore paws and covered by sweeping soil over them with the muzzle.

Maned wolves are extremely shy and solitary, and males and females come together only during the breeding season, which occurs during November and December, the South American spring. Two to four cubs are born about two months later and are probably reared solely by the mother.

MANGABEYS, Old World monkeys closely related to baboons, from which they differ by their long tails, shorter faces with deep fossae under the eyes, and white eyelids. They are essentially forest-living, arboreal baboons; they are robustly built, large monkeys not much smaller than many baboons, 18-24 in. (45-61 cm) long with a tail as much as half as long again. The generic name for mangabeys is *Cercocebus*.

The various species of mangabey fall into two groups: crested and uncrested. The uncrested group includes those with short, speckled coats, blackish hands and feet, striking white eyelids, and white or yellow underparts. The included species are the gray-black sooty mangabey, *Cercocebus atys*, from west Africa (Senegal to Dahomey); the white-collared mangabey, *Cercocebus torquatus*, with its white neck band and maroon-red crown, from the lower Niger River to equatorial Guinea; the brown golden-bellied mangabey, *C. chrysogaster*, with its bright golden-yellow underparts and robust, macaquelike build, from south of the Congo River; and the smaller, gray-speckled agile mangabey, *C. galeritus*, from Gabon, equatorial Guinea and the forests north of the Congo River, and the gallery forests of the Tana River, in Kenya. The more distinctive crested group includes just two species: the gray-cheeked mangabey, *C. albigena*, which has a long black coat, gray underneath, and an untidy floppy crest on the head; and the black mangabey, *C. aterrimus*, which is jet black with an upright, pointed coconutlike crest on the head and fan-shaped cheek-whiskers, which are either gray or black. Both these species have thin faces and sunken cheeks and are rather gargoylelike; the gray-cheeked is found from Cameroun and Gabon east to Uganda, always keeping north of the Congo River, while the black mangabey is found south of the Congo River. Except for a slight overlap between the white-collared and agile species in Equatorial Guinea, within each species group the different species replace one another geographically, but the two groups as a whole overlap widely over the whole central African forest region: as will be seen, they differ considerably in their ecology and way of life.

In Equatorial Guinea, Jones and Sabater Pi found that gray-cheeked mangabeys live in humid primary forest and also in secondary forest and are common in the vicinity of rivers, whereas collared mangabeys live both at lower levels in the trees and in swamp forest and near human habitation. Collared mangabeys often come down to the ground and raid plantations, which gray-cheeked never do. The troops of the collared mangabey are larger and bisexual; the gray-cheeked lives in troops of about 10— only half as large—with only one adult male in each, although in Uganda, where they live in areas where there are no uncrested mangabeys, they have larger troops and more than one male per troop. Mangabey troops occupy home ranges that overlap, but the troops rarely meet because of the males' loud spacing calls.

Collared mangabeys, like most other monkeys, sit on branches, supported from below by their hands and feet; gray-cheeked sit on just their rumps with their limbs sprawled out on either side in front of them. They also stand and hold their tails in different ways, and gray-cheeked mangabeys may wrap their tails around a branch for support. When alarmed, collared mangabeys stop calling, and keep quiet, then make off through the trees or on the ground, making for swampy areas; on the other hand, gray-cheeked mangabeys call with increased gusto, and finally run into the tallest trees.

Mangabeys have quite a tight-knit troop, especially among the females, who groom one another frequently. Males are rather aggressive toward one another. Females also groom males, but rarely vice versa: a baboonlike characteristic of mangabeys. Another baboonlike trait is that female mangabeys swell up in the perineal

African mantis with a conspicuous eye-spot on each wing.

region in estrus, and at this time there is an increased frequency of copulations. The young, which clings to its mother's belly, is a focus of attraction for other troop members, even the adult males.

MANNIKINS, a group of usually rather small, heavy-billed and dull-colored seed-eating birds of the waxbill family. They are widely distributed from Africa, through Arabia and southern Asia to Japan, and south to Indonesia and Australasia. They are mainly adapted for living and feeding in grassland.

MANTIS, long narrow carnivorous insects usually found in the tropics and never occurring where the winters are very cold. There are many species of praying mantis, although for most people in Europe the term applies to a particular species, *Mantis religiosa*, common in southern Europe and north Africa. The name stems from the habit of these insects of sitting motionless with the fore legs raised and held together almost as if they were praying. Far from this being so, they are actually sitting waiting for their prey. Mantids feed on other insects and sit motionless, waiting for something to come within striking distance of their front legs. Because the eyes of insects are particularly good at seeing movement, but not so good at making out detail, many insects will be quite unaware of the waiting mantis until it is too late. The mantis watches its prey, turning its head slowly to follow movements, then, when the insect is close enough, the front legs are shot out at great speed to catch it. These legs are long and specially modified for grasping the prey and bringing it back to the mouth, where it is held and systematically chewed up. Mantids are not very selective in their choice of food; they will take almost anything that comes their way, including other mantids. This makes the business of mating rather hazardous for the male, which is often smaller. His craft is usually sufficient to enable him to mate successfully, but often he falls a victim to the female's greed as soon as this process is complete.

MARABOU, *Leptoptilos crumeniferus*, a very large species of stork, characterized by a massive pointed bill, its almost featherless head and neck (the latter with a long distensible pouch of pink skin), slaty upperparts and long legs. It is confined to Africa, although the adjutant, *L. dubius*, of southern Asia is similar. It breeds in colonies, building bulky nests in large trees, often in villages. It feeds on carrion and offal but will also catch small animals. Marabou soar majestically, with a wingspan of up to 8½ ft (2.6 m), stalk on open ground and also perch, rather grotesquely, on the topmost boughs of trees. It stands about 4 ft (1.2 m) high.

MARKHOR, *Capra falconeri*, a species of wild goat with compressed spiral horns. Males stand 38-41 in. (95-100 cm) high and weigh 80-100 lb (35-45 kg). The color is gray-brown, old males becoming nearly white. Females are less than half the weight of males and are dark fawn. In winter, males develop long fringes, and both sexes are clothed in long silky hair.

Markhor inhabit some of the ranges of Kashmir (Astor, Pir Panjal, Kaj-i-Nag, Baltistan, Gilgit, Chitral) south through the Suleiman range to the trans-Indus hill ranges in the

The unlovely marabou, an African stork.

USSR. The horns vary in size and shape from one locality to another. In Astor and Pir Panjal, the horns have an open corkscrew 65 in. (165 cm) long; in the Suleiman range the horns are tightly twisted like a stick of licorice, 48½ in. (123 cm) long; around Kabul the spiral is again more open, but the horns are only 39½ in. (100 cm) long; in the Soviet Union they are similar but still shorter, rarely above 28 in. (70 cm) long. Nine races have been described from this comparatively restricted area, but probably only about three will prove to be valid.

MARLINS, large tropical oceanic fishes related to the sailfish and the swordfish and included under the general heading of billfishes. The bony snout is often as long as, or longer than, that in the sailfish but does not reach the length of the "sword" in the swordfish. Great strength is given to the vertebral column, especially in the tail region, by overlapping processes on the vertebrae. Marlins are great sporting fishes. They are cunning and wily when hooked and are tremendous fighters. Because of the difficulty of preserving such large fishes for study there is still considerable uncertainty regarding the number of species and their identification.

The black marlin, *Istiompax marlina*, is variously referred to as the silver or white marlin. It is characterized by its pectoral fins which cannot be folded back against the body (true of the sharks also). It is widespread in the tropical Indo-Pacific area and grows to 11½ ft (3.5

m), and can weigh up to half a ton (500 kg). The blue marlin, *Makaira ampla*, of the Pacific, also known sometimes as the black marlin, grows to 1,400 lb (630 kg) and often has stripes on the body that causes some confusion with the striped marlin, *M. audax*. The blue marlin is of great commercial importance and is caught on long lines off Pacific islands. The striped marlin of the Pacific is a solidly built fish with definite vertical stripes on the body. It grows to 500 lb (225 kg) and is also of commercial importance. A related species in Australian waters, *M. zelandica*, is also referred to as the striped marlin, while *M. australis* is called the black marlin in Australia. One of the Atlantic marlins, *Tetrapturus albidus*, is occasionally caught to the southwest of the British Isles. A related species from the Pacific, *T. angustirostris*, has a very short snout, hardly longer than the lower jaw.

MARMOSETS, a family of New World monkeys, the Callitrichidae. They differ from the other New World monkeys, family Cebidae, in several features. They are all very small with claws instead of flat nails on their hands and feet (except for the great toe). They have simple brains almost unfissured and small orthognathous faces. They are the only higher primates with carpal vibrissae (i.e., tactile hairs on the wrist). Most marmosets lack the third molar, upper and lower, but this is not a very important difference as not only does the somewhat aberrant genus *Callimico* possess only a

The alpine marmot has a wide (but discontinuous) range in mountains, from the Swiss Alps to the North American Rocky Mountains.

MARMOTS, ground-living, burrowing rodents forming a distinctive genus, *Marmota*, of the squirrel family. Marmots are found throughout the north temperate region, especially on mountains and open plains. Only one species is a woodland animal, namely, the woodchuck, *Marmota monax*, of North America.

Marmots are large, heavily built and ponderous by comparison with other members of the squirrel family, measuring 1-2 ft (30-60 cm) in length and weighing up to 15 lb (7 kg). The legs are short but powerful, and the tail is usually about a third the length of the body and only moderately bushy. The color is of various shades of brown without any conspicuous pattern.

Whether living on mountains or plains, marmots are sociable animals, living in colonies based on a system of deep burrows that may extend for several yards underground. Like most members of the squirrel family, they are diurnal, spending the day feeding or just sunning themselves, but never far from an entrance to the burrow. Any animal that suspects danger emits a sharp whistle that sends the whole colony scuttling into their burrows with a remarkable turn of speed.

Marmots hibernate, probably more completely than any other ground squirrels. They feed on a great variety of green vegetation, and they lack the cheek pouches found in most other seed-eating ground squirrels. They do not store food, but they become very fat in autumn. A new nest of dry grass is made deep in a burrow, and an entire family may sleep together for up to six months, depending on the region. The winter bedding is changed when the animals emerge in spring, and breeding soon begins, the litter of two to four being born after a gestation period of about six weeks. The young grow slowly and are not mature until they are two or even three years old.

The Alpine marmot, *Marmota marmota*, of Europe is a familiar animal on alpine pastures above the tree line in the Alps, and the same, or very similar, species is found throughout northern Asia and western North America.

MARSH DEER, *Blastocerus dichotomus*, which stands about 44 in. (112 cm) high at the shoulder, is the largest of the South American deer, and occurs throughout much of southern Brazil, Paraguay and northeastern Argentina, and perhaps still in Uruguay as well. Local names for this deer are *pantanos* and *guazu pucu* deer. Eight is the usual number of points on marsh deer antlers, but heads of ten or twelve points, or even more, are by no means rare. This deer, as its name implies, is fond of marshy ground and is seldom found far from water. Its color is a deep rich red.

MARSUPIAL CAT, name given to a number of carnivorous marsupials that superficially resemble the true cats. They are included in the family Dasyuridae, which also includes smaller carnivorous forms (marsupial mice) and

small third molar, but many cebids, especially spider monkeys and sakis, commonly lack this tooth as well. There has been a great deal of debate over whether marmosets should be regarded as primitive survivals of the ancestral New World monkey. But, although they are peculiar in many respects, it is most likely that they are not especially primitive, for example their claws are more likely to be an adaptation for a small-sized animal that cannot easily grasp a branch, than a survival of early clawed forms.

Marmosets live in tropical forests and the south Brazilian hardwood forests, keeping to larger branches where their claws can dig in and provide a firm hold. They tend to eat insects, leaves and fruit, but the black tamarin is said to eat fruit only, and the pinché feeds on seeds. Different species live at different levels in the forest.

Marmosets live in family groups, of a male, female and one or two offspring. The group is territorial, and the female is more aggressive than the male in maintaining the territory against intruders, marking branches and trunks with her genital glands and urine. Many species indulge in long bouts of mutual grooming with hands, teeth and tongue; a marmoset grooming itself, on the other hand, generally restricts itself to scratching.

Marmosets move by scurrying, rather like a squirrel, and most species frequently jump from branch to branch or from tree to tree in the wild, but the pied tamarin, which is longer-legged than other species and more lankily built, does not jump. The pinché tamarin often stands bipedally when excited. Marmosets sleep curled into a ball.

The genus *Callithrix* contains the most familiar types of marmosets. They are smallish animals weighing 3-6 oz (175-360 gm). The head and body length is 6-8½ in. (16-22 cm), with a tail of 10-14 in. (25-37 cm). There are three species, each of which varies geographically to a considerable extent. The common marmoset, *Callithrix jacchus*, is marbled black and gray or black and brown; the tail is ringed with black and gray; the ears have white, yellow or black tufts growing either in front of the pinna (and usually above and behind it too) or from its inner surface. The face is pale and hairy, and there is a white blaze on the forehead that in one race extends back to the vertex (the highest point on the head). Common marmosets are found along the Brazil coast from the river Paraiba in the north to the river Ribeirao in the south, and inland from the coast of Bahia to the Brazilian Highlands. The coastal distribution explains why the species is so commonly seen in captivity. Darker-colored races tend to live in more humid climates.

The pygmy marmoset, *Cebuella pygmaea*, is very small, being only 5-5½ in. (13-14 cm) long with a tail of about 8 in. (20-21 cm). It has a rounded skull with a very small face and is found on the upper Amazon.

The Emperor tamarin, a marmoset with very long mustaches.

the larger carnivores (marsupial wolf and Tasmanian devil). The Dasyuridae are distinguished from other marsupials by having four incisor teeth on each side of the upper jaw and three incisors on each side of the lower jaw. The canine teeth are well developed in both jaws. The marsupial cats are distinguished from most of the remaining dasyures by having only two, instead of three, premolar teeth in each jaw. Four molar teeth are present. Each fore foot has five toes. The little northern marsupial cat, *Dasyurus hallucatus*, the black-tailed marsupial cat, *D. geoffroyi*, and the tiger cat, *D. maculatus*, have five toes on the hind foot, but the eastern marsupial cat, *D. viverrinus*, has only four toes on the hind foot. The smallest marsupial cat is the little northern, which has a head and body length of about 11 in. (28 cm); the largest is the tiger cat, which is about 2 ft (61 cm) long excluding the tail. In all species the tail is about ¾ of head and body length and is hairy. The characteristic spotted pattern of the pelage is continued onto the tail only in the tiger cat.

The smaller marsupial cats are largely insect

The American marten or sable looks and behaves like a large squirrel when in the trees, speeding from bough to bough. It is hardly less agile on the ground.

feeders, but they also feed on small birds, mammals and reptiles. The tiger cat feeds largely on mammals and birds. All marsupial cats are to some extent arboreal, the most arboreal being the tiger cat and the least the eastern, marsupial cat.

MARSUPIAL FROG, several species of tree frog in which the female carries her eggs in a large pouch on her back.

MARSUPIAL MICE, name given to about 40 species of small, generally insectivorous, marsupials of the family Dasyuridae with shrewlike snouts and ranging in size from that of a small mouse to a rat. *Planigale ingrami* has an adult body weight of only about ⅛ oz (5 gm), whereas Byrne's marsupial mouse, *Dasyuroides byrnei*, is about the size of a house rat. All have carnivorous dentition similar to that of the marsupial cats.

Marsupial mice, of one or other species, are found in all habitats, ranging from extreme desert to tropical rain forest in Australia and New Guinea.

MARSUPIAL WOLF, *Thylacinus cynocephalus*, doglike in appearance and size, it is the largest of the carnivorous marsupials of the family Dasyuridae and may now be extinct. A general tawny yellow-brown animal, with 16-18 dark brown bars across the back, its overall length is about 5 ft (1.7 m), the head and body being about 3 ft (1m) long. There are

four upper and three lower incisor teeth, the same number of canines, three upper and lower premolars and four upper and lower molars. The marsupial wolf resembles some extinct South American carnivorous marsupials of the superfamily Borhyaenoidea, but such resemblances are examples of parallel evolution and are not indicative of close relationship. In historical times the marsupial wolf was confined to Tasmania, but it inhabited New Guinea 10,000 years ago and parts of the Australian continent 3,000 years ago. It has not been positively identified in Tasmania for more than 30 years. In 1863 it still inhabited the remoter parts of Tasmania, including the tops of mountains at about 4,000-ft (1,200-m) altitude. According to naturalist Ronald Campbell Gunn, "It would eat only what it killed and that fresh so that after killing a sheep it would never (or very rarely) return to the dead carcass, but kill another." Very little is known about the breeding, but it is thought that births of three or four young occur in the summer. The male has a pouch, which encloses the testes and is thought to prevent the pendulous scrotum from swinging during fast movement.

MARTENS, cat-sized carnivores, among the world's most valuable fur bearers. They are widely distributed in the forested areas of North America and Eurasia south to Malaya, dividing into six species along the way. Members of this

genus (*Martes*) have a darkly colored, lustrous fur with a white, yellow or orange neck patch or chest bib. Sometimes the bushy tail and limbs appear to be darker, almost black, in comparison to the hair on the back, where the dense underfur forms a lighter background. They resemble other members of the family Mustelidae with their short legs and elongated body, but the head is more triangular and the round ears are large, giving them a distinctive appearance. While the tail serves as a balancing rod, the small paws, with haired soles and semiretractile claws, are also great assets to these arboreal animals that seem to leap from branch to branch effortlessly, making them among the most agile and graceful of the mustelids. Their elevated pathways, marked with scent and urine, are patrolled with great regularity by males, which may wander more widely than the females, (up to 10 miles [22 km] in one night), thus giving the false impression that they lead a solitary existence. Magpies' or squirrels' nests and hollow trees are used for shelter. Martens are a highly adapted group, each species having evolved a slightly different set of habits to suit its habitat.

The pine marten, *Martes martes*, the range of which extends over Europe to the Caucasus Mountains, measures 24.8-31.5 in. (63-80 cm) and weighs from 1.5-3.8 lb (0.8-1.6 kg). Typically found in dense pine forests, it avoids populated areas. The breeding season is in June or July, and, after an eight month gestation, two to five cubs are born the following April. Consequently, females have a litter only every two years. The young are naked and blind at birth, opening their eyes two weeks later. Males reach sexual maturity at two and a half or three years. Pine martens make a huffing-puffing sound when excited, rather like a polecat, a moaning growl to threaten and a purr interspersed with chirps during the breeding season. The bulk of their omnivorous diet consists in squirrels, rodents, birds, insects, fruit and eggs.

The stone marten, *Martes foina*, is smaller than the pine marten, rarely exceeding 20½-29½ in. (52-75 cm) in total length. It has a shorter neck with a pure white chin and chest patch. Less arboreally inclined, it can usually be found on rocky slopes. This marten often lives in close proximity to man, settling in empty barns or attics, quite unafraid. Consequently it is known in German as the "house marten." Delayed implantation occurs only in northern females. For instance, in North America there is an eight-month gestation, whereas at the other end of the extended range, in India, it is only nine weeks long.

The sable, *Martes zibellina*, famous for its silky soft fur, lives in Russia. Mainly terrestrial, it can be recognized by its stouter limbs and larger ears. The yellow-throated martens, *Martes flavigula* or *Martes gwatkinsi*, are perhaps the most attractive species with their extensive gold throat and chest patches. The tail is two-thirds of the head and body length, which adds to the graceful appearance. Found in the forested belt of the Himalayas and in Malaya, they live in small groups, uttering characteristic chuckling contact calls as they move from tree to tree.

The American marten, *Martes americana*, like the pine marten prefers quiet conifer woods.

The eight-to-nine-month gestation period, due to delayed implantation, can be shortened by three months in warmer southern areas where the young, born early in January, are more likely to survive. Cubs are weaned after seven weeks and reach adult weight of about 4½ lb (2 kg) three months later.

The fisher or pekan, *Martes pennanti*, is the largest member of the group, measuring 32.7-40.7 in. (83-103 cm) and weighing 14-18 lb (6½-8½ kg). The name is derived from its reported habit of raiding baited traps and "fishing out" the contents, rather than a predilection for fishes. Surprisingly enough, porcupines seem to be a favored food. After carefully stalking its prey, the fisher flips the porcupine over with a deft stroke of the fore paw and pounces on the unprotected abdomen. It also feeds on squirrels, carrion and fruit. Two or three cubs are born in the spring after an almost year-long gestation. More terrestrial than the pine marten, fishers were nearly exterminated for their valuable fur, unable to escape the numerous traps or multiply fast enough to meet the demand. Now they are making a slow recovery in the northeastern United States.

MATA MATA, *Chelys fimbriata*, one of the South American side-neck turtles, is most bizarre and weird looking, as its broad, flattened head and neck is covered with fringes of skin giving it a mossy appearance. A fish wandering too close to its jaws may be engulfed by a sudden forward lunge of the head and inrush of water down the turtle's throat. Narrow, hard-rimmed jaws can snap open to a circumference equal to the wide, tubular throat. An elaborate hydraulic-action by use of a powerful hyoid apparatus not only pulls water into the throat but also snaps the head forward to engulf the prey. Forward movement of the head is apparently stabilized somewhat by horizontal skin flaps over the huge tympanum.

MAYFLIES, aquatic insects of riverbanks and lake shores, familiar because of their summer swarms. There are 1,300 living species fairly

closely related to the dragonflies that similarly cannot fold their wings back. Mayflies rest with the wings raised vertically above the body and at right angles to its main axis. Traditionally, adult mayflies are supposed to live only for a day. Certainly they have very short adult lives, for their mouthparts are vestigial as adults so they cannot feed and can live only long enough to mate and lay eggs.

Mayflies are soft-bodied insects with short bristlelike antennae, two pairs of wings—the hind pair being very much the smaller—and three long bristlelike "tails" projecting from the end of the abdomen. They have aquatic larvae and are therefore never found far from fresh water. They are widely distributed throughout the world.

In due course the aquatic nymph rises to the surface, and the skin splits along its back. The insect that then emerges has perfectly developed wings and the form of the adult, but it is dully colored, its wings are opaque and it cannot fly well. Within a short time, usually only an hour or so, this "subimago," or "dun," as it is known to fishermen, molts again and emerges as a complete, active and fully colored adult ("imago") with shining translucent wings. The subimago stage followed by the molt to the imago is unique among insects.

MEERKAT, *Suricata suricatta*, or suricate, a small mongoose that has developed separately

The mayfly *Ephemera danica*, found on the banks of trout rivers.

from the other species. It has the same inquisitiveness that typifies the mongoose, but its method of feeding is different and it has developed more precise use of the fore paws. It is confined to southern Africa, mostly south of the Orange River.

Slender of build, it is only 10-14 in. (25-35 cm) long in head and body with a tapering tail 7-10 in. (18-25 cm) long. Its coat is a light grizzled gray with a thick, dark reddish underfur. Its back is marked by eight or ten black transverse stripes. The head is tapered and almost white, and the snout extends well beyond the lips. The ears are black and the tail yellowish with a black tip. Unlike the other mongooses, the meerkat has only four toes, and the fore feet are well equipped with powerful claws for digging.

Meerkats live mainly in the dry sandy plains where there is little vegetation other than low grass and also in some rocky areas. They usually live in colonies in burrows they dig out for themselves or in crevices among the rocks. An average colony will consist of 24 or more individuals. They are powerful diggers, and their burrows have numerous entrances leading down to passages that may go as deep as 10 ft (3 m) underground.

In the wild, the meerkat feeds mainly on insects, spiders and centipedes and will also catch lizards, birds, rats and mice and will eat the eggs of birds and reptiles. It will occasionally kill moderate-sized snakes. It relies on its very keen sense of smell, not only to locate its food, but also to identify it before consumption.

MERGANSER, various species of saltwater diving duck or sea duck, characterized by a long slim bill with serrations for gripping slippery prey. There are six species of merganser spread over Europe, Asia and South and North America. The only South American species is the Brazilian merganser, *Merganser octosetaceus*, living on streams in forests. The most familiar species of merganser is probably the red-breasted *Mergus serrator*, which is very widely distributed through northern North America, northern Europe and northern Asia. It lives on lakes, rivers, estuaries and sea coasts—anywhere where fishes live in fairly clear water. The prey is captured beneath the surface, mergansers being specialized for underwater swimming. Mergansers are highly streamlined and have legs placed well back on the body with a well-developed musculature that enables the webbed feet to give a powerful thrust.

The red-breasted merganser drake is a handsome bird with a green-black double-crested head separated from a cinnamon breast by a broad white collar. The back is black, the belly cream and the sides vermiculated in black and white. The bill and legs are red. There is also much white in the wing that, when the bird is in post-breeding eclipse plumage, distinguishes it from the female, which has a much more somber plumage. She is basically gray with a brown head, and this is typical of all female mergansers.

Another well-known species is the similar and closely related American merganser, *M. merganser*. The European race is known as the goosander. This species has the green/black head, but the crest is a single backward sweep, changing the apparent shape of the head. The back is black, but the underparts are entirely cream. It is found much more frequently in wooded areas than the previous species.

Two smaller but perhaps even more handsome species are the smew, *M. albellus*, and the hooded merganser, *M. cucullatus*, of Eurasia and North America respectively. Both are found primarily on inland waters that have good woodland cover around them and nest in tree holes and crevices. They take fishes and also frogs and various invertebrates.

The drake smew is largely white with a black face patch and a black band running from the side of the head to the nape. Other black markings on the body and wings are more noticeable in flight, when the bird appears boldly pied. The hooded merganser is also largely black and white, but has rufous-brown sides, vermiculated darker, and a green sheen on the black of the neck and head. Its most outstanding feature, however, is its strikingly beautiful crest, which extends upward from the top and back of the head in a black-bordered white fan. The crest is depressed in flight and when diving, and is fully erected in display.

MERLIN, *Falco columbarius*, a very small, long-winged falcon; males $9\frac{1}{2}$ in. (24 cm), females $12\frac{1}{2}$ in. (31.8 cm). The males have slate-blue upperparts, streaked with black but lack mustachial stripes. They have a broad rufous-buff nuchal collar, and their underparts are also rufous-buff, boldly streaked blackish brown. The females are browner above and more creamy-white underneath. The bill is bluish-horn, yellow at the base, and the legs and feet are yellow. The eyes are dark brown. Immature birds resemble adult females. There are a number of races, varying slightly in size and color.

Holarctic in distribution, they migrate south in the winter. They inhabit open wet or dry lowland areas and upland moors, and have traditional breeding territories, nesting on the ground or in trees using old nests of other species. The eggs are light buff-spotted red-brown, the clutch varying from two to seven. Incubation is by the female, while the male hunts, preying chiefly on small birds, such as larks and pipits, but also taking small wading birds. The prey is captured by a dashing aerial chase. Merlins are frequently trained for falconry.

MESITES, a family of three species of birds related to the rails but with some superficial similarity to songbirds, such as the babblers. They are wholly confined to Madagascar. They are 12-14 in. (30-35 cm) long, with short rounded wings and fairly long tails. The legs are strong and well developed. Although the wings appear functional, all the evidence suggests that mesites never fly; and while it has been suggested that the two species of *Mesoenas* might use the wings if closely pursued by mammalian predators, the one species of *Monias* seems to be incapable of flight. They walk and run well, the head nodding in rhythm with the steps in pigeon fashion.

MEXICAN BEADED LIZARD, *Heloderma horridum*, one of the only two known venomous lizards, in the family Helodermatidae, restricted to North America. This lizard was known to the Indians of pre-Columbian times, and was described first by Hernandez in 1651. Even at that time, although the lizard was known to be venomous, it was described as seldom producing a fatal bite. It attains a length of nearly 3 ft (1 m), a third longer than its closest relative, the Gila monster. There are at least three subspecies, ranging from northern Mexico to beyond the Guatemalan border. It is mottled with yellow on a black ground.

MIDWIFE TOAD, *Alytes obstetricans*, of western Europe, in which the male collects the

Giant millipede of Kenya, with mites.

eggs, as they are laid in strings by the female, around his legs. Thereafter he takes care of them, visiting water periodically to keep the eggs moist and finally entering water as the tadpoles hatch.

MILLER'S THUMB, *Cottus gobio*, also known as bullhead, a rather grotesque freshwater fish found in clear, fast-running and shallow streams in Europe. The head is broad and flat, the body round and lacking scales except along the lateral line. The pectoral fins are large. The color varies to accord with the bottom on which the fish rests. It spends most of its time under stones, darting out only to catch a passing fish or insect larva. The eggs are guarded by the male, which will display by expanding its gill covers to an intruder. Its common name derives from the shape of a miller's thumb, which, used in testing flour by rubbing it between the thumb and forefinger, often became rather spatulate.

The Alpine bullhead, *C. poecilopsis*, is found in northern and central Europe, and can be distinguished from the miller's thumb by its long and narrow pelvic fins, which reach as far as the vent.

MILLIPEDES, slow-moving arthropods well adapted for defense. They are well armored; their cuticle is thick and calcified like the hard shell of a crab. When disturbed they "freeze" or curl up into a tight spiral or sphere, and many have a row of stink glands down both sides of the body. Millipedes, therefore, contrast markedly with centipedes, which are adapted for swift running and offense.

There are over 8,000 species throughout the world. The flat-backed millipede, *Platyrrhacus pictus*, of South America is 5 in. (13 cm) long and ¾ in. (2 cm) broad. The African snake millipede, *Graphidostreptus gigas*, is 11 in. (28 cm) long and nearly ¾ in. (2 cm) in diameter. The European *Macrosternodesmus palicola* is probably one of the smallest species, $\frac{3}{16}$ in. (3.5 mm) long.

On the head there is a pair of slightly elbowed and clubbed antennae, which probe the ground immediately ahead, and two pairs of jaws. Just behind the antennae there may be a group of simple eyes or ocelli.

The trunk consists of many articulating rings, most of which carry two pairs of legs.

Most millipedes tend their eggs. In the snake and the flat-backed millipedes an elaborate dome-shaped nest is built with moistened soil and excrement.

Most millipedes living in temperate regions are nocturnal and secretive. They live in leaf litter and soil, under loose bark, under fallen logs or in dead wood. Some leave these places at night and roam over the surface or climb trees and walls. However, in damp conditions in temperate latitudes and frequently in rain forests, millipedes wander about in the open by day.

Millipedes eat dead and decaying vegetation, playing a role equivalent to that of the earthworm. Some are suspected of damaging young seedlings, but probably they merely aggravate damage to living plants initiated by some other agent.

MINIVETS, slender, arboreal birds of the cuckoo-shrike family with brightly colored plumage. They are mainly birds of the tropics,

European midwife toad *Alytes obstetricans* with eggs. Male carries eggs on his back.

their distribution extending from northeastern China and Japan to Burma and India, and southward to Malaysia, Indonesia and the Philippines. The most northerly forms, the ashy minivet, *Pericrocotus divaricatus*, which has a range extending to Manchuria and Amurland, and the rosy minivet, *P. roseus*, which occurs in China and the Himalayas, migrate southward in winter.

Minivets are 6-8 in. (15-20 cm) long, with long tails that taper toward the tip, and slightly large heads. The legs and feet are relatively small and slender. The bill is usually finely and sharply hooked at the tip of the upper mandible for effectively gripping insect prey.

Although minivets are birds of the treetops, their vivid plumage makes them conspicuous.

The scarlet minivet, *P. flammeus*, has a typical pattern, the male being jet black on head, mantle, wings and central tail, while the greater part of the body and parts of wings and tail are bright scarlet. The female is less intensively black on wings and tail, pale gray on the back and top of the head, and the remainder of her plumage is yellow. In some species the male also may be gray rather than black, and the brighter colors vary from red to orange and yellow. On Jerdon's minivet, *P. erythropygia*, the plumage is black and white, the other color being reduced to a pink patch on the breast and a reddish rump; however, in some parts of its range the rosy minivet loses the reds and yellows progressively to produce some populations with gray and white plumage. The ashy minivet is gray, black and white only.

MINK, semiaquatic, like martens in size and appearance, except for the small ears, dark fur and bushy tails, their fur being coveted by women the world over. New World mink, *Mustela vison*, distributed from Alaska south to the northern United States, are more widespread than the smaller European palearctic species, *M. lutreola*. Deep chestnut in color, the fur has long soft guard hairs, responsible for its lustrous quality. White patches may sometimes be present on the lips, chin and underparts of the European species. Males are usually larger than females, measuring from 17-29 in. (43-73 cm), the tail accounting for nearly half the length, and weighing up to 3.8 lb (1.6 kg). Even though the coat is not water-repellent and the small paws with naked

The American or New World mink has escaped from mink farms in Britain and become feral.

soles and semiretractile claws are only partially webbed, minks were once known as "marshotters" because of their amphibious habits. Exploiting both streams and woodland habitats, this versatile carnivore is both a proficient swimmer and terrestrial hunter.

In Scandinavia and Great Britain, North American mink which have escaped from local fur farms adapt quite well to their new surroundings and may compete with the dwindling otter populations as well as with their European counterpart. This is particularly true in Russia, where American mink were deliberately introduced and have now spread widely. Otters prey on an assortment of mussels, frogs, crayfishes, eels and nongame fishes with seasonal diet variations, whereas mink tend to catch small fishes all the year round, sometimes hoarding and killing more than they can eat, which the otter seldom does. The mink's greatest drawback is no doubt its predation on eggs and fledglings during the waterfowl nesting season. Bold and fearless, like other members of the genus *Mustela* (which includes the polecat, stoat and weasel), mink can kill animals of similar size, such as rabbits or even muskrats, during droughts. In fact, they often choose to live in old muskrat huts or abandoned otter holts with submerged entrances, so as not to be too far from water, but in dry areas, hollow logs or burrows are also used.

MOCKINGBIRDS are mainly thrush-sized and characteristically have loud, prolonged and varied songs. The northern mockingbird, *Mimus polyglottos*, is conspicuous not only for the loudness and variety of the song but

also because it can be heard for most of the year and not merely in the period just before nesting. In the breeding season also its song may be heard at night when other songbirds are silent. The name implies that these birds are regular mimics, but this has been somewhat exaggerated, for, although mimicry of other birds does occur, it forms only a minor part of the repertoire of the northern mockingbird and seems to be even less apparent in other species.

Tower mockingbird *Nesomimus parvulus*.

A character of the northern mockingbird's song is the repetition of most phrases several times in succession, usually delivered from a conspicuous perch and often uttered in flight. The alarm calls are loud and harsh.

The northern mockingbird feeds extensively

on insects taken from the surface of the ground, but mockingbirds as a whole also take fruit. It has a habit of flicking the long tail and may also momentarily spread its wings. The reason for this wing flashing is unknown, although it has been suggested that it may disturb insects that would otherwise remain hidden. Mockingbirds are relatively unsociable, both single birds and pairs maintaining territories through most of the year, and they are as aggressive in defending the territory against other mockingbirds as they are in attacking potential predators.

The nests of mockingbirds are large and cup-shaped, usually built in the cover of foliage. The eggs are pale greenish or bluish and usually heavily spotted or blotched. Both parents assist at the nest and boldly attempt to drive away potential enemies, including man. Several broods may be reared in a season.

MOLE, COMMON, *Talpa europaea*, a small insectivore adapted to an underground digging existence. It is a glossy black animal with very large "hands" projecting sideways from the body. The muzzle is long, mobile, sparsely covered with hair and pink at its tip. The eyes are extremely small and usually hidden in the fur. There is no external ear. The neck is short so that the head is hunched back into the shoulder region. The tail is short and bristly. Body sizes vary according to locality and season, but the head and body together usually total about $5\frac{1}{2}$ in. (14 cm), the tail is $3-3\frac{1}{4}$ in. (3-3.5 cm) and the animal weighs $2-2\frac{1}{4}$ oz (80-120 g). Males are bigger than females.

Anatomically the mole is uniquely modified as a digger. Its fore limbs are massive and operated by enormous muscles. The humerus is a short, very solid bone of peculiar shape and points upward. The fore feet are large, flat and turned outward from the body, and the wrist contains a bony modification that broadens and strengthens the hand. The shoulder blade and pelvis are long and thin, and the latter is solidly fused to the vertebral column. The skull is flat and narrow, and the jaws bear 44 teeth.

Mole fur is soft and velvety, with all the hairs approximately the same length. These hairs are nondirectional and will lie equally well in any direction, thus allowing the animal to go backward or forward in its tight-fitting burrows without difficulty. The fur is black, appearing grayish sometimes. Color varieties do occur (mainly cream, piebald and albino), but these are uncommon and affect less than one in a thousand animals.

The mole's snout is extremely sensitive, being studded with numerous tiny sensory capsules called Eimer's organs. Moles are not blind, because they can be taught to perform tests involving discrimination between light and dark, but the eyes are very small, simple in structure and usually hidden by the skin and fur of the face. Inexplicably, the mole's eye has a unique cellular lens.

The mole's range extends from the extreme south of Scandinavia, to north Spain and Italy and eastward to Siberia. It is absent from many islands and is rarely encountered above the tree line in mountainous areas.

The mole's burrows are oval in cross section and usually excavated near the surface of the soil, but some go down to a depth of 40 in

The small mongoose of India, one of the most widespread of mongooses.

(1 m). Digging is done with the fore feet, loose soil being collected together and pushed upward to form a heap on the surface (the molehill or tump). The mole can push upward a mass of earth 20 times its own weight using the fore limbs. The snout and head are not used in digging, and in this it differs from the golden moles. The speed and extent of digging varies with soil conditions, but in light soil a mole might manage $7\frac{1}{2}$ in. (20 cm) of burrow in a minute; one produced a 13-lb (6-kg) molehill in 20 minutes. In springtime "mole fortresses," which are very large molehills overlying the breeding nest, are built.

The nest itself is made of leaves and grass collected from the surface, and in it the young are born in April or May. Litter sizes average four (range two to seven) in Britain, and second litters are rare. Newborn young are pink and naked, but hair becomes evident after two weeks, and they leave the nest when about five weeks old to lead solitary lives thereafter. Moles feed on soil invertebrates, mainly worms (about 90% of their diet) collected underground. They often store worms that have been immobilized by mutilation of the front end; one such store contained 1,280 worms. Excursions on the surface may be made, especially at night or when subterranean food is scarce due to drought. Moles are active around the clock with a three- to four-hourly rhythm of alternating rest and activity. They do not hibernate.

MOLE, GOLDEN, subterranean insectivore superficially like true moles but of the family Chrysochloridae, in which there are five genera, totaling 20 species. They have dense fur with long iridescent hairs. The ears and eyes are tiny and buried in the fur; the tail is not evident externally. They are between 3-9 in. (7-24 cm) long and have a pointed snout with a leathery pad at its tip much used during digging. The limbs are held below the body, not out to the side as in the common mole, and the fore feet are armed with two very strong claws for digging.

Golden moles live in sandy soils and cultivated areas of Africa from Uganda south to the Cape. They live in underground burrow systems, feeding on invertebrates. The young are born two to a litter during the wet season in a grass nest, deep underground.

MOLOCH, *Moloch horridus,* the "thorny devil" of the arid regions of Australia, is a lone genus in the large family Agamidae. Slow-moving, introverted and harmless, this lizard forms a complete antithesis of its name. About 8 in. (20 cm) long, colored orange and brown and covered with spines, it feeds almost exclusively on ants. Another lizard, the horned lizard or "horned toad" of the American southwestern deserts, although of an entirely different family, has evolved along parallel lines both in diet and in its spiny appearance.

There is no ready explanation why the moloch is spined, although this would provide ample defense against a number of would-be predators. Molochs are egg layers.

MONGOOSES, a group of carnivores with a reputation for killing snakes and stealing eggs. There are about 48 species living in the Mediterranean region, Africa, Madagascar and southern and southeast Asia, all much alike in

Yellow monitor *Varanus flavescens* of southern Asia.

form and most of them similar in habits. They are long-bodied with short legs, a sharp muzzle and a long, tapering, bushy tail. Their short ears are almost hidden in their long, coarse fur, which is usually a speckled gray or brown but sometimes striped or banded as in the banded or zebra mongoose, *Mungos mungo,* and the broad-striped mongoose of Madagascar, *Galidictis striata.* Most species have five well-clawed toes on each foot. The largest mongooses are $3\frac{1}{2}$-4 ft (1.05-1.2 m) long, of which just under a half is tail. In their general form they show their relationship with the civets and genets, but they lack the retractile claws and the scent glands.

One of the larger and best-known mongooses is the common or Egyptian mongoose, sometimes known as the ichneumon, *Herpestes ichneumon,* 2 ft (60 cm) long with a tail of 18 in. (46 cm). In Africa it ranges from the Cape to Egypt, while in Europe it is found in southern France and in Spain. Its body is a grizzled iron-gray with a black tip to the tail. The ancient Egyptians regarded it as sacred, and it is thought that they called it "Pharaoh's cat." It is known mainly for its prowess as a killer of snakes. Another large species, the crab-eating mongoose, *Herpestes urva,* of southeast Asia, Nepal to southern China, Burma and Malaya, is 4 ft (1.2 m) long. It is gray with a longitudinal white stripe on either side of the neck. The smallest is the dwarf mongoose, *Helogale parvula,* of Africa south of the Sahara, which is sometimes as much as 18 in. (46 cm) long but usually smaller. The small Indian mongoose, *Herpestes auropunctatus,* is one of the most widespread of mongooses. It is gray and about 2 ft (60 cm) long and ranges through Persia, Nepal, northern India, Assam, Burma, Thailand and Malaya. It has been introduced also into the West Indies and Hawaii to keep down the snakes and rats. This has proved a doubtful blessing, as wherever it has

been introduced it has become a pest, attacking small native animals and poultry.

All mongooses are alert and very active with lightning-fast reflexes, particular assets for those species that kill snakes sometimes as large as 7 ft (2.1 m) long. They are highly immune to snake venom, and when attacking a snake they erect the hair all over the body, which probably helps to disconcert the reptile. Another favorite food is eggs. Some mongooses will break them by throwing them through their hind legs onto a wall or rock or will rise semierect holding the egg in the fore paws and crash it on the ground to break it. The main diet consists of lizards, birds, small mammals and insects. Some species eat fruit or green leaves. The crab-eating mongoose lives mainly on shellfish and crabs but will also dive after fish, as does also the water mongoose. This large brown mongoose is remarkable in the method it employs to crack open the freshwater mussels it collects. It holds them in its fore paws and then smashes them against rocks to crack them open in much the same way as other mongooses break eggs. The dwarf mongoose drinks by dipping its fore paws into water and licking it. It has been noticed that a tame dwarf mongoose will eat such soft foods as egg custard in just this same way.

MONITORS, lizards belonging to the genus *Varanus,* of which 25 species are known. They range from small to very large, the biggest being the Komodo dragon, *V. komodoensis,* which may reach a length of 9 ft (3 m) and a weight of 165 lb (75 kg). Monitors are very snakelike with slender bodies and long necks and tails, and a forked tongue, which can be extended well beyond the mouth. On the other hand, no monitor has degenerate legs.

Other characteristics in which monitors resemble snakes include their ability to swallow large prey and with this a solid bony sheath around the brain protecting the brain itself from pressure as the large prey is being engulfed. The bones of the jaws are movable on each other and give a wide gape to the mouth, the temporal arch is completely ossified and the two halves of the lower jaw are joined by a ligament, much as in snakes. On the other hand, no monitor has poison fangs, and the teeth are relatively long, single-pointed, slightly curved and sometimes serrated on the rear edge, enabling the lizards to grasp and seize their prey and also, in some species, to tear it to pieces for swallowing. In the more primitive monitors the rounded nostrils are on the end of the snout, but in the more specialized species they form slits lying back near the eyes. The forked tongue can be stretched far forward, and it carries back into the mouth odorous particles to be tested in the Jacobson's organ in the roof of the mouth. The body scales are without underlying bony bases, or, if these are present, they are only slightly developed. Each of the four strong legs bears five toes armed with strong claws. The tail is muscular and long and can be used to beat off enemies, and as there are no fracture points in the tail vertebrae no part of the tail can be discarded and regrown as in true lizards.

Monitors are found in Africa, southern Asia and, more especially, the islands of the East Indies and Australia.

All monitors are predators on any animals they can overpower. The smaller kinds eat mainly insects, the larger species catch lizards, small mammals and birds. Some are egg eaters, and the Nile monitor is famous for its ability to find the nests of crocodiles and gorge itself on

the eggs (see ichneumon). The Komodo dragon also eats carrion, and *V. exanthematicus* and the Nile monitor are remarkable because of the change in feeding habits as they grow, the older individuals eating hard-shelled food such as large snails and crabs, and the form of their teeth changes to thick rounded crowns suitable for cracking such prey.

The female deposits her eggs in a pit and covers them with sand and vegetable material, which, warmed by the sun, act as an incubator.

MOORHEN, *Gallinula chloropus,* common dark-colored water bird of the family Rallidae, otherwise known as the waterhen or, in America, the common gallinule. There are a number of other closely related species that replace it in certain parts of the world.

The moorhen proper, *G. chloropus,* is an almost cosmopolitan species, being found in all parts of the world with the exception of the extreme northern and southern latitudes and Australasia. In Australia it is represented by *G. tenebrosa,* the very closely related dusky moorhen.

The moorhen is a familiar bird of the water's edge, brownish black, some 13 in. (33 cm) long, with a pattern of white streaks on the flanks and a short tail that is frequently "flirted" to display the white undertail coverts separated by a black central band. The underparts are grayer in tone than the back, the legs are green with a red garter above the tarsal joint, the frontal shield on the head and the base of the bill are red and the bill tip is yellow. As in other rails the head is jerked back and forth when the bird is walking or swimming.

Moorhens frequent a considerable variety of

freshwater habitats, as long as there is cover close at hand. Thus they may be found around or on small field and garden ponds as well as lakes and slow-moving rivers. They also wander over fields or parkland adjacent to water in search of food. They have a varied diet, taking many kinds of small animals and a larger proportion of vegetable material—leaves and fruits. Food is not often taken from beneath the surface of water, though moorhens may dive when attacked or in display. Flight is strong once the bird is on the wing, with the feet trailing behind the tail.

The nests of moorhens may be found on or near the edge of still or slowly moving fresh water from swampy rain forests to isolated desert pools, and from sea level to over 13,000 ft (3,960 m) in the Andes. The nest is largely made of dead reeds, sedges and similar plants, though sometimes the birds will build above ground or water level, occasionally using old nests of other species. Usually 5-11 eggs are laid, and normally there are two broods in a season, frequently three. Extra nests are built and are used by the young, which may be fed by the young of an earlier brood.

MOORISH IDOL, a highly colored, tropical reef fish allied to the surgeon fishes. The derivation of the name is obscure, the more so since of all religious groups the Muslims have perhaps been the least prone to idolatry. In both shape and color, the Moorish idol is so distinctive that any written description is superfluous. There appears to be only one species in the family, *Zanclus canescens,* a fish that grows to about 7 in. (18 cm) in length. It is found on reefs and rocky areas throughout the Indo-Pacific region. In larger fishes a "horn" develops on the forehead. It would make a most attractive aquarium fish but unfortunately has proved hard to keep in captivity.

MOOSE, *Alces alces,* a very large, long-legged deer of cold climates. It is known as elk in northern Europe and Asia and moose in North America and has a wide distribution in both eastern and western hemispheres. In the former, its present range extends from Norway in the west to eastern Siberia and Manchuria in the east. Within this range, two subspecies are represented, *A. a. alces* in the west, and *A. a. cameloides* in the east, the approximate line of demarcation being about longitude 90°E. In the western hemisphere the moose has a wide distribution north of about latitude 40°N ranging from Novia Scotia and New Brunswick in the east to Alaska in the west. Within this range, four subspecies are represented, the largest of which is the Alaskan form, *A. a. gigas.* In North America the wapiti is generally known as elk, but this is a misnomer. The moose, deep brown in color, is distinguished by its large size, up to 7 ft 9 in. (2-3 m) at the shoulder, long legs, short tail and broad, overhanging muzzle the nose of which is covered with short hair except for a small bare triangular patch between the nostrils. Its rump is noticeably lower than its shoulders. The neck is comparatively short, and from it hangs a dewlap or "bell," which is well developed in the North American form but short in the elk. Adult males, particularly of the Alaskan moose,

The moorhen, found over almost the whole world. In America it is called the common gallinule.

develop extremely large palmate antlers, which may measure up to 6 ft (183 cm) in span. Outstanding bulls from Alaska may weigh as much as 1,700-1,800 lb (733-816 kg).

The gestation period is about eight months, and the young calf is born unspotted, one of the few deer to be so. It will weigh about 25-30 lb (11-14 kg) at birth with twin weights averaging considerably less than single calves. Twins are fairly common, and even triplets have been recorded, but these must be comparatively rare.

MOSQUITOES, two-winged flies with some 3,000 known species. The females of many species feed on vertebrate blood. They have their mouthparts drawn out into slender stylets, which, except when feeding, are enclosed in a sheath forming a long proboscis. During blood feeding the sheath is drawn back, and the bundle of stylets is inserted through the skin into a small blood vessel. This type of feeding is known as capillary feeding. It contrasts sharply with the mode of feeding of other groups of biting flies. These have shorter mouthparts, which are used to excavate a small pit, or sump, into which the blood flows and from which it is imbibed, a mode of feeding known as "pool feeding."

Four larval stages and a pupal stage are passed through in the course of development. The larvae are found in water of almost every imaginable kind. Besides absorbing oxygen through the general body surface, they take in air at the surface of the water through a pair of breathing pores or spiracles situated at the tip of the abdomen. In the tribe Anophelini the spiracles are flush with the surface of the abdomen. Anopheline larvae consequently rest parallel with the surface film to which they adhere by means of rosettelike hairs. In the larvae of the other tribes (Toxorhynchitini, Sabethini, Culicini) the spiracles are situated at the tip of a tubular respiratory siphon. In these, only the tip of the siphon is flush with the surface film from which the larvae hang downward at an angle.

The pupae of the different tribes differ only in detail. They breathe in all cases, by means of paired respiratory trumpets situated on top of the thorax. The tip of the abdomen is furnished with a pair of paddles, used, in most cases, like a lobster's tail to propel them through the water. The larvae mostly swim by means of a jerky side-to-side movement of the body, but some culicines employ a curious vibratory motion presumably adaptive to progression through the thick sludges in which they breed. The larvae of most species have a pair of dense hair tufts, known as mouth brushes, at the front of the head. These are used for sweeping food into the mouth, but they can also be used for propulsion with the body held rigid. This type of locomotion is seen particularly in certain species breeding in tree holes or plant pitchers and with the body covered with dense rosettes of stiff spines thought to be protective against predation. Finally, the larvae of some sabethines, breeding in water collecting in the leaf bases of broad-leafed plants, can crawl rapidly over moist surfaces and thus are able to migrate from one leaf base to another.

Only the females feed on blood. Male mosquitoes feed on sugary substances such as the nectar secreted by flowers. Such substances

A male mosquito *Culex pipiens,* just emerged from the pupa at the surface of water.

The blue-crowned motmot, noted for perching for long periods and for its monotonous call.

also form an important part of the diet of the females of many species, the blood meal serving mainly to provide additional resources for developing the eggs.

MOTMOT, common name for eight tropical, forest-dwelling birds. They are confined to the New World, occurring from Mexico south to northeastern Argentina, but most of them are found in Mexico and in northern Central America.

Motmots vary in length from 6½-19¾ in. (16½-50 cm) and have short and rounded wings, a long tail and a large bill, which is broad and somewhat decurved with serrated edges. In most species the tail is graduated, and in typical motmots the two central tail feathers are much longer than the others and in the subterminal parts the barbs are loosely attached and fall off when the bird preens them. This results in part of the vane becoming naked while at the end of the vane the barbs remain intact, leaving a racquet or spatula tip.

Most motmots are greenish, olive or rufous brown with some bright colors on the head and

Mousebird, of Africa south of the Sahara and Ethiopia, eating flowers of the kaffirboom.

a dark spot on the breast. They live in dark forests and, because of their blending plumage, are difficult to locate among the foliage even when they call constantly. They feed mainly on insects and berries, which are snapped off in flight.

The blue-crowned motmot, *Momotus momota*, has the widest distribution of all, occurring from northeastern Mexico south through Central and South America to Argentina. It is green above and brownish below. Its crown is black, the forehead cobalt blue with an ultramarine band behind the crown and a black area around the eyes. On the breast is a small black spot edged with pale blue. This is the only species of motmot found in the Guianas, where it is not uncommon in dark forests. It often sits motionless for long periods, not very high up on a branch of a tree, uttering its peculiar and rather loud call—a rhythmical and staccato "jootoo, jootoo." Two birds often sing in duet. When sitting, they sometimes swing their long tails like a pendulum from side to side.

Motmots nest in holes that they dig with their strong bills in the ground, either in a bank or in the level ground of the forest floor. The entrance pipe can be long and even curved, and at the end is a chamber where the eggs are laid. Usually the nests are very difficult to find. The eggs are roundish and pure white. The best-known species lay three or four eggs. Both sexes incubate, and the incubation period is from 21-22 days. The nestlings are naked when hatched and are brooded and fed by both parents.

MOUNTAIN GOAT, *Oreamnos americanus*, one of the most peculiar hoofed mammals; not a goat, but a goat-antelope and a distant relative of the European chamois. It is an animal specialized to live on steep, wet, and usually snow-covered mountains, where chilling snowstorms are not absent even in summer and where winter reigns for up to eight months of the year. The mountain goat has massive, muscular legs that terminate in large, broad hoofs. It is a methodical climber, not a jumper, and resembles a bear when moving. Only exceptionally is it found away from steep, broken cliffs. Its white coat is made up of thick, long underwool and long guard hairs. There is a hair ridge on the neck, withers, and rump, and long hair parts on hind and front legs. The underwool tends to be greasy to the touch. Males and females are almost identical in appearance, and both have large chin beards, narrow pointed ears and short, recurved, black horns. The males carry large glands behind the horns, which swell in size during the rutting season.

Mountain goats are distributed from Wyoming to southeast Alaska and the Mackenzie Mountains of northern Canada. They are most abundant in the wet, cold mountain ranges such as the coastal ranges, Selkirks or the western slopes of the Rocky Mountains. Here they reach much larger sizes than in dry mountains. Large males in wet mountains may weigh over 260 lb (118 kg), but will be only about 140 lb (64 kg) in the dry southern Rockies. Little is known about the goat's life expectancy, its parasites or diseases. We do not know how important predators are in checking the growth

of goat populations. The steep, dangerous terrain favored by goats, and the readiness of this rather nervous though apparently phlegmatic animal, to ascend high cliffs if a wolf howls, appear to protect it sufficiently.

MOUSE, wood mouse, *Apodemus sylvaticus*, or long-tailed field mouse, as it is often called, is probably the most abundant and ubiquitous of all the mammals in Europe. In winter, especially, it sometimes enters houses, where it is frequently mistaken for the house mouse. It is, however, a much more attractive animal than the house mouse, with large bright eyes, larger ears, a rich yellow-brown fur above and pale silvery gray below with a little streak of yellow in the center of the chest. Its extreme agility enables it, when it goes foraging at night, to travel over open ground away from the protection of grass and dense vegetation, to which slower and more diurnal species like the voles and shrews are confined. The wood mouse, for example, is the only species that will be found in woodland with a closed canopy and no undergrowth, but is equally at home in gardens and hedgerows, shrubby sand dunes or on rocky hillsides. In spite of its alertness and agility, the wood mouse is far from immune from predation, and it forms a large proportion of the diet of the tawny owl. The versatility of the wood mouse extends to food as well as habitat. Seeds, buds and insects are all relished, and nuts are gathered in autumn for use in winter.

MOUSEBIRDS, a family of small, long-tailed and crested arboreal birds. They are similar to songbirds but assigned to an order of their own. They are found only in Africa, in areas south of the Sahara and in Abyssinia, where they occur in the more open tree savannas and forest edge habitats. The six species are finch-sized, and all have tufted crests and long, slender, graduated tails. The plumage, particularly that of the head and neck, is soft and lax. It is generally dull in color, overall brown or gray, at times with some fine transverse barring. The only distinctive coloring is on the head, which is red-cheeked on *Colius indicus*, white on *C. leucocephalus*, and blue-naped on *C. macrourus*. There is bare skin around the eyes and nostrils, which may be red, blue or gray; and the bill may be black and white or red in different species.

In general appearance mousebirds are squat and large-headed, with a stout, short and curved bill. The legs are short, and the feet are strong, with long sharp claws. The hind toe can be reversed so that all four toes may point forward, and the bird will frequently hang from a twig, tail downward, with the toes hooked over the perch; but on a flat surface both the outer toe and hind toe may point back to give a zygodactyl foot. The birds tend to rest back on the whole tarsus when moving or resting, but can shuffle and clamber about the branches of trees in a variety of acrobatic postures, and hop or run on level surfaces. The shafts of the tail feathers are stiff and lend some support. The birds clamber rapidly about on trees and bushes, feeding mainly on fruit, but also eat other parts such as leaves, and occasionally take insects.

The nests are shallow cups, built of a great variety of material, coarser outside and finer within. The eggs, up to four in a clutch, are cream-colored or white with brown streakings in different species.

MUD SKIPPERS, a family of small gobylike fishes living in brackish waters and mangrove swamps of the Indo-Pacific region. Their common name refers to their habit of leaving the water and skipping across the exposed sand or mudflats at low tide. Their overall shape is similar to that of the gobies, with a steep forehead and tapering body, but the eyes bulge from the top of the head and the pectoral fins have become modified with a fleshy base so that they can support the body. In some species the pelvic fins are also modified in this way, although in others they are joined to form a sucker.

One of the most widespread species is *Periophthalmus koelreuteri* of the Indian Ocean. It spends a great deal of its time, when the tide has receded, perched on the edge of small pools in mangrove swamps with the tip of its tail just in the water. When disturbed it will skip to the next pool, often with jumps of 2 ft (60 cm), rarely missing its target. These leaps are made by curling the body and then straightening it suddenly. It can also leap along the surface of the water. This species grows to about 5 in. (12 cm) in length, but certain of the Indo-Malayan species grow to about 12 in. (30 cm) and are dug up from holes in the mud by local fishermen.

The limblike pectoral fins are provided with extra strong muscles and can be used for "walking." They are thrust forward, the rest of the body being dragged after. The mud skippers are able to absorb oxygen in two ways. The gill cavities can be filled with water before the fish emerges, but exchange of gases can also occur in the mouth and throat, which are well supplied with fine blood vessels. Thus, gill breathing probably takes place for the first part of their stay out of water, but after this the gills and throat must still be kept moist for air breathing.

MULE DEER, *Odocoileus hemionus*, which is found over a vast expanse of western North America and in a variety of habitats, from high mountains to plains and deserts. Its range is confined almost entirely to the western half of the United States extending northward from central Mexico to southern Alaska, and the Great Slave Lake of Northwest Territories in Canada. 11 subspecies are recognized, of which the typical deer, *O. h. hemionus*, has the greatest range. The type inhabiting the Northwest Pacific coastal areas, *O. h. sitkensis*, is generally referred to as the black-tailed deer or sitka deer. The smallest mule deer is *O. h. peninsulae*, found in lower Baja California, while the two insular types, *O. h. sheldoni* and *O. h. cerrosensis*, from the islands of Tiberon and Cerros respectively, are not much bigger.

Typical mule deer antlers consist of a main beam with a number of even forks sprouting from it as compared with the typical white-tailed deer antlers which consist of a main beam and simple upright spikes sprouting from it. In some heads, particularly those of immature animals or of malforms, it is, however, sometimes difficult to distinguish the antlers of the two species.

The mule deer is fairly uniform in color throughout its range, and distinctions between sub-

species are not too well defined. Generally speaking, the winter coat is a brownish-gray which changes to a rusty tan or red in summer.

MUNTJAC, *Muntiacus*, five species and 17 subspecies of deer from southern and southeastern Asia. Other names for it are barking deer and rib-face deer. The latter name has arisen because the antlers, which consist of a short brow tint and an unbranched beam measuring about 3-7 in. (7.6-17 cm) in length dependent on subspecies, are supported on long skin-covered pedicles which continue down the forehead as converging ridges. The cry of this deer is a loud, short bark, similar to that of a dog, and this is often repeated many times, thus earning it the name of barking deer.

A feature of the muntjacs is that both sexes have canine teeth in the upper jaw, those of the bucks extending to about 1 in. (2.5 cm). These canines are used for fighting.

An adult male Indian muntjac, *Muntiacus muntjak*, of which there are a number of subspecies, measures 22-23 in. (56-58 cm) high at the

shoulder, the body color being a deep chestnut in summer, slightly darker in winter. The muntjac of southern India, *M. m. malabaricus*, is the largest of the Indian muntjacs, while the muntjac of Burma, *M. m. grandicornis*, produces the more massive antlers.

In central, eastern and southern China three species are represented. Reeves' muntjac, *Muntiacus reevesi*, which is sometimes referred to as the Chinese muntjac, has the widest distribution. It is smaller than the Indian muntjac, measuring only about 16-18 in. (42.5 cm) at the shoulder. One of the rarest of the Chinese deer is the black or hairy-fronted muntjac,

Muntiacus crinifrons, which occurs in Chekiang Province. It is a large muntjac, with a shoulder height of about 24 in. (61 cm).

MUSK DEER, *Moschus*, represented by three species, and the Chinese waterdeer, *Hydropotes inermis*, a typical feature of both being that the males are completely devoid of antlers. Instead, they are armed with long upper canine tusks, 2¾-3 in. (7-7.6 cm) in length. On the does they are much shorter.

An adult musk deer stands 20-22 in. (53 cm) high at the shoulder, and although there is some variation in color, generally it is a rich dark brown, mottled and speckled with light gray above and paler beneath. Its principal habitat is forest and scrubland at elevations of about 7,000-11,000 ft (about 2,120-3,350 m).

This deer is much hunted for its musk which is a brownish waxlike substance secreted from a gland on the abdomen of the bucks, used in the manufacture of perfume and soap. About 1 oz (28 gm) of it can be obtained from a single male. Unlike all other species of deer, the musk deer possesses a gall bladder. It is a

Mud skippers *Periophthalmus chrysospilos*, small bogy-like fishes of Malaya which live in brackish water, on a muddy beach at low tide.

solitary deer, and seldom are more than two seen together.

The distribution of the three species of musk deer is as follows: *Moschus moschiferus*, of which there are two subspecies, in northern India, from where it extends eastward into Transbaikalia. Here it is replaced by *M. sibiricus*, likewise divided into two subspecies, one being an insular form, *M. s. sachalinensis*, restricted to the island of Sakhalin or Karufuto in the Sea of Okhotsk. The third species, *M. berezovskii*, is an inhabitant of Szechwan, China.

MUSK OX, *Ovibos moschatus*, heavily built member of the family Bovidae from the Arctic

Indian hill mynah.

of North America; not a true ox, but more closely related to sheep and goats. It is 44-60 in. (110-150 cm) high, weighing at least half a ton (500 kg). The long guard hairs are blackish, and the soft light brown underfur is dense, and is shed in patches in the summer. The stocky legs are white. There is a hump on the shoulder. For all their clumsy appearance, musk-oxen are agile. They are highly aggressive.

For most of the year, musk-oxen form herds of four or five or up to 100 or more. Adult bulls are often solitary, young bulls herd together. The cow is sexually mature at four years, the bull at five or six. The rutting season may start as early as mid-July, but does not reach its peak until September. The bulls give off a strong odor at this season, butt each other, bellow and try to mate with the cows. Finally a few dominant bulls emerge; each of these stands with his fore feet on a rock or mound, preparing to defend his cows; often a bull manages to cover only one cow, but sometimes large numbers. The smaller bulls, usually not yet full-sized, quickly learn to avoid the dominant ones. Calves are born in April or May, a single calf to each cow; it weighs on average 15½ lb (7 kg) and stands 18 in. (46 cm) high. The mother protects her calf fiercely. Suckling continues for nine months; hence a female will calve only every other year. The calf moves along with the herd from its first day. Calves are heavily predated by wolves, and the herd's defense mechanism is to form a circle of adults around the calves, facing outward showing a well-nigh impenetrable array of horns. Musk-oxen live for twenty years.

Musk-oxen have always been heavily hunted for their fur. Nowadays the fur is farmed, and musk-oxen have been introduced into Norway and Spitsbergen. In 1930, after the Thelon Game Sanctuary in Northwest Territories (Canada) was set up, there were 9,000-10,000 in Canada, mostly on the arctic islands, and perhaps the same number in Greenland. Today there are at least twice that number, and most musk-ox populations are increasing rapidly.

MUSKRAT, *Ondatra zibethica,* or musquash, the largest of the voles, found throughout North America wherever the habitat is suitable. There is a second species, *O. obscura,* which is restricted to Newfoundland. Its total length is up to 25 in. (63 cm) including a scaly, rudder-like tail up to 10 in. (25 cm) long. It weighs about 2 lb (0.9 kg). Its coat ranges in color from silvery brown to almost black and is composed of a thick waterproof underfur overlaid with long glistening guard hairs. Its feet are broad and flat, the hind feet being webbed. The muskrat is aquatic, living in fresh water or salt marshes, or by streams and rivers. In summer it lives among the water plants, and in the winter it builds a house of stems and other vegetation, which projects above the water line. Sometimes it tunnels into the riverbanks and builds its nest there. It feeds mainly on water plants but will also take fish, frogs and freshwater mussels. It mates in the water and has 3-5 litters a year with an average of 5-7 young in each. The gestation is 19-42 days.

The muskrat is valued for its fur and also because it helps to keep waterways clear of water plants. It has been introduced into Europe where it is kept on ranches for its fur. Many, however, have escaped and gone wild and caused serious damage to riverbanks.

MUSSELS, bivalve mollusks mainly living in the sea but with a few freshwater species. The best known of the marine mussels is the edible mussel, *Mytilus edulis,* which often occurs in dense communities near the mouths of estuaries or elsewhere on the shore where there is a suitable surface for attachment. The tissues are completely enclosed by the black-blue shell, which may reach 3 in. (7.7 cm) in length, and which is attached by means of a series of threads, the byssus threads, to the substratum. Each of the shell valves is the mirror image of the other and can be closed very tightly by means of an anterior and a posterior adductor muscle that join the two valves transversely.

The shells of mussels that have grown rapidly are either smooth black or variegated ("pitch pine" mussels), while those living under crowded conditions have shells from which the outer periostracum has been worn away. This gives them a bluish color, and often they are encrusted with barnacles. In dead or dying mussels the shell gapes because the adductor muscles are no longer able to overcome the elasticity of the ligament that joins the valves at the hinge region.

MYNAH, the Indian name for birds of the family Sturnidae, of which the common starling is perhaps the best-known member. This common name is shared by seven birds in the same family; the common mynah, *Acridotheres tristis,* and the hill mynah, *Gracula religiosa,* are the best known of these birds.

The common mynah was originally found only in Afghanistan, India, Pakistan, Nepal and Indochina, but, like the house sparrow and starling, it has been successfully introduced into a number of countries—Malaya, Natal, Australia, Hawaii, Seychelles, New Zealand and many other islands. It is a little larger than a starling with a black head and neck, brown body and white underparts. The bill, legs and a patch of bare skin on the face are yellow, and conspicuous white patches on the wings and tip of the tail show in flight.

Mynahs are well known wherever they are found because they live in association with man and make plenty of noise. They are as common a sight on the roadsides of pastoral New Zealand as they are in the dry hills of India. The food consists of fruit as well as insects, which makes them unpopular with fruit growers, though they are certainly blamed for a good deal of the damage done by less conspicuous birds.

In autumn and winter mynahs return at night to communal roosts, much in the manner of starlings, but apart from this they are very territorial birds. Both members of a pair can be found on the territory for most of the year. The song and calls consist of a variety of rather loud raucous, chattering and whistling notes, some parts of which are quite melodious. The nest is an untidy cup placed in a niche in a tree or building, and from four to seven pale blue eggs are laid. Incubation takes about a fortnight, and the young spend as long again in the nest before fledging.

The hill mynah or "grackle" (not to be confused with the New World grackles, family Icteridae) is best known as a cage bird and is perhaps the best mimic of them all. It is similar to the common mynah, but has prominent yellow wattles pointing backward from behind the eye. Its natural habitat is the forests of India, Ceylon, Pakistan, Burma, the Malay peninsula and Indonesia, where it feeds in noisy flocks in the canopy of trees. The food is said to be insects, fruit and berries, with nectar in season. Though they make excellent mimics in captivity they do not seem to imitate other sounds in the wild, but have a fantastically varied and noisy vocabulary of their own. The usual bulky mynah nest is placed in a hole in a tree, and the clutch is two or three greenish-blue eggs with brown spots (see starlings).

NARWHAL, *Monodon monoceros,* a toothed whale, probably the prototype of the fabulous unicorn. Even its name is strange, being of Scandinavian origin and meaning "corpse whale." The teeth are completely absent in both sexes except for the single specialized left-sided tusk in the male. Sometimes the tusk may be oh the right side, and rarely there are two tusks, one on each side of the head, but the spiral always turns the same way regardless of which side the head it is on. The body may be some 15 ft (5 m) and the tusk half as long. The function of the tusk is quite unknown. The narwhal is an arctic species that feeds on a wide variety of food that it catches effectively in spite of its toothless mouth. All the toothed whales have asymmetrical skulls, and that of the narwhal shows this feature to a considerable extent. This is not only on account of its tusk but is particularly associated with asymmetry in its nasal passages, probably related to needs of sonar sound production.

One strange feature of the narwhal is the large amount of vitamin C that it appears to store in its skin. This is reported to be of the order of 31.8 mg per 100 gm of skin, the same amount as in many vegetables and fruit. This fact has been long appreciated by Eskimos, who are otherwise short of this vitamin in winter and who have hunted the narwhal especially for its skin, which they chew. Like that of the white whale, the skin makes excellent leather, a rare feature among Cetacea.

NEWTS, tailed amphibians, including in Europe the smooth newt *Triturus vulgaris* and the warty newt, *T. cristatus.* They are terrestrial during the greater part of the year and become aquatic during the breeding season, when the male develops a prominent crest on back and tail. The crest is nonmuscular, sensory and is usually brightly colored. It is used to attract the female during the elaborate courtship displays by the male prior to breeding. The male's sperm are deposited in the water in a structure called a spermatophore produced by special glands in the wall of the cloaca. This is picked up by the female with her cloaca and the sperm then leave the spermatophore and swim to a specialized portion of the female reproductive system, the spermatheca or receptaculum seminis, and are stored there. The female generally lays 200-450 eggs, which are fertilized inside her by the sperm. Laying usually starts three to ten days after the spermatophore is picked up, and the female will not accept further spermatophores until the eggs are laid. These may be deposited singly or in small clusters, and they are usually attached to the stem or leaf of a water plant or to a small rock. Laying usually takes place in the spring, and metamorphosis is usually complete by the end of the summer, when the young adults leave the water. They then remain on land until they become sexually mature three or four years later, when they return to water to breed for the first time. Hibernation during the cold months of the year also takes place on land.

The fire salamander, or spotted salamander, *Salamandra salamandra,* also extends into north Africa. In ancient times it was thought this salamander was able to live in fire. The myth probably arose from people seeing the animal emerging from a log, in which it had sheltered, when this was put on the fire. The fire salamander mates on land usually in July, and about ten months later the female enters the water to bear live young. Each litter contains 10-15 young about 1 in. (2.5 cm) long and possessing external gills that are lost during metamorphosis, when the animals become terrestrial and acquire the orange-yellow patches characteristic of this species. The skin of the adult is kept moist by secretions of the dermal glands, which also produce poisonous substances that afford some protection from predators. In

Alpine newts often develop neotenously in the cold waters of high level lakes.

Two male warty newts leaving their breeding pond in the autumn at the end of the breeding season.

laboratories the fire salamander has been known to survive for at least 12 years.

Newts of the genus *Pleurodeles* occur in Spain, Portugal and northern Africa and have long pointed ribs, which may even pierce the skin. The Waltl newt, *Pleurodeles waltl*, is said to live for 20 years in captivity.

The red-spotted newt, *Diemectylus viridescens*, of the eastern United States has a terrestrial stage, the "red eft" which lasts two to three years. After this individuals reenter the water and become permanently aquatic. They lose their bright red color and assume a dull green appearance.

The California newt, *Taricha torosa*, lives in the coastal mountain ranges of California, is usually aquatic and develops a crest during the breeding season. Newts also occur in China and Japan, and the Japanese newt, *Cynops pyrrhogaster*, is commonly kept as a pet.

Development in newts often involves neoteny. This is especially well documented in the alpine newt, *Triturus alpestris*. Members of those species which inhabit freshwater areas in the low-lying plains of France and Italy metamorphose in the normal way and are not neotenous. In the cold lakes at higher altitudes, development is retarded and neoteny is common.

Neoteny should not be confused with overwintering, which also occurs in newts, when there is an early winter or a late spawning and the larvae, unable to metamorphose in time, are forced to hibernate as larvae. These tend to grow unduly large, so resembling the partially neotenic forms, but will metamorphose in the normal way the following spring or summer.

Newts are voracious feeders and will eat worms, slugs, snails and insects when on land, and aquatic larvae, small crustaceans, mollusca and even frog's spawn when in the water. Food is detected by sight and smell. Swimming is by the use of the tail, the limbs being usually held alongside the body.

Newts, like most tailed amphibians, can regenerate amputated parts such as the tail, limbs and even some parts of the head. The power of regeneration decreases with age.

NIGHT HERONS, wading birds closely related to, but smaller than, other herons. They are mainly active at night. The typical race of night heron, the black-crowned night heron, *Nycticorax nycticorax*, has a range extending across southern Europe, Asia and Africa, with small colonies farther north. It is replaced in the Americas by geographical races making it one of the most cosmopolitan of birds. Three other species are American, and a fourth, the Nankeen night heron, *N. caledonius*, is found in Australia, the Philippines and Polynesia.

The black-crowned night heron is about 24 in. (60 cm) long. Juvenile plumage, present for two years, is entirely different from that of the adult, having upper surfaces chocolate-brown streaked with buffish-white, and longitudinally streaked gray underparts. It is easily distinguished from a bittern, however, as it habitually perches in trees. The crown and back of adults are black with a greenish gloss, the forehead and throat are pure white and the rest of the plumage is dove-gray or white tinged with gray. The sexes are similar in appearance. On assuming adult plumage they acquire pale cream plumes, usually three, which fall from the nape and may measure 8 in. (20 cm). Except when wind-blown, or at a nest, these are held together and resemble a single plume. The large eyes have crimson irises. The bill is black and shorter than that of a typical heron, while the yellow legs, which may redden at nesting time, are comparatively short, giving the bird a characteristically hunched appearance when perching. Throughout the daylight hours the birds perch gregariously in trees, becoming active soon after sundown, when they fly to the marshes looking, with their rounded wings, very owllike. They feed on small fishes, amphibians and a variety of other animals, including small mammals.

They breed in large colonies, sometimes consisting of several heron species. Said to be very tame, those nesting with little egrets in stone pines in the Rhône delta remained on the nests until approached very closely by the author. The nests of sticks are small compared with the size of the bird and are sometimes so flimsy that the contents are visible from below. Many colonies are in low bushes—alders, tamarisks and sometimes even in reed beds.

NIGHTINGALE, *Luscinia megarhynchos*, bird of the thrush subfamily Turdinae, renowned for its outstandingly beautiful song. The nightingale is a rather ordinary-looking bird, some 6½ in. (16½ cm) long, warm brown in color, lighter beneath, and with a chestnut-brown tail. It has a southwestern palearctic distribution, including much of Europe, and its preferred habitat is deciduous woodland with dense undergrowth and a rich humus layer on the ground in which it searches for insects and other invertebrate food. It is frequently found in damp places. It nests near the ground, usually laying four or five eggs which are incubated for 13-14 days.

Some other closely related species are also known as nightingales, particularly the thrush nightingale, *L. luscinia*, which overlaps with the nightingale in western Asia and which also has an outstanding song.

NIGHTJARS, birds also known as goatsuckers, a term favored in the Americas and derived from a myth that these birds drank milk from nanny goats. There are some 67 species, usually nocturnal, insectivorous and well-camouflaged, covering most parts of the world excepting high latitudes, New Zealand and some oceanic islands.

In America several of them have some general name that is derived from the bird's call, for example the whippoorwill, *Caprimulgus vociferus,* of North and Central America, chuck-will's-widow, *C. carolinensis,* of the eastern United States and the poorwill, *Phalaenoptilus nuttallii,* of the west.

The name "nightjar" is derived from the churring or jarring song of several species, particularly the European nightjar, *C. europaeus.* This species is some 10½ in. (26 cm) long with a gray-brown plumage that is mottled and barred to provide extremely good camouflage. Nightjars nest on the ground and rest on the ground or on branches, typically along the branch rather than across it. The rapid churring of the European nightjar may continue for several minutes, rising and falling, and is normally heard only at night.

The nightjars are almost entirely insectivorous, although some will eat other small invertebrates. Most of them have large eyes, being nocturnal or crepuscular, "hawking" for insects on the wing. As an adaptation for their mode of feeding the mouth has a very broad gape, and most species have a fringe of stiff, bristlelike feathers around the mouth to increase the catchment area. As in the European nightjar, the plumage normally matches the ground on which they nest. The eggs also are highly cryptic, marbled and blotched with shades of brown, gray and purple, on a pale background. Eggs and young are cared for by both sexes.

The poorwill is apparently unique in the depth of dormancy shown, at least in the winter. It may retire among rocks and become so torpid, with a temperature drop from 106°F (41°C) to 66°F (19°C), that it must be regarded as being in hibernation. Its loss of weight during the winter period, while it is in this state, is minimal. In this way it survives the winter period when flying insects are very scarce.

NILE PERCH, *Lates niloticus,* the largest of the freshwater fishes in Africa. It is found in the Nile, in some of the African lakes and in the larger west African rivers such as the Congo and the Niger. Its distribution reflects an ancient drainage system that linked some of the present rivers of west and central Africa with the Nile system. Most specimens of the Nile perch are 4-5 ft (1.2-1.5 m) in length, but giants of 6 ft (1.8 m) and weighing more than 250 lb (112 kg) are by no means rare. These powerful fish make a splendid adversary for the angler and have been introduced into other parts of Africa both as a food fish and for sport. Nile perch have now been introduced into Lake Victoria, where they have some effect on the important fisheries for *Tilapia*.

The Nile perch was well known to the ancient Egyptians, who drew accurate pictures of it on the walls of their tombs. These fishes were not infrequently embalmed and placed in tombs. At Esneh on the Nile in Upper Egypt, the Nile perch seems to have been worshiped as an important god, and this town was later renamed Latopolis, the City of Lates. Some of the fishes were so well preserved that even the fin membranes are intact.

Investigations into their biology has shown that they spawn in relatively sheltered conditions in water of about 10 ft (3 m). The eggs are pelagic and contain a large oil globule, which gives them buoyancy. The adults are fish eaters and are sometimes cannibalistic. In certain parts of Africa they are important in local fisheries.

NILGAI, *Boselaphus tragocamelus,* a large antelope of the hilly grasslands of peninsular India. Its closest living relative is the four-horned antelope, but the abundant fossil remains from the Siwalik Hills (India and Pakistan) show that it is a living relic of the group that gave rise to the cattle and buffaloes. Bull nilgai reach 52-56 in. (130-140 cm) at the shoulder and may weigh 600 lb (270 kg); females are much smaller. The horns are short, smooth and keeled, averaging 8 in. (20 cm) in length; they are found only in males. The build is fairly robust though less so than in cattle, and the withers are higher than the rump. Males are iron-gray, looking blue in some lights (hence the alternative name of "blue bull"), but females and young are tawny. Both sexes have a white ring above each

Night heron, easily distinguishable from gray herons by its black plumage and short neck and legs. It is found over Europe, Asia and Africa and occasionally visits the British Isles.

Nilgai or blue bulls have a white ring above the hoof.

Old World nutcracker. The strong bill is used for opening nuts and digging for seeds and insects.

hoof, two white spots on each cheek, and white lips, chin, inner surfaces of ears, and underside of the tail. There is a dark mane on the neck, and the male has a tuft of stiff black hair on the throat.

Nilgais are found from the base of the Himalayas to Mysore, but on the peninsula only, not in East Bengal, Assam or West Pakistan nor on the Malabar coast. Small groups of four to ten are seen together (sometimes more), consisting of cows, calves and young bulls. Adult bulls live alone or associate together in bachelor groups. In the rutting season, which varies from place to place but in northern India is usually March to April, the bulls fight, dropping to their knees and locking their foreheads, pressing down with their necks. The gestation period is between eight and nine months. The herds have no territories, but have much-used core areas in their home ranges, with habitual places for resting, defecation, drinking and so on. The herds may mingle while grazing, which they do in the morning and evening.

NOCTILUCA, a marine protozoan of relatively large size (up to 2 mm), remarkable for being highly luminescent. It often occurs in enormous numbers in surface waters and is responsible for the "phosphorescence" of the sea at night and its reddish tint by day.

NODDIES, five species of birds of the family Sternidae. They are small terns restricted to tropical seas. Two species are dark colored, two are intermediate and one, the fairy tern, *Gygis alba*, is pure white. Some species lay their single egg in an untidy nest in bushes or on the smallest of rock protuberances. The fairy tern has the most unstable of any bird's nest site—the egg is commonly laid in a crevice or depression on a bare tree branch. In some areas the common noddy, *Anous stolidus*, nests at less than annual intervals and is one of the few seabirds to breed and molt at the same time.

NORWAY LOBSTER, *Nephrops norvegicus*, also known as the Dublin Bay prawn, is lobster-like, 3 in. (7.5 cm) long in the body with claws of nearly equal length, and is a beautiful orange-red. It lives in depths of 60-300 ft (20-100 m) on a soft mud bottom. It used to be fished in the Irish Sea but is now more commonly taken in the North Sea.

NUTCRACKER, name for two species of smallish crowlike birds, the Old World nutcracker, *Nucifraga caryocatactes*, found at high altitudes and latitudes in coniferous forest throughout the palearctic region (with eastern thin-billed and western thick-billed forms) and Clarke's nutcracker, *N. columbiana*, which occupies similar habitats in western North America.

The Old World nutcracker is some $12\frac{1}{2}$ in. (32 cm) long, dark brown with noticeable white spots, white undertail coverts, a white tail tip and white beneath the tail at the edges. The tail is otherwise very dark, as are the wings. The flight is rather heavy and undulating. On the ground, progress is by strong bounding hops. The bill is black, long and strong and is used for digging in the ground for a variety of food—particularly fruits, seeds and invertebrate animals—and for hacking open nuts or pine cones. Clarke's nutcracker is very similar in general form and habits, but its plumage is light gray and it has conspicuous white patches in the wings and tail. It is occasionally found as far east as the Great Lakes or beyond.

The nutcrackers are outstanding examples of birds that store food. Many other members of the crow family spend a certain amount of time carrying and storing food, but in the nutcrackers this has become a fundamental part of their annual cycle of activities.

Nutcrackers nest in conifers, the nest being made of twigs, moss, lichens and earth, lined with grass and hairy lichen, and typically placed near the main trunk, 15-30 ft (4½-9 m) from the ground. Three or four eggs are laid and incubated by the female, who is fed on the nest by the male from his throat pouch. The young are fed by both, largely from this pouch.

NUTHATCHES, a family of typically small, dumpy birds, usually to be found foraging on tree trunks for insects and spiders. They are found almost throughout the forested regions of Asia, Europe and North America.

Most members of the family are about 5 in. (12 cm) long, though the largest, the giant nuthatch, *Sitta magna*, of Burma is about 9 in. (23 cm) long. Apart from three very brightly colored southern Asian species, the upperparts are normally gray-blue with the underparts white, gray, red-brown or chestnut, often more richly colored in the male. The wallcreeper resembles the nuthatches in build, but the bill is slender and curved and the feet are weaker. The toes and claws are long to aid climbing on vertical surfaces. Unlike woodpeckers and treecreepers the tail is short and not stiffened for use as a prop when climbing. Instead, nuthatches climb obliquely, hanging from one foot, supported by the other. They are equally capable of progressing down a tree trunk head first, they and the treerunners being the only birds able to do so. The bill is strong and daggerlike, and may be used for hammering open seeds that are first wedged in crevices in the bark. The English vernacular name is derived from the ability of the European nuthatch, *S. europaea*, to open hazelnuts in this way. Some populations of the brown-headed nuthatch, *S. pusilla*, in America are able to use flakes of bark as tools to prize off other flakes in the search of hidden prey.

All but two nuthatches nest in holes in trees, some of the small species excavating the hole themselves. Many of the Old World species reduce the size of the entrance hole by plastering it with mud, which may deter some of the larger potential nest-site competitors. Two species of *Sitta*, the rock nuthatch, *S. tephronota* and *S. neumayer*, have left the forest environment and inhabit rocky hillsides of southwestern Asia. These are the most industrious plasterers, walling up a rock cavity and constructing an entrance entirely of mud that may project 6-8 in. (15-20 cm) from the rock face. None of the New World species uses mud, though the red-breasted nuthatch, *S. canadensis*, smears the edge of its hole with pine resin. Four to ten eggs, white with reddish spots, are laid in the nest, which may consist simply of bark flakes or be lined with moss or hair. Incubation lasts two weeks and is carried out solely by the female, but the male assists in feeding the young, which remain in the nest for a further three weeks before fledging.

European nuthatch bringing insect food to its nestlings.

O

OARFISH, *Regalecus glesne,* a relative of the dealfishes, long, ribbonlike and unusual. If any one fish could be held responsible for stories of sea serpents it would undoubtedly be the oarfish. The body, which may reach 20 ft (6 m) in length, is thin, and the color of polished silver. The dorsal fin is almost as long as the body. Starting just over the eye, the anterior rays are elongated into long plumes, and these and the rest of the dorsal fins are bright red. The name of the fish derives from the scarlet pelvic fins, which are thin and elongated but expanded at their tips like the blades of an oar. The anal and caudal fins are lacking in the adult. The body is naked, and the shining silver color comes from guanine crystals deposited in the skin. Very few of these fishes have ever been seen alive, but it is said that when one swims at the surface it lies on its side and undulates its body. Because the snout is short and the face a little horselike, there are all the ingredients for a real sea serpent—a bright red mane, shining body, large size and humps (undulations when swimming).

In some areas this fish is known as the "king of the herrings" (a name also given to the John Dory). Pelagic eggs and some young stages have been found in the Straits of Messina, but almost nothing is known of the biology of the oarfish.

OCEAN SUNFISHES, large disc-shaped oceanic fishes apparently lacking a tail, which has earned them the alternative name of headfishes. The body is greatly compressed, the head large and the dorsal and anal fins prolonged into paddlelike structures. The most striking feature of these fishes, however, is the abrupt termination of the body behind the dorsal and anal fins so that they seem to be all head. During development, the rear end of the vertebral column atrophies and there is a rearrangement of the small bones of the tail, somewhat as in a normal fish when the tail has been amputated at an early stage. The result is that the muscles of the body that should attach to the base of the tail no longer do so, and instead they are attached to the bases of the dorsal and anal fins. This increases the power of these fins, and they are the principal means of locomotion, the fish gently flapping them as it propels itself slowly along. Pelvic fins, as well as the swim bladder, have been lost. The skin is leathery and up to 3 in. (7.5 cm) thick. The mouth is unexpectedly small, and the teeth are fused together into a beak. The ocean sunfish *Mola mola* is the largest

in this family. It can reach 11 ft (3.6 m) in length and weigh up to a ton. It is found in all oceans and even comes as far north as the British Isles.

With their thick skins, they have few predators, and they make little effort to escape when caught; the skin has been known to ward off rifle bullets.

Five species of sunfishes are known, placed in three genera. The truncated sunfish, *Ranzania truncata,* rarely grows to more than 3 ft (90 cm) in length. The mouth in this species is apparently unique among fishes in that it closes as a vertical slit and not horizontally. The tailed sunfish, *Masturus lanceolatus,* is a rare species in which there is a small pointed tail.

Young sunfishes are nothing like the adults. They hatch at about $\frac{1}{16}$ in. (2.5 mm) and at first resemble the larvae of any other fish with a normal tail.

OCELOT, or painted leopard, *Panthera pardalis,* a medium-sized South American representative of the Felidae. It is closely related to the margay, *Panthera tigrina,* and both are in the unfortunate position of having skins that are sufficiently attractive to be of interest to the fur trade. The ocelot is a small but highly patterned cat of the general leopard type. It has a shoulder height of about 20 in. (50.8 cm) and an overall length of some $4\frac{1}{2}$ ft (1.4 m) including the tail, which accounts for approximately one third of the total length of the animal. Its coloring is more varied than in the majority of cats, as the ocelot is a golden sand color on the head and down the dorsal region, silvery on the sides of the body, and shading paler on the belly. The head and neck are marked with black longitudinal stripes, while the body line is broken by the presence of black spots that are arranged roughly in rows.

The ocelot is found from the southern part of Texas, at the north of its range, through the isthmus of Panama and right down into Brazil and Bolivia. It is capable of pulling down and killing prey up to the size of a fawn of the smaller deer species, but for the main part it subsists on smaller animals and birds, and has developed the habit of living and hunting near villages and farms, where the domestic animals provide an easily obtainable supply of food.

Ocelots in Texas produce their young in the autumn, while those in Mexico breed in January. The period of gestation is approximately 115 days, at the end of which time two kittens are born. The den usually consists of

Boldly masked ocelot. The name is from the Mexican tlalocelotl or field tiger.

The common octopus is eaten in many parts of the world and is also widely used as a laboratory animal.

and recorded by Aristotle in the 4th century B.C. It is eaten in several Mediterranean countries, including Italy, Greece, Spain and Tunisia, and also in Japan. The male is recognizable by two external features. One is the hectocotylus or specialized third right arm, on the underside of which there is a groove running the length of the arm to the terminal spadelike portion. The spermatophores, or bundles of sperm, pass along the groove and are deposited in the mantle cavity of the female. The other feature is the presence, on the second pair of arms, of some especially large suckers that can be exposed as a display. Mating occurs close inshore from March until October in different parts of the world. As many as 50,000-180,000 eggs are laid in long strings and attached to rocks. The eggs are guarded and cared for by the female for between three and a half to nine weeks. Meanwhile, the mother eats little or nothing, and once the eggs have hatched she dies.

The octopus moves by jet propulsion, achieved by the expulsion of water through the funnel by contraction of the mantle muscles. Movement by this means is usually backward—that is, with the body in front of the head and the arms close together and streamlined behind the head. Forward movement is possible by ejecting the water toward the hind end of the body—indeed, considerable accuracy in steering is possible by deflection of the funnel.

OKAPI, *Okapia johnstoni,* with a build like a giraffe, but with foreshortened legs and neck, is found only in the dense rain-forest areas of west central Africa, which explains why it was unknown to zoologists until 1901. Before that time okapi were hunted and trapped by Pygmies, but no other people were sure of its existence. Sir Harry Johnston, its discoverer, at first believed that this animal, with a 6 ft (2 m) head and body length and a tail of 17 in. (43 cm), must be a type of horse, as he had seen pieces of its hide striped with black and white that reminded him of a zebra. However, all the other known members of the horse family lived in open areas, not in forests. He realized that it was a relative of the giraffe when he obtained several skulls. These possessed the lobed canine teeth and permanent horns partly covered by skin that are characteristic of the family Giraffidae. In the okapi, the pointed horns are present only in the males.

When people see the few captive okapi near giraffe in zoos, they seldom appreciate their close relationship, as superficially there are striking differences. This is because the environments of the two species are so unlike. The forest habitat of the okapi has affected it in many ways. Its coat is a rich black shade with white markings on the thighs, legs and throat. This coloration serves to camouflage the animal in the dark forest. Its ears are large, to enable it to detect danger that may be close but unseen in the thick vegetation. It lives a relatively solitary life, because there is little advantage in herds when vision is so restricted; indeed, the noise that a number of individuals would make might drown the sounds of their chief predator, man. Like the giraffe the okapi is not mute, but it seldom makes any vocal noises, at least in captivity, which is the only place where it has been

a small cave in the rocks or of a hollow log. The young are fully marked at birth, but are somewhat darker than their parents. Food is supplied by both the parents, and the pairs do not separate after the litter has been born. When hunting or near the den, the adult animals communicate with each other with a mewing sound.

OCTOPUS, a mollusk whose most striking feature is the eight arms that encircle the mouth at the front of the head. On either side of the head are two large eyes enabling the animal to see all around itself. Behind the head is a saclike body containing the viscera. Movement is either by "walking" on the arms or by jet propulsion. Octopuses can change from almost white to dark reddish-brown very rapidly, providing almost perfect camouflage. They have large brains and show considerable capacity for learning.

At the center of the bases of the arms the mouth is found. There is a circular lip surrounding the chitinous beak, the upper half fitting into the lower one. Inside the mouth is a tooth-bearing ribbon, the radula. From the back of the mouth the esophagus goes through the

brain to the crop and stomach in the visceral mass.

The viscera are enclosed by the mantle muscle, which is free and forms a cavity below the viscera that opens just below the head. This is the mantle cavity into which the gills project. A locking mechanism, which consists of a cartilaginous stud or ridge fitting into a socket on each side of the mantle, seals the mantle cavity, while the exhalant jet of water is expelled through the funnel. During inspiration, the lips on the funnel are drawn together so that water enters the mantle cavity and passes over the gills. The octopus can expel water forcibly through the funnel from the mantle cavity, and when it does so a locomotory jet is produced.

The common octopus, *Octopus vulgaris,* is widely distributed, being found in tropical and temperate seas throughout the world. In the west Atlantic it can be found along the coast from Connecticut in the United States south to Brazil. It is found around the south and southwest coast of England and southward as far as Africa. The common octopus is fished in the Mediterranean, where it was observed

studied. Like the giraffe too, the okapi is entirely a browser.

OLM, *Proteus anguinus,* aquatic salamander living in a permanent larval or neotenous form and retaining three pairs of external gills throughout life. The single species from deep caves in Carniola, Carinthia and Dalmatia in Europe grows to 12 in. (30 cm) and is a uniform white except for the gills, which are bright red. It lives in perpetual darkness, and the eyes are concealed under the opaque skin. It seems that the olm retains some sensitivity to light since, when it is kept in the light in captivity, it turns black. The olm is said to keep very well in captivity provided the temperature of the water is kept at about 50°F (10°C). The limbs are of moderate size, and there are three fingers and two toes. Fertilization is internal, and the eggs, laid singly, are fastened to the undersurfaces of stones and take about 90 days to hatch into larvae about 1 in. (2.5 cm) long. It has been reported that sometimes the female retains the fertilized eggs in her body and the young are born alive.

OPOSSUMS. The name "opossum" has also been applied to Australian marsupial phalangers, which are herbivorous and arboreal. The latter are now usually called "possums" to distinguish them from American forms. The opossums are the most primitive living marsupials and are distinguished from the phalangers by having the substantially unmodified carnivorous dentition of their remote ancestors. They are described as being polyprotodont because of the large number of teeth—five upper and four lower—on each side of the jaws. The opossums are also didactylous, having the second and third digits on the hind feet separate, whereas the phalangers are syndactylous, with the second and third digits fused. The opossums thus most resemble, among the Australian marsupials, the dasyurids (marsupial cats), which are also polyprotodont, didactylous and carnivorous.

All opossums are quadrupedal with five digits on each foot, the first toe of the hind foot usually being opposable to the remainder. The tail is usually prehensile. In size the opossums range from mouselike to forms a little larger than domestic cats. They are crepuscular and nocturnal, occurring from northern U.S.A. to Patagonia in South America and on some West Indian islands.

The pouch, or marsupium, is well developed in some kinds of opossums, but vestigial or absent in others. The Virginian opossum, *Didelphis marsupialis virginiana,* has a well-developed pouch usually containing thirteen teats, although there may be as few as nine or as many as seventeen. The teats are usually disposed in a horseshoe pattern with the open end toward the pouch opening. The occurrence of an uneven number of teats is unusual in marsupials; all Australian species have an even number consisting of one, two, three or four pairs. The mouse opossums, *Marmosa,* of South America are pouchless and with seven to fifteen teats. Mostly, the teats are arranged on the posterior region of the undersurface

(abdominal and inguinal regions), but in some species some teats are farther forward, in the pectoral region. All opossums produce a large number of very small young often in excess of the number of teats present so that some new-born young, which cannot find teats, die.

The Virginian opossum breeds at least twice in each year throughout its extensive range. After conception, the fertilized eggs, often 25 or more in number, develop for 13 days in the uterus. There they are nourished by a placenta consisting of vascularized and non-vascularized parts of the yolk sac. The young at birth are minute and weigh $2\frac{1}{2}$ gr (0.16 gm). They are suckled in the pouch for about 80 days before they can ride the mother's back.

The Virginian opossum is the only form occurring in North America, to which it migrated from South and Central America in comparatively recent geological times. Within recorded history the Virginian opossum has rapidly extended its range and is now found in parts of southern Canada. It is about the size of a house cat with grizzled gray pelage, white head, black legs, unhaired ears and scaly, naked, prehensile tail. The clawless great toe of the hind foot is held at right angles to the other toes. Opossums live in woodchuck burrows, natural cavities in rock piles, tangles of low vines and, more rarely, in natural cavities in trees. Although it is an adept climber, the opossum's natural habitat is on the ground, where it feeds on fungi, various parts of green plants and fruit, insects and other invertebrate animals, frogs, reptiles, birds and small mammals. Climbing is the opossum's first means of escape, but it has a remarkable facility for "playing possum" or appearing to be dead by entering a state of tonic immobility. The nest is made of dried leaves and grasses gathered by mouth and passed under the body to be grasped by the curved tail and carried to the nest site.

Virginian opossum, the only member of the family to live in North America, where it is spreading northward into Canada.

There are about 60 species of opossums of various sorts native to Central and South America. Allen's opossum, *Philander opossum,* has spots of light-colored fur above the eyes and is often called the four-eyed opossum. About 40 species of small opossums, included in the genus *Marmosa,* are arboreal animals with grasping digits and prehensile tails, which eat insects and fruit. *Marmosa robinsoni* is about the size of a house rat but is distinguished from this rodent by its large mouth, large eyes, naked ears, prehensile tail and feet with five distinct toes. Other species of *Marmosa* are well known as animals frequently included accidentally with a cargo of bananas on boats leaving South America.

OPOSSUM SHRIMPS, small, transparent crustaceans mostly less than 1 in. (2.5 cm) long, with some species attaining a length of about 6 in. (15 cm) and colored red. All species live in marine or estuarine (brackish) water except *Mysis relicta*, which lives in freshwater lakes, and two species of *Antromysis*. The body is similar in many ways to that of the crayfish and lobster.

Species of opossum shrimps occur in most sandy bays and among the seaweeds on rocky shores throughout the world. They often occur in large shoals in brackish-water areas and are most easily found at the water's edge at low tide. The incoming tide spreads the populations throughout the intertidal area, and the outgoing tide concentrates them at low water. Most mysids live close to the bottom, and nets on sledges that are towed over the seabed are usually required to catch species that live in deeper areas. Many opossum shrimps migrate vertically from the seabed to the surface of the sea at night but swim back down to the seabed at sunrise; many species of plankton perform this migration, the phenomenon being known as a diurnal vertical migration. The freshwater species *Mysis relicta* lives in lakes in North America and Europe, including Ireland and Lake Ennerdale in England. The two species of *Antromysis* are restricted to caves in Yucatan and crab holes in Costa Rica.

ORANGUTAN, *Pongo pygmaeus*, large anthropoid ape related to the chimpanzee, gorilla and man. Two subspecies have been described: the Borneo orangutan, *Pongo pygmaeus pygmaeus*; and the Sumatran orangutan, *Pongo pygmaeus abelii*, but differences between them are not great enough to make it easy, even for an expert, to identify a captive live orangutan whose origin is not known, as coming from Sumatra or Borneo. Members of the Sumatran subspecies tend to be larger and lighter-colored than those from Borneo. The name "orangutan" is derived from two separate Malay words: "orang," meaning man, and "utan," meaning jungle. In west Borneo, the native term "maias" is nearly always used, as it is by Malays.

The orangutan is a large animal with marked sexual dimorphism, males growing to double the weight of females. The coat has coarse, long and shaggy hair, especially over the shoulders and arms, where it may grow up to 18 in. (0.5 m) long. It comes in shades of orange to purplish or blackish brown and is generally darker with increasing age. Infants and subadults have a shock of hair standing up from the crown of their heads, which becomes short and flattened in adults, falling in a slight fringe over the forehead. The face and the gular pouch are bare, except for a fringe of hair on the upper lip and chin, growing strongly orange-colored in adult males. The skin is tough and papillated,

dark brownish to black in color with irregular, wide patches of blue and black shining through the hair, especially on the abdomen. A large gular pouch and prominent cheek flanges of fat and tissue placed at the side of the face are the most striking features of heavy adult males—but many males have them poorly indicated, especially if they are lean or in poor health. The face is somewhat concave, like a dish, with projecting, mighty jaws and slightly pronounced eyebrow ridges. The eyes

and ears are small, the latter pressed close to the skull. The arms, a principal asset in arboreal locomotion, are extremely long—the longest and strongest of all apes—with spans ranging between 7 ft and 8 ft (2.15-2.40 m). Similarly, the hands are longer than those of either the gorilla or chimpanzee and are extremely powerful, except for the thumb, which is small. The legs are relatively short and weak. Nails on hands and feet are strongly curved. The nail of the first or big toe is frequently

Marine opossum shrimp, showing the brood pouch where eggs are fertilized and embryos develop.

The female orangutan lacks the fleshy cheek flanges of the male but has a gular pouch like a monstrous double chin.

missing in Bornean orangutans. There is no tail.

The impression of huge size is gained mainly when an adult male is seen with arms extended, moving arboreally. However, owing to its short legs, when standing up the animal may only reach about 4 ft 6 in. (1.37 m) in height; a female is about 3 ft 6 in. (1.15 m). Males may weigh between 165-220 lb (75-100 kg), females between 75-100 lb (35-45 kg), depending on their age and state of health, and the food supply in their range, which varies with the seasons.

The animal lives in confined tropical rain-forest areas, usually within river boundaries or mountain ranges over 6,500 ft (2,000 m) high, which it is unable to cross.

Like all the great apes, the orangutan is essentially a quadrupedal climber. Though able to move on the ground, it is also the only truly arboreal ape, spending most of its life in trees. It walks up trunks using irregularities in the bark to give a firmer grip for fingers and toes, and usually proceeds in the middle stories of the forest, cautiously and silently. The orangutan walks quadrupedally along branches or bipedally with arms holding on above, its powerful, far-reaching "hook grip" bearing its main weight. It progresses, balancing and swinging slowly, using its body weight on trees and branches to bend them in a desired direction. Occasionally the orangutan brachiates for short distances, legs freely dangling, especially in moving from one tree

to another and in a quick flight. It does not jump. Descending a tree is a reversed climb or a jerky glide.

While moving and at rest in trees, the orangutan continually manipulates vegetable and animal matter within its reach, testing them as food. It specializes in many kinds of jungle fruit and eats or chews an infinite variety of buds and leaves, flowers, bark, epiphytes, canes, roots, even mold and humus. It forages at leisure, alternately squatting and climbing, nibbling, masticating and rejecting through its lips the shells and seeds of fruit and portions of the vegetables tested. It picks and plucks its food with cupped hands; if the material is small it uses the side of the index finger in an inter-digital grip. It may use its feet in a similar way or lock one or both of them into vegetation in order to secure its arboreal position. Honey of wild bees, the eggs of birds and occasionally soil are also eaten by orang-utans. Their favorite fruit is the durian, a spiky, smelly fruit with soft, pulpy flesh the size of a football. This particular fruit has brought the animal into conflict with man, as in many areas it is cultivated in orchards.

The nests of the orangutan are conspicuous tangles of branches, and they are constructed in a few minutes at any place where there is sufficient support and a suitable quantity of small branches, especially in crowns and forks. Branches are bent or broken toward the animal, assembled and laid across each other and finally lined with additional, smaller twigs. All

Male of the European garden spider courting the much larger female.

is then patted down with hands and feet into a rough circle of about 2-3 ft (0.6-1 m) diameter. Nests may be placed at between 10 and 100 ft (3 and 30 m) above ground and sometimes even higher. The orangutan sleeps in its nest from dusk to dawn, and may occupy one and the same nest for several nights if it happens to be in an area with plenty of fruit near. But as a rule, a new nest is built each night, and additional nests may be constructed for daytime rest and sleep. Some animals make their nests larger and more elaborate than others, even including a leaf "roof." Nests used during bad weather, sickness and parturition are usually large and sturdy.

Small infants share a nest with their mothers and drop off to sleep on her lap, even while she is moving. Youngsters start building their own nests at about two years old, but may also share a nest with another youngster. Some individuals rest and sleep without first building a nest. When on the move, a group normally stops to rest in the late morning and during the hot hours of noon.

ORB-WEAVING SPIDERS. The crowning achievement of spiders' engineering triumph is the orb web of the Argiopidae. A circular design was encouraged by the habit of random wandering from one fixed base while trailing a silk thread. Space can be bridged by a drifting thread, and a center is established by hauling downward and fixing a loose horizontal thread, thus forming a triangle with a vertical thread from its apex. Radial threads like spokes are followed by spirals, which alone are coated with gum.

Genera and even species have adopted variations in design. Most hubs are meshed, but *Meta* has an open ring. *Zygiella* omits one segment. *Argiope* and *Cyclosa* spin wide silk bands, or other designs, to camouflage their own presence at the center. Many have a home line down which they can slide quickly from their retreats.

All this knowhow is an inherited instinct as the mothers abandon their eggs. It is also remarkable that an apparently unrelated cribellate family, Uloboridae, has independently evolved similar orb webs.

The Argiopidae do not repair webs, but they normally remove the viscous spirals every night and spin new ones. Insect prey is usually touched, bitten and securely wrapped in silk. An oily coating to the legs prevents the spider sticking to its own threads.

The thick golden threads of the largest tropical species of *Nephila* were once used for making silk garments, but the spiders ate one another and were too difficult to rear to make the experiment a success. Despite their fertility, the largest British species, *Araneus quadratus*, lays nearly 1,000 eggs, which weigh twice as much as the female laying them.

ORIBI, *Ourebia ourebia*, a small tawny-colored African gazelle. See Dwarf antelopes.

ORIOLES, medium-sized arboreal songbirds, often brightly colored. In several species the males have bold yellow and black patterns. There are 26 species comprising the genus *Oriolus*, the two figbirds, *Specotheres* spp., and the aberrant tylas of Madagascar. Orioles occur through Africa and southern Asia, south through the Philippines and Indonesia to Australasia, while one species, the golden oriole, *O. oriolus*, extends up into the palearctic as far as northern Europe.

Oriolus spp. are starling- to thrush-sized but tend to be more slender-bodied, like the former. The bill is slender and a little elongated. The wings are long and pointed. The legs are somewhat short, but legs and feet are strong. The males are mostly brightly colored: yellow, red or maroon combined with varying amounts of black, or more rarely green. The females are usually duller, olive-green and often streaked beneath, and the young resemble them. These are birds of the treetops, usually of forest or forest edge, but also in more open spaces where scattered tall trees are present. They usually occur singly or in pairs, feeding on insects and fruits, and are troublesome in orchards in some regions. Some calls are harsh, but others, and the song, consist of loud liquid or fluty notes. The nest is a woven structure of flexible strands of plant material, suspended in a fork with the edges bound to the twigs. The eggs are yellowish or white with some sparse dark spots. The female incubates, the male feeding her at the nest, and both help to feed the young.

The two figbirds of Australasia are larger and more heavily built. The males are yellow and green with black on the head and bare red skin

around the eyes. The females are brown, streaked on the underside. The bills of these species are heavier than those of the typical orioles. The figbirds are more sociable than other orioles, often occurring in parties or small flocks. They are fruit eaters and, as the name suggests, feed extensively on wild figs, a fruit to which other orioles are also very

Female golden oriole at nest. The male is brilliant yellow with black wings and tail but is very secretive. Glimpses of green woodpeckers sometimes give rise to reports of golden orioles.

partial, and the seeds of which they help to disseminate. They have harsh calls. The nests are similar to those of other orioles, but the eggs are greenish and heavily marked in red-brown and purple.

Tylas eduardi of Madagascar is now regarded as an oriole, although earlier it was put with the bulbuls. It is about 6 in. (15 cm) long, of plump build, gray above and buff below, with a black head. The plumage may show a greenish tint. The black bill is fairly long and though stout is like that of an oriole. It is a sluggish bird of the upper parts of trees, where it takes insects from the branches.

OROPENDOLA, the common name for 12 species of social birds of the family of New World orioles. They have a very wide distribution in the Americas from Mexico, throughout Central America and south to northern Argentina.

Oropendolas are rather large birds, somewhat larger than a jackdaw, and in all species the male has brighter plumage and is considerably larger than the female. They live the whole year round in flocks, feeding, nesting and roosting in groups. They are to be found among the trees in forests, forest edges, along forest-fringed rivers, and some species at least are also common in plantations and in groups of trees in open country.

Among the best known is the crested oro-

pendola, *Psarocolius decumanus*, which has a very wide distribution, occurring from Panama south through Colombia, Venezuela, Ecuador, Peru, the Guianas, Brazil, Bolivia, Paraguay and northern Argentina. It is largely glossy black, the lower back, rump and lower underparts are chestnut, and in the rather long tail the middle feathers are black and the outer ones yellow. The strong bill is pale yellow and the eyes a very fine pale blue. Males vary in weight from 10-10½ oz (280-290 gm), but females are only 5-6 oz (148-167 gm). The male has few long and narrow feathers on the head forming a crest. It feeds partly on fruits, and, as it takes cultivated fruits, such as mangos, papayas and citrus, it can do a lot of damage in cultivated areas. Insects also form a large part of its diet, and it captures flying termites on the wing.

Crested oropendolas nest in colonies, preferring trees that tower over the surrounding vegetation—for instance the huge cotton trees or palms, although colonies in quite low isolated trees are also known. The nest is a wonderful structure, a long, pendent purse about 3 ft (1 m) in length with the entrance right at the top. The nest chamber is at the end, and the bottom is lined with dry leaves on which the eggs rest. It is entirely constructed of long stripped leaf fibers and is fastened at the very end of a branch, very often at a considerable height. Nests move freely in a strong wind, and during storms they sometimes break off and fall to the ground. It takes from 9-25 days to complete the nest. Owing to their inaccessibility, the contents of the nests are difficult to examine without first cutting off the branch to which they are attached. Nest building, incubation and the

A group of gemsbok, one of the antelopes known as oryxes, found widely in the open parts of Africa.

feeding of the nestlings is confined to the female.

The size of a colony varies from two to more than 40 nests. Isolated nests are also known but are apparently never used. The eggs, one or two in number, are pear-shaped and white with varying numbers of purple dots, blotches and lines. The incubation period is from 17-18 days, and the nestlings stay in the nest for a period of 31-36 days. The female roosts in the nest, covering the nestlings for 23-26 days. The nesting season is rather extended, and in Surinam the earliest date for young birds on the wing is February 8 and the latest July 20. After the nestlings have left the nest they wander about in noisy flocks, constantly uttering their begging cry and frequenting fruit-bearing trees, where they are fed by the accompanying females for what seems to be a long period.

ORTOLAN, *Emberiza hortulana*, a bunting found in Europe and western Asia, formerly regarded as a delicacy. It was netted alive in very large numbers and sometimes fattened for the table. In France the term "ortolan" has been used for all buntings and the word has become extended in other parts of the world to cover all small birds considered to be delicacies, such as the bobolink, *Dolichonyx oryzivorus*, of North America.

ORYX, a medium- to large-sized antelope, one of the most typical inhabitants of Afro-Arabian desert. The nearest relatives are the

sable and roan antelope and the addax. The Arabian or Beatrix oryx, *Oryx leucoryx*, is 40 in. (1 m) high. It is white with brown limbs, blackish frontal and nasal patches and eye stripes that expand below and unite under the lower jaw, being continued back as a throat stripe. The horns in this species are often at least as long as the animal is tall. Formerly found all over the Arabian peninsula and north into Syria and Iraq, this species has been drastically reduced in numbers and in range. It has always been hunted because of the believed connection between killing an oryx and virility. Recently oil-rich sheikhs have led large hunting parties into the desert, which have mown down great numbers with automatic rifles and machine guns from fast cars and even from airplanes. An expedition from Qatar into the Rub' al Khali in 1963 accounted for over 300—at that time thought to represent more than three quarters of the surviving stock. Accordingly, in 1964 the Fauna Preservation Society mounted "Operation Oryx" under Ian Grimwood, which captured two males and a female to form the nucleus of a breeding herd in Phoenix Maytag Zoo, Arizona, against the probable extermination of the species in the wild. This expedition focused world attention on the plight of the species. The London Zoo contributed its female, and both the Sheikh of Kuwait and the King of Saudi Arabia revealed that they possessed private herds in captivity, out of which each

generously donated animals to the Phoenix pool. A number exist in a zoo in Ta'iz, Yemen, and David L. Harrison has recently obtained evidence of the animal's continued existence in the wild. The Phoenix herd has now increased to 34 (as of 1972), and there is every hope that the species has been saved.

Arabian oryx, the most desert-adapted species, retire into the dunes when alarmed. In the heat of the day they dig into the sand for shade. They are less aggressive than gemsbok, and they run with a slower, more cumbersome canter, without the twists and turns that gemsbok make.

OSPREY, *Pandion haliaetus*, a large, fish-eating bird of prey, the only member of its family, but found regularly throughout the world except for South America where it occurs only on migration. In North America it is known also as the fish hawk. Measuring up to 24 in. (61 cm) long, the osprey is unusual among birds of prey in being almost wholly dark above and white beneath. It has a white head with a dark streak through the eye, a barred tail, and angled wings with a dark patch beneath at the angle.

The osprey feeds very largely on fish and occupies both freshwater and marine areas where there is sufficient food. It hunts by cruising above the water at heights up to 200 ft (60 m), and takes its prey by plunging in a shallow dive. The talons are brought forward to strike the prey just as the osprey hits the

The hearing of birds sometimes surpasses that of man. The tympanic membrane shows clearly on the semibald head of an ostrich.

An ostrich squatting. Note the almost bare neck and legs.

Ostriches are found in the open country of eastern and southern Africa, where they feed on a variety of plants and animals.

water, and it sometimes completely submerges. The feet are very strong, the claws long and sharp, and the toes bear spiny tubercles on their undersurfaces to give a good grip on a slippery fish. Furthermore, the outer toe is large and can be moved to face backward as in the owls. The grip following a good strike is so efficient that the bird may not be readily able to let go, and there is at least one record of an osprey striking a very large fish and being dragged under the water. After a successful strike the osprey rises from the surface, shakes the water from its plumage, arranges the fish head forward and carries it to its young or to a suitable feeding place. In addition to fish, ospreys take mammals, birds, sea snakes and even large sea snails. In some regions they are frequently robbed of their prey by eagles, such as the American bald eagle, *Haliaetus leucocephalus.*

Ospreys usually nest in trees or on cliffs, though they also nest on the ground. In some regions, particularly eastern North America and northeast Africa, they nest in large colonies. The nests usually command a clear view of the fishing grounds, and are built of sticks, grasses or any other available material. They are often used year after year and may have material added to them annually. Ospreys usually mate for life, which may be 20 years or more. Two to four, usually three, eggs are laid, and incubation, largely by the female, begins with the first egg. The eggs hatch after about 35 days. For five or six weeks the young are fed by the female with food brought by the male. Then, when the young can deal with the food themselves, it is dropped to them by both parents. The losses of eggs and young are high, but successful young leave the nest after 8-10 weeks.

OSTRICH, *Struthio camelus,* the largest living bird, is flightless and is the only member of the family Struthionidae. At one time found in large numbers in many parts of Africa and southwest Asia, the ostrich is now common in the wild only in parts of east Africa. There are also considerable numbers in ostrich farms in South Africa, and domesticated birds have become feral in South Australia.

The ostrich is well adapted for a terrestrial life, being able to run well. The legs are very long and strong, and with the long neck make up a considerable part of the ostrich's height, which in a large male may be 8 ft (2½ m). Such a bird would weigh up to 300 lb (135 kg). Typically, the color of the male is black with white wings and tail and that of the female gray-brown. The female is smaller, but in both sexes the head appears small in proportion to the rest of the bird. The head and most of the neck are almost bare, being sparsely covered with down and bristlelike feathers. The legs also are almost bare, and the skin of the neck and legs is grayish or reddish according to subspecies. Well-developed eyelashes are present.

The ostrich has only two toes on each foot—the original third and fourth digits. The third is much the larger. This is an adaptation to a predominantly running and walking mode of life, giving greater strength and thrust to the foot, as in the reduction of the horse's foot to one strong digit. The ostrich, with its long neck and keen eyes, is able to see long distances, and its long, strong legs make it capable of speeds up to 40 mph (64 kph) so that it can outpace most pursuers. It is commonly found, therefore, in open country with little cover and is very difficult to approach except in nature reserves, where the birds have become accustomed to vehicles. In some areas, however, vehicles have brought about the ostrich's downfall, for in combination with high-powered rifles they make it possible for the birds to be hunted successfully by man. It is almost certain that the reduction in numbers of the ostrich in many parts of its range has been because of excessive human predation.

Ostriches are omnivorous, though the bulk of their food is usually of plant origin. They take a variety of fruits and seeds and also the leaves and shoots of shrubs, creepers and succulent plants. A variety of invertebrates and smaller vertebrates, such as lizards, are also eaten. Quantities of grit and stones are swallowed, aiding digestion by assisting in the grinding of resistant foods. The succulents provide a certain amount of moisture, and they obtain more from some fruits and animals, which has probably given rise to the supposition that ostriches can survive for long periods under desert conditions without water. This is not the case; they must either obtain water from their food or else have access to drinking water.

They also bathe when they have the opportunity.

All ostriches are polygamous. Varying numbers of females are recorded as forming the harem of a single male, but this may be the result of differences in the availability of females rather than racial differences. The nest is a shallow pit dug in sandy soil, and in this the eggs of all the females of a particular male are laid. Varying clutch sizes are recorded, from 15-60 eggs. It seems that each female lays from six to eight eggs, usually one every other day. The eggs vary in surface texture in the different subspecies, some smooth, some pitted, and there is also a certain amount of variation in size. The average egg is around 6 in. (15 cm) long and 5 in. (13 cm) wide, with a weight only 1.4% of that of the laying female. This is an unusually low figure for such a large bird.

The incubation period is around 40 days, rather short, possibly a development resulting from the considerable amount of predation to which ostrich nests are subject. The male bird incubates at night, but one or more of the females takes over for much of the day. A large proportion of eggs fails to hatch. An injury-feigning distraction display may be performed by either or both of the sexes if danger threatens the eggs or young, and, as in other birds, this is more frequent around the time of hatching. The young are precocial and can follow the parent as soon as they are dry. They can run as fast as the adult when they are only one month old.

Ostriches are usually seen in small groups, a male being accompanied by the females of his harem and a number of young. Groups of 5-15 are common, though single birds may be seen and also parties of 50 or more. In some areas ostriches may accompany mammals such as zebra and wildebeest. It may be that, as in other cases of bird-mammal association, the ostrich is better able to detect danger visually, while the mammal may pick up a dangerous scent before the danger is visible. The association is thus to the animals' mutual advantage.

Ostriches have been farmed for their plumes in Cape Province in South Africa since the 1850s. In the early part of this century there were over 700,000 birds in captivity, and ostrich farming enterprises had also been started in north Africa, the U.S.A. and various European countries, as well as South Australia. But the First World War almost eliminated the industry, and now there are only about 25,000 birds in captivity in South Africa, principally for the production of high-quality leather.

Accounts of the ostrich's proverbial habit of burying its head in the sand to escape danger, on the principle that if it cannot see it cannot be seen, date back to classical times. In the version recorded by Pliny the ostrich pushes its head into a bush. The probable explanation is that when an ostrich is sitting on its nest its reaction to disturbance is to lower its head with the neck held out straight. The head may then be hidden behind a hummock or clump of herbage, while the body, although inconspicuous, is still visible. The ostrich's habit of swallowing indigestible materials, from nails

Short-clawed otter eating a fish. This is the smallest otter.

to bottles of beer, is also exaggerated.

OTTER, a carnivore belonging to the weasel family and living a semiaquatic existence.

Otters are well adapted to their semiaquatic life. Covered with a close, waterproof underfur and long guard hairs, their body is muscular and lithe, built for vigorous swimming. The limbs are short, and the trunk cylindrical. The paws, each with five digits and non-retractile claws, are generally webbed, and the fore feet are shorter than the hind ones.

The fully haired tail, thick at the base but quite flexible, tapers to a point and is flattened on its under surface. Below its base are situated two scent glands (except for the sea otter, in which they are absent), which give the otter a characteristic mustelid odor, sweet and pungent. Adults may use these glands when suddenly frightened, but do not direct the scent toward the aggressor like a skunk, *Mephitis.* Considerable color variation exists, partly geographic (tropical species are usually

Seiurus aurocapillus, one of the many ovenbirds, a group that has a wide variety of nesting habits. The species shown above does not make an oven nest.

271

lighter than their northern counterparts), partly seasonal (the fur is paler before the molting season), ranging from black-brown to a pale gray hue. Stiff vibrissae, or whiskers, are numerous around the snout, reminding one of a seal, and smaller ones appear in tufts on the elbows. Used for tactile purposes in dark or muddy waters, they are very sensitive, detecting the slightest turbulence. When swimming underwater, the nostrils and small, round ears are kept shut, and the otter must rely on its facility in swimming, on its sight, whiskers, and in some species manual dexterity, to detect and catch its prey.

The European otter, *Lutra lutra*, is not in fact confined to Europe but extends from the British Isles across Russia to Japan and is also found in north Africa. In Asia, a number of subspecies are recognized, the most southerly occurring in Sumatra. Smaller than the Canadian otter, *Lutra canadensis*, the European otter measures 36-48 in. (91-122 cm) and weighs 10-25 lb (4.3-11.4 kg). It is a territorial animal, the male's home range covering approximately 10 miles (16 km) of stream or lake. Generally solitary and nocturnal, they rarely vocalize and are the shyest members of the otter group. Pairs or small family groups can be found only at certain times of the year. One to five cubs, but more usually two or three, are born at any season after a nine-week gestation.

The Canadian otter is restricted to North America south to Texas. The genus, however, spreads down through South America to Tierra del Fuego, separating into several other species along the way. The specific differences are based on the shape of the nose pad and skull variations, but these otters are very similar in appearance and behavior. For instance, the Canadian and South American otters share a common vocal repertoire, which includes a low chuckling greeting call unique to otters of the New World. An adult may measure 36-50 in. (91-127 cm) and weigh 11-35 lb (5-13.7 kg). Females are smaller and lighter, especially in tropical forms.

OVENBIRD, 220 species of small to medium-sized South American perching birds, usually with brown or chestnut-brown plumage. The ovenbirds are a very diverse group of dull-colored birds found from the high Andes to the seacoasts of South America. Most of them are drab birds with plumage of some shade of brown, some have rufous or chestnut upper-parts and a few have white undersides or white wing markings. Only a very few species have any bright colors, though some of the genus *Asthenes* have a small patch of reddish-chestnut on the throat, and the white-cheeked spinetail, *Schoeniophylax phryganophila*, has a patch of bright chrome yellow color on the chin. The true ovenbird or pale-legged hornero, *Furnarius leucopus*, is also one of the more brightly colored species, being a bright chestnut-brown above and white below.

Ovenbirds vary from about 4-9 in. (10-23 cm) in length, but they show considerable diversity in the proportions of the various body parts. This family also shows a great diversity in nesting habits. Some of the ground-living species nest in holes, either natural ones in banks, rocks or stony ground, tunnels dug

Specter of the bird world, the barn owl (seen here with its fluffy chicks).

by the birds themselves (as in the common miner) or in burrows excavated by mammals. Some, such as the thorn-tailed rayadito, *Aphrastura spinicauda*, and the tawny tit-spinetail, *Leptasthenura yanacensis*, nest in tree holes, and a number of species related to these nest in cracks behind the bark peeling off dead trees or the disused, domed nests of other birds. In striking contrast to these, the wrenlike spinetail, *Spartonoica maluroides*, and a few other species build neat cup-shaped nests in bushes and trees. Most remarkable, however, are the huge domed nests built by the true ovenbirds, the name "ovenbird" having originated in the resemblance that these nests bear to an old-fashioned stone oven.

The white-throated cachalote, *Pseudoseisura gutturalis*, builds a gigantic nest. The bird is only 8 in. (20 cm) long, but the cavity inside the nest is big enough to contain a turkey, and it is said that a man can stand on the domed roof of the nest without causing any damage.

OWLS, soft-plumaged, short-tailed, big-headed birds of prey, the nocturnal equivalent of the Falconiformes. They have large eyes directed forward surrounded by facial discs. The bill is hooked and the claws sharp. Owls vary in size from the sparrow-sized pygmy owl to the huge eagle owls. There are about 132 species, and four others have become extinct in historical times. All owls are rather similar, and taxonomists have chosen to take the degree of ear development and asymmetry as a means of classification. Owls are divided into two families: the Tytonidae and Strigidae. The former includes the barn and grass owls, *Tyto*, and the curious bay owls, *Phodilus*. The nine species of barn owls are characterized by big heart-shaped facial discs and long tarsi, and by having the middle claws on each

foot expanded into serrated "combs." *Tyto alba* must be one of the most widespread of birds since it is found in the New World as well as in Africa, Europe, southern Asia and Australia. The majority of the *Tyto* species are found in the Australian New Guinea-Celebes region with isolated species in Madagascar and southern Africa. Two Afro-Asian bay owls are intermediate between the barn owls and the remaining species.

The majority of the Strigidae belong to the subfamily Buboninae, which generally have reduced or flattened facial discs—i.e., are "frown-faced," and less well-developed external ear flaps than those of the other subfamily, the Striginae.

The most widespread of the bubonine genera is *Bubo*, the 12 kinds of eagle owls occurring in the New World and across to the Philippines. The hawk owls, *Ninox*, replace the eagle owls in Australasia, as they do the eared, scops and screech owls, *Otus*, which stretch in a long chain of species from the Americas through Eurasia and Africa to the southwest Pacific islands.

The second subfamily, the Striginae, includes six genera characterized by their well-developed facial discs and sophisticated hearing. Best known are the "earless," chiefly black-eyed, wood owls, *Strix*, most of which have feathered feet and rounded wings. They are chiefly found in temperate woodlands in the New and Old Worlds, excepting Australia and New Zealand. The tawny owl, *Strix aluco*, is found over most of Europe and parts of Asia and north Africa. Six *Asio* owls are all "eared," and the long- and short-eared owls, *A. otus* and *A. flammeus* respectively, are both widely distributed circumpolar species. The second of these, being found even in temperate

The white-faced scops owl *Otus leucotis (left)*, about 9 in. (23 cm) long, is common in the African bush and also penetrates far into the Sahara and other desert areas. The oriental hawk owl *Minox scutulata (right)*, of eastern and southern Asia is 9 in. (23 cm) long. It hunts insects in forest edges and jungle clearings.

South America, must be one of the most southerly of all owls.

The owl family has not undergone so much adaptive radiation as the diurnal hawks. The least pygmy owl, *Glaucidium minutissimum*, of Mexico and the Amazon valley is the smallest at 4½-5 in. (12 cm), and by comparison the 27-in. (67-cm) eagle owl is a giant. Between these two extremes, there is a range of size with smaller species tending toward insectivorous diets and increasingly large prey being taken by the bigger owls. As hunters, owls tend to sit, watch and pounce, or else to drop onto prey when patrolling at a low height. They are not basically interceptors, although some, like the screech owl, *Otus asio*, catch insects with their bills, flycatcher fashion. The hawk owl, *Surnia ulula*, is the nearest equivalent to a small falcon or hawk. It is diurnal, has "hard" plumage, comparatively small eyes, while the facial discs and sense of hearing are not as well developed as in the nocturnal species.

Once the sun has set, illumination is reduced to a very low level, but owls make the best use of the available light. They seem to be capable of resolving as much detail from the night scene as we can by day, having vision estimated to be 35-100 times more sensitive than our own.

The sense of hearing of most owls is also extremely well developed, particularly in the Tytonidae and the strigine owls. This is indicated by the enormous semilunar ear openings and their attendant flaps, which fringe the facial discs. The part of the medulla of the brain concerned with hearing is also well developed, containing 95,000 nerve cells in the barn owl, compared with only 27,000 in a crow twice the weight.

The sensitivity to high frequencies is exploited by the tawny owl and long-eared owl in hunting, because their rodent prey produces noises of these frequencies, either vocally or by scampering across dry leaves and twigs.

Further adaptations to living by night may be found in owls' communication behavior. Animals usually carry their own recognition marks, which have been compared with national flags, but brightly colored patterns are of limited use at night. However, when seen against the night sky, owls have quite distinctive silhouettes, and in 57 species these are enhanced by the presence of ear tufts—groups of long feathers that project from the scalp. These vary enormously in size from species to species.

Owls are also vociferous, and their hoots and screams make up a well-developed language. On the whole, there is a relationship between the size of the owl and the basic frequency of the voice.

Owls are basically hole nesters, laying their eggs in crevices, open cavities, or inside the vacated homes of woodpeckers. A cryptic site like this must be advertised and shown to the mate, so that there can be no doubt about its location. Voice plays an important role in this: Tengmalm's owl calls repeatedly from the nest entrance, and little owls scream a great deal while flying around the nest. Eagle owls, too, have a nest-site display, and it is during these ceremonies, when the sexes are close together, that visual displays may come into play. In particular, movement of the throat when hooting often reveals white patches; thus visual signals synchronize with the vocalization. Like many hole-nesting species, most owls have pale-plumaged chins and throats.

A few owls have taken to nesting on the ground, like the snowy owl and the short-eared owl.

Like the eggs of other hole-nesting birds, those of owls are white.

Except in the pygmy owl, hatching is asynchronous, an adaptation to a varying food supply. By the time the last egg has chipped, the oldest chick may be as much as two weeks old. Should food be plentiful then the parents will be able to bring enough to satisfy the demands of several owlets, but in the event of shortage, the oldest and strongest chick will dominate the supply and the others will die.

OXPECKERS, two aberrant starlings adapted for feeding on hides of large grazing animals. They are found only in Africa, where they have probably evolved in the environment provided by the extensive herds of large grazing animals on the African grasslands.

The yellow-billed oxpecker, *Buphagus africanus*, is widespread over western and central Africa except in the Sahara and the Congo Basin, while the red-billed oxpecker, *B. erythrorhyn-chus*, is confined to the eastern third of Africa from the Red Sea to Natal. In the central regions their distributions overlap and the two species may be found together without apparent interaction.

In their manner of scrambling over the bodies of large animals the oxpeckers have evolved a rather woodpeckerlike manner, and they superficially resemble them in other features. They are about 7-8 in. (18-20 cm) long, with a rather slender body. The longish, pointed tail is well developed, with stiff-shafted tapering feathers which can help to prop the bird as it clings upright to a vertical surface. The legs are fairly short and stout with strong feet equipped with large sharp claws to give a firm grip. The wings are long and pointed. The bill is rather heavy, and broad at the base with an arched culmen to the upper mandible, while the lower mandible is deep, with flattened sides sloping outwards towards the cutting edge and with a sharp angle at the front. Oxpeckers are relatively dull in color. Both species are a dull coffee-brown above, with a duller, paler brown on the head. The belly is light buff, grading into a dull olive-brown on the breast. The yellow-billed oxpecker is a little larger than the other species and has pale buff on the rump and upper tail coverts. It has a bright chrome-yellow bill, the lower mandible being bright red towards the tip. The red-billed oxpecker has a wholly red bill and also has a conspicuous ring of bare yellow

skin around the eye. The bill of the first species is proportionally larger than that of the second, and the lower mandible is very deep and broad toward the base.

Oxpeckers tend to live and feed entirely on the skin of animals such as buffalo, rhinoceros, giraffes, zebras and various antelopes. Where human occupation has introduced domestic cattle, or replaced other grazing animals with these, the birds transfer their attention to the domestic stock. The oxpeckers usually occur in small groups. They shuffle over the bodies of the animals with considerable agility, but prefer to perch woodpecker-fashion, with head uppermost, and in this posture will move sideways or drop back or downwards in short jerky hops. The host animals appear to show little or no reaction while the birds move around the body and neck, up and down the limbs, and even over the face. They even tolerate the birds' attention to open sores. However, where the birds have been introduced to animals without previous experience of them, these show alarm on the first occasion that the oxpeckers settle on them. The principal activity of the birds is feeding on the blood-gorged ticks which they remove from the animals' bodies. They also take bloodsucking flies and remove any blood that shows. Experiments with captive birds have shown that the bills are well adapted to rapidly removing every trace of blood from a surface using a sideways scissoring movement. They also remove scar tissues and, while their attention is mainly beneficial in removing the ticks, they may also do some harm in enlarging the wounds during this cleaning-up process. The

extent to which these birds are dependent on ticks for food is shown by the way in which, when dependent on domestic cattle, oxpeckers disappear from areas where cattle are treated for external parasites.

OYSTER, primitive bivalve mollusks whose structure and life history can best be described with reference to the European flat oyster, including the British "native," *Ostrea edulis*. The body, exposed after removal of the flat upper valve and the mantle flap that forms this, is organized around the central adductor muscle. This consists of two parts, a "catch" muscle of smooth fibers that closes the shell for, if necessary, prolonged periods if exposed to the air or to unfavorable conditions in the sea; and a "quick" muscle of striated fibers responsible for repeated sudden contractions needed for the ejection of sediment and other waste that collects within. These muscles act in opposition to the elastic ligament, that part of the shell which is uncalcified and unites the two valves at the hinge. The heart, displayed after opening the cavity in which it lies above the adductor, consists of two auricles, which receive blood from the gills and pass this into the muscular ventricle. Above this again lies the visceral mass containing the alimentary system, including the stomach and intestine with the style sac in which a gelatinous rod, the crystalline style, rotates and serves both to stir the contents of the stomach and also assist in their digestion by dissolving at the tip to release enzymes.

The pair of extremely voluminous gills, the "beard" of the oyster, lies in a half-circle around the adductor muscle. Each consists of two flaps, or lamellae, made up of great numbers of parallel filaments, attached from place to place and so forming a highly complex latticework covered with different series of highly active ciliary hairs. Those that line the sides of the filaments create a powerful inhalant current of water, which passes through the latticework to emerge above and behind the gills and then leave the shell as a posteriorly directed outflowing current. It carries with it waste products from the anus and from the kidneys.

Particles contained in the inflowing water are retained on the surface of the gills where, under the action of frontal cilia on these surfaces, mixed with mucus, they collect into streams along the base and free margins of each half gill (demibranch), where a further series of ciliary currents convey them to the mouth. The entrance to this is guarded on each side by a pair of fleshy lips or labial palps. The inner, opposed faces of these are ridged and even more elaborately ciliated than are the gills, their function being to control the quantity of mucus-laden material that passes into the mouth. There is also, although it is a little difficult to determine precisely how this is accomplished, some selection for quality. The food of oysters, like that of other bivalves, consists essentially of the largely microscopic plant life of the surrounding seawater, and this is what almost exclusively passes into the alimentary system, which is highly adapted for dealing with such finely divided material of plant origin. What is rejected by the palps, and to some extent by the gills as well, is

The cosmopolitan oyster catcher at its nest where it lays three well-camouflaged eggs. The strong red bill is used for smashing open shellfish.

dropped onto the surface of the mantle lobes, carried by cilia to their margins where it collects forming "pseudofeces" which are expelled from time to time by sudden contractions of the quick muscle. It is the extreme efficiency of the feeding and digestive systems that is largely responsible for the great success of oysters. Where the water is extremely turbid, with high concentrations of suspended silt, this system may break down.

The reproductive organs ripen and spread over the two sides of the body during the summer months. Although at any one time an individual is either male or female, there is an alternation of sex, the speed of succession depending on temperature. Thus, at the northern end of its range, *Ostrea edulis* may function as male and then female in alternate years. In British waters it may often spawn in both capacities in the one year, whereas in southern France oysters may spawn as females with intervening activity as males several times in the same season. Spawning begins when the sea temperature reaches about 59°F (15°C). Only in the male phase are the sexual products, the spermatozoa, released. In countless numbers they are carried out in the exhalant current and almost immediately drawn in by the inflowing current of individuals in the female phase, the eggs of which are fertilized as they issue from the oviducts. Surprisingly, the fertilized eggs then pass through the meshwork of the gills, that is, against the flow created by the cilia lining the sides of the filaments, to be incubated in the mantle cavity, forming first a creamy mass, when the oyster is said to be "white sick" and then, as pigment accumulates in the digestive organs, becoming gray or "black sick." After the initial discharge of sperm by perhaps only a single individual in the male phase, the presence of sperm in the water leads to a chain reaction, other males as well as females spawning so that the whole oyster bed is speedily in a state of intense reproductive activity.

Up to 1 million larvae may be incubated for a period of up to two weeks, depending on temperature. When liberated, during the "swarming" process, they are shelled and have a velum or ciliated sail used for swimming and for the collection of the most minute members of the plant plankton for food. For anything from one to two and a half weeks, these active larvae are members of the animal plankton at the mercy of water movements. Settlement involves change in both habits and structure with the loss of the velum and the temporary appearance of a foot and sense organs, including an eye spot. These enable the larva to find a suitable substrate, when it turns over on the left side and attaches itself by a spot of cement. In general, such settlement occurs in shady places and usually where other oysters have settled, so that oysters tend to aggregate in large numbers. Following this "spat fall," the young assume the appearance of the adult with development of gills and palps. Shell growth proceeds by a series of sudden marginal extensions known as "shoots."

OYSTER CATCHERS, seashore wading birds, about the size of a pigeon. There are usually considered to be four species and 21 subspecies, of which seven are completely black while the remainder have pied plumage. The pied forms, including the European oyster catcher, *Haematopus ostralegus*, are very distinctive, having a black head and upperparts, white belly and rump, white tips to the primary wing feathers and patches on the secondary wing feathers forming a broad wing bar. The feet and legs of all species are stout with three toes, and the color varies from bright orange-pink to a pale pink, depending on age and subspecies. The bill is long, laterally compressed and about 3½ in. (9 cm) long. It is bright orange in the breeding season, but becomes duller during the rest of the year. The iris is bright scarlet. The sexes are very similar, but females are slightly larger and the bill size can be used to sex individuals in the hand.

Immature oyster catchers have brownish tinges to the plumage and also differ from the adult in having darker legs, a black tip to the bill and a brown eye iris. They have a white half-collar on the front of the neck, which is also common to the adult bird outside the breeding season.

Oyster catchers are found on the shores of every continent except Antarctica. Most of them are separated geographically from their neighbors, but in certain areas two species exist together. In areas such as South America, the Falkland Islands, New Zealand and Australia, one of the pair of species is pied, the other black. The pairs also differ in their feeding and breeding habits, a circumstance which helps to maintain the species separation.

They are adapted to feeding on marine shellfishes and worms and are hence restricted to coastal areas in winter. Most of them remain on coasts to breed on sand dunes, among rocks and shingle or on small islands, but some move inland to breed, notably in parts of Europe, USSR and New Zealand.

Most forms are migratory, and in early spring the oyster catchers arrive on their breeding grounds where they set up territories. Each territory, comprising part of a mudflat, mussel bed or sand beach, enables access to a good supply of food and also includes an area that will, subsequently, be used for nesting purposes. Nests are simply scrapes in the ground, crudely lined with small flat stones or pebbles, pieces of seaweed, fragments of shell—any suitable material near at hand. Scraping appears to be of great importance, and a large number of scrapes may be prepared before the final choice is made. The first eggs appear in May, about eight weeks after the birds arrive. A clutch of three is usual, but two or four are not uncommon. The eggs are incubated by both parents for 25-28 days, and the chicks are fed on shellfishes and marine worms, again by both adults, until they fly at about five weeks of age.

Edible oyster *Ostrea edulis*, exposed by a low spring tide. It is famed as food. Many of the original beds have been destroyed by overfishing and pollution.

P-Q

PACA, *Cuniculus paca*, a rodent found in tropical America, from Mexico to southern Brazil. It belongs to the same family as the agoutis and acushis, but is rather larger, up to 2 ft (60 cm) in length. Like its relatives, it has almost no tail, but it is distinguished by a pattern of bold white spots and lines on the back and sides. The skull of the paca is unique among mammals in having the cheek bones grossly enlarged and inflated, with a peculiar honeycombed texture, perhaps acting as resonating chambers. Pacas are vegetarians and sometimes damage crops.

PACARANA, *Dinomys branicki*, a South American rodent superficially resembling the paca, although it is placed in a different family, of which it is the sole member. Like the paca, it is a heavily built animal about 22 in. (56 cm) long, and it has a similar pattern of rows of white spots on a background of dark brown. It differs from the paca, however, in having a conspicuous tail, about 8 in. (20 cm) long. The pacarana is confined to the lower slopes of the Andes from Colombia to Bolivia. It is a poorly known animal and appears to be becoming very rare. Its survival is not helped by its temperament, being unusually docile and slow moving for a rodent.

PADDLEFISHES, primitive fishes related to the sturgeons and sharing with them a skeleton of cartilage and a spiral valve in the intestine. There are only two species of paddlefishes, one from the Yangtse in China (*Polydon gladius*) and one from the Mississippi basin (*P. spathula*). The body, although naked, resembles that of the sturgeons except for the prolonged snout, which is extended into a flat sword. This "paddle" is sensitive and easily damaged and is not used, as might be expected, as a probe or digger in feeding, since these fishes tend to swim with their mouths open and gather their food in that way (small planktonic organisms). The barbels that are characteristic of the mouth of sturgeons are reduced to small protuberances under the paddle. The Mississippi paddlefish rarely grows to more than 150 lb (68 kg). The Chinese paddlefish is reported to reach 20 ft (6 m) in length. Paddlefishes spawn in turbulent waters, and it is only in recent years that their larvae have been identified and studied. Like the sturgeons, the paddlefishes have become rarer owing to pollution of rivers.

PAINTED SNIPE, a family of marsh-frequenting snipelike birds, one species of which is found from Japan to Australia and west to Africa and the other in South America. The first is the painted snipe, *Rostratula benghalensis*, with the Australian population subspecifically differentiated from that in Africa, southern Turkey, India, China and Japan; and the second is the American painted snipe, *Nycticryphes semicollaris*, found from southeastern Brazil to northern Argentina. The two species are rather similar in plumage but differ in a few minor characters such as the amount of webbing between the toes, and the conformation of bill and tail.

In size and shape painted snipe are comparable with true snipe, but have rather shorter legs and a shorter bill, only slightly longer than the head, and a little decurved toward the tip. Like snipe, the tip of the bill is pitted and sensitive to the movement of the worms, insects, crustaceans and mollusks for which the bird probes in mud. The eyes are set at the sides of the head, so that painted snipe have monocular vision through the entire 360° field, and probably have binocular vision in front of and behind the head. Like snipe too, the plumage is cryptically patterned, olive above with creamy lines over the forehead and crown, through the eyes and over the shoulders, and buff spots and bars on the wings and tail. But here the similarity ends, for painted snipe are more richly colored, especially the female, which is a little longer than the male and takes the initiative in courtship although playing a minor role in nesting.

In the Old World painted snipe it is the female who displays toward the male, and she may

Giant panda, adopted by the World Wildlife Fund as a symbol of endangered animal species.

be polyandrous. In a spectacle that few have recorded, the wings are spread and brought forward to display to best advantage to the male their rich pattern of buff circles on an olive ground, and to offset the cinnamon breast and face with its white spectacles. The male builds a simple nest on marshy ground, concealed by tall sedges or a tangle of thorny shrubs, and incubates the clutch of four or more glossy whitish eggs with handsome black or purple blotches. He rears the downy young, while the more aggressive female defends the territory. In the American painted snipe the reversal of the roles of the sexes has not proceeded so far, and the female incubates.

PALOLO, the South Sea islanders' name for a marine bristle worm, *Eunice viridis,* which swarms at the surface of the sea at certain times of the year. The adult worm lives on the sea bottom in muddy sand, among rocks or in crevices in coral. It is superficially like a ragworm, but the head end bears five short tentacles or antennae and two additional tentacles just behind the head. The body is long with many segments and with gills arched over the back. As the worm becomes sexually mature, the rear half of the body becomes loaded with ova or sperm, and it looks then quite different from the front half. At the spawning season the worms leave their shelters on the seabed, the rear half of their body breaks away and swims to the surface, where it splits, shedding its ova or sperm into the water. This mass of eggs is netted by the people of Fiji and Samoa, taken ashore and cooked. The season of spawning is precisely determined, the worms maturing under hormonal influences governed by a combination of factors related to the lunar cycle and to light. This ensures that all the worms that normally live independent lives on the sea bottom release their genital products at the surface at the same time, giving maximum opportunity for the ova to be fertilized.

The average date of the appearance of the great swarms was November 27 in data gathered up to 1921. In fact, swarming occurs seven, eight or nine days after the full moon. Very careful analysis of the timing of the breeding behavior during the day shows that this also is remarkably regular.

Swarmings of other kinds of worms, in the warmer parts of the Atlantic as well as other parts of the Pacific, are sometimes called palolo, although wrongly, as for example Japanese palolo.

PAMPAS DEER, *Ozotoceros bezoarticus,* of which three subspecies are recognized, the typical form *O. b. bezoarticus* in Brazil, *O. b. leucogaster* in Paraguay and adjacent areas of Argentina, Bolivia and Brazil and *O. b. celer* which is restricted to the pampas of Argentina. Standing about 27 in. (69 cm) at the shoulder, the pampas deer is, without doubt, the most elegant of all the South American deer, and in many respects is rather similar to the roe deer, particularly in the formation of the antlers, which normally bear six points. The general color of its short, smooth hair is yellowish-brown, with the insides of the ears and underparts white. The upper hairs on the tail are dark brown to black.

Pampas deer are generally found in small groups of 5-15, except during the fawning season when the doe becomes more solitary. As the name suggests, this is a deer of the open plains and it avoids, as far as possible, woodlands and mountainous country.

PANDA, GIANT, *Ailuropoda melanoleuca,* one of the rarest, most puzzling and yet very popular mammals known today. Since the first live black and white panda reached a zoo in the western hemisphere 32 years ago, only 14 others have been kept in captivity outside China. The giant panda's inaccessible and restricted range is no doubt one of the factors responsible. It lives in a most inhospitable environment: densely vegetated mountains that are either covered in snow or shrouded in mist, while deep gorges and ravines form such efficient barriers that it takes weeks for a man to contour them. The beishung, as the giant panda is known locally, lives in the bamboo and rhododendron forests of Yunnan and Szechwan in an area approximately 500 miles (800 km) wide. Because of these formidable odds, the panda was discovered only in 1869 by Père David, a French zoologist-explorer. Ever since then, taxonomists have been at a loss to decide its correct classification, whether it belongs to the bear family, Ursidae, or the raccoon family, Procyonidae, which includes kinkajous, coatis, olingos and cacomistles. The giant panda weighs 160-400 lb (75-180 kg) and measures 44.5 to 63 in. (1.3-1.6 m) including the diminutive tail. The fur is white or yellowish, contrasting with black limbs, shoulders, ears and eye patches.

Bamboo forms the major part of its diet in the wild. Young shoots in particular are eaten, whereas the hardened stems are broken off and discarded. While foraging, the giant panda sits on its haunches, legs spread, and holds the delicate branch with one fore paw, using a remarkably strong "pseudothumb" to grip it. This unusual sixth digit is in reality a modified wrist bone under the pad of the foot that has evolved to a thumblike size and flexibility, permitting precise plucking movements. Giant panda cubs are born after a short 120-140-day gestation, weighing 5 oz (142 gm), approximately $\frac{1}{800}$th of the adult weight.

PANDA, LESSER, *Ailurus fulgens,* also known as the red panda, discovered by Hardwicke in 1821, was described by Frédéric Cuvier only four years later, in 1825. It remained the only panda known to science for nearly 50 years. Like the giant panda it favors remote bamboo forests between 7,000 and 12,000 ft (2,300-4,000 m), but it is more widespread and may be found from Yunnan and Szechwan southward to Sikkim and westward to Nepal. The "fire fox," as it is aptly called in China, has an elongated body covered in soft, dense fur of a brilliant chestnut red. The tail, 2 ft (60 cm) long, is bushy, faintly ringed with creamy orange and russet. In contrast with the body, the triangular face and large pointed ears are usually a light fawn color. A dark stripe circles each eye and runs down to the chin. The lesser panda's broad paws with hairy soles and semiretractile claws are well adapted to the rigors of its environment, being useful in both climbing trees and gripping icy boulders. Although heavy limbed in appearance, this medium-sized mammal weighs only 7-12 lb (3-5 kg) and measures 32-44 in. (80-112 cm) including the tail. The "red fox" is crepuscular and arboreal, resting during the daylight hours either curled up in the fork of a tree or sprawled along a branch. Descending to browse in the bamboo groves, it eats young shoots, fruit and, should it stumble across them, an occasional

Lesser panda, or red panda, was the first of two species to be given the Nepalese name "panda." The latter name was later also applied to the giant panda of China.

A golden-headed South American parakeet, *Aratinga* sp., of the kind sometimes referred to as conures.

egg or nestling. This small panda feeds in the same manner as its larger relative, plucking bamboo leaves with either fore paw, which has a special callosity on the pad allowing certain prehensile movements.

PANGOLINS, mammals with the body covered with scales that overlap in an imbricated manner, like slates on a roof. Only the abdomen, the inner sides of the extremities, the throat and part of the head are free from scales. These are covered with hair. The scales are outgrowths of the true skin and are coated with epidermis. The head is conical, the eyes are small and the outer ear is also small. Since anteaters have no teeth, they feed on termites and ants by means of the very long, wormlike extensile tongue. They have short, sturdy legs and the front paws have claws. The terrestrial types have a strong, muscular tail, approximately the same length as the body, which is completely scaly. In arboreal types the slender tail is longer than the body with a bare patch at the tip of the tail (about the size of a finger) and is used as a prehensile limb when climbing. The seven living species are as follows. The Indian pangolin, *Manis crassicaudata*, is found in India and Ceylon, with a head and body length of 2 ft (60 cm) and a tail 18-20 in. (45-50 cm) long, the scales being pale yellowish-brown, and the skin brownish. The Chinese pangolin, *M. pentadactyla*, found on Taiwan and in southern China, with a head and body length of 20-24 in. (50-60 cm) and a tail 12-16 in. (30-40 cm) long, the scales being blackish brown and the skin grayish white. The Malayan pangolin, *M. javanica*, inhabits the Malay peninsula, Burma, Indochina, Laos and the

islands of Sumatra, Borneo, Java and others, its head and body length being 20-24 in. (50-60 cm) with a tail 20-32 in. (50-80 cm) long. The scales are amber to blackish brown, and the skin is whitish. These three Asiatic types differ, among other things, from the African species by the presence of hair between the scales. On the African continent four types are distinguished. The small-scaled tree pangolin, *M. tricuspis*, of the tropical rain forests from Sierra Leone to the central African trough fault, has a head and body length of 14-18 in. (35-45 cm) and a tail 16-20 in. (40-50 cm) long. The scales are brownish gray to dark brown and the skin white. The long-tailed pangolin, *M. tetradactyla*, also inhabits the rain-forest areas of Africa, but is rarer than the small-scaled tree pangolin. Its head and body length is 12-14 in. (30-35 cm), and its tail 24-28 in. (60-70 cm) long. The scales are dark brown with yellowish edges and the skin dark brown to blackish. The giant pangolin, *M. gigantea*, is also an inhabitant of the tropical rain forests. Its head and body length is 30-32 in. (75-80 cm) and its tail 22-26 in. (55-65 cm) long. The scales are grayish brown and the skin whitish. The Temminck's pangolin, *M. temmincki*, lives in the east African savannas, from Ethiopia to the Cape. Its head and body length is 20 in. (50 cm) and the tail 14 in. (35 cm) long. The scales are dark brown and the skin light with dark hairs.

PAPER NAUTILUS, *Argonauta argo*, a cephalopod mollusk related to the common octopus, which it closely resembles in its general structure. It has a rosette of eight sucker-bearing arms around a beaked mouth,

a rather large brain, which is typical of all modern cephalopods, prominent eyes and a baglike body containing the digestive and reproductive organs. Below the main part of the body there is a pouch—the mantle cavity—containing the gills and the openings of the reproductive and digestive systems and the kidney ducts. The animal lives in the surface waters of the Mediterranean and other warm seas. Exceptionally large specimens may have a shell of 10 in. (25 cm) or more in diameter. It is a very remarkable mollusk in a number of ways. Thus, it has a shell that is not a shell in the normal molluskan sense. Instead of being a rigid calcareous structure attached to the rest of the animal by connective tissue and muscles, it is thin, almost papery as its name implies, and the octopod inside can crawl out and leave at will. This papery false shell is secreted by glands on the edge of the webs bordering the first pair of arms, and appears to act both as protection (the animal can withdraw inside) and as a float (since nearly all specimens captured seem to have air trapped in the upper part of the shell). The unchambered shell also serves as a cavity in which to hang the eggs. *Argonauta*, like other octopuses, broods its eggs until they hatch.

This last use of the shell introduces a second oddity of the paper nautilus, because the shelled animals are all females. The males of the species are minute—1-2 in. (3-5 cm) long—shell-less and at first sight a quite different kind of animal. They are clearly octopods and characteristically bear an enormously enlarged third right arm that acts as a receptacle for the sperm packets produced by the animal. This arm is inserted into the mantle cavity of a female, where it breaks off. The male then swims away to regenerate another arm.

PARADISE FISH, *Macropodus opercularis*, one of the labyrinthfishes found in China and southeast Asia. It grows to about 4 in. (10 cm) long and has a browny-blue body with a dozen vertical, thin orange-red stripes. The tail fin is also red, and there is a red-edged black spot on the gill cover. Samuel Pepys wrote in his famous diary, toward the end of the 17th century, of "a prettily marked fish living in a glass of water" that was then a recent import. This has often been cited as the first mention of a goldfish in England, but it has been suggested that it was a paradise fish because of the marks. The Chinese had been breeding these fishes for a long time, and, as they are fairly hardy, this interpretation is feasible.

PARADOXICAL FROG, *Pseudis paradoxa*, a remarkable frog, of Trinidad and the Amazon basin, in which the tadpole, which is 7.5 cm long, is more than three times the length of the adult.

PARAKEETS, a loose name given to small or medium-sized parrots, usually those with long tails. There is no natural group of parakeets, the name having been given to many species that are not closely related to one another. Examples from different genera are: the now extinct Carolina parakeet, *Conuropsis carolinensis*, of North America; the green parakeet, *Myriopsitta monachus*, of Argentina; the rose-ringed parakeet, *Psittacula krameri*, of the Old World tropics; the ground parakeets, *Pezoporus*, of Australia; and the crested parakeets,.

Cyanoramphus, of New Zealand. To add to the confusion, the familiar budgerigar is sometimes called a grass parakeet, and the diminutive lorikeets, which have quite short tails, are alternatively called hanging parakeets.

Here only the members of the genus *Psittacula* are described. All have long, finely pointed tails. Most *Psittacula* parakeets have a distinctive and colorful pattern on the head, which relieves the overall green of the body and wings. One of the most beautiful is the Alexandrine parakeet, *Psittacula eupatria*. The head of the male is banded with red, purple, black and turquoise. The body and wings are bright green with a red flash on each shoulder and the long tail is azure blue.

Indo-Malaya is the home of most of the

Hyacinth macaw or blue ara, *Anodorhynchus hyacinthus*, a Brazilian parrot.

Psittacula species, but one of them, the rose-ringed parakeet, *Psittacula krameri*, has an extremely wide distribution: from Senegal in the far west of Africa to Indochina. Like most of its relatives, the rose-ringed parakeet lives in semiarid country where, however, it keeps largely to the trees. Only a few species inhabit heavily forested regions, and these seem to prefer the more open parts. The long-tailed parakeet, *Psittacula longicauda*, of the Malayan peninsula and the Indonesian archipelago is a common bird of the lowlands, which were once completely clothed in jungle.

For much of the year parakeets are gregarious, flying around in noisy flocks, but when the breeding season begins they pair off and nest singly or in small colonies. Mating is preceded by elaborate and comical courtship displays in which the male struts along a branch toward the female and the pair engage in rhythmic swaying and "necking" exercises. Like most other parrots, parakeets are hole nesters. The natural site is a hole in a dead tree or branch, but where parakeets live in towns and villages, as in India, temples, bridges and other man-made constructions provide many good nest sites. Two or three dull white eggs form the usual clutch and are laid at the bottom of the cavity with scant or no nesting material.

The natural food of parakeets is largely fruits, nuts, buds and flowers. A tree bearing ripe fruit may attract hundreds of birds, yet the observer may have great difficulty in seeing a handful of them, so well are they camouflaged among the foliage.

PARAMECIUM, the best known ciliate, formerly called the slipper animalcule, and one that has been studied intensively not only at the school and university level but also in the wider fields of cell and molecular biology. Eight species of *Paramecium* are known, but some authorities recognize more. All are free-living in fresh or stagnant water, and one occurs in brackish water. The ciliates are elongate and round in cross section. To one side is the conspicuous mouth lying in a deep groove called an oral groove. The ciliate is covered with a uniform layer of cilia, and those in the region of the mouth are essentially no different from those covering the rest of the body, although they beat more strongly. The ciliate swims by revolving about its axis and does not feed while swimming. When stationary, the ciliate creates a vortex by beating the cilia in the oral region, and this actively draws fine particles into the oral groove and thence into the mouth. *Paramecium* feeds on bacteria and particles of plant material.

PARROTFISHES, a family of colorful tropical marine fishes related to the wrasses. Parrotfishes are moderately deep-bodied and have a fairly long dorsal fin but shorter anal fin. The fin spines are rather weak. The teeth in the jaws are fused to form a "beak," and this, together with their very bright colors, earns them their common name. Parrotfishes feed on coral, biting off pieces and grinding them very thoroughly with the pharyngeal teeth in the throat. This is then swallowed, the food extracted and the calcareous matter excreted, often in regular places where a small pile of coral debris accumulates. Parrotfishes are, in fact, responsible for most of the erosion that

occurs on reefs. Many species show strong homing instincts, returning to the same spot after foraging for food, and schools of parrotfishes have been seen regularly following the same route through the coral landscape. At night, some species, such as the rainbow parrotfish, *Pseudoscarus guacamaia*, secrete a tent of mucus around themselves. This may take up to half an hour to produce and as long to break out of in the morning. This mucous envelope would seem to be a protective device that perhaps prevents the odor of the fish from reaching predators.

Parrotfishes usually reach 2-3 ft (60-90 cm) in length, but a Tahitian species has been reported to attain 12 ft (3.6 m). In some species, such as *Scarus coeruleus* of the Atlantic, the larger individuals develop a curious hump on the forehead. The extraordinarily bright colors of the parrotfishes (often vivid patches of varying green and red) have made identification of species difficult. The young fishes are often quite different from the adults, and the latter may show striking sexual differences, sometimes greatly complicated by sex reversal with or without the appropriate change in color. In the surf parrotfish, *Scarus fasciatus*, of the Atlantic this sexual dichromatism is so pronounced that the male and female can hardly be recognized as the same species. The species known as *S. taeniopterus* and *S. croicensis* were long thought to be distinct species until it was realized that only males of the first had ever been found and only females of the second.

PARROTS, a family of about 320 species, widely distributed in the tropics and southern temperate regions. Parrots are 4-50 in. (10-130 cm) in total length and are mostly brightly colored, although there are some drab gray, brown, green or black species. The family is sharply set apart from other groups of birds, although they are probably more closely related to the pigeons than to any other group. The hooked bill, bulging cere at the base of the bill, more or less rounded wings, short legs, zygodactyl feet and other characters, as well as general appearances enable all parrots to be easily recognized as such.

The family is divided into four subfamilies (more in older classifications) of very unequal size. The first subfamily, the Strigopinae, includes only the peculiar kakapo or owl parrot, *Strigops habroptilus*, of New Zealand. This is a large flightless bird which lives in burrows among tree roots during the day, emerging at night to feed on the leaves, flowers and berries of low growing plants.

The 17 or so species of cockatoo in the subfamily Cacatuinae are distributed from Australia to the Philippines. They are mainly medium-sized to large with rather short tails.

Over 200 species of parrot are classified in the subfamily Psittacinae. The members of this large group are 4½-40 in. (11-100 cm) or more long, and are found in the Americas, Africa, Asia and Australasia, as well as on many islands in the Pacific and Indian Oceans and the Caribbean. All American parrots belong to this subfamily, and the total of approximately 130 species is greater than the total number of parrot species inhabiting any other continent. American parrots vary in size from the huge, long-tailed macaws to the sparrow-sized Andean parakeets of the genera *Bolborhynchus*, *Amoropsittaca* and *Forpus*. Many of the larger American parrots, for example the genera *Ara*, *Anodorhynchus* and *Amazona*, including the macaws and Amazons, feed mainly on fruit and seeds from forest trees. Some of the smaller species eat fruit pulp, small seeds and nectar and pollen collected from flowering trees, as for example trees of the genera *Touit* and *Brotogeris*; and some species inhabit open country in temperate South America or on the Andean plateau, feeding mainly on the seeds of low-growing plants, for example the genera *Amoropsittaca* and *Bolborhynchus*.

So far as is known all American parrots nest in holes in trees but sometimes in holes in cacti, ants' nests, rocks or buildings. The hooded parakeet, *Myiopsitta monachus*, is unique among parrots in building huge com-

Green-cheeked Amazon parrot, *Amazona viridigenalis*, of northeastern Mexico.

munal nests of sticks in the branches of trees, each pair having a separate nest chamber within the main structure.

The extinct Carolina parakeet, *Conuropsis carolinensis*, was the only parrot with a range extending into the northern temperate regions of North America. It was formerly very abundant in the Gulf and middle United States but was wiped out by human persecution. Several macaws and Amazon parrots on the islands in the Caribbean have also become extinct in the past few hundred years. Some of these are known only from skins and other identifiable remains, while others are known only from contemporary descriptions and paintings, so that it is not clear how many different species there were. A trade in young parrots for the pet industries of Europe and North America has flourished for many years in the West Indies, so that other island species of parrot may well be threatened with extinction if the trade continues unchecked.

Africa has surprisingly few species of parrot, perhaps because of the aridity of large areas and the relative scarcity of fruiting forest trees in the wetter areas. The African gray

parrot, *Psittacus erithacus*, a medium-sized, short-tailed species, is found throughout tropical Africa, and the lovebirds *Agapornis* are also peculiar here.

Many species of the subfamily Psittacinae are found in southern Asia and Australia. The eclectus parrot, *Eclectus roratus*, and the New Guinea hawk-headed parrot, *Psittrichas fulgidus*, are large, forest-inhabiting species found in the Papuan region, both of which eat mainly fruit from forest trees.

The fourth subfamily Loriinae includes the lories and lorikeets, the pygmy parrots, the fig parrots, the kaka, *Nestor meridionalis*, and kea, *N. notabilis*, of New Zealand and the Australian broad-tailed or platycercine parrots. Various of these groups have been considered as separate subfamilies in earlier classifications. The six species of pygmy parrot, *Micropsitta*, are 4-4½ in. (10-11 cm) long, so that they are

the smallest of all parrots. They are found from New Guinea to the Bismarck archipelago and the Solomon Islands. All six are forest birds.

The pygmy parrots are probably unique among birds in feeding on fungi. They scrape slimelike fungi from decaying wood with their stumpy, weak bills and suck it into their throats through a tubelike tongue. Termites, lichen and seeds are also eaten at times, although little is known of the diet of several species.

The five species of fig parrot, *Opopsitta* and *Psittaculirostris*, are small, stout parrots with brilliantly colored plumage, brilliant shades of red, orange, yellow, green, blue, black and white being arranged in sharply contrasting patterns in the different species. These heavy-billed parrots are forest dwellers which live mainly on the fruit and seeds of forest trees.

The Australian broad-tailed or platycercine parrots are a group of about 31 species, most of them confined to the Australian mainland, but some occurring in New Zealand, New Caledonia and even the Society Islands. The budgerigar, *Melopsittacus undulatus*, a familiar cage bird, is the smallest of the platycercine

parrots and it is found only in arid areas of Australia.

PARTRIDGES, game birds distributed throughout Europe, Asia and Africa. They include the little stone partridge, *Ptilopachus petrosus*, which is found in the rocky parts of west and east Africa, and the even smaller "bush quails," *Perdicula*, of India, which are really dwarf partridges. The see-see partridges, *Ammoperdix*, are sandy colored desert birds to be found in the Middle East and western India.

Most well known of all is the gray or common partridge, *Perdix perdix*, 12 in. (30 cm) long, with a chestnut horseshoe on its breast. This famous game bird is monogamous, and coveys usually consist of one pair with their offspring from the previous summer. The coveys break up early in the new year, when the birds pair off and take up their breeding territories. Unlike many game birds, the cock partridge is a devoted parent and helps the hen in the care of the chicks.

The gray partridge is also found in Europe, east to the Caspian area, but is replaced in Turkestan, Mongolia and northern China by the closely related *Perdix barbata*, and in central Asia and the Himalayas at high altitudes by *Perdix hodgsoniae*. The partridge population of the British Isles and other areas has declined in recent years because of changing agricultural methods, including the use of pesticides and the grubbing up and removal of hedgerows, which provided shelter and nesting cover.

The red-legged partridge, *Alectoris rufa*, is a more colorful bird than the gray partridge. It originated on the continent of Europe but has been introduced as a game bird to the British Isles. It is more at home in wooded and rocky country as well as in mountains. The chukar, *Alectoris graeca*, is very similar to the typical red-legged partridge in appearance. It is found from the Alps eastward to Manchuria, while its near ally, *A. barbara*, inhabits north Africa, and the giant *A. melanocephala* lives in Arabia.

PATAGONIAN HARE, or Patagonian cavy, two species of which, *Dolichotis patagona* and *D. salinicola*, are found in Argentina and Patagonia. These short-tailed rodents look like long-legged rabbits or hares, the hind limbs being long each with three toes bearing hooflike claws. *D. patagona* is much larger than *D. salinicola*, having a head and body length of 27-29½ in. (69-75 cm) with large individuals weighing 20-35 lb (9-16 kg). The coat is dense, the upperparts grayish and the underparts whitish, with some yellow-brown on the legs and feet. Patagonian hares are diurnal, sleeping during the night in burrows, where their 2-5 young are also born. They are fond of basking in the sun and feed on any available vegetation.

PEA CRABS, so called for their smooth, rounded bodies and because the males of many species are no larger than a pea. They readily form associations with other animals, and the best-known species are those that live inside the mantle cavities of bivalve mollusks. The female of such a species may grow to such a size that she can no longer escape from the interior of the mollusk shell, but the males

Common mussel gaping, exposing a pea crab living within its mantle cavity.

has a more easterly distribution and occurs in Burma, Thailand, Indochina, the Malay peninsula and Java.

The Congo peacock belongs to the genus *Afropavo* and is the only pheasant originating outside Asia, being found in the rain forests of the east-central Congo basin in Africa.

PEARLY NAUTILUS, a shelled member of the Cephalopoda, the group of mollusks that includes octopuses, squids and cuttlefishes. It is the sole surviving representative of a group of mollusks that dominated the seas in Paleozoic and Mesozoic times. There are six living species, all placed in the genus *Nautilus*, the best known being *N. pompilius*. All are found in deep water around coasts in the southwestern Pacific and Indian oceans.

The shell is coiled, chambered and 6-8 in. (15-20 cm) in diameter in adult specimens. The animal lives in the last chamber, further partitions being laid down behind it as it grows. The partitions are never quite complete, and a long thin prolongation of the abdomen connects through to earlier sections of the shell. This is known as the siphuncle and is richly supplied with blood vessels, and it has the capacity of absorbing salt from the fluid—mostly seawater—that is left in the abandoned chambers as the animal grows. Removal of the salt creates an osmotic gradient between the blood of the animal and the water in the shell chambers, with the result that much of the water is drawn out, to be replaced by gas coming out of solution in the bloodstream. As a result, the partially gas-filled shell counterbalances the weight of the animal inhabiting it, so that this otherwise rather bulky creature is able to float delicately just above the bottom, propelling itself with jets from a funnel slung below the main part of the body.

Very little is known about the habits of *Nautilus*. In aquaria, they are mainly active at night. The eyes are large but simple, lacking the lens and cornea of more recent cephalopods, and although the animals are plainly able to see a little (they will, for example, approach the light of an electric torch shone through the walls of their aquarium), they do not appear to recognize their surroundings or their food by sight. Their chemical sense, in contrast, is obviously very acute. The pearly nautilus appears to be a scavenger, and if food, such as dead fish or crustaceans, is placed in a tank with a *Nautilus*, the cephalopod extends its many tentacles and circles to search the area. Having found the source of the smell, the pearly nautilus grasps the food by the cephalic tentacles, which are very mobile and extensible, and transfers it to the mouth to be bitten off in chunks by the massive beak. The arms are numerous, and each consists of a leathery sheath and a slender tentacle that can extend far beyond the sheath. These bear no hooks or suckers. The top of the head is developed into a tough shield that closes the mouth of the shell when it has been disturbed.

PECCARY, a piglike mammal with long slender legs. The collared peccary, *Pecari tajacu*, is grayish to black, paler underneath than on the back, and with white annulations on the bristles. An erectile mane extends from the head to the scent gland on the rump at the dorsal midline. An incomplete whitish collar

remain small and so can enter the shells and mate with the trapped females.

Pinnotheres pisum, which lives inside the mantle cavities of the mussels *Mytilus* and *Modiolus*, obtains its food from that collected by the host. The crab positions itself at the edge of the gills, where a large ciliary tract conveys food toward the bivalve's mouth. The pincers of the crab are hairy, and are used to remove mucus and entrapped food from the ciliary current. The pincers are then wiped across the mouthparts, and these work the food backward into the mouth.

Other pea crabs have somewhat bizarre associations with other animals. One species lives inside the rectum of a sea urchin, and another lives inside the respiratory trees of sea cucumbers.

PEAFOWL, relatives of pheasants, the males having a long train of up to 150 feathers formed from the tail coverts, which can be erected to make a showy fan. For centuries, peacocks have figured in mythology and folklore. They have been admired for their beauty and hunted for their flesh, and in more recent times they have become a popular and graceful addition to parks and gardens. There are two species, the Indian peafowl, *Pavo cristatus*, which comes from Ceylon and India, and the green peafowl, *Pavo muticus*, which

Peacock displaying to a peahen.

crosses the neck and extends obliquely upward and backward. The young are reddish brown with a lighter collar and usually a blackish dorsal band. There is a puffy scent gland, devoid of hair, and about 4 in. (10 cm) in front of the base of the abortive tail, which is only about 1-1.5 in. (2.5-3.7 cm) long. Males and females are of a similar size, the length of the head and body of the male averaging 36.9 in. (93.7 cm) and of the female 36.1 in. (91.7 cm). The height at shoulders of 20 males measured averaged 19.8 in. (50.3 cm). For females this height was 20.5 in. (52 cm). The adult male weighs 44 lb (20 kg) and the female 45 lb (20.6 kg).

Collared peccaries range from southwestern U.S.A. as far south as Rio Platte in Argentina. Some breeding takes place throughout the year, but at the northern edge of the range most of the young are born in July and August. The gestation period is 142-148 days, and there are usually two in each litter. The young are precocial and weigh about 1 lb (0.45 kg).

They inhabit mountainous canyons and brushy thickets and are vegetarian, eating fruits, green plants, acorns, cactus pads, cactus fruit and roots. Peccaries are highly social animals, traveling in bands of up to 20 individuals. Mothers with their young follow the herd, which remains in a home-range area of about $\frac{1}{2}$ sq mile (about 1.3 km²). When alarmed, the animals exude musk from their dorsal gland, causing a strong odor. When friendly animals meet, each rubs its cheek over the scent gland of the other. They also mark rocks, trees and other landmarks within their home range.

The only other living member of the family Tayassuidae is the white-lipped peccary, *Tayassu pecari*, which is larger than the collared peccary, having a head and body length of 43-47 in. (100-120 cm). This species has long bristly hair and a scent gland on its back like the collared peccary, but unlike the latter its cheeks, nose and lips are white. It inhabits dense forests of southern Mexico, Central America and South America. It is a highly social animal, traveling in large bands and living off nuts, fruits, green vegetation and roots. Little is known of its breeding. It has a reputation for ferocity when cornered or wounded.

PELICANS, large or very large aquatic birds highly adapted for swimming but ungainly on land. They have long bills provided with an expansible pouch attached to the flexible lower mandible. The short powerful legs are set far back on the body, ideally suited to rapid swimming but making the traversing of land difficult. Oddly enough, all are magnificent fliers, capable of sustained soaring flight over great distances. The whole body, beneath the skin, and even in the bones, is permeated with air spaces, probably assisting buoyant flight. Pelicans weigh 10-25 lb (4½-11 kg), the larger species being among the largest flying birds, with wing spans of 9 ft (2.7 m). They are usually white or mainly white, with areas of gray, brown or black plumage. The American brown pelican, *Pelecanus occidentalis*, is mainly brown. In the breeding season the colors of areas of bare skin, beak, pouch and legs, are intensified, and several species grow crests at

The collared peccary, of tropical and subtropical America, is so named for its white neck band.

Brown pelican preening.

this time. In the American white pelican, *P. erythrorhynchos*, a strange hornlike growth develops on the bill, and in the great white pelican, *P. onocrotalus*, the forehead becomes swollen. Males are considerably larger, and can also be distinguished from the smaller females by the color of bare skin and other features, such as the length of the crest. Despite their somewhat grotesque shape, pelicans in breeding dress are often beautifully colored; in *P. onocrotalus*, for instance, a rosy flush derived from a preen gland above the tail tinges all the silky-white plumage. Brown and yellow colors in this plumage are derived from staining in the waters in which the pelicans swim.

All seven species of pelicans belong to one genus, *Pelecanus*, which divides conveniently into two, perhaps three, superspecies or groups. In the first group there are four very large species, the American white pelican, the great white pelican, the Dalmatian pelican, *P. crispus*, and the Australian pelican, *P. conspicillatus*. All these breed in large colonies on the ground, roost on the ground and rarely perch in trees or bushes. The second group consists of three smaller species, the brown pelican, the African pink-backed pelican, *P. rufescens*, and the Asian spotted-billed pelican, *P. philippensis*. These all breed in smaller, looser colonies in trees, but occasionally on the ground. They readily perch and roost in trees. The brown pelican is rather different from the other two, and should perhaps be placed in a superspecies or group of its own, because of its more specialized fishing habits. Contrary to general belief, the bill is not used for storage or holding fish, but is simply a catching apparatus, resembling a scoop net in function.

All pelicans breed gregariously, in colonies varying from about 50 to tens of thousands. The largest colonies reported in recent times have been of the great white pelican in southern

The Gentoo penguin lives on the South Shetlands and other Antarctic islands.

Tanzania, where up to 40,000 pairs have been sighted.

Newly hatched young are ugly, being naked and pink at first, turning black or gray, then growing a coat of gray or blackish down. In ground-breeding species they collect in groups or "pods" when they can walk, after about three weeks. Even so, each parent recognizes and feeds only its own young in these pods. Feeding is by regurgitation at all times. The newly hatched young is fed small quantities of liquid matter dribbled down the parent's bill, and the young of the great white pelican peck feebly at the red nail on the end of the bird's bill at this time—perhaps the origin of the medieval belief that the pelican fed her young on drops of blood from her own breast. Larger

chicks reach into the parent's bill and gullet to obtain food, and feathered chicks recognize their own parent and run to meet it on arrival. A parent feeding a large chick must face a violent struggle, the chick's head being thrust far down the parental gullet, and the chick struggling and gyrating in its efforts to obtain the food.

PENGUINS, the most highly specialized of all aquatic birds, with 17 species restricted to the southern hemisphere.

The distribution of penguins is by no means limited to the Antarctic. In fact, only two of the 17 species actually cross the Antarctic Circle. Five more nest in regions with a varying ice cover; six species belong to the south-temperate zone; and four species are tropical or subtropical. The distribution of some of the species is circumpolar, so that around the Cape Horn-Antarctic region can be found seven species; in the south Indian Ocean four species; and around the New Zealand region eight species. This distribution has been achieved by dispersal from the Antarctic region of origin by means of ocean currents, the birds either swimming or being carried by ice floes. Thus they are now found not only in the Antarctic and subantarctic but also around the south coasts of Australia, Africa, and South America, and north along the west coast of South America as far as the Galápagos on the equator.

The size of penguins seems to have become reduced the farther they have spread from their original center of dispersal. Thus the largest species is the emperor of Antarctica and the smallest the Galápagos penguin, *Spheniscus mendiculus*, and the little blue penguin of Australia. The Peruvian penguin, *Spheniscus humboldti*, which extends northward along the coast of Peru, seems to be an exception, for it is larger than the little blue penguin but extends farther north. However, it is the exception that proves the rule, for the major factor influencing penguin size has probably been heat loss—the larger the animal the smaller its surface area relative to bulk and the smaller its heat loss. The Peruvian penguin seems to have pushed northward up the west coast of South America together with the cold Humboldt Current. This current even influences the Galápagos and must have been the means of transport of the Galápagos penguin. This species, with a very small population of possibly under 2,000 individuals, would probably not be able to survive on the Galápagos if it were not for this current.

Penguins nest on the ground, usually on the surface, but some, such as the little blue penguin, nest in burrows or crevices. An exception is the emperor, which breeds on floating sea ice and is, therefore, the only bird never to touch land. The nest is usually made of pebbles, grass, sticks or bones, depending on what is available. The emperor and king penguin make no nest but carry their single eggs on their feet, covered with a flap of skin. Most species nest in colonies or "rookeries," sometimes with hundreds of thousands of nests packed together, each sitting bird being just out of pecking range of its neighbors. The

Adélie penguins on Michelson Island.

White pelicans at the nest.

The European perch has given its name to a large order of spiny-finned fishes.

clutch consists of two, sometimes three, eggs, except in the king and emperor penguins, which lay a single egg.

Penguins are long-lived, and they are usually faithful to mate and nest site throughout life. The more southerly species undergo a long fast during the incubation period. The Adélie penguin, *Pygoscelis adeliae*, for example, fasts for 2½-3½ weeks while establishing its territory and nest, and the male then continues to fast for another 2-2½ weeks while incubating, until relieved by the female, who has been to sea to feed. In the emperor penguin, the male incubates and fasts for 64 days. Both parents collect food for the chicks, and even when half-grown chicks have collected in "crèches," the parents still pick out and feed only their own chicks. Their food consists of crustaceans, such as krill, fish and squid.

Most rookeries are near the sea, but rookeries of the emperor penguin and Adélie penguin may be many miles from the open sea when the birds take up territory, for unbroken sea ice extends a considerable distance from the land. As the ice breaks away later in the season it would be unsafe for the birds to nest on it. Thus, Adélies may travel 30-40 miles (50-65 km) or more over the ice to reach their nests, and the emperor perhaps twice as much. The tracks of an emperor have been found 186 miles (300 km) from open water, and those of a chinstrap penguin, *Pygoscelis antarctica*, 250 miles (400 km) inland. It is possible that these birds wandered off course during overcast weather, for it is known that penguins can navigate using celestial clues, as other birds can.

The breeding habits of the Adélie penguin have been studied in the greatest detail and can be considered as typical of penguins as a whole. The penguins arrive at the rookery in spring,

sometimes traveling over the frozen sea from their winter feeding grounds farther north. They immediately occupy their nest sites and commence courtship. Nest building starts as pebbles are exposed by the melting of the snow. Both sexes assist in building the nest, collecting the pebbles or small stones from near the rookery or stealing them from neighboring nests. After laying two eggs the female goes back to the sea to feed, leaving the male to incubate. By the time he is relieved he will have lost about 40% of his body weight, but he returns to the nest just before the eggs hatch. At first the chicks are brooded, but they later gather in crèches of 100 or so.

The problem facing the two large species, the king and emperor penguins, is to raise a very large chick within the short space of the Antarctic summer. The emperor penguin, *Aptenodytes forsteri*, has overcome this problem by laying during the winter so that the egg hatches in early spring. The male incubates the egg for 64 days and feeds the young chick on a secretion from its crop. The king penguin, *A. patagonica*, has a different solution. Eggs are laid in the summer, and the chicks lay down a large supply of fat, then stay in the nest throughout the winter, when they lose about half their weight. The following spring brings plentiful food, and their development is completed. It is then too late for their parents to lay again that season, so king penguins lay only in alternate years.

Another species that differs from the more typical penguins in a number of ways is the yellow-eyed penguin, *Megadyptes antipodes*, which breeds only in New Zealand and neighboring islands. First, it is sedentary and may be found in the breeding area at any time of the year; second, it does not breed in large colonies, and the nests are somewhat

isolated from one another; and third, it breeds in forest areas.

One of the problems that penguins have to face is the molt, which in these birds is unusual in that sizable patches of feathers come away in one piece, rather as in a reptile sloughing its skin. During this period, which may last for a month or more, the birds are very lethargic. As they are no longer waterproof, they cannot feed, and they therefore may lose 40% of their body weight.

PERCHES, a family of spiny-finned fishes containing the darters, the pike perches and the true perches.

The European perch, *Perca fluviatilis*, is common throughout Europe as far as the Soviet Union. It is a deep-bodied fish with two barely separated dorsal fins, the first spiny with a prominent black spot near the rear. The back is olive brown, the flanks yellowish (often brassy) with about six, dark vertical bars, and the belly white. The tail and the lower fins are often tinged with red. The perch prefers slow and sluggish waters but can live almost everywhere. When in fairly fast streams, these fishes form small shoals in the eddies. Perches are predators, feeding throughout life on small fishes and invertebrates. Beloved of anglers, because of the ease with which they take the bait, perches have been caught up to 6 lb (2.7 kg) in England and up to 10 lb (4.5 kg) on the continent of Europe. They spawn among weeds in shallow waters in late spring. The males first congregate at the spawning grounds, and when the females arrive several males will accompany one female as she lays long strings of eggs entwined in weeds. Perches make excellent eating.

The American perch, *P. flavescens*, is closely allied to the European species and is found over large parts of central and southern United States, in some regions reaching as far north as 60°N. The body is greenish-yellow, and it is slightly smaller than its European counterpart, attaining a weight of 4 lb (1.8 kg).

PÈRE DAVID'S DEER, *Elaphurus davidianus*, a species which no man, living or dead, has ever recorded seeing in the wild state. It was in 1865 that the French explorer, Armand David, first saw this animal inside the walled Imperial Hunting Park of Nan-Hai-Tsue near Peking. Before the close of the century, specimens of this rare deer had reached some of the European zoos, as well as Woburn Park in England. And not a moment too soon, for by about 1900, due to flooding in the Imperial Park, the wall was breached and the escaping animals were either killed off by the starving peasantry or by troops during the Boxer Rebellion. By 1910, not only was the animal extinct in China, but also in the European zoos, and the only beasts that remained alive were those at Woburn. From these few survivors a thriving herd of over 300 deer has now been built up.

Its physical attributes are unique. Standing about 48 in. (122 cm) high at the shoulder, it has an extremely long tail, wide-splayed hoofs somewhat similar to reindeer, and strange-looking antlers which appear to be worn back to front, with all major tines protruding to the rear. On rare occasions, two sets of antlers have been produced in a year.

Flat periwinkle, small sea snails common on European shores. They derive their name from the obliquely flattened apex of the spiral shell.

During the summer, the general body color is a bright red which in winter changes to a dark iron-gray with fawn shading. Gestation lasts about ten months.

An old Chinese name for this deer was ssu-pu-hsiang which means literally "not like four," i.e, like, yet unlike the horse; like, yet unlike the ox; like, yet unlike the deer; like, yet unlike the goat. Another name is milou.

PEREGRINE FALCON, *Falco peregrinus*, the most widespread of the large falcons, being found on all continents but Antarctica and on many oceanic islands.

Peregrine falcons are the most valued of all falcons used in falconry. Relatively easily obtained and trained, they are docile but large enough to kill game birds in spectacular style. Their speed in the diving attack (stoop), variously estimated at 100-275 mph (160-440 kph), coupled with their readiness to "wait on" above the falconer until he flushes the quarry, makes them superior to larger falcons such as gyrs or sakers. Peregrine eyries were jealously guarded in the Middle Ages, and records show that some are still in use.

Peregrines usually inhabit mountains or sea cliffs, but some live in tundras or in boggy areas among conifer forests. Cliffs are usually necessary for nesting, and peregrines sometimes make use of manmade "cliffs" in cities. The most famous of these, the Sun-Life falcon of Montreal, bred on the building of that name for 16 years.

Peregrines normally feed on birds up to the size of wild duck, caught in the air and either struck dead with the foot or seized and carried to the ground. Their favorite prey is pigeons, but in smaller races smaller birds are more usually taken. They eat an occasional small mammal, and in Alaska one has been known to catch a fish.

Peregrines probably pair for life and return annually to the same cliff to breed. They lay between three and six eggs in a scrape on a ledge, or sometimes in an old nest of a crow or other raptor. Some north European peregrines breed on the ground, and nesting in trees is known. The smaller male (tiercel) feeds the female (falcon) during courtship and incubation, and provides for the whole brood until they are half-grown. Thereafter the female assists him in catching prey.

Usually the female incubates the rich red-brown eggs alone for about 28 days, but males may play some part. The new hatched young are covered in white woolly down. They are feathered and can fly at six to seven weeks, but are still fed by the parents near the nest site for some time after their first flight.

PERIWINKLES, small sea snails living mainly on the shore and possessing a horny operculum and breathing by means of a single gill that has leaflets along one side only. They feed on seaweeds, rasping these with their horny tongue, or radula.

Common species include the crevice-dwelling small periwinkle, *Littorina neritoides*, which occurs at high-tide mark on exposed rocky coasts in northwest Europe (but not in the southeast of England) and in the Mediterranean. The shell rarely exceeds $\frac{1}{5}$ in. (5 mm) in length and is dark black-brown in color. Aside from living in crevices, large numbers are often aggregated among lichens and in dead barnacle shells on the upper shore. A larger species is the rough periwinkle, *Littorina saxatilis*, which occurs from the middle to the upper shore throughout much of northwest Europe and on both east and west coasts of North America. Its shell commonly reaches $\frac{1}{2}$ in. (12.5 mm) in length and is marked with fine spiral lines. The sutures between each whorl of the shell are

deep. This winkle is easily confused with small specimens of the common or edible periwinkle, *Littorina littorea*, but may be distinguished by the tentacles having longitudinal dark stripes in the rough periwinkle and transverse concentric stripes in the edible periwinkle. The latter is a widespread intertidal and estuarine winkle, which is commonest on the middle shore and below but may extend much higher than this in some areas. It occurs throughout northwest Europe and on the east coast of North America. The final European species is the flat periwinkle, *Littorina littoralis* (or *obtusata*)—so called because the apex of the spire is obliquely flattened. It occurs predominantly on the bladder wrack, *Fucus vesiculosus*, but can also occur on other seaweeds such as the knotted wrack, *Ascophyllum nodosum*. The flat periwinkle is an active animal found browsing when uncovered by the tide; the shell may be brown-orange, black or yellowish and is commonly $\frac{1}{2}$ in. (12.5 mm) long. Other littorinids include *Littorina punctata*, which occurs in west Africa, and *Littorina knysnaënsis*, which occurs on rocky shores in South Africa.

PHALAROPES, three species of fairly small semiaquatic wading birds, similar in build to the sandpipers and stints—that is, with thinnish, though rather long necks and small heads. Since all three species swim a great deal, their morphology shows several adaptations that parallel those found in water birds—dense plumage on their breast, belly and underparts, to provide both waterproofing and buoyancy; legs with an oval cross section so that the width of bone is much smaller in the direction of movement than at right angles to it, thereby cutting down resistance to the water flow without losing leg strength; and toes broadened or lobed and slightly webbed at the base. They have different plumages in summer and winter. In the breeding plumage, females are more brightly colored than males, an unusual feature found also in the dotterel, a mountain plover. Two of the species are of arctic origin. The gray has a patchy circumpolar breeding distribution in the tundra zone, and the red-necked is found usually to the south of the gray, in both tundra and boreal climatic zones between about 60° and 70°N. However, in several parts of Canada the two species breed alongside each other. It is not yet known whether they have any differences in food preference in these areas. In Britain, the red-necked phalarope breeds in small, but decreasing numbers to the south of its normal range, while it is reported to be extending its breeding range northward in other parts of the world. Both species reach their arctic breeding grounds in June, but stay there only a few months. After breeding, they move first to coastal waters close to the breeding areas, where they gather in large concentrations on the sea to molt, in the same way as many ducks. Later, they migrate to tropical and subtropical oceanic regions where "upwellings" occur, i.e., where persistent offshore winds bring cold water laden with nutrients to the sea surface, so that phytoplankton, and consequently zooplankton, are abundant.

PHEASANTS, game birds in which the male is usually far more brightly colored than the female. All pheasants spend a great deal of their time on the ground searching for food. They scratch the earth with their feet like domestic chickens in search of seeds, worms and insects. They have long and powerful legs and can run far and fast, so much so that many pheasants prefer to run into cover rather than take to the air when alarmed. When they do fly up, they rise almost vertically on their short but broad wings.

Most pheasants nest on the ground, making a scrape under cover of a bush, tussock of grass, or even among dead leaves on the forest floor. A few species prefer to lay their eggs in an elevated position, and in these cases they usually take over an old nest of some other bird such as a pigeon. The tragopans and the Congo peacock adopt this method, but it may well occur among other species. Some pheasants are polygamous or even promiscuous, but in at least seven of the 16 genera this is not the case. As sporting birds they have been widely introduced throughout the Old and New Worlds.

There are numerous races of the common pheasant, *Phasianus colchicus*, stretching across Asia from Transcaucasia in the west to Taiwan in the east. A number of different forms have been introduced as sporting birds to Britain and elsewhere in the world, so that the so-called common pheasant is a mixture of several races. The green pheasant, *Phasianus versicolor*, is similar in form but is confined to the islands of Japan.

The blood pheasant, *Ithaginis cruentus*, resembles a partridge in shape and size and is among the most delicately colored game birds. The general color of the male is bluish-gray with pale greenish below and crimson markings on the throat and beneath the tail. Blood pheasants live at a higher altitude than any other in the mountains of central Asia from Nepal through Tibet to northwest China. They always live near the snow line, usually between 9,000 and 15,000 ft (2,750-4,500 m) according to the season.

Koklas, *Pucrasia macrolopha*, are medium-sized pheasants of fairly dull coloration. The male has a long crest of dark green and rufous-brown feathers, glossy green-black head and upperparts of silvery gray streaked with brown. The female is brownish. It is found in the Himalayas from Afghanistan to central Nepal and in northeastern Tibet and eastern China, and is confined to mountain forests between 4,000 and 13,000 ft (1,200-4,000 m). Large and powerfully built, the males of the

three species of monals, *Lophophorus*, are the most brilliantly colored of all pheasants. The iridescent metallic hue of their plumage is rivaled only by the hummingbirds. Monals are mountain birds whose distribution extends from eastern Afghanistan along the Himalayan range to the mountains of western China. They spend a great deal of time searching for grubs, insects and roots, digging with their powerful bills and never scratching with their feet.

The ten gallopheasants, *Lophura*, are upright in stature and have longish legs armed with sharp spurs. The tail is compressed and ridge-shaped. There are large wattles of bare skin covering the face around the eye, blue in two species and red in all the others. Gallopheasants are forest birds living at low or moderate altitude from the Himalayas through southeast Asia to Taiwan and Borneo. The silver pheasant is the best-known member of this genus and the most frequently kept in captivity. Swinhoe's pheasant is also popular as an aviary bird, though rare in its native land of Taiwan, where it is threatened with extinction. In 1967, the Pheasant Trust sent 15 pairs of Swinhoe's pheasants that had been bred in captivity back to Taiwan for release to help build up the depleted wild population.

There are three distinct species of eared pheasants, *Crossoptilon*: the white-eared, brown-eared and blue-eared. All are large distinctive birds adapted to life at high altitudes and obtaining much of their food by digging with their powerful bills. The general color of both sexes is implied by the name of the species, and the so-called ears refer to the elongated white ear coverts common to all of them. The eared pheasants are confined to China, Tibet and Mongolia.

The sexes of the cheer pheasant, *Catreus wallichi*, are rather similar in coloration, both being dull gray and brown. The cheer pheasant inhabits the western Himalayas as far east as Nepal, living in forests and scrub between 4,000 and 10,000 ft (1,200-3,000 m).

There are five well-defined species of long-tailed pheasants, *Syrmaticus*, including Elliot's pheasant from eastern China, the bar-tailed pheasant from Burma, the Mikado pheasant from Taiwan, the copper pheasant from Japan and Reeves pheasant from China. The last named is the best known and most often kept in captivity. The Mikado pheasant is confined to Taiwan and is in danger of extinction in the wild. The Pheasant Trust is engaged in a project to reinforce the wild population with young birds, of which no less than 140 were bred in the Trust's collection in 1969.

The two species of the ruffed pheasant, *Chrysolophus*, the golden and the Lady Amherst's pheasants, are among the most beautiful and certainly the most popular of all game birds. They get the name "ruffed" pheasant from the males' large ruff of wide feathers, which are spread like a fan across either side of the head and neck during display. Both species come from the mountains of central China.

Peacock pheasants, *Polyplectron*, are small,

The common pheasant spread across Europe (and has since been taken to other parts of the world) as a sporting bird.

The golden pheasant *Chrysolophus pictus* is one of the most handsome pheasants.

dainty birds with loose plumage of intricate design. They are related to the argus pheasants and, like them, lay only two eggs in a clutch. There are six species, all inhabiting the tropical forests of southeast Asia from the eastern Himalayas south to Sumatra and east to Borneo and Palawan.

The great argus, *Argusianus argus*, is one of the most highly specialized of the pheasants, and though the male's plumage may look less colorful than that of some species, his display, which is one of the most remarkable in the bird world, more than compensates. The secondary wing feathers are very broad and of tremendous length with a line of beautiful ocelli running the length of each feather, while the two central tail feathers are even longer and twisted toward the tip. During his display the male argus faces the hen, bends forward and spreads his wings, twisting them so that they meet in front of his head. The tips of the two central tail feathers project above the circle of feathers. The great argus lives in the tropical forests of the Malay peninsula, Sumatra and Borneo. The male makes a display or dancing area to which he attracts the hens by his loud calling. The male of the crested argus, *Rheinartia ocellata*, also called Rheinart's crested argus, is remarkable for the length and breadth of its central tail feathers, which are over 5 ft (1½ m) long and 6 in. (15 cm) wide. Its habits are similar to those of the great argus. This species inhabits the tropical forests of Vietnam, Laos and the Malay peninsula.

PICHICIAGO, the smallest armadillo. There are only two species belonging to two genera. Pichiciago menore or the fairy armadillo, *Chlamyphorus truncatus*, measures 6 in. (16 cm) at the most in length, and its spatula-shaped tail is less than 1 in. (25 mm) long. In this smallest armadillo the many-banded head and body armor is scanty and is anchored in only two places, on the skull over the eyes and by a narrow ridge of flesh down the animal's back. Pichiciago menore is the only armadillo in which the dorsal armor is almost separate from the body. The squared-off rump, however, is covered securely with a large anal plate attached firmly to the pelvic bones. Pichiciagos are said to use this rear plate to plug their burrow entrances. The tail projects from a notch at the bottom of the anal plate and cannot be raised because of it. The shell in pichiciago menore is pale pink (hence they are sometimes called pink fairy armadillos), and the rest of the body is covered with soft, fine white hair that hangs down over the legs and feet. Supposedly, the babies are hidden beneath this protective curtain when they are very young.

Fairy armadillos are found only in the sandy arid plains of west-central Argentina, where they burrow in the hot, dry earth where cactus and thorn bushes grow in abundance. Pichiciago menore is a very rapid digger, supporting itself with the stiff tail and using all four feet. In soft soil it is reported to be able to disappear from sight before a man can dismount from his horse. At dusk, fairy armadillos emerge from their burrows to feed principally on ants and occasionally on worms, plant tops and roots. Burmeister's armadillo or pichiciago mayroe, *Burmeisteria retusa*, closely resembles the fairy armadillo and is only slightly larger.

PIDDOCKS, marine bivalve mollusks that bore either into soft rock or into wood; those boring into wood are distinguished from the shipworms by the absence of calcareous tubes and long extensions of the siphons. The piddock's shell is cut away anteriorly, leaving a permanent gape through which the rounded foot projects. The end of the foot forms a sucker that anchors the piddock to the end of the burrow while rasping movements are made. Such movements of the shell valves are brought about by the alternate contraction and relaxation of the anterior and posterior adductor muscles that rock the valves about a central area of articulation or fulcrum. There is one such area in all the rock-boring piddocks belonging to the family Pholadidae, but in the wood piddock, *Xylophaga dorsalis*, there are two fulcra (a dorsal and a ventral one as in the shipworm). A common feature of all the true piddocks is the presence of accessory shell plates, which lie between the valves on the upper surface.

The largest European species is the common piddock, *Pholas dactylus*, the shell of which reaches 3 in. (7.7 cm) in length; there are four accessory plates, three anterior and one longer posterior one.

The wild boar has been wiped out in parts of Europe, but introduced into many other parts of the world.

The bizarre Nicobar pigeon *Caloenas nicobarica* of the East Indies, most highly adorned of pigeons.

Feral pigeons, descendants of the domesticated rock dove, in Trafalgar Square, London.

PIG, DOMESTIC, also known as hog or swine, has been used by man for the last 5,000 years, since the Neolithic period. Probably, Asiatic pigs were domesticated first, and there is evidence that these were first brought to Europe, although the European wild boar was later domesticated. By the nature of its behavior, the pig was unlikely to have been domesticated before human communities became agricultural. Nomadic tribes would have difficulty in moving about with a relatively slow and refractory animal, and it seems likely that the use of hogs spread from one settled village to another.

Although primarily a prolific source of meat and cooking fat, pigs were put to other uses. They have at times been employed as draft animals, and in ancient Egypt were used to tread the corn, their hoofs making an imprint in the ground of just the right depth at which wheat should be sown. Individual pigs have at various times been used for rounding up cattle, for retrieving game and for detecting truffles, the fungus delicacy that grows 1-2 ft (30-60 cm) below ground. Their main use, apart from being suppliers of meat, was in clearing the ground. Pigs root for food, turning over the soil with their snout, feeding on acorns, beechmast and other fallen fruits, roots and tubers, and in the process destroying seedlings, even uprooting small saplings as well as bushy undergrowth. Grass then covers the ground, and from Neolithic times onward they seem to have been used for ground clearance by the farmer, to convert open woodland into arable and pasture land. The swineherd was then an important member of the community.

PIGEONS, a family of some 255 species of birds spread over most of the world except for polar and subpolar regions and some oceanic islands, varying in size from that of a lark to that of a hen turkey, from 6-33 in. (15-84 cm) in length. They also vary considerably in plumage, some of them being among the most brightly colored of birds, others rather drab. The most typical plumage is some pastel shade of gray, brown or pink, with contrasting patches of brighter colors. The feathers are soft and often loosely inserted in the skin, but are nevertheless strong and dense. The wings and tail show much variation in size and shape, but the legs are usually short, being rather long only in some of the terrestrial species. The body is compact, the neck rather short and the head small. The bill is usually small, soft at the base but hard at the tip, and at the base of the upper mandible is a fleshy "cere," a naked area of skin that in some species is much swollen. This swelling is much more common in domesticated varieties. Some species have a noticeable crest. In most species the male is rather brighter than the female, but in a few species the sexes are similar and in others they are very different.

Most species of pigeons perch readily and regularly in trees, but some are terrestrial, others cliff-dwelling and some have taken to nesting on buildings in towns and cities. The feral pigeons that are so common in towns are all "escapes" or descendants of such, from stocks of domestic pigeons. All domestic pigeons are derived from the rock dove,

Pika *Ochotona pusilla*, one of the mouse hares or calling hares, a short-eared tailless rabbit.

Columba livia, of Europe, which in the wild form nests on rock ledges, so feral pigeons take naturally to breeding on buildings. Most species are gregarious, at least outside the breeding season, and some of them may be seen in large flocks. In the passenger pigeon, *Ectopistes migratorius*, now extinct, flocks of literally countless millions were common. Most pigeons are very strong on the wing.

The food of pigeons is very varied, including berries, nuts, acorns, apples, seeds of many kinds—for example, weed seeds and cultivated grain—and also buds and leaves. Many species also take animal food, such as snails, worms and caterpillars. Food is stored temporarily in a crop, which may be capacious. The distended crop of a wood pigeon, *Columba palumbus*, for example, after a successful day's feeding may be seen clearly as the bird flies home to roost. Most pigeons have a large, muscular gizzard that with the enclosed grit deliberately swallowed, grinds up even the most intractable food. Pigeons build a simple, rather unsubstantial and usually platformlike nest of twigs, stems or roots, in a tree or bush, on a cliff or building ledge, or sometimes on or in the ground or in a tree cavity. Two pale, unmarked eggs form the usual clutch, and both sexes incubate. The young are helpless when first hatched and sparsely covered with a filamentous down. They are fed by both parents for the first few days on "pigeons' milk," a curdlike material secreted by special cells lining the crop. This is scooped up by the broad, soft bill of the young, inserted deeply into the parent's mouth to obtain the regurgitated material. Gradually it is supplemented with food partially digested by the parents.

PIGMY POSSUMS, dormouselike marsupials with prehensile tails of the family Phalangeridae, found in Australia, New Guinea and Tasmania. Unlike other phalangers, the pigmy possums have teeth adapted for an insectivorous diet, chromosomes alike in number and morphology to those of marsupial cats and some American opossums, and they give birth to up to six young at one time. These features indicate that the pigmy possums are the most primitive of the phalangers and are perhaps like the stock from which the phalangers arose. Broom's pigmy possum, *Burramys parvus*, previously known only as a Pleistocene fossil, was found alive in Victoria, Australia, in 1966.

PIKA, like a small, short-eared tailless rabbit. It weighs only 5 oz (140 gm). The body is short and cylindrical, the ears short and rounded, the tail not apparent externally, and the hind legs not much longer than the front ones, permitting a scampering run rather than the leaping gallop typical of other lagomorphs. The color of the 14 species, with many subspecies, varies from dark brown or dark slaty-gray to pale sandy or ash. Twelve species are distributed from eastern Europe to Japan and from the Himalayas to Siberia. .Two North American species range from Alaska southward down the Rocky Mountains. Although predominantly a high-altitude animal living above the tree line among the rocks and crevices of mountain slopes, a few species inhabit plateaus and open grasslands down to sea level. All species live in a cold climate, and all dig in the soil. Breeding occurs from late spring to summer. Gestation lasts 30 days, the newborn are furred, with closed eyes and are dropped in burrows.

Pikas live in colonies, which are spaced at distances from one another in accordance with the availability of food in different areas. Defended territories are marked with the secretion from cheek glands. Pikas use characteristic sounds frequently and loudly, presumably for intraspecific communication, hence one of their common names, "calling hare."

Pikas occupy marginal areas where the supply of food is drastically reduced during severe winter conditions. They do not hibernate but have developed the habit of food-hoarding, involving not only storing it under rocky ledges but also preliminary drying and turning it as in haymaking. They feed on a variety of vegetation, the hay piles often being brush piles consisting of grass, wood twigs, and also pine cones, clumps of moss and sprigs of conifer needles. The stacks can be quite sizable, reaching a few pounds in weight. In some countries, the pikas' enterprise is exploited by herdsmen who feed their sheep on the "haystacks" in winter.

The Mount Everest pika, *Ochotona wollastoni*, occurring on Mt. Everest and in northern Nepal, is a member of the south Asian *roylei* group of pikas. It lives at altitudes up to 20,000 ft (6,000 m), thus having one of the highest vertical distributions among the mammals.

PIKE PERCHES, fishes of the genus *Stizostedion* that are true members of the perch family but show a similarity in general form to the pike. The European pike perch, *S. lucioperca*, is an elongated, pikelike fish widely distributed across central Europe and introduced into southern England and southern Scandinavia. A population exists in the Baltic that migrates into the lagoons from the sea in winter. This species is also known as the zander. There are two dorsal fins, the first with stout spines. The mouth is armed with large caninelike teeth interspersed with many smaller ones. The back and flanks are greenish-gray with vertical dark bars in the young, and there are longitudinal rows of dark spots on the two dorsal fins. The young feed on small aquatic ani-

The European pike-perch, also known as the zander, is a true perch that is pikelike.

mals, but the adults are greedy predators feeding chiefly on fishes. The pike perch grows to 4 ft (120 cm) and a weight of 22 lb (10 kg). The flesh is good to eat.

PILCHARD, *Sardina pilchardus*, a marine fish of the eastern North Atlantic and Mediterranean belonging to the herring family and widely known in its marketed form as both pilchard and sardine. The pilchard somewhat resembles the herring in its cylindrical body and fairly smooth belly, the series of small serrations along the belly (scutes) being poorly developed. It differs from the herring in having the last two anal rays longer than the rest, the coloring of the back rather greenish as opposed to bluish and the presence of a dark spot behind the gill cover, followed by a series of small spots along the flanks (sometimes absent). The pilchard grows to 8 in. (20 cm) in length and is found from north Africa to the southern coasts of Norway and also in the Mediterranean and adjoining seas. The fishes of the north African coasts and the Mediterranean have slightly more gill rakers and are considered to belong to a distinct subspecies, *S. pilchardus sardina*. The species spawns in autumn and winter in the southern part of its range, but in late summer in the north.

The pilchard is of considerable economic importance to several European countries. Adult fishes are caught off the coasts of Cornwall, England, and although the fishery is rather small, the canned products are of excellent quality. Larger fisheries exist along the French coasts of Brittany and the Bay of Biscay. Off the coasts of Spain, southern France and in the Mediterranean, the juvenile fishes, of about a year and a half, are caught in enormous numbers and are referred to as sardines. In Basque country the sardines are caught by ring nets, the fishes being enticed into compact shoals either by casting *rogue* (salted Norwegian cod's roe) onto the water or by the use of powerful incandescent lamps at night.

PILOTFISH, *Naucrates ductor,* a small fish habitually associated with sharks and larger fishes and related to the mackerels and tunas. This pelagic fish is found all over the world in tropical and temperate oceans. The name comes from its alleged habit of not only accompanying sharks and sometimes other large fishes but of actually leading them to their prey. The pilotfish feeds on the scraps of food left by its host and thus deserves the description "commensal" (literally, "feeding from the same table"). Pilotfishes swim around sharks, making brief sorties and returning, but it is doubtful if they act as pilots. V. V. Shuleikin calculated that sharks swim three times as fast as a pilotfish. He suggested the pilotfishes are carried along by the shark's boundary layer, that is, the layer of water over its surface that, as with any body moving through water, travels at the same speed as the shark.

Young pilotfishes are often found sheltering among the tentacles of jellyfishes, a habit shared with the young of the horse mackerel and the Portuguese man-o'-war fish.

The pilotfish has a mackerellike body with the first dorsal fin reduced to a few low spines. The second dorsal fin and the anal fin are moderately long and are opposite each other. The body is dark blue on the back and silver on the sides with about six vertical dark bars on the flanks that fade with age. The fish grows to about 2 ft (60 cm) in length.

PILOT WHALES, several species of large dolphins with a worldwide distribution. The North Atlantic species, *Globicephala melaena*, is also called the Ca'aing whale or blackfish. Adults are about 25 ft (8 m) in length, and they

are found in schools of several hundreds. As the scientific name implies, pilot whales have a rounded forehead that bulges forward of the lower jaw. They are black overall but for a white patch below the jaw. The flippers are distinctive, being relatively long and narrow, about onefifth of the body length.

Pilot whales are highly social, yet nervous, and as a result of this fall an easy prey, particularly in the Faroes, where the economy of the Faroese has been to a considerable extent dependent upon them. When a school approaches the islands, the boats congregate and form a line to the seaward of the school, which is then driven toward a bay. A few are lanced and driven onto the beach, and the remainder follow, and they can be dispatched at leisure. The remarkable feature is that although some may escape to sea the social sense is so strong that they return to the remainder.

The North Atlantic pilot whale is found as far south as Scotland and New Jersey. There is evidence of some migration, but how far is not certain. A very similar animal is found around the Cape of Good Hope and New Zealand and is said to be the same species. The other species are very similar externally but have skull differences. *G. scammoni* is the pilot whale of the North Pacific, and *G. edwardi* lives off the Indian and South American coasts. *G. macrorhyncha* has a somewhat shorter flipper and is found on the north coast of America. All three are entirely black.

PIN WORM, *Enterobius vermicularis,* a nematode parasite, that lives in the large intestine of humans. It is particularly common in children in warm countries, and infection rates of 40% have been recorded. The adult parasite attaches itself to the wall of the gut of its host, and when the female is ready to lay her eggs she migrates to the anal region of the host, where the eggs are deposited. The irritation that results often causes the host to

dislodge the eggs by scratching, and these may be ingested again from contaminated fingers. If no reinfection occurs, the initial infection dies out in about a month.

PIPEFISHES, a family of highly elongated and rather specialized fishes related to the trumpetfishes, shrimpfishes and sea horses. They have a worldwide distribution and, although mainly marine, include some freshwater species. The long, thin body is completely encased in bony rings, but it is surprisingly flexible and prehensile. The fish is well camouflaged for a rather secretive life among weeds. The pelvic fins and the tail are often lacking. The prehensile body is developed to its greatest extent in the sea horses, the head being bent at an angle to the body. There are pipefishes that approach this arrangement but still retain the head in the normal position.

The pipefishes show a most interesting method of caring for their young. In the most primitive forms (e.g., *Nerophis*), the fertilized eggs are stuck together on the underside of the male. In the next and more advanced group, the eggs are embedded singly in a spongy layer that develops along the belly of the male. Finally, in forms like *Doryrhamphus*, the bony plates encasing the body are enlarged to form a groove in which the eggs are placed. This is carried to the extreme in the sea horses, where the male has a distinct brood pouch. One of the largest is the great pipefish, *Syngnathus acus,* a species that reaches 18 in. (45 cm) in length. It has a small anal fin and a small tail. As in all pipefishes the rays of the dorsal fin are soft and flexible, and each ray can be moved independently. This is important, because, like the sea horses, the pipefishes swim by undulations of the dorsal fin. The small worm pipefish, *Nerophis lumbriciformis,* rarely grows to more than 6 in. (15 cm). The body is dark brown, and the body plates are not easily seen. It is often found on the shore under stones and weeds. In July the males can be

Most pipefishes, a family of highly elongated and specialized fishes, live in the sea. A few, like this *Syngnathus pulchellus* of the Congo, have become adapted to a freshwater life.

The male golden pipit of east Africa, unlike most other species of pipits, is not particularly somber in his coloring.

found with eggs stuck to the underside of their bodies.

Most pipefishes are rather drab in color, but the male of the straight-nosed pipefish, *Nerophis ophidion*, has a greenish body with blue lines along the abdomen. Some of the pipefishes are quite small. *Doryrhamphus melanopleura*, from coral reefs of the Pacific region, reaches only 2½ in. (6.5 cm) in length. It is bright orange-red with a longitudinal bright blue band from the snout to the tail, the latter being orange at its base, followed by blue with a white margin.

Most pipefishes live among corals or weeds in shallow water, but a few live in burrows, and *Corythoichthys fasciatus*, of the Indo-Pacific, inhabits the intestinal cavity of bêche-de-mer. *Syngnathus pelagicus*, on the other hand, is pelagic and lives among the sargassum weed floating at the surface in the Atlantic.

Piranhas, South American river fishes, are credited with unusual ferocity; yet some species, including the red piranha, are kept as aquarium fishes.

PIPITS, small, somberly colored terrestrial birds, brown, streaked with black, above, and buffish white or yellow, with or without streaks, below. The sexes are usually alike. The tail is relatively long and is often bobbed up and down in a wagtail fashion. The bill is fine and pointed, the legs long and slender and the hind claw often very long. Although frequently quoted as being exceptional among passerines in possessing only nine pairs of primary wing feathers, pipits do in fact possess the normal quota of ten, the tenth pair being very much reduced.

Pipits feed on insects mainly. They are very much birds of open ground, inhabiting grassland, steppe and savanna country, usually in well-watered areas, alpine meadows in mountainous regions and arctic tundra, although a few species prefer areas with scattered bushes or trees. They walk or run but never hop. Most pipits occasionally perch on trees, particularly when disturbed, but few species do so habitually. The name "pipit" is derived from the thin, twittering call uttered by many species in flight. The general similarity of pipits to larks has in some regions earned them the name "titlarks."

Almost all pipits of the genus *Anthus* are brown above, buffish-white below and more or less heavily streaked, with prominent white outer tail feathers. One of the most widespread, Richard's pipit, *A. novaezeelandiae*, occurs from New Zealand and the neighboring subantarctic islands through Australia and southeast Asia to India and also over much of Africa. Although absent from the Middle East and Europe as a breeding species, it has occurred irregularly on migration as far west as Britain. The tawny pipit, *A. campestris*, prefers drier habitats than most pipits, occurring on sandy wastes, arid pastures, and barren rocky ground in southwest Asia, southern Europe and north Africa. Two of the commonest pipits in Europe, the tree pipit, *A. trivialis*, and the meadow pipit, *A. pratensis*, although very similar in appearance, show quite different habitat preferences. The tree pipit is a bird of wood edges, forest clearings and heaths with scattered trees. It seems that elevated perches are essential to tree pipits during the breeding season, as the erection of a line of telegraph poles across an otherwise bare heath may lead to their colonizing the area. The meadow pipit prefers rough grassland, moors and grassy tundra. It is replaced ecologically farther east by the red-throated pipit, *A. cervinus*. This is very similar in appearance to the meadow pipit in winter, but in summer it has a brick-red throat and breast, which makes it quite distinctive.

The food of all pipits is mostly insects, although a few seeds may be taken in winter. Several African species are particularly fond of termites. The rock pipit has a more varied diet of small worms, sandhoppers and periwinkles, in addition to a wide variety of insects.

PIRANHAS, or caribes, small but very ferocious freshwater fishes from South America belonging to the family Characidae. They are renowned for their carnivorous habits and are among the most infamous of all fishes. Travelers' tales relate cases where large animals and even men have been attacked and the flesh

Blue-winged pitta, of southern and eastern Asia. Pittas are considered by some people the most beautiful of all birds.

picked off their bones in a very short time. At river crossings in South America a lookout is kept for the shoals of piranhas so that people fording the river can be warned. There are several species involved, the largest growing to about 15 in. (38 cm). The jaws are short and powerful and are armed with sharp cutting teeth. Their main diet is fish or mammals, but they are reputed to be strongly attracted to the smell of blood so that a single bite will draw hundreds of other members of the shoal to the same spot.

The white piranha, *Serrasalmus rhombeus*, of the Amazon is one of the largest species. The body is olive to silver with irregular dark blotches. The red piranha, *Rooseveltiella nattereri*, also from the Amazon, grows to 12 in. (30 cm) and has an olive-brown back, light brown flanks and numerous bright silver spots on the body. The belly and the fin bases are, appropriately enough, blood red; the dorsal and anal fins are black.

PITTAS, beautifully colored ground-living birds that are very plump, with short wings and tail, and powerful legs and feet. Because of their gorgeous coloration, they have been called "jewel thrushes." The allusion to thrushes is ill-chosen, however, for they are not conspicuously thrushlike in appearance, nor related to them. The pittas are, without a doubt, among the most beautiful birds in the world. Few others have so many varied colors combined in their plumage. The male green-breasted pitta, *Pitta sordida*, of southeast Asia, for example, has a brown and black head, green back and breast, red undertail coverts, blue rump and wings patterned boldly with blue, black and white. Other species display even more gaudy arrays of color, but one or two, like the banded pitta, *Pitta guajana*, of Malaysia, have a relatively somber pattern of browns, relieved only by small splashes of bright color. The females of most species are

much duller than the males, and a few are quite drab-looking.

Pittas are characteristically birds of the forest, but they are not confined to any one vegetation type, being found in lowland and mountain rain forest, bamboo thicket and coastal mangroves. When feeding, they move about quickly on the forest floor by hopping, but can fly rapidly if they have to. Their food is apparently very varied: spiders, ants and other insects, snails and seeds have all been found in the stomachs of shot specimens. Though they probably obtain all their food on the ground, they do fly into trees on occasion, and roost on branches at night. Generally speaking, pittas are shy birds, more often heard than seen, despite their showy appearance.

Their nests are large, untidy constructions of sticks, dead leaves and roots, built on or near the ground. The young remain in the nest for several weeks and are fed by both parents.

PLAICE, *Pleuronectes platessa,* perhaps the most popular of all the edible European flatfishes. The common name derives from an Old French word for "flat." The plaice is one of the most easily identified species because of the irregular orange spots on the upper surface that persist after the fish is dead (similar spots in the flounder soon disappear when the fish is out of water). The blind side is a translucent white. Unlike the flounder, there are no rough scales on the head or at the start of the lateral line, but there are small tubercles between the eyes. In contrast to most of the flatfishes, the plaice is a sedentary species, found mainly over sand or gravel. It is, therefore, very susceptible to overfishing, and the maintenance of a fishery depends on strict control of catches to balance the natural replacement each year. Experiments have been conducted in the rearing of young plaice for restocking depleted areas and also in the transferring of plaice from British North Sea

coasts, where they are numerous, to the eastern shores of the North Sea, where growth conditions are better. Plaice grow to 33 in. (1 m) and may reach 15 lb (6.7 kg) in weight.

PLAINS WANDERER, *Pedionomus torquatus,* a small quaillike bird closely related to button-quails, but placed in its own separate family. It is confined to open grasslands of southeastern Australia. The role of the sexes is reversed, the female being bigger, more brightly colored, the male incubating the eggs and caring for the young. This species differs from button-quails in possessing a hind toe, in the hairy texture of its feathers and in having a clutch of four large, pear-shaped eggs. It also tends to move in a very upright posture, raised on the toes, unlike the crouching postures of button-quails.

PLATYPUS, *Ornithorhynchus anatinus,* Australian four-legged amphibious animal, which, with the echidnas, comprises a distinct subclass of the Mammalia, the Prototheria or egg-laying mammals. In size it is considerably larger than the land mole. The eyes are very small. The fore legs, which are shorter than the hind, are observed, at the feet, to be provided with five claws, and a membrane or web, that spreads considerably beyond them; while the feet on the hind legs are furnished not only with this membrane or web, but also with four long, sharp claws, that project as much beyond the web, as the web projects beyond the claws of the fore feet. The tail of this animal is thick, short, and very fat; but the most extraordinary circumstances observed in its structure is its having, instead of the mouth of an animal, the upper and lower mandibles of a duck. By these it is enabled to supply itself with food, like that bird, in muddy places, or on the banks of the lakes in which its webbed feet enable it to swim; while on shore its long and sharp claws are employed in burrowing, nature thus providing for it in its double or amphibious character. These little animals have frequently been noticed rising to the surface of the water and blowing like the turtle.

The webbing on the hand projects beyond the ends of the claws, being carried by five long leathery extensions of the digits. When walking on land, this extraneous web is folded back under the digits. The claws on these are used for the construction of the burrows to which they retire after swimming and feeding. All feeding is done in the water, and the prey consists of various crustaceans, mollusks, aquatic insect larvae, and even large flying insects like the cicada, *Melampsalta denisoni,* which may fall into the water. The platypus lives only in freshwater lagoons, lakes and pools in small and large rivers. It is a beautiful little animal with a streamlined body and a tail like that of a beaver. The body is covered with a dense, very fine fur about 0.75 in. (1.5 cm) long somewhat concealed with coarser long hairs, but the tail is covered with coarse, bristlelike hairs, densely on the dorsal surface and very sparsely on the ventral surface.

An adult male platypus is about 20 in. (51 cm) long (a length of 26 in. or 65 cm has been recorded) and weighs 4.2 lb (1.9 kg); adult females are much smaller, weighing 2½-3 lb (1.1-1.3 kg). The legs are short and stout, the ankles of the males bearing a curved spur about 0.75 in. (1.5 cm) long. This spur is hollow, and communicates with a duct that emerges from a gland situated on the dorsal side of the upper hind leg. The gland secretes a poisonous substance during the breeding season: the function of the apparatus is unknown. There is no scrotum, the testes being internal as they are in reptiles. The eyes are situated dorsally on the broad, flattened head, and immediately posterior to the eye is the external opening of the ear. There is no external pinna, and both the eye and ear are situated in a groove or fold of fur. Both the eye and ear openings are closed by the apposition of the lips of this groove when the animal is under water. Very young platypuses have molariform teeth. In the adult, the teeth are replaced by horny plates situated on a flattened area of the lower jaw just anterior to the coronoid process and on a lateral projection of the posterior portion of the maxillary bone. The horny plates serve to crush the food, some of which is then stored temporarily in cheek pouches: doubtless, larger items pass direct to the stomach since the cheek pouches are small. The stomach, as in the echidna, has no gastric glands, so presumably all digestion takes place in the intestines.

The platypus is found in fresh waters throughout Tasmania and eastern Australia. Platypuses spend little of their time in the water—maybe a total of three or four hours a day in the winter, less in summer; nobody knows for sure, since identified animals have not been studied; maybe observation of tagged animals will provide an answer to this question. The platypus spends most of its time in burrows, which it digs in the soft earth of the banks of the waterways, or in sunning itself in the open. The entrances to the burrows are said to be above water level. There are two types of burrow: one used for shelter and another for breeding. The latter is constructed and inhabited by a pregnant female only and is not shared with any other platypus, male or female. In 1884, W.A. Caldwell demonstrated that platypuses do not produce their young as other mammals do, but lay eggs. Copulation takes place in the water, and fertilization is followed by an unknown period of gestation in the left uterus only, the right ovary and oviduct being nonfunctional. The female then retires to the complicated nesting burrow where she has excavated a brood chamber containing a nest of grass, leaves, reeds and so on. Generally two eggs are laid, sometimes three; the eggs adhere to one another, their shells being sticky when laid. Nobody knows how the eggs are incubated (there is no pouch as in the echidna), nor for how long: equivocal evidence suggests that the period may be between seven and ten days. After hatching, the tiny young, about 0.65 in. (17 mm) long, are suckled by paired mammary glands that open at a pair of milk patches or areolae on the ventral surface. These milk patches are not well defined like those of echidnas but are hidden by thick fur; consequently scientists find them very hard to detect, but the newly hatched platypus presumably has no such difficulty; possibly sense of smell helps it. The young are surprisingly like newborn marsupials with their enormous,

strong fore limbs and rudimentary hind limbs, except that an egg tooth and caruncle are present on the head, which are used for breaking out of the keratinized egg shell. Doubtless the relatively great fore limbs of the newly hatched are used for clinging to the fur over the milk patches. As the young grow, the mammary glands become very large, reaching almost from the armpits to the pelvis longitudinally and up around the flanks laterally. The glands are made up of alveoli and ductules as they are in other mammals. Analysis of a sample of milk shows that, in its fatty acid content, it is like that of marsupials, insectivores, and echidnas, with the difference that there is about 8% lauric acid in the platypus milk, whereas it is scarcely detectable in the other types of milk mentioned.

As far as is known, copulation takes place in the months of August and September, and the young emerge from the breeding burrow for the first time in December and January and are about 12-14 in. (30-35 cm) long, i.e., not much smaller than their mothers. They growl, squeak, and play like puppies.

PLATYS, live-bearing cyprinodont fishes from fresh waters of the New World, placed in the genus *Xiphophorus* (formerly *Platypoecilus*, from which the common name was derived). The genus *Xiphophorus* also contains the swordtails. These fishes are attractive and easy to keep in an aquarium. The platy *X. maculatus*, which grows to $2\frac{1}{2}$ in. (6 cm), is found in the rivers of the Atlantic drainage of Central America. In the wild, the normal color is an olive green with a pair of black spots near the tail, but it is a very variable species, and red and black varieties are also found. These characters have been developed by breeders to produce stable red, orange and yellow varieties with assorted black markings. The wagtail platy was the result of a cross between a wild platy and the golden variety. Wagtails have orange and red bodies with a black dorsal fin and tail. The variegated platy, *X. variatus*, comes from southern Mexico. It too is very variable in its colors but usually has bands or zigzagging rows of spots along the flanks. Many color varieties have now been bred by aquarists. The green swordtail, *X. helleri*, which is greenish in the wild, has been crossed with the red variety of the platy to produce the common red swordtail.

Platys are of interest not only to the aquarist. Their ability to cross with other species and the range in color forms that can be produced have been of considerable interest in studies on genetics, and a great deal can be learned from them of the mechanism of inheritance.

PLOVERS, small- to medium-sized, plump, wading birds ranging in size from 6-16 in. (15-40 cm). Their wings are long, pointed in the true plovers but rounded in the lapwings. The bill proportions are fairly uniform throughout the family: straight, hardly tapering, fairly stout and of moderate length compared with head width. The neck appears short and thick when the bird is feeding or resting, but can be extended considerably when the bird is on the lookout for predators or is patrolling its breeding territory. In many species the eye is large, as expected of birds that are active by night as well as by day. Leg length varies from species

to species, and plumage is also very variable. A few features are common to most species. The tail, which is rather short, has a band of a dark color close to the tip, and another dark band or patch of color occurs on the breast or belly. Head patterning, which is usually prominent, varies within the family, but most species have a white band on the nape, a feature shared by the chicks. This band is hidden when the chicks crouch to escape predators, and the colors of the rest of their down blend in well with the background on which they are found. In contrast, the adults rely on disruptive coloration, with broad bands of dark and light colors "breaking up" the outline of their bodies, to escape detection.

The majority of the true plovers belong to the genus *Charadrius*, which is represented in all parts of the world. In North America, where no lapwings occur, one of the larger plovers, the killdeer, *C. vociferus*, breeds on agricultural land, and may be considered as the ecological replacement of the larger lapwing, *Vanellus vanellus*, of Europe. The killdeer has a transequatorial breeding range, from Canada to Peru. To the north is found the semipalmated plover, *C. semipalmatus*, a chiefly shore-living species that should probably be regarded as conspecific with the ringed plover, *C. hiaticula*, of Europe, as hybridization has been shown to occur in the area of overlap. The most recent suggestion is that they are different forms of a polymorphic species, as in the herring gull. In South America, despite the presence of three species of lapwings, several plovers breed at both ends of the continent: the killdeer, mentioned above, the winter plover, *C. modestus*, in Patagonia and the Falkland Islands and the southern dotterel, *Eudromias ruficollis*, on high ground at the southern tip of South America. The last two migrate north toward the equator after the breeding season.

In northern Europe and Asia, the ringed plover replaces the Kentish plover as the typical seacoast species north of about latitude 50°N. It also nests inland in the tundra and shrub tundra. Farther south in Europe and Asia it is replaced as an inland breeding species by the little ringed plover, *C. dubius*, which has spread gradually northward in Britain this century, colonizing gravel workings almost as far north as the Scottish border. The movements of ringed plovers provide an example of "leapfrog" migration, the most southerly breeding population (that of Britain) being almost sedentary, while the most northerly (Arctic) breeding birds winter farthest south, on the western coasts of South Africa. The European populations of little ringed and Kentish plovers also winter in Africa, whereas the tropical populations are chiefly sedentary. Another plover, the dotterel, *Eudromias morinellus*, is found in the arctic regions of Europe and western Siberia and on high mountain ranges farther south, with even a few in the Austrian Alps and the mountains of central Italy. This species has a probable counterpart in the southern Rockies of North America in the mountain plover, *C. montanus*. Dotterels migrate in autumn to the sandy and muddy shores of the Mediterranean and Persian Gulf.

The plover family also includes a unique species, the wrybill plover, *Anarhynchus fron-*

talis, of New Zealand. This bird has its bill tip turned to the right and feeds by poking it under stones on beaches for insect food. It breeds only on South Island, but moves to North Island in the nonbreeding season.

Plovers lay a single clutch of usually four eggs in a scrape in the bare soil, mud or sand that form their chosen habitats. The chicks leave the nest soon after hatching and are then supervised (but not actually fed) by the parents, who warn them of approaching danger. The adults frequently draw away potential predators from the vicinity of the chicks by "broken-wing" and other distraction displays. After the young are able to fly, most plovers congregate in large flocks for the winter months, or before migration. Plovers feed largely on insects and small Crustacea living on or in

climb less readily than stoats or weasels and swim only when the need arises. Hunting more by smell than by sight, the polecat traverses obstacles or high grass in short leaps, stopping now and then to rear up on its hind paws like so many other short-limbed mustelids. Rabbits used to be a favored prey before the myxomatosis epidemic, but now rodents, insects, eggs and fruit are eaten. The thick, closely knit fur with the underlying layer of fibro-elastic tissue makes polecats almost impervious to the bite of enemies, whether fox, dog, or snake. They possess such a loose-limbed agility that they often give the impression of being able to bound backward as well as forward. This lithe muscularity and speed are no doubt instrumental in making them successful in encounters with larger predators. If

the same summer. The four to eight cubs are blind and naked at birth and weigh no more than 0.3 oz (10 gm). Their eyes open after 20 days, and a gray, downy fur with white muzzle and ear tips appears. For seven weeks, the young explore near the leaf-lined nest, and the female weans them gradually, bringing back live frogs or insects. The male does not participate in rearing and is excluded forcibly from the nest area by the female. Full size is reached in a year, but sexual maturity is not attained until the end of the second year.

Frequently confused with the polecat, the ferret, *Mustela eversmanni furo*, is a smaller, domesticated version of the Asian polecat. Yellowish white or albino in color, it is still used by European farmers to flush rabbits from their burrows.

Platypus, one of the two egg-laying mammals, swimming in a tank at Healesville Sanctuary, Victoria, Australia.

soil, mud and sand, but they also take a small proportion of vegetable matter in their diet.

POLECATS, terrestrial carnivores, larger than a stoat and smaller than a marten, best known for the pungent odor secreted by their anal scent glands. They measure 17.3-25.2 in. (44-64 cm) and weigh 1 lb 7 oz-2 lb 11 oz (0.65-1.2 kg). Females are much smaller and lighter than males. Certain individuals of the European polecat, *Mustela putorius*, are very dark, almost black, in fact, because the thick yellowish underfur is more or less masked by the guard hairs, which may be either completely black or black-tipped. A dark brown band across the eyes contrasts with the creamy white of the throat, forehead and edges of the ears.

Depending on the species, polecats are found in a variety of habitats in the northern hemisphere, from open grassland to woods or thickets. The nonretractile claws are used to excavate the animal's den, but empty rabbit warrens or fox holes are also used. Polecats

suddenly alarmed, they will snort and constrict the scent glands, sending a jet of milky white fluid 20 in. (50 cm) backward, which acts as a pungent deterrent. By contrast, when attacking, the polecat hisses and keeps its head low, the back humped into an inverted U shape, lunging forward to bite the prey on the neck. The victim is then shaken violently from side to side. The orientation of this neck bite is a learned behavior pattern that the young acquire while pouncing and biting during play-fighting bouts.

Neck grasping also occurs during mating. Females come into estrus in March, and the copulations that follow may last for over an hour at at time. During that period, males scream loudly and chase each other out into the open in broad daylight, taking little notice of man. Gestation usually lasts from 40 to 45 days, although some authors claim 63 days as a maximum. Should the litter not survive, the female will immediately come into estrus again, and another litter will be born later in

POLLACK, *Pollachius pollachius*, a codlike fish found in the eastern Atlantic as far north as Norway, but absent from the Mediterranean. It resembles the cod but lacks a barbel on the chin, has smaller pelvic fins and has elongated, light-colored smudges on the flanks. The pollack feeds on small fishes, especially sand eels, as well as worms and crustaceans. It reaches 24 lb (11 kg) in weight and is of some commercial value.

POND SKATERS, familiar aquatic bugs, often seen in groups skating with their long legs over the surface film of still water. Others are found in flowing water, and some live in the sea. These are slender insects although the body is not as elongate as that of the water measurers, another group of bugs that occur on the surface of still water. Within the same population of pond skaters can be found several different kinds of individuals, some with wings, some without and intermediates with very short wings. The front legs are very short, and the rear two pairs only are used to support

In spite of the fact that porcupine fishes blow themselves up and erect their spines, they are sometimes swallowed by sharks. Nobody knows what effect this has on the sharks.

the insect, the middle pair being used for propulsion and the rear pair for steering. Pond skaters feed mainly on small animals that fall onto the water. They lay their eggs in groups on submerged plants.

PORBEAGLES, large sharks related to the mackerel sharks and the great white shark. They are heavy-bodied but streamlined fishes with two dorsal fins (the second quite small), awllike teeth in the jaws and a symmetrical tail. They can be distinguished by the presence of two keels on the slender part of the body in front of the tail and by an additional small cusp beside the main slender cusp of the teeth. Two species are known, the porbeagle *Lamna nasus* of the Atlantic and Mediterranean and the related *L. ditropis* of the Pacific, both species growing to about 12 ft (3.6 m) in length. Along American Atlantic coasts the porbeagle is sometimes known as the blue shark from the dark blue of the back, which changes abruptly to white on the belly. The porbeagles are chiefly found in temperate waters and reach as far north as Alaska, Newfoundland and the British Isles. They are fished for by anglers but do not give quite the same spirited fight as the mackerel sharks. As in other members of this family, the young are born alive after hatching in the uterus of the female. Both species are potentially dangerous to man.

PORCUPINE FISHES, a family of tropical marine fishes in which the teeth in each jaw are fused completely together to form a beak ("Diodontidae," meaning two teeth). They are related to the puffer fishes and are best described as puffer fishes with well-developed spines on the body. The spines, which are modified scales, normally lie against the body. As in the puffer fishes, the porcupine fishes are capable of inflating the body with air or water as a means of defense, the fish presumably being too large, in that condition, to be swallowed by predators. With inflation, the spines on the body become erected and provide an additional means of defense.

The most frequently illustrated species is *Diodon histrix,* one of the several forms commonly sold to tourists dried and inflated as a curio or even as a lampshade.

PORCUPINES, large, spiny rodents belonging to two quite distinct families, one confined to the Americas (family Erithizontidae) and the other to the tropics of the Old World (family Hystricidae). Both families belong to a group of rodents known as the Hystricomorpha.

Old World porcupines include about a dozen species found throughout Africa and southern Asia. They are among the largest of rodents, and the entire body is covered with spines although these vary greatly in length and thickness in different parts of the body and in different species. On the bodies of some species and on the feet of the others, the spines are no more than rather stiff, bristly hairs, whereas in the crested porcupines the spines of the back may reach 14 in. (35 cm) in length. The best known of the Old World porcupines are the crested porcupines, genus *Hystrix,* which occupy the savanna and steppe zones from southern Africa to India. These are large, weighing up to 55 lb (25 kg) and measuring up

to 34 in. (85 cm) long. The black-and-white banded spines on the back are very long and when erected, along with the enormous crest on the head, greatly increase the apparent size of the animal as well as provide a formidable weapon of defense. The quills on the tip of the tail are modified to form a rattle. Each quill is hollow, and the end breaks off as soon as the quill is fully grown, forming an elongated, goblet-shaped structure. These clash together with considerable noise when the tail is shaken. *Hystrix cristata* occurs all around the edge of the Sahara and south through east Africa to Tanzania. It is also found in Italy, but it seems very likely that this represents an introduction, although perhaps dating back as far as Roman times. Rather similar species occur in southern Africa (*Hystrix africaeaustralis*) and from Turkey and Arabia to India and Ceylon (*Hystrix indica*).

The question is often raised as to how porcupines mate. The smart answer, but in fact also the correct one, is "very carefully"—the mating position is not unorthodox as is often suggested. The gestation period is about four months, which is unusually long for an Old World rodent, and the two or three young are well developed at birth. In this they closely resemble the South American hystricomorph rodents.

Crested porcupines are nocturnal, spending the day in deep burrows excavated in a bank, or in caves or crevices in rock. Several animals may share an earth or den, but little is known of their social structure. They are strictly terrestrial, feeding on a variety of vegetable matter but mainly on roots, tubers and bark. They can do considerable damage to crops, especially root crops and melons. The consolation, for those farmers who can catch them, is that they make excellent eating.

Brush-tailed porcupines, genus *Atherurus,* are a rather distinctive group of Old World porcupines replacing the *Hystrix* group in the rain forests of Africa (*Atherurus africana*) and southeast Asia (*Atherurus macrourus*). These are smaller and more slender animals than the crested porcupines, with longer tails. The rattle quills are quite different, being long and ribbonlike, with a number of segments flattened alternately in different planes. Brush-tailed porcupines can and do climb trees to some extent, but are nevertheless predominantly terrestrial like their relatives on the plains.

The American porcupines, family Erithizontidae, comprise about a dozen species. Most of them live in South and Central America, but one species, and by far the best known, *Erithizon dorsatum,* occurs in North America from Mexico to Canada and Alaska. The North American porcupine is about the same size as the larger Old World species but is not so heavy. It is covered with an armor of formidable spines, but when they are relaxed they are almost concealed in a coat of long brown hair (especially well developed in the animals in eastern Canada). There is also a dense woolly underfur. The spines have finely barbed points and are readily detached when they penetrate an adversary, but they cannot be shot at an attacker as is often supposed. This is a woodland species occurring throughout the western states of the U.S.A. and

The North American porcupine climbs well, and its quills are largely hidden in fur. The animal spends a large portion of its time in trees.

the whole of Canada except for the barren tundra. The courtship of this porcupine is noteworthy for the fact that the male sprays the female with urine before mating. As in other porcupines the gestation period is very long, about seven months, and a single, very well developed youngster is born in the spring. It is a remarkable fact that a newly born porcupine is larger than a newly born black bear, although the adult bear is many times larger than the porcupine.

The North American porcupine is less arboreal than the more southern species, but more so than any of the Old World porcupines. It can climb well and spends a considerable time in trees. It is predominantly nocturnal and spends the day in a hollow tree, rocky den or burrow. In summer a variety of vegetation is eaten, but in winter the porcupine becomes more of a specialist, feeding largely on the bark of trees and to a lesser extent on the needles of conifers.

The North American porcupine would seem to be well protected against predators. When it is threatened, it crouches low with the head drawn down between its legs and gnashes its teeth. If the attacker persists, the porcupine leaps round to present a bristling and thrashing rear, which can very quickly plant hundreds of spines in a predator's nose. In spite of these tactics, one species of predator in particular seems to have mastered the art of tackling porcupines. This is the fisher, *Martes pennanti*, a large member of the weasel family, which can dart in to attack the unprotected underside of the porcupine and is at the same time agile

enough to avoid the porcupine's attack.

The Central and South American porcupines mostly belong to the genus *Coendou* and are much more strictly arboreal than the North American species. *Coendou prehensilis* is the best-known species. The spines are short but are not concealed by hair. The tail is very long with the terminal part naked and prehensile like that of the cebid monkeys found in the same habitat. This species is found in rain forest throughout Brazil. They are slow-moving animals, in contrast to the less well-armored monkeys, and live almost entirely in trees where they feed on fruit, buds and leaves.

PORPOISES, small toothed whales of the family Phocaenidae. They are fairly easily distinguished from the dolphins, being usually under 6 ft (2 m) in length, rather tubby in appearance and having a rounded head and no projecting beaklike mouth. The teeth, which are found in both jaws, are spadelike instead of being conical as in the other toothed whales. The dorsal fin is characteristically triangular, whereas in many of the dolphins it is falciform (sickle-shaped). The fore flippers are quite small and spadelike.

The porpoises are divided into three genera, of which one, *Neomeris*, is finless, but the other two, *Phocaena* and *Phocaenoides*, both have triangular fins.

The common porpoise, *Phocaena phocaena*, is the common porpoise of the North Atlantic and neighboring seas. It is by far the commonest cetacean species around the British coasts and most of the North Sea and Baltic, but is found in smaller numbers in the Mediterranean. On

the American coast it is common between Davis Strait and New Jersey. To what extent it is found in the mid-Atlantic is not known. It is said to be essentially coast-loving, but if this is completely true probably the only contact between east and west animals will be via the populations around the coasts of Iceland and Greenland. The common porpoise is also numerous along the Pacific coast of North America from Point Barrow in Alaska to California, where it is known as the harbor porpoise.

It is an attractive, rotund little animal, rarely reaching 6 ft (2 m) and is a very fast swimmer. It is black on the back, shading to white on the belly. The flippers and tail flukes are always black. The small, rounded black flipper is attached to the white part of the trunk, but a black line runs from the angle of the jaw to its root. Occasionally there are small tubercles on the leading edge of the flipper.

The common porpoise feeds mainly on such fishes as herring, whiting, sole, pilchards and mackerel. Fishermen usually complain that the arrival of porpoises means a rapid disappearance of these shoaling fish from the area for several days. Other seafood is also taken. Porpoises breed in the summer, and the young are born a year later. There is a tendency for populations to collect at certain coastal areas for these activities, and in the past this has led to an easy porpoise-hunting season for the local fisherman.

PORTUGUESE MAN-OF-WAR, *Physalia physalis*, a colorful jellyfish of the kind known as siphonophores. It consists of a colony of four kinds of polyps, the most obvious of which is a gas-filled bladder 12 in. (30 cm) long and 6 in. (15 cm) in diameter, which carries a high crest and is colored blue to purple. This normally floats on the surface, and on its underside hanging down in the water are many polyps of three kinds, those concerned with feeding, those concerned with reproduction, and the long trailing tentacles armed with stinging cells that may be up to 40 ft (12 m) long.

Normally an inhabitant of the warmer seas, the Portuguese man-of-war may be carried into temperate latitudes by persistent winds and there cast up on the shore in thousands. The contractile float is hollow and lined with chitin. There is an apical pore with a sphincter muscle that allows the escape of gas when the animal sinks below the waves as it does when the surface of the sea is turbulent. The float can be refilled with gas secreted by a gas gland at the bottom of the float. Analysis of the gas in the float shows it to be a mixture of nitrogen and oxygen together with carbon monoxide; the latter may be up to 13% of the total. Carbon monoxide accumulation is seldom found in animals, and the cells of the gas gland contain high concentrations of folic acid. It is probable that folic acid is involved in the secretion of carbon monoxide into the float. This secretion of carbon gas may serve to inflate the float, and it is slowly replaced by air through diffusion and exchange once the float is above the surface of the water.

POTTO, *Perodicticus potto,* a clumsy-looking, heavily built mammal with forward-facing eyes and powerful, grasping hands and feet. It

The potto, a "half-monkey" that fights with its backbone.

has thick, woolly fur, which is brownish-gray to rufous-brown on the back and paler on the belly. It weighs about 3 lb (1.2 kg) and is the heaviest of the family Lorisidae (see loris, angwantibo). The head and body measure 16 in. (40 cm), and there is a very short tail reaching only 3 in. (8 cm) long.

The potto is completely arboreal and nocturnal, occurring at medium height in the trees of the west central African rain forest. It is more commonly seen than the angwantibo, a closely related mammal which occurs in the same areas. It is omnivorous, like the angwantibo, and similarly relies largely on animal food. Its basic diet probably consists of insects, with fruit and other animal food to make up the total. The canines are large and stout, and seem to be used both in killing fairly large prey and in defense against predators. Another striking device that is used against predators and in butting members of the same species is a row of prominences along the back of the neck, formed by spiny processes of the vertebrae. Using its stout limbs and powerful grasping hands and feet, the potto can cling tightly to a branch and butt a predator until it falls off or moves on. As an adaptation for powerful branch-grasping, the index finger is reduced to a small knob, and the thumb is widely opposed to the remaining fingers.

There is normally one baby at birth, and this crawls straight onto the mother's fur. It is then carried around most of the time on the mother, though it is left hanging on a branch from time to time while the mother is feeding. The gestation period is about 5 months, and the baby is fairly well developed when it is born. Growth is slow, and there is a long period of maternal care before the offspring becomes self-sufficient. Like the lorises and the angwantibo, the potto is solitary or pair-living. It does not make a nest and sleeps curled up in a branch crotch or in a hollow tree during the daytime.

PRAIRIE CHICKEN, the name given to two species of North American grouse, *Tympanuchus cupido* and *T. pallidicinctus*, weighing about 2 lb (1 kg). They were once widespread in eastern and midwestern North America but are now uncommon and only locally distributed in areas of virgin grasslands and prairie. Initially they were reduced by overhunting by the early settlers, but the decline continues, despite some protection, because of the pressures of farming and ranching. The species are probably doomed in many areas but might survive elsewhere if methods of land-use more suited to the prairie chickens' needs could be adopted.

Pratincole *Glareola pratincola*: small ploverlike bird that hunts insects like a swallow.

PRAIRIE DOGS, in spite of the name, ground squirrels of the genus *Cynomys*. They are social animals, very characteristic of the open plains of North America.

PRATINCOLES, widely distributed, small, elegant birds living near water. Pratincoles feed in flight. The genus *Galachrysia* has two species in Africa and one in Asia; and *Glareola* has one species in Madagascar and another, the pratincole *G. pratincola*, with a wide breeding range from the Mediterranean to China, India and parts of Africa.

At rest they resemble small plovers, 7-9 in. long (17-23 cm), with small heads and short bills, long wings and rather long legs. But in the air they are like large swallows, the wings being pointed and the tail deeply forked, the flight buoyant and graceful. Despite the diminutive bill, the gape is very wide, enabling them to catch small insects on the wing. Pratincoles will also pursue locusts, and sometimes pick insects from the ground. The plumage matches the earth on which they nest, being gray or dun brown. The rump, the base of the tail and the belly are white. There is a whitish wing bar and generally some chestnut in the plumage. In all species the bill is black with a carmine base, and the legs are brown or red.

A depression on the ground serves as a nest. The two to four eggs are laid on sun-baked mud in *Glareola pratincola*, on sand in the west African gray pratincole, *Galachrysia cinerea*, and the Indian little pratincole, *G. lactea*, of southeast Asia, or even on bare rock in the collared pratincole, *G. nuchalis*, which inhabits rock-strewn rivers in Africa. Except for *Galachrysia*, pratincoles breed socially, but all are gregarious at one season or another.

PRONGHORN ANTELOPE, *Antilocapra americana*, the only horned animal that sheds its horn sheath (this it does every year), and the only one with branched horns. It differs from the true antelopes (Bovidae) in this shedding of the outer sheath. Males can be distinguished from females by their larger horns and by a conspicuous black spot beneath the ear. They are larger than females. The height at the shoulder varies from 32-41 in. (81-104 cm); head and body length from 39-59 in. (1-1.5 m); tail length 3-4 in. (75-100 cm); and weight from 79 to 134 lb (36-60 kg). The long white hairs of a large rump patch can be erected to form a heliographic disc that may flash for several miles in bright sunlight, sending warning signals from one group to another across open grassland and semidesert shrubland habitats. The brown coat is interrupted by white on the undersides extending halfway up the sides and up the ventral side of the neck, where it is banded by a partial lower band of brown, a complete band halfway up the neck, and a dark brown collar. White cheeks and lips contrast sharply with a black rostrum down the center of the face. This disruptive pattern of brown and white produces a fine camouflage in the wide-open spaces in which the pronghorn lives.

The pronghorn's range extends through the semiarid lands with a rainfall of 7-14 in. (18-35 cm) of western North America from southern Canada to northern Mexico and Baja California, usually at elevations of 3,000-8,000 ft (1,000 to 2,500 m).

Breeding occurs from mid-August to mid-October, at which time bucks defend groups of females. Does are sexually mature at 16 months, and normally give birth to twin fawns after a gestation period of about eight months. Conservation efforts have restored the pronghorn from an estimated low of about 30,000 in 1924 to nearly 400,000 at present.

PTARMIGAN, species of game bird belonging to the grouse family. They are typically birds of the Arctic and subarctic. The rock ptarmigan—in Britain known simply as the ptarmigan—*Lagopus mutus*, and the willow ptarmigan or willow grouse, *L. lagopus*, are circumpolar in

The white-tailed ptarmigan of western North America.

distribution. In northern Britain the willow ptarmigan is represented by the red grouse, sometimes given specific rank as *L. scoticus*. This form is adapted to less extreme environmental conditions and does not molt into a white plumage in winter. Its ecological equivalent in western North America is the white-tailed ptarmigan, *L. leucurus*.

PUFF ADDERS, African snakes of the genus *Bitis*, highly venomous, so called because of their loud hissing or blowing air from the lungs—that is, they puff. The name is also one of several common names for the North American hog-nosed snake, which, although nonvenomous, also tends to blow air loudly, as if in warning, when disturbed.

PUFFBIRDS, New World birds varying in size from that of a sparrow to a thrush and having a thick soft plumage, a large head, rounded wings, short feet and, usually, a strong hooked bill. They comprise 30 species and occur from southern Mexico south through Central and South America to southern Brazil, Bolivia, Paraguay and Argentina.

They live in trees in forests and at forest edges, where they sit motionless and silent in an upright position on a branch. They feed by suddenly darting after flying insects, im-mediately returning to their lookout. They also take beetles and caterpillars from leaves or branches while in flight.

Among the best-known species is the white-necked puffbird, *Notharchus macrorhynchos*. It has a wide distribution from southern Mexico all over Central and South America as far south as eastern Paraguay and northeastern Argentina. The smaller pied puffbird, *Notharchus tectus*, ranges from Panama southward to northern Brazil. The white-necked species is about the size of a thrush whereas the pied puffbird is sparrow-sized. Both are black above and white below with a broad black band across the breast, and both nest in holes that they make with their strong bills in termites' nests in trees. As in all puffbirds, the eggs are white and the clutch is of only two. The newly hatched nestlings are wholly naked, a feature again common to all members of the family.

PUFFINS, stubby, large-headed seabirds of the auk family. The best-known species is the Atlantic puffin, *Fratercula arctica*, which breeds on most coasts around the North Atlantic. It is 12 in. (30.5 cm) long, black above and white beneath, with a white face and a much enlarged laterally-compressed bill. At the beginning of the breeding season this bill is enlarged by the development of a sheath striped in red, blue-gray, and yellow. Also, the skin at the angle of the mouth is enlarged and develops a yellow hue, as does the inside of the mouth, and a small triangular blue-gray plate is developed above the eye. The total effect is somewhat clownlike, but the function of these features is a serious one, being connected with the breeding displays. The feet and legs of the adults are colored vermilion in the summer, fading to yellow in winter.

The Atlantic puffin nests in colonies on grassy cliff slopes on or near the edge of the sea, often on offshore islands. It is the only Atlantic auk that digs a burrow for its nest, but it will

Common auk: the puffin, readily recognized by its colored bill.

appropriate the burrow of a rabbit or shearwater or use a natural cavity. The burrowing activities of a large colony may so undermine an area as to cause it to collapse. One large whitish egg is laid, and incubated largely by the female. The young are fed by both parents on fish, which is first presented in a partly digested state. The puffin carries several small fish arranged laterally in a row in its strong bill, having caught them by underwater pursuit. After about their 40th day the young are deserted by their parents, and after a few days find their own way to the sea. They cannot fly at this stage but make their way out to sea by paddling and dive when attacked by gulls. They apparently have to learn to catch fish for themselves without any parental assistance.

PUMA, *Felis concolor*, often known in the United States and Canada as the cougar. Probably no animal has received so many common names: mountain lion, catamount and painter, to give only a few. As many as 30 subspecies have been described, 14 of which are to be found in North America. The puma ranges over practically the whole of America from western Canada in the north to Patagonia in the southern half of South America. It can live quite happily in mountains, swamps, savanna and forest from sea level to as high as 13,000 ft (4,000 m).

At first glance it resembles a very slender and

The puma, in the United States and Canada called cougar and also known as the mountain lion, has an ill-deserved reputation for ferocity.

The California crested quail lives in mixed woodlands and in the city parks in western North America.

sinuous lioness, although the size varies considerably from region to region. The maximum total length recorded for a male was $9\frac{1}{2}$ ft (3 m), of which about a third was tail, and 260 lb (120 kg) weight, but it may be as little as 4 ft (1.2 m) total length and 46 lb (21 kg) weight. The females are generally smaller than the males. The subspecies found in the tropical zones are generally accepted to be smaller and more brightly colored than those from the farther ends of the range. The head is particularly worthy of note, as it is very much rounder than is usual in cats, and the ears are more rounded and somewhat shorter. The shoulders are very powerful, and the large paws have long retractile claws with terrific ripping power. The color of the short close fur varies considerably from yellowish brown to red, sometimes being darker in the winter. The throat, chest and belly are white. The ridge of the back and the tail is usually marked by a darker line, and the tail is tipped with black, but has no tuft. There have been some records of dark, almost black, pumas, but these do not appear to be common.

The puma leads a solitary life, keeping very much out of sight. It is outstanding for its stamina and strength. It is able to cover up to 20 ft (6 m) in one bound, and a leap of 40 ft (12 m) has been recorded. It can also leap upward to a height of 15 ft ($4\frac{1}{2}$ m) and has been seen to drop to the ground from a height of 60 ft (18 m). It will travel 30-50 miles (48-80 km) when hunting, but its usual range is restricted to an area of about 12 sq miles (40 km²).

It is generally accepted that a puma screams in a bloodcurdling manner. When contented it will purr like any domestic cat. Its screaming may have contributed to an undeserved reputation for ferocity, but records show that attacks on human beings are very rare.

The favorite food of the puma is one type or another of deer, although an amazing variety of food has been recorded, from slugs and snails to porcupines and very rarely moose and buffalo. It sometimes attacks domestic stock such as sheep and goats, as well as horses and cattle, and this is the main reason why the puma has been wiped out or its numbers seriously reduced in parts of its range. 1-6 kittens are born in a den among rocks or dense thicket at any time of the year after a gestation of 90-96 days. At birth the kittens are blind but well furred, spotted and with a ringed tail. They are up to 1 ft (30 cm) long and weigh up to 1 lb (0.45 kg). Their eyes open in 7-14 days, and they are weaned between 1-3 months, but these periods seem to vary considerably over the range. As they mature, the kittens lose their spots and the rings on the tail. The mother takes them out hunting when they are 9-10 weeks old, but they do not become truly independent and leave the mother until they are about two years old. The life span of the puma in the wild is up to about 18 years, but it may live longer in captivity.

PYTHONS, the Old World equivalent of the New World boas, and like them bearing small spurs that represent the vestiges of hind limbs. These two snakes are clearly the closest living relatives of the ancestral snake type. There are seven genera in the subfamily Pythoninae, inhabiting the warmer regions from Africa to Australia. The largest and best known, the true pythons, belong to the genus *Python*. They all

have bold color patterns, mainly in browns and yellow. Three live in Africa: the African python, *P. sebae*, which reaches a length of 32 ft (9.9 m); the ball python, *P. regius*, which when molested rolls itself into a tight ball with its head inside; and the Angolan python, *P. anchietae*. Several other species are found from India to China and the East Indies. The largest species, the reticulate python, *P. reticulatus*, reaching a length of 33 ft (10 m), ranges from Burma to the Philippine Islands and Timor. Although so large, it has been found to be remarkably inoffensive in the wild, and most accounts of its attacks on humans are exaggerated or invented. The Indian python, *P. molurus*, with a length up to 20 ft (6 m) is found from India to China and on some of the islands of the East Indies.

The pythons live in a wide variety of habitats. Some live near water like the Indian python, which is almost semiaquatic, others prefer jungles and climb trees, whereas the African python lives in open country. The reticulate python has a preference for living near human settlements. They kill their prey by constriction, small mammals being preferred, but birds, reptiles, frogs and even fish are also taken. The larger African pythons also take small antelopes. A large python can swallow prey weighing up to 120 lb (54 kg), and one 18-ft (5.5-m) African python is known to have eaten a leopard, but this is exceptional.

All the pythons lay eggs usually 3-4 months after mating. The number varies considerably from eight to over a hundred. The female pushes them together in a heap and coils herself around them, brooding them for 2-3 months. Most pythons merely guard their eggs, but some, including the Indian python, actually incubate them by keeping their body temperature several degrees above that of the surrounding air.

QUAILS, two distinct groups of game birds, the Old World quails and the New World quails. Old World quails include the genus *Coturnix*, and birds sometimes placed in the genera *Excalfactoria* and *Synoicus* as well as the rare mountain quail, *Anurophasis*, of New Guinea and the pretty *Margaroperdix* of Madagascar. There are a number of species found in Africa, *Coturnix africana*, *C. delagorguei* and *C. adamsoni*, and in Asia, *C. coturnix*, *C. coromandelica* and *C. chinensis* as well as in Australia, *C. pectoralis*, and *C. chinensis*. The larger and duller swamp quail, *C. ypsilophorus*, is confined to Australia, New Guinea and the Lesser Sunda Islands.

Old World quails are small, rounded birds with dainty bill and legs and no visible tail. They vary in size from the comparatively large mountain quail, which is only a little smaller than the common partridge, *Perdix*

perdix, down to the pretty little painted quail, *C. chinensis*, which is no bigger than a finch.

The most widespread and well-known species is the common quail, *C. coturnix*, which is found throughout most of Europe, Asia, parts of Africa and the Atlantic islands. There are five recognized races or subspecies, of which one, *C. c. japonica* of Japan, has been domesticated for centuries, being kept in small cages to produce both eggs and meat. By means of selective breeding these birds now mature very quickly, and females have been known to lay eggs only 12 weeks after they themselves were hatched. In recent times the commercial farming of the Japanese race has spread to the United States and Europe, the quail being kept in large numbers in small battery cages. The call of the male Japanese quail is much harsher than that of the nominate race.

The common quail may be found in any open country provided there is enough cover available such as tall grass, lucerne, clover and cereals. When disturbed, quail prefer to run and hide rather than fly, and for this

reason they are difficult to see in the wild. Indeed, the first indication of the birds' presence is usually the males' characteristic and often-repeated callings. This has often been described as sounding something like "wet-my-lips," but if the observer is close enough to a calling male at the height of the breeding season he may hear a soft "va-va" immediately before the louder notes.

The nest of the common quail is a grass-lined hollow scraped in the ground under the cover of dense vegetation. The clutch consists of 7-15 eggs, which are pale buff, spotted and blotched with dark brown. The chicks hatch after 21 days and leave the nest as soon as they are dry, following their parents.

The common quail eats grain, seeds, shoots and insects, the young feeding on insects alone for the first few days.

The tiny painted quail is a popular aviary bird and is one of the most colorful quails. The male has a distinctive black and white pattern on his throat, a slate-blue breast and chestnut-red belly. The female is more somber and is patterned with buff and darker brown streaks. It inhabits grassland and low vegetation in tropical regions from southern Asia to Australia.

The American quails differ from Old World quails in anatomical features, and they are usually larger and rather more colorful, but their habits are similar. There are some 36 species, of which the best known is the bob-white, *Colinus virginianus*, one of the most popular game birds in the United States. There are a number of races found from southern Canada to the border of Guatemala. It lives on open plains and gets its name from the male's loud call, which sounds something like the words "bob white."

QUETZAL, *Pharomachrus mocino*, a large bird closely related to the trogons and distinguished by the dazzling beauty of the male. The breast and upperparts of the plumage are of shimmering, iridescent green, contrasting vividly with the crimson and white belly. The female is much duller in appearance. When the male comes into breeding condition, four of the upper tail coverts grow into gorgeous metallic-green plumes, the central pair extending 2 ft (0.6 m) beyond the tail.

The quetzal is found only in the dense mountain forests of Central America, from southern Mexico to Costa Rica. Like other trogons, quetzals nest in holes in trees. When the male enters the nest hole to incubate, he turns about, so that, as he sits facing the entrance, the long tail plumes are bent over his back, their ends trailing outside.

QUOKKA, *Setonix brachyurus*, or short-tailed scrub wallaby living in thickets and freshwater swamps in remnant colonies in Western Australia, on Rottnest Island near Perth and on Bald Island near Albany. These coarse-haired marsupials are brownish-gray in color with a total length of $28\frac{1}{2}$-$37\frac{1}{2}$ in. (72-95 cm). When moving quickly they hop on their hind legs. They feed mainly at night on grasses and other plants. The female usually has only one young, which stays in her pouch for about five months. As in some other marsupials and phalangers, delayed birth is a feature in the quokkas.

The resplendent quetzal has been adopted by the Guatemalans as their national symbol.

RABBIT, a small- to medium-sized terrestrial herbivore, usually having long ears and a very short tail. Members of eight genera, *Pentalagus, Pronolagus, Romerolagus, Caprolagus, Lepus, Sylvilagus, Nesolagus* and *Oryctolagus* are called rabbits. The European wild rabbit, *Oryctolagus cuniculus*, is the best known and most widely distributed. Although originally confined to the western Mediterranean region, it has spread with the advance of agriculture over western and central Europe. It has been introduced in recent times to most parts of the world, notably Australia and New Zealand and numerous islands, where it has attained plague proportions. It has a high degree of adaptability and a high rate of reproduction, which assure its continuing success. It is the only fully domesticated member of the order Lagomorpha, and numerous domestic breeds exist. The Angora rabbit was recorded in Roman times. The normal color in the wild state is agouti, but occasional sandy, black, gray, white, and piebald individuals are found. Bucks and does do not vary greatly in size, having a head and body length of about 18 in. (45 cm) and weighing up to 5 lb (2.2 kg).

Oryctolagus attains maturity at three months of age and is capable of breeding at monthly intervals practically all the year around. It is an opportunist breeder, taking advantage of suitable conditions. Gestation occupies 28 days. Litters are deposited in separate breeding chambers lined with grass and the mother's fur. The entrance to the nest chamber is sealed with an earth plug and is marked with urine and feces to discourage other rabbits from entering. The young are born naked and are blind for the first ten days. They are suckled once a day for three weeks.

Oryctolagus lives in discrete social groups each with its own linear order of dominance based on aggression, and separate for each sex. Territory is respected by strangers and defended by all members of a group; it is demarcated by the scent of the secretion of the anal gland coating fecal pellets. Regular defecation sites, maintained mainly by males, are important in this demarcation. The chingland secretion and urine are also used in marking. All of these sources of odor, together with inguinal gland secretion, are used in interspecific communication. Stress arising from social subordination affects the physiology of rabbits generally and the reproduction of the does particularly. Compared with dominant does, subordinates show a higher incidence of intra-uterine mortality, and their progeny are less healthy, grow more slowly and have a lower survival rate. Despite this population self-control mechanism, the numbers attained are not tolerated by modern agriculturalists. Shooting, ferreting, trapping, snaring, digging out, poisoning and gassing, as well as biological control methods, are used to reduce numbers. Myxomatosis, an epidemic disease caused by a virus, was introduced into Australia and later England and France to control rabbit numbers.

RACCOON, probably the best-known American mammal, this carnivore belongs to the family Procyonidae, to which coatis, kinkajous and pandas also belong. The stout body is roly-poly and bearlike, an impression enhanced by the short fore paws and pigeon-toed gait. The famous black "mask" across eyes and cheeks and the five to eight black bands on the tail immediately identify it. Although the overall coloration is grayish, the long guard hairs are buff, tipped in black, and the thick underfur is creamy white. Several island species exist which are similar in appearance, but the two mainland species are quite distinct; the raccoon *Procyon lotor*, distributed from Canada south to northern Central America, and the rarer crab-eating raccoon, *Procyon cancrivorus*, ranging from Panama to Brazil, which looks like a larger, short-haired version of the northern species. 'Coons, as they are familiarly known, measure from 24-40 in. (60-100 cm) and weigh from 4.4-44 lb (2-20 kg), depending on the area and abundance of the available food supply.

No webbing is present between the elongated, spindly fingers, which can therefore be spread widely apart and are capable of an almost monkeylike dexterity. The claws are non-retractile, and the digit extremities have a well-developed network of sensitive touch nerve fibers. Appropriately called "wasch-bären" (wash bear) in German, the raccoon has the habit of "washing" any object or food item between its fore paws. Although this is a common behavior pattern in captivity, it probably does not occur in the wild, where raccoons usually search for crayfish, turtles, mussels, frogs or small fish in shallow pools,

The rabbit has everywhere become a pest.

using both fore paws simultaneously, and then swallow the prey forthwith without such meticulous preamble. Raccoons are omnivorous, feeding on almost any vegetable or animal they happen to come across and also scavenging in rubbish heaps. Considered intelligent by some, raccoons are actually more ingenious and persistent and are considerably aided in their endeavors by their manual dexterity. Unlike other small North American mammals, 'coons are not on the decline at present, having managed to adapt quite readily to the impact and proximity of man. Considerable numbers have been captured by fur trappers over the past two centuries, and the 'coon cap with the tail dangling down the back was a favorite among early settlers.

Descending from their tree perches at nightfall, raccoons forage alone or in small family groups over their restricted, 1-sq-mile (2.2-sq-km) home range. Young raccoons can climb with ease and agility, but increasing weight may hamper the bulkier adult males who are more slow and deliberate in movement and carefully slide down from trees tail first. Nor do they dive when swimming, dog-paddling instead, using their bushy tail as a rudder. In the colder regions, raccoons sleep for most of the winter in hollow logs or underground dens, living on stored body fat until February, when the thaw heralds the start of the breeding season. In the southern United States, where these carnivores remain active throughout the year, the breeding season can begin in early January, sometimes resulting in two litters a year. The young are born after a 60-70-day gestation in a tree nest. Litter size varies from one to six cubs, more usually four, which are blind and small, 2.5 oz (70 gm) at birth. They leave the nest at seven weeks and can follow the parents on hunting forays a month later.

RACCOON DOG, *Nyctereutes procyonoides*, so called for the patch of black around and under each eye resembling the "robber mask" of the raccoon. It is a short-legged wild dog with a total length of about 2 ft (61 cm). Its long fur is yellowish brown with dark hair on the shoulders, the tip of the bushy tail and on the legs. Raccoon dogs are native to the forested areas of eastern Asia, where they are hunted for their valuable pelt. They have been introduced into parts of eastern Europe and are now rapidly spreading west.

RADIOLARIANS, single-celled animals possessing a skeleton usually siliceous but sometimes of strontium sulphate. They are all marine, and the majority of them are pelagic and occur in large numbers within the plankton. The cytoplasm is divided into an ectoplasm and an endoplasm, and the two layers are separated by a membrane called a central capsule, which is characteristic of the group. The form of the skeleton varies within the different suborders. In some species it is absent or very much reduced, but in the majority it is spherical and possesses numerous spikelike outgrowths. The elaborations on this simple theme are many, and radiolarian skeletons possess an amazing variety of spines and hooks embedded in some kind of latticework. When these animals die their skeletons sink to the bottom and there, over the years, build up into thick oozes covering large areas of the ocean

The raccoon of North America in a characteristic pose, feeling in the water for its food.

The purple gallinule *Porphyrio porphyrio* on a marshy lake in Kenya is also a species belonging to the rails.

The ratel or honey-badger of Africa.

bed. The skeletons can be separated from these oozes by soaking them in acid, to dissolve out the organic matter mixed with them. In the living animals the ectoplasm is very vacuolated, due to the presence of gases. This is a flotation device, and in rough weather the bubbles burst and the organisms sink. Later, the vacuoles re-form and the organisms again rise to the surface when the weather has improved. The endoplasm, lying within the capsule, contains the nucleus or nuclei (for some forms possess several nuclei) and a number of other inclusions such as oil droplets, granules of pigment and algae. The outer surface of the ectoplasm is very fluid and can be thrown out into filamentous projections that may fuse to form a feeding net beyond the skeleton. The whole of this outer surface is used for trapping particles of food, which are drawn into the ectoplasm to be digested in the region just to the outside of the capsule. Radiolarians reproduce in a variety of ways. Some, which do not possess skeletons or have only a few skeletal elements, are able to reproduce by simple division. In others, the contents of the central capsule divide up into flagellated structures known as swarmers, and these leave the parent to grow into adults on their own.

RAILS, term used for certain species of water or marsh-dwelling birds. The type genus *Rallus* contains a number of species that together inhabit most of the world. The water rail, *R. aquaticus*, is the only palearctic species. It is 11 in. (28 cm) long and has darkly streaked olive-brown upperparts, flanks barred with white and near black, and slate-gray underparts from the breast forward and up onto the sides of the head. It has long brownish feet and legs and a long red bill. It is discontinuously spread across the whole of the palearctic from Iceland to Japan, wintering around the Mediterranean and in southern Asia. On migration it is frequently attracted to lighthouses, under certain heavily overcast conditions, and is killed in large numbers.

The plumage, form and general habits of the water rail may be taken as representative of most crakes and rails, specific differences deriving mainly from the occupation of different ecological niches. The water rail's nest, for example, is of a fairly generalized type, built of dead reeds and other plants and hidden among the dense vegetation of reedbeds, marshes or the edges of lakes or rivers. There are usually 6-11 eggs in a clutch, sometimes more or less. Both sexes take part in the incubation of the eggs and care of the young. The nestlings are covered with a dark down and leave the nest soon after hatching.

The rails make a great variety of sounds—some of which are most unbirdlike—including various screams, cackles and squawks. The water rail has a particularly distinctive call, a "singular discordant noise generally beginning as a kind of grunting and ending as a squeal, rather reminiscent of the squealing of pigs, and sometimes rising to a regular scream." As the rails are secretive and crepuscular, such unusual vocalization is presumably a substitute for visual communication.

In North America the equivalent of the water rail is the very similar Virginia rail, *R. limicola*. Other North American *Rallus* species are the king rail, *R. elegans*, a larger bird of the freshwater marshes, and the clapper rail, *R. longirostris*, inhabiting salt marshes.

RATEL, or honey badger, *Mellivora capensis*, a badgerlike mammal, heavy-set and powerful. The ratel's aposematic coloration is reminiscent of the grison: gray or white above and black on the limbs and ventral surface. Completely black, presumably melanistic, specimens are found in certain forest areas. Males may reach 32 in. (80 cm) in length and weigh over 30 lb (15 kg). It is widely distributed from southeast Asia, throughout most of Africa down to Cape Province.

A strange commensal association has developed between the ratel and the honey-guide bird *Indicator*. If a ratel is in the vicinity when the honey guide gives a series of call notes, it will follow the bird to the bee nest, rip it open, and both will then feast on the combs and larvae. Usually crepuscular or nocturnal, the ratel can be found in the open steppe of arid regions or in humid primary forests. Terrestrial, it may climb trees when searching for nestlings or honey. Like many other mustelids, it is omnivorous, feeding on an assortment of berries, lizards, snakes, eggs and even carrion when hard pressed. Ratels will hunt singly or in pairs, covering up to 20 miles in a single night's foraging with a tireless jog trot. Strong claws used for digging, nauseating scent glands and muscular jaws serve its

White-necked raven of eastern and southern Africa.

Although the brown rat is the predominant pest in Europe and temperate America, on a worldwide basis the dominant villain is undoubtedly the species that in Britain is called the black rat, *Rattus rattus*. The black rat occurs in a bewildering variety of races and varieties, few of them black, especially in tropical Asia, and it has been argued that the brown rat is no more than a subspecies of the ubiquitous black rat. This group of rats originated in southeastern Asia, where many forms, and also many other distinct species of *Rattus*, still live an outdoor life independent of man. The spread of rats from this ancestral area probably began as soon as man himself began to use vehicles for transport. The black rat was the dominant rat in European towns until the 16th century, when it began to be replaced by the brown rat, which arrived in North America toward the end of the 18th century. In these temperate areas the brown rat can live outdoors, on farms, in fieldside ditches, on the seashore and in rubbish dumps, whereas the black rat is now almost completely confined to cities, especially seaports. Even in the tropics most forms of black rat are found in and around houses and tend to be replaced by other species in the fields.

These rats of the genus *Rattus* are very unspecialized members of the family, and their structure differs little from that of a house mouse. The tail is about the same length as the head and body and almost naked, and the fur tends to be harsh. The black rat has a more slender build than the brown rat, with a longer tail and more prehensile feet, and it is a much more able climber, being commonly found in the upper stories of buildings, whereas the brown rat is much more at home burrowing in the foundations or in the drains. The brown rat is a capable swimmer and in Europe is frequently mistaken for the water vole, *Arvicola amphibius*.

Both species are prolific breeders. The gestation period is short, 21-24 days, the litter size may be very large, up to 12, although six or seven is more usual, and litters may follow one another in rapid succession. Rats have a strong exploratory urge, and this, combined with their ability to feed on almost anything edible, makes them versatile and persistent pests. They are major pests to stored food of all kinds throughout the world; in addition to their ability to transmit a great number of dangerous diseases, this makes their control of major economic and medical importance. Black rats, by way of their fleas, were responsible for transmitting the outbreaks of plague that decimated the population of Europe during the Middle Ages, and rats also transmit typhus, tularemia and *Salmonella* food poisoning. Control of rats is now usually carried out by means of poisoning by Warfarin, but a recent problem is the development, apparently quite independently in many areas, of populations that are immune to the usual dosages of Warfarin.

The largest rats in Africa are the giant pouched rats, *Cricetomys*. These measure up to 18 in. (45 cm) long without the tail, and the terminal half of the tail is white. In spite of their size they are docile, inoffensive animals and are considered good eating. They have recently

reputation as a fearless fighter. Its thick skin, hanging like a loose coating of rubber, makes it apparently impervious to tooth, fang or sting, and no adversary seems formidable enough for it. Rushing from its burrow, a ratel will charge an intruder as large as an antelope or buffalo and is reported to be a match for a pack of dogs. Once it has bitten, it never relaxes its grip, snarling and shaking its head until the victim drops from exhaustion or shakes it loose. A tongue-clicking sound is given in greeting, while an unmistakable growl serves as a warning. Two cubs are born in

the underground den or hollowed-out tree trunk after an unusually long gestation of six months. The ratel may live to the age of 24 in captivity.

RATS, a vast number of species of rodents belonging to a number of distinct families. The smaller members of all these groups tend to be called mice, and there is no sharp distinction between mice and rats. The best-known rat is the brown rat, *Rattus norvegicus*, which is equally familiar as a farmyard and warehouse pest or, in its albino form, as the white rat that is so widely used in laboratory experiments.

been considered to belong to the family Cricetidae rather than to the Muridae on the basis of their tooth structure and the presence of internal cheek pouches that are used for carrying food.

The Philippine Islands are the home of some of the most remarkable of all the murid rats. The bushy-tailed cloud rat, *Crateromys schadenbergi*, for example, has long black fur and a long bushy tail that would look more in place on a tree squirrel. Like the tree squirrels, it is strictly arboreal. In contrast, the shrewlike rat, *Rhynchomys soricoides*, of the Philippines has a long slender snout like a shrew and only two rather degenerate cheek teeth in each tooth row.

In New Guinea and Australia there is a group of water rats that forms a rather distinctive subfamily of the Muridae, the Hydromyinae. The Australian water rat, *Hydromys chrysogaster*, for example, is a large rat with the terminal half of the tail white and the fur closely resembling that of an otter or other highly adapted aquatic mammal, having dense woolly underfur and sleek, flattened guard hairs.

RATTLESNAKES, named for the rattle on their tail, belong to two genera *Crotalus* and *Sistrurus*. They are unique to the Americas. There are about 30 species and 60 subspecies, ranging from Canada southward through South America. In the United States, they are the most respected and feared of snakes. All species north of Mexico have a hemotoxic or blood poison, while many south of there have a mixture of blood and nerve poison. Fortunately, a single polyvalent antivenine is now available for all rattlesnake bites.

The rattle, the most outstanding feature of

the hiss of escaping steam or the rapid crackling of frying fat. When wet, the rattle makes no noise.

Rattlesnakes are pit vipers. The pits, one on each cheek, between the eye and the nostril, are extremely sensitive heat detectors. Rattlers have movable fangs, which fold parallel to the roof of the mouth when not in use and are shed about every three weeks. New fangs move in behind the old ones just before they are shed. Rattlers have eyes equipped for both day and night vision. A vertical elliptical pupil dilates after dark to become round, admitting more light. All rattlers give birth to living, poisonous young.

The first rattler encountered by the white settlers in North America was the timber or banded rattlesnake. Early accounts led it to be named eventually *Crotalus horridus*, for obvious reasons. It averages 4 ft (120 cm) in length and may attain 6 ft (182 cm). The black or brown bands on a brown or yellowish body camouflages it too well on the forest floor.

The next important snake the settlers had to contend with was *C. adamanteus*, the eastern diamondback rattlesnake. *C. adamanteus* is one of the truly dangerous snakes of the world and inhabits the southeastern area of the United States. Although accounts vary, it is quite certain that 8-ft (243-cm) specimens have been captured. Such a snake would weigh well over 15 lb (6.8 kg) and be 15 in. (38 cm) in circumference. An adult diamondback could have fangs ¾ in. (19 mm) long, capable of being driven through almost any boot and of injecting well over a lethal dose of venom into a victim.

Much of the western great plains are infested with the little prairie rattler, *C. viridis*. It is too

too like mice but, unlike the *Crotalus*, will also eat frogs.

RAVENS, several large dark crows of the genus *Corvus*, including the raven or common raven, *C. corax*, the African white-necked raven, *C. albicollis*, the Australian raven, *C. coronoides*, the thick-billed raven, *C. crassirostris*, the American white-necked raven, *C. cryptoleucos*, the fan-tailed raven, *C. rhipidurus*, and the brown-necked raven, *C. ruficollis*. These do not form a natural group of related species, but a rather haphazard assemblage of the larger crows, the smaller forms being called crows, jackdaws or rooks. Being large crows they have proportionately heavy bills, especially the African white-necked and thick-billed ravens.

The ravens are all mainly black in color. The African white-necked raven and the thick-billed raven have a white half-collar on the back of the neck, while the brown-necked raven and the fan-tailed raven have a brown patch in the same position. The Australian raven and the American white-necked raven both have black napes, but the feather bases are white.

The common raven is found in North America, Europe and Asia, most commonly in open or rather wild country. The Australian raven is confined to Australia, and the American white-necked raven to the U.S.A. and Mexico. The other three species are found in Africa north of the equator, the fan-tailed raven being characteristic of desert areas, the brown-necked raven of desert, subdesert and wooded savannas and the African white-necked raven of both arid areas and savannas with scattered trees.

All of the ravens are omnivorous, eating animal carrion, other birds' eggs, seeds, fruit, fishes, insects, small reptiles and amphibians. Some species catch small mammals and even fully grown birds at times. All species usually nest either in tall trees or on ledges of cliffs or rocky crags. The nests are substantial cups or baskets of sticks, lined with grass, hair, dead leaves or feathers. Ravens lay from three to seven pale blue or green eggs with dark brown or gray spots and freckles. Only the female incubates (period 19 to 22 days), living mainly on food that the male brings to the nest and regurgitates for her. The young are fed by both parents (for about 40 to 50 days) while they are in the nest, and for some weeks after fledging.

RAZORBILL, *Alca torda*, black and white seabird of the North Atlantic, named from the shape of its laterally compressed bill. It breeds in colonies in crevices and clefts on rocky coasts, principally in the Old World. Its food consists largely of fish, which it catches underwater, using its wings as flippers for swimming. It is somewhat similar in its habits to the guillemots and lives in the same areas, but competition is avoided by differences in food and nest site preferences.

RAZOR SHELLS, known as jackknife clams in America, are bivalve mollusks with elongated shells and a large powerful foot specialized for deep burrowing into sand. The slightly curved, oblong shell is up to eight times longer than broad, and of a constant width; shaped, in fact, like an old-fashioned cutthroat razor. In the strict sense the name applies to the family

Two views of the pod razor shell: side view *(above)* and dorsal or hinge view *(below)*.

these snakes, is composed of a material similar to our fingernails. The baby rattlesnake is born with a button at the end of the tail, and the first time it sheds its skin the piece next to the button remains, and each time the skin is shed, normally three or four times a year, a new segment is added to this, so building up the rattle. Each new segment is convoluted to fit loosely into the previous segment, and it is the clicking together of these segments that produces the characteristic rattle. The rattle is vibrated so fast that it may appear as no more than a blur, and the noise it makes is like

small normally to cause death to a treated human, but untreated horses and cattle bitten on the nose while grazing occasionally die as the resulting swelling blocks the air passage. *C. viridis* seldom exceeds 4 ft (120 cm) in length and averages about 2 ft (60 cm). It is gray-green with a series of dark blotches on the back. Its range extends from Canada to Mexico.

The genus *Sistrurus* embraces the little rattlers of the eastern United States. These are dangerous because their tiny rattles sound like insects and because their dark colors hide them in the swampy or wooded areas where they live. They

Red deer stag roaring to warn off others. Note blood on antlers.

Solenidae. They are closely related to the sunset shells, family Tellinidae, and some members of this second family have adopted a similar mode of life, and their structure and habits are very similar to those of the Solenidae.

The shell of a razor shell gapes at both ends. At the front end, which always lies at the bottom of the burrow, the large, powerful foot may be protruded. The hinge and ligament are also very near the front end, instead of more or less central as in most bivalves. At the hind end the siphons always protrude. These are rather short and stout, joined for most of their length, separated only near their openings, and water enters by one and leaves through the other after passing over the gills within the mantle cavity. When undisturbed, particularly when covered by water, a razor shell will lie near the top of its burrow with the siphons just protruding from the key-hole-shaped aperture. At the slightest disturbance the powerful foot is distended by a flow of blood into the cavities enclosed by its muscle strands and anchors in the sand. Then the shell is pulled down after it by the contraction of the foot retractor muscles. In seconds, the mollusk lies safely at a depth of well over a foot (0.4 m), penetrating the sand with the help of its streamlined shape.

Razor shells are found on sandy beaches and in shallow sublittoral waters throughout the world. The common European pod razor, *Ensis siliqua*, is one of the best known and has a wide distribution from Norway to Morocco.

RED DEER, *Cervus elaphus*, or wapiti, a deer with a rich red-brown summer coat widely distributed in the northern hemisphere north of about latitude 30°N in North America, where it is known as the wapiti, and 25°N in Europe and Asia. In Britain the name "red deer" is used.

In Europe the largest deer are found in the Carpathians in the eastern part of their range, where a good stag will stand about 54 in. (137 cm) at the shoulder, which is some 12 in. (30 cm) higher than the average stag from Scotland.

The general color of the European red deer is a rich reddish brown in summer which becomes grayish brown in winter. Color abnormalities include white, cream and albino animals.

During the autumn rut, when each stag collects a harem of hinds, all challengers are greeted by a lionlike roar, and fights between rival males often occur, and on occasions will end fatally for one of the contestants. On rare occasions antlers have become locked together, which has resulted in the death of both participants from starvation.

Good specimens of western-type red deer antlers should bear at least 12 points, the terminal points clustering in the form of a crown. This crown formation, however, is lacking in true wapiti antlers, the two or three terminal points pointing upward in the same plane as the beam. It is rare for wapiti antlers to exceed 16 points, but a 17th-century German red deer head once ran to 66 points. Stags

without antlers at any season do occur, and are known as hummels.

Compared to the red deer of western Europe, the wapiti of North America is a much larger animal, with good bulls weighing up to 1,000 lb (454 kg) live weight, and standing close on 60 in. (152 cm) high at the shoulder. In color, this deer resembles the red deer, except that in summer it is not quite so red. A typical feature of wapiti coloring is the light rump patch, which is more prominent than in the red deer. The rutting habits of the wapiti are similar to those of the red deer, except that instead of a lionlike roar, the wapiti utters an undulating bugling noise that is prolonged before dropping into a series of grunts. When both deer are present, as happens in South Island of New Zealand, where both were introduced during the latter part of the last century, interbreeding will take place.

REDSHANK, *Tringa totanus*, common wading bird of Europe and Asia, named for its red legs. The spotted redshank, *T. erythropus*, of northern Eurasia, is closely related.

REEDBUCK, a delicate, medium-sized antelope related to the waterbuck. It has slender, ridged horns that are simply curved back and then up, 8-10 in. (20-25 cm) long. It is further characterized by its stiff gray coat and the large bare patch, possibly glandular, below each ear.

There are three species of reedbuck. The common reedbuck, *Redunca arundinum*, is 30-36 in. (76-91 cm) high, with a large bare muzzle. It is light gray-fawn, becoming tawny on the neck and whitish below. The horns are simple and concave forward. The Bohor reedbuck, *R. redunca*, is smaller, 27-31 in. (67-78 cm) high, with a smaller bare area on the muzzle. The horns are strongly hooked at the tips and longer than the head. The color is yellowish, white below as before, but with the head and neck coloration not contrasting with that of the body. Finally, the mountain or Rooi reedbuck, *R. fulvorufula*, is the same size as the Bohor, but the horns are only slightly hooked at the tips, and shorter than the head. The tail is more bushy, and there are distinct reddish tones, especially on the head and neck. The common reedbuck is found from Cape Province north to Lake Victoria in the east, and southern Gabon in the west, in savanna areas. The Bohor reedbuck is found in northern savannas, from Senegal to the northeastern Congo, and as far north as 15°N in the Sudan and central Ethiopia. Its range overlaps that of the common reedbuck throughout Tanzania. The mountain reedbuck has a disjunct distribution, being found only in hilly country in three areas, in each of which is a distinct subspecies.

The common reedbuck is not highly gregarious, being usually found in pairs, or solitary and never in numbers above six. When alarmed, the group runs off, its members staying together with tails held bolt upright. This contrasts with other reedbuck, which scatter when alarmed. Bohor reedbuck hold their tails pressed down between their legs when running, bounding along with their limbs extended. All species have a shrill alarm whistle.

Mountain reedbuck are much more gregarious, associating in herds of up to 20. They occur

Females and young of Chanler's reedbuck.

on mountainsides among bushes and dry grass. When alarmed they run off, down the mountain or around it obliquely, going for 300-400 yd (275-365 m) before stopping to look back. They come down from the hillsides at night to drink, returning in the morning.

Bohor reedbuck travel in small groups, usually one buck with three or four does. They favor swampy valley bottoms with dense grass. Young males live together in small groups. When disturbed, the reedbucks stand with one hind foot in advance of the other, broadside on to the intruder with the heads turned, sniffing the air and watching.

REINDEER, *Rangifer tarandus*, a large ungainly-looking deer, the males and females both

Dorsal view of a remora, showing the large oval sucker on top of the head.

bearing antlers. It has a wide distribution in both eastern and western hemispheres, being referred to as reindeer in the east and caribou in North America.

REMORAS, or shark suckers, curiously specialized fishes in which the first dorsal fin has become modified into a sucking disc by which the fish attaches itself to sharks, other large marine creatures or even to ships. The body is elongated and flattened on top, and the sucker lies over the head. The sucker is one of the most remarkable examples of the adaptation of a body part for quite a different function from its original one, since this disc is in fact a highly modified fin. The finrays have become flattened and deflected alternately to the left and to the right to form a series of ridges, sometimes with serrated edges. The rim of the sucker is raised, and the platelike finrays can be adjusted to form a strong vacuum. The grip of a remora is remarkably strong, especially on a slightly rough surface, and it is possible to tear the disc from the fish before the latter will relinquish its hold. The various species are distinguished chiefly by the number of plates or lamellae in the disc, which range from ten in *Phterrichthys* to 20-28 in *Echeneis*. Some species are more elongated than others, but in all the belly and the back are the same general color. This, coupled with the fact that the lower jaw is longer than the upper, has earned these fishes the Spanish name of "reverso," the fish appearing to swim upside down.

It is not certain what advantage is derived by the remora from attachment to another animal. The arrangement may simply be one of phoresy or the transport of one organism on the body of another. It is also possible that the remora feeds on the scraps left by a shark. There have been many records of remoras entering the mouths of Manta rays and several species of large sharks, as well as some of the larger billfishes. In this, they seem to resemble the cleaner fishes, but there is no evidence that they undertake the cleaning duties (the removal of parasites) of the former. In at least one instance, fairly large remoras (*Echeneis naucrates* in this case) have been recorded from the stomach of a sand shark. Remoras are good swimmers and often leave their host to forage and then return into the water. When the remora has fixed itself to some large fish or turtle, the line is then hauled back to the boat. It is remarkable that this fishing method should have been evolved independently by fishermen in the western Atlantic and in the Indo-Pacific region.

RHEA, two species of large, flightless, running birds of South America, related to the ostrich within the general group of "ratites." Rheas stand up to 5 ft (1.5 m) tall and weigh up to 55 lb (25 kg), which makes them the biggest birds in the New World. They have a floppy loose-webbed plumage and rather larger wings than the other ratites. The legs are long and powerful, and the feet bear three toes only. They run very swiftly and frequently keep company with bush deer, *Dorcelaphus bezoarticus*, or even cattle. The larger species, the common rhea, *Rhea americana*, is a gray-brown bird with white tail feathers, but a white variant is not uncommon. It breeds from northeastern Brazil to central Argentina. Farther west it is replaced by Darwin's rhea, *Pterocnemia pennata*, which is smaller and spotted with white. The sexes are similar.

The rheas live in grassland and open bush country, often near rivers or swamps. They feed on a variety of plant materials including seeds, roots, grasses and other leaves, and also take animal food such as insects—particularly grasshoppers, mollusks, and small vertebrates. They are gregarious, living in flocks of 20-30 or more. They will crouch, or run, dodging, to escape danger, and they also swim well.

Rheas have quite well developed voices, and the males utter mammallike roars in their breeding displays. They also posture with wings spread. They are polygamous, and successful males will acquire harems of 6-8 females. The hen birds lay up to 20 eggs each, but many are wasted, several being deposited by most females before a nest has been built to receive them. The nest is a scrape in the ground made by the male, who also adds a lining of some dry vegetable material. The nest site is usually a concealed one, though birds seem to clear a small area surrounding the nest by biting off the herbage. Possibly this acts as a kind of fire break. All females of a particular harem lay in the same nest. The eggs are a golden yellow or deep green and are incubated by the male, the full clutch being very variable in size, frequently of 20-30 eggs. Incubation takes 35-40 days, and the young are gray with dark stripes, giving a disruptive camouflage.

They are able to run soon after they are dry, and they leave the nest with the male. Those that survive grow rapidly and reach full size in about 5 months, but they do not breed until they are 2 years old.

RHEBOK, *Pelea capreolus*, a small South African antelope, related to the dwarf antelopes or to the reedbuck, depending on the authority followed. 28-32 in. (70-80 cm) high, weighing 44-50 lb (20-23 kg), rhebok are distinguished by their woolly, soft, rabbitlike fur, which is brown-gray in color. The face and lower parts of the legs are yellowish and the underparts white. The build is very slender with long legs, and long pointed ears. The muzzle is bare and slightly swollen, glandular. The lateral hoofs are jointed across the midline. The horns are short, nearly straight but slightly back-curved. Rhebok live in hilly, grassy country above 4,000 ft (1,200 m) in South Africa and extreme southeastern Botswana, north to 24°S. They spend the day on the moorland plateau, descending at night to lower ground. They live in family groups, which occasionally combine to form groups of 30 or more with a single adult male. Other males live solitarily. Rhebok are very alert, the sentinel giving a warning cough at approaching danger, causing the group to bound away with a jerky motion.

RHINOCEROS, with the exception of the elephant the heaviest land mammal, characterized by its nasal horn or horns. Five species of rhinos still occur. These are *Ceratotherium simum*, the white or square-lipped rhinoceros of Africa, *Diceros bicornis*, the black or hook-lipped rhinoceros of Africa, *Rhinoceros unicornis*, the great Indian rhinoceros, *Rhinoceros sondaicus*, the Javan rhinoceros and *Dicerorhinus sumatrensis*, the Sumatran rhinoceros, which is found in Sumatra, Malaysia, Burma and perhaps still in Borneo.

The horns of rhinos are in structure and position quite different from the hornlike organs of other ungulates, i.e., the horns of Bovidae and the antlers of Cervidae. They consist of a tight mass of horny fibers that are continuously built up by a special tissue covering a hump on the nasal bones. *Ceratotherium*, *Diceros* and *Dicerorhinus* have two horns, both species of *Rhinoceros* only one. The horns are weapons used in intraspecific conflict and also against enemies such as large carnivores and man. The African rhinos have no other weapon but their horns, whereas in the Asian species the almost tusklike incisors (known as tushes) are also used in fighting and can inflict deep wounds.

In comparison to the majority of ungulates, especially the Equidae and most ruminants, who can survey the entire surroundings with their eyes, the rhinos have poor eyesight. In contrast, their sense of smell is very keen and their hearing quite acute. As can be seen from the list of weights below, all rhinos, with the exception of the Sumatran species, reach or exceed the size of the large Bovidae.

Species	Weight reached by large bulls
Square-lipped rhino	over 3 tons (3,000 kg)
Great Indian rhino	$2\frac{1}{2}$ tons (2,500 kg)
Black rhino	$1\frac{1}{2}$ tons (1,500 kg)
Javan rhino	$1\frac{1}{2}$ tons (1,500 kg)
Sumatran rhino	$\frac{1}{2}$ ton (500 kg)

Rheas (South American "ostriches") drinking.

The black rhinoceros of Africa is dark gray and can be recognized by its prehensile, hooked upper lip.

The white rhinoceros is slate-gray; a preferable name for it is square-lipped rhinoceros.

Rhinos are bulky animals, yet their trot and gallop are both powerful and elastic. They can cope with very rough country. The Indian and the Javan rhino can cross deep morasses, the latter and the Sumatran species climb very steep and sometimes slippery slopes. All push through dense and thorny vegetation like armored cars. They are all "pachyderms," their thick skin serving as an armor in intraspecific fighting, against lion or tiger, and when moving through thorny thickets. In fact, in the Indian and Javan rhino the skin resembles the armor of a medieval warrior, consisting of armorlike plates with deep folds as flexible joints between them. Plates and folds show characteristic differences in the two closely related species.

The skin of all rhinos but the Sumatran is hairless, except that the tip of the tail has a hairy brush and the ears have fringes of hair in all species. Rough, long and scanty hair grows on the body of the Sumatran rhino. In the adult animals, it may be rubbed off on the back and the sides.

All five rhino species are on the verge of extinction, the Asian being in a worse position than those in Africa. They all survive only in small islands of their former wide range.

The gestation period is not yet known for all five rhino species, and observations made in zoos are conclusive only for the Indian and the black rhino, with approximately 480 and 450 days respectively.

As in other ungulates, the newborn rhino—which in the Indian species weighs 130-160 lb (60-70 kg)—is able to rise within an hour of birth. It then searches for the nipples in the angle between legs and body of its mother and drinks. In these first hours after birth, some kind of reciprocal imprinting between mother and calf occurs. They know each other from then on and form an intimate social unity. With time the calf, guided and protected by its mother, becomes familiar with the large excursion range. Also the reaction to man, which is different in rhinos from different regions and depends on past experience, is transferred from mother to calf.

RHINOCEROS VIPER, *Bitis nasicornis,* a species of puff adder, up to 4 ft (1.3 m) long, gaudily colored and with large pointed and erectile scales on the tip of the snout. It is sometimes called the nose-horned viper, but, in fact, its near relative, the Gaboon viper, has similar hornlike scales on the snout.

Horned vipers are two distinct species, *Cerastes cerastes* and *C. cornutus,* living in the sandy deserts of northeast Africa and southwest Asia. They have a hornlike scale over each eye. They bury themselves in the sand, and it is supposed that the "horn" keeps sand grains out of the eyes when the head is almost completely buried.

RIBBONWORMS, elongated, soft-bodied invertebrates, also known as proboscis worms or nemertineans. There are nearly 600 species, mainly marine although there are freshwater species and a few living on land. Some species are commensal in sea squirts, sponges and bivalve mollusks, taking their food from their hosts' feeding currents. Those living in the sea may shelter under rocks or among seaweeds on or near the shore, although some live as deep as 4,500 ft (1,500 m) burrowing in mud, and many are free-swimming or free-floating, from surface waters down to 9,000 ft (3,000 m), the swimmers having fins on the sides or hind end, the floating forms being gelatinous. Most ribbonworms are around 8 in. (20 cm) long, but some are only 2 in. (5 cm) or less, while the bootlace worm, *Lineus longissima*, may extend several yards, a length of 60 yd (55 m) having been recorded.

Ribbonworms feed mainly on annelid worms, or nematodes and turbellarians in the case of freshwater species, usually swallowing them whole, but they also eat small crustaceans, mollusks or fish, the larger prey being sucked in a piece at a time.

RIFLEMAN, *Acanthisitta chloris*, the smallest New Zealand bird, sharing an endemic family with the "New Zealand wrens." It is only 3 in. (7½ cm) long, and is greenish brown, the male having a green back and the female a brown one. They are forest birds and feed like tree creepers, searching bark, epiphytic mosses and lichens for insect food. The members of a pair or family group call continually with a high-pitched "zit zit."

The loosely woven nest is constructed in a hole or crevice, and four to five white eggs are laid. Both sexes incubate and feed the young, and there are usually two broods a year.

RIGHT WHALES, so called because in the early days of whaling they were literally the right whales to catch. They are relatively slow and timid and so were more easy prey for the rowing boats equipped with hand harpoons of the early whalers. When killed, they floated, and hence were to easy to handle. Then their large plates of baleen, as well as the oil, offered great financial reward. The Greenland right whale, *Balaena mysticetus*, was once abundant

around the coasts of Greenland, but the activities of whalers reduced it almost to extinction. It is a northern form that does not migrate far from the ice. It extended from Greenland, around Spitzbergen and Jan Meyen in the east, to Baffin's and Hudson's Bay in the west. An essentially identical animal known to the American whalers as the bowhead was found in vast numbers in the Bering Straits and Okhotsk Sea.

The Greenland right whale is a large animal growing to around 60 ft (20 m), and most adults reach 50 ft. It has an enormous head and mouth with large lips some 15-20 ft (5-6 m) in length and 5-6 ft (up to 2 m) in height. The head is about a third of the animal's length. The baleen plates are the largest of any whale, with an average length of 10-11 ft (3.5 m) and some reach 15 ft (5 m). With some 300 plates on each side of the mouth, this gave well over a ton of baleen in a large whale. Baleen was used for ladies' corsets, umbrellas and parasols, chairs, sofas, trunks, fishing rods and even carriage springs in view of its remarkable strength and elasticity.

When added to the value of the oil, the proceeds from the baleen meant that one whale would more than cover the expenses of a whaler's trip and any others caught were clear profit. When whalers first moved into these northern waters in about 1611 there were vast numbers of right whales. But they became so rare that they are virtually unknown in modern times, and the best report is that published in 1820 by William Scoresby, a remarkable biological study. Fortunately, with protection, the number of right whales is increasing in the Hudson's Bay and the Bering Sea.

The Biscayan right whale, as the North Atlantic form is often called, has been known for many centuries in the Bay of Biscay. Even

in the 12th century it was an important fishery. The pygmy right whale, *Caperea marginata*, is smaller than the others, reaching only 20 ft (6.5 m). It inhabits only southern seas and is known around New Zealand, South America and South Africa. As it has virtually no economic importance very little is known of it. Although called a right whale, it is only distantly related to the other species, the body being less bulky and the baleen short; the 250 or so plates are only about one-tenth of the body length though very tough and flexible.

ROACH, *Rutilus rutilus*, one of the commonest of the carplike fishes of Europe but also found right across Asia to the Amur basin. The roach is a fairly high-backed but slender fish, the upperparts being gray or blue-green, fading to silvery on the flanks and belly, and the fins red. It is found in both rivers and lakes, the young fishes shoaling and readily taking any bait offered but the larger adults becoming more solitary and difficult to catch. Roach grow to 16 in. (40 cm) and weigh over 3 lb (1.4 kg). The young feed on algae and plankton but later turn to worms, insects, crustaceans, small fishes and fish eggs, as well as aquatic plants and mollusks. The roach is very variable in form, depending on the type of water mass that it inhabits. Lake forms are usually deeper-bodied than those from flowing water, and this was once considered a good basis for recognizing several species, one of which was from brackish water. All are now placed in the same species.

ROBBER FLY, so called because the adults of both sexes feed exclusively by catching other insects and sucking them dry. Usually the prey is chased and captured while it is in flight, and robber flies have become highly adapted to this predatory habit. The eyes are large, the proboscis is powerful and sharp and the head and legs are equipped with exceptionally strong bristles, which both help to hold the prey and protect the robber fly from counter-attack. The flies are quick to notice movement, but do not see detail well and often chase anything that moves. In the tropics some robber flies are huge, up to 3 in. (7.5 cm) or more in length, and may capture and eat the largest and fiercest bees and wasps.

ROBINS, several species of the bird family Muscicapidae, sharing the common characters of rather small size and a red breast.

The common European robin is a small brown thrush with the upper breast orange-red and the lower breast light gray. A dumpy little bird, it is a familiar sight in gardens in Britain, but shyer and more confined to its natural woodland habitats on the Continent. It feeds on insects, worms, spiders, food scraps provided by man and sometimes small seeds. Both sexes defend territories in the autumn and winter with a sweet, mellow song. The male alone defends the breeding territory in the spring and early summer. The nest is most often built in a hollow in a bank, but also in a variety of other sites. Clutches of two to six eggs are usually laid and incubated by the female, but both parents bring food to the young. The incubation period is about 13-14 days and the fledging period about 12-14 days. Young robins have a spotted breast, showing their relationship with the thrushes.

European robin, about to land. It is braking with its wings and its legs are ready for touch-down.

The five-bearded rockling, a relative of the cod, is abundant between tide-marks on the coasts of Europe.

ROCKLINGS, small and elongated codlike fishes of shallow waters and especially rock pools, from which they derive their common name. They are found off the shores of Britain and the European continent. The general body form is similar to that found in the other codlike species, but in many rocklings there is more than one barbel around the mouth. The three-bearded rockling, *Gaidropsarus vulgaris*, has two barbels near the nostrils and one under the chin; it is the largest of the eastern Atlantic species, growing to 20 in. (51 cm). The four-bearded rockling, *Rhinonemus cimbrius*, has an additional barbel near the upper lip, and the five-bearded rockling, *Ciliata mustelus*, has a total of five barbels. The young of all these species are silvery and pelagic and look quite unlike their brownish and spotted parents. The difference between young and adults is so striking that the young were at one time placed in a separate genus, *Couchia*.

ROE DEER, *Capreolus*, with three subspecies, are widely distributed in Europe and the Middle East and northern Asia, the typical deer, *C. capreolus capreolus*, being present in Scotland and England, and in almost every country of Europe as far east as the Ural Mountains in Russia, beyond which it is replaced by the larger Siberian roe deer, *C. c. pygargus*. The former also occurs in Anatolia, Transcaucasia and northern Iran. In the Far East, the Siberian roe deer is replaced by *C. c. bedfordi* of China and Korea.

The European roe deer stands 25-29 in. (64-74 cm) high at the shoulder, and an adult buck may weigh from 38-50 lb (17-23 kg), with exceptional beasts from Poland weighing as much as 80-90 lb (about 38 kg).

The full head of a buck should be six-pointed, although multipointed heads occur. The tail is not readily visible, although a small one is present. During the winter the doe develops a prominent anal tush—a tuft of long hair—which is sometimes mistaken for a tail. The summer pelage of a rich foxy red changes to a grayish fawn in winter with a most marked white rump patch. In the Netherlands and elsewhere melanistic roe deer occur.

The rut takes place during late July and early August. Due to delayed implantation, no development of the embryo within the uterus is visible until December. Thereafter development is fairly rapid and the young, frequently twins and occasionally triplets, are born from late April to early June.

ROLLERS, a small group of solitary jay-sized birds centered on the Old World tropics, deriving their common name from their tumbling courtship flights. Some species extend to temperate Eurasia, and these, together with a few of the African and oriental forms, are well-known birds because of their tendency to breed around human habitations.

The typical rollers of the genus *Coracias* and family Coraciidae are handsome thickset birds, with shining azure-blue wings and chestnut shoulders. The body is blue, olive, pink and brown in delicate combination. The bill is stout, the shape of a jackdaw's and the head is rather large and flat, the neck thick, legs short but strong and the wings and tail moderately long. In the single European species, the roller *Coracias garrulus*, the outer tail feather is a little attenuated and lengthened, and the Abyssinian roller, *C. abyssinica*, of northern tropical Africa has these outer feathers elongated into 6-in. (15-cm) streamers, whereas they are long and racquet-shaped in the African *C. spatulata*. Outside the breeding season, rollers are rather stolid, keeping watch for long periods from a shrub or overhead wire for beetles, grasshoppers, small lizards and similar prey captured on the ground at the end of a rapid glide; or from time to time flying to a new vantage point with shallow flicking wing beats. But during the breeding season their presence is made known not only by continuous raucous calls (hence the name *garrulus* for the European species), but also by the vigorous chasing of the female by the male. During courtship they climb high in the air and swoop in an erratic tumbling flight, twisting or rolling about their own axis, and apparently even somersaulting. Such dives also feature in territorial disputes, and human intruders into the territory of a nesting pair are dived upon fearlessly.

Rollers nest in holes in trees, buildings and termite hills. The two to four eggs are white, and the nests are unlined cavities. The nestlings' feces are not cleaned away so that, like all other coraciiform birds, the nest quickly becomes noxious. The young grow quickly, however, and fledge after about five weeks in a plumage similar to that of the adults.

ROOK, *Corvus frugilegus*, a member of the crow family Corvidae, widespread in Britain, Europe and parts of Asia. Also in the genus is the Cape rook, *C. capensis*, of South Africa. Both species are rather lightly built crows with long bills. Adult Eurasian rooks have a bare patch of skin at the base of the bill and nest in colonies of several thousands in the tops of tall trees, hence the term "rookery." Cape rooks lack the bare face patch and nest singly in trees. Both give harsh cawing calls. The story goes that rooks sometimes gather in a circle around a forlorn and dejected member of their species and, after a great deal of cawing, set upon it and peck it to death. One suggestion is that this "rooks' parliament" is a court of justice in which an offender is tried, sentenced

The pugnacious lilac-breasted roller *Coracias caudatus* is reported to capture other birds, even snakes, at times.

and executed, while another is that the rooks are carrying out a mercy killing. The explanation of a sick bird being put out of its misery is the more likely, but there are so few accounts of this strange social behavior that it is impossible to reach any firm conclusion about its significance. Most eyewitness accounts describe the rooks as forming one or two circles around the condemned bird, and there may be several hundred taking part. They caw loudly, then fall silent before cawing again and finally setting on the victim, who in several cases has been found to be emaciated and badly parasitized. Support for the theory that it is sick birds that are attacked comes from two accounts of turkey parliaments. In both cases the hen turkeys formed a circle around the sick bird and, after a ritual of bowing and gobbling, the cock advanced and struck it with its spurs, after which each hen in turn pecked the victim.

The female ruff, known as the reeve, with its chick.

RUDD, *Scardinius erythrophthalmus*, a carplike freshwater fish from still waters of Europe. It closely resembles the roach but has the beginning of the dorsal fin set behind a vertical line from the base of the pelvic fins. The rudd is an attractive fish with a fairly deep, silvery body, red fins and a red eye. It grows larger than the roach, reaching 4 lb (1.8 kg) in weight but, like the roach, feeds on most small aquatic animals. It hybridizes readily with the roach, which the offspring strongly resemble.

RUFF, *Philomachus pugnax*, wading bird of freshwater marshland and meadows showing extreme sexual dimorphism, males being much larger than females, which are called "reeves." In breeding plumage, males acquire colored eartufts and a "ruff" of feathers around the neck. It is rare for two males in the same breeding area to have identical plumage. Ruffs breed throughout the north temperate zone of the palearctic from Britain (where they are rare) eastward to eastern Siberia. In autumn they migrate both west and south, some of the Siberian birds wintering in western Europe while the majority winter away from the seacoasts in Africa. Large concentrations have been recorded in winter and on passage from the vicinity of Lake Chad. In spring a pronounced northward passage occurs through Italy. On the breeding grounds communal display and mating occur at a lek—that is, an open (often grassy) area on which several males display from fixed positions (residences) about 3 ft (1 m) apart. Females visit the leks and choose residences at which to be mated. Males with residences at the center of the lek usually

mate with more females than do peripheral males, but each female may visit several residences or even several leks. Hence the mating system is best described as promiscuous. In spite of the extreme plumage diversity shown by the males, two classes may be distinguished: those with white ruffs and white or colored (but not black) ear tufts, and all others (mainly those with colored or black ruffs). These two classes show behavioral differences at the lek, the former acting as "satellite" males to the latter, which defend the residences. Satellite males are tolerated on the residences, since they show no aggressive behavior toward the owners, whereas other males are attacked and driven off.

RUSA DEER, *Cervus timorensis*, the most widespread species of deer in the Indonesian archipelago in which some eight insular races are represented. It is by no means certain, however, that all these are indigenous to some of the islands, for it is known that a considerable importation of deer has taken place.

The largest member of this species is *C. t. russa* from Java, which has a shoulder height of about 43 in. (110 cm), while *C. t. floresiensis*, one of the smaller representatives, is found on Flores. The typical deer *C. t. timorensis* also occurs on Flores, as well as on Timor, Hermit and Ninigo.

Dark brown in color and with the largest antler seldom exceeding 27 in. (68.5 cm) in length, this is a deer of the grassy plains, though persecution will make them resort to cover. This deer has been introduced to New Guinea and New Zealand (North Island).

SABLE ANTELOPE, *Hippotragus niger*, a large African antelope closely related to the roan antelope and blaauwbok. Sable stand 51-57 in. (127-143 cm) high and may weigh 500 lb (230 kg); they are shorter-bodied than roan and blaauwbok. The horns are much longer than either, being 28-70 in. (70-173 cm) with 35-59 rings. They are slender and strongly curved, elongated oval in section with the lateral side flattened. The ears are only half as long as the head. There is a mane on the neck but not on the throat, and the color is black in males, golden-brown to black in females, with lips, muzzle, upper throat, underparts and a line from eye to muzzle white. There are a number of small white tufts in the region below the eyes.

In northern Tanzania and Kenya is found the short-horned race, *H. n. rooseveltí*, in which cows are light brown. The best-known and most beautiful race is the giant sable antelope, *H. n. variani*, restricted to an area of Angola bounded by the rivers Luando and Dunda in the north and east and the Cuanza and Luasso in the south and west; a few occur farther north, in the Cangandala area. Giant sable are slightly larger than other races, with a longer, narrower face, more vertical horn bases and much longer horns. Bulls are shining black, cows a brilliant golden chestnut. In both sexes the white eye stripe stops short of joining the white of the muzzle, being virtually restricted to the white eye tufts. South of the Zambesi, the race *H. n. niger* has long horns, and both sexes are black. North of the Zambesi, horns are shorter and the cows are brown (*H. n. kirki*).

Giant sable antelope, discovered by Varian in 1916, are rare today; only a few thousand exist. As of 1970, around 2,000 were in the Luando Reserve. They live in herds varying in size from 8 to 21.

Other races of sable have been reported in larger herds of 10-40 head, sometimes assembling in groups of a hundred. The usual pattern is one bull with several cows and calves. Herds without a bull are also encountered. Unlike roan, sable are grazers; they visit the waterhole regularly in the evenings, and other antelopes yield place to them. They drink quickly and leave rapidly, unlike other species. Sable antelope have a gestation of 270 days. The breeding season is not restricted, but the peak of births comes in January and February. All over its range this magnificent antelope is somewhat of a rarity, and its conservation is a matter of urgency.

SAIGA, a unique sheeplike antelope of central and western Asia.

The saiga stands 30-32 in. (75-80 cm) high, and weighs 88-110 lb (40-50 kg), although females are often as little as 66 lb (30 kg). Only males have horns, which are 11-15 in. long (28-38 cm), nearly straight with a very slight lyration, a pale waxy color and semitransparent. They bear 18-20 ridges. The coat is woolly, buff in summer and whitish in winter. The most striking feature of the saiga is its nose, which is swollen. The nostrils are terminal, downward-looking and very mobile, and in each nostril is a sac with mucous membranes that warm and moisten the inhaled air. According to Bannikov, this sac is an adaptation to fast and prolonged running in herds in a semidesert environment. The dust is kept out of the nostrils by their low position and mobility, and by the low carriage of the head while running.

The above description applies mainly to the common species, *Saiga tatarica*, which is found in the USSR, from the Kalmyk ASSR to the Chinese border, and slightly north into Dzungaria. A second species, the Mongolian saiga, *S. mongolica*, lives in the Gobi Desert, being isolated from the first by the Gobi Altai range. It is smaller, being only 24-27 in. (60-67 cm) high, with thin, delicate horns under 9 in. (22 cm) long, and a duller coat, gray-sandy in summer and gray-brown in winter. The horns are poorly annulated.

Saigas walk with a shambling gait and with their heads low, but they can also run very fast, up to 60 mph (95 kph), for quite long periods. They also, with the exception of the Mongolian saiga, make considerable leaps. Saigas live on the open steppe, from sea level to 5,000 ft (1,600 m). There are no trees in this environment; the dominant vegetation is wormwood and cespitose. Every year, mass migrations take the herds south in the autumn and north again in spring. The males are generally the first to move. The migration of the herds west of the Volga may spread over 200-220 miles (300-350 km), moving at anything from 2 mph (3 kph) to 12½ mph (20 kph).

The magnificent sable, the most handsome of the antelopes.

The saiga, Asian antelope, is distinguished by its inflated nostrils.

SAILFISH, *Istiophorus platypterus*, a large oceanic scombroid fish with a saillike dorsal fin. This fish was formerly referred to as *I. gladius*, and several other species were recognized, but recent studies have shown that all belong to a single, worldwide species occurring in all tropical and subtropical seas. The sailfish has the powerful torpedo-shaped body of fast oceanic swimmers. The dorsal finrays are enormously extended to form a "sail" along the back. Pelvic fins are present (as in the related swordfish but not in the marlins). The tail is crescent-shaped, and there are small finlets behind both the dorsal and the anal fins. The upper jaw is extended into a "bill" or "sword" which is longer than the head and is rounded in cross section (flattened in the swordfish). In the juveniles the upper and lower jaws are the same length and are provided with pointed teeth, but by the time that the fish is 2 in. (5 cm) long the upper jaw has outgrown the lower and the teeth have disappeared. Just in front of the tail there are two keels on the side of the body (a single keel in the swordfish). The body and the dorsal fin are blue or blue-black, often ornamented with small dark spots.

The sailfish is an exceedingly fast swimmer, the huge dorsal fin being folded down at high speeds. It has been said that they can reach 60 mph (100 kph), and although this is by no means impossible, no accurate measurements have been made. On calm days the sailfish basks at the surface with the dorsal fin fully erect, giving rise to the perhaps correct assumption that this fin is actually used as a sail. They feed on fish and squids.

SALMON, highly palatable anadromous fishes of the northern hemisphere belonging to the genera *Salmo* and *Onchorhynchus*. Of all the fishes in the world, the Atlantic salmon, *Salmo salar*, is perhaps the most famous. It has been eaten for at least 2,000 years, has had books and poems devoted to its virtues and habits, has made the owners of stretches of good salmon rivers wealthy, and artists and photographers have spent a lifetime trying to capture the salmon's triumphant leaps as it makes for its spawning grounds. Cavemen eagerly sought the salmon, and the drawings and models of this fish found in caves in the Pyrenees suggest that they well appreciated the nobility of the salmon. The thrill of spearing a salmon with a sharpened length of wood is now gone, and regretfully the salmon is to many people a fish that is bought by the can or caught only by the wealthy.

The Atlantic salmon lives in the North Atlantic and breeds in the fresh waters of Europe and North America. It is the only salmon in this whole region. In the North Pacific, however, there are several species belonging to the related genus *Onchorhynchus*—namely, the Chinook salmon, *O. tschawytscha*; the coho, *O. kisutch*; the sockeye, *O. nerka*; the pink or humpback salmon, *O. gorbuscha*; and the chum or dog salmon, *O. keta*. These are all confined to the western side of Canada and the United States. A further species, *O. masu*, is found on the eastern seaboard of northern Asia. The largest of these is the Chinook, which can weigh over 100 lb (45 kg), and the smallest is the pink salmon, which rarely exceeds 10 lb (4.5 kg). In their general biology, these species of *Onchorhynchus* resemble the Atlantic salmon in ascending rivers to spawn, the young descending to the sea either shortly after hatching (pink salmon) or months or years later. One principal difference between these species and the Atlantic salmon is that the former do not return to the rivers to spawn a second time but die on completion of their first spawning.

The adult salmon is an elongated, powerfully built yet graceful fish. The fins are soft-rayed, the tail is slightly emarginated and a small adipose fin is present. In color the fishes are silvery with small black "freckles" and a darker back, but when the breeding season approaches the male becomes suffused with a reddish tinge and its jaws become curiously hooked. The adult fishes feed in the sea for one or two years before they return to the rivers and are in good condition for the climb ahead of them. Quite often they will approach the mouth of a river but lie off it for another year before making the ascent. The salmon run up the rivers in August to September, and the major spawning period is in the early winter. In some areas a second run of salmon occurs in spring. During the actual spawning run they need all the energy reserves they have built up, as they do not feed in fresh water. It is not known why salmon, although fasting, will take the flies and spinners of the fisherman.

It has often been said that salmon return to their natal stream to spawn, and the individual marking of fishes with numbered tags has shown that, at least for the recaptured fishes, this is perfectly correct. Over 400,000 young salmon were marked from one Canadian stream before they migrated to the sea, and of these 11,000 were recaptured in their natal stream and none in any other stream.

Most rivers are fairly easy to ascend, but if weirs have been built the salmon display extraordinary tenacity in trying to leap over them, trying again and again until they either succeed or fall back exhausted. A large salmon can make a leap of 10 ft (3 m), which is normally quite sufficient.

The journey upstream continues into the smallest brooks and streams, often where there is only just enough water to cover the back of the fish. Here the nest or "redd" is built, the male making a large trough in the pebbles by lashing with its tail. Several males will often make their redds within a few feet of each other. The female then lays the eggs in the redd (about 800 eggs per lb weight of fish), and the male fertilizes them and covers them over with gravel.

The newly hatched alevins are about $\frac{1}{2}$ in.

Atlantic salmon have dark bands on the flanks known as parr marks.

(1.3 cm) long, and they remain among the pebbles living on the food in their yolk sac. When about 1 in. (2.5 cm) long, they leave the nest, and those that survive lead a secluded life in shallow waters feeding on small insect larvae. They reach about 4 in. (10 cm) in the first year and 6 in. (15 cm) in the second year. During this time they bear the parr markings of 8-10 dark oval blotches on the flanks with a red spot between each oval. Usually in the second year (occasionally in the first or the third), a silvery pigment develops over the parr marks, and the fishes are then termed smolts (if the silvery pigment is scraped away the parr markings are visible below). The smolt then migrates to the sea, spends a year or more feeding, returning as an adult to struggle back upstream to its birthplace.

SAMBAR, *Cervus unicolor,* one of the largest and most widespread of the deer of southern Asia, extends from India in the west, through Burma and southern China, to Indonesia and the Philippines in the east. It has also been successfully introduced to the North Island of New Zealand. Throughout its range, 16 subspecies are recognized, the largest of which, *C. u. niger,* is found in India, this animal being similar, but slightly larger than the typical sambar, *C. u. unicolor,* of Ceylon. A large sambar stag from the Central Provinces will stand about 52-56 in. (132-142 cm) high at the shoulder, and weigh about 600 lb (272 kg). With antlers normally bearing six points and measuring up to 50 in. (127 cm) in length, this deer is a uniform dark brown, the calves, which are never spotted at birth, being the same color.

The sambar of South China, *C. u. equinus,* is also a large animal, but the one found on Borneo and adjacent islands, *C. u. brookei,* is a slightly smaller beast. Eight insular races are represented in the Philippines, the smallest of which, *C. u. nigricans,* has a shoulder height of only about 24-26 in. (60-65 cm) and occurs on Basilan Island.

Sambar occur at all altitudes from sea level up to about 10,000 ft (about 3,000 m), for their habitat is very variable, ranging from coastal forest and swamp land to agricultural fields and mountains. They are seldom far from water.

SAND EELS, small, silvery eellike fishes, in no way related to true eels, found around the coasts of the Atlantic Ocean. They spend a lot of their time buried in sand. They have pointed jaws well adapted for burrowing, a long dorsal fin and no pelvic fins. Their main significance lies in their being an important item of diet for many commercially valuable fish like cod and halibut.

SAND FLIES, small two-winged flies belonging to the blood-sucking genus *Phlebotomus,* but in some places this name is given to the biting midges (Ceratopogonidae) as they are also minute and blood-sucking, and may be common around estuaries and sandy river banks. In Australia, the blackflies (Simuliidae) are also known as sand flies.

True sand flies (*Phlebotomus*) are tiny, fragile flies found throughout the tropics and subtropics, generally in damp shady places and particularly at dusk and during the night. Their flight is weak, and they are easily disturbed when feeding on their vertebrate hosts. It is

usually only the female that sucks blood, and it is therefore on account of the females that sand flies are medically important as vectors of various diseases, notably "three-day fever." They are also a considerable nuisance, not least because they are so small that they can easily get through the meshes of all but the finest mosquito nets.

The eggs are laid in damp crevices in rocks, drains and damp earth, and the larva feeds on any organic detritus in the damp environment that is essential for its survival. Such breeding places are common around human habitations, while the humans themselves provide ample food for the adult flies.

SAND GROUSE, desert-living, plump, ground birds, pigeonlike in size and shape and with a general resemblance to game birds.

They have short feathered legs and toes and walk with a mincing gait. Their wings are long and pointed and their flight is strong and direct. In some species the central tailfeathers are elongated to form a pin-tail. The males are brighter than the females but both have cryptic coloration, the plumage being predominantly in browns and grays, often dappled and barred on the back with chestnut, buff and black. They inhabit deserts and steppes in Eurasia, India and Africa but they are peculiar among

is rarely situated in the shade, which means that the female sits, without relief, in temperatures of up to 104°F (40°C) and over, for several hours daily. Overheating is prevented by rapid movements of the loose skin of the throat, known as gular fluttering. Heat regulation in this way, by evaporation, is rare among desert birds and is only made possible by daily drinking, whereas most desert birds rely on behavioral adaptations to avoid overheating. The chicks leave the nest immediately after hatching (nidifugous) and feed on dry seeds. This means that they also require water daily, and this is brought to them trapped in the specially adapted breast feathers of the male. The young "strip" the water from these feathers.

Pallas's sand grouse, *Syrrhaptes paradoxus,* is famous for its occasional irruptions out of the steppes of central Asia.

SANDHOPPER, *Talitrus saltator,* a small crustacean that burrows in sand on European shores. It differs from other amphipods in that it does not lie on its side but rests in an upright position on the claws of the thoracic legs, which are well developed and can withstand abrasion. The last three segments of the abdomen are shortened and partially tucked under the body. The usual method of locomo-

The pin-tailed sandgrouse ranges across Asia to Europe and Africa.

dry-country birds in that they eat no succulent green or animal food, relying entirely on seeds. This means that they must drink water daily which restricts their distribution to areas within about 20 miles (32 km) of water, although nests of Burchell's sand grouse, *Pterocles burchelli,* in the Kalahari Desert may be up to 50 miles (80 km) from water.

The nest is a simple scrape or unaltered hollow such as a dry hoof mark, in which usually three cryptic ellipsoid eggs are laid. Both parents incubate, the male at night and the more cryptic female during the day. The nest

tion is by jumping, whence its name. It can also swim and walk. In jumping, the abdomen is suddenly flexed, pushing against the substratum and giving a jump of up to 3 ft (1 m) in distance, and to a height of about 1 ft (30 cm), a greater distance than that jumped by any other animal of similar size.

When disturbed, it jumps around in a random succession of jumps until it finds shelter. In swimming, it gives an initial thrust with a flexure of the abdomen and swims with the aid of the pleopods and with repeated blows of the abdomen against the water. In walking,

The sargassumfish lives among the floating weed of the Sargasso Sea, beautifully camouflaged.

some 450 miles (724 km) up the Amazon and is also found in Lake Nicaragua. Other species are known from fresh waters throughout the tropical world, and in Thailand a specimen of 26 ft (7.8 m) has been caught and a monster of no less than 46 ft (14 m) has been reported. Although sawfishes have been reported as docile, considerable respect is shown for them by local fishermen, both in the water and when landed.

SAWFLIES, the more primitive members of the Hymenoptera, which includes the better-known ants, bees and wasps. Sawflies, of which there are at least 2,000 species throughout the world, have two pairs of membranous wings, often with the venation much reduced and with the fore and hind wings interlocked by means of a row of small hooks. The Hymenoptera (bees and ants) are most easily recognized by having a "waisted" appearance, due to very narrow first and second segments of the abdomen. In the sawflies, however, this is not apparent; the abdomen is broad, and there is no constriction between it and the thorax. Unlike the rest of the Hymenoptera, the sawflies have fully legged larvae, which look very like caterpillars of butterflies and moths except that they generally have three pairs of thoracic legs and at least six pairs of abdominal legs, whereas true caterpillars have only four such pairs.

Sawflies are entirely plant feeders. Their larvae have varied feeding habits. Some feed inside plant galls and stimulate their formation, whereas others feed in the twigs of plants or in the hard wood of trees and shrubs. The majority, however, live on the leaves of plants much in the manner of caterpillars of moths and butterflies. In consequence, many sawflies are well known as defoliating pests of timber and arable crops.

The large and strikingly colored wood wasps or horntails (family Siricidae) have wood-boring larvae. The adults can be up to 1.7 in. (42 mm) long, are black or brown and yellow, or in some species a bright metallic blue. The long ovipositor of the female projects beyond the end of the abdomen and gives the false impression of a ferocious sting. It is in fact used for boring holes into wood for egg laying. The European *Sirex gigas*, the yellow and black species, attacks coniferous wood, and *S. noctilio*, the steel-blue species, is found especially in larch and silver fir plantations. The adults are flying from June to early autumn, and egg laying commences in July. The female drills a hole beneath the bark of decaying trees and especially resinous felled logs. The larvae, which live for 2-3 years, tunnel extensively in the superficial layers of the wood and finally pupate close to the surface. Owing to the large size and extent of the tunnels, such infested logs are useless commercially, and the wood wasp has become a considerable pest, especially from timber imported into Australia.

When egg laying, the female sawflies move the serrated blades of the ovipositor alternately to cut a deep slit in leaves or stems. Sawfly caterpillars can cause considerable damage to their host plants, and many of them, such as the common currant sawfly, *Nematus ribesii*, and the larch sawfly, *Diprion sertifer*, are well-known defoliators. Pine, spruce and larch

it pushes the body forward over the sand with movements of the abdomen.

SARGASSUMFISH, *Histrio histrio*, a small fish 6-8 in. (15-20 cm) long, belonging to the family of frogfishes, which lives among the floating weeds of the Sargasso Sea. Like the angler fishes, it has a "fishing rod" or illicium modified from the first dorsal fin ray. It is very beautifully camouflaged in blotches of yellows and browns to match the Sargassum weed and is found also in other tropical areas where this weed occurs. It moves slowly among the fronds, using its pectoral fins rather in the manner of arms, and can modify its colors to match its surroundings. The body is stocky and appears clumsy, but if dislodged from the weed it can move surprisingly quickly in its efforts to regain the safety of the sargassum. To complete its camouflage, the outline of the fish is broken up by small, irregular flaps of skin.

SAWFISHES, members of a family of flattened cartilaginous fishes somewhat resembling the guitarfishes but with the snout greatly elongated and bearing a series of 16-32 teeth on either side. Like the guitarfishes, the sawfishes have a rather sharklike body, with small pectoral fins that are not used for propelling the body forward, the motive power for swimming being derived from sinuous movements of the body, as in sharks. The sawfishes, however, have the gill slits on the underside of the head and are thus clearly allied to the rays and not to the sharks. They are found in tropical marine and brackish waters but also occur in some tropical fresh waters.

The curious "saw" is a flattened blade of cartilage, calcified to give it rigidity, and the teeth along its two edges are implanted in sockets. The saw has always been assumed to be a weapon for defense and for attacking prey (other fishes), but observations have shown this to be only partially true. A specimen at the Lerner Marine Laboratory in Bimini (Bahamas) was seen to strike sideways at a fish, impale it on one of its saw teeth, and then retreat to the bottom where the impaled fish was rubbed off and eaten. The proverbial use of the saw to cut up larger fishes, and attack whales and even boats, is, however, unfounded. The common sawfish, *Pristis pectinatus*, is found in the Atlantic and Mediterranean and is known to reach a length of 18 ft (5.4 m). The related *P. perotteti* has been caught sawflies are serious pests of their host trees. A few of the tenthredinid sawflies stimulate gall formation by their host plants. The willow sawfly, *Pontania*, stimulates the willow leaves to produce gall very quickly after the egg is laid. The growth of the gall is said to be stimulated by a secretion injected by the sawfly female during oviposition.

SAW SHARKS, sharks with a bladelike extension of the snout armed with a series of teeth. The saw sharks somewhat resemble the sawfishes but have relatively short bases to the small pectoral fins, and gill slits are placed at the side of the head and not underneath; these

appearance is similar to that of a Roman comb, which is why the principal genus was called *Pecten* by Pliny. Perhaps the best known of the scallops is the great scallop or the St. James's shell, *Pecten maximus*, which occurs on the Atlantic coast of Europe and was the emblem of pilgrims visiting the tomb of St. James the Apostle in northwestern Spain. The second major type belongs to the genus *Chlamys*. There is an enormous number of species, including the tiger scallop, *Chlamys tigerina*, the hunchback scallop, *Chlamys bistorta*, and the variegated scallop, *Chlamys varia*. One of the commonest European forms, however,

particles are further sorted by ridged and ciliated structures called the labial palps or lips, and are then ingested.

One of the truly remarkable features of scallops is the great swimming ability shown best by the queen scallop but also exhibited by the great scallop and by several other species. The mobility of scallops has been known for a long time. The young animals are found near the low-water mark on the shore, but as they grow older, as adults they undertake substantial migrations into deeper water. This migratory ability is thus quite distinct from the escape reactions shown by some cockles, although

A female sawfly, *Tenthredo mesomelas*, photographed while laying her eggs in a grass stem.

two features serve to place the saw sharks with the true sharks and not with the rays. The evolution of a distinctive and specialized feature in two unrelated groups is not uncommon in nature. The saw sharks have a fairly slender body with two dorsal fins, and there are the usual five gill slits in species of *Pristiophorus* but six in species of *Pliotrema*. The blade of the "saw" is cartilage strengthened with calcified tissue, and there are two long barbels trailing from it about halfway down its length. Unlike the teeth in the sawfishes, those of the saw sharks are alternately long and short. Saw sharks reach a length of about 4 ft (1.2 m) and are found in South African waters (*Pliotrema warreni*) and throughout the Indo-Pacific region (three species of *Pristiophorus*). They are live bearers, the young hatching within the uterus of the mother. During the birth process the teeth are said to lie flat against the blade so they do not damage the mother.

SCALLOPS, bivalve mollusks, of which there are almost 300 species throughout the world. The shells of all have a generally similar shape, one or both of the valves being convex and of rounded outline, often with a series of ribs radiating across the surface. The general

is the queen scallop, *Chlamys opercularis*. Other scallops include *Chlamys islandicus*, which occurs off the coasts of northern Europe and North America, *Chlamys nodosus*, which occurs off the southeastern United States and the Caribbean, as well as a vast number of strictly subtropical forms. All the species of *Chlamys* are easily distinguished from *Pecten* by the fact that in the former both shell valves are convex, whereas in *Pecten* the left valve is convex and the right one is flat. The great scallop is also much larger than the queen scallop and related species.

As in most other bivalves, scallops feed by sieving particles suspended in an inhalant stream of water, which is created by the activity of lateral cilia on the gill filaments. The gills are delicate paired structures, one on each side of the foot, and fill much of the mantle cavity. Each consists of a large number of gill filaments alternating on each side of a central gill axis. Each filament is extremely long and is reflected into a U shape, so that the whole gill in section is W-shaped. Food particles are trapped on the frontal surface of the gill filaments and are transported by frontal cilia to food grooves that lead to the mouth. Here

scallops are also able to escape rapidly from their principal predator, the common starfish. In effect, a scallop swims by snapping its shell. Turning movements can be made by the great scallop, which can, in addition, right itself if the shell has been overturned. All of such movements depend on the expulsion of powerful jets of water from the mantle cavity by the rapid contraction of the adductor muscle. Such movements are most clearly seen in the queen scallop. In this animal prior to swimming the valves are opened more widely than usual, so that the mantle becomes filled with water. Then the adductor muscle contracts vigorously, and jets of water are forced out from each side of the hinge where the mantle margins do not meet. The valves are then opened again and the process rapidly repeated while the animal is up in the water.

SCARAB, common name for some of the more than 20,000 beetles of a family that includes the dung beetles, chafers, Japanese beetle and dor beetles. Most of the species are either scavengers on decaying organic matter, especially dung, or feed on the foliage or roots of growing plants. Many are large insects, $\frac{1}{2}$–1 in. (12-25 mm) long, but a considerable number are small enough to be easily overlooked. The adult beetles are more or less convex, with the abdomen projecting from under the end of the wing cases. For breeding, some scarabs come together in pairs, each pair combining to mold a ball of dung, which they then roll to a selected site and bury. The female, working underground, now molds it into a pear-shaped mass, tamping it hard except for the neck of the pear, which is left hollow. In this one egg is laid, the robust, whitish, C-shaped larva feeding on the rest. Each female lays 2-4 eggs in one season.

The sacred scarab, *Scarabaeus sacer*, of the ancient Egyptians, is of special interest. The notched protuberances on the front of the scarab's head and the spiny projections on the legs were regarded by the early Nile civilization as symbols of the sun's rays, and the spherical dung balls made by the beetle were symbols of the earth itself. It seems that the Egyptians took the view that just as the scarab beetle made its ball of dung revolve, so some gigantic celestial scarab kept the earth revolving. The sacred scarab was also considered symbolic of the moon, as it was believed that the beetles did not start to feed on their hidden treasure (the buried dung balls) until the 28 days of a lunar month had passed. The scarab was, therefore, frequently represented in the carvings of the ancient Egyptians, and its likeness often appeared on their tombs, rings and charms.

SCORPIONS, with spiders, among the best known of the terrestrial arachnids, although their distribution is limited to the warmer parts of the world.

The Scorpiones is a remarkably homogeneous group of over 600 species. In general appearance, one species of scorpion shows much of a resemblance to any other, although there are variations in size, color, the number of eyes, the development of the limbs and certain minor characters. As a group they are relatively large arachnids, just under 8 in. (20 cm) in length, the largest being the west African *Pandinus imperator*. In most groups the body color is brown, occasionally green, bluish or black. The division of the body into an anterior prosoma and a posterior opisthosoma is well marked, for the upper surface of the prosoma is covered with a single large rectangular shield, or carapace, whereas the opisthosoma is clearly segmented. This posterior part of the body is subdivided into a broad "pre-abdomen," consisting of seven segments, and a much narrower cylindrical tail, or "post-abdomen" of five segments, terminating in a curved and sharply pointed sting. In the normal posture, the tail is reflexed over the rest of the body so that the sting is directed forward. This menacing attitude is enhanced by the formidable palps, two massive clawlike limbs inserted on either side of the mouth and held extended in predatory fashion. These two claws are used for capturing prey, and in this they are sometimes, but not always, assisted by the stinging tail. The scorpion's diet consists mainly of other arthropods, such as grasshoppers, crickets, moths, flies, ants, termites, beetles and spiders, although small vertebrates, such as lizards and mice, are sometimes taken. Once secured in the firm grip of the palps, the prey is held against

the mouth where the chelate jaws, or chelicerae, tear it to pieces.

The mouth of the scorpion is very small, and feeding consists of the soft tissues and juices of the prey being sucked up into the digestive tract by the pumping action of the pharynx; this method of feeding is slow, and a scorpion may take several hours to consume a small insect. Like many other arachnids, scorpions can go for several months, even up to a year in some cases, without feeding.

Scorpions are mainly confined in their distribution to warm, dry regions of the world.

The mating habits of scorpions have attracted considerable attention, particularly since they include a courtship "dance" which has certain parallels in human behavior. The details of this intricate behavior pattern vary from genus to genus, but the overall effect is to maneuver the female into a position suitable for taking up the sperm, which is deposited by the male in a cylindrical stalk, or spermatophore. The male, with his tail held vertically, approaches the female until they are face to face, and grasps her palps with his own. Thus embraced, the two partners move backward and forward, sometimes sideways, in a ritual which may last for several hours, and during which the tails of the two may become interwined, or the male may move over the body of the female, stimulating her with his walking legs. In some species, only one palp is used to link the partners, who then promenade side by side, while in others the link is with the jaws instead of the palps. At some point in the dance, the male extrudes the spermatophore from the genital

Scorpions *Euscorpius flavicaudis* on the island of Elba, about to take partners for a courtship dance.

aperture and cements it at an angle to the vertical on the soil surface. He then retreats, jerking the female forward, sometimes rather violently, over the spermatophore. The tip of the spermatophore is loaded with sperm and as this touches the pectines on the underside of the female, the plates guarding her genital aperture open and the sperm mass is maneuvered inside. In *Euscorpius carpathicus* the spermatophore is equipped with an ejector apparatus and a trigger, which causes the sperm to be fired into the genital tract of the female.

Scorpions do not lay eggs, but give birth to living young; the fertilized eggs develop into embryos within brood chambers in the mother's body. Each embryo is attached by a tubular umbilicus to the maternal intestine, from which nutrients are obtained. Immediately after birth, the young scorpions climb onto the mother's back and adhere by means of suckers at the tips of their legs. Here they remain for several days or weeks, relying for sustenance on their own supply of embryonic yolk. Shortly after their first molt, they leave the mother and disperse, gradually maturing to the adult stage through six or seven molts extending over the period of a year.

SCREAMERS, three species of South American gooselike birds, the closest relatives of which are the ducks, geese and swans of the family Anatidae. The screamers differ, however, in having unwebbed feet, in general body proportions with longer legs and in a number of other peculiarities.

The crested or southern screamer, *Chauna torquata,* is found from central Argentina and southern Brazil to Paraguay and eastern Bolivia, the closely related black-necked or northern screamer, *C. chaviara,* is found in western Venezuela and north Colombia, and the horned screamer, *Anhima cornuta,* in tropical South America from Colombia to Venezuela, southern and central Brazil and eastern Bolivia. All three species are the size of a large goose. On the leading edge of the long, rounded wings are two spurs 1 in. (2½ cm) or more long, that are used in fighting. Screamers rise heavily from the ground but fly well, frequently soaring and gliding like eagles. The crested and black-necked screamers are both about 28-32 in. (71-81 cm) long and are mainly gray, but with a ruff of gray or black feathers, a paler head and a crest of elongated feathers on the back of the crown. The horned screamer is about 30-36 in. (76-91 cm) long (the size of a turkey cock), and its plumage is mainly black, with white on the neck, head, wings and breast. A "horn" of cartilage about 3 in. (7½ cm) long projects from the bird's forehead, bending forward over the base of the bill.

They are mainly ground-living birds that live near water and marshes, sometimes walking on floating mats of vegetation to collect food and occasionally swimming. The crested screamer is also often found on the open pampas, where it walks about and feeds among herds of cattle and sheep.

Screamers are so called because of their loud, bisyllabic trumpeting call. This is audible up to 1 mile (1.6 km) away, and is usually preceded by a much softer rasping, rattling noise that appears to be caused by the inflation of the air sacs under the bird's skin. It is the presence of these that enables the trumpeting call to be so loud.

SCULPINS, name given to marine relatives of the bullheads or miller's thumbs. This widespread group of shore living fish are like larger versions of the miller's thumb. They are cryptically colored, difficult to see against a natural background. Some species change color with the season. All have large heads, and spines are usually present in front of the gill cover. They rarely exceed 2 ft (61 cm) in length and are of almost no economic importance.

SEA ANEMONES. Probably the most familiar animals of rocky shores, they are worldwide but are more abundant in warmer seas, although a considerable number of species live in colder seas. Named for the flowerlike appearance of the expanded oral disc, sea anemones have only the polyp present in the life history, the medusa seen in Hydrozoa and jellyfish being absent. There are over 1,000 species.

Sea anemones are all solitary and do not form a skeleton. The polyp has a relatively short, cylindrical body, attached to a rock or other substrate at the basal region or pedal disc, and flattened at the oral region to give a wide disc with numerous tentacles arranged in rings or cycles around the mouth, which opens into a deep throat or pharynx, which then opens into the enteron. The enteron, or "stomach," is subdivided by mesenteries or septa that project from the body wall. These partitions are composed of a central axis of mesoglea surrounded by endoderm, and bearing at their free ends digestive filaments. In some families these can be pushed through the body wall when the anemone is disturbed. At the oral end the mesenteries fuse into the throat, making pocketlike regions. These mesenteries are known as digestive mesenteries and always occur in pairs. Anemones may be hermaphrodite or dioecious, and gonads develop in the mesenteries near their lower ends.

SEA BREAMS, marine perchlike fishes not closely related either to Ray's bream or to the freshwater breams, found mainly in warm and tropical waters. The body is usually deep with a single long dorsal fin, the first part of which has spines, and three spines in the anal fin. There are well-developed teeth in the jaws, sharp in front, often rounded and molarlike behind. Many species are known from the eastern Atlantic and the Mediterranean, and a few penetrate north as far as British shores. The common or red sea bream, *Pagellus bogaraveo,* a reddish fish with a black blotch on the shoulder, is the most frequently caught around British coasts, although Neolithic sites in Scotland have shown that the fish was evidently much more common in those times. The black sea bream, *Spondyliosoma cantharus,* is occasionally caught off Cornish coasts, where it is known as the "old wife." It has an iron-gray back, silver flanks with dark horizontal bands, and reaches 18 in. (46 cm) in length. The largest of the Mediterranean species is the dentex, *Dentex dentex,* which reaches 4½ ft (1.4 m) in length and is highly prized by anglers for its fighting qualities. Its name derives from the two large caninelike teeth on either side of the jaws, which are

The snake-locks anemone *Anemonia sulcata* has long tentacles.

The sea cucumber *Cucumaria saxicola*, 4 in. (10 cm) long, hides its white body in burrows excavated in rocks by other marine animals. From there it stretches out its sticky tentacles to trap small organisms for food.

feet the ends of which are suckerlike. There are five double rows of tube feet altogether but those of the sides and upper surfaces have bluntly rounded tips and are mainly sensory. Most sea cucumbers are bottom dwellers, some concealing themselves in crevices or under stones, or among seaweed, others lying in burrows in sand or under mud. They may form very dense populations especially in the deep seas. In one deep sea trench they formed 50% of the living forms at 13,120 ft (4,000 m) and 90% at 27,880 ft (8,500 m).

A more unusual adaptation is found in those sea cucumbers which have structures called Cuvierian organs branching from the bases of the respiratory trees. When the animal is irritated these are extruded from the arms, and since they are very sticky any attacker is likely to become trapped among them. The British species *Holothuria forskali* can perform in this way, and is commonly called the "cotton-spinner" as a result.

Those sea cucumbers without Cuvierian organs may defend themselves by evisceration. When irritated they simply split the body wall and eject the gut and respiratory trees. These, it seems, provide a meal for any attacking predator, while the animal itself moves away and later regrows the missing parts.

There are three sources of food: plankton, detritus, and the organic matter in the bottom sediment. Some species are mainly planktonic feeders, extending their branched and sticky tentacles to catch small organisms, either from the open water or swept from the surface of the sediment. Sea cucumbers without branched tentacles shovel detrital material into the mouth with their tentacles or they may burrow through the sediment and simply ingest it. A single specimen may swallow over 100 lb (45 kg) of sediment per year. Thus it has been estimated that in a small bay in Bermuda of only some 1.7 sq miles area *Stichopus* species pass between 500 and 1,000 tons of sediment through their intestines each year. As such, sea cucumbers are responsible for causing significant changes in the composition of the sea floor.

In the Indo-Pacific region species of *Holothuria* are marketed under the name of trepang or bêche-de-mer. The smaller ones are eaten raw in salads, but the larger ones are split open, gutted, boiled, and then dried or smoked before sale and transport, then either soaked and minced to make soup or roasted and eaten whole.

SEA FANS, relatives of sea anemones but made up of colonies of polyps. Each colony forms a plantlike growth with a short main stem and lateral branches in one plant, the whole strengthened by a central axis of horn or gorgonin. They are found mainly in tropical and subtropical seas and are classified with the horny corals.

SEA GOOSEBERRY, *Pleurobrachia*, the best known and also the most typical of the Ctenophora or comb jellies. They are pelagic and often occur as swarms in the plankton. They are globular, gelatinous animals shaped rather like a hen's egg with a mouth at the narrower

followed by a single row of cónical teeth, sometimes hooked. Unlike most sea breams, there are no flat molar teeth in the jaws. The dentex is also found off the Atlantic coasts of southern Europe and occasionally reaches the British Isles. The gilt-head, another large Mediterranean species.

SEA BUTTERFLIES, marine mollusks that have two large flaplike extensions to their mantle called parapodia, which are used to propel them through the water. They are mostly carnivorous and are found only in offshore surface waters.

SEA CUCUMBERS, small, slimy, sausage-shaped marine animals related to starfish and sea urchins. On the front of the body is a mouth surrounded by a ring of tentacles. The skin is leathery and strengthened by variously shaped calcite spicules, such as rods, crosses, hooks, anchors or wheels. The color of sea cucumbers is usually dull: gray, brown, black or purple, only rarely red or orange. The earliest known fossil sea cucumbers seem to have had a continuous skeleton rather like a sea urchin. Most sea cucumbers crawl sluglike on their lower surface using two double rows of tube

The European sea horse *Hippocampus guttulatus* swimming in typical upright position.

end. Two long, retractile tentacles armed with adhesive cells or colloblasts are spread out either side, rather like fishing lines. Sea gooseberries are carnivores, and planktonic crab and oyster larvae and fish eggs are caught by the sticky tentacles and conveyed to the mouth. Rapid digestion of the prey takes place in the long pharynx, and food particles are circulated through a complex canal system around the body. Sea gooseberries swim using the eight rows of comb plates of fused cilia, which "flap" in an orderly fashion, from the mouth backward, so that these animals swim mouth forward. Coordination of the beat of the comb rows is brought about by a sense organ at the opposite (or aboral) end to the mouth. In the dark, sea gooseberries luminesce, the light apparently coming from the outer walls of the digestive system, which lie beneath the comb plates. These animals are able to regenerate damaged parts, for example, the comb plates.

SEA HARE, a sluglike marine mollusk with a pair of tentacles at the front end and, behind these, a pair of tentacles called rhinophores, which carry many sensory organs and are rolled inward and shaped like hare's ears, whence the common name. A reduced platelike shell is present, but concealed by mantle lobes that fuse along the back of the animal leaving only a small opening to the mantle cavity. Flaps, known as the parapodia, are developed at the sides of the foot, with which the animal can swim clumsily. The common European sea hare, *Aplysia punctata*, may have been the species described by Pliny, who was the first to record its common name and some of its habits. The sea hares of the genus *Aplysia* are worldwide but particularly abundant in subtropical and tropical waters, and are all very similar in appearance and habits. They lay very large numbers of eggs on shore in tangled pink or orange ribbons. One giant Californian sea hare, weighing 5 lb 12 oz (about 2.6 kg), laid approximately 470 million eggs in 18 weeks. The larvae swim in the plankton and then settle as young adults below low-tide level.

SEA HORSES, small highly specialized marine fishes related to the pipefishes. They are unique among fishes in having the head set at right angles to the body. In many ways the sea horses represent an end point in several trends already apparent in the pipefishes. The body is entirely encased in an armor of bony plates or rings, but the tail fin is absent and the hind part of the body is prehensile and can be twined around seaweeds to anchor the fish. Swimming is accomplished by wavelike vibrations in the dorsal fin, the fish progressing in a characteristic upright position. Care of the young by the male has reached a point where an enclosed pouch is present, formed by the bony plates of the body. Camouflage reaches bizarre proportions in such members as the Australian leafy sea horse, *Phyllopteryx foliatus*, in which fleshy leaflike appendages decorate the body, simulating seaweed. This grows to 12 in. (30 cm) which is large for a sea horse.

The behavior of the living sea horses is just as interesting as the ancient legends. These fishes can show remarkable color changes to match their surroundings. Their eyes can move independently of each other, and, using their heads, they can clamber about the weeds in which they live, "chinning" themselves from one strand to another. Not infrequently, several adults come together and twine their prehensile tails, a charming "dance" in the adults but possibly lethal in the juveniles, who are sometimes unable to free themselves and thus die of starvation. When breeding, the male and female of most species wrap their tails around each other, the eggs then being passed from the female by a cloacal appendage into the brood pouch of the male along his belly. The brood pouch is lined with soft tissue into which the eggs sink in little compartments, while blood vessels in this tissue enlarge transforming the pouch into a spongy "womb." A parallel can be drawn between the vascular tissue surrounding the eggs and the placenta found in mammals. The male may be visited by a number of females. Eggs that do not find a pocket in the pouch fail to develop, but those that do succeed hatch within the pouch and remain in their pocket until the yolk is used up. The "birth" or ejection of the young seems to be exhausting for the male. Firmly grasping some support with the tail, he rubs the pouch against shell or rock until the young and assorted pieces of tissue are ejected. There seems to be no truth in the contention that the young thereafter resort to the pouch for shelter when danger threatens. A large male may give birth to as many as 400 young, each a little replica of the adult.

Sea horses live in shallow warm temperate and tropical seas around the world but have a patchy distribution, being absent from large stretches of the coast of west Africa, and in the Indo-Pacific area may be present in one region but absent in another. There are about 100 species known. They feed on tiny pelagic organisms, which they suck up with their long snouts.

SEA LIONS, eared seals differing from the fur seals in having blunter, heavier snouts, a coat with a very small number of underfur hairs and hind flippers with the outer digits longer than the three inner ones.

All the sea lions are big animals, the males ranging in length from about 7 ft (2.1 m) to about 10 ft (3 m), the smallest and perhaps the best known being the Californian sea lion, *Zalophus californianus*. This is the animal that is seen in most zoos, and performs in so many circuses. Sea lions live along the coast of California and Lower California, occurring mostly on the offshore islands from San Francisco to Cedros Island, and there are also populations on the Galápagos Islands and off the coast of Japan.

The other sea lion of the northern hemisphere is the big northern or Steller's sea lion, *Eumetopias jubatus*, of the North Pacific. This sea lion is perhaps the biggest of them all, the males reaching over 10 ft (3 m) in length and up to a ton (1,016 kg) in weight. Its range is wide, reaching from Hokkaido in the west to California in the east. It is found on the Kurile Islands and in the Sea of Okhotsk, off Kamchatka and the Alaskan coast, and on the islands off British Columbia and California. But the center of abundance is on the islands of the Aleutian chain, where there may be over 100,000 animals. Both males and females are light brown, and the male develops a heavy mane around his neck and shoulders. During the breeding season, which starts in May, harems are formed and territories defended.

Male sea lion of the Galápagos Islands.

The pups are born in June and are fed by their mothers for at least three months. In earlier times this sea lion was exploited by the Aleutians, who used the skins for waterproof clothing, boots and boat coverings. The meat was eaten and the fat used for fuel.

The southern sea lion, *Otaria byronia*, has almost the same distribution as the South American fur seal. It occurs on the coast of South America south of Rio de Janeiro, on the Falkland Islands and around Cape Horn and also along the coast of Chile and Peru. Except during the breeding season, when they are more pugnacious, the sea lion and fur seal tolerate each other's presence, but normally they tend to keep away from each other, the sea lion preferring the sandy beaches while the fur seal lives in more rocky areas. The general behavior and harem formation is much as in the northern sea lion, but in the southern hemisphere the pups are born at the beginning of January.

SEAL, TRUE. Although any member of the highly adapted aquatic mammal order Pinnipedia may popularly be called a seal, the true seals, otherwise known as phocids or earless seals, belong to the family Phocidae. They have no external ears, cannot bring their hind flippers forward underneath the body, and their coat is composed mostly of guard hairs.

The classification of the true seals is almost coincident with their natural division into those of the northern and southern hemispheres. The family Phocidae is divided into two subfamilies: Phocinae and Monachinae. The Phocinae includes all the northern seals such as the bearded seal, *Erignathus barbatus*, the hooded seal, *Cystophora cristata*, the gray seal, *Halichoerus grypus*, the common or harbor seal, *Phoca vitulina*, the Caspian seal, *Pusa caspica*, the Baikal seal, *Pusa sibirica*, the ringed seal, *Pusa hispida*, the harp seal, *Pagophilus groenlandicus*, and the ribbon seal, *Histriophoca fasciata*. The Monachinae includes the monk seal, *Monachus schauinslandi* and *M. tropicalis*, the elephant seal, *Mirounga angustirostris* and *M. leonina*, the Weddell seal, *Leptonychotes weddelli*, the crabeater seal, *Lobodon carcinophagus*, the leopard seal, *Hydrurga leptonyx*, and the Ross seal, *Ommatophoca rossi*. In earlier classifications the elephant seal and the hooded seal are grouped together in the subfamily Cystophorinae because they are the only seals that have an inflatable nasal region. A more detailed comparison of their skulls and skeletons has shown that they are not so closely related to each other, and that the hooded seal is closer to other northern seals, while the elephant seals are closer to the other antarctic seals. The characters separating the two subfamilies Phocinae and Monachinae involve various details of the skull and skeleton, among which may be mentioned the reduced spine on the scapula, the first metacarpal which is noticeably longer than the others, and the reduced claws on the hind flippers in the Monachinae. In the Phocinae there is a well-developed spine on the scapula, the metacarpals of digits 1 and 2 are approximately the same size, and there are large claws on the hind flippers as well as on the fore flippers.

The most northerly seal of the northern hemi-

Laysan monk seal *Monachus schauinslandi*, the best known and most common of the monk seals.

Cow of gray or Atlantic seal, a species living on both sides of the North Atlantic.

sphere is the ringed seal, which lives along the circumpolar arctic coasts. It may be found as far north as the Pole, living in the water of the gaps in the fast ice and along fjords and bays, but not usually among floating pack ice or in the open sea. Two forms of this seal live in the freshwater lakes of Saimaa and Ladoga in Finland, and these, together with other geographical forms, have been given separate subspecific names.

Ringed seals are small animals, growing to about 4 ft 10 in (1.4 m) in length and weighing about 200 lb (90.7 kg). The coloring of the fur is rather variable but usually consists of black spots surrounded by a light ring, giving the animal its name, on a light gray background.

Two other little seals closely related to the ringed seal are the Caspian seal, *Pusa caspica*, and the Baikal seal, *P. sibirica*. The Caspian seal, as its name suggests, lives in the Caspian Sea, moving around this lake during the year, the greatest numbers always being found in the cooler parts. The Baikal seal lives only in Lake Baikal, a very large freshwater lake in eastern Russia. Both these seals are about the same size as the ringed seal, and grayish yellow in color, the Caspian seal being irregularly spotted with black.

Another arctic circumpolar seal is the bearded seal, *Erignathus barbatus*. This species does not occur as far north as the ringed seal and prefers shallow water near coasts. It is not gregarious

Southern elephant seal *Mirounga leonina* threatening the photographer. Once nearly exterminated, its numbers are now increasing.

and is rarely found in large numbers in any one locality. Bearded seals are about 7 ft 6 in. (2.3 m) in length and weigh about 600 lb (272 kg). They are gray with a slightly browner area down the middle of the back. The very profuse set of whiskers gives the animal its name, and these are curious in that when they dry they curl into tight spirals at their tips. The pups, 4 ft (1.2 m) long and gray in color, are born out on the ice floes in April or May and are said to remain with their parents for some time. One pup is born every second year, an unusual occurrence among the phocids, where one pup a year is more usual. Bearded seals eat animals living on the ocean floor, such as shrimps, crabs and sea snails, and are said to have a special liking for whelks. In spite of this, shells are never found in the stomach. In this they resemble the walrus. Because of their scattered distribution, these seals are of little commercial importance, but great use is made of them by the Eskimo. The skin is too thick to be used for clothing, but is very useful for boot soles, kayak covering and heavy ropes. The flesh, intestine, blubber and flippers are eaten, but again the liver is avoided.

One of the few phocids to have a distinctly patterned coat is the banded or ribbon seal, *Histriophoca fasciata*. This little-known seal, about 5 ft 6 in. (1.7 m) long, lives along the eastern coast of Russia, in the northwestern Bering Sea and in the Sea of Okhotsk. Although both males and females are similarly patterned, the females and young are much paler. The adult males are chocolate brown in color with wide creamy ribbonlike bands around the neck and hind end of the body, with a big circle on the side of the body around the junction of each fore flipper. The pups are born in the spring and have a coat of long white hair.

The harp seal, *Pagoplilus groenlandicus*, is another arctic seal, but is much better known than the banded seal. It is known from the northern coasts of Europe and Asia from Severnaya Zemlya to Spitsbergen, Greenland, and the northeastern coast of Canada from the Gulf of St. Lawrence to Baffin Island, but does not normally find its way into Hudson Bay. This seal is about 6 ft (1.8 m) in length and light gray in color, the male having a black horseshoe-shaped band running along the sides and across the back. This marking is present in the female, but is less distinct. The newborn pups, about 2 ft 6 in. (0.8 m) long, have a coat

of thick white wool that is shed after a month for a gray coat spotted with darker gray and black. There is enormous variation in the patterning of the coat of these immature animals, which are known as "beaters," and the pattern gradually changes over several years to the adult one. The pups are born at the end of February, and are suckled for only 10-12 days, putting on a thick layer of insulating blubber, which then acts as a reserve food supply until the pup learns to catch its own food. This is a very abundant seal: an estimate, based on counting the individuals seen in many aerial photographs, has put the numbers at 5 million. These seals are of great commercial value, the young pups, up to ten days old, are known as "whitecoats," and are much in demand, but the soft thick coat of the beater may be even more valuable. Older immature seals and adults are taken for oil and leather, the oil producing about half of the total revenue from these animals. International agreement controls the numbers of seals killed each year.

The hooded seal, *Cystophora cristata*, occurs from Newfoundland and Baffin Island to Greenland, Iceland and Spitsbergen, coinciding approximately with the western part of the range of the harp seal. Hooded seals occur farther out to sea than the harp seals, but produce their pups at the same time. Consequently, the same sealing trip is used to catch both kinds of pups, and it is almost impossible to distinguish between them in references in the older literature on sealing. Hooded seals are mostly solitary animals, but during the breeding season in March they collect in widely scattered groups of families, the chief concentrations being in the Newfoundland area and near Jan Mayen. Before it is born the pup sheds its first coat of long hair for a very beautiful coat that is grayish blue on the back and creamy white ventrally. These "bluebacks" are much in demand for making fur coats. Adult hooded seals are about 10 ft (3 m) long and weigh about 900 lb (408 kg). They are gray with large irregular black spots on the back and smaller spots ventrally. The most characteristic feature of the hooded seal is the large inflatable hood, an expansion of the nasal cavity found in the adult male. When not inflated it is slack

and wrinkled and hangs down in front of the mouth, but when blown up it sits like a cushion on top of the head. Another interesting feature of hooded seals is their ability to blow from one nostril a curious red balloon-shaped structure about the size of an ostrich egg. This is formed from the wrinkled and very extensible internasal septum, which can be blown out of one nostril. The reason for this behavior is still unknown, as both excited and perfectly quiet seals have been seen to do it, both in captivity and in the wild.

In slightly more temperate waters, the gray seal, *Halichoerus grypus*, lives on either side of the North Atlantic. On the western side it is found on the shores of the Gulf of St. Lawrence and Newfoundland, and on the eastern side in Iceland, the Faroe Islands, the British Isles, the Norwegian coast and in the Baltic. The largest breeding colonies are on the British coasts on North Rona and the Orkneys. Other colonies exist on the Shetlands and Hebrides and at many suitably rocky places, princi-

pally along the western side of Great Britain, but also on the Farne Islands on the eastern side. Many young animals have been tagged over recent years, and although there is no real migration there is dispersal after the breeding season, the young animals in particular wandering widely, but tending to return to their birth site each year. Gray seals are not quite as big as hooded seals, reaching about 9 ft 6 in. (2.9 m) in length and about 650 lb (294 kg) in weight. Although there is enormous variation in the color, all shades of dark and light gray and brown being found, it has been established that in males the background is darker with lighter spotting, while in the females the background is the lighter color on which are the spots of the darker color. As well as the color difference, the adult males have a pronounced "Roman" nose.

The pups are born at different times according to the locality of the colony. The Baltic and St. Lawrence pups are born in February and March, whereas in Britain they are born

Australian fur seals *Arctocephalus doriferus*, living along the southern coast of Australia. Little is known of their habits.

between September and December. The newborn pup is about 2 ft 6 in. (0.8 m) in length, and like so many phocids is covered in long white fur, which it keeps for about three weeks before shedding it for a short blue-gray coat. Lactation lasts for two or three weeks, and the daily weighing of a captive mother and pup has shown how much change in weight there is. At 3 days old the pup weighed 43 lb (19.4 kg), increasing to 92 lb (41.7 kg) by the time it was 18 days old, at which time it was weaned. The mother, who fasts during lactation, lost 95 lb (43 kg) during this period, her weight falling from 371 lb (168 kg) to 276 lb (125 kg). Male gray seals establish and defend territories during the breeding season.

The common or harbor seal, *Phoca vitulina*, in spite of its name is not so common around the British coasts as the gray seal. It lives in estuaries and on sandbanks uncovered at low tide. It is widely distributed, and various subspecific names have been given to the various geographical groups.

Monk seals live in warmer waters than many phocids. There are three species: the Mediterranean monk seal, *Monachus monachus*; the West Indian monk seal, *Monachus tropicalis*, and the Laysan monk seal, *Monachus schauinslandi*. It has been suggested that they are called monk seals because of the cowllike effect of the rolls of fat behind their heads. The Mediterranean monk seal is almost certainly the seal that has been known longest to man within historic times. Living in the Mediterranean, they were well known to the ancient Greeks and Romans.

The remaining phocids live in the cold waters of the Antarctic. They are all circumpolar, but living in different areas and with different food habits, they do not interfere with one another. The leopard seal, *Hydrurga leptonyx*, occurs farthest north, living on the outer fringes of the pack ice. Most of the animals are migratory and move north in winter, and at this time leopard seals are seen on most of the subantarctic islands, such as Macquarie, Kerguelen and South Georgia. Occasional stragglers reach the southern beaches of Australia and New Zealand, and one has even been seen as far north as Rarotonga in the Cook Islands. Young animals move farther afield than the adults. Leopard seals are large, solitary animals, reaching about 10-11 ft (3-3.3 m) in length and weighing about 600 lb (272 kg). Adult females may be up to 2 ft (0.6 m) longer than adult males. These seals are dark gray dorsally and light gray ventrally, variably spotted on their sides and throats. They have very long slim bodies, and the head seems disproportionately big, with a wide mouth armed with large three-pointed teeth. Its appearance has probably been responsible for its reputation of extreme ferocity, but unless annoyed it is normally dangerous only to penguins and the fish and squids on which it feeds. Like other pinniped penguin-eaters, the leopard seal shakes the bird vigorously until it is flung out of its skin, and penguin skins turned completely inside out may be found on the ice. As it lives on the pack ice not much is yet known about its general life history, but it is believed that the pups are born between November and January.

The crabeater seal, *Lobodon carcinophagus*, is perhaps the most abundant seal of the Antarctic. It is found on the circumpolar drifting pack ice, moving north with the ice in winter and south in summer when the heavier ice breaks up. It is a gregarious animal, being found in large concentrations in the southern summer. Very occasionally animals may reach Australia and New Zealand. Crabeater seals reach about 8 ft 6 in. (2.5 m) in length and are a silvery brownish gray in color, but this color fades during the year, and also with age, so that older animals may be almost white, even when newly molted. Again, few details are known about its habits, but what information there is suggests that the pups are born at the end of September. The name of the crabeater is misleading, as it feeds not on crabs but cn the shrimplike crustaceans known as "krill." It swims into a shoal of krill with open mouth and then sieves the krill from the water through its cheek teeth, the complicated cusps of which interdigitate to form a sieve.

The Ross seal, *Ommatophoca rossi*, is perhaps the least known of the Antarctic seals. It occurs in the heavier pack ice around the edge of the Antarctic continent, but is rarely seen. The Weddell seal, *Leptonychotes weddelli*, lives farther south than the pack ice, usually within sight of land or fast ice. It is the most southerly of the Antarctic seals and does not migrate, though occasional stragglers reach such places as Macquarie Island, Falkland Islands and New Zealand. Adults are about 9 ft (2.7 m) in length, females being slightly larger than males. They are dark brown or black dorsally, fading to white ventrally with variable white streaks and splashes on the dark ground color. The coat fades in summer, but not so much as that of the crabeater. Fish is the usual food of this seal, though many invertebrates, such as squid, crustaceans and sea cucumbers, may also be taken. The seals themselves are much used for dog food for Antarctic expeditions, and, occurring in areas where man also lives, they have been studied, and more is known about them than other

The sea slug *Coryphella lineata* reminds us that these marine gastropods include some of the most exquisitely colored of all the lower animals.

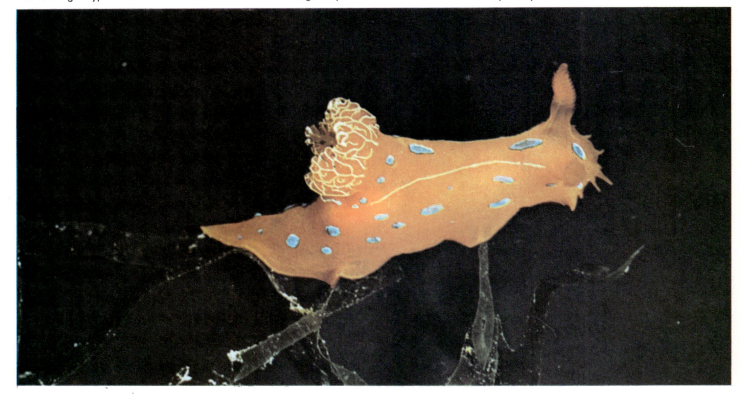

Antarctic seals. Diving depths and the production of underwater sound have been particularly studied. Depth recorders attached to Weddell seals have shown that they frequently dive to 980-1,310 ft (300-400 m), but rarely go deeper. The time spent under water usually is about 20 minutes, and although submersion for 40 minutes is possible, it seems that this may be near the limit for this seal. Many underwater sounds have been recorded, used by the seals for communicating with one another and for locating food and breathing holes in the ice. The pups, which are nearly 5 ft (1.5 m) at birth, are born during September and October and have a grayish woolly coat.

SEA MOUSE, *Aphrodite*, is a polychaete worm found below low-tide level on sandy bottoms. It is often brought up by fishermen in the dredge. There are various species, ranging in length from 1-8 in. (25-200 mm). The body is covered with a fine "felt" of silky chaetae, and it is to this apparently furry appearance that its common name is due. The body is more or less shuttle-shaped, being relatively short and stocky and tapering at each end. Sea mice live for the most part just beneath the surface of the sand or mud. They are rather lethargic, though they can scuttle quite rapidly for a short distance when disturbed. If the matted felt over the back is removed, a series of large disc-shaped scales will be seen overlapping over the back. The sea mouse is in fact nothing more than a very large scaleworm, in which the characteristic scales are concealed by the matted felt of hairlike chaetae. The segmented nature of the body is revealed on the underside when the animal is turned over.

These worms feed on other worms or moribund animals on the sand. When picked up, they often discharge a brownish fluid from the anus. In the water this makes a sort of "smoke screen," which might have a protective function. Sea mice respire by drawing water into the space between the dorsal felt and the back; and the tip of the abdomen is raised just above the surface of the sand as a kind of snorkel.

SEA PENS, colonies of polyps living on a central axis and looking like quill pens. In contrast to the other soft corals, sea pens are motile colonies generally living in warm coastal waters, but limited to soft muddy bottoms.

The sea pens are known for their luminescence, which comes from a slime exuded on stimulation. Some are luminous at any time, whereas others are luminous only at night or after being kept in the dark. A local stimulus will bring about a flash of light, and on stronger stimuli waves of luminescence will pass outward from the point of stimulation.

SEA SLUGS, unlike land slugs, are among the most beautiful of invertebrates. One group is exemplified by the sea lemon, *Archidoris pseudoargus*, common in European waters, and includes forms flattened in the vertical plane with a pair of sensory retractile tentacles on the front of the back which otherwise bears only small warts until the mid-dorsal anus is reached. This is surrounded by a ring of feathery gills which can be completely retracted. The sea lemon and related sea slugs feed mostly on sponges, swallowing large lumps. The yellowish sea lemon feeds on the

The edible sea urchin *Echinus esculentus* of European coasts, with its tube-feet extended. The shell is cut open and the roes extracted for food.

yellowish crumb-o'-bread sponge, *Halichondria panicea*, and the red sea slug, *Rostanga*, feeds only on red sponges, both matching the colors of their prey.

Some sea slugs have rows of small fingerlike processes running the length of the back. They feed on sea anemones and sea firs and it is said that the stinging cells from these are not digested but move from the sea slug's stomach up into the processes on the back and these are used by the sea slug in defense.

SEA SNAKES, poisonous snakes that live permanently in the sea and are fully adapted to that way of life, having the body flattened from side to side and the tail flattened and paddle-shaped. They swim with a sculling action of the tail, keeping mainly near the surface but able to submerge, closing the valvular nostrils on the top of the snout. They are, however, fully air breathing and must surface periodically or drown. The head is characteristically small and the front half of the body slender with the hindquarters more fully rounded. Some sea snakes reach a length of 10 ft (3 m), but the majority are 4-5 ft (1.2-1.6 m) long. All have venom glands with cobralike fangs, hollow and permanently erect in the front of the mouth. The 50 species are confined to the tropics, especially in the eastern Indian Ocean and western Pacific, mainly living in inshore waters. Some sea snakes swarm in the cracks and crannies of rocks in the intertidal zone. A few are known to enter tidal rivers up to 100 miles (160 km) inland. One species only, *Hydrophis semperi*, is confined to fresh water, in the Philippines. One of the commonest and most widely distributed is the yellow-bellied sea snake, *Pelamis platurus*, which is black on the back and yellow or pale brown on the belly, the two colors being sharply separated. The tail is dappled black and yellow. Some species,

living among seaweeds, are cross-barred with brown on a dun or gray background.

The slender front part of the body serves as a long neck, allowing the snakes to strike downward or sideways at prey, killing them with their venom, which is powerful enough to immobilize small fishes in a matter of moments. Some sea snakes feed exclusively on eels, but those that habitually lie among seaweeds or under floating objects eat the small fishes that go there to shelter. Crustaceans are also eaten.

SEA SPIDERS, marine arthropods only distantly related to spiders, and often known as pycnogonids. However, they differ in so many respects from all other arthropods that they are sometimes placed in a separate subphylum, which may have evolved as an offshoot from the early arachnid stock. Grotesque in appearance, they seem to be all legs and no body, and it is from the long legs, varying in number from 4-9 pairs and radiating in spider fashion from the inconspicuous body that they derive their common name. The body, which consists almost entirely of the trunk region, or prosoma, is usually narrow, occasionally disc-shaped, rarely more than a few millimeters in length, and totally inadequate to house some of the larger organs such as the stomach, which extends into the more voluminous legs. The abdomen is greatly reduced, in most cases, to a short stump at the end of the trunk; in a few species the abdomen is a long, spiny tube. At the anterior end of the trunk a tubular proboscis carries the mouth at its tip.

SEA URCHINS, spiny, spherical to somewhat flattened marine animals with the radial five-fold symmetry, an endoskeleton of calcite plates, and water vascular system that gives rise to avenues of tube-feet, so characteristic of the echinoderms to which sea urchins belong, on the surface.

Secretary bird playing with a tussock of grass. It throws it into the air with its feet and leaps after it.

The basic structure of modern sea urchins is that of a sphere made up of 20 columns of plates fused one to another to provide a rigid, usually boxlike, skeleton, known as the test, pierced at the top by an opening for the anus and at the center of the underside by the mouth. Sea urchins can, however, be divided into two groups. There are those in which the test has the original pronounced radial symmetry; there is a set of well-developed jaws, the Aristotle's lantern, in the mouth, named after the famous Greek scientist and its similarity in shape to a lantern. The body is covered with spines which vary considerably in size and shape between species. They may be large or small, long and pointed or short and club-shaped. Each spine is on a ball and socket joint, and the spines can be moved and used for protection against wave action or enemies, and they can be used for locomotion. In some species the spines are poisonous.

Some sea urchins live in shallow water or even between tide marks, but the greater number live between 650 and 3,250 ft (200 and 1,000 m), some living as deep as 13,120 ft (4,000 m). They are numerous in the Indian Ocean and western Pacific and in the Antarctic Ocean.

SECRETARY BIRD, *Sagittarius serpentarius*, a long-legged, diurnal, terrestrial bird of prey, confined to Africa south of the Sahara. The secretary bird stands almost 4 ft (1.2 m) high, and spans 6-7 ft (1.8-2.1 m). It is mainly gray, with black wing quills and thighs, and bare orange skin on the face. The head is adorned with a black-tipped crest, the central tailfeathers are very long and the gray legs are long with short stubby toes. The name comes from the gray and black clerical livery and from a fancied resemblance of the crest to quill pens stuck behind the ear.

Secretary birds inhabit short grassland 15-18 in. (38-46 cm) high, preferably shorter. They avoid long grass and extreme desert but sometimes adapt to cultivated areas. They roost on low thorny trees usually in pairs, and early in

the morning jump to the ground and begin hunting. They walk steadily through the grassland, nodding the head backward and forward like hens. Now and again one makes a quick dart to catch an insect or mouse in the bill, or breaks its pace to stamp rapidly for half a minute or so, presumably to disturb prey. The food consists mainly of rodents, insects and snakes, in that order of preference. Secretary birds kill far fewer snakes than they are usually credited with.

Normally terrestrial, they can fly well and can soar like a vulture or an eagle. They soar mainly in nuptial display, but sometimes to move from place to place, and have been recorded at 12,000 ft (3,650 m) above the ground. They are probably nomadic, becoming more common in certain areas in response to rodent or insect plagues, but some pairs are permanently resident, occupying a territory of about 4,000 acres (1,600 ha).

Secretary birds nest in low, very thorny trees, usually acacia or the desert date. They make a large, flat nest of sticks, lined with grass, and use it for several years. Two to three greenish white eggs are normally laid and incubated by both sexes for about 45 days.

When the chicks hatch, their development resembles that of other large falconiforms: helpless and downy at first, feathering by about 30 days and leaving the nest at 60-70 days. They do not fight with one another, and two are often reared. They are at first fed by the parent with regurgitated liquid matter, probably predigested insects. Later, the parent regurgitates a mass of grasshoppers, rats, snakes, etc. onto the nest, and the young feed on it. Since the secretary bird has short toes, it cannot carry prey in its feet, but brings all food for the nestlings in the bill or the crop.

SERIEMAS, birds resembling small brown cranes. There are two species: the crested seriema, red-legged seriema or crested cariama, *Cariama cristata*, and the burmeister's seriema, black-legged seriema or burmeister's cariama,

Chunga burmeisteri. Seriemas have long, slender legs, long necks and rather long, heavy bodies. The crested seriema stands about 30 in. (76 cm) tall and is about 36 in. (91 cm) long, whereas burmeister's seriema stands about 25 in. (63 cm) tall and is about 30 in. (76 cm) long. The bill is short, stout and hooked at the tip, the wings short and rounded and the tail long with dark and light bars. The crested seriema has an orange bill and gray-brown plumage finely barred and vermiculated with dark brown and black. The wings and tail have broad, contrasting bands of black and white. Burmeister's seriema is darker in color, with the underparts heavily streaked.

The crested seriema is found from central and eastern Brazil, through Paraguay to Uruguay and northern and northwestern Argentina. Burmeister's seriema is known only from the Cordoba to the Jujuy in northwestern Argentina and the western Chaco of Paraguay.

Seriemas are found on open ground, areas of open thorny woodland and scrub. Occurring in pairs and small groups, occasionally in larger flocks, they are usually seen running away in the distance, as both species are shy, being hunted as game. They can run very rapidly, but the flight is weak, with the legs trailing behind. Because of their shyness they are often most easily found by their calls, which include yelping and screaming notes in the crested seriema, and high-pitched yelpings in burmeister's seriema. In both species several birds often caterwaul in chorus. The crested seriema is often taken when young and reared among chickens by the South American Indians. The Indians do this because the noisy calls of the seriemas warn the fowls of the approach of predators long before they would notice them themselves.

Besides eating berries, seeds and leaves, they take insects, reptiles and small mammals. Snakes are often eaten, and it was considered that seriemas were immune to snake venom until recently, when an American research worker killed some by injecting snake poison into their circulatory systems.

SEROW, a medium-sized goat antelope, related to the goral, with long, mulelike ears, ranging from the eastern Himalayas through China to Japan and southeastward through Burma to Malaya.

SERVAL, *Felis serval*, unusual-looking member of the cat family because of its very long legs and large erect ears. It is entirely confined to Africa south of the Sahara, where it lives in regions of medium cover in the broken country of the foothills.

The serval is of slender build, standing some 20 in. (50.8 cm) at the shoulder, with a total length of about 4 ft (1.2 m) of which about 1 ft (30 cm) is tail. It weighs about 34 lb (15.4 kg). The ears are very large for a cat, in some ways resembling the shape found in certain species of bat, and give the serval acuity of hearing beyond the normal cat's. They are so wide at the base that they almost meet on the top of the head. The smooth, short hair is tawny on the back and sides, marked all over with a conspicuous

The serval of Africa, one of the cat family, has an unusual build.

Chionis alba, scavengers of the Antarctic, feed on dead animals, excrement and refuse, as well as on small shore animals.

pattern of bold black spots and stripes. The undersides are almost white. The ears are black on the outside with a prominent white spot in the middle. The upper part of the inner surface of each fore leg is marked with two black horizontal bands. The servaline, which used to be classified as a distinct species, is now accepted to be only a second color phase of the serval. It is tawny gray shading to off-white on the underparts and marked all over with closely scattered black spots giving a dusky or brownish hue to the whole body. There is also a melanistic form, which appears to be found more frequently in the higher parts of the range.

SHEARWATERS, a group of some 15 species of fairly large seabirds of the family Procellariidae, which also includes the typical petrels. They are tube-nosed birds, having nostrils opening through horny tubes lying on top of the upper mandible. The function of these tubes is not clear. They may be connected with the elimination of the salt solution produced by the salt excretion glands situated on the skull; they may be a means of protecting the nostrils when the bird is feeding in the water; they are perhaps connected with the well-developed sense of smell; and it has even been suggested that they help to spread oil over the plumage during preening. Shearwaters, like other tube-

nosed birds, produce large amounts of oil in the gut and discharge it through the mouth. The oil is secreted by special glands lining the part of the stomach preceding the gizzard and is used in a number of ways. In some species it is used defensively, being vomited in emergency and directed at an intruder. This habit would seem to be a simple development of the common defense reaction of vomiting and defecating. Also, the oil is used in the feeding of young and in courtship feeding. Chemically, it is a wax very rich in vitamin A. It is therefore probably also an excretory product, since vitamin A, accumulated in large amounts by many animals that feed in the sea, is toxic in high concentration. In some tube-nosed birds the oil is also discharged through the nostrils during preening and becomes spread over the plumage.

Shearwaters' bills are generally rather long, deeply grooved, being covered with separate horny plates, and are hooked for the retention of prey. The feet are webbed, all species swimming well and some capturing their food beneath the surface of the water. The plumage is dense and strongly waterproof, for shearwaters spend all their time at sea when breeding. Its coloration is dark—usually black—above and white beneath, or else uniformly dark. Shear-

waters breed colonially in burrows on offshore islands, which most species visit only at night. Some nest in hilly inland areas, and one central Pacific species nests in the open. As the general adaptations of shearwaters are connected with an efficient oceanic existence, their legs are placed far back for efficient swimming whereas their wings are long and narrow for efficient high-speed gliding flight. This means that on land they have difficulty in taking flight in windless conditions; they have to push themselves along on their bellies until they have climbed some eminence from which they can launch themselves. Thus they are very vulnerable to predators when on land, which is the main reason for their nocturnal habits.

The Manx shearwater is some 14 in. (36 cm) long. The closely related little shearwater, *Puffinus assimilis*, is some 3 in. (7.5 cm) shorter and has a more fluttering flight. The sooty shearwater is only a little larger than the Manx shearwater but has a wholly dark plumage. British colonies of the Manx shearwater seem to fly considerable distances to feed. Welsh birds, for example, flying to feed in the Bay of Biscay on the abundant sardine crop. Post-breeding movements are also considerable, at least some of the British birds wintering off the east coast of South America. Even the birds of the year take part in this migration, recently fledged young from Britain having been found on the coast of Brazil. This species seems to make use of the trade winds to assist its flight to the coastal waters of the Argentine, where it passes through its winter molt. It would seem that some birds get carried past their target by the prevailing winds, for one British-ringed bird, at least, has reached Australia.

SHEATHBILLS, aberrant birds showing both plover- and gull-like affinities and having the appearance of heavily built white pigeons. Both *Chionis alba* and *Chionis minor* are also sometimes called sore-eyed pigeons, paddies and kelp pigeons.

Their uniform white plumage covers a dark gray down, and the stout bill has a characteristic saddlelike horny sheath covering the base of the upper mandible. There is a cluster of pink caruncles around the base of the bill and a bare patch below the eyes, which have a bald ring around them. Sheathbills in their first year lack caruncles and have a white tip to the bill that is later tipped with black. The legs are dark gray with unwebbed toes and strong claws. The sexes cannot be reliably distinguished on the basis of external characters. Adult *C. alba* weigh about 1½ lb (0.6-0.7 kg).

In *C. minor*, the smaller of the two sheathbills, the bill and bare skin of the face are dark, and the front of the sheath is raised like the pommel of a saddle. This species is found in the south Indian Ocean on Prince Edward Islands, Crozet Islands, Kerguelen Islands and Heard Island, where each population is sedentary and is regarded as a different subspecies.

C. alba is more wide-ranging, occurring in the South Atlantic from Patagonia and the Falkland Islands south to the Antarctic peninsula, breeding regularly at South Georgia, South Sandwich, South Orkney and South Shetland Islands and the Antarctic peninsula.

Sheathbills have been much maligned because of their habits of scavenging on refuse tips, stealing penguin eggs and feeding on excrement and seal afterbirths. Also, when handled, they discharge the contents of their large rectal caecae, which has surprised many an unwary ornithologist. In fact, they keep their white plumage remarkably clean and should be looked upon as resourceful birds that have to take advantage of all available food supplies in their harsh environment.

SHEEP, DOMESTIC, *Ovis aries*, short, stocky animals with curly horns and a dense coat of wool. They are primarily grazers and are therefore normally kept on open grassland, often in enormous flocks.

The wild sheep, *Ovis vignei*, of central Asia, is probably the ancestor of our domestic forms, which have proved to be such useful animals that they have been introduced to many distant places. Sheep were taken to America at the time of Columbus, to Australia in the 18th century, and vast numbers are also kept in Europe, New Zealand, South Africa and Argentina.

The strong herd instincts of sheep make them excellent ranch animals, as they keep together in tight and easily managed flocks and do not disperse widely all over the available land, where they would be difficult to protect and almost impossible to round up.

Wool has been of great economic importance for thousands of years, and to this day, the Lord Chancellor sits upon the Woolsack in the House of Lords, as a symbol of the wealth of the English sheep-rearing industry in the Middle Ages. Wool is a normal attribute of many animals, but in domestic sheep its development has been enhanced and exaggerated by selective breeding. Most mammals have two types of hair: primary hairs, which are long, stiff and straight; and between these, dense masses of short curly, fuzzy secondary hairs, or wool. It is the latter that is responsible for body insulation, helping maintain the mammal's high body temperature. In sheep, selective breeding has encouraged the development of animals that grow only a dense coat of secondary hairs, and this coat or "fleece" is clipped off once or twice a year and used for spinning and weaving. Before the advent of man-made fibers, wool was the only real alternative to skins in the large scale manufacture of warm human clothing for use in temperate and cold regions.

A top wool-producing breed of sheep, such as the Corriedale of New Zealand, may yield up to 20 lb (10 kg) of high-quality fleece in a year. Some breeds have only 100 lambs per 100 ewes in a year, but others, notably the Dorset, may have many twins and triplets, resulting in 140 or more lambs per 100 ewes. However, a prolific breeder may produce only 6 lb (2.7 kg) of wool in a year, or perhaps a fine mutton-sheep breed, like the Southdown, has the disadvantage of being very small.

Merino sheep, a breed developed in Spain, were, and still are, renowned as the basis of the great wool industry. They produce very fine wool with long fibers and little tendency to

The Soay sheep are semiwild and have been on the St. Kilda islands off western Scotland since prehistoric times. They are believed to be descendants of sheep kept by Neolithic farmers.

shrink, two features of prime importance. The breed is very adaptable and can thus be kept in a variety of different types of country. It is particularly favored in Australia.

Karakul sheep, from central Asia, are the source of "Persian lamb" skins. The fleece itself is of poor quality and is not clipped. Instead, the whole skin is taken as a pelt from very young lambs. It is not true that the ewes are sacrificed in order to get the lambs before they are born. "Persian lamb" skins have glossy, dark, tightly curled fur, which is periodically much in demand as a fashionable fur trimming for clothes. Their value fluctuates widely, subject to changes in fads and fashions. In addition, Karakul do not produce many lambs or good meat, so are not a very attractive commercial proposition, though they are very hardy and can survive the most arduous conditions.

SHEEP, MOUNTAIN, successful Ice Age mammals of Eurasian origin that arose between 2 and $\frac{1}{2}$ million years ago, prior to the major ice ages. The living species include mouflons, urials, argalis, snow sheep, thinhorn sheep and bighorn sheep (desert sheep).

Mouflon is the smallest and most colorful sheep, native to Corsica and Sardinia, but successfully transplanted to the European mainland, the United States and Argentina. Unfortunately, the transplanted populations have often been crossed with urials or domestic sheep to produce better trophy animals and may differ considerably from native mouflons. These sheep differ from urials in having a distinct rump patch, a broad dark tail, out-curling horns and a white saddle across the back. The Cyprus urial should be considered to be a mouflon.

Argalis include the giants among sheep. They are found from the foothills of the Pamir throughout the mountain ranges of central Asia to northern China and central Siberia south of Lake Baikal. The westernmost races are small and uriallike, but the easternmost are huge animals. Large Siberian argali males carry a skull and horns up to 60 lb (27 kg) in weight, about 13% of their live weight and almost twice their remaining skeletal weight. The longest horns are grown by argalis from the Pamir and Tien-Shan mountains. These are the famous Marco Polo sheep. They are only a little larger than Rocky Mountain bighorns in body size, but their horns may reach 75 in. (190 cm) in length. The large argalis, excepting the Himalayan race, differ from urials in lacking neck manes and cheek beards and in growing relatively larger, curling horns. In other respects they are in essence overgrown urials, adapted to great cold and often high altitudes, with many similarities to the American bighorns. There are nine races.

Snow sheep are the primitive, Asiatic representatives of American-type sheep. The most primitive race appears to be the one in Kamchatka. They have small, thin horns and small rump patches, although in body size they almost rival the Rocky Mountain bighorn. Snow sheep are found on mountains beyond the Arctic Circle, where they attain a body size similar to thinhorn sheep. These northern snow sheep appear to be very similar in color and general appearance to the gray northern Stone's sheep from the Yukon Territory.

Thinhorn sheep are intermediate in horn, skull, and rump patch characteristics between snow sheep and bighorn sheep. Thinhorn sheep include the pure white Dall's sheep and the gray to black Stone's sheep. The distribution of the former is in Alaska, the Yukon and Northwest Territories, while Stone's sheep are found from the central Yukon, throughout northern British Columbia. Like snow sheep, Dall's sheep penetrate beyond the Arctic Circle in the Brooks Range of Alaska. In the St. Elias range, these sheep live in the sight of huge glaciers, and are here found in the largest numbers. Stone's sheep are a little larger than Dall's sheep and very variable in color. Three races of thinhorn sheep are recognized.

Bighorn sheep is the most diverse of American sheep. The smallest is the Nelson's bighorn, which is no larger than the urial and shows strong affinities with thinhorn sheep. This, as well as five of the seven living races, is a desert sheep, characterized by large ears, short wool, and extended breeding and lambing seasons. The largest bighorns are the cold-adapted, short-eared, massive-horned Rocky Mountain and California bighorns from Canada. These races approach a medium-sized argali in size. In fact, the bighorn females tend to be larger than the argali females. Large bighorn rams weigh about 250 lb (113 kg), and large females

lean body weight in some individuals, which compares favorably with the Siberian argali. Bighorn sheep have been eradicated over much of their former range in the United States and exist there mainly in small, relict populations. The bighorn ranges of Canada, excepting those of the California bighorn in British Columbia, appear to be largely intact.

SHELD-DUCK, various species of duck, some of which are rather gooselike in appearance and are therefore called sheld-geese, but all of which are quite closely related to the typical dabbling ducks. The sheld-ducks and sheld-geese comprise the tribe Tadornini.

The common sheld-duck, *Tadorna tadorna*, breeds on the coasts of northern and western Europe and eastward through Eurasia, where it frequents the shores of inland lakes and seas as well as ocean coastlines. It nests in holes in the ground, sometimes in the burrows of mammals such as the rabbit and the steppe marmot. It may also be found in agricultural areas and sometimes even nests in trees.

As in most species of the tribe, the sexes of the common sheld-duck are similar in appearance, though the male is larger. The head and neck are dull green, but the body is white with a chestnut band around the breast and black above the wings. The feet and bill are red, and in the male the latter has a large frontal shield

X-ray photograph *(left)* of a piece of shipworm-riddled wood, showing the animals' burrows. Shipworms and their burrows exposed *(right)*.

exceed 180 lb (81 kg) and may even be 200 lb (90 kg). The sexual dimorphism here appears to be less than that of argalis. Typically, bighorn rams break off their horns in fighting once the horns have formed about three-quarters of an arc. Compared to thinhorn sheep, bighorns are more aggressive, clash more and are apparently rougher in their combat and courtship behavior. Bighorn rams have relatively, and absolutely, larger skulls, horns, and rump patches than thinhorn sheep. The weight of dry upper skull and horns exceeds 13% of

in the form of a knob on the forehead. Both sexes have a duller postbreeding plumage, corresponding to the male eclipse plumage seen in many other ducks. They are in full plumage from October to July. The male is about 24 in. (60 cm) long.

Other species of sheld-duck are the muddy sheld-duck, *T. ferruginea*, of southeast Europe and central Asia, the very similar Cape sheld-duck, *T. cana*, of South Africa, the New Zealand sheld-duck or paradise duck, *T. variegata*, the Radjah sheld-duck, *T. radjah*, of the East

Indies and Australia and the Australian sheld-duck, *T. tadornoides.*

SHIELD-TAILED SNAKES, forty species of primitive snakes related to the pipe snakes and showing many adaptations to burrowing. There are usually two lungs, but the left is only of a moderate size and the skull is a rigid compact structure with peculiar features in the supporting axis and atlas vertebrae. There are no specialized enlarged belly scales. The eyes are minute and are hidden under the head scales. The tail shows varying degrees of modification in different species. Some simply have a large scale at the end of a short tail, but in others this is embellished with two or three spines. Most, however, have the scales replaced by a single circular spiny shield at the tail end. It is believed that the shield is thrust into the soil to provide a firm anchorage while the small pointed head is pushed forward, but it has also been suggested that it is used to plug the entrance to the snake's burrow. Shield-tailed snakes are mostly 1 ft (30 cm) or less in length. The largest on record was 29 in. (73.8 cm). These quiet, inoffensive snakes have a limited distribution in peninsular India and Ceylon, being found burrowing in the soil and under stones and logs from sea level to 7,000 ft (2,500 m) in the forested areas. They give birth to between three and eight young at a time. With their rigid skulls, shield-tailed snakes seem unable to eat large items, and their food consists of soft-bodied grubs and worms.

SHIPWORM, a bivalve mollusk, despite its name, notorious for burrowing in the timbers of piers and wooden ships. The common shipworm, *Teredo navalis,* makes burrows of up to 18 in. (40.3 cm) long and $\frac{1}{4}$ in. (7 mm) wide, lined with a white calcareous material given out by the shipworm. Most bivalve mollusks are entirely enclosed by the shell valves, but in the shipworm these are modified to a tiny pair of abrasive plates at the head end. The body is long and wormlike and consists mainly of an elongated tubular mantle that connects the end of the burrow with the seawater outside. Water is drawn into the tube through an inhalant siphon and is pumped out through an exhalant siphon, as in many other bivalve mollusks. The siphons normally project from the surface of the timber but can be withdrawn and the hole closed by a pair of calcareous plates called pallets. The shipworm is reputed to be able to bore tunnels as long as 1 ft (30.5 cm) in one month. Neighboring worms avoid boring into each other's tunnels, so bored timbers have large numbers of parallel white burrows. These scarcely show on the surface since the only opening to the exterior is the tiny hole made by the larva when it first entered. The shipworm is hermaphrodite, both male and female gametes being ripe at the same time in any one animal, and the larvae are retained within the mantle cavity and later released as miniature adults. Dense and destructive populations may therefore be rapidly built up by recruitment from one adult.

SHOEBILL, *Balaeniceps rex,* or whale-headed stork, a heavily built gray swamp bird standing nearly 4 ft (1.2 m) high. It has a long neck, long black legs and a massive head. The bill is grotesque, almost as broad as it is long and sharply hooked at the tip.

Shoebills are shy, retiring birds and spend most of their time singly, or in pairs, in the permanent swamps of eastern Africa (Congo, Zambia, Uganda, Sudan). Large sluggish fishes feature in their diet, which also includes frogs, turtles and probably small mammals too. Two eggs form the usual clutch and are laid close to the ground on a flattened platform of swamp grasses or papyrus.

SHORE CRAB, *Carcinus maenas,* a species common on European shores and also those of North America, where the crab is known as the green crab, or Joe Rocker. The green color is found mainly in the males; the adult female tends to be more orange in color. Young crabs show a very wide range in coloration and pattern. Some are almost black, others green or white with red or black patterns. This variety of coloration is lost as the animals become adult.

The shore crab is abundant on rocky shores, but also extends a considerable distance into estuaries, and may be found in muddy areas and salt-marsh pools. When living in water of low salinity, the shore crab can maintain the concentration of its blood well above that of the surrounding water down to a concentration of about one-sixth that of seawater. Below this

value the crab can live for short periods, but not permanently.

The shore crab is most active at night and at high tide. In the summer the crabs move upshore with the tide, and remain on the shore when the tide ebbs, provided there is sufficient shelter. In the winter not so many remain on the shore, and most move down again as the tide retreats.

SHREW, small mouselike mammal with short legs, a long pointed nose and long tail. Shrews form the biggest family in the Insectivora, with over 200 species. They have a worldwide distribution, though absent from the polar regions, Australia, New Zealand and most of South America. Their distribution in time is equally great; fossils are known dating back to Oligocene times. The family is divided into three subfamilies. The Soricinae are the most common and widespread and include familiar species such as the European common shrew, *Sorex araneus,* and the short-tailed shrew, *Blarina brevicauda,* of North America. All soricine shrews are characterized by having bright red enamel on the crowns of their teeth. The subfamily Crocidurinae have all-white teeth and the head is usually more triangular in shape with bigger ears. This group is absent from America, but very abundant in Africa.

The ungainly looking shoebill of the East African swamps.

In Europe they are less widely distributed than the soricines and are represented by such species as the musk shrew, *Crocidura russula*. The Crocidurinae includes both the largest and the smallest of all shrew species. The third subfamily, the Scutisoricinae (armored or hero shrews), consists of just two large species found in central Africa. They look quite ordinary from the outside, but internally are notable for their extraordinary vertebrae which have massive, interlocking processes and unique dorsal and ventral spines. The spinal column of these creatures is said to be so strong that the shrew will survive a man standing on it.

Shrews all look rather similar with a body length of less than 8 in. (20 cm) (usually less than 4 in.). Shrews weigh up to 1.2 oz (35 gm). Some genera (e.g., *Neomys*, the European water shrew) are adapted to a semi-aquatic life and have bristle-fringed tails and feet to aid in swimming. The same sort of bristle-feet are seen in certain Asiatic shrews, where the stiff hairs are said to enable the animals to run about on soft, loose sand without sinking in.

Shrews are nervous, highly active creatures with a pulse rate sometimes nearing 1,000 beats per minute. They are on the move at all hours of the day and night resting for only brief periods. They scurry about very rapidly, often emitting a low twittering noise and stopping to sniff the air or investigate objects with the long, inquisitive nose. Shrews do not hibernate in winter. They are insectivorous in their diet, but also behave as small-time carnivores and eat a variety of invertebrate and even small vertebrate animals. The semi-aquatic species also have access to crustaceans and other water creatures not normally available to ordinary shrews. In addition many shrews eat small quantities of seeds and other plant material. They are voracious feeders and soon starve if deprived of food. They may eat more than their own weight of food in a day, partly

A litter of babies of the common shrew.

due to their great and constant activity, resulting in high energy consumption, and partly because the shrew's diet is not very nutritious for the most part, consisting mainly of water and indigestible materials.

Most shrews breed during the summer (March-November) in the northern temperate lands, but in the tropics they often breed throughout the year. The gestation period varies from species to species, but is between two and four weeks. Up to four litters may be produced in one season. The young are born naked and blind in a breeding nest, usually built under leaf litter or a log or stone for shelter. The young are weaned after about three weeks and

may remain together as a family for up to a month.

SHRIKES, aggressive and predatory perching birds. They kill insects, birds and mammals by striking with their hooked bill. Because they impale their victims on thorns they are called butcherbirds, from the fancied resemblance to a butcher's shop.

The true shrikes (Laniinae) are mostly between 7-10 in. (18-25 cm) long and are characterized by a strong hooked bill, resembling a falcon's with a toothlike projection behind the hook on the upper mandible and a notch on the lower; a bill unique among passerine birds. The feet are strong and the claws are sharp and are

Female red-backed shrike feeding nestlings.

used for grasping prey. The shortish rounded wings have ten primary feathers, while the rounded tail has 12 main tail-feathers. There are well-developed bristles around the mouth. The sexes are generally alike and there is a single molt in autumn.

One of the largest, the great gray (northern) shrike, *Lanius excubitor*, is 9½ in. (24 cm) long and breeds widely in Europe, temperate Asia and North America, although only a winter visitor to Britain. This, and its near relative the loggerhead shrike, *L. ludovicianus*, are the only New World species. The great gray shrike is pale grayish with a black mask, wings and tail, and during the winter in the north feeds largely on birds and small mammals. The rather similar lesser gray shrike, *L. minor*, breeds in southern Europe and central Asia and winters in east Africa, and feeds largely on insects. Other species tend to be various shades of brown and gray. The red-backed shrike, *L. collurio*, breeds in southern England, Europe and temperate Asia, wintering in tropical Africa and southern Asia.

The name shrike is derived from the shrieklike calls which are harsh and discordant. The songs, however, often contain musical warbling as well as harsh notes, and some species are excellent mimics.

SHRIMPS, a term applied to those decapod crustaceans that swim using their abdominal limbs, instead of crawling like lobsters and crabs. A typical shrimp has five pairs of walking legs, of which at least two pairs bear pincers of some sort. These pincers may be quite large and powerful, or they may be very long and delicate. The body is usually elongated and more or less cylindrical with a distinct tail fan. The two pairs of antennae are generally very long and two-branched, at least one branch is carried reflexed over the body, reaching just beyond the tail fan. This enables the shrimp to walk backward into crevices, and also gives warning when touched by a predator approaching from behind. A sudden flick of the abdomen is the usual response to the approach of a predator, and serves to jerk the shrimp well away from harm. The eyes are stalked, and in some deep-sea forms are very large, but in other deep-sea forms and some cave-dwelling species they may be small or absent. There is great variation in the body length of shrimps, from ½ in. (1 cm) to 1 ft (30 cm).

Some shrimps are scavengers, others are predators and some have developed special techniques for collecting small particles. Some use brushes on their pincers to sweep up detritus and pass it to their mouthparts. Others have large pincers and will prey upon small fish if the opportunity arises.

A typical inshore shrimp, such as *Crangon vulgaris* of European shores, spends much of its time during the day buried in sand in shallow water. The sand is cleared away from under the body by movements of the limbs, so that the crab sinks downward. During the night the shrimp walks about and preys upon worms and any other small creatures that it can catch in its pincers. Shrimps often abound in estuaries. Around the coasts of Europe *Palaemonetes varians* is often abundant where the sea is diluted by the inflow of fresh water. A closely related species, *P. antennarius*, is

Verreaux's sifaka, of Madagascar, is credited with traveling through the trees with leaps of up to 40 ft (12 m).

found in some of the lakes in northern Italy. Freshwater shrimps are most abundant in the tropics, particularly members of the families Atyidae and Palaemonidae. In some large tropical lakes the adult live around the edges, and the larvae are planktonic in the middle of the lake.

SIFAKA, a magnificent white herbivorous lemur with orange, maroon or black markings and a terrierlike muzzle. Two species are recognized: Verreaux's sifaka, *Propithecus verreauxi*, and the diademed sifaka, *P. diadema*. The first is white with maroon or black patches on the head, back, belly or limbs, according to the subspecies; the second is larger, also white, but has patches of orange to deep brown and black. The name "sifaka" is from the Malgache name (pronounced "shifak") for Verreaux's sifaka. The diademed sifaka is referred to as the simpona (pronounced "shimpoon") by local villagers. In both species, the face and ears are always black. The head and body length is 18-22 in. (45-55 cm) with a tail of roughly the same length. The two species are confined to Madagascar, and are both completely arboreal and diurnal.

Their food is leaves, buds, flowers, fruits and

the bark of large trees. Sifakas are vertical clinging and leaping forms, with extremely long and powerful hind legs. When moving fast, they can make leaps of 15 ft (4-5 m) or more, taking off and landing in a vertical position. The animal pushes off with the hind legs and lands on its feet. The thumb and big toe are opposable, so that vertical supports can be firmly grasped in squatting and leaping. Verreaux's sifaka can even leap around in the spiny forests of the south of Madagascar, characterized by tall, cactuslike trees with spiral arrays of spines. Apparently, the sifaka's leap is so accurate it can land even on these trees without injury, for the skin of the hands and feet does not appear to be tough enough to prevent penetration of the spines when the sifaka lands after one of its enormous leaps.

Gestation is four to five months with usually a single young at birth. Twins occasionally occur. Immediately after birth, the infant climbs onto the mother's fur. It is then carried around by the mother, and later by other adults. At about one and a half months, it starts to move away from the mother in the trees and to feed independently. However, at the slightest sign of danger it returns to cling onto her fur.

The silver bream is a carplike fish of north and central Europe.

The sifaka is placed in the family Indriidae together with two other vertical clinging and leaping forms: the indri, *Indri indri*, and the avahi, *Avahi laniger*, all of which are alike in diet and habits. The first is much larger than the sifaka with a head and body length of 28 in. (70 cm), while the second is smaller, with a head and body only 12 in. (30 cm) long and a tail 16 in. (40 cm) long. The indri, diurnal like the sifaka, has an extremely short tail, only 1 in. (2½ cm) long. The avahi is nocturnal. Both are specialized herbivores, eating leaves, buds, flowers, fruit and bark of a very small number of trees.

The indri can leap farther than any other lemur, and with great accuracy, although it does not even have a tail to steer it during the jump. Like the sifaka, it lives in small social groups, probably family units with two to four adults, which seem to be territorial. The indri is well known for its eerie howls, which carry over great distances and are answered by howls from neighboring groups. A group can be induced to produce this howl by playing a tape-recording of their own calls. Like the sifaka, the indri also utters short, intermittent grunts as a mild alarm call. There is a single offspring, which is carried on the mother's fur.

The avahi, in spite of its nocturnal habits, also makes fairly large leaps. It also forms small groups (probably family units) of two to four individuals, but is most often encountered singly at night. As with all nocturnal lemurs, its eyes reflect a bright red glow, and it can easily be spotted with a torch in the dark. High-pitched, whistling calls are produced at night, and these may serve as territorial signals. A single young is produced toward the end of the dry season and the infant is carried on the mother's fur.

SIKA DEER, *Cervus nippon*, of eastern Asia, its range includes Japan (six subspecies), Taiwan, eastern China (three subspecies), Manchuria and Korea (two subspecies) and northern Vietnam and Annam.

The sika deer are animals of medium size, vary-ing in shoulder height from about 25½ in. (65 cm) in one of the Japanese forms to about 43 in. (109 cm) in Dybowski's deer from Manchuria. Antlers, usually eight-tined, seldom exceed 26 in. (66 cm) in length.

The summer pelage, apart from melanism in the small Kerama sika deer, *C. n. keramae*, from the Ryukyu Islands, varies from a rich chestnut red to a more yellowish-brown hue, dependent on the subspecies, and liberally spotted, which has given rise to the name spotted deer. In winter these spots, in some animals, are barely discernible. A typical feature of the sika group, except those of mela-nistic strain, is the white caudal disc which is fanned out when the deer is alarmed.

The breeding habits are similar to the red deer with which this deer will interbreed on occasions. The rut call of the stags is a peculiar whistle which may change into a high pitched scream. The deer is fond of cover.

SILKWORM, the caterpillar of a moth, *Bombyx mori*. Like many other caterpillars it spins itself a cocoon of silk in which it pupates, and this cocoon is the source of virtually all our commercial silk. *B. mori* was originally a native of China, where the value of its silk was realized in about 1800 B.C. Since then it has been introduced to many countries and is now no longer known in the "wild" state at all. The main silk-producing countries are Japan, whence comes two-thirds of the world's silk, China, India, Russia, Italy and Brazil, but the annual production now is significantly less than 30 years ago because of competition from synthetic fibers such as nylon and rayon. In 1939, about 50,000 tons of silk were produc-ed, compared with only 41,500 tons in 1970. Most of the world's silk comes from eastern countries, because the moth breeds continuously so that there may be six or more generations in a year. By contrast, in Europe there is only one generation a year, and the eggs remain in a state of diapause throughout the winter. The eggs of the continuously breeding race hatch in 8-10 days.

The caterpillar feeds only on the leaves of the mulberry tree, and it normally takes a month to reach full development. Then it starts to spin its cocoon. The silk is made in special glands, which in many other insects produce saliva. It is a mixture of proteins and is formed into a thread as it is forced through a small opening just behind the mouth. The caterpillar makes a continuous thread commonly 1,000 yd (900 m) long, and sometimes as much as 2,000 yd; by unraveling this thread, in some places by hand, we obtain our silk. Normally, when the adult insect emerges from the pupa it damages the silk. To prevent this happening the pupae are killed by heating to a high temperature for a short time.

The caterpillars of some related moths belonging to the genera, *Antheraea* and *Attacus* (family Saturniidae), also produce large silken cocoons. Some commercial use is made of their silk, but this is insignificant compared with that of *Bombyx*.

SILVER BREAM, *Blicca bjoerkna*, also known as the white bream, a carplike fish of north and central Europe that is also found in south-east England. It closely resembles the common bream, but its body feels hard and less slimy. Also, it has only 19-24 branched rays in the anal fin, whereas there are 23-29 in the bream, and it is a rather more silvery fish with reddish tints to the head and the bases of the pectoral and pelvic fins. The two species frequently hybridize. The silver bream grows to about 15 in. (35 cm). It is of little interest as a sporting fish and is rarely eaten. The Pomeranian bream, common in Cambridgeshire and once consid-ered a distinct species, is now known to be a hybrid of the silver bream and the roach.

SILVERSIDES, small marine or freshwater fishes with a bright silvery band down the flanks, related to the flying fishes. They are almost worldwide in distribution, being found in the fresh waters of Australia, where they are known as rainbow fish. They have two dorsal fins, the first with rather weak spines and the second soft-rayed. The eggs have long filaments sprouting from them, which help anchor them to weeds or seaweeds. The common sand smelt, *Atherina presbyter*, found off the coast of northwest Europe, grows to about 6 in. (15 cm) and is common in harbors, bays and estuaries. These fishes are gregarious and feed on small crustaceans and other invertebrates. With their bright silver band on the flank they make excel-lent bait for trolling and spinning. The second European species is Boyer's sand smelt, *A. mochon* (formerly *A. boyeri*), a less common species, which can be distinguished by its slightly larger eye and fewer rays in the anal fin (13-15 as against 15-19 in the common sand smelt).

There are several freshwater species on the western side of the Atlantic. The brook silver-side, *Labidesthes sicculus*, from a wide range of waters in the eastern parts of the United States, grows to only a few inches in length and is semitransparent and pale green with a violet iridescence. It lives in enormous shoals, and when alarmed by predators these scatter rap-idly, the fishes skittering over the water, which has earned them the name skipjack.

Along the shores of Lower California is found the grunion, *Leuresthes tenuis*, which grows to

Reminiscent of the magic carpet, the thornback ray moves gracefully through the water using wavelike movements of its large pectoral fins.

6 in. (15 cm) and normally lives in shallow water. It has remarkable breeding habits. It spawns at night high up on the beach at the time of high spring tides, the fish allowing themselves to be washed as far up the beach as possible by the surf and then wriggling forward over the wet sand. The female digs herself into the sand, and the male wraps himself around her. The eggs are shed and fertilized about 2 in. (5 cm) below the surface. The fishes then wriggle back into the surf. The eggs remain in the sand until the next high spring tide, when they hatch within a very short time of first being wetted,

the larvae making their escape to the sea. Since the spawning of the grunion is so closely correlated with the tides, it is thought that these fishes are able to appreciate the phases of the moon.

SKATES, a family of flattened cartilaginous fishes belonging to the order of rays but differing from all other raylike fishes in producing eggs and not giving birth to live young. Typically, the skates are highly flattened fishes with large winglike pectoral fins that spread evenly from the snout and then taper more or less abruptly. The tail is slender, equal to or shorter

than the body, and there are two dorsal fins set far back toward the end of the tail. The eyes are on top of the head, followed by a pair of conspicuous spiracles through which water is drawn to aerate the gills. The body, and sometimes the pectoral "wings," are often lined with spines or bucklers (modified denticles). The gill slits are on the underside, as is the mouth, which bears a series of flat, pavementlike teeth in each jaw. In some species the snout is produced into a point or rostrum. The skates lay large eggs enclosed in a horny rectangular capsule with short pointed tendrils at each corner; these are commonly found on beaches and are known as "mermaids' purses." The young remain within the egg capsule for at least four months and sometimes as long as 14 months. Skates are worldwide in their distribution, mostly in shallow waters but a few in depths down to over 7,000 ft (2,200 m). Some species have weak electrical organs developed along the side of the tail, and these are used possibly for echo location of prey, mates, obstacles and predators. Color is very variable among species and among individuals and can to some extent be controlled to match the background. The majority of skates are fairly small, reaching 1 or 2 ft (30-60 cm) in length, but a few may reach 8 ft (2.4 m). The skates are more important commercially off European coasts than in the western North Atlantic but are of local importance to fisheries in other parts of the world. The common skate of Europe is *Raja batis.*

SKIMMERS, ternlike birds of tropical rivers and shores, deriving their name from their habit of skimming low over the surface of water when feeding. There are three species, one in the Americas, one in Africa and one in southern Asia. They are very similar, differing only in minor respects of bill color and plumage pattern, and a description of the salient features of one species will serve for all. Much still remains to be discovered about these birds, with their graceful aerial acrobatics and unique manner of feeding.

A relative of the gulls and terns, the African skimmer, *Rynchops flavirostris* (more aptly called scissorbill, or even shearwater or razorbill were not these names already in use for quite different birds) is 18 in. (45 cm) long and has the proportions of a large tern. The underparts are white, in sharp contrast with the dark blackish-brown upperparts, and the forehead, sides of the tail and trailing edge of the wing are white. In winter the hind part of the neck becomes white also. As in terns, the tail is forked and the wings very long, the tips projecting well beyond the tail when the bird is at rest. The short legs and shallowly-webbed toes are vermilion. Quite the most remarkable character is the bill; both mandibles are laterally flattened, and the lower is about 1 in. (2-3 cm) longer than the upper. It is as thin as a penknife blade, but fairly flexible. In this species the bill is yellow, in the other two orange or red and black.

Skimmers feed by flying gracefully back and forth over still water, the bill wide open and the lower mandible slicing the surface for yards at a time. Although the wing beat is full, the tips never touch the water, partly because the body is somewhat angled during skimming,

The thick tail of the West African skink *Riopa fernandii* contains a store of fat.

head down and tail up. There has been considerable debate about how the food is procured, and it has been held that the principal function of skimming is to create plankton phosphorescence, the light of which attracts fishes, captured by the bird on its return flight. More likely, skimming is merely an unusual and effective means of catching small fishes and crustaceans, which rise to the surface of lakes and the seas in twilight. The lower mandible is doubtless very sensitive to touch, and striking a small particle during skimming provokes an immediate closing of the upper mandible—with an audible snap—together with a tilting downward and backward of the bill. If the prey is caught, the bill is lifted clear of the water, the food deftly swallowed, and skimming resumed with hardly a check in flight.

SKINKS, with over 600 species, comprising one of the two largest lizard families. They are generally smooth-scaled and cylindrical, with conical heads and tapering tails, short legs or none at all, overlapping scales and a protrusible tongue. The majority are under 1 ft (30 cm) in length, and most are less than 8 in. (20 cm), the smallest being barely 3 in. (7 cm) long. They are mostly ground dwellers and burrowers, but many climb about in bushes or trees, although only one has a prehensile tail and thereby shows some specialization for arboreal life. This, the Solomon Islands giant skink, *Corucia zebrata*, is also the largest of the family and just exceeds 2 ft (61 cm) in length. No skink is truly aquatic, and there are no real runners among them; they only crawl or scamper. They are secretive, and many species spend much of their lives underground. Adaptations for burrowing are evident in the body shape, in the shovel-shaped snout and in the reduction in size or even absence of the limbs in some species. Another feature is the degenerate eye, which is covered by a scaly lower lid or by a transparent nonmovable disc. The ear opening is closed, and the inner ear degenerate.

Skinks are the most abundant of the lizards in Africa, the East Indies and Australia. In Australia the number of species of skinks exceeds that of any other family of reptiles. Half or more of the species occur in southeastern Asia and the East Indies, and only about 50 species are found in the western hemisphere. However, some skinks are found on every continent except Antarctica. The smaller species of the skink family feed mainly on insects, the largest skinks are wholly or partly vegetarian.

One of the best-known skinks is the Australian blue-tongue skink, *Tiliqua scincoides*, 12-15 in. (30-40 cm) long, frequently kept as a pet. The name refers to the light or deep blue tongue, which is slowly but constantly flicked in and out whenever the animal is alert or on the move. The young of this species are heavily banded, the bands tending to break up as the animal matures. Another attractive Australian skink is the shingle-back lizard, *Trachysaurus rugosus*, up to 18 in. (45 cm) long, also known as stumpy-tail, bobtail, double-headed lizard, boggi, and pine-cone lizard. Unlike the other Australian skinks, the shingle-back generally has only two young at one time, only occasionally triplets. It is a sluggish animal, and feeds on flowers and fruit, as well as snails and slugs.

SKUAS, seabirds related to the gulls and terns, which have been described as "gulls turned into hawks." They are best known for their piratical attacks on other birds, forcing them to disgorge their food, and for the vigor with which they defend their territories against trespassers. The shape of bill and feet are gull-like, except that the bill is strongly hooked and the feet bear curved claws. The great skua, *Catharacta skua*, is about the size of a herring gull but is heavier. The plumage appears uniformly dark brown, but, close up, the feathers can be seen to be streaked with rufous brown or white. Maccormick's skua, *C. s. maccormicki*, of the Antarctic, has a much paler body than the other subspecies. The great skua is readily identified by the white patch at the base of the primary feathers, which is conspicuous in flight and during the characteristic display with the wings raised.

The *Stercorarius* skuas or jaegers are smaller and more lightly built. The central two tail feathers of the adults are elongated; those of the long-tailed skua, *Stercorarius longicaudus*, may be $\frac{1}{3}$ of the total body length, those of the pomarine skua, *S. pomarinus*, are twisted, while the tail feathers of the arctic skua, *S. parasiticus*, are comparatively short. There are dark, light and intermediate color phases.

The great skua is unusual in having a bipolar breeding distribution.

Skuas are basically fish-eaters but are best known for their piratical and predatory habits. In many places skuas are the scourge of other seabirds. Gannets, kittiwakes, terns and others are chased on their way back from feeding and forced to give up their prey to escape the skua's harassing. Eggs and chicks of kittiwakes, guillemots and prions are stolen, and the adults of many birds, such as snow petrels, fulmars and puffins, are killed. Even herons are attacked by great skuas, and the arctic skua feeds on small birds such as wheatears and buntings. In the tundra, lemmings are a very important food, and the breeding success of jaegers depends on the abundance of lemmings.

SKUNKS, carnivorous mammals renowned for the foul smell they produce. The word should not be confused with "skink," which refers to a group of lizards. There are about ten species of skunks, distributed throughout the Americas, belonging to the weasel family, Mustelidae. The best known is the striped skunk, *Mephitis mephitis*, found throughout the United States and the adjacent parts of Canada and Mexico. It is about the size of a small cat, about 18 in. (45 cm) in length with a tail of about equal length. The body is more thickset than in true weasels, but less so than in badgers. The pattern is unique: black with two bold white stripes on the back converging in front to meet on the crown of the head. The tail is a mixture of black and white, and the hairs are several inches long, producing a great plume that is raised high in the air at the first sign of danger.

All members of the weasel family have scent glands under the root of the tail, with which they mark their territory. In the skunk these glands are highly developed for purposes of defense: the secretion not only smells pungently but it causes severe irritation on a victim's skin, and it can be shot out in fine jets to a distance of about 10 ft (300 cm). This weapon is used only if the intimidation display fails, but it can be repeated several times in rapid succession if necessary. The

The striped skunk, the best known of the skunks of North America.

animal arches its body before taking aim, and can be remarkably accurate.

The striped skunk is an abundant and versatile animal in North America. It is nocturnal and omnivorous. At most times insects constitute a large part of the diet, although mice, eggs, frogs and fruit all add variety. Activity is greatly reduced in winter, especially in the north of its range, although it does not undergo a true hibernation. Breeding takes place in early spring, when a litter of about five young is born in an underground nest. The young follow their mother for much of the summer before dispersing in the autumn.

The spotted skunk, *Spilogale putorius*, is a

smaller species, boldly patterned with white lines and spots on a black background. When performing its threat display, with its long, white plumed tail held high, the spotted skunk may make up for its smaller size by doing a handstand, a unique piece of behavior among mammals.

In South America are found several species of hog-nosed skunks, *Conepatus*, and one of these, *C. leuconotus*, reaches the southern states of U.S.A. Apart from their piglike noses, which give them their name, the hog-nosed skunks are distinguished usually by totally white upperparts, including the whole tail, the rest of the body being black. This unusual pattern closely resembles that of the honey badger or ratel of Africa, and, like the bold pattern of all the skunks, is a good example of a warning pattern that enables potential predators easily to learn to recognize the animal as obnoxious.

SKYLARKS, *Alauda arvensis*, small songbirds about 6 in. (15 cm) long, well known for their magnificent song, characteristically delivered while hovering high in the air. The plumage is streaky brown above and white below with dark spots on the throat and breast. There is a small erectile crest on the crown. It lives in open country, never occurring near trees or in small fields with tall hedgerows. Its range extends across the whole of Europe and Asia, except for the extreme north, and into northwest Africa. A small introduced population exists on Vancouver Island, British Columbia. Its food is seeds and insects taken from the ground. Three or four white eggs, thickly spotted with brown, are laid in a nest on the ground. Two or three broods are usual.

SLOTHS. The modern tree sloths, family Bradypodidae, are uniquely adapted to an arboreal existence. The active part of their lives is spent upside down in trees, and this characteristic, coupled with their extreme slowness, makes them curious mammals indeed. Sloths rarely descend to the ground. Because of the anatomy of their limbs, they cannot stand or walk, as do most mammals, but must sprawl awkwardly on a flat surface and almost drag their bodies along. The arms are longer than the legs, and the actual digits (both fingers and toes) are closely bound together by tissue, covered with skin and hair, and terminate in long, strong, permanently curved claws. The palms of the hands and the soles of the feet are leathery pads and together with the claws form efficient hooks that can easily suspend the body from branches. Sloths can hang by all four limbs or even two, but more often they prop themselves in forks of trees or support their backs against branches while resting—always, however, with at least one foot securely hooked to a bough. A sloth climbs with a slow, deliberate (but graceful) hand-over-hand motion. Surprisingly enough, sloths are good swimmers, using an overarm stroke while in a right-side-up position.

Although peaceful and inoffensive creatures, sloths can defend themselves by biting and slashing out with their arms and formidable hooks. But perhaps their best protection is their very slowness and concealing coloration. Remaining motionless for long periods (sloths may sleep and doze 18 hours out of every 24) in a shaggy but compact mass high in a tree,

a sloth is far from conspicuous to predators. During the rainy seasons, algae growing on the fur give a greenish cast to the sloth's thick wiry coat, increasing its protective camouflage. (With its head down on its chest, a sleeping sloth can strongly resemble a bunch of dried leaves or even a wasp nest.) In addition, being able to curl into an almost impenetrable ball and having thick, tough skin also helps it to weather attacks from its principal enemies—jaguars, ocelots, other carnivorous mammals that climb well, big tree snakes and large birds of prey, such as harpy eagles. Sloths are very tenacious of life and have been known to survive severe injuries, doses of poison, and shocks from electrical wires that would kill other mammals.

Despite being classed as toothless mammals, sloths do have teeth, about 18 in number and all molars. The teeth are a primitive type, lacking enamel, and have cupped grinding surfaces. They grow throughout life. There are two basic types (genera) of tree sloths—the three-toed *Bradypus*, and the two-toed *Choloepus*.

The three-toed sloth or ai has three digits with their corresponding hooked claws on both hands and feet. The general color is grayish brown, often tinged with green when algae are growing on the hairs. The males have a pronounced dark brown or black marking on the back in the shoulder region. Adults usually weigh 9 or 10 lb (4-4.5 kg) and measure 19.5-23 in. (50-60 cm) in head and body length. Three-toed sloths have a short stubby tail about 2.5 in. (6.5-7 cm) long. Whereas the usual number of cervical (neck) vertebrae in mammals is seven, these sloths have nine, permitting great flexibility in rotating their heads.

There are several species of three-toed sloths, the exact number being in doubt, but the typical species is *Bradypus tridactylus*. All are solitary by nature and inhabit forests from Honduras south through Brazil and Paraguay to northern Argentina.

There are two species of two-toed sloth: the two-toed sloth, *Choloepus didactylus*, and Hoffmann's sloth, *Choloepus hoffmanni*. Both have three toes on the hind feet but only two fingers on each fore foot. Fur coloration ranges from a uniform blond to dark brown. Frequently the hairs have bands of light and dark color, producing a frosted or grizzled effect, and often the color varies on different parts of the body. Two-toed sloths are tailless and may weigh up to 20 lb (9 kg). The head and body length is about 23-25 in. (60-64 cm), and the claws may be almost 3 in. (7.5 cm) long. Two-toed sloths have only six cervical vertebrae. The first set of molars in each jaw is advanced from the rest of the teeth, and these four, triangular in shape, resemble canines. With them the animals can bite viciously if molested.

Two-toed sloths are found from Nicaragua south through Venezuela, the Guianas and Brazil to Peru.

SLOWWORM, *Anguis fragilis*, limbless and smooth-scaled lizards with eyelids, without lateral grooves, rather stiffly cylindrical, the head, trunk and tail being not easily distinguishable, and with the ability to cast their tails. The young are silver-gray or light bronze, with a blackish middorsal stripe and underside.

Adults, up to 20 in. (50 cm), are darker and duller olive-gray or bronze. The females retain the "juvenile" stripe, whereas males, which have relatively shorter trunks and longer tails, are usually less heavily marked and often show bluish markings dorsally.

Slowworms are found over temperate Eurasia west of the Caucasus Mountains, but not in Ireland and some other islands. In parts of north Africa and the Middle East to Iran they are found in sheltered nonarid situations; elsewhere their habitat is usually rocky or wooded. In Britain, pairing takes place soon after emergence from hibernation and before dispersal for the summer. Three to 23 or even more young are born ovoviviparously or viviparously, late in the summer. The young are independent from birth. Slowworms eat many slugs, earthworms and insect larvae, so it is regrettable that these lizards, harmless to man and his livestock, are so often killed in mistake for venomous snakes.

SLUGS, shell-less, or nearly shell-less mollusks living mainly in the soil. They have an elongated mucus-covered body, which in many species is covered at the anterior third by a loose flap of tissue, the mantle. Within this is the air-breathing lung, which opens to the outside by a small hole, the pneumostome. Gaseous exchange occurs in the lung and also all over the moist body surface. The head bears

Land-slug *Arion* with its two pairs of sensory tentacles (one pair of which bears eyes) extended.

two pairs of tentacles and can be retracted under the mantle. One common group of tropical slugs, of the genus *Vaginula*, have no mantle cavity, and the gut is unusual as it ends at the posterior of the body and not, as in most slugs and snails, near the pneumostome, on the right-hand side of the body.

The most obvious single feature that distinguishes slugs from snails, their nearest relatives, is the apparent lack of a shell. In fact, in many species, a reduced internal shell is present, and one rather unusual family, the Testacellidae, have a small external shell under which is a very much reduced mantle cavity, an arrangement identical to that of snails.

Slugs are protandrous hermaphrodites—that is, the male system becomes mature, followed by the female system, the animals then being true hermaphrodites, and after this the male system becomes senescent and the animals are solely female. They mate on the soil surface after a long circuitous dance. Slugs lay their transparent eggs in the soil. These hatch into small slugs about 1.5 mm long, which probably feed on fungi and decomposed plant matter before becoming large enough to feed directly on the green plant. Like snails they have a rasplike radula and a chitinized jaw, which scrapes small pieces of food into the alimentary canal. The undigested portion passes out of the body as a compacted fecal rope often containing plant tissue almost unaltered, suggesting that in some cases the digestive enzymes are not very powerful.

All slugs crawl by means of muscular waves passing over the surface of the foot. They secrete mucus while crawling, which lubricates the tissues of the foot and so prevents excessive abrasion. The secretion of mucus (hydrated muco-polysaccharide) accounts for a significant decrease in the body weight of these animals while they are active, and this water loss may limit the extent of surface activity.

SMELT, small estuarine fishes related to the salmons. The European smelt, *Osmerus eperlanus*, is an elongated, compressed fish somewhat resembling a small trout but with a silvery body and slight blue-green tinges on the fins. A small adipose fin is present, and the lateral line is short, not extending beyond the pectoral fins. The mouth is large, with fine teeth in the jaws, and there are conical teeth on the roof of the mouth and several large fanglike teeth at the front of the tongue. Smelt live in large shoals in the estuaries and coastal waters of Europe and are commonly used by fishermen for bait. They are anadromous fishes, migrating up into fresher water to spawn. A peculiar feature of the smelt (whence it may have derived its common name) is the odor of the flesh, which resembles that of cucumbers. In the days before many of the large European estuaries became so polluted that fish were rarely found in them, fishermen were said to be able to detect the presence of smelt by the smell of cucumbers.

The smelt rarely grows to more than about 8 in. (20 cm) but is considered to be a great delicacy. Connoisseurs claim that smelt should be eaten within an hour of capture, should not be washed at all but lightly fried or grilled.

SNAILS, gastropod mollusks with a spirally coiled shell. They may occur on land in fresh water or in the sea, and it is usual to qualify the term with that of the habitat in which the snail is normally found: pond snail, water snail, land snail and sea snail. Some of these are given separate entries. Snails without any qualification may be normally taken to be land snails. Land snails, together with some aquatic snails, may be divided into two groups: operculate or nonoperculate; the latter are also called "pulmonate." Operculate snails are those with a disc, usually horny, on the foot, which closes the mouth of the shell when the snail withdraws. Pulmonates are those that have lost their gills and breathe air.

Snails mostly crawl by means of waves, the pattern of which varies with the species, passing across the undersurface of the muscular foot. But some crawl by the beating of cilia in the mucus secreted by the pedal glands; species showing this mode of crawling are mostly restricted to very moist ground. All mollusks secrete mucus—the well-known, silvery slime trails—while crawling, and this lubricates the

foot and so prevents abrasion of the epithelium while the animal is moving. The animal loses large amounts of water while on the surface, and the amount of this water loss may govern the extent, and so the duration, of surface activity.

Most snails are vegetarians and feed by rasping plant material into their mouths with the radula. The fragments are mixed with saliva and passed into the gut, where they are acted upon by digestive enzymes including cellulases, which digest the cellulose. Some snails are carnivores, and these usually have a modified radula, more powerful proteases (protein-digesting enzymes) and a shorter gut.

Land snails conserve water by having a kidney that can resorb water and produce insoluble uric acid. But they also have retained the characteristic unwieldy shell. In water snails, the shell probably gives protection mainly from enemies. On land, it acts as a major water conservation device, protecting the snail from drying up. Moreover, land snails can seal the mouth of the shell with a mucus, or sometimes calcified door, the epiphragm, which is secreted when the animals hibernate or estivate, especially in summer when conditions are too hot and dry for them or in winter when it is too cold. Thus, snails can exploit periods of favorable conditions and then retire into a protected microhabitat, the shell, when the climate becomes unsuitable.

Where the giant African land snail has been introduced, it is now regarded as a serious agricultural pest. This has led to the search for natural predators, and in some islands a predatory pulmonate snail, *Gonaxis kibweziensis*, has been found to produce an effective control. In Europe some snails are pests of vines and soft fruit, and in most parts of the world they will attack crops. They are less harmful than slugs, however, as they feed only above ground and can be destroyed by collecting them from areas where salad crops are being grown or by using baits, such as metaldehyde and bran, which will kill them.

SNAPPING TURTLES, widespread and abundant in the fresh waters of North America, and named for their aggressive disposition. Instead of retreating into its shell when approached, the common snapping turtle, *Chelydra serpentina*, up to a foot or more (30 cm) long, turns to face the intruder, even advancing to the attack, snapping and biting.

The head of the snapping turtle ends in a strong, hook-shaped, down-turned beak. Three keels run along the dorsal carapace, made up of single knobs on each scute. The tail is about the same length as the carapace and has knobby, horny scutes on the upper side. The plastron has degenerated to a small, cruciform scale on the center of the belly. These turtles can barely swim and usually lie quietly on the bottom of shallow pools. Since, in addition, they leave the water only in order to lay their eggs, they need no special protection on the underside. If they are disturbed, they are extremely irascible and are even said to be capable of biting through a cane with their knife-sharp jaws.

The common snapping turtle, also known as the loggerhead snapper, has a carapace of up to 15 in. (43 cm) long and weighs up to 85

lb (38.6 kg). It eats mainly fish, invertebrates and a large proportion of vegetation, and also, carrion. Loggerhead snappers are disliked because they also attack waterfowl, and if disturbed by bathers may easily snap off fingers or toes.

SNIPES, long-billed birds of the subfamily Scolopacinae with flexible bill tips that can be opened below ground to grasp food items. As might be expected in birds that are active at low light intensities, snipes and woodcocks have large eyes, set well back in the head, giving almost all-around vision. Their bodies are rather dumpy and their legs fairly short. They have cryptic plumage and keep very still until disturbed at very close quarters, when they rise sharply, often with erratic changes in flight direction.

The common snipe, *Gallinago gallinago*, breeds throughout the northern hemisphere, and also in Africa and South America, where the birds are chiefly sedentary. Geographical races of the species have been named, for example Wilson's snipe of North America, which winters in Central America. The western palearctic populations of common snipe migrate southwest in autumn to the British Isles and, in cold winter weather, to Iberia; other populations reach tropical Africa. The pintail snipe, *G. stenura*, breeds in Siberia. It boasts 26 tail feathers compared with the 14 carried by the common snipe; it migrates south to India in winter. In the central parts of the palearctic, the great snipe, *G. media*, breeds in marshy woodland in the boreal zone, as far west as Scandinavia (where it is decreasing rapidly) and east to central Siberia, where it is replaced by Swinhoe's snipe, *G. megala*. The great snipe migrates south to winter in tropical east Africa.

The common snipe produces a resonant drumming (or bleating) sound by vibrating its outer pair of spread tail feathers. Other species of *Gallinago* also produce noises by tail vibration, with the exception of the great snipe, which displays communally and very vocally in early summer at a lek (similar to that of the ruff) in the twilight of the arctic night.

SNOW LEOPARD, *Uncia uncia*, or ounce, one of the most beautiful of the big cats. It has been placed in a separate genus from the true leopard because of its short muzzle and the differences in the skull. It is found mainly in the Altai, the Hindu Kush and the Himalayas, at heights of 6,000-18,000 ft (1,800-5,500 m), living in the coniferous scrub.

The snow leopard has thick long fur of a soft shade of gray with a pattern of dark rosettes on the upperparts. The ears are white with an edging of black, and there is a thick black line down the back. Its total length is about 6 ft 5 in. (1.9 m), standing about 2 ft (0.6 m) at the shoulder. It is nocturnal and shy, so little is known of its habits except that it feeds on wild sheep, mountain goats, marmots and domestic stock and usually has 2-4 young.

SOAP FISHES or soapies, fishes of the marine family Leiognathidae, which, when handled, secrete a slippery mucus that makes them difficult to hold. Presumably this is used by the

Two apple snails, of Europe, mating. This species is also known as the edible snail, the escargot of the gourmet.

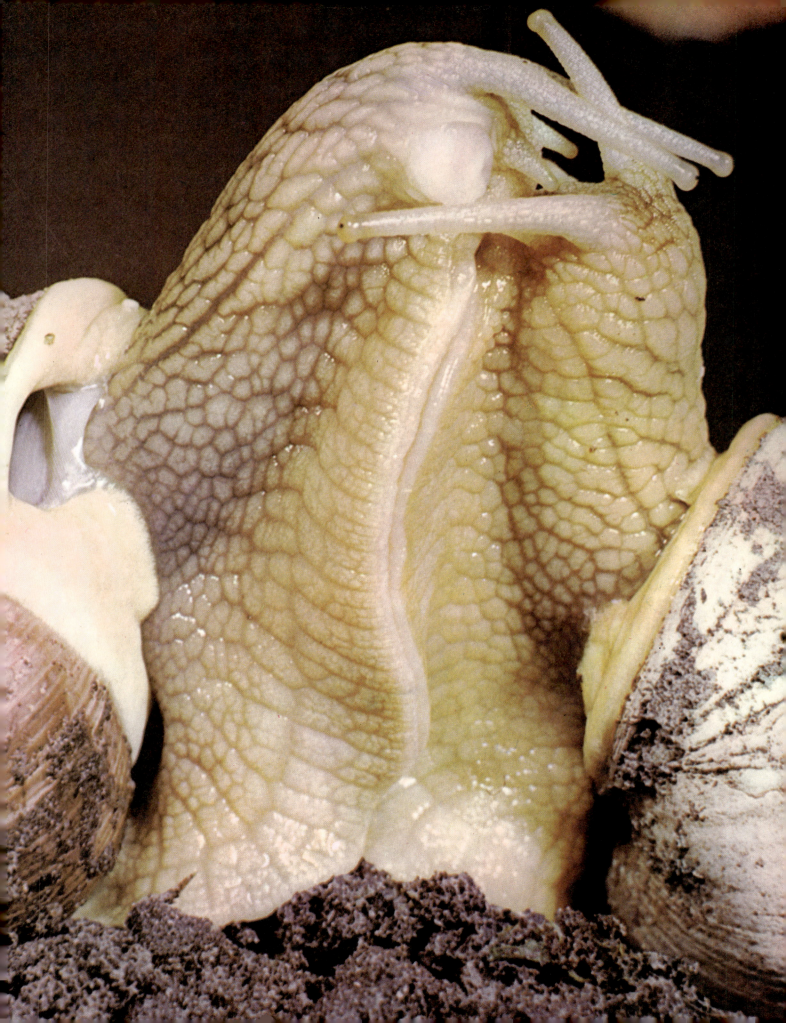

Representative of an African race of the common snipe of Europe, by Lake Naivasha, Kenya.

fish as a defense mechanism against predators. In America these fishes are also called slip-mouths, and in Australia they are referred to as ponyfishes. The soap fishes have deep and very compressed silvery bodies with the dorsal and anal fins beginning at the deepest point of the body, so that the profile of the fish is almost disclike. A characteristic feature is the protrusile mouth, which is usually extruded when the fish is taken from the water, giving it a horse-like appearance and earning it the name of ponyfish. A protrusile mouth enables a fish to seize or suck in small food particles from the bottom, although in the genus *Secutor* the mouth is in fact protruded upward. The soap fishes are small shoaling fishes that often enter brackish lagoons, especially as juveniles. In spite of their small size and relatively little muscle, they are so numerous in some areas that they make an important contribution to protein resources and are caught and sun-dried in large numbers.

SOLDIER BEETLE, a term applied to members of the families Cantharidae and, in the United States, Meloidae. The first of these families is large, with 3,500 species, and is closely related to the glowworms. These Cantharid beetles often show warning coloration and are distasteful to predators. Their bright red color being a somewhat similar color to the soldier's uniforms of the 17th and 18th century may have led to their common name. They are carnivores both as larvae, which live in the soil, and adults, which can often be seen on flower heads and leaves.

The second of these families is widely distributed, with about 2,000 species, and with other members of the group are known as blister beetles.

SOLENODON, an insectivore, about the size of a small cat with a long, slender snout, giving it the appearance of a large, slow-moving shrew. The length of the head and body is 11-13 in. (28-32 cm) and the length of the tail is 7-10 in. (17-25 cm). The fur is mixed gray to blackish and buff to deep reddish brown with a rectan-

gular spot of white on the nape. The several areas of the body that are naked of fur include the feet, tail and the rump around the base of the tail and around the anus. These naked areas may relate to the reduced role of the mouth and fore paws in cleaning and scratching.

Two species of solenodon, *Solenodon paradoxus* and *S. cubanus,* are the only surviving members of a primitive family, Solenodontidae. They are to be found on two neighboring islands: Haiti and Cuba. Solenodons sleep in burrows, caves, hollow trees and logs. Their diet consists of a variety of fruits, insects and small vertebrates such as reptiles. They are slow and rather clumsy and have survived on the islands only in the absence of many predators. Since the introduction of cats, dogs and mongooses, their survival has been threatened. Many governments and zoos of the world have agreed to reduce the importation of solenodon.

SOLES, a family of rather elongated flatfishes of considerable economic importance. The eyes are on the left side, the mouth is small and the snout projects well beyond the mouth to give the fish its characteristic appearance. The Dover sole, *Solea solea,* has a long oval body, and the dorsal fin begins over the head and reaches almost to the tail. The eyed side is dark brown with darker blotches. It lives in shallow water in the Mediterranean and along European shores as far north as the coasts of Britain, but becomes rarer in the north. Occasionally it enters estuaries, and in winter it migrates into slightly deeper and warmer waters. It is caught by trawl, especially in the southern North Sea and the Bay of Biscay.

The thickback sole, *Microchirus variegatus,* is very much smaller and thicker. It has a more southerly range and is of no economic importance. Of all the European soles it is the most easily identified, having transverse black bands on the body. A fairly rare Atlantic sole is *Bathysolea profundicola,* a deep-water species found down to nearly 4,000 ft (1,290 m) in the North Atlantic and Mediterranean. The naked sole, *Gymnachirus williamsoni,* from Florida coasts is less elongated than the European species and has a pretty pattern of dark chocolate bands on a reddish background.

SONG SPARROWS, among the most familiar North American land birds. They resemble the Old World sparrows only superficially.

The song sparrow, which derives its name from the male's persistent but musical song, breeds commonly in a wide range of suitable habitats, preferring damp areas with low and dense vegetation. Its breeding range extends from Mexico north to the outer Aleutians and across North America from Pacific to Atlantic. The song sparrow is a very variable species, for throughout this range there are over 30 subspecies. Before the complete range of the song sparrow was known, a number of these subspecies seemed sufficiently distinct to be classed as separate species. Later, intermediate populations were found, and it gradually became apparent that the different populations were subspecies of one and the same species.

The nominate race, *Melospiza melodia melodia,* known as the eastern song sparrow, breeds in eastern Canada and the United States west to the Appalachians, in typically damp habitats along streams, beside ponds and in shrubby

wet meadows or swamps. Along the central Atlantic coasts it is replaced on the beaches and barrier islands by the Atlantic song sparrow, *M. m. atlantica,* a race rarely found far from salt water. The Mississippi songsparrow, *M. m. euphonia,* became one of the best-documented races following the classical studies of a population at Columbus, Ohio, by Mrs. M. M. Nice between 1928 and 1938. In her study of the eastern song sparrow Mrs. Nice found that the males advertise their territories by singing and defend them by fighting. Any female that enters a territory is first attacked but later courted if she is persistent. She builds a nest and lays 4-5 eggs, which she incubates for 12 days. The chicks are fed by both parents for the 10 days they are in the nest and for three weeks after they can fly. Three broods are raised each season, but many nests are parasitized by cowbirds, the cowbird chick being raised with the song sparrows.

SOUSLIKS, ground squirrels of the genus *Spermophilus* found in eastern Europe and Russia. They are small, colonial, short-tailed squirrels, either yellowish brown (*S. citellus*) or spotted with white (*Citellus souslicus*).

SPARROWS, small seed-eating birds forming the subfamily Passerinae of the weaverbird family Ploceidae.

The house sparrow, *Passer domesticus,* has become one of the most widespread and familiar birds in the world. This species almost certainly evolved with man in the Middle East as the practice of agriculture spread, and is thus adapted entirely to living in man-made environments. In North America, where it arrived long after the main immigration of human populations had occurred it spread, in under 30 years, from a few introductions on the east coast, the 3,000 miles (4,750 km) across the continent.

Like the true weavers, sparrows tend to be gregarious breeders, but the degree of association is very variable. Tree sparrows, *Passer montanus,* nest almost entirely in holes, very rarely building their own free-standing nests, and so must nest where holes are available, not necessarily close together. In the social weaver, *Philetairus socius,* on the other hand, many pairs cooperate in the building of an enormous thatched communal nest structure under the shelter of which are the separate nests of the individual pairs. The chestnut sparrow, *Passer eminibey,* is social, but possibly only because it usurps the newly built nests of other sparrows and weavers, themselves colonial. A colony of gray-headed social weavers, *Pseudonigrita arnaudi,* may contain 10% chestnut sparrows which devote the time other sparrows spend nest-building to displaying for hours on their stolen property.

The displays of sparrows are related to those of weavers. The nest-site is advertised by the male sitting on or near it and chirping: a female is attracted to the site, and if satisfied that it is a good one, a pair is formed. Pair formation in this way may take place well before the breeding season in species that live near the nesting places all the year round. The male chestnut sparrow has a display on its stolen nest very similar to the intensive wing-shivering displays of the *Ploceus*-weavers. In sparrows both sexes build the nest after the site has been settled and the pair formed, in contrast to the weavers, the males of which build the nest before attracting a mate. The house sparrow has, in addition, a curious sexual display reminiscent of the "pursuit flights" of *Anas* ducks. A number of males actively chase a female, flying furiously around. After a few seconds they all land in a bush, and the males hop around the female, chirping excitedly before dispersing. Tree and house sparrows have a short incubation period of 11-12 days, and the young can sometimes leave the nest successfully after a further 11 days, though 15-17 is more usual. This rapid turnover, and their readiness to feed the young on virtually any kind of insect food available, allows them to have two, three, or even four broods in European summer.

After the breeding season all sparrows are gregarious, forming foraging flocks that have led to several species being branded as pests, as considerable damage to cereals can be caused.

SPERM WHALE, *Physeter catodon,* is by far the largest of the toothed whales. Also known as the cachalot, it is the best known among whales because of its enormous squarish head, which makes up about a third of the animal's length, some 60 ft (20 m) for a large male, the female being rather more than half that length. Moby Dick in Melville's famous novel of that name was a white sperm whale. The front of the head contains an enormous reservoir, the "case," filled with clear spermaceti oil and below that the junk, a lattice of fibrous and elastic tissue containing oil. Spermaceti, although fluid in the living animal, solidifies when exposed to air in cooler temperatures to form a "wax" originally used for candles and cosmetics.

The lower jaw is much shorter than the upper and appears as a small underslung unit. Nevertheless, this is more an illusion due to the enormous size of the head, for it is powerful enough. Teeth are found only in the lower jaw. These are variable in number, ranging from 18-28 on each side and often different on the two sides. They are large and powerful, reaching as much as 8 in. (20 cm) in length and conical in shape. The teeth enter strong fibrous sockets in the upper jaw when the mouth is closed and so give effective grip to the food. The teeth have traditionally been whalers' perks and used for scrimshaw work, superb carvings and scratched pictures having been made from them.

Sperm whales are darkish gray on the back, lightening in the flukes and belly to pale gray or white. There are often white scars and circles particularly in the head region, as a result of fights with giant squid, which form a main item of diet.

Sperm whales are found in all oceans, but breeding populations tend to stay in the warmer waters though migrating nearer the poles in summer. Those in the polar seas are usually males, who migrate there during or after the breeding season. The sperm whale is a deep-water species and is rarely found in shallow or "green" water with depths less than 330 ft (100 m). Most are to be found where deep oceanic currents produce water mixing and hence good supplies of food. They dive deeply to 1,500 ft (500 m), and one is reported to have broken a cable at twice this depth. The deep diving is presumably associated with hunting for giant squid, for most of the squid found in sperm-whale stomachs has been of deep-water forms. The ambergris, which has considerable value as a fixer for perfumes, is found in the stomachs and intestines. It appears to be a secretion caused by irritation of the gut wall by squid beaks in the food.

The females and young and the attendant males form large schools of up to several hundreds; one count of well over 1,000 has been reported. Bulls not having a harem group and particularly those in polar seas tend to travel alone or in small numbers.

SPIDER CRABS, characterized by their long legs and triangular bodies, the body having a point in front. The back is usually provided

Spider crab *Stenorhynchus seticornis*, of the tropical Atlantic.

with hooked or angular spines or hairs, and many species attach pieces of seaweeds, hydroids and sponges to these hooks, and so camouflage themselves. The old camouflage is removed with the pincers, and a more appropriate decoration is put in its place. The triangular shape of the body allows the pincers to reach over the back. This is a feat of which most other crabs are incapable.

One of the heavily camouflaged species, *Hyas coarctatus*, has pink patches on its pincers, which it waves about among the seaweeds, so attracting small fish. When the fish draws close enough, it is seized and eaten. Some of the deep-sea species have extraordinarily long legs, which seem to be an adaptation for walking over the soft ooze of the deep-sea bottom. The largest living arthropod belongs to this family: *Macrocheira kaempferi* from the deep sea near Japan has a body a foot (30 cm) across and legs that span over 8 ft (2.5 m).

SPIDER MITES, plant-feeding mites including many serious agricultural pests occurring in orchards and greenhouses. Their common name reflects their habit of spinning a fine silk web around the leaves on which they are feeding. They use this to anchor their eggs to the leaf surface, but it probably also serves as a protective canopy under which colonies of mites can feed undisturbed.

Spider mites comprise a large group of forms infesting a wide range of host species. They are not host specific; any one species may occur on a range of hosts, and it is not uncommon to find more than one species of spider mites on the same host plant. Economically important species include the fruit-tree red spider mite, *Panonychus ulmi*, which occurs on apple, pear, plum and damson, the glasshouse red spider mite, *Tetranychus urticae*, which is a pest of cucumbers and tomatoes in greenhouses and may also be found outdoors on blackcurrant and ornamental plants, the citrus spider mite, *T. telarius*, of California and Florida, which may transmit a virus disease to cotton, and *Bryobia praetiosa*, the clover mite, which is a serious pest of clover, peas, almonds, peaches and alfalfa in the United States. This last-named species often appears in large numbers in human dwellings, and its wide range of distribution includes Europe, where, however, it is not a serious pest. Although many of these forms are often collectively referred to as "red spider mites," there are complexes of related species that the experts sometimes find difficulty in separating. Furthermore, the body color may vary from green or yellow to brown, orange or red. The glasshouse red spider mite is probably not one species but a complex of related forms, some of which are green in summer and red in winter, while others are reddish-brown throughout the year. In some cases at least, it has been shown that these two color forms can hybridize, although the progeny are either sterile or all males.

The life cycle of spider mites varies from species to species and also with climatic conditions. In *Panonychus ulmi* two different kinds of eggs

Spoonbills feed by sweeping their bills from side to side, picking up anything edible in the water.

are produced—namely, winter eggs that are deposited in sheltered crevices on the stems and branches of the host, and that go into a resting "diapause" condition, and summer eggs that are laid on the leaves. Winter eggs hatch in spring and early summer; during the summer, the life cycle is completed in about a month, and there may be as many as five generations, possibly more, produced during this season. The population build-up is often staggering at this time, which is also the peak period for feeding activity. Toward the end of the summer the production of summer eggs ceases and winter eggs appear, which will enable the species to survive until the following year. In northern climes, the active stages in the life cycle are killed off by the winter cold, but in warmer latitudes adults may be active throughout the year, and up to 15 generations may be produced.

The mouthparts of spider mites are sharp stylets that puncture the leaf tissue of the host and enable the mite to feed by sucking out the cell contents. At the height of the feeding activity, the damage produced may kill the leaf, and large-scale defoliation results. Control of these pests can often be obtained by introducing other mites, notably gamasids of the family Phytoseiidae, which prey upon them.

SPOONBILLS, graceful storklike birds, related to the ibises but with long straight bills broadening to spatulate tips. The four species of *Platalea*, with mainly white plumage, between them cover most of the tropical and subtropical parts of the Old World. *P. leucorodia* has a northern outpost in the Netherlands, where it is a summer visitor. The roseate spoonbill, *Ajaia ajaja*, is the New World representative. They nest in trees, bushes or reedbeds, or on the ground. Gregarious at all seasons, they frequent the neighborhood of fresh water, where they feed on small aquatic animals.

SPRING HARE, *Pedetes capensis*, or Cape jumping hare, one of the most bizarre of all the rodents. It is found in the short-grass plains of eastern and southern Africa and is the sole representative of the family Pedetidae, which has very uncertain relationships among the rodents. The spring hare is a large rodent, measuring up to about 16 in. (38 cm) with a bushy tail of about equal length. It is bipedal, in some respects resembling the jerboas that replace it in northern Africa and Asia, but much larger and more closely approaching the size of some of the Australian wallabies. As in all bipedal mammals the hind legs, and especially the hind feet, are very long and the front feet disproportionally short. The long ears, large bulging eyes, prominent whiskers and deep muzzle with large protruding orange teeth combine to give the head a uniquely quaint appearance. The pelage is of uniform yellowish brown except for the brush of the tail, which is black.

The front feet bear five long, curved claws with which the spring hare can very quickly dig a burrow. The burrows may be quite complex but are usually inhabited by only one animal or a pair. Spring hares are strictly nocturnal, emerging after dark to feed on bulbs, rhizomes and other fleshy plant material. They are slow breeders. The litter consists of one, or occasionally two, young, which are fairly well developed at birth and open their eyes within two days.

SPRINGTAILS, also known as Collembola, are a group of insects that at no stage in their life history have wings and seem to have no close evolutionary relationship with any other insects. They are grouped together with the Thysanura or bristletails, the Diplura and the Protura, as the Apterygota or wingless insects. About 2,000 species of springtail are known, but there are doubtless many more to be discovered. They are quite small insects. Few are bigger than $\frac{1}{4}$ in. (6-7 mm), and most are much smaller, but the ecology of the Collembola is important because they occur in extremely large numbers in the soil or at the soil surface. Populations of up to 100,000 per square meter of soil are common in temperate woodland and grassland soils; big populations are also found in the soils of extreme environments such as the polar regions. In fact, it is likely that the springtails are the most numerous of all insect groups. The only environments in which they are not abundant are those with dry soils, such as deserts. Springtails cannot survive in dry places because they have little ability to withstand desiccation. Despite this, many species have very wide geographical distributions, and some are known to occur in all continents: this is probably the result of their great evolutionary age, which has given them plenty of time for dispersal. The oldest known fossil insects which are about 320 million years old and come from the Devonian period, are Collembola and are very like modern species.

Most species of Collembola are between 1 and 3 mm long. Some are white or colorless; others have simple or more elaborate color patterns. The head carries a pair of four-segmented antennae and generally a number of simple eyes or ocelli, though these are absent in many forms. Externally the mouthparts are barely visible because they are hidden in a pouch formed by downgrowth of the sides of the head, a condition called entognathy and found in some other wingless insects, for example, the Diplura.

SQUID, the common name used to describe a number of cephalopod mollusks that have in common a more or less streamlined fishlike shape. The group, properly called the Teuthoidea, includes the largest and fastest, as well as some of the most beautiful invertebrates. All are predators, and the great majority are powerful swimmers, competing successfully with fish. A few, *Loligo vulgaris*, *L. pealii* and their close relatives, live in coastal waters, but the group is more typical of open water, living at the surface and down to great depths in all oceans.

Loligo, though not quite typical in habitat, is the best-known species, and it shows most of the developments typical of this group of mollusks. It is a smallish, streamlined squid, running up to about 2 ft (60 cm) in length, with lateral fins and ten arms, facing forward. It can swim backward or forward with equal facility. The eyes are prominent, and, like other cephalopods, *Loligo* is distinguished among invertebrates by the possession of an unusually large brain (see cephalopods). The arms are used to seize prey—fish for the most part, and other small animals living in midwater. Two of the tentacles are longer than the rest, and in *Loligo* (though not in most of the other squids) these are kept coiled away, hidden within the circle of the remaining eight arms. When the animal strikes at its prey, the long tentacles are shot out with great speed and accuracy to grab the victim and pull it in toward the mouth. *Loligo* has suckers on its arms and tentacles; other squids also have clawlike hooks, a combination of offensive weapons ensuring that the squid very rarely misses its prey. So far as the smaller inhabitants of the ocean are concerned these highly intelligent predators must be very dangerous animals indeed.

Enclosed in the elongate hind part of the body is a space, the mantle cavity, which opens to the sea outside through a funnel at the front end below the head. This cavity encloses the gills and houses the openings of the excretory, sex and digestive systems, as in other mollusks (see mollusks). In squids the wall of the mantle cavity has become specialized as an organ of propulsion. It is thick and muscular, and an unusual arrangement of nerves ensures that the whole wall can be contracted simultaneously. The resulting blast of water, directed forward or backward by the mobile funnel at the exit from the mantle, can produce rocketlike accelerations. Some, such as *Onchiteuthis*, are even capable of leaping clear of the water to glide for a few seconds on their outstretched fins. Squids, like other cephalopods, have an impressive capacity for rapid color change. Some are red, others pale greenish or semitransparent. The switch from one pattern to the other takes a fraction of a second. Combined with the ability to discharge a cloud of black "ink" from a gland in the mantle cavity, this provides an effective escape mechanism; the would-be predator is not only confused by the ink but is also left searching for an animal of quite different color from the squid, which is by now rapidly removing itself from the scene of the action. The spread of the color change is caused by the structure of the chromatophores, small pockets of bright color each expandable by a ring of radial muscles; the muscles are under direct control of the central nervous system. Many squids, the deep-water forms in particular, have in addition to chromatophores light-producing "photophores." These are pits containing luminescent bacteria. Structurally, the photophores may be of considerable elaboration, with lenses to focus the emitted light and shutters, derived from chromatophores, that can close off or alter the apparent color of the luminescence. Nobody knows what use squids make of these photophores. In some instances they differ from one sex to the other, which suggests that they may play a part in the sex recognition of deep-water forms, equivalent perhaps to the color patterns displayed by some of the cephalopods living in better-lit environments. An alternative suggestion is that they act as lures to attract prey.

SQUIRRELS, a term usually meaning tree squirrels, but the squirrel family is one of the largest families of rodents and, besides the numerous species of tree squirrels, includes a great variety of ground squirrels, including the marmots and woodchuck, and flying squirrels. Tree squirrels are found in most forested parts

Flying squirrel of the eastern United States.

of the world and range in size from under 3 in. (7.5 cm) in the pygmy African squirrel, *Myosciurus pumilio*, to 18 in. (45 cm) in the giant squirrels, *Ratufa*, of southeastern Asia. Tree squirrels typically have long bushy tails, usually about equal in length to the head and body, short muzzles and large, bright eyes. They are predominantly diurnal, and, like other diurnal mammals such as the monkeys, they tend to have patterns and colors bolder than the more typical nocturnal rodents.

There are few squirrels in the temperate zone, but they have been intensively studied. In temperate Eurasia there is only one widespread species, the red squirrel, *Sciurus vulgaris*. Most red squirrels are indeed a rich reddish brown, with attractive tufted ears, but coloration is extremely variable. In some areas, e.g., in Switzerland, they are predominantly black and white, and in many parts of Europe both red and black individuals may be found in the same area, i.e., the population is dimorphic.

Red squirrels live equally in coniferous and deciduous forest, but in Britain, where the American gray squirrel, *S. carolinensis*, has been introduced, it is noticeable that the gray squirrel has, on the whole, replaced the red squirrel from the deciduous woodland but not from the coniferous. Their food consists largely of the seeds of conifers and nuts such as acorns and beechmast. But when the seed crop is poor in coniferous woodland they will concentrate on buds, especially in winter, and may cause considerable damage to the trees. During the summer, like most northern animals, red squirrels are much less specialized in their feeding habits, and animal food such as insects and birds' eggs may be included in their diet.

Red squirrels do not hibernate, but they do store food and are less active in winter. They have been seen to collect toadstools in autumn and to store them by jamming them in forks of twigs where they dry and provide an addition to the winter diet.

The nest is usually in a fork or on an exposed branch of a tree, and it is used at all times of the year. Two litters may be produced during the summer, usually of four or five young.

North America is richer than Eurasia in squirrel species. The North American red squirrel, *Tamiasciurus hudsonicus*, is smaller than its Eurasian counterpart. It, and especially its western relative, *T. douglasi*, is often known as the chickaree on account of its chattering alarm call. The American gray squirrel is a more southern species and lives predominantly in deciduous woodland. Both red and gray squirrels will gather a proportion of their food on the ground, but whereas the red will always carry the food up a tree before eating it, the gray will often eat on the ground where it spends a larger proportion of its time.

STAG BEETLES, so named because the males have enormous jaws with projections resembling the antlers of a stag. Even within a single species there is great variation in the size of the jaws; the larger males have disproportionately larger jaws. The common stag beetle, *Lucanus cervus*, is the largest British beetle, but many tropical species are much larger. *Odontolabris dux* reaches a length of 4 in. (10 cm), and *Cladognathus giraffa* is similar in size, with jaws equal in length to the rest of the body.

The larva of the common stag beetle has a strong, heavily armored head with powerful jaws, but its body is thin-walled, white and swollen. It is commonly found eating rotten

wood buried in the ground. Larval development takes about four years.

STARFISHES, star-shaped marine animals with a radial, usually five-rayed, symmetry, a skeleton of meshlike calcite plates buried within the body tissues and a water vascular system that gives rise to avenues of tube-feet on the lower surface, by which the animals move about. They are related to sea urchins, sea lilies and sea cucumbers, and these together make up the phylum Echinodermata (literally, spiny skins).

A starfish consists of a central disc surrounded by five or more radiating arms. Five is the commonest number, but species with up to 50 are known, as in *Heliaster*. Typically, the arms are two or three times as long as the disc is broad, but there is a wide range from species with long slender arms, such as *Brisinga* to the compact cushion stars. The total size range is also large: from about $\frac{2}{5}$ in. (10 mm) to 2 ft (61 cm) across. The color varies from gray, green and blue, to vivid red or orange, sometimes mottled or with banded patterns.

The upper surface is somewhat rough to touch. The supporting skeleton of calcite plates varies from a network with fleshy areas between to a continuous hard pavement of overlapping plates. In some starfishes the surface is bordered by rows of large marginal plates. Others have plates that bear erect columns with expanded tops set with little tubercles or spinelets. These mushroom-shaped structures are called paxillae, and form a formidable defensive surface. Other defensive or offensive structures are the pedicellariae, spines modified to pincers. Four types of pedicellariae are recognized: two have jawlike calcite plates set on a stem, in one case the plates crossed, in the other straight; a third type, called sessile, is formed of three spines set close together on the surface of the starfish; the fourth type, known as alveolar, has the

Female stag beetle, *Lucanus cervus*, lacks "antlers."

The male stag beetle, with large "antlers."

349

jaws sunk in a groove in the body wall. They function as grasping organs, used to deal with small prey or to pick off detritus that would otherwise foul the body surface.

One of the most remarkable features of starfishes is their ability to generate damaged or lost parts of the body. Whole arms when broken off by accident may be regrown again entirely. Moreover, species such as *Linckia* can actually grow a new disc and arms from a single severed arm. *Linckia*, indeed, may break off one of its arms for no apparent reason, and the resulting pieces then grow into two separate individuals. Such asexual reproduction is found in a few other starfishes, such as *Nepanthia* and *Coscinasterias*, but these split in two across the disc, and then each half regenerates the arms and part of the disc that it lacks. Specimens in which such regeneration has taken place naturally show a very unequal development of the arms until the process is completed.

STARLINGS, a family of 105 species of medium-sized to large songbirds having, in general, slender bills, a rather upright stance and smooth, often glossy, plumage. They occur naturally in the Old World only, but a few species have been introduced into other parts of the world, where they have usually proved to be pests.

The typical starlings have slender, straight and tapering bills, and when feeding in turf, among plants or leaves, they thrust the closed bill into the ground or vegetation and open it, making a hole. The eyes are set close to the base of the bill, and the bird looks along the opened bill, down into the hole, and seizes any food that the probing has revealed. These close-set eyes give the starlings their characteristic appearance. Starlings of the genus *Sturnus* have elongated spiky feathers on the head and throat and are soberly colored in black-gray, white, buff and, more rarely pale pink, with some iridescence on the darker plumage. They are mainly terrestrial in their feeding, and walk and run with ease. Their flight is rapid and direct, with constant beating of the tapered wings. The relatively similar proportions during flight of the tapering wings, tapering head and bill, and short tail, gives these birds the "starlike" outline that accounts for their original name. Outside the breeding season they tend to flock together for feeding and roosting. Some of the roost flocks are of huge size, and the rapid coordinated maneuvering of such flocks over the roost is typical of starling behavior. At times the numbers may be such that the weight of the birds breaks the branches, and the accumulated droppings kill trees in plantations where they roost. They have taken to flying into the centers of many cities to roost on the ledges of buildings.

The food of starlings consists of insects and invertebrates, fruits, some seeds and parts of green plants. The typical probing starlings specialize on insect foods to a greater extent than do some of the others. They will also take ripe fruit, however, and may do so to an extent that makes them unpopular with fruit growers. The common starling, *Sturnus vulgaris*, has been introduced into North America and eastern Australia and has established itself and spread on both continents. The rosy pastor, *S. roseus*, is a bird of the Asiatic steppes and,

Large cushion-shaped sea star *Protoreaster linckii* of the tropical Pacific.

Superb glossy starling *Spreo superbus* of East Africa.

New Zealand smooth stick insect *Clitarchus hookeri*, in its shape and color as well as its habit of remaining still, bears a deceptive resemblance to a twig.

like other species of those parts, may periodically show large increases in numbers, and then in an unfavorable period spread out over wide areas, at such times often appearing in western Europe. The Asiatic mynahs, small brown or gray species of the genus *Acridotheres* or the larger glossy-plumaged *Gracula* species, are more stoutly built and have the shorter, heavier bills of more generalized feeders. They tend to have bristly or crested foreheads, bare skin around the eyes and yellow wattles on the sides of the head.

Starling nests tend to be rather crudely made and bulky. Those built in the open in trees are usually rounded structures with a side entrance, rarely a cup; but when a natural hole is used a cup-shaped nest may be built within. Some African species, such as *Spreo superbus*, may use old nests of the larger weaver species. The common starling, using a hole in a tree or house, may also retain this as a year-round roost and with successive nestings may accumulate a very considerable amount of material. This species occasionally nests in thick creepers or shrubs on the walls of houses.

STICK INSECTS, often called "walking sticks" in America, get their name from the shape of the body which is very long and slender, and usually green or brown, so that they look like slender twigs. The deception is enhanced by the fact that if they are disturbed they remain quite motionless. With too much disturbance they fall to the ground without showing any sign of life, with the long thin legs held close against the body so that they look just like twigs among the litter of leaves under plants. The largest stick insects may be up to 9 in. (23 cm) long, but the species commonly kept in schools and laboratories, *Carausius morosus*, is not more than about 4 in. (10 cm) long at most.

Stick insects are usually without wings even as adults, and even in those species that have wings the fore wings are usually very small. Since the fore wings offer so little protection, the front edges of the hind wings are leathery and cover the membranous parts of the wings when they are folded up. Stick-insect eggs are hard and rounded, about an $\frac{1}{8}$ in. (3 mm) across. They drop to the ground as they are being laid and often remain for a long time before hatching. The young, when they hatch, are about $\frac{1}{2}$ in. (12.5 mm) long, but otherwise resemble the adults. They feed on leaves, and in Australia, where stick insects are particularly abundant, they may completely defoliate eucalyptus trees and so are of some economic importance.

STICKLEBACKS, a family of very common freshwater and marine fishes of the northern hemisphere, characterized by the series of sharp spines in front of the soft-rayed dorsal fin. The three-spined stickleback, *Gasterosteus aculeatus*, is found in almost every brook and pond in England and is common throughout Europe, across Asia to Japan and in North America. There are, as the name suggests, three spines before the soft dorsal fin, and the pelvic fins are each reduced to a single spine.

In spring, the male three-spined stickleback changes to a bright blue with red on the chest. He then constructs a nest from plant strands, which are stuck together with a sticky secretion from the kidneys. The nest is ball-like, and the male enters it and makes a large central chamber. The nest may be placed in a hollow on the bottom, but nests have also been found in old tins lying in the water. The male defends the nest from other males or from intruders but entices gravid females to enter and deposit their eggs. After the eggs are fertilized, the male guards the nest and aerates the eggs by fanning movements with the pectoral fins, carefully removing any dead or infertile eggs. Two other species are found in Europe and

Asia. The ten-spined stickleback, *Pungitius pungitius*, is less common in England than the previous species and is often found in brackish water. It is, however, widespread across Europe and occurs in North America. Other species of *Pungitius* are found in Asia. The marine fifteen-spined stickleback, *Spinachia spinachia*, grows to about 9 in. (23 cm) in length and is commonly found around European coasts. It is more elongated than the freshwater species and leads a solitary life among weeds. It has a brown, well-camouflaged body and also constructs a nest.

In North America there are two species of sticklebacks, the four-spined stickleback, *Apeltes quadracus*, and the American brook stickleback, *Culea inconstans*, which usually has five spines and is the more northerly of the two. The latter species may have four or six spines but more usually five. Some variation also occurs in the other species, the three-spined occasionally having two or four spines and the fifteen-spined sometimes having fourteen or sixteen.

STILTS, very long-legged, wading birds which, with avocets, form the family Recurvirostridae. They are distinguished by their legs, which are longer in proportion to body size than in any other wader, and by their small heads and thin, dark, almost straight bills (in contrast to the up-curved bills of the avocets). Their plumage is chiefly black and white, but the banded stilt, *Cladorhynchus leucocephala*, also has a chestnut breastband. The latter is confined to Australia whereas the stilt, *Himantopus himantopus*, has a worldwide distribution, five geographical races having been named by different qualifying adjectives describing the plumage—for example, the black-winged stilt of the southern palearctic. They are catholic in choice of habitat, feeding in both salt- and freshwater lagoons and flooded grassland. The banded stilt breeds beside temporary salt lakes, feeding on the shrimps found there.

STINGRAYS, a family of raylike fishes with a venomous spine at the base of the tail capable of inflicting a painful or even fatal wound. The greatly flattened body and winglike pectoral fins vary in outline from round to triangular or diamond-shaped, followed by a thin and whiplike tail that may be longer than the body. The eyes are on top of the head, and close behind are the spiracles through which water is drawn to aerate the gills, the latter being on the underside behind the mouth. On the upperside and near to the base of the tail (not at its tip as is commonly thought) is the sting, a sharp spine with a pair of grooves down the hind edges, in which lie the glandular cells that secrete the venom. The spine, which is usually 3-4 in. (7.5-10 cm) long but may be up to 15 in. 38 cm) in a large fish, is sometimes followed by one to four additional spines. The sting is used solely in defense, the fish lashing the tail from side to side or up and down, sometimes with sufficient force to drive the spine deep into a plank or through a limb. There are about 90 species of stingray, ranging in size from less than 1 ft (30 cm) across the disc of the body to 6-7 ft (1.8-2.1 m) in the case of Captain Cook's stingaree, *Dasyatis brevicaudata*, of Australasian waters and *D. centroura* of the western North Atlantic. A common species off the

The stingray is fairly common in shallow waters, especially in summer, off the coasts of Europe.

Atlantic shores of Europe and in the Mediterranean is the stingray *D. pastinaca*, a species that was well known to Pliny, who repeated the fable that the sting would wither a tree if driven into the trunk. Stingrays lie on the bottom and are often extremely well camouflaged, so that great care should be taken when wading on sandy or muddy beaches where these fishes are known to occur. Some 1,500 accidents with stingrays are reported in the United States every year, mostly of a minor but nonetheless very painful nature, although some fatalities have been recorded when the sting has penetrated the abdomen causing paralysis of the muscles of the heart. In many parts of the world the spines are used as tips for native spears. The stingrays are bottom-living fishes that feed on shellfish, crustaceans and fishes, the food being crushed or ground up by a pavement of teeth in the jaws. All species are ovoviviparous, the young hatching within the uterus of the female.

STINKBUG, North American name for bugs of the large family Pentatomidae. The pentatomids, in common with many other bugs, possess glands secreting substances which are pungent-smelling or definitely obnoxious in some species. The English name, shield bug, which is applied to the Pentatomidae and to a few other closely allied families, seems a more generally applicable name for the group. The pentatomids are moderately large bugs having flattened, shield-shaped bodies which are often brightly and distinctly patterned. One of the dorsal plates of the middle segment (metathorax) of the thorax is enlarged to form a prominent triangular shield which may cover much of the wings. The pentatomids do not occur in Antarctica, and are common in Africa, South America and the Indo-Australian region.

Most species of pentatomids lay eggs in masses on various plants and their larvae (also known as nymphs) resemble the adults except they are wingless and are often differently colored. Some female pentatomids, for example the birch shield bug, *Elasmucha grisea*, guard their eggs and young larvae by covering them with their body until they are fully active. The great

Scale insects of Nigeria.

majority of the family are plant feeders but a few species are either entirely predatory on other insects or may feed on both plants and animals. Most pentatomids are of little economic importance but some are serious pests of cotton, cabbage, tomato and fruit trees. One of the most important is the harlequin bug, *Murgantia histrionica*, which is exceedingly destructive to cabbage and related plants in the southern United States.

STOAT, or ermine, *Mustela erminea*, a small carnivore closely related to the weasel but slightly larger and readily distinguishable by the longer tail with the characteristic black tip. Although the sexes are of similar appearance there is a striking size difference: while the male may measure 17.3 in. (44 cm) and weigh 15.5 oz (445 gm), the female is rarely over 10.2 in (26 cm) long and may weigh less than 6 oz (170 gm). Confusingly, a small female stoat may be the same size as a large male weasel but the tuft of black hair on the tail immediately identifies the stoat. The back is brown or russet and the underside white or cream colored with an even line separating the two.

It has a north temperate range encompassing the forest and tundra zones of Eurasia and North America. In the northern parts of this range, ermines usually take on an overall white color in winter, after molting, save for the black tail tip. The ventral surface will often be much yellower than the dorsal, due to the secretions from the anal glands used to mark territorial boundaries. This change in pelage color can be induced experimentally at any time of the year by lowering the temperature and shortening the number of daylight hours and it would appear that these are the two triggering factors concerned. The stoat lives in woods, hedges, or wherever undergrowth is thick enough to provide sufficient cover. Mainly terrestrial, it may climb and swim when hunting. Stoats rarely come out into the open, preferring to skirt a field or a clearing rather than risk falling prey to a kite or a buzzard while crossing it.

These small carnivores feed mostly on voles and mice but large insects, earthworms, shrews and moles may also be eaten. Young rabbits are only occasionally killed by stoats. Standing motionless on its hind legs, neck arched, the stoat hunting alone will sight its prey and then chase after it with surprising speed and agility. An unerringly accurate neck bite swiftly kills the victim which may be left on the spot or dragged back to the den. Often crepuscular, even diurnal in certain places, the stoat can be seen disappearing under a bush or racing across the road with the typical mustelid humped-back gallop.

Mating takes place in June followed by delayed implantation which retards the birth until the following spring. The blastocyst does not become implanted in the uterine wall until a month before parturition. The young may be born in a hollow log or under a rock pile. The cubs open their eyes 9 to 12 days later and are able to hunt alone when two months old.

STONE CURLEWS, an almost cosmopolitan family of large wading birds, some of which are alternatively called "thick knees." The family includes nine species of medium-sized to rather large birds, 14-20 in. (35-51 cm) long, classified in three genera, two of which contain only one species. They are found in Europe, temperate and tropical Africa, Australia and Asia and in tropical America. The northern species migrate to warmer areas for the winter. Birds of open country, they are found on stony or sandy ground, seashores, along sandy riverbeds and on grassy savannas.

Stone curlews bear a superficial resemblance to bustards, another group of birds found in open country, but their real relationships are with the waders, plovers and gulls included in the varied assemblage of the order Charadriiformes. They have long unfeathered legs, long wings and large eyes. They hunt at dusk or by night, and they remain crouched in concealment during the daylight hours. When disturbed, they run swiftly, and their flight is strong, though it is usually brief and infrequent. The plumage is of varied shades of concealing gray, brown and buff, with broad stripes on the side of the head. The bill is short and ploverlike in the seven species of the genus *Burhinus*; longer than the head, heavy and swollen in the beach stone curlew, *Orthorhamphus magnirostris*, from Australia and islands nearby, and similarly

ally takes a larger share. In some species she receives little or no help from the male. The young are very active within a few hours of hatching and soon leave the nest area. Both parents tend them for a number of weeks. The water thick knee, *Burhinus vermiculatus*, often nests on the breeding grounds of crocodiles on the banks and muddy islands of African rivers. It would appear that nesting among these large, dangerous reptiles protects their eggs and young from other predators.

STORKS, large, sometimes very large, birds of heavy build, with long legs, long necks, wings that are both long and broad and long, stout, pointed bills, which may be either straight or turned up or down. The plumage is commonly black and white, in a bold pattern, and some species have bright red bill and legs.

The family is widely represented in the tropics and subtropics of the Old World, and two species breed in the temperate zone of Europe and Asia, performing long migrations from there. A single species is found, not exclusively, in Australia but there is none in New Zealand. Only three species belong to the New World, and none of them to its higher latitudes.

Storks are adapted, in general, to wading

seen. Various display attitudes can be observed. In a characteristic greeting ceremony between mates at the nest, both birds bend the neck backward until the head touches the top of the body. Most species are vocally silent, or nearly so, but a noisy clattering of the mandibles is common.

Storks most commonly nest in trees, but sometimes on cliffs or buildings, breeding in colonies if the availability of sites permits. The nest is usually a bulky platform of sticks. The eggs are white, or almost so, and there are usually three to six in a clutch. The young are hatched almost naked, but a plumage of down is soon grown, and it is a feature of the group that this is succeeded by a second down plumage before this, in its turn, gives way to true feathers. The young take a few years to reach breeding maturity. Both sexes share in incubating and rearing the young.

The white stork, *Ciconia ciconia*, of which there are several geographical subspecies, is the most familiar member of the group and also the one that has been most closely studied scientifically. It nests chiefly on buildings, and it finds much of its food on cultivated land. Being generally esteemed as beneficial to man, it shows little fear; and it is so conspicuous that it is easily observed. It thus makes an almost ideal subject for population studies. Unhappily, these confirm the impression that its numbers are greatly diminishing in most European countries. It is still found as far north as the Netherlands and Denmark, and eastward through Germany and Russia to Asia. It was never a regular inhabitant of the British Isles, and it has ceased to be one in Sweden. Southward, it breeds in Spain and Portugal, north Africa and also in southeastern Europe and again eastward, but not in Italy.

The white stork is a notable migrant, and in many places the movements can be readily seen in progress—large birds flying by day in flocks, and often making use of the upward air currents over the warm land. The species has also been ringed, mostly nestlings, on a large scale. The European population winters in Africa south of the Sahara, many of the birds going as far as Cape Province. Of the latter, a few sporadically remain to breed, and immature birds may also linger in winter quarters. The more easterly populations migrate to southern Asia.

STURGEONS, primitive, often large fishes from temperate waters of the northern hemisphere, descended from the ancient paleoniscids The dense bony skeleton of the paleoniscids has, however, been replaced by cartilage, the scales have been lost and the dermal armor is represented by a series of bony plates or bucklers. The biting mouth is now an underslung, protrusile sucking mouth. The sturgeons are, however, larger than their ancestors, some reaching well over 20 ft (6 m) in length. Although not related, the sturgeons have a rather sharklike appearance that is heightened by the steeply rising tail.

Sturgeons are found throughout most of the cold and temperate waters of the northern

The common Atlantic sturgeon lives in the sea but enters rivers to spawn.

long and massive in the great stone curlew, *Esacus magnirostris*, of India, but in this species it is slightly up-turned at the tip. All of the species that have been studied feed on large insects, small reptiles and amphibians, crustaceans, mollusks, worms, nestling birds and small mammals.

A clutch of one or two rounded, cryptically marked eggs is laid either directly on the ground or in a small unlined hollow. The eggs are incubated by both parents, though the female usu-

in shallow water or walking in marshes; the long legs and slightly webbed feet being adaptations for this. They mostly find their food in such situations, but some species feed on drier ground. The food consists largely of freshwater animals and of large insects, but three species feed mainly on carrion. Storks fly strongly, with extended neck and trailing legs in most species. Some are notably adept at soaring in thermal air currents. They tend to be gregarious at all seasons, and large flocks may be

The white stork winters in Africa and nests on buildings in Europe in summer.

European swallows at the nest in a barn.

hemisphere, some species living in the sea and migrating up rivers to spawn, whereas others live permanently in rivers or are landlocked in lakes. There are about 25 species, all rather slow-moving, that browse on the bottom. Fleshy barbels surround the mouth and are used to detect prey (usually bottom-living invertebrates). Sturgeon in the sea also feed on fishes. There is a spiral valve in the intestine, a primitive feature also found in the sharks.

The largest of the sturgeons is the beluga, *Huso huso*, of the Volga and the Black and Caspian seas, for which a length of 28 ft (8.4 m) and a weight of 2,860 lb (1,300 kg) have been recorded. A related species, *H. dauricus*, from the Amur basin and the Far East, is smaller. The Atlantic sturgeons all belong to the genus *Acipenser*. The largest from the New World is the white sturgeon, *A. transmontanus*, of the Pacific coasts of North America, which now grows to about 300 lb (135 kg) but in the past has been known to reach over 1,200 lb (540 kg). These fishes rarely go to sea until they are almost mature, the younger individuals living in rivers and migrating up and down stream each winter and spring.

The common Atlantic sturgeon, *A. sturio*, reaches a weight of 700 lb (315 kg), although the males are smaller. They live in the sea and migrate into rivers to spawn. Formerly widespread and occurring in most European rivers, Atlantic populations now survive only in the Guadalquivir in Spain, the rivers of the Gironde in France and Lake Ladoga in the Soviet Union.

SUNBIRDS, tiny, brilliantly colored birds belonging to the family Nectariniidae, which also includes the spiderhunters. They are the Old World equivalent of the American hummingbirds, although they cannot match the latter in beauty, powers of flight, or minuteness, and are not related to them. The smallest of the hundred or so species of sunbirds is $3\frac{3}{4}$ in. (10 cm) long, the largest over 8 in. (21 cm). All have lightly built bodies, slim, delicate legs and strong feet. The bill is fine and long, and in many species it is strongly curved downward. The tongue is tubular along part of its length as an adaptation to nectar feeding. Most male sunbirds are brilliantly colored with green, purple or bronze metallic colors on the breast and upper parts and a nonmetallic, though often colorful, belly. In a few species, such as the widespread beautiful sunbird, *Nectarinia pulchella*, of Africa, the central tail feathers are greatly elongated. Female sunbirds are usually dull brown or olive. In some species the males molt into a dull plumage outside the breeding season and then resemble the females. Many male sunbirds, and some females too, have small patches of bright red, orange, or yellow feathers (called pectoral tufts) on the sides of the breast. Though normally hidden, these tufts can be erected during display.

The greatest variety of sunbirds (66 species) is found in Africa (Kenya alone boasts 15). There are four species in Arabia, and one reaches Israel. Sunbirds are widely distributed in the warmer parts of Asia, and a small number have reached New Guinea and Northern Australia.

Sunbirds are found in many kinds of vegetation: lowland and montane tropical forests, mangrove swamps, savanna and even semidesert, wherever there are trees or shrubs providing sufficient nectar and insect food. A surprising number of sunbirds are in fact montane dwellers. Several live between 4,000-9,000 ft (1,200-2,750 m) in Africa and the Himalayas, and one, the scarlet-tufted malachite sunbird, *Nectarinia johnstoni*, has been seen above 12,000 ft (3,650 m) in the Ruwenzori Mountains. Some species are very tame and live, sometimes in large numbers, in the heart of towns and cities where they feed on ornamental trees and garden flowers. The males, especially, are very noisy, pugnacious birds, chasing one another a great deal.

SUN BITTERN, *Eurypyga helias,* South American wading bird resembling a small heron. Sun bittern is the common name for the family Eurypygidae, in which there is only the single species. It has a wide distribution, occurring from Mexico south throughout Central and South America to eastern Peru, southeastern Brazil and Bolivia.

Sun bitterns are about 18 in. (45 cm) long, have long thin necks, daggerlike bills, long legs, broad tails and broad wings. The plumage is full and soft, almost like that of an owl, and is barred and mottled gray and brown. The crown is black with a distinct white streak above and below the eyes. When the bird spreads its wings a pattern of black, chestnut and yellow becomes visible. On the tail there are two bands of black and chestnut. The eyes are red. The upper mandible is black and the lower one yellow. The feet are orange-yellow.

Sun bitterns live along overgrown creeks and rivers in the tropical rain forest and are often to be seen wading with slow movements through the shallow water of a creek, stopping suddenly, looking intensely into the water and with a quick dart of the bill seizing prey. Their food consists to a large extent of water insects and their larvae, creatures such as water beetles, flies and the larvae of dragonflies, but they also eat small snails, crabs and fishes.

They spend a large amount of time on the ground, but perch in trees when disturbed. They can swim and will cross creeks in this way.

SUNFISHES, or centrarchids, a family of common freshwater fishes from North America containing the crappies, bluegills and black basses. The two dozen species are perchlike, with a spinous anterior portion to the dorsal fin, and in some species the body is deep and compressed. These fishes are nest builders, the male scooping out a hollow and guarding the eggs once the female has deposited them. The black basses (species of *Micropterus*) have been dealt with elsewhere. The crappies (species of *Pomoxis*) from the northeast of the United States, which can grow to 21 in. (54 cm), are popular sport fishes that have now been introduced into fishing waters elsewhere in the country. The white crappie, *P. annularis,* prefers rather still and turbid waters, whereas the black crappie, *P. nigromaculatus,* is generally found in clear waters. The bluegill sunfishes or bluegills are fairly deep-bodied and are characterized by a bony projection from the upper corner of the gill cover, popularly referred to as an "ear flap." The pumpkinseed bluegill, *Lepomis gibbosus,* is one of the most colorful and best-known species. The back is dark green to olive, the undersides yellow to orange, and there are red, blue and orange spots arranged irregularly on the flanks. There is a brilliant scarlet spot on the "ear flap," from which this species derives its common name. It reaches 9 in. (20 cm) and is found in the maritime provinces of Canada and in the neighboring areas of the United States. The green sunfish, *L. cyanellus,* grows to the same size but is less colorful. The bluegill, *L. macrochirus,* has been introduced into all the states except Alaska. Reaching 4 lb (1.3 kg) in weight, the bluegill can be recognized by the bluish color of the lower jaw, lower part of the cheek and gill cover. The red-ear sunfish, *L. microlophus,* a

smaller species, is dull olive with a red band on the gill cover. It is also known as the shell-cracker, because it has strong teeth in the throat (pharyngeal teeth) with which it crushes snails on which it feeds. The spotted sunfish, *L. punctatus,* is reputed to linger beside tree stumps and half-submerged logs waiting for insects or frogs to settle above it, thereafter charging the log to knock its prey into the water. This has never been verified, however. The Sacramento perch, *Archoplytes interruptus,* is the only species of sunfish found naturally in the rivers and streams of the western part of the United States.

SUN SPIDERS, also called wind scorpions or jerrymanders, are terrestrial arachnids of hot dry desert regions. The common name "sun spider" relates to certain South American species that are active during the daytime, although most species are nocturnal. They are relatively large arachnids, up to nearly 3 in. (7 cm) in length, superficially similar in appearance to large hairy spiders, but distinguishable from them by the segmented abdomen, or opisthosoma, the absence of spinnerets and the truly massive pair of jaws at the front end of the body. They are very hairy and move across the ground with lightning speed, and this habit, coupled with their ability to bend the abdomen upward in the manner of a scorpion, has also given them the common name "wind scorpion." During the day, and in winter when they hibernate, sun spiders hide in burrows, which they dig in the desert soil by means of their legs and mouthparts. They also possess a remarkable ability, for so large an arachnid, to climb vertical surfaces, and this they do with the aid of suckerlike pads at the tips of the pedipalps, a pair of leglike appendages attached to the body on either side of the jaws. In addition to the jaws and the pedipalps, the anterior part of the body (cephalothorax or prosoma) also bears four pairs of legs, of which the first pair is long and slender, and is usually extended forward as feelers, while the remaining three pairs are used for walking and digging. The last pair of legs carry a unique type of sensory hair, shaped like a miniature tennis racquet, and called a "racquet organ."

Sun spiders are voracious carnivores, feeding on insects, wood lice, spiders, scorpions, lizards and even small mice and birds. The pincerlike jaws, which may be as large as the entire prosoma, are used to seize the prey, which is macerated into a soft pulp and sucked into the mouth. Feeding will often continue until the gut is so full that the animal can hardly move. Travelers in the desert often come across these arachnids, for they invade tents and sleeping bags, and an encounter with the large hairy body surmounted by massive, widely opened jaws can be a daunting experience. There are many records of these animals biting people, and the effect is painful although probably not poisonous, despite popular belief.

SUNSTAR, a starfish aptly named, its shape resembling a conventionally drawn sun, with a large disclike body surrounded by 8-15, most usually 10-12, radiating arms or rays. In America the name is more usually applied to starfishes of the multiarmed genus *Heliaster.* The color too is fitting: usually the central disc is purplish red, the arms whitish with a broad

transverse red band, when viewed on the upper or dorsal side: more rarely this side may be purplish all over; the lower or oral side is whitish. The total diameter may reach 14 in. (34 cm) but is usually less, the arms being about as long as the disc is broad. The skeleton of the upper surface consists of an irregular network of narrow calcite bars enclosing fairly large patches of tissue in which little blisters of the body wall known as papulae are found. The skeleton itself bears large broom-shaped groups of rather long, slender spines, the groups sparsely distributed over the upper surface but conspicuously large at the margin.

The common sunstar, *Solaster papposus,* is widely distributed over most of the north Atlantic, its southern limit being the English Channel and, on the American coast, approximately latitude 40"N.

SURGEON FISHES, a group of marine fishes characteristic of coral reefs. Their name derives from the little bony keels, often extremely sharp and bladelike, on either side at the base of the tail (i.e., on the caudal peduncle). In some species these little "knives" are hinged at the rear and can thus be erected to point forward, so that care should be taken when handling live specimens. In the unicorn fish there are several bony keels on each side of the peduncle. The surgeon fishes are deep-bodied and compressed, with small terminal mouths bearing a single row of cropping teeth in each jaw. They feed by scraping algae and other organisms from rocks and coral.

Some variation in color is found in the surgeonfishes. In the yellow surgeon, *Zebrasoma flavescens,* there are yellow and gray-brown color phases, and the common surgeon fish of the Atlantic, *Acanthurus bahianus,* has been observed to change when chasing another member of the species, from its normal blue-gray coloring to white anteriorly and dark behind. The five-banded surgeon fish, *A. triostegus,* an Indo-Pacific species that reaches 10 in. (25 cm), has a dark, apple-green body with dark brown vertical bars. Also known as the convict fish, it has been involved in cases of fish poisoning. During development, the young surgeon fishes pass through a curious larval or acronurus stage in which the body has vertical ridges and does not closely resemble that of the adult.

SURINAM TOAD, *Pipa pipa,* a tongueless frog of curiously flattened appearance and living wholly in water in the Amazon and Orinoco basins.

SWALLOWS, a distinctive and successful cosmopolitan family of some 78 species of perching birds. They are all rather small birds, varying from 3¾-9 in. (9.5-22.8 cm) in length, and in a number of species much of this is taken up by the long, forked tail. The plumage is generally dark; black, brown, green, or blue, often with a metallic sheen. Several species show white in the spread tail, and many species are paler on the underside of the body. The neck and legs are short, and the feet small and weak. Swallows perch readily on wires, branches and other vegetation, but because of their leg and foot structure they are clumsy on the ground. In several species the whole of the legs, sometimes even the toes, are feathered. The general characteristics of the swallows are centered around their adaptations for feeding on the aerial plankton, the insects and other invertebrates, which are carried by, or fly weakly in, the air currents. For this reason one of their most outstanding features is the short, broad and flattened bill, which can be opened to a very wide gape, forming a highly efficient insect trap. It also acts as a trowel for scooping up and carrying the mud with which many of the species build their nests.

Swallows have a worldwide distribution, some individual species being found in both Old and New Worlds. Only the extreme latitudes and some oceanic islands are without one or more species of swallow. Most species are gregarious, and all are migratory. In temperate climates especially, birds that depend on the aerial plankton are forced to migrate for that part of the year when the insects on which they feed are not flying. Some of these migrations are of very great length. The European swallows, *Hirundo rustica*, for example, may fly 7,000 miles (11,000 km) from northern Europe to South Africa. And, if it survives long enough, it will undertake this journey twice a year.

Occasionally some species of swallows will feed on berries. Tree swallows, *Iridoprocne bicolor*, for example, are sometimes caught by an early autumn cold snap in Cape Cod, Massachusetts, and may then be seen feeding on bayberry fruits.

Most species breed more or less colonially. The nest may be made in a hole in a tree or rock face, or even a building; or a tunnel may be excavated in the ground—usually in a bank—with the actual nest perhaps several feet underground; or a mud structure may be built, either cup-shaped or enclosed, with an opening in the side, and placed on a branch, beam, rock surface or fixed to a rock face. Buildings make good substitute cliffs, and some species, such as the house martin, *Delichon urbica*, have taken to applying their mud nests to the walls under the eaves of buildings. In the species in which the nest is made of mud it is common for the female to do the building while the male carries the material. The female also seems to play the major role in incubation. The clutch contains three to seven eggs, according to species, which are white, sometimes speckled. The newly hatched young are helpless and almost naked, and the developmental period is an extended one, as the young must be able to fly well on leaving the nest. The fledgling period may thus be longer than three weeks, both parents feeding the young throughout this period. In spite of the long development, some species may rear two or three broods in a season, even in temperate countries. However, at the northern limits of a species' range there may be time for only one brood in the brief period between winter snows.

SWAMP DEER, *Cervus duvauceli*, included among the endangered species of deer, having only a limited distribution in eastern and northern India, and in southern Nepal and Assam. In some parts of India it is known as gond, while in others as barasingha, a name that is also used to describe the hangul or Kashmir deer. There are two subspecies: *C. d. duvauceli* in the area north of the Ganges, and *C. d. branderi* in the central parts of India.

They stand about 47-49 in. (119-124 cm) high at the shoulder and are generally reddish-brown in color, while the calves are spotted at birth. There is considerable variation in the pattern of the antlers, which usually bear 10 to 15 points, or even more, the majority of which project above the main beam from about two-thirds of the way up. During growth the antlers are covered with a beautiful red-colored velvet. During the rut, which in India generally takes place about the end of the year, each stag collects a harem of hinds in much the same way as does the European red deer stag. The stags utter a peculiar braying sound comparable to that of a donkey.

While the northern race is a dweller of swamp lands, the favorite habitats of the southern race are large grassy plains or maidans, on which the deer can graze and rest, preferring to live in, or on the edge of, these plains rather than to penetrate far into the jungle-clad hills. While resting, the deer are constantly attended by the Indian mynah birds, which search for ticks in their coats.

SWANS, a small group of large, long-necked aquatic birds. There are only eight species, but their size and, at times, aggressiveness assured them of a place in the economy and folklore of man in earlier times. Of the eight species, seven are placed in the genus *Cygnus*. The remaining one, the coscoroba swan, *Coscoroba coscoroba*, may not be a true swan, for it has characteristics that suggest that it provides a link between the swans and geese and another group, the whistling ducks. The coscoroba, a white bird with black wingtips, has a shorter neck than the other swans, pink legs and feet and a bright red bill. It breeds in the southern part of South America and migrates north to warmer areas in the southern winter. It derives its name from its call.

Two of the seven species of *Cygnus* are found in the southern hemisphere. The black swan, *C. atratus*, is naturally confined to Australia, though it has been introduced elsewhere and now flourishes in New Zealand. It is black with white primaries and a red bill. The black-necked swan, *C. melanocorypha*, comes from southern parts of South America. It is, as its name suggests, a white bird with a black neck and head. It has a red bill and knob and is the smallest of the genus. The other five species are all white in adult plumage, though they have bills of different colors, and are all found in the northern hemisphere.

The trumpeter swan, *C. buccinator*, breeds in northwestern U.S.A. and southwestern Canada. Largely resident, it has been close to extinction, but strict protection has enabled it to increase in numbers to its present level of around 1,500 birds. The whistling swan, *C. columbianus*, breeds in the high Arctic of western Canada and Alaska and winters largely on the eastern and western seaboards of the United States. Bewick's swan, *C. bewickii*, also breeds in the high Arctic, across Russia and Siberia, and migrates south for the winter, many then being seen in Europe. The whooper swan, *C. cygnus*, breeds largely to the south of the Bewick's breeding areas and in Iceland. A few pairs breed as far south as Scotland, but most of these also have to migrate to milder areas to the south for the winter—the time when they can be seen most easily in Britain. The last species, the mute swan, *C. olor*, breeds largely south of the whooper. It is mainly resident in Europe, but in Asia and in the cooler parts of Europe it may have to leave for the winter. The natural range of this last species is not so well known, since much of its present distribution may be due to man's interference. From the Romans onward the birds have been moved around to establish new stocks. They were much valued for food, and from the 13th to the 18th century all swans in Britain were the property of the Crown and the ownership of herds under license was a much valued possession. During this time the complex series of bill markings,

Male swordtail with young females. This male started life as a female and, after giving birth to several broods, underwent a change of sex and became a fully functional male.

which identified ownership, evolved. This has now largely ceased, though the Vintners and Dyers Companies still retain the right to mark the birds on the lower reaches of the Thames. There has even been discussion as to whether this species is a true member of the British avifauna, since it may not have bred there before its introduction by the Romans.

SWIFTLET, the name usually reserved for birds of about 20 species of the genus *Collocalia*—small spine-tailed swifts distributed through southeast Asia and parts of the western Pacific. Some nest in huge colonies in caves. The nests of some are used as the basis of "bird's-nest soup." Some species find the way to their nests by echo location in a manner similar to that used by bats.

SWIFTS, small, fast-flying birds with an almost worldwide distribution. They share the order Apodiformes with the hummingbirds. Usually, they are placed in the suborder Apodi and the hummingbirds in the Trochili, but some authors consider the similarities between the two groups as being caused by convergent evolution (i.e., evolution of the same modifications by animals of different groups because both animals live in the same way) and so place them in different orders. Both hummingbirds and swifts are striking in aerial behavior, although the modes of flight are very different. Both groups have very small feet (hence the name Apodiformes—"no feet") and legs. Another shared characteristic not shown by any other bird is the extremely short humerus (upper arm), short ulna (lower arm) and extended hand bones. In all birds the major flight feathers are attached to the ulna and the hand, but the swifts and hummingbirds have a peculiar wing in which the hand feathers (primaries) predominate and only a small area is taken up by the feathers in the inner part of the wing (secondaries).

The wing of a swift is long and very narrow, so they are extremely fast fliers and are among the most aerial of all birds. They catch their food on the wing, and at least one species spends the night on the wing. They have small, weak bills but large, wide gapes with which they catch insects. The birds are very streamlined, even the eyes being slightly sunk into the side of the head. It is often said that they are unable to take off from flat ground (due to their very short legs and high stalling speed), but this only applies to weak or injured birds (and most grounded birds are sick or injured). However, the European common swift, *Apus apus*, is unable to hover and cannot fly for long in small circles. Swifts are poor at maneuvering and need open spaces for flying.

SWORDFISH, *Xiphias gladius*, a large oceanic fish with the snout produced into a powerful, flattened sword. Swordfishes are worldwide, mainly in tropical oceans but also entering temperate waters. They have been recorded off the coasts of northern Europe and occasionally stray as far as Iceland. Rather solitary fishes, they grow to a weight of 1,500 lb (675 kg) and are chiefly found in open waters, often at the surface with the high but short dorsal fin cleaving the water like the dorsal fin of a shark. Where common they are exploited commercially, usually being caught by harpoon, and are also much sought after by anglers. Swordfish

Mute swan *Cygnus olor* preening.

The black-necked swan of South America and the Australian black swan artificially join company.

feed on fishes and squids, and examination of their stomach contents shows they also penetrate to depths and feed on deep-sea fishes.

The "sword" is reputed to be used to thrash among shoals of fishes, the swordfish feeding at leisure on the injured fishes. The fish has been known since ancient times both because of its size and because of instances when it has rammed wooden vessels. In the British Museum (Natural History) is preserved a piece of timber from a ship that has been penetrated to a depth of 22 in. (56 cm) by the sword of a swordfish. It was also reported that H.M.S. *Dreadnought* was punctured by a swordfish on its return voyage from Ceylon to London, the sword passing right through the copper sheathing of the hull. Swords are occasionally found broken off and embedded in the blubber of whales, and it has been suggested that swordfishes mistake ships for whales. It is possible, however, that the swordfish, a fast and powerful swimmer of the open seas, credited with speeds up to 60 mph (98 kph) has on occasions

been unable to divert its course in time to prevent a collision.

SWORDTAIL, *Xiphophorus helleri*, a live-bearing cyprinodont fish from Mexico and Guatemala. The lower caudal fin rays in the male are prolonged into a "sword," which is quite soft and used in sexual display. The swordtail has many remarkable features, which make it a most instructive aquarium fish. In some instances the female, having given birth to up to 100 young, changes into a male, losing her dark "pregnancy mark" as the anal fin changes shape and the male sword develops. Such males, once the transformation has been completed, are fully capable of fathering another 100 or so young. In some strains, up to 30% of the females change sex. The change from male to female has never been recorded.

In the wild, the swordtail is the "green sword," but the species is variable. A cross with a reddish individual of the closely related species *X. montezumae* (the Montezuma sword) has produced the red swordtail.

TAHR, a close relative of the goat. Males have a characteristic odor, a little different from that of the "billy goat." There is no beard and the horns are short, highly compressed bilaterally and keeled in front; they are simply back-curved. The Himalayan tahr, *Hemitragus jemlahicus*, is 36-40 in. (90-100 cm) high. Big males weigh 200 lb (90 kg), but females may weigh only 80 lb (35 kg). It has a heavy body, narrow ears and coarse shaggy hair that forms a mane on the neck and shoulders, reaching to the knees. It is brown with a dorsal stripe. It is found along the Himalayas from Kashmir to Sikkim, on precipitous cliffs, in scrub and forest, from 10,000-12,000 ft (3,050-3,660 m).

The Nilgiri tahr, *H. hylocrius*, is slightly bigger, 39-42 in. (100-110 cm), with a short yellow-brown coat, old males becoming deep brown with a distinct light "saddle patch" on the back. Females have only one pair of teats; other species have two. It is found from the Nilgiris to the Anaimalais, south along the western Ghats at 4,000-6,000 ft (1,220-1,830 m), on scarps and crags above the forest.

The Arabian tahr, *H. jayakari*, is small, only 24-26 in. (60-65 cm) high, slenderly built, sandy with a dorsal crest but, like the Nilgiri tahr, with no mane. It has long shaggy hair, however. It is restricted to the mountains of Oman. Male tahr live solitary lives most of the year. Unlike females, they never, at least in the Himalayan species, graze in open clearings, but emerge from the oak and cane forests only in the evenings. They migrate up and down the slopes according to season and snowfall. The rut takes place in the winter, and rams are often killed by falling during the fierce fights. Little else is known about the breeding of these animals.

TAILORBIRDS, two genera of warblers. The nine species of *Orthotomus* are found in southern Asia, from India to the Philippines. They are common garden birds and are named after their method of nest building, in which leaves are stitched together with fibers to form a pouch. The Australian tailorbird, or golden-headed fantail-warbler, *Cisticola exilis*, makes a similar nest.

TAIPAN, *Oxyuranus scutellatus*, a slender elapid snake and the largest and deadliest of Australasian snakes, growing to a little over 11 ft (3.4 m) in length. It has a large head distinct from its neck, a relatively slim fore body and tapered tail. Its fangs are large, and its venom one of the most potent neurotoxins known; death is usually caused by paralysis of the nerve centers controlling the lungs and heart. It is reputed to be the world's deadliest snake. Australian taipans are rich brown above and cream below, while those from New Guinea are usually blackish with a rusty-red stripe along the middle of the back.

The taipan is found throughout many parts of northern and northeastern Australia, ranging from coastal rain forests to the drier inland regions. In New Guinea it is found largely in the savanna woodlands along the southern coasts. About 16 eggs are laid in a clutch. The taipan is a timid, retiring snake that may become very aggressive when provoked. It may be seen in weather conditions that are too hot for other snakes, and although generally diurnal, it may move about at night if the weather has been excessively hot. It feeds upon small mammals and reptiles.

Specific and polyvalent antivenenes have been developed for the taipan, without which the chances of recovering from a bite are slender. The name "taipan" is a Cape York aboriginal name for this snake.

TAKAHE, *Notornis mantelli*, a large, flightless bird of New Zealand. Subfossil remains show that it was recently widely distributed, but in the 19th century European settlers found only five specimens. It was thought to be extinct until its dramatic rediscovery in a remote, high, tussock-grassland valley in 1948.

Extensive study since then has shown it to be in danger of extinction, with only 300 birds left in an area of 200 sq miles (320 sq km). Numbers seem to have been shrinking before the arrival of Europeans, suggesting that the Polynesians were probably responsible for the decline. The takahe does not breed successfully in captivity, but attempts are being made to establish it in other parts of New Zealand.

TAKIN, *Budorcas taxicolor*, an ungainly looking goat antelope related to the musk-ox. Takin, 42 in. (110 cm) high, weighing 500-600 lb (230-275 kg), have a convex face, heavy

The takin of the wooded mountains of southern Asia, one of the goat-antelopes.

The taipan of Australasia, reputed to be the most dangerous of all snakes.

muzzle, thick neck, short thick legs, humped shoulders and an arched back. The color is golden to dark brown or black on the flanks and haunches (according to race); the withers are always lighter toned. Calves are black. The horns, found in both sexes, are thick and triangular in section. They are at first upright, then turn out, then up at the tips. Takin are found along the flanks of the mountains from Bhutan through Szechwan to Shensi. They inhabit steep, thickly wooded slopes, most characteristically the dense bamboo and rhododendron jungle at 7,000-10,000 ft (2,135-3,050 m), but in the Mishmi hills of northern Assam they go as low as 3-4,000 ft (9-1,200 m). During the rut, which in Szechwan takes place in July to August, the males, which are always aggressive, are extremely dangerous and fight fiercely. Gestation lasts about 200 days, and calves are born in early April. The cow enters the forest to give birth.

TAMANDUA, *Tamandua tetradactyla,* an anteater intermediate in size between the giant and the tiny silky species, which like them, is completely toothless and belongs to the the mammalian order Edentata along with the sloths and armadillos.

Tamanduas have short coarse hair on the body and on top of the tail near its base, but most of the grasping tail is naked and blotched with irregular markings. Tamanduas vary greatly in coloration, but typically they have a black band encircling the middle of the body, and this joins, on the back, a black ring or "collar" around the neck. Head and legs are tan or cream-colored, so the animals look as if they are wearing a dark vest or waistcoat. The ears are rounded but more conspicuous and proportionately larger than those of the giant anteater. Fox-sized, the tamandua has a head and body length of 21-22.5 in. (54-58 cm) with a 21-in. (54-cm) tail. Tamanduas have the characteristic anteater spoutlike snout (ideal for poking into rotting logs and termite nests) and long extensile tongue, no bigger in diameter than a lead pencil. The nostrils and the tiny mouth are at the tip of the snout. The nails of the hands are long and sharp, particularly those of the middle fingers, and tamanduas have great strength in their arms and tearing power in their claws.

The tamandua ranges from southern Mexico to Argentina and is found in tropical forests and occasionally savannas. Little is known of its reproductive biology except that the single young apparently is carried on the mother's back from birth.

The tamandua is equally at home in the trees and on the ground, walking on the edges of its palms with claws turned in, but nevertheless-it seems to spend most of its time in trees, being an excellent climber, and its prehensile tail suggests a long history of arboreal living.

TAMARAO, *Bubalus* (*Bubalus*) *mindorensis,* a small species of buffalo from the island of Mindoro in the Philippines, closely related to the water buffalo, and by some authorities, such as H. Bohlken, classed as a subspecies of it. It differs considerably from the water buffalo in its small size, being only 40 in. (1 m) high, in its much stouter and more robust build, its short and very thick horns, and its color, which is jet black with white bands above the hoofs only instead of white shanks. Tamarao weigh 600-700 lb (270-320 kg), and the horns are only 14-20 in. (35.5-51 cm) long and slightly in-curved. In breeding it is very similar to other borids.

TANAGERS, small colorful birds of the family of Thraupidae. There are about 200 species all confined to the New World, ranging from southern North America, through Central America and the West Indies to the tropical parts of South America. The largest number of species lives in the tropical areas.

Tanagers vary in size from that of a tit to that of a finch. Many of them, especially those living in the tropics, are brilliantly colored. Among these are the members of the genera *Euphonia* (20 species) and *Tangara* (43 species). The euphonias are the smallest, and in most of them the males have glossy blue upperparts and yellow or orange underparts. The females are quite different, having a dull green plumage that makes them difficult to distinguish in the field. They live in trees and shrubbery at forest edges and feed mainly on the berries of mistletoes. Euphonias make dome-shaped nests with a side entrance, sometimes between the leaves of tree orchids. The nests are built by both sexes, and the female lays three to four eggs, which is more than most other tanagers.

TAPACULOS, a family of about 26 species of small to medium-sized perching birds found in South and Central America. They vary from about 4-10 in. (10-25 cm) long and have soft, fluffy plumage in which gray-brown or black usually predominates. They are dumpy birds that often hold the tail cocked over the back. Their legs are strong, correlated with a terrestrial habit, and the bill is straight, thin in the small species and stouter in the large ones. The wings are short and rounded, and the tail varies from a short tuft to a long flag of stiff feathers.

Tapaculos live in thick cover, whether in dark woodland thickets or dry thorn scrub on the pampas. They creep or run about among the thick vegetation, only rarely taking to the wing. In most of the species the male and female share the same dull coloring, but in a few, such as the slaty bristlefront, *Merulaxis ater,* the male is more brightly colored than the female. Little is known about their feeding habits, but both insects and seeds have been found in the stomachs of several species. They feed by picking food up from the surface of the ground or off vegetation and by scratching in the litter of dead leaves, some species having very stout feet and claws.

Brazilian tapir on the bank of the Amazon.

TAPEWORMS, internal parasites, so named because they are long and flat. The whitish body (called the strobila) is marked off into many segments, or proglottids, each containing a complete set of male and female reproductive organs. The proglottids are connected by muscles, two lateral nerve cords and the two pairs of excretory ducts connecting the flame cells. The proglottids are budded off continually from a region just behind the minute scolex, or head region, by which the long tapeworm is attached to the intestinal mucosa of the vertebrate host. At first the proglottids are very small and the reproductive organs embryonic, but farther down the strobila the proglottids become progressively more mature. In each proglottid the male organ matures first, so ensuring cross fertilization; the female system too becomes mature until in the terminal proglottids the branching uterus is packed with fertilized, shelled eggs, and the gonads and eggshell-forming glands have degenerated.

The large tapeworms of man include the pork tapeworm, *Taenia solium*, and the beef tapeworm, *T. saginata*. As their names suggest, these use pigs and cattle respectively as their sole intermediate hosts, and man acquires the infection by eating measly pork or beef. The proglottids expelled in the stools are extremely active, unlike those of the pork tapeworm, and may even themselves creep out of the anus. In conditions where sewage disposal is primitive, eggs contaminate pasture grazed by cattle so that these become infected. Oncospheres are then liberated from the egg, penetrate the gut wall and make their way via the lymph ducts or blood vessels to the muscles, mainly the jaw muscles and heart. Here the oncosphere larva loses its six larval hooks and develops into a bladderworm or cysticercus (plural cysticerci),

in which the scolex is not merely retracted but actually invaginated into the body, which is hollow, bladderlike and filled with fluid. The cysticerci of the beef tapeworm may measure up to $\frac{2}{5}$ in. (1 cm) across. When man eats undercooked measly beef, these cysticerci escape, the scolex evaginates and the worms attach to the gut wall and become adult. These tapeworm infections of domestic animals long associated with man probably explain why pork is forbidden to Moslems and Jews and beef is not eaten by Hindus. The pork tapeworm is particularly dangerous because, if the eggs of this worm are accidentally ingested by man, they can develop into bladderworms in what would normally be the definitive host. If these reach the brain, they may produce epileptic fits and paralysis.

Larval multiplication. Some tapeworms reproduce asexually by budding while in the intermediate host, thus increasing the chances that infection of a definitive host will occur. This larval multiplication (or polyembryony) is most highly developed in a small tapeworm, only three proglottids long, called *Echinococcus*, which lives as an adult in the intestine of dogs, foxes, wolves, rats and cats. The usual intermediate hosts of *Echinococcus granulosus* are sheep and cattle, and in them a kind of "bladderworm" develops that produces internally many daughter or brood cysts containing secondary scolices. This mother cyst is termed a hydatid cyst because it is filled with fluid. It can become very large and may reach the size of an orange and produce millions of scolices. After 10-20 years the hydatid cysts may reach an enormous size and can contain over three gallons of liquid. Man can inadvertently serve as an intermediate host for *Echinococcus*, and hydatid cysts develop as

a result of infection with eggs originally liberated in dog feces but that may be picked up from the fur or tongue of the dog after this has licked itself. Hydatid disease of man is complicated by the fact that, should the cyst containing brood capsules be ruptured, these can take root and grow in other tissues of the body. In view of the fact that the life cycle of *Echinococcus* relies on contact between domestic ruminants and, for instance, dogs, it is hardly surprising that human hydatidosis is prevalent in sheep- and cattle-raising areas such as parts of Africa, New Zealand and Australia.

TAPIR, a large brown or black and white ungulate with many similarities to the rhinoceros.

They are plump, thick-skinned creatures, the males weighing about 400-800 lb (180-360 kg) and the females 200-400 lb (80-180 kg). The fore feet bear four toes; the hind feet, three, but the outer toes of the fore feet are small and on hard ground may not leave an imprint, so that tapir tracks are sometimes mistaken for those of small rhinos. The dental formula of the tapir is $\frac{3143}{3143} = 44$, but, as in the rhinoceros, P_1^1 may be unrepresented in the permanent dentition. The lower canine is well developed, but the upper is smaller than $1\frac{3}{3}$ The cheek teeth are low-crowned, with no cement, as tapirs are browsers. The chief external differences from rhinos are the absence of horns, the elongation of the snout into a short proboscis with terminal nostrils, the long neck, the absence of skin folds, and the presence of at least a sparse covering of hair. The skull is distinctly different, with a high narrow crown and short, high-placed nasals. The four living species, all placed in the single genus *Tapirus*, are not very closely related, but the Malay tapir is the most distinct and can be placed in the subgenus *Acrocodia*. The three New World species all belong to the nominate subgenus, and, according to Hershkovitz, who places them unnecessarily in three different subgenera, they each represent a separate invasion into South America.

The Malay tapir, *Tapirus indicus*, is the largest species, weighing as much as 800 lb (360 kg). It is black, with a white body and haunches. The white begins behind the fore legs and extends over the rest of the body except for the hind legs and tail. There is no mane on the neck, the proboscis is longer than in other species and the build is heavier and stouter. Malay tapirs live in lowland, especially swamp forest, in Sumatra, Malaya, Tenasserim, southern Thailand.

The South American tapir is found throughout the tropical forest and subtropical hardwood forest zones, from Colombia through the Guianas and Brazil into Paraguay and northern Argentina. In the Andes it ascends to about 4,000 ft (1,200 m). It is very similar ecologically to the Malay species, being fond of lowland forest and especially swampy regions. It bathes and wallows a great deal, and it walks with its snout close to the ground. The name "tapir" comes from the Tupi language (Brazilian Indian).

The social life of the South American tapir is unknown. In the San Diego Zoo, the captive group forms a structured herd, with dominant and subordinate animals of both sexes. The dominant male and female make what is called the "sliding squeal," less than a second in duration. On hearing this sound the others make a "fluctuating squeal," which is longer and quavers rather than merely decreasing in pitch. This is also uttered when a dominant individual approaches, apparently as an appeasement call and as a sign of pain or fear. Tapirs also utter a challenging snort, and a click made with the tongue and palate, perhaps as a species identification.

TARPONS, powerful silvery fishes related to the tenpounders. They are strong swimmers and are renowned for their fighting powers when hooked. There are two species, the Atlantic tarpon, *Tarpon atlanticus*, and the Indo-Pacific tarpon, *Megalops cyprinoides*. The two are outwardly so similar that many authors place them in the same genus. Recent studies have shown, however, that there are fundamental differences in the skull that suggest that the two tarpons are very different fishes. The body is fairly compressed with large silvery scales, and there is a single dorsal fin with the last ray prolonged into a filament. Although primarily marine, the tarpons are not infrequently found in fresh water and sometimes even in foul waters. The Indo-Pacific tarpon reaches 3 ft (90 cm) in length, but the Atlantic tarpon is a huge fish, growing to 8 ft (2.4 m) and weighing up to 300 lb (135 kg). Like the ladyfish and the tenpounder, the tarpons begin life as a ribbonlike leptocephalus larva resembling that of the eels.

TARSIER, a peculiar nocturnal mammal related to lemurs and also to monkeys, apes and man as indicated by its large, forward-directed eyes, its opposable thumb and big toe and its relatively large brain. It is about the size of a rat, with a head and body length of

about 5 in. (13 cm) and a tail 8 in. (20 cm) long, remarkable for being hairless apart from the tip, which has a featherlike arrangement of fine hairs. The tarsier always rests and jumps in a vertical position, typically moving around in bushes and high grass. It can rotate its head through a half circle and, like an owl, look directly backward over its shoulder in order to sight a suitable landing point for its next jump.

The tarsier ranges over the Malay archipelago, from Sumatra to Borneo, Celebes and the Philippines. There are three species: the Philippine tarsier, *Tarsius syrichta*, the spectral tarsier, *T. spectrum*, of Celebes and neighboring islands, and Horsfield's tarsier, *T. bancanus*, of Sumatra and Borneo. All are mainly insectivorous, but eat some fruits.

TASMANIAN DEVIL, *Sarcophilus harrisii*, a carnivorous marsupial included in the family Dasyuridae, which also includes the native cats, banded anteater and marsupial wolf. The Tasmanian devil is a fox-terrier-sized animal of powerful build with a widely gaping mouth and strong teeth. The muzzle is short

and broad, and the ears short and rounded. The head and body are about 28 in. (70 cm), and the tail about 12 in. (30 cm) long. There are five toes on the fore feet and four on the hind feet, each with a strong claw. The general body color is black, but there is a white band across the chest and sometimes a white band across the rump.

The jaw gape of the Tasmanian devil is wide, and the teeth are large and strong. As in other dasyurid marsupials there are four upper and three lower incisor teeth. The canine teeth are powerful and the premolars sharp and reduced to two in each jaw. The four molar teeth of the upper jaw bite only partly on the four lower molars, a powerful crushing action being obtained by the uppers sliding past to the outside of the lower molars.

The Tasmanian devil is now confined to Tasmania, where it has been described as rare. Recent studies have, however, shown it to be very abundant over at least some of its former range, and it has been implicated in the killing of some domestic animals. Local eradication or control measures are sometimes under-

Usually represented as highly ferocious, the Tasmanian devil looks more dangerous than it is.

taken. Tasmania is, however, but a remnant of the former range of the animal. It occurred in the Northern Territory of Australia, at about 12°S latitude, 3,000 years ago, in various other parts of Australia at about the same time and during the Pleistocene epoch. In Victoria, Australia, at about 38°S latitude, it was present 500-700 years ago.

The Tasmanian devil is a usually solitary, terrestrial, nocturnal carnivore that feeds on mammals and birds, a variety of insects and invertebrate animals and on carrion. The litter size is up to four, and the young are born in the autumn of each year.

TASMANIAN WOLF, *Thylacinus cynocephalus*, also called thylacine or marsupial wolf, the largest carnivorous marsupial, once numerous over most of the continent of Australia but persecuted to the point of extinction. It now probably survives only in the wilder regions of western Tasmania.

TAYRA, *Eira barbara*, a South American forest-dwelling member of the weasel family (Mustelidae) renowned for its swiftness and climbing ability. The single species is distributed over southern Mexico south to Argentina and Paraguay. Tayras have relatively long, slender legs and tails, which set them apart from the other usually short-legged mustelids. The very short, coarse fur is sepia or black, save for the brown head and light yellow throat or chest patch, which may be speckled with white. Individuals with an overall buff coloration have been reported occasionally. The broad head with flat, rounded ears has such a human resemblance that Mexicans refer to the tayra as "cabeza de viejo" (head of an old man), and the subspecific name is appropriately *senilis*. Adult males are markedly larger and more muscular than females. Consequently, overall length varies from 38.5-45.3 in. (98-115 cm), of which the tail accounts for at least 16 in. (40 cm), and the weight varies from 9-13 lb (4.2-5.8 kg).

Like the marten or fisher, tayras are equally at home in trees as on the ground and have often been observed in the wild chasing one another in the loftiest treetops. This marked arboreal tendency allows them quickly to outdistance such terrestrial predators as man or dog. The strong, nonretractile claws and naked soles are great assets in climbing, and, although the tail is not prehensile, it serves as a balancer. Wild tayras have been frequently seen eating bananas and fallen cecropia fruit, but they also prey on a variety of animals such as tree squirrels, agoutis, mouse opossums, snakes and birds, which are swiftly dispatched with a neck bite and a shake of the head. When chasing a victim, a tayra may swim but does not do so willingly. The precise length of the gestation period is not known; two to four cubs have been found, however, in makeshift grass nests in May. The young, blind and helpless at birth, are covered in sparse black fur, the brown head fur appearing only six months later.

TEGU, *Tupinambis nigropunctatus*, one of the largest of the lizards in the family Teiidae, an inhabitant of tropical South America, about 3 ft (1 m) long. It is terrestrial and lays its eggs in the large spherical nests of certain termites. It is omnivorous and disliked for its

tendency to take chicks and eggs around farms. The bulk of its diet, however, consists of frogs, insects, other lizards, and even leaves and soft fruits.

TELESCOPE EYES, deep-sea fishes with tubular eyes of the family Giganturidae and distantly related to the whalefishes. They are also called giant tails. These fishes have cylindrical bodies and reach 9 in. (23 cm) in length. The body is naked, the pectoral fins are set behind the head and very high up on the flanks and the lower lobe of the tail is elongated into a long banner. They live in all oceans at depths down to 12,000 ft (3,600 m). The tubular eyes give these fishes excellent binocular vision, an attribute that is rare in fishes but one that enables them to make accurate estimates of distances when pouncing on prey. Like many deep-sea fishes, the telescope eyes have elastic stomachs and are capable of swallowing fishes larger than themselves.

Gigantura is a relatively solid-looking fish and is unusual amongst deep-sea forms in having a silvery body. Its relative, the Pacific telescope-eye fish, *Bathyleptus lisae*, is much more fragile in appearance, but both fishes have long, fanglike teeth. Almost nothing is known of the biology of these fishes, and only about 12 specimens of *Bathyleptus* have ever been caught.

TENCH, *Tinca tinca*, perhaps the most easily identified of all the carplike fishes of European fresh waters. It is a stocky fish with very small scales, heavy and rounded fins and a body that is dark olive-green shot with gold. The belly is light gray or reddish gray with a violet sheen. One variety is green shot with gold, with the mouth red. The tench, usually a rather sluggish fish, spends most of its time on the bottom rooting around the mud in ponds or slow-flowing waters. It hides away in winter, sometimes in the company of other tenches, almost in a state of hibernation, but it will stir on very warm days. A really large tench can weigh up to 10 lb (4.5 kg), and any angler who has caught these large fishes will vouch for the fact that they can be anything but sluggish.

Old legends refer to the tench as the doctor fish because of its alleged habit of allowing injured fishes to rub their wounds on its healing slime. These healing properties were supposed to confer some kind of immunity from attack on the tench, but this is certainly not true since both perch and pike will readily eat this fish. The tench is also reputed to remove leeches from carp in much the same way as the cleaner-fishes remove parasites from other and larger species.

TENRECS, related to moles and shrews. Half the genera have spines, and *Setifer* and *Echinops* closely resemble hedgehogs. *Oryzorictes* has developed fossorial habits and anatomical adaptations. *Limnogale* is semiaquatic, and *Microgale* parallels the shrews in its way of life. Tenrecs range in size from only 2 in. (5 cm) long up to rabbit-sized species of 16 in. (40 cm) in length. Some are tailless, but *Microgale* has a very long tail supported by 47 vertebrae, more than in any other mammal, apart from some of the pangolins. The skull is narrow and long (especially the nasal part), but is not constricted

in the orbital region. Many tenrecs have spiny skins, but are unable to roll up into a spiny ball as effectively as the hedgehogs because their skin musculature is less well developed. They are found in Madagascar and the adjacent Comoro Islands, where they inhabit a variety of biotopes including montane forests, grassland, scrub and marshes. Most live in burrows and are either crepuscular or nocturnal. They are basically terrestrial animals, feeding on a variety of plant and animal foods, but mainly on ground-dwelling invertebrates. *Tenrec* hibernates in the cold dry season, but *Hemicentetes*, at least, is active all the year around. Some tenrecs are among the most prolific of mammals, rivaling even the rodents in having up to 25 young per litter. Some of the larger species of tenrec are hunted by the local people as a source of food.

TERMITES, insects closely related to cockroaches, having similar biting mouthparts and an incomplete metamorphosis. They are the most primitive insects to have developed a social system, and all termites live in well-regulated communities, there being no solitary forms. Approximately 2,000 species have been recognized, and these are divided into six families.

Termites are polymorphic, and each community is made up of several distinct castes. First in importance come the reproductives, the king and queen, or in some cases several pairs together in one nest. Then there are the workers and soldiers and juvenile forms in various stages of development. At certain times of the year there are also present numbers of fully winged young adult termites waiting to swarm.

Winged termites vary in size according to species with wingspans of ½-3½ in. (12-87 mm). The head is round or oval with large eyes, two small ocelli, a pair of long antennae made up of 15-32 beadlike segments and mouthparts of the biting type. Like the cockroach, the head is held at right angles to the body. Thorax and abdomen are chitinized yellow to dark brown, and there is no narrow waist as with bees and ants. The membranous wings have a unique line of fracture close to the thorax, known as the humeral suture, which enables them to be discarded rapidly once male and female termites have found each other during swarming. The two sexes are very similar in appearance.

Worker and soldier termites are sterile individuals of either sex whose development has been arrested at an early stage. Workers have rounded heads and mandibles similar to those of the adult. The head is chitinized, but the body remains white and nymphlike. Soldiers undergo two special molts and develop large, hard heads with mandibles of a more aggressive type. Only the soldiers and workers of harvester termites have functional eyes, and this is associated with their habit of feeding in daylight on plants in the open.

A queen termite develops an enlarged abdomen in order to be able to supply eggs in increasing numbers as the community grows. While the

The tayra, a South American member of the weasel family.

dry-wood queen only becomes a little larger, others grow a great deal and in the case of some African *Macrotermes* reach 7 in. (17.5 cm) in length. Such queens are capable of producing one egg every two seconds. The king retains his original size.

Termites are to be found in all the warmer regions of the world, broadly speaking between latitudes 47° North and South, but their numbers and variety are greatest in the tropics. The mass exodus of flying termites from the nest is known as swarming.

Two things in particular are commonly associated with termites—their mounds and the damage they do to the woodwork of buildings. Large mounds are a feature of the tropical landscape away from the darkness of jungle or rain forest, in the savannas of South America, the woodlands and grassy plains of tropical Africa and the eucalyptus scrub of Australia. Small mounds come in a variety of shapes, some resemble large mushrooms, others are conical while others leave the ground altogether and are like footballs attached to the trunks and branches of trees. In Northern Australia the tall, slender, wedgeshaped mounds of *Amitermes meridionalis* always point north and south, giving it the common name of "compass termite," and whatever advantages this may have, no other termites appear to have adopted this system. In jungles, many small mounds built up against the trunks of trees have a series of steep, overhanging roofs to shed the heavy rain.

TERNS, resembling gulls, but mostly smaller and more lightly built, often with long forked tails. They are short-legged birds with webbed feet and a bill that is long and tapers to a point in all but one species. They all have long, pointed wings, and most have deeply forked tails, hence the old name, "sea swallows." Most species are white with a gray back and wings and a black cap on the head in the breeding season, though this is often lost in the winter. Some have black upperparts in the breeding

A termitarium, housing a termite colony and rising like some bizarre ruin of a building, is built by the cooperative labors of its thousands of inhabitants.

season, a few species also having a black breast.

The common tern, *Sterna hirundo*, the arctic tern, *S. paradisaea*, and the roseate tern, *S. dougallii*, are representatives of this group. They are mainly coastal birds that feed on small fishes caught by plunging into the water from a height. The arctic tern is remarkable in that it probably enjoys more hours of daylight each year than any other bird. It breeds as far north as about 82°N and migrates south in the autumn to winter in antarctic and subantarctic waters. This group also includes a freshwater species, the smaller black-bellied tern, *S. melanogaster*, of southern Asia. The Sandwich tern, *S. sandvicensis*, is the only species to breed in Britain; it migrates to

South Africa and the surrounding seas for the winter. The largest of all the terns, the Caspian tern, *S. tschegrava*, belongs in this group; it is about the size of a herring gull and breeds in scattered places in Europe, Africa, Asia, North America, Australia and New Zealand. Other rather smaller crested species are the crested or swift tern, *S. bergii*, the elegant tern, *S. elegans*, and the royal tern, *S. maxima*.

Then there is a small group of little terns, the most common of which is the widely distributed little or least tern, *S. albifrons*. The eggs are usually gray or brown in color with dark brown and pale gray-brown spots and blotches. Incubation periods vary from about 19 to 26 days, and both sexes sit. The chicks are active and alert within a few hours of hatching and are covered with thick down that is cryptically marked in most species. Fledging periods vary from about four to eight weeks.

Most terns feed mainly on fishes that they catch by plunge-diving into the sea from some height, but the marsh terns pick food from the surface of water and hawk insects in the air. The gull-billed tern hawks insects and also catches insects and other small animals on the surface of grassland.

THAMIN or ELD'S DEER, *Cervus eldi*, of which three subspecies are recognized: *C. e. eldi* from Manipur; *C. e. thamin* from Burma and Tenasserim; and *C. e. siamensis* of Thailand, Vietnam and also Hainan Island, all of which are rare.

Standing about 45 in. (114 cm) high at the shoulder, the general color of the thamin is reddish brown, lighter in summer than in winter. The most noticeable features of this deer are the antlers in which the brow and main stem (beam) of the antler form a more or less continuous bow-shaped curve, with a number of small jags sprouting from the upper surface of the outer tine of the terminal fork. Another peculiarity of this deer is that the foot has been modified, enabling it to walk on swampy ground, which is its principal habitat in the Manipur valley. The deer cannot tolerate heavy forests or hills, and to a great extent it is a grazing animal, feeding largely on wild rice, as well as browsing on certain trees.

THRESHER SHARKS, *Alopias vulpinus*, and two very similar species, large sharklike fishes characterized by the enormous upper lobe of the tail, which may be equal to the length of the rest of the body. The thresher sharks have a worldwide distribution in tropical and temperate seas, mainly living a pelagic life in the upper waters but with at least one species found in deep water. In general form they resemble the gray sharks, but their enormous tail easily identifies them. Stories of threshers using their tails to lash at whales, while swordfishes execute a *coup de grâce* with their swords, are quite without foundation, but there are reliable accounts of the thresher using its tail to stun or kill fishes and birds at the surface, which it then eats. There is still some doubt, however, whether threshers really encircle shoals of fishes and with their tails beat them into a compact mass before charging in and feasting. The threshers reach about 20 ft (6 m)

Winged adults, nymphs and eggs of the termite *Neotermes*.

A pair of common terns with chicks.

in length and may weigh up to 1,000 lb (454 kg). The young are hatched within the uterus of the female and when born are 4½-5 ft (1.2-1.5 m) long.

THRUSHES, slender-billed songbirds of small to medium size. The plumage is most often gray or brown but is sometimes chestnut, blue, green black or pale buff. Some thrushes have long pointed wings, some short rounded wings, with many intermediates. The tail is usually rounded or square, occasionally rather strongly graduated, and it is held erect in some species.

The thrush subfamily, the Turdinae, is distributed throughout the world, with the greatest number and variety in Africa, Asia and Europe. There have been several invasions of thrushlike birds into Australia and New Guinea and into the New World. Thrushes have become isolated on islands for example on Tristan da Cunha and Hawaii, where they have developed into distinctive forms. There are about 300 species, combined into two groups, the "true thrushes" centered on the genera *Turdus* and *Zoothera*, and the large varied group of chats.

The genus *Turdus* includes about 63 species. European species include the blackbird, *T. merula*, the song thrush, *T. philomelos*, the mistle thrush, *T. viscivorus*, and the ring ouzel, *T. torquatus*. It also includes the American robin, *T. migratorius*, a common garden bird in the United States. Thrushes of the genus *Zoothera* are closely related to those in the genus *Turdus* but they have more rounded wings and white bases to the wing feathers. Most are found only in tropical Asia, but

White's thrush, *Z. dauma*, reaches Europe and several species are found only in the New World, including the varied thrush, *Z. naevia*, of the western United States and the Aztec thrush, *Z. pinicola*, of South America. The forest thrushes of the Caribbean islands, *Cichlherminia*, and the Tristan da Cunha thrush, *Nesocichla eremita*, are offshoots of this group, as are the American genera *Catharus* and *Hylocichla*. These last are two genera of small, slender thrushes found in woodland, scrub and gardens. The hermit thrush, *C. guttata*, the veery, *C. fuscescens*, and the wood thrush, *H. mustelina*, are familiar birds in North America.

The chats are confined to the Old World. In general they have weaker, less melodious songs, and more slender legs than the true thrushes, and many of them are smaller European chats.

The brightly colored bluebirds of North America are chats, with bright blue plumage. They are found mainly in open country. Other distinctive groups are the shortwings, *Brachypteryx*, and the genus *Zeledonia*. The former are sedentary round-winged birds found in the dense jungles of southern Asia. The latter has only one species found in western Panama and Costa Rica in Central America.

Most thrushes and chats defend territories in the breeding season. True thrushes sing from perches and chats sing from perches and during display flights. Some species defend territories through the winter, as well as in the breeding season, although winter and summer territories may be different. The territorial instincts of

some chats are so strong that they defend territories for even a few days while they are resting on migration. Migrant wheatears are sometimes grounded in large numbers by adverse weather and they can be seen in rows on the tops of stone walls, each bird singing only a few yards away from its neighbor.

Nearly all the "true thrushes" nest in bushes or trees, building rounded cup-nests of grass, leaves and other materials. The chats have more varied nesting arrangements: some nest in tree or rock holes, some in partially enclosed hollows and some in the open but surrounded by thick vegetation. The color of the eggs varies in relation to the kind of nest site that is used. The hole nesting species such as the redstart and wheatear lay pale blue or white eggs without markings, those nesting in partially enclosed sites lay pale-colored eggs with spots and blotches and those nesting in the open lay darker eggs that are more heavily marked. The eggs of the nightingale are so heavily blotched that they appear a uniform dull green, matching the dark background where they are laid.

In those species that have been studied, the female undertakes all the incubation and she is given food at the nest by the male in a few species such as the European robin. Incubation periods vary from 13-15 days and the young of most species hatch with a covering of down, which is dark-colored in the species nesting in the open. The young are fed by both parents and leave the nest after a period of 12-16 days. After they fledge the young are often fed by their parents for several weeks. Most of the chats eat mainly insects but also

Tiger taking its ease in the shade.

take spiders, tiny reptiles, worms, snails and other animals. Some of the true thrushes subsist mainly on worms and snails but many of them take fruit and berries in the autumn and winter.

TICKS, annoying parasites of man and domestic animals, the importance of which in disease transmission did not become apparent until just prior to the opening of the present century. Studies then that implicated ticks as vectors of a protozoan disease of cattle, Texas cattle fever, focused world attention on the potential danger of this group of arthropods.

The subclass Acari, which contains both mites and ticks, differs from most arthropods since the body is not divided. A true head is lacking, and the thorax and abdomen are fused, producing a saclike appearance. Ticks should not be confused with the wingless insects such as the sheep ked, often referred to as the "sheep tick." The suborder Ixodides, which comprises the ticks, differs from the suborders of mites by their larger size, the presence of a pair of breathing pores, or spiracles, behind the third or fourth pair of legs, and by having a unique type of sensory organ (Hallar's organ) situated on the distal segment of each of the first pair of legs.

In place of a true head, ticks have a gnathosoma bearing the mouthparts. It consists of a basal portion, a pair of four-segmented palps and a rigid holdfast organ, usually toothed, called the hypostome, which serves to anchor the parasite to its host. In addition, a pair of cutting organs or chelicerae permit the tick to cut the skin for the penetration of the hypostome.

The 700 species of ticks are all blood-sucking, external parasites of vertebrates, including amphibians, reptiles, birds and mammals. Because of this specialized way of life they have several special adaptations, one of which is a powerful sucking pharynx. After the chelicerae have broken the skin of the host, the hypostome is inserted into the wound and the tick commences feeding. The salivary glands of some ticks produce secretions that prevent blood from coagulating, and the blood is pumped in by the pharynx and forced back into the esophagus, stomach and diverticula of the stomach. The body of ticks is covered with a leathery cuticle capable of great distension as the blood is being forced into the diverticula of the stomach. Unengorged ticks of different species vary tremendously in size (1-30 mm). Some species are capable of ingesting hundreds of times their weight in blood, and after engorgement they may attain a size many times greater than when unengorged.

Ticks have four stages in their life cycle: the egg, a 6-legged larva, and an 8-legged nymph and an adult.

TIGER, Panthera tigris, the major feline of the Asian continent. A number of attempts to separate the races of tiger into different species have been abortive, and it is now accepted that all tigers belong to one species. There is a close link with lions, and the two will interbreed in captivity to give fertile young. The cubs are known as "tiglons" or "ligers," the first syllable of the name coming from the male parent. The cubs take on some characteristics of each parent.

Tigers are the largest of the cats, but they are slimmer and narrower in the body than lions, and only an expert can distinguish between the skeletons of the two species. That of the tiger provides support for the powerful muscles of the hindquarters for the spring, and strength to the fore limbs for gripping and dragging down prey. The coat provides a natural camouflage against the patterns of light and shade in the natural surroundings, and it is said that a running tiger in the jungle looks gray against the background. Regardless of race, the coat usually has the same basic coloring, the ground color being a shade of reddish fawn, broken at intervals by dark vertical stripes. The belly is white, and there are patches of white on the face. The claws are hooked and retractile, and the teeth are used for the holding, tearing and cutting of flesh, the staple diet. In terms of both armament and camouflage, the tiger is one of the best equipped of all the carnivores.

Fossils show that the tiger originated in Siberia and the New Siberian Islands and then spread south to occupy most of Asia and parts of the Malaysian archipelago. The northernmost race still lives in Siberia, Manchuria and North Korea. Tigers shed their coats seasonally and are intolerant of great heat, indicating that they originated in a colder climate.

They become mature at three years of age, and for about the next eleven years a tigress will mate approximately every two to two and a half years. The interval is devoted to the rearing and training of the previous litter. In captivity, tigers will breed more frequently, as there is no need to train the cubs. The gestation period is 105-115 days, and a litter consists of two to six young, but only one, or perhaps two of the cubs will ever reach maturity. The tigress herself controls the size of the litter to some extent by "weeding out" the sickly and injured young. The cubs are blind at birth, but have a fully marked coat and weigh from 2-3 lb (1-1.5 kg). By the time they are six weeks old, the cubs will have been weaned, and at the age of seven months, they will have started to kill for themselves. In spite of the fact that the mortality rate in an individual litter is so high, a tigress may still leave up to 20 descendants. The males are polygamous, and play little or no part in the rearing of the young, although they may hunt with the mother, thus indirectly contributing to the supply of food. For the most part, tigers are solitary, although they may occasionally hunt as a pair, one animal driving the prey down to the ambush formed by the hunting partner. Although tigers belong to the so-called roaring cats, they do not roar as much as lions, doing so only briefly when charging or threatening. A surprised tiger will give a "whoof" of alarm before making off. Another sound has been described as "belling" because of the similarity to the noise made by the sambar stag. This sound indicates alarm or uneasiness. Finally, there is a moaning or mewing sound made when the tiger is moving through cover. This appears to be the same as the purr of the domestic cat and may indicate contentment and satisfaction.

Tigers have been classed as mainly nocturnal cats, but this is not necessarily so. They avoid the heat of the day, and will often lie in water to keep cool. They are excellent swimmers. Preferring to hunt in the cool of the day, they will take a variety of prey from wild pig to buffalo. In times of food shortage, they have been known to eat lizards, frogs and even crocodiles. The tigers of India also appear to have a fondness for the durian fruit. All tigers will eat carrion and will return to a kill a number of times. A full-grown tiger will eat 40-60 lb (18-27 kg) of meat at one feed in the wild. A probable average intake would be about 20 lb (9 kg) per day, involving the killing of 45-50 deer a year.

It is from the Bengal tiger, *P. tigris tigris*, that the species receives its name. Found in India from the Himalayas to the south, it has shorter hair and is more brightly colored. A male can weigh 420 lb (190 kg) and measure 9.5 ft (2.67 m). The females, as is the case with all tigers, weigh about 100-150 lb (45-68 kg) less.

TIGER BEETLES, popular name for a group of beetles, which is well deserved since they are formidable predators with large, sharply toothed jaws, probably the most voracious and ferocious of insects, especially as larvae. Their eyes are large, their legs are long, and they run and fly quickly. Their habit is to run speedily when disturbed, then take to the wing and fly for about 50 ft (15 m) before again alighting on the ground. There are about 2,000 species, typically 1 in. (2.5 cm) long, mostly in the warmer parts of the world. They are often common in sandy areas near the sea or on lake shores. Some species are found on riverbanks and others live on dry heathland. Many of the species are most active in the middle of the

day in the hot sun, but there are also a few nocturnal species. Their colors are often bright, particularly the upper surface of the abdomen, which is exposed in flight.

The larva has a large flattened head, and lives in a vertical shaft in the ground where it waits for passing prey that it seizes with its large jaws.

TIGERFISHES, African freshwater fishes of the genus *Hydrocyanus*, deriving their common name from their striped coloring and voracious appetite. The body is elongated and powerful, and the mouth is armed with curved, daggerlike teeth. These fishes prey chiefly on other fishes, and the largest species, *H. goliath*, which reaches 125 lb (56 kg) in weight, has been suspected of attacking cattle and even human beings. Fish eaten by the tigerfishes either are swallowed whole or have the soft parts cleanly bitten off.

Young tigerfishes shoal in shallow waters, but larger fishes are mostly solitary, living in deeper water. Because of the importance of freshwater fishes as a source of protein in Africa, the tigerfishes have been the subject of various investigations, both as a source of food and because of their effect as predators on other useful species. It has been found that *H. vittatus* reaches its peak of efficiency as a predator on other fishes when it is 12-18 in. (30-45 cm) long. Fishes smaller than this catch less relative to their body weight, while larger fishes also catch less because their bodies are fatter and their fins relatively smaller and they are thus less efficient swimmers.

The tigerfishes are found in the Nile and in those rivers and lakes that once had a connection with the Nile.

TIGER SNAKES, a group of closely related Australian elapid snakes of the genus *Notechis*. All are restricted to the coast ranges and wetter parts of the interior of southern Australia, including Tasmania.

They are relatively heavy, thick-bodied snakes with broad, rather massive heads. The common tiger snake of southeastern Australia, *N. scutatus*, averages only a little over 4 ft (1.2 m) and varies considerably in color and pattern. It ranges through various shades of gray, brown, reddish or olive to almost black. Typically it has numerous light yellowish cross bands along its length, but these are often indistinct or absent.

Tiger snakes are normally shy and inoffensive, but react savagely when provoked. The neck is then flattened like that of a cobra, while the first quarter or so of the flattened body is held off the ground in a long, low arc. All are highly venomous, their venoms being among the most potent known. Although they have relatively short, immovable fangs they cause a high proportion of the average of only four deaths that result from snake bite in Australia each year. Before the development of an antivenine in 1929, the first to be developed for any Australian snake, about 45% of tiger snake victims died. Like that of other dangerous Australian elapids, the venom of tiger snakes acts largely on the central nervous system, causing death by respiratory paralysis. The potency of tiger snake venom varies considerably in different populations, the most potent yet recorded being that from the black tiger snake, *N. ater*, from Reevesby Island in

Spencer Gulf, South Australia. It was found to have venom more than twice as deadly as that of tiger snakes from the mainland of southeastern Australia.

Tiger snakes produce living young, an average litter numbering about 50. The young measure about 8 in. (20 cm) at birth, and are usually much more strongly banded than the adults.

TILAPIA, perchlike fishes of the genus *Tilapia*, found principally in Africa but occurring also in Lake Tiberias and other water masses connected with the northern extension of the African Rift Valley. Species of *Tilapia* are found in lakes, streams and rivers, but some will penetrate into brackish waters. Typically, the body is fairly deep and compressed, with a long dorsal fin (the anterior rays being spiny) and a moderate anal fin. In some species a black spot occurs on the hind part of the dorsal fin, occasionally edged in yellow. This "Tilapia mark" is not always retained in the adults. Members of this genus have successfully invaded a wide range of habitats, but are principally found in lakes where they form the basis for large local fisheries. Certain dwarfed species, such as *T. grahami* of Lake Magadi, Kenya, have become adapted to highly alkaline hot springs, while others, for example *T. spilurus*, are typically found in rivers. In good growing conditions, many species will reach at least 12 in. (30 cm), and in Lake Rudolf, Kenya, specimens of *T. nilotica* weighing 14 lb (6.3 kg) have been recorded.

Many species of *Tilapia* are mouth brooders. At the onset of the breeding season the male becomes more highly colored and establishes a territory, usually in shallow water over sand or mud. The male then excavates a nest, a shallow circular depression 12 in. (30 cm) or more in diameter, taking material in the mouth from the center and spitting it out beyond the rim of the nest. When the nest is complete, the male guards it from intruding males or members of other species by a series of postures and swimming antics, but will go through a special ritual if a female of its own species approaches the nest. When the female has laid the eggs at the bottom of the nest, the male discharges sperm over them, and the eggs and sperm are taken into the mouth of the female. In some species it is the male that takes up the eggs. The female mouth brooder then leaves the nest, or is chased away, and thereafter broods the eggs until they hatch, which happens after about 5 days. Even then the fry are retained in the mouth of the parent for some days, and even when they venture forth in a little cloud they will swim back into the safety of the mouth if alarmed. In some species it has been noticed that the fry will "freeze" at the sudden appearance of a predator while the parent deals with the danger. While brooding the young, the parent is unable to feed.

In a few species of *Tilapia* the eggs are left within the nest to incubate.

TILEFISH, *Lopholatilus chamaeleonticeps*, one of the largest members of the family Pseudochromidae, reaching 3 ft (90 cm) in length. It is related to the sea perches. There is a sharp crest at the back of the head. The body is fairly slender, and the anal fin and single dorsal fin are fairly long. The discovery and the subsequent sudden disappearance of the tilefish present a rather curious story. The fish was first discovered in 1879. It was found at the bottom of the Gulf Stream slope off the shores of New England. The water mass in which the tilefishes were found was at that time warm since it was composed of Gulf Stream water that had traveled northward from the Gulf of Mexico. The fishes were found in large numbers and a fishery subsequently developed. Three years later, after some extremely severe gales, the course of the Gulf Stream altered, and the area in which the tilefishes lived was invaded by a much colder body of water from the Labrador Current. For some reason the tilefish is exceedingly sensitive to changes in water temperatures, and this sudden cooling was enough to kill off the fishes in millions. In March 1882 an area of some 15,000 sq miles

some other parts of the world, to which it has been transported through commerce, the best-known timber-beetle species and, economically, the most important, is the house longhorn beetle, *Hylotrupes bajulus.*

Some species, such as *Macrotoma heros* and *Titanus giganteus*, are among the largest known insects, with a body size of about 6 in. (15 cm) long. On the other hand, some species are very small, and a few are even minute. The most characteristic feature of the longhorns, shared by all but a few species, is the extreme length of the antennae, which are usually as long as, or longer than, the body. In *Batocera kiebleri* they reach a length of 9 in. (22.5 cm). The eyes are usually large and are frequently bow-shaped.

TINAMOUS, some 50 species of neotropical birds, pheasantlike in plumage and like guinea fowl in form. They fly weakly.

after hatching. They run well and, like the adults, crouch when danger threatens and merge with their surroundings. Some species breed throughout the year, while others have a well-defined breeding season.

TITS, term normally used to cover the small birds of the family Paridae of which there are some 50-60 species distributed throughout the world with the exception of the land masses in the southern hemisphere—Australia, New Zealand, Madagascar and South America. The word has been misleadingly used in different parts of the world as a name for birds in many different families, for example bearded tit, tree tits, wren tits, shrike tit, etc. The word "tit" is an abbreviation of the earlier "tit-mouse," though the latter is still used to describe one group of these birds in North America, while "chickadee" is used for the others, on that continent.

All the birds are small, only the sultan tit, *Melanochlora sultanea,* weighing much over $\frac{3}{4}$ oz (20 gm). While the North American and European species include some that are well known (indeed the great tit must be among the most studied of all wild birds), the tits of Central Asia and the tropics are not well known, being small and rather inconspicuous members of large avifaunas. However, the great majority of the species are resident, or undergo only short migratory movements. All nest either in holes or in domed nests that they build themselves. They have large clutches, often of eight or more eggs, which are white with red spots.

The family is currently divided into three subfamilies, the Parinae or true tits, the Aegithalinae or long-tailed tits and the Re-mizinae or penduline tits. There is some doubt as to whether all these families are really closely related and some taxonomists divide them into three separate families. It has been suggested that the Parinae are more closely related to the Sittidae (nuthatches) than to the other "tits." It is only recently that the bearded tit, *Panurus biarmicus,* now sometimes called the reedling, was removed from this family and put with a family of the flycatcher group (Muscicapidae). Even if the three families are related it is likely that one or two of the species still included in them will be shown eventually to be more closely related to birds of other families.

The penduline tits fall into three groups. The penduline tit, *Remiz pendulinus,* is patchily distributed throughout southern Asia and its range extends into southern and eastern Europe, where it has been increasing in recent years. Titlike in behavior, it has a longish tail, is reddish chestnut above and buff beneath and the pale gray head and neck with the big black mask is diagnostic. It is a bird of marshy scrub and reeds. In the summer it feeds itself (and its chicks) on tiny insects which it collects off the herbage with its needle-shaped bill. In the winter it supplements its insect diet with small seeds from reeds, etc. One of the most striking features of this species is its nest, which is a very soft pouchlike structure with a small tubular entrance at the top. The nest is suspended from small twigs in a bush (hence the name of the species) and when it is built of the down from the seeds of the reed-mace, or

Tinamous are running birds, living mainly in South America, that are believed to be related to ostriches.

(40,000 sq km) was strewn with dead tilefishes. For the next 20 years no tilefishes were caught, and it was presumed that the species was extinct. Then they slowly made a reappearance in the warmer waters of the Gulf Stream, and their numbers gradually built up once more.

TIMBER BEETLES, members of one family, the Cerambycidae, one of many families of beetles whose larvae and adults are known to damage timber. About 20,000 species have already been named, and they are to be found throughout the world wherever trees or bushes grow, and wherever timber is transported or used. Although the longhorns are usually rather more than medium size for beetles, a number, especially from South America and, to a lesser extent, from Africa, are of gigantic proportions, and often they are of remarkable form and they show an infinite variety of color and pattern.

The great majority of the larvae infest some part of the woody tissue of woody plants. A few species, however, consume the roots, and others the stems, of herbaceous plants. Many species of longhorns attack only one species of plant, but some are polyphagous, attacking a large number of plants. One, *Stromatium barbatum,* is known to feed upon 311 different tree species, while, on the other hand, 38 species of timber beetles are listed as attacking the single tree species *Shorea robusta* (a member of the genus from which the timbers meranti and seraya are converted). In Europe and in

Tinamous are efficient running birds and have a well-camouflaged plumage of browns and grays, streaked, spotted or barred. They range widely over South America from southern Mexico to Patagonia and from the tropical rain forests to 14,000 ft (4,200 m) up in the Andes. They vary in length from 8-21 in. (20-53 cm). The body is compact, with a very short tail and the wings are short and rounded. The legs are strong, and the hind toe is elevated or missing entirely—a tendency followed in many other running birds. The neck is quite long, and the head is small but often strikingly marked in black and white, or crested. The bill is rather long and is decurved and pointed for feeding on a variety of plant materials and small animals picked up from the ground. The sexes are rather similar in appearance, but the female is usually larger.

Most species are solitary, except during the breeding season when they are polygamous —some polygynous and some polyandrous. The nest is a depression in the ground, poorly lined. The number of eggs in a nest is anything up to 12, the larger clutches probably being laid by more than one female. The coloration of the eggs is remarkable. They are always unmarked with a surface sheen resembling polished metal or porcelain and may be green, blue, yellow, purple, black or chocolate. Incubation is carried out by the male alone and lasts for about 20 days. The young leave the nest with the male in the first day or so

A male bearded tit among the reeds, its usual habitat. The tendency today is to classify this species with the flycatchers, family Muscicapidae.

similar material, it is strikingly soft and pliable. The true tits of the subfamily Parinae include the best-known species. They are small, active, arboreal birds, usually fairly tame and easy to see, often quite noisy. The great majority of them have the same general appearance and are easily recognized as tits, once the observer is familiar with one or two of the species. The majority have a "capped" appearance. Usually the cap is black or brown and this is enhanced by white or pale cheeks. A black or dark brown bib is also often present, though not in most of the titmice of North America. Some, including this last group and the European crested tit, *Parus cristatus*, have quite conspicuous crests. The sexes are always similar and, in most species, cannot be distinguished in the field. However, in the two most common European species, the great tit, *Parus major*, and the blue tit, *Parus caeruleus*, the males are generally brighter in color. In the great tit the male has a broader, blacker and shinier band down the center of the yellow underparts.

As far as is known they are all omnivorous, taking a mixed diet of insects, other small animals and seeds. Their short, stout bills are very strong and a blue tit can break open the scales of pine cones to remove the insects inside, while the larger great tit will hold down an acorn or a hazelnut with its feet and hack it open with woodpeckerlike blows. The great tit has also been recorded killing young

birds and adults of some of the smaller species such as goldcrests. It is probably a fair generalization to say that most species feed on seeds and a few insects in the winter and almost exclusively on insects in the summer. Most of the European species feed their young on caterpillars off trees.

Although most species are resident, some of the northernmost tits may migrate south for the winter.

These tits are predominantly woodland dwellers, but they are very agile and versatile and one of the reasons why they have been popular with people is that they have found it easy to live in garden habitats and to take food from bird feeders in winter.

All species are hole-nesters—unlike the species in the other two subfamilies. Many species need to find a ready-made hole. The clutches are large, usually of seven to twelve eggs, though some of the tropical species may have smaller ones. The blue tit not infrequently lays more eggs than this.

The young usually hatch in about 13 days from the start of incubation. Normally most young hatch within a 24-hour period. The large number of young need a lot of food and the parents have to work hard to collect it. At the point in the nestling period when the young are needing the most food, the parents not infrequently bring a caterpillar a minute to the nest for 14 or 16 hours a day. A brood not infrequently receives 10,000 caterpillars

during the time that it is in the nest (about 18-21 days).

TOAD, a name referring strictly only to members of the family Bufonidae, but the terms "frog" and "toad" are the only common names available to describe the 2,000 species of the order Anura.

Toads are nocturnal, spending the day concealed in holes in walls, drainpipes or similar places. They emerge at night and feed on any small animal that moves. They are sometimes found sitting near a lighted window or lamp, snapping up the insects attracted by the light. Like other anurans they are able to recognize food only if it moves, but the feeding action of a toad is more deliberate than that of some anurans. If a worm is placed a few inches away from a toad, for example, it walks rapidly up to it and looks down at it for several seconds, its nervous energy being shown by a twitching of its toes. The worm is then snapped up, the toad's tongue flicking rapidly out and dragging the worm into the mouth.

Toads hibernate on land, burying themselves in loose soil, either alone or in groups. On emerging from hibernation, usually in March, they migrate to the breeding pond, where the males attract the females to them by their calling. They usually call while raising themselves on their front legs in shallow water. The male clasps the female behind the arms and the pair swim about in amplexus until the female comes into contact with some waterweed. Oviposition then commences. The female extrudes the eggs in the form of a long string while the male ejects sperm over them. This continues at intervals for several hours, the female swimming around so that the long string of eggs, about 7-10 ft (2.1-3.0 m) long, is wrapped around waterweed. A total of about 4,000-5,000 eggs is laid. The tadpoles hatch after about 12 days, but the time taken to develop into the adult varies considerably, depending on the temperature. About three months is an average period, although in warm areas it may be two months and in some cold areas the tadpoles have been known to hibernate and metamorphose the next year.

The American toad, *Bufo americanus*, is found throughout the eastern half of North America in almost any habitat, from gardens to mountains, the only requisites being shallow breeding pools and moist hiding places. It is 3-5 in. (7.5-12.5 cm) in length and usually a reddish brown, although there is a great variation in the markings. Like several species of *Bufo*, the American toad has a pattern of bony crests on the head.

The southern toad, *B. terrestris*, is more restricted in its range, occurring in the southeast coastal plain from North Carolina to the Mississippi River. It is 1½-3½ in. (3.7-8.7 cm) long, and the bony crests have two pronounced knobs behind the eyes, which give the toad a horned appearance. It is found in most open situations but is particularly abundant in sandy areas. There are about 16 species of *Bufo* in North America, although the number of subspecies is uncertain.

TOP SHELL, marine snails, related to periwinkles and named for their resemblance to the children's whip top of the 19th century. They belong to the most primitive group of prosobranch gastropoda, having two auricles to the heart, two kidneys and only one gill. The head of the snail bears a pair of sensory tentacles for touch and taste, and each has an eye near its base. Along the sides of the foot are 3-6 pairs of longer tentacles, and at the rear of the foot is a circular cover, or operculum, which closes the entrance to the shell when the mollusk withdraws into it.

There are over 50 genera of top shells, ranging from $\frac{1}{25}$-6 in. (1-150 mm) across the base, with the lower part of the shell usually marked with distinct spiral ridges or oblique bands of color. The painted top shell, *Calliostoma zizyphinus*, of Europe, is pink or yellow, flecked with crimson or red-brown. It is 1 in. (26 mm) high and slightly less across the base. The brick-red top shell, *Astraea inaequalis*, is found on the Pacific coast of North America. One of the largest top shells, measuring up to 5-6 in. (13-15 cm) across the base, is the button shell of the shallow waters of the Great Barrier Reef, the northern Australian coasts and the Indo-Pacific. Its shell is white, marked with wavy red-brown bands, and, because of its mother-of-pearl interior, it is collected for the manufacture of buttons, and a thriving button export industry has been built up in Australia.

Many species of top shell are found between tide marks, but most of them live offshore and down to depths of 600 ft (182 m) being most numerous in tropical and subtropical seas. Most of the species feed by rasping off minute particles from very small seaweeds by means of a ribbonlike radula that has many rows of unspecialized teeth.

TORTOISES, slow-moving, heavily armored reptiles that first appeared some 200 million years ago and have remained relatively unchanged for 150 million years. The body is enclosed in a box or shell, which in many species is rigid and into which the head, tail and limbs can in many instances be withdrawn. The top of the shell, known as the carapace, is formed from overgrown, widened ribs. The lower part of the shell, called the plastron, is also made up of bony plates. Both carapace and plastron are covered with horny plates or shields known as scutes. The males are usually smaller than the females and often have a longer tail, and the plastron may be concave.

They have no teeth, but the jaws are covered with a horny bill that can be used for tearing food apart. They have movable eyelids that are closed in sleep. Their external ear openings are covered with a membrane, and it is doubtful if the Testudines can hear airborne sounds, but, like snakes, they probably can pick up vibrations through ground or through water. They all lay eggs, and the aquatic species must come ashore to lay them. In temperate climates land tortoises hibernate in the ground and freshwater species under mud at the bottoms of ponds. As a group they are noted for their long life spans, the longest recorded being in excess of 158 years for one of the giant tortoises, but even the garden tortoise has been recorded as living 50 years or more.

Tortoises mate in the spring; the male usually makes squeaking noises during copulation, which often takes hours.

Tortoises live in areas that extend from dry, hot desert regions to the humid jungles; this includes the tropics, subtropics and warmer temperate regions in all the continents except Australia. During the intolerable heat of the day they hide in holes that they dig in the ground and that consist of a long passage and a terminal, spacious living chamber. They come out at dusk when the air has cooled down and usually eat cacti. The epiglottal shields on the underside in the male are extended very much forward and are used, when fighting for the female, to lever the rival onto his back and put him out of action. It is questionable whether these extensions are also used as spades when digging the burrow.

In the New World, the genus *Testudo* is found only in South America; to the north of this the only tortoises are the species of *Gopherus*. The wood tortoise, *Testudo denticulata*, and the coal tortoise, *T. carbonaria*, inhabit the humid, tropical rain forest. *T. chilensis* is found only in Argentina and, despite its scientific name, is not found in Chile.

The carapace of the giant tortoises, *T. elephantopus*, of the Galápagos Islands and *T. gigantea* of the Seychelles is up to 4 ft (1.2 m) long. The Galápagos types lack the uneven nape scute on the front margin of the carapace, which is developed in the Seychelles species. Several strains of both species inhabit the small islands of the two archipelagos, but some have died out or are threatened with extinction. In some of their subspecies the front part of the carapace curves upward in the shape of a saddle. This enables the animals to stretch the head and neck up farther than other tortoises and thus to feed on plants higher up. These island tortoises originally had no natural enemies, and their shield has become relatively thin.

TOUCANS, a family of about 37 species of South American birds, with heavy bodies and long and often bulky bills. They vary from 12-24 in. (30-60 cm) in length, most species having a bold pattern of black, yellow, orange, red, green, blue or white plumage, in one combination or another. The most arresting feature of the larger species of toucans is the huge bill. This may be 6 in. (15 cm) long, 2 in. (5 cm) deep and colored in gaudy patterns, often of orange, red or yellow. The smaller species tend to have duller plumage patterns and smaller bills. The sexes are alike in plumage in nearly all species, and both sexes of most of them have a patch of bare skin around the eye. All toucans are heavy-bodied birds with rounded wings, a rather long graduated tail and short strong legs, with stout feet of the zygodactyl pattern—that is, with two toes pointing forward and two back.

They are birds of the woodlands and forests of Central and South America, found from Veracruz southward to Paraguay and northern Argentina. All of them are primarily fruit-eaters, though some species have been recorded taking large insects, small snakes and the nestlings of other birds. The huge bill may serve one or more functions: it is probably used to reach fruit that is far out on thin twigs; it may help in intimidating other birds that might rob toucans' nests were it not for this defense; and it may be used in display.

Nests are in tree holes, either natural ones caused by decay or those excavated by other birds. Two to four white eggs are laid and incubated by both the male and female. Incubating toucans are very restless for birds of their size, seldom staying on the eggs for more than an hour at a time.

TRAPDOOR SPIDERS, members of several families of mygalomorph spiders that dig deep burrows, the entrances to which are closed with a hinged door carefully camouflaged with moss or debris. This may be thin like a wafer or thick like cork and formed of earth and silk. Some burrows have a branch burrow closed by a second trapdoor. The doors serve as protection from rain and dust, from excessive heat and dryness and from the entrance of enemies (though some pompilid wasps have mastered even this defense). The spider holds the door firmly closed during the day but darts out to attack passing insects, chiefly at night.

Trapdoor spiders are found in all tropical areas and the warmer temperate regions, including southern Europe. *Atypus affinis*, the only mygalomorph spider in Britain, has wrongly been called the trapdoor spider. A closed silk tube extends from its burrow, which has no lid.

Trapdoors have also been devised independently by some ateneomorph spiders, especially certain species of the family Lycosidae.

TREECREEPERS, a family of small, dull-colored birds which obtain their food by climbing along the trunks and branches of trees, usually in an upward direction.

There are six typical treecreepers of the genus *Certhia*. Three are confined to southeast Asia around the Himalayan region. The fourth and fifth together have a circumpolar range, the common treecreeper, *C. familiaris*, through the temperate forests of Eurasia and *C. americana* replacing it in similar regions of North America; the two apparently differing only in voice. The sixth is the short-toed treecreeper, *C. brachydactyla*, of southern Europe which overlaps the range of the common treecreeper, *C. familiaris*. They differ in both feet and voice and where they overlap the former tends to occupy deciduous forest, the latter conifers. The North American species occupies both habitats.

Typical treecreepers are small birds with large feet and fairly long, slender, decurved bills for probing in crevices for insects. The tail feathers have strong shafts and spiny tips and, pressed against the tree, act as a prop to support the bird. The upper plumage is brown and streaky, the underside white. The treecreeper moves head upwards, backing down if it wishes to return to a spot it has passed. When it has climbed one trunk it flies to the bottom of another and spirals up it.

Treecreepers roost in hollows in the bark of tree trunks, sometimes scraping their own hollows in the spongy bark of a Sequoia. A nest of moss, twigs and bark is built in the narrow cavity behind a loose piece of bark. The newly emerged young climb with agility when they can barely fly.

TREE DUCK, term applied to certain ducks which commonly perch in trees. Two groups of ducks are involved: the whistling ducks of the tribe Dendrocygnini and the perching ducks of the tribe Cairinini. An alternative name for the whistling ducks is in fact "tree ducks," but they perch in trees much less regularly than the Cairinini and some of them do so hardly at all.

The whistling ducks form a distinct group of gooselike ducks which are placed with the geese and swans in the subfamily Anserinae. They have long legs and walk well and in a more upright posture than most ducks. There are eight species, covering much of the world excepting Europe.

The perching duck tribe includes a wide variety of forms but that with which we are most concerned here is the North American wood duck or Carolina duck, *Aix sponsa*. This species breeds in eastern North America, frequenting secluded woodland pools and streams where it feeds along the banks and also in the woods, taking a variety of foods including some of the hardest fruits and seeds. The male is strikingly beautiful with bold bright plumage including an iridescent green crest. Even the female is more brightly colored than is usually the case in ducks.

TREEFROGS, strictly only those frogs which belong to the family Hylidae. The habit of living in trees from which they are named has in fact been adopted by many species belonging to other families of frogs, while some members of the Hylidae do not have this arboreal habit. The Hylidae is a large family, containing about 34 genera, but, with two exceptions, *Nyctimystes* and *Hyla*, they are only found in the New World. *Nyctimystes* occurs in New Guinea while *Hyla* is almost worldwide, but is absent from the Arctic and Antarctic and most of Africa.

There are several hundred species of *Hyla* but they are all rather similar in habits to the European treefrog. There is, however, considerable variation in the general shape and color. *H. versicolor*, the common or gray treefrog of North America, is a short squat frog abou 2 in. (5 cm) long, with a rough warty skin. Although, like most treefrogs, it is able to vary its color quite considerably, it is usually gray or brown with irregular dark patches on its back and difficult to see when sitting on the lichen-encrusted bark of a tree. The green treefrog, *H. cinerea*, also of North America, is also about 2 in. (5 cm) long, but is a very slim, long-bodied frog with a completely smooth skin and long legs. Its color is usually bright green but may occasionally be gray.

Although treefrogs live in trees, usually some distance from water, most of them have to return to water to breed and have the pattern of development characteristic of most frogs, with a free-swimming tadpole stage. Some species, however, have developed breeding methods in which the vulnerable eggs and tadpoles are protected to a certain extent from predators.

Although most genera of hylids are arboreal in their habits there are exceptions. The chorus frogs belong to the genus *Pseudacris* which is found only in North America. Most of them are small and delicate, about 1 in. (2.5 cm) long, patterned with brown or green stripes or spots. They do not climb very much and the adhesive discs on their fingers and toes are small. Their toes are only slightly webbed and they are poor swimmers. They only call during

The South American treefrog *Phyllomedusa pulcherrima* is nocturnal, as shown by its large eyes, and slow-moving, in contrast to the more typical treefrogs. Treefrogs generally are characterized by having suction discs at the tips of the toes and forwardly directed eyes.

the breeding season, which is in the spring in the north but during the winter rains in the south. After that they are rarely found due to their small size and the effective camouflage of their coloration among vegetation.

The cricket frog, *Acris gryllus*, is a representative of a genus which has progressed even further from the arboreal habits of most hylids. It is small, ½-1¼ in. (13-31 mm) long, with a pointed snout. Its toe discs are very small, and it resembles a true frog (*Rana*) more than a treefrog. It is very variable in color and may be gray, brown, reddish tan or green. It never climbs but lives always on the ground among the grass bordering streams and swamps. Its name refers to its chirping call which is heard in chorus during the spring. Unlike most frogs it is active during the day.

TREE HOPPERS, leaf-sucking insects related to spittlebugs, cicadas and leaf hoppers. The

adults and nymphs are, like aphids, often attended by ants, and they exude "honeydew." They lay their eggs in slits that they cut in twigs, and they guard their young when the eggs hatch. They are powerful jumpers.

TREE-SHREW, *Tupaia glis*, one of several small, squirrellike mammals with a head and body length of 9 in. (22 cm) and a bushy tail of 8 in. (20 cm). The Malay name "tupai" is used for both tree-shrews and squirrels, underlining their close resemblance.

The common tree-shrew is basically an arboreal animal, constructing nests with leaves and mosses in hollow trees, but it also spends much time searching for fruit and insects in the leaf-litter of the forest floor. All members of the tree-shrew family (Tupaiidae) are restricted to the rain forests of southeast Asia, where the various species occupy a number of different ecological niches.

All tree-shrews have a small number of offspring (one to three) in each litter, and birth takes place in a convenient hollow in a tree or an abandoned rodent burrow, which is lined with leaves a few days beforehand. Laboratory studies show that the common tree-shrew female gives birth in a different nest from the one used as a refuge. This is linked with the fact that she apparently visits her young only once every two days to suckle them. Unlike most other small nest-living mammals, the female does not clean her offspring, and they are left to keep themselves warm. Nevertheless, the young keep a constant body temperature of 99°F (37°C), undoubtedly aided by the fact that the maternal milk has

fishes. The common name derives from the triggerlike action of the enlarged first dorsal spine, which can be locked in the upright position by the much smaller second dorsal spine. With the spine locked, the fish is both difficult to remove from rock crevices and difficult for a predator to swallow. The body is deep and compressed, and is covered by small bony plates. The mouth is terminal and small. The triggerfishes are often brightly colored with grotesque or even absurd color markings that contrast with their slow and rather dignified movement around the reefs. There are about 30 species, the largest rarely exceeding 2 ft (60 cm) in length. They have powerful teeth with which they crush mollusks

vertically or at times tilted slightly forward below the perch. The squatting appearance is due in part to the small and comparatively weak legs and feet. The foot is peculiar in that the inner toe is turned backward to give two toes forward and two back. In other families it is the outer toe that is moved in this way. The large dark eyes are capable of good vision in poor light. The bill is short, very broad, with serrated edges to the curved culmen. Trogons have a thick soft plumage and very thin skin.

The coloring is usually bold. The breast, belly and undertail coverts of males are frequently bright red, pink, orange or yellow, and these areas are similar but sometimes paler or more subdued on females. On males the rest of the plumage tends to contrast with the underside; the back from head to tail, and the upper breast and head, being glossy, sometimes black, but more often glossed with purple, blue or green, the last often vivid and metallic and sometimes with gold or bronze tints. The wings tend to be black, marked with white, or so finely vermiculated as to appear gray. Females have a duller plumage with more gray, brown or buff tints. The only exceptional plumages are those of *Pharomachrus* species, the male quetzal, *P. mocino*, in particular. They have spiky crown feathers forming a rounded crest and very elongated feathers of the upper tail coverts, which are several times the bird's length, curving down to form a train of metallic green feathers.

TROUPIALS, colorful, small and medium-sized birds belonging to the genus *Icterus*, and the name "troupial" is also applied without qualification to *I. icterus*. Plumages of the true troupials range from the nearly all black epaulet oriole, *I. cayensis*, to the very striking yellow or orange and black of the majority of species.

Icterus species are essentially birds of the broad-leaved tropical forests of the New World, but some, such as the Baltimore oriole, *I. galbula*, breed as far north as southern Canada and migrate south in winter. In South America true troupials occur only as far south as northern Argentina.

Although true troupials are adapted to fruit-eating and nectar-drinking, the bulk of their food is made up of insects when these are plentiful, as in the northern summer. There are two groups of true troupials. The fruit-eaters feed by thrusting the closed bill into the soft fruit then gaping and lapping up the juices released by the pressure of the opening mandibles with the brush tongue. Nectar-drinkers have thinner bills and weaker jaw muscles by comparison, and their brush tongues are much better developed. They feed by probing into the open flower or by piercing its base.

The nests of the true troupials are deep hanging cups beautifully woven from grasses and fibers. The two sides are completed first, hanging from adjacent twigs at the very tip of a branch, and these are then joined to form a purselike structure narrower at the top than the bottom. The nest is well hidden by some species but daringly exposed by others. In the latter, intending nest robbers are deterred because they have to expose themselves either to the irate parents or to predators. Tropical

The triggerfish *Rhinecanthus aculeatus* is known in Hawaii as the humuhumu-nukunuku-a-puaa.

a high proportion (25%) of fat. It seems likely that this unusual maternal behavior is common to all tree-shrews.

All tree-shrews have a gestation period of about seven weeks, producing typical nest-living offspring (eyes and ears closed, teeth not yet erupted).

The relationships of tree-shrews to other mammals are not yet clear. They were at first classified with the Insectivora and later with the Primates, but recent studies indicate that they should be placed in a separate order, characterized by the retention of many primitive characters.

TRIGGERFISHES, marine fishes of warm seas related to the trunkfishes and puffer-

and crustaceans. Their flesh is reported to be poisonous, but it is not yet established that this is not simply a form of ciguatera poisoning. In Hawaii the species of *Rhinecanthus* are called humahuma and have been immortalized in song.

TROGONS, a family of brightly colored, arboreal birds of tropical forests. There are 34 species, all similar in appearance and varying from about the size of a thrush to that of a dove, about 9-13 in. (23-33 cm). They are found throughout the tropics. They are stout, heavy-bodied birds with fairly large heads and distinctive long, broad tails. Their characteristic posture is relatively upright on a perch, squatting close to it with tail hanging down troupials lay 2-3 eggs, the northern ones 5-6, the ground color being pale blue or whitish with darker spots or streaking and fine scribbling. Incubation is by the female alone, and both parents feed the young.

TROUT, members of the salmon family found in the fresh waters of the northern hemisphere. The brown trout is closely related to the Atlantic salmon, and the parr of the two species are very similar. The tail is less deeply forked in the brown trout juveniles, and there is always an orange or reddish tinge to the adipose fin. The natural range of the brown trout is from England to the Pyrenees and eastward across to the Urals, with a very small population in north Africa. Because of their sporting qualities, however, brown trout have been introduced into the coastal regions of North America, Chile, Argentina, southern India, highland areas of east Africa, South Africa, New Zealand and Australia. In warm countries trout have been limited to the upper reaches of rivers and have not interfered with the populations of local species farther down. In the Drakensberg Mountains of South Africa, however, trout introduced in the last century have thrived to such an extent that, a small carplike fish, *Oreodameon quathlambae,* known only from that area, is now extinct. The danger of introducing a species to a new area cannot be too highly stressed.

The brown trout is a solid, powerful fish, usually spotted but with great variability in the number, size and color of the spots. The mouth is large and toothed, and the tail is emarginate. These are swift, active fishes that favor highly oxygenated and cool waters. In Europe they can live at sea level, but in Africa the best conditions for their survival may well lie at 5,000-6,000 ft (1500-1800 m) above sea level. In small mountain streams only a few inches deep, brown trout will thrive, feed and breed but will never reach more than a few inches in length. The great lake trout, on the other hand, will tip the scales at 30 lb (14 kg), showing that trout, like many other fishes, are extremely sensitive to the amount of food and living space available.

The breeding habits and time of spawning of brown trout are similar to those of the Atlantic salmon. Sea trout and forms living in estuaries migrate up rivers to breed. Lake trout move into the feeder streams, and riverine forms merely run farther upstream. Like the salmon, the trout female constructs a shallow nest, the redd, into which the eggs are deposited and there fertilized. The young trout pass through stages similar to those of the salmon (alevin, parr, smolt), the comparison being closer in the case of sea trout. Throughout the whole of the northern hemisphere spawning occurs from September to April, although the actual time depends on water temperatures.

TRUMPETERS, a family of three species of chicken-sized birds of the forests of the tropical areas of the New World. They have rather long legs, long necks, short tails, rounded wings and short, somewhat curved bills. The plumage is mainly black with purplish reflections on the foreneck. The lower upperparts are either gray, white or brown, according to the species.

Typical forest birds, which live in flocks on the forest floor, they seldom fly but will roost in trees. Very little is known about their life in the wild, but they are often kept in captivity, becoming very tame. They nest in large holes in trees and lay roundish gray-white eggs. The incubation period is still unknown, but the nestlings are believed to leave the hole immediately after hatching. They feed on insects and fruits.

TSETSE FLY, one of the greatest scourges of tropical Africa, the only important agent capable of transmitting human sleeping sickness from person to person. Tsetse flies are true, two-winged flies and in general appearance and in structure are very similar to house flies, except that the mouthparts are highly adapted for piercing the skin of man and other mammals and sucking blood. The mouthparts form a prominent, forward pointing proboscis consisting of a lower lip or labium with a tip for rasping and piercing, which bears in its grooved surface the needle-like labrum-epipharynx and hypopharynx. The first conveys blood from the host animal into the insect's gut, and the hypopharynx carries the saliva that, as in other blood-sucking insects, is an anticoagulant, which prevents clogging of the delicate mouthparts. When the tsetse is at rest, the wings are folded one on top of the other over the back, whereas most other similar flies rest with the wings spread laterally. Tsetse fly species range in size from about the house fly up to larger than blowflies. They occur in Africa south of the Sahara and north of South Africa, within the Tropics of Cancer and Capricorn, except that a single species has a foothold in southwest Arabia.

Tsetse-fly larvae are born fully grown and ready to change into pupae. The female ripens only a single egg at a time, and after fertilization this comes to rest in the uterus. When the larva hatches it is nourished by a secretion from special nutritive glands, which it imbibes directly from a papilla or nipple. The female deposits her fully developed larva (a maggot) in the shade of a tree, shrub or log, depending on the species and the climate of the area. Immediately after this, the whitish larva burrows into dry, loose soil or under ground litter and changes into a dark brown pupa with two prominent knobs on the posterior end. The adult flies may live for upward of 70 days, but the average length of life is only about four weeks. Although a female tsetse fly may deposit a maximum of only twelve larvae, frequently less, in the course of her life, the task of eliminating this most harmful species has, up to the present time, proved impossible in spite of massive research over half a century.

African sleeping sickness is caused by microorganisms known as trypanosomes, which live as parasites in man and many big game of Africa, as well as in the tsetse fly *Glossina.* The presence of trypanosomes in the bodies of man and other vertebrates is known as trypanosomiasis. Two species of trypanosome, *Trypanosoma gambiense* and *T. rhodesiense,* have been named as the cause of human African sleeping sickness, and both parasitize wild game and domestic animals, and domestic pigs and antelopes, especially wildebeest, are important reservoirs of trypanosomes. Other species, such as *T. vivax* and *T. congolense,* are highly pathogenic to domestic animals, especially cattle, and are also transmitted by tsetse flies. It is important to note that where trypanosome and mammalian host have long been associated, disease symptoms are either completely lacking or are very slight, whereas in new associations the microorganism is usually highly virulent toward its host. Thus, the native game animals of Africa harbor trypanosomes without any trace of nagana, the animal form of tsetse-borne trypanosomiasis, and in areas where human infections are of long standing, the disease in indigenous populations tends to be relatively mild. On the other hand, newly exposed domestic animals and Europeans have little resistance to trypanosomiasis, and effects are usually fatal if untreated.

TUATARA, *Sphenodon punctatus,* belonging to the otherwise extinct reptilian order Rhynchocephalia. The rhynchocephalians are characterized by the presence of a beaklike upper jaw, and first appeared in the Lower Triassic some 200 million years ago. The group was virtually extinct by the Lower Cretaceous about 100 million years later. The members of the group formerly had a widespread distribution, and fossils have been found from all the continents except North America. The order is now represented only by a single genus with one species, the tuatara, which is restricted to approximately 20 islands off the coast of New Zealand. The ancestry of the tuatara may be traced back to the fossil reptile *Homoeosaurus* from the Upper Jurassic, and the similarities between the two forms are striking. It seems, therefore, that the structure of the tuatara has remained virtually unchanged for some 130 million years. The existence of the tuatara is just as astonishing as the discovery of a large dinosaur would be—perhaps even more so, since the tuatara is the sole survivor of a group that reached its peak about 180 million years ago, whereas the dinosaurs were at their peak about 140 million years ago in the Jurassic and Cretaceous.

The tuatara is lizardlike in general appearance but is distinguished from the lizards by several skeletal features involving the skull and the ribs. The generic name *Sphenodon* means "wedge tooth," and this refers to the chisellike teeth on the upper and lower jaws, which are fused to the jawbone, not set into sockets. The Maori word "tuatara" means "peaks on the back," and this describes the triangular folds of skin that form a conspicuous crest down the back and tail of the male. The female has only a rudimentary crest. Tuataras vary in color from black-brown to dull green, while some may have a reddish tinge. The upper part of the body is covered with small scales that may have small yellow spots. The feet have five toes each with sharp claws and are partially webbed. A vestigial parietal "eye" is found on the top of the head in very young animals, but soon becomes covered over, and is invisible in adults. The presence of this "eye" is considered to be a very primitive feature since it also occurs in the ancestors of the rhynchocephalians. It is usually further reduced or even absent in modern lizards. The parietal eye retains some traces of a lens and a retina, but there is doubt about its function. It may be that it acts as a register of solar radiation and controls the amount of time the creature spends in the sunlight. This is important, since the tuatara, like all reptiles, has a body temperature that is affected by the temperature of its surroundings, and is said to be ectothermic. The body temperature may therefore be controlled to some extent by basking in the sun or seeking the shade. Nevertheless, the tuatara spends most of the daytime in its burrow, leaving it only occasionally to sunbathe, mainly in late winter and spring. It is therefore largely nocturnal and is active at temperatures that are much lower than those favored by lizards. Available reports indicate that the tuatara may be active at temperatures as low as 45°F (7°C) and, further, that even in winter it hibernates only lightly. Allied to its low body temperature is the fact that it has a very low metabolic rate. This means that it requires very little energy to keep the vital body processes, such as excretion and digestion, ticking over. The tuatara is reputed to grow very slowly and probably does not breed until it is 20 years old. Growth may continue until the age of 50 or beyond, and the animals are said to be extremely long-lived. Estimates of the life span vary from about 100 to 300 years.

The mating habits of the tuatara are also remarkable in that the male has no copulatory organ, and mating is accomplished by the apposition of the male and female urinogenital openings (known as cloacal apposition). Pairing usually takes place in January, but the sperm is stored within the female until October-December. She then scoops out a shallow nest in the ground and lays 5-15 eggs with soft white shells. These remain in the nest for a further 13-15 months before hatching, the longest incubation period known for any reptile.

At the present time the tuatara is found only on some small islands off the east coast of the North Island of New Zealand and in Cook Strait between the North and South Islands. These small islands also have large colonies of birds such as petrels and shearwaters, which nest in burrows.

TUCO-TUCO, a nocturnal burrowing rodent whose common name is derived from its bell-like call note. The 50 species in the genus *Ctenomys* range from tropical to subarctic regions of South America. It is stockily built, 9-13 in. (23-33 cm) in total length, its fur varying in color from dark brown to creamy buff.

Tuco-tucos live in colonies and excavate extensive burrows. The litter of 1-5 young are born at the end of one of the long tunnels. The breeding season is usually in the wet season, when food is abundant. Food consists entirely of plant material such as grass, tubers, roots and stems.

TUFTED-DEER, *Elaphodus cephalophus*, derives its name from the crest of long and dense bristly hair crowning the summit of the head around the antlers, sometimes completely concealing them. Tufted-deer also have small tusks similar to muntjac. Three subspecies are recognized, extending from Burma to southern China, including eastern Tibet.

Generally brown in color, this small deer, which stands 22-25 in. (55-63 cm) high at the shoulder, according to subspecies, inhabits mountainous country, where it may be found at up to 15,000 ft or more (about 4,570 m). When feeding, tufted-deer carry their tails high, and when bounding off their tails flop with every bound, displaying the white underside in much the same manner as the white-tailed deer.

TUNA FISHES, or tunnies, large oceanic members of the mackerel family. Almost every feature of the tunas seems to be adapted for their life of eternal swimming. The body is powerful and torpedo-shaped, the dorsal and pectoral fins fold into grooves and the eyes are flush with the surface of the head, all of which help to reduce the drag caused by turbulent eddies as the fish cleaves the water. In addition, the scales are small and smooth and are reduced merely to a corselet around the pectoral fins. The tail is crescentic, the ideal shape for sustained fast swimming, and the fin rays are closely bound to the end of the vertebral column so that the tail is no longer a flexible appendage of the body but an integral part of it. Behind the second dorsal and the anal fins are small finlets that may serve to control the formation of eddies, while the sides of the caudal peduncle, or base of the tail, bear little keels for further streamlining. A swim bladder is absent, but a considerable amount of oil is present, which aids in buoyancy. One remarkable feature of these large fishes is that the energy expended in fast swimming warms the blood to a few degrees above that of the surrounding water—an unusual feature for a "cold-blooded" vertebrate. The tunas are carnivorous fishes that feed on pelagic organisms and especially on squid. They are mostly tropical in distribution, but some of the larger species are found in the colder northern waters. There are six species of great tuna in the world: the albacore, *Thunnus alalunga*; the yellowfin, *T. albacares*; the blackfin, *T. atlanticus*; the bigeye, *T. obesus*; the bluefin, *T. thynnus*; and the longtail, *T. tonggol*. Because of the size of these fishes and the difficulty of preserving specimens for comparative studies in museums, it is only recently that the confusion of names, both scientific and popular, has to some extent been resolved. Some species are worldwide in their distribution and have received a variety of common names, some of which are used for quite different species elsewhere. The names used here are those that are now generally agreed by fishery workers.

TURACOS, a family of 20 species of active and often highly-colored arboreal birds, sometimes known as plantain-eaters or louries. They occur in Africa south of the Sahara in all habitats containing trees, from dry thornbush and gallery forest to dense evergreen forests. They are related to the cuckoos, and are fairly large, 14-18 in. (35-45 cm) long, with fairly slender bodies, rather long tails and short rounded wings. The legs are rather long, with a reversible outer toe, and the birds run and leap among the branches of the trees with great agility, in contrast to the poor powers of flight. The head is often crested, the crest varying from hairy tufts or rounded domes to slender tapering structures rising to a point at the front, or to a coronet of feathers as in the great blue turaco, *Corythaeola cristata*. The bill is stout, somewhat broad, and slightly decurved and may be very deep, with a high arched culmen, and boldly colored in yellow and red in some species, black in others. In some it extends back to form a forehead frontal shield. The nostrils may vary in shape and position from one species to another. In most species the feathers of head and breast are hairy in texture, as are those of the crest, and tend to be less glossy than wings and tail. The plumage of *Tauraco* and *Musophaga* species is usually green, violet or dark blue, and on such birds the inner webs of the flight feathers, and sometimes the crest, are vivid red, showing as a scarlet flash when the bird flies. The green and red pigments of these plumages are peculiar to the turacos. The great blue turaco is entirely blue and green, and a group of species of the genera *Crinifer* and *Corythaixoides* are only a sober gray ornamented with touches of black or white. In general, the duller colored species occur in the more open, drier country. The head and crest are often ornamented with patches of white, red or pink, and there may be small red wattles around the large dark eyes.

The turacos are fruit-eaters, feeding the young on regurgitated fruit, but they will also eat buds and seeds and some insects and invertebrates.

TURBOT, *Scophthalmus maximus*, a large flatfish from the North Atlantic. The turbot is one of the flatfish species that rests on its right side, the left side being pigmented and bearing the eyes. It is a shallow water species found in the Mediterranean and all along European coasts. The body is diamond-shaped, with large symmetrical jaws lined with sharp teeth. There are no scales on the body but the eyed side is covered by warty tubercles. The turbot is found mostly on sandy bottoms in water of 10-200 ft (3-60 m). Turbot are avid fish-eaters and tend to concentrate in areas where food is plentiful, such as shallow banks near the mouths of rivers. They lie in wait for their prey and, like most flatfishes, are well camouflaged, the upper side being brown or gray with lighter and darker specklings.

The turbot is one of the best flavored of British fishes. It grows to over 40 lb (18 kg) in weight and is extremely prolific, a female of 17 lb (7.5 kg) having been recorded with 9 million eggs in the ovaries.

TURKEYS, two species of large game birds. The common turkey, *Meleagris gallopavo*, of North America occurs in open woodland and scrub. This is the domestic turkey. Its brown, barred plumage is glossed with bronze. The head and neck are naked and colored red and blue, with wattles and a fleshy caruncle overhanging the bill. A tuft of bristles hangs from the breast. In display the wattles swell and the color intensifies, the tail is erected and fanned, the body feather ruffled and the wings drooped until the tips sweep the ground. The male's call is the noisy "gobbling," whereas

A turaco *Tauraco livingstonii* shows the pigment that has been the cause of erroneous ideas. Unique in the animal kingdom this red pigment has been said repeatedly to be washed out by rain, which is untrue.

the female has a softer sharp note. The wild turkey is a strong flier. It feeds on a wide variety of seeds, nuts and small creatures. It is polygamous, females laying in shallow scrapes in well-concealed situations.

The ocellated turkey, *Agriocharis ocellata*, occurs in Yucatan and Guatemala. It lives in jungles and is brighter in color than the common turkey. The grayish body feathers are tipped with large iridescent ocelli of green and bronze, while the head and neck are bright blue with scarlet and white caruncles. The display and behavior are similar to that of the common turkey, but the voice is different and it flies more frequently.

TURNSTONES, wading birds usually found on rocky seashores. They are members of the subfamily Arenariinae and are normally placed in the family Scolopacidae, but sometimes in the Charadriidae. They are dumpy, short-billed and fairly short-legged birds, less than 12 in. (30 cm) long, and the sexes are alike.

During most of the year, they are found chiefly on the coast, though the surfbird, *Aphriza virgata*, nests inland (in the Alaskan mountains). The other two species in the subfamily belong to the genus *Arenaria*: the (ruddy) turnstone, *A. interpres*, has a circumpolar, mainly coastal, breeding distribution, chiefly to the north of the Arctic Circle, while the black turnstone, *A. melanocephala*, replaces it in Alaska, on the coast of the Bering Sea. Although the turnstone is among the world's most northerly breeding birds (to at least 83°N), it also occurs in glacial or postglacial relict populations on the Scandinavian islands and shores of the Baltic. The breeding range of the turnstone has decreased in Europe this century, possibly associated with an improved climate and higher mean summer temperatures. It no longer nests on the Baltic coast of Germany, where it was common in the 19th century. Its preferred breeding habitats are low-lying stony coasts and islands adjacent to tundra, occasionally inland on moss or lichen tundra and, in the Baltic area, even on the grassy edges of conifer-clad islands.

TURTLES, MARINE, tortoises adapted to life in the sea. Like the land tortoises their body is encased in a shell, which consists of a dorsal part, the carapace, and a ventral part, the plastron, joined together on either side by the bridge. From land tortoises and terrapins, they differ in the shape of the limbs, which have developed into flat flippers. These make turtles good swimmers, but make their movement on land very cumbersome. The head and neck cannot be completely withdrawn within the shell.

Seven species of turtle are known. The largest is the leathery turtle or leatherback, *Dermochelys coriacea*, the carapace of which may reach a length of 6 ft (1.80 m). The bony carapace, which is in no way joined to the vertebrae and ribs, consists of a mosaic of bony platelets covered with a leathery skin. The species is easily recognized by the presence of seven ridges, often notched, running lengthwise over the back. The plastron consists of four pairs of bony rods, arranged to form an oval ring; more superficially, six rows of keeled platelets are present. The leathery turtle is blackish above with numerous scattered,

The hawksbill turtle yields the valuable tortoiseshell of commerce.

small, irregular whitish or pinkish spots; below, it is white with black markings. It is found in all tropical and subtropical seas, and from there it wanders far to the north and south into temperate regions. It is a fairly regular visitor to British and French waters; in Norway it has been found up to 70°N. The food of the leathery turtle consists mainly of jellyfish and salps. Although it breeds all through the tropics, only a few nesting beaches are known, where large numbers of females come ashore to deposit their eggs, for example, on the east coast of Malaya and in French Guiana.

In the rest of the sea turtles the bony shell is of a more solid construction. The carapace consists of bony plates, which are firmly joined to the vertebrae and the ribs, and of a series of smaller bones around the margin; the plastron consists of nine flat bones, which leave some openings between them. Both the carapace and the plastron are covered with horny scutes. On the carapace, the horny scutes are arranged in three longitudinal rows, one row consisting of the nuchal scute (or precentral) and a number of vertebral scutes (centrals) along the middle of the back, with a series of costal scutes (or laterals) on either side, and with a series of small marginal scutes along the border of the shell. The number of scutes in the various series are used to identify the species.

The genus *Chelonia* contains two species: the green turtle, *Chelonia mydas*, occurring in all tropical and subtropical seas, and the flat-backed turtle, *Chelonia depressa*, which is found only along the north and east coasts of Australia. Both species have four pairs of costals, and a single pair of prefrontal shields on the snout. The green turtle is the larger of the two; the carapace may reach a length of 55 in. (1.4 m). It is this species that is in demand for preparing turtle soup; for this one uses not only the meat, but also the gelatinous cartilage ("calipee"), which fills the openings between the bones of the plastron. The *Chelonia* species are mainly vegetarian, with a preference for sea grass. However, the hatchlings are carnivorous, and adult green turtles, in captivity, can be fed

with fish. Because of the unlimited harvesting of eggs, and to a much lesser extent, the killing of adult turtles, the populations have declined. The hawksbill turtle, *Eretmochelys imbricata*, like the green turtle, has four pairs of costal scutes, but the scutes of the carapace overlap, like the tiles on a roof. The carapace may reach a length of 36 in. (0.92 m). Moreover, this species has two pairs of prefrontal shields on the snout. The hawksbill is the species that yields tortoiseshell. In some areas, as in the Caribbean, it is also much appreciated as food, but in other areas, for example, New Guinea, the meat is known to be highly poisonous. The hawksbill is carnivorous, feeding on various kinds of small marine animals. It is believed never to move far from the nesting beaches. It is found in all tropical and in some subtropical seas, for example, the Mediterranean. Those from the Indian and Pacific oceans are more darkly colored than those from the Atlantic. The loggerhead turtle, *Caretta caretta*, has five pairs of costal scutes: the snout is covered with two pairs of prefrontals, which often have one or more scales wedged in between them. Its general color is reddish brown above, yellowish below. The loggerhead occurs in all oceans, also in the tropics, but it is more common in the subtropics, where it breeds. It is often found far from land in midocean, and on its wanderings it comes to temperate seas; it is a fairly regular visitor to the Atlantic coasts of Europe, and it even has been found at Murmansk in northern Russia. The logger-head feeds on a variety of marine invertebrates, such as shellfish, squids, goose barnacles and jellyfish. The carapace may reach a length of 40 in. (1,017 m).

The Olive Ridley, *Lepidochelys olivacea*, and Kemp's Ridley, *Lepidochelys kempi*, are characterized by minute openings (pores) on the hind borders of the scutes that cover the bridge. Both species have the snout covered by two pairs of prefrontal shields. Kemp's Ridley has five pairs of costal scutes. It is a small turtle, the carapace reaching a length of only 27½ in. (0.69 m). Its only known nesting beaches are on the Gulf coast of northern Mexico. There, the females may arrive in large

"arribadas," hundreds and sometimes thousands coming ashore at the same time.

TYRANT FLYCATCHERS, small birds of a very large and varied family restricted to the New World where they replace the Old World flycatchers or Muscicapinae.

Generally they are 3-9 in. (8-23 cm) long with the exception of a few species with long tails, but are so diverse in plumage, shape of bill and other characters that it is impossible to give a general description. Few species are really brightly colored, most being in shades of green, brown, yellow, gray, white and almost black. Red is unusual, though the vermilion flycatcher, *Pyrocephalus rubinus*, is one of the best-known and most widely distributed species, and blue is almost unknown. Perhaps the most striking and common character is a crown patch of yellow, orange, red or white, which is usually opened in display, and partly or entirely concealed at other times.

The family ranges from Canada to Tierra del Fuego and reaches the Galápagos Islands, but is best represented in the tropics. In higher latitudes most species migrate and almost all North American species such as kingbirds, phoebes and pewees winter in the Caribbean and northern South America. Similarly many Patagonian species move north during the southern winter.

Breeding habits are as diverse as other aspects. Most species make open cup-shaped nests, but these range from small felted cups (*Elaenia* spp.) to loose untidy twig nests (*Tyrannus* spp.). Some make domed nests with a side entrance (*Tolmomyias*, *Camptostoma*) and others nest in holes. Nests are usually placed in trees, but many open country species nest on the ground and some in reeds over water. Species of *Tolmomyias* and *Camptostoma*, and perhaps of other genera, often nest close to aggressive wasps. The eggs are generally white, cream or buff, usually spotted and blotched with dark colors and the clutch size varies from two to four according to the latitude. Some species are known to lay at an interval of 48 hours (*Camptostoma* spp.). The incubation period varies between 14 and 23 days and the nestling period between 13 and 24 days.

VANGAS, a family of specialized shrikelike songbirds peculiar to Madagascar. The variation in bill shape is particularly pronounced, but in spite of this there appears to be little variation in the feeding habits. Vangas are arboreal, feeding mainly on insects and small invertebrates that they take from the branches, twigs and leaves of the trees, but the larger-billed helmet bird, *Aerocharis prevostii*, and the hook-billed vanga, *Vanga curvirostris*, also take small reptiles and amphibians.

The tiny red-tailed vanga, *Calicalicus madagascariensis*, is like a small grayish, rufous-tailed flycatcher with a stubby bill. The vangas of the genus *Leptopterus* are finch- to starling-sized, with a thrush-type bill with a small hook at the tip. Chabert's vanga, *L. chabert*, is white below and black, glossed with dark blue, above. The blue vanga, *L. madagascarinus*, is a vivid overall sky-blue above and white below tinted with buff on the female, and the white-headed vanga, *L. viridis*, is white with a green-glossed black back, wings, and tail. Bernier's vanga, *Oriolia bernieri*, is similar in size to the last but with a light brown plumage with fine black barring. The sicklebill, *Falculea palliata*, has a plumage pattern like the white-headed vanga, with white head and body and blue-glossed black on back, wings and tail, but it is half as large again and has a long slender down-curved bill about 2½ in. (6 cm) long.

The rufous vanga, *Shetba rufa*, and *Xenopirostris* species have a bill of proportionally similar length to *Leptopterus* species, but it is very deep along its whole length and laterally flattened, with a slight terminal hook. The former is white below and chestnut-red above, and Lafresnaye's vanga, *X. xenopirostris*, is white and gray, both being black on the head. The hook-billed vanga has a black, white and gray patterned plumage, and the bill has a sharp, vicious-looking hook. The larger-billed helmet bird has a bill as big as its head, thick and with an inflated and high, curved upper mandible that rises higher than the crown of the head. The adult is black with bright chestnut on back and midtail; the young bird is dull brown.

VICUÑA, *Lama vicugna*, a member of the camel family living in the western High Andes, is regarded by some people as the original form of the domesticated alpaca. It is the smallest, most graceful and most agile of the four *Lama* types, having a shoulder height of 2½ ft (75 cm). Vicuñas have a uniform, short-haired coat, reddish in color, and a white blaze on the chest. The lower incisors, in contrast to all the others

of the camel family, have an open root—as in rodents—and thus grow continuously. In captivity, for lack of wear, they often project considerably from the muzzle.

In the last few centuries the range of the vicuña has contracted and at the present time is between latitudes 10° and 30° south. The preferred habitats are the tablelands of the central Andes, those broad, high plateaus between the chains of the Andes extending from southern Ecuador to Argentina and loosely covered with grasses, mosses, resinous bushes and cushion plants hard as stone. The vicuñas are the highest climbers of all the New World camelids, up to 16,600 ft (5,500 m) and prefer the regions just below the snow line, in the "bofedales" fed by rain and melting snow. The hemoglobin of their blood has a particularly high affinity to oxygen, enabling them to live at high altitudes.

The mare is sexually mature at the age of one year. During the rutting season from April to June there are fierce battles between the "kings" of the herd. In addition to the fighting habits typical of all the New World Camelidae, the vicuñas have the peculiarity not only of kicking backward but also sideways at the opponent. In doing this they take up an attitude crossways on to the foe, drive him broadside on into a corner and then attack him at lightning speed. This close-in position also occurs in mating. In captivity vicuña stallions are always belligerent. They not only frequently attack the females and their young, but even inanimate objects, and stallions that are kept in isolation even attack themselves! In the domestic types the preliminary mating ceremony (driving the mare and snapping at and around the legs), just as in the case of the wild guanacos, is more violent and longer than the actual mating.

The mares foal after 10 months. At the first signs of sexual activity the young stallions are driven off by the mares and join large leaderless herds. In the rutting season they each try to get a herd of their own, the size of which fluctuates all the time and which has no permanent territory.

Both in the case of vicuñas and alpacas, the dung always lies in an extremely small space, since several animals do not defecate side by side, as in the case of guanacos and llamas, but only one after the other. Vicuñas have a remarkable ritual for this: after thoroughly sniffing the dung, they stamp and scrape in it, turn round and evacuate exactly over the heap. A similar behavior may be observed in the oribi antelopes, for example. It has been as-

sumed to be marking behavior, since oribis in this way impregnate the dung with the secretion of the glands between the toes. Vicuñas also have glands of this kind, and the ritual of evacuation might therefore have the same function.

Garments of vicuña wool, woven by priestesses of the Temple of the Sun, were reserved for the use of the Inca kings. The vicuña gives the finest wool, with a diameter of 5/10,000th of an inch. It is half the thickness of the best Australian merino wool. One animal yields little more than 18 oz (500 gm), and 10-12 animals are needed for one yard, so that vicuña wool coats are among the rarest and most expensive in the world. The vicuña, however, were not only hunted for their fine fleece but also for the bezoar stones found in the stomach, which were sought after as remedies, particularly as antidotes to poisoning. The hunting and export of vicuñas has been forbidden by law for several decades.

VINEGAROON, *Mastigoproctus giganteus*, a whip scorpion living in the southern part of the United States and Mexico. It is large for a whip scorpion, 3 in. (7.5 cm) long. When attacked, the vinegaroon flexes its tail forward over the body and sprays its assailant with a fine cloud of acetic acid, which has the characteristic odor of vinegar. Vinegaroons are often encountered in human habitations and although reputed to be poisonous, at most they can inflict a sharp, but not serious, bite with their mouthparts and a mild irritation from the acid spray.

VIPER FISHES, slender-bodied deep-sea fishes. Those of the family Stomiatidae are active pelagic fishes found in greatest numbers at about 6,000 ft (1,800 m); at nighttime they rise to the surface waters. They have large heads and a mouth well equipped with vicious dagger-like teeth. To swallow large prey, the mouth can open to an extraordinary extent. To achieve this, the head is jerked upward, a most unusual movement in fishes since the skull is normally very firmly attached to the anterior vertebrae. The skeleton is considerably modified to accommodate this unusual movement. In *Eustomias brevibarbatus*, for example, the first few vertebrae are widely spaced and are linked by a loop of cartilage that allows for the movement of the head.

The body in viper fishes tapers evenly from the head, the dorsal and anal fins being just in front of the tail. The pectoral and pelvic fins are reduced to a few long rays. Along the lower half of the flanks are rows of luminous organs giving the appearance of portholes. There is usually a luminous gland under the eyes that is larger in the males than in the females and can be shut off by rotating it. Most of the stomiatoids have a barbel under the chin. In species such as *Melanostomias spilorhynchus* the barbel is fairly short, but in *Ultimostomias mirabilis* the barbel is no less than ten times the length of the body. In some species the barbel ends in a slight swelling, and in others it has been described picturesquely as resembling a bunch of grapes, strings of beads or branches of a tree. The swellings usually bear luminous organs, which presumably serve to lure the prey nearer the mouth. Sir Alistair Hardy has noted that the tip of the barbel in *Stomias boa* resembles very strikingly a small red copepod. Possibly the barbels serve also as means of species recognition, while the more elaborately branched barbels may have a sensory function. From the bathysphere, William Beebe noted that at the slightest disturbance of the water near the barbel the fish immediately threshed around, snapping with its jaws.

VIPERS, a family of snakes with a highly developed venom apparatus, comparable with that of the nearly related pitvipers. The true vipers are found in Africa, where they are most numerous in species, Europe and Asia. The best-known species is the European viper or adder, *Vipera berus*. The smallest is Orsini's viper, *V. ursini*, of southern Europe, less than 1 ft (30 cm) long. The largest is the Gaboon viper, *Bitis gabonica*, up to 6 ft (2 m) long and 6 in. (15 cm) diameter.

Most vipers are short and stoutly built, with a short tail, and are typically terrestrial. Some, like the mole vipers, burrow in the ground or, like the horned vipers, burrow in sand. Few climb, although some of the puff adders climb into bushes and the tree vipers are arboreal and have a prehensile tail. Characteristically, vipers do not pursue their prey but lie in wait for it.

They strike, wait for a while then track down the victim that has crawled away to die. Their prey is mainly lizards and small mammals. The dead prey is tracked by flickering movements of the forked tongue, which picks up molecules of scent in the air, testing these by withdrawing the tongue into the mouth and placing the tips of the fork into the taste-smell organ in the roof of the mouth, known as Jacobson's organ. The fangs of a viper are hollow, efficient and large, and are typically folded back when not in use. They are automatically erected as the mouth is thrown wide open for the strike. The poison flowing through the hollow fang comes from a venom gland, a modified salivary gland, at its base. This can be very large and vipers typically have broad heads to accommodate them. In some species the venom glands extend behind the head, to as much as one-fourth the length of the body in night adders.

Species additional to the European viper which rarely exceeds 2 ft (60 cm) in length, with a record of 32 in. (80 cm), are the asp viper, *V. aspis*, Lataste's viper, *V. latasti*, and the sand viper, *V. ammodytes*, all of southern Europe, and all slightly larger than the adder. They have the dark zigzag line down the back as in the adder. In *V. palestinae*, the Palestinian viper, the zigzag has become a continuous line of dark diamond-shaped markings.

VIREOS, 20 or so species of small nondescript perching birds, 4-7 in. (10-18 cm) long, mostly gray or brown above and white or yellow below. They have rather heavy, slightly hooked bills. Most vireos are forest-dwelling birds preferring the shrub level, although a number are arboreal, haunting the crowns of broad-leaved trees. The gray vireo, *Vireo vicinior*, lives in scrub where there are no trees, and a few others inhabit thickets with little well-grown timber.

Vireos are widespread in Central and North

America, occurring as far north as subarctic Canada. The northern populations are migratory, wintering in Central and South America south to central Argentina. Only a few vireos breed in South America, and these are all widely distributed in North America.

Most vireos feed in a leisurely fashion, moving slowly through the tree or shrub and picking insects in a deliberate manner from the undersides of leaves. They often hang from a branch or stretch out to search the more inaccessible places. The black-capped vireo, *V. atricapillus*, is exceptional in this respect, moving restlessly through the branches like a warbler. It is also the only vireo with an immediately recognizable plumage that its name suggests.

During the breeding season vireos are territorial, and the males are persistent but unmusical songsters. The nest is a fairly deep cup hanging from the fork of two twigs. It is normally well hidden and constructed from strips of bark, grasses and fine leaves held together with cobwebs and lined with fine grasses. The two to four eggs are white, ranging from unmarked to heavily spotted, depending on the species. Both parents attend the nest for incubation, brooding and feeding.

VISCACHAS, South American rodents closely related to the chinchilla and more distantly to the guinea pig. The two groups of viscachas, the plains viscacha, *Lagostomus maximus*, and the mountain viscachas, *Lagidium*, are superficially very different and occupy quite different habitats, as their names indicate, but they resemble each other in many details of structure and way of life.

The single species of plains viscacha is perhaps the most familiar rodent in Argentina. It is a large, heavily built animal about 2 ft (60 cm) long, with a short tail about a quarter of the body length. The most striking feature is the disproportionately large head, accentuated by bold horizontal black and white lines on each cheek. The rest of the pelage is gray above and white below, and the texture is coarse.

Plains viscachas are highly colonial, living in large and conspicuous colonies known as *viscacheras* on grassy plains or open scrubland. About 20 individuals may inhabit an extensive network of burrows with a number of entrances, and the whole warren is made more conspicuous by the viscacha's habit of collecting any hard object it can find and depositing it outside the burrow. These accumulations may include all kinds of man-made trash as well as natural objects such as stones, twigs and bones.

Mountain viscachas inhabit almost the entire chain of the Andes from Peru to Patagonia and are represented by four species. They look remarkably like rabbits, except for the ears, which are rather shorter, and the tail, which is very long and rather coarsely haired in contrast to the soft pelage of the rest of the body. They live only on dry rocky mountainsides at altitudes up to 15,000 ft (5,000 m), making their nests in crevices in the rocks. They are active and agile animals, in contrast to their rather lumbering relatives of the plains, but in many respects their way of life is similar: they are colonial, diurnal and vegetarian, feeding on the toughest of montane plants.

VOLES, generally distinguished from mice by their shorter tails, less pointed snouts and long,

The red-eyed vireo *Vireo olivaceus*, one of the commonest birds of the deciduous forests of the eastern United States.

dense coats, but one of the most important differences is not so easily visible and concerns the molar teeth. These number three in each row, as in the majority of mice, but in most species of voles molars, like the incisors of all rodents, grow throughout life without developing roots.

Voles are therefore well equipped to deal with tough herbage, and they are one of the dominant groups of herbivorous animals throughout the northern hemisphere. There are about a hundred species, and one or more species can be found, often in great abundance, in almost any piece of grassland, marsh, heath or woodland with dense ground vegetation, from the northernmost tundra south to north Africa, the Himalayas and the mountains of Mexico. Voles are prolific breeders. They are capable of breeding at the age of about two months, and since the females can become pregnant while still suckling and the gestation period is only about three weeks, several litters, often of about five or six young, can be produced during one summer.

VULTURES, comprising two groups of large diurnal birds of prey (Falconiformes), not closely related to each other. The New World vultures, found only in the Americas, are a primitive family, the Cathartidae. The Old World vultures of the warmer parts of Europe, Asia and Africa are one branch of the large family Accipitridae, their closest true relatives being some species of eagle.

All vultures share several features, presumably through convergent evolution. Their heads and necks are partly or wholly naked, probably because they feed on and inside dead animals, those that feed mainly on large carcasses having longer, more naked necks. Since, with few exceptions, they do not kill their own prey, they lack powerful grasping feet. The claws are short and blunt, more adapted to walking than to killing. Several have specialized tongues, to feed rapidly on soft flesh or perhaps extract bone marrow. All are large birds, adapted for soaring. New World vultures include condors, the largest flying birds, but some of the Old World species are nearly as large, weighing more than 15 lb (6½ kg) and spanning 8 ft (2½ m). The smallest, but not the weakest, is the Egyptian vulture, *Neophron percnopterus*.

New World vultures differ from the Old World species in having "pervious" nostrils, opening through the bill at the point. The hind toe is rudimentary, placed above the three front toes on the leg. They have large olfactory chambers and, apparently, some sense of smell, though this is not used to locate prey. In America leaks in gas pipelines are sometimes located by circling vultures, attracted by a small amount of vile-smelling chemical added to the gas.

The seven species of the New World vultures include the two huge condors, *Vultur* and *Gymnogyps*, the extraordinary king vulture, *Sarcorhamphus papa*, brown and white in adult plumage, with brilliant orange, red, blue and purple bare skin on head and neck, the turkey vulture, *Cathartes aura*, widespread in North America, two other *Cathartes* species and the rather small, but locally abundant black vulture, *Coragyps atratus*. All eat flesh, but the black vulture is also said to eat decaying or overripe vegetable matter.

Old World vultures include 13 species in eight genera, all but one being monotypic. Two species, the Egyptian vulture and the lämmergeier, *Gypaetus barbatus*, have more heavily feathered heads and more powerful feet than is typical. Both are remarkable, specialized birds, the Egyptian vulture being a tool-user, and the lämmergeier, the avian equivalent of the hyena, feeding largely on bones. The hooded vulture, *Necrosyrtes monachus*, common in most of Africa south of the Sahara, is relatively small, but still a large bird.

The feeding habits of New World vultures are less fully described than those of the Old World. All, however, find their prey mainly or exclusively by sight, not smell. Their eyesight is probably acute, but not so acute as that of some raptors that feed upon small, moving animals. Vultures can see a large carcass, or each other, from several miles. They also observe other birds, and even lions and hyenas, in their search for a meal. Since they locate prey by sight they cannot find it in heavy woodland or forest and, since they have little sense of smell, a covering of branches or grass will usually conceal a dead animal, even in open country. Normally they are exclusively diurnal feeders, but in India both black vultures and white-backed vultures, *Gyps bengalensis*, have been known to feed by moonlight on tiger kills.

Usually one of the more handsome vultures, these lämmergeiers have moulted their crests.

Long-tailed field mouse of Europe, also called European wood mouse.

W

WAGTAILS, small birds deriving their name from their habit of wagging their long tails rapidly up and down (or side to side in one species—the forest wagtail—*Dendronanthus indicus*) as they stand or walk.

The white wagtail, *Motacilla alba* (and the British race of this, the pied wagtail) has a boldly patterned black and white plumage, as do the African and Indian pied wagtails. The six or more other species have much yellow in the plumage, particularly on the underparts. The beak is finely pointed in all wagtails.

The wagtails are an Old World group, though one race of the yellow wagtail extends across the Bering Strait, into Alaska. One or two species are resident in the tropics, and some of the palearctic breeding species migrate into Africa and tropical Asia (some individuals even reach Australia) in winter.

The most intriguing feature of the group is the perplexing geographical variation some species exhibit. This reaches such a complexity in one species, *Motacilla flava*, that taxonomists differ widely in the number of species they recognize. Some treat each form as a geographical race (i.e., a subspecies) while others give them species rank. Many of the 20 or so different races (as they will be termed here) are markedly different from one another in appearance and therefore in their vernacular names, which refer to the males in breeding dress. The race confined to Britain, and a few small areas across the English Channel, is the yellow wagtail, *Motacilla flava flavissima*, which is yellow all over. On the European continent there are several other races in which the male's head is distinctively colored: the blue-headed wagtail, *M. f. flava*, in France, Germany and adjacent areas; the gray-headed wagtail, *M. f. thunbergi*, in northwest Europe; the black-headed wagtail, *M. f. feldegg*, in the Balkans, and several others. In Asia, too, there are many races, one of which, the white-headed wagtail, *M. f. leucocephala*, of central Asia, is particularly striking. The main reason for treating such morphologically distinct populations as geographical races, and not as distinct species, is the frequency of interbreeding where the races come into contact with one another. A second reason is the tendency for aberrant birds to appear in the breeding range of a particular race, the aberrant individuals having the characteristics of a different race. In the winter quarters in tropical Africa and Asia, individuals belonging to several races may live together in mixed flocks and show few, if any, differences in their general behavior. However, when spring comes, they finally separate, the individuals belonging to races breeding in warmer parts of the breeding range setting out on their northward migration first and leaving the rest behind. The last to leave are generally those belonging to the most northerly races, the breeding grounds of which do not offer suitable conditions for nesting until early summer.

The nest is an open cup of grass sometimes placed among grass and herbage but often in cavities in walls, banks or trees. The clutch is of four to six eggs, generally incubated by the female alone. Incubation lasts two weeks, and the young leave the nest two weeks later, but are fed by the parents for a few more days.

WALLABIES, a large and diverse assemblage of about 30 species of kangaroolike marsupials of generally smaller size than the true kangaroos, of the subfamily Macropodinae of the family Macropodidae which also includes the kangaroos. The wallabies have large hind feet, strong hind limbs and a long tapered or untapered tail. They are herbivorous and adopt a bipedal method of locomotion when moving quickly.

The Dorcopsis wallabies, two species of *Dorcopsis* and two of *Dorcopsulus*, are ground dwelling but show more resemblances to the tree kangaroos than to the remaining wallabies. They have a tapered tail which is proplike.

Red-necked wallaby, one of several species killed for its fur.

Behind the body it extends parallel to the ground for about one-third of its length then bends downwards to touch the ground at its tip. The Dorcopsis wallabies have functional canine teeth in the upper jaw as do the tree kangaroos. The larger species weigh 12-15 lb (5.5-7 kg) but the smallest species, *Dorcopsulus vanheurni*, weighs 4.5-6.5 lb (2-3 kg).

The pademelons, *Thylogale* (4 species), and rock wallabies, *Petrogale* (about 6 species), are medium- to small-sized wallabies weighing 10-25 lb (4.5-11 kg). The pademelons have a tapered tail whereas the rock wallabies have an untapered tail. The hind feet of rock wallabies are equipped with pads and granulations and the claws are short, adaptations to rock haunting habits which are absent in the forest dwelling pademelons. The quokka is a stockily built gray-brown wallaby with a short tail and weighs 5-10 lb (2-4.5 kg).

The scrub wallabies are a group of larger wallabies weighing up to 45 lb (18 kg). Bennett's wallaby, *Macropus rufogriseus fruticus*, is a large gray wallaby with a reddish tinge in the fur on shoulders and rump, an indistinct white cheek stripe and naked muzzle. It is confined to Tasmania.

The scrub wallabies contribute significantly to the Australian fur trade. Up to 45,000 skins of agile, red-necked, swamp and whiptail wallabies have been sold each year since 1954 from the State of Queensland. Over two million Bennett's wallaby skins were sold from Tasmania during the period 1923 to 1955. The pademelons are not generally exploited for the fur trade except in Tasmania where two and a half million skins were sold during the period 1923 to 1955. Shooting for skins does not, however, appear to have caused a decline in numbers of the exploited species. These remain relatively abundant whereas a number of species which have never been extensively hunted are now extremely rare.

WALL LIZARD, name used for several European species of typical lizards, the best known being *Lacerta muralis*. Their natural habitat is among rocks, but when living in gardens they use walls instead.

WALRUS, similar to a sea lion in appearance but heavier, more wrinkled and having distinctive tusks. Their hind flippers can be brought forward underneath the body, but walruses resemble true seals in lacking a visible external ear. Studies of chromosomes have shown that those of walruses are similar in some respects to those of both otariids and true seals. There is one species of walrus, *Odobenus rosmarus*, with two subspecies, the Atlantic walrus, *O. r. rosmarus*, and the Pacific walrus, *O. r. divergens*. The differences between them are not great, the Atlantic walrus having shorter tusks and a narrower facial part of the skull.

Walruses are perhaps one of the most spectacular and best known of all the pinnipeds, and their heavy wrinkled bodies and whiskery, tusked faces are easily recognizable. Adult males are about 12 ft (3.6 m) in length and weigh up to 3,000 lb (1,360 kg), and females are slightly smaller, about 10 ft (3 m) in length and 1,800 lb (816 kg) in weight. The head looks rather square in side view and has small blood-shot eyes, rows of thick whiskers and long tusks that are present in both sexes. These

tusks, which are formed from the upper canine teeth and erupt when the animal is about four months old, grow to about 4 in. (10 cm) long at two years, about 1 ft (30 cm) by 5-6 years, and may reach a length of over 3 ft (1 m). The skin is rough and wrinkled, and adult males have many warty tubercles about their necks and shoulders. The thickness of the skin increases with age and in an adult animal may reach over 1 in. (2.5 cm) on the body and about 2½ in. (7 cm) on the neck. Under the skin is a thick layer of blubber that may also be 2½ in. (7 cm) thick and weigh over 900 lb (408 kg). The hide is a light gray, but when basking in the sun the blood vessels dilate and the animals appear rust red. Young animals have a scanty coat of reddish hair, but old animals have a practically naked skin.

A single pup is born in April or May after a gestation period of almost a year, but in any one year only about half the adult females produce pups. The newborn pups are about 4 ft (1.2 m) in length, 100 lb (45.3 kg) in weight and grayish in color, though they soon molt to become a reddish color. They are suckled by their mothers for over a year, but become more independent during the second year, supplementing the milk with small invertebrates. Even after weaning they remain with their mothers for another year or so, possibly because with their very short tusks they are unable to dig up enough food for themselves and have to rely to a large extent on animals stirred up by their mothers.

and fish are also taken. A full stomach of a bull walrus has been shown to contain about 85 lb (38 kg) of mollusk feet. Such stomachs, full of the feet of the mollusk *Cardium*, are considered a great delicacy by the Eskimo. The long tusks of the walrus are used to stir up the gravel, and the lips and whiskers sort out the food material and convey it to the mouth. It is a curious fact that although mollusks form the main food, it is only very rarely that pieces of mollusk shell are found in a walrus stomach. Mussel shells are frequently found in quantities near walrus breathing holes, and it has been suggested that the walrus may hold the mussel on the ground with its whiskers and then suck the fleshy parts out of the shell. Certainly the mollusk feet and siphons that are found in walrus stomachs are entire, and not crushed, so it is very probably that they feed in this way. Occasionally walruses will eat young bearded and ringed seals, and even young walruses, but it is believed that they do this only when other food is scarce, though an occasional rogue walrus is seen with tusks and skin stained with grease from habitual seal eating.

WARBLE FLIES, known in North America as grubs, lay their eggs in the skin of cattle, the larvae from these causing swollen skinboils or warbles. The flies look like small bumblebees but have only one pair of wings, so are true flies. The two cattle warble flies, *Hypoderma lineatum* and *H. bovis*, are common throughout the temperate zone of the northern hemisphere, but

A lacertid of the islands off Eastern Spain, *Lacerta pityuensis*.

Mating is believed to take place shortly after the birth of the pups, but as this happens far out on the ice floes it is difficult to get reliable information.

Walruses have few enemies, man, killer whales and polar bears being their only predators. Walruses are found chiefly in shallow waters of 40 fathoms (72 m) or less, where there is abundant gravel that supports a large fauna of mollusks. Bivalve mollusks form the greater part of the food eaten, although echinoderms

do not occur farther south. They are usually very host-specific, but serious conditions occasionally arise in other animals, and in man, for example, eye infection with a maggot, following chance contact with egg-laying flies. During summer the adult *Hypoderma* females irritate cattle by flying around them and darting down to lay their eggs on the legs and underside of the body. This can cause entire herds of cattle to panic and run, known as "gadding" (hence the alternative name of gadfly for the warble fly). The eggs are attached to the hair, rather like lice eggs, although they are placed well down in the coat and so not usually very obvious. The name *lineatum* refers to the lines of 5-12 eggs laid on individual hairs by this species.

The eggs hatch in 3-4 days, and the $\frac{1}{25}$-in. (0.75-1.00 mm) first stage larva crawls down the hair to the skin, which it penetrates by means of digestive juices and its two tiny cutting mouth hooks. It then travels along underneath the skin.

WARBLERS, small, perching birds, comprising the Sylviinae, a subfamily of the Muscicapidae. Almost all warblers have thin, pointed bills (though there are exceptions, such as the thick-billed warbler, *Phragmaticola aedon*) and are mainly insectivorous. They have ten primaries (flight feathers) and, with a few exceptions, for example Rüppell's warbler, *Sylvia rüppelli*, the sexes have similar or identical plumages. Where there is a sexual difference, the male is the more conspicuously colored and, in these cases, the juvenile plumage resembles that of the female. The plumage is usually similar at all seasons, with no special breeding plumage.

The English name (warbler) refers to the pleasant and melodious song of many of the species, especially some of those best known in western Europe. The blackcap, *Sylvia atricapilla*, a common woodland warbler, for instance, has a rich warbling song rivaling, in many people's opinions, that of the blackbird, *Turdus merula*, and nightingale, *Luscinia megarhynchos*. Other species, such as the grasshopper warbler, *Locustella naevia*, however, have monotonous and relatively unmusical songs and some species even have harsh and unpleasing songs (e.g., whitethroat, *Sylvia communis*).

While some (mainly tropical) species are brightly colored, most are various shades of brown, green or gray and identification is often not easy, and frequently depends on behavioral differences.

Most species of warbler build their nests at or close to ground level, in grass, reeds or low bushes. The nest varies from a shallow cup to a domed nest or even, in the case of some of the grass-warblers (*Cisticola* spp.), a bottle-shaped nest. In some species, such as the blackcap, the male may build several "cock's nests" of which only one is eventually used. Various species lay from three to ten eggs, but most of the familiar western European species lay from four to six.

These are usually pale-colored, often with faint spots or streaks. In some species, for example the whitethroat, both sexes may incubate, whereas in others the female incubates alone (e.g., chiffchaff), but both usually share the task of feeding the young. Most species breed in isolated pairs, but some (e.g., reed warbler) form loose colonies.

Being mainly insectivorous, warblers are almost constantly on the move, searching for food.

WARTHOG, *Phacochoerus aethiopicus,* a large gray-brown African hog with smooth or rough hide sparsely covered with hair, except around the neck and on the back. As the name suggests it has warty bumps on the face beside the eyes and on the sides of the face. The dental formula is: incisors $\frac{1}{3}$; canines $\frac{1}{1}$; premolars $\frac{3}{2}$; molars $\frac{3}{3} \times 2 = 34$. The lower canines are highly developed, curving to the top of the snout and being up to 24 in. (61 cm) long in some specimens. The feet are typical of the hog family with four toes on each foot. The length of the head and body of 12 males measured averaged 52.3 in. (133 cm). The same number of females averaged 46.3 in. (117 cm). The height at the shoulder of the males averaged 30.5 in. (77 cm) and the females 26.4 in. (67 cm). The tail length of the 12 males averaged 18 in. (46 cm), whereas that of the females was 15.7 in. (40 cm). Males are heavier than females, the average weight of the 12 adult males studied being 191.4 lb (87 kg), and that of the females was 132 lb (60 kg).

The warthog is found over almost the entire continent of Africa, but principally south of the Sahara.

Warthogs inhabit savannas, open plains and bushy edges. They are not found in dense forest. Usually they form small groups, but they may sometimes be solitary. They are almost entirely diurnal, spending the nights in burrows, usually those made by ant bears or porcupines. When retreating into burrows they back in, presumably in order to face any enemy. They run with tails erect and can reach speeds of 20-30 mph (32-48 kph).

They are vegetarian, feeding largely on roots and rhizomes of grasses and other plants. The warthog is a frequent visitor to waterholes or "pans" and is fond of mud wallows.

WART SNAKES, unique in being covered with a skin that is like sandpaper to the touch. The two species are the elephant's trunk snake, *Acrochordus javanicus*, and the file snake, *Chersydrus granulatus*. Both are harmless and highly specialized for aquatic life. Adaptations to aquatic life include a small head with nostrils facing upward and tiny eyes. The nostrils can be closed when the snake is submerged by a flap of cartilage in the roof of the mouth. There are no enlarged belly scales, and the skin consists of small almost uniform granules. The scales do not overlap each other as is usual in snakes but instead lie side by side, sometimes with skin showing between them. Each body granule is wartlike and has a central tubercle, which gives the snake a granular appearance and abrasive texture. The skin of the elephant's trunk snake is flabby, and this and the snake's girth account for its common name. Occasionally, the elephant's trunk snake grows to 6 ft

Wood warbler *Phylloscopus sibilatrix*, of Europe.

Walrus bulls hauled out on a beach.

(1.8 m) long, with a girth of 1 ft (30 cm). The file snake grows to only half this size and has a shorter and more compressed, rudderlike tail. The elephant's trunk snake occurs in southeast Asia from Cochin China to New Guinea and also in northeastern Australia. It is found in streams, pools, canals and estuaries. The file snake is almost entirely marine and more widespread; it is found around coasts and estuaries from Ceylon and India across southern Asia as far as the Solomon Islands.

WASPS, typically stinging insects, banded black and yellow, related to bees and ants. The majority are solitary. That is, male and female come together only for mating, after which the female alone provides for her offspring. A small minority are social, which means they live in a colony of thousands of individuals, the workers, with a queen that devotes most of her active life to laying eggs. Examples of solitary wasps are the velvet ants and ichneumon flies, which have their separate entries.

During the long winter months, queen wasps produced during the previous summer undergo a period of hibernation, for which they are well supplied with fat, and with a low metabolic rate, so are able to survive their six months' sleep. Hibernation sites are generally in well-insulated places such as outhouses, dense leaf litter and crevices in bark and wood, but large aggregations of queens have also been observed hibernating in the original wasp nest.

When the queens leave their winter quarters in April they begin searching for nesting sites. These are often the disused burrows of small mammals, but almost any cavity will do, including attics, roofs and cavity walls of houses, disused beehives or even old cardboard boxes. In the search for limited nest sites, fighting between competing queens often occurs, sometimes resulting in the death of one or both combatants. The nest begins as a blob of wasp-paper fixed to the roof of the nest site from which small umbrellas of paper are built. These "envelopes," which surround the first comb of cells, insulate the nest and permit a high and constant temperature to be maintained. The initial comb is composed of 30-40 hexagonal cells in which the queen lays her eggs and the larvae develop. Throughout this period, the queen behaves in the same way as the solitary wasp, which practices progressive-provisioning, but with the emergence of the first daughter wasps—the workers—the queen becomes more restricted to egg-laying activities, and soon ceases to leave the nest as the workers take over foraging duties.

The larvae are fed on protein, which is collected in the form of insect prey (including flies, caterpillars and bees), carrion and even meat and fish from shops and markets. When the prey is caught the wasp normally kills it by biting it in the neck—the sting is used principally for defense. The wasp then cuts off the head, legs and wings before carrying the carcass back to the colony. The adults feed themselves on carbohydrate, such as nectar, honeydew of aphids and other sources such as

jam. This is stored in the wasp's crop and, on the return to the colony, is distributed to the adult occupants.

Nest building is one of the major duties performed by the workers and involves the collection of massive quantities of chewed-up wood fibers—scraped from fences, posts and dead trees. The wood is mixed with saliva to form a thick paste, which is drawn through the mandibles and applied to the nest in a strip. When the load of pulp is used up, the completed strip is allowed to dry—a process essentially similar to the paper-making techniques of man. Each strip represents a single load, and the different colors illustrate the source of wood pulp.

The typical mature wasp nest has the size and shape of a football—a sphere representing the most economical shape for building—although the site can impose severe restrictions on nest construction and influence its ultimate shape and size. As development proceeds, more combs are added below the original queen-built comb, and existing combs widened. Each comb is supported by a large number of stout pillars made of wasp-paper, and separated from one another to afford sufficient space for adult wasps to crawl between them. By the end of the season, most nests have 8-9 combs. Surrounding the combs, the original envelopes of the embryo nest are replaced by a thick, cellular wall with excellent insulation properties. The temperature in wasp nests is remarkably constant, especially in mature colonies with their thick walls and large populations. Heat is produced through the activity of adults and larvae, and if the temperature rises above the optimum, wasps in the nest entrance fan vigorously with their wings to produce a stream of cooling air. If the temperature continues to rise, water is brought to the nest and allowed to evaporate on the combs and walls.

Building accounts for much of the activity of workers until early August, when colonies of 10,000 cells or more can be found. As each cell is completed or vacated by an emerging adult, the queen lays an egg in it. The queen, whose ovaries enlarge so that the abdomen is almost entirely filled with eggs, attains an oviposition rate of 300 eggs per day at the height of the season.

With the recruitment of emerging workers to the population, the number of adults increases, reaching a peak toward the end of August of up to 5,000. By the end of the season, colonies produce up to 40,000 workers. All cells in a colony are used at least once, and about 75% of them used to produce a second generation by the end of the season. The reuse of cells makes possible the high rate of productivity in the wasp colony.

WATER BEETLES, several beetle families having aquatic members but the two most important being the Dytiscidae and the Hydrophilidae, both widely distributed in the fresh waters of the world. The dytiscids are mainly predators; they swim strongly with synchronous oarlike movements of the hind legs, and catch small aquatic invertebrates as food. The largest species, of the genus *Dytiscus*, will even attack small fish. The larvae of this family are also fierce predators with hollow needlelike jaws that pierce the prey and act as ducts

through which digestive juices are passed into the prey, and through which the liquefied contents of the prey are sucked into the gut.

The adult hydrophilids are in the main herbivorous, although a few of the larger forms are predators. They are generally more feeble swimmers than the dytiscids and use the hind legs alternately. A few of the smaller species are found in salt marshes and even in rock pools on the seashore. The larvae of hydrophilids are carnivorous and deal with their prey in a similar way to the dytiscid larvae.

WATER BOATMEN, aquatic bugs of two distinct families. Because of the confusion of common names given to the two families it is best if the carnivorous and predatory Notonectidae are known as back swimmers, although they are also called water boatmen, and the plant-feeding Corixidae are called water boatmen, although they are often called lesser water boatmen to distinguish them. The Corixidae tend to be smaller than the Notonectidae, and the common British species *Corixa punctata* is only ½ in. (12-13 mm) long.

The Corixidae have been found to resemble the color of the floor of ponds in which they live. E. J. Popham noticed that immature corixids or young adults retain the color of their background after molting and that a pale floor inhibited the formation of dark pigment in their bodies. Popham also discovered that corixids, if given a choice, generally came to rest on a background with which they merged. In a series of experiments, Popham showed the advantage of this camouflage. In a laboratory experiment, small fish (rudd) destroyed about three times the number of corixids that were not well camouflaged compared with those that were. Popham also observed that the proportion of corixids living in fishless ponds that matched their background significantly increased if minnows were then introduced.

WATERBUCK, large African antelope related to the reedbuck but larger, with more spreading horns, and lacking the bare patch below the ear. All of the antelope which are loosely described as waterbuck belong to the genus *Kobus*, but the five species involved fall into three subgenera: *Kobus* for the true waterbuck, *Adenota* for the kob and puku, and *Hydrotragus* for the lechwe. Like reedbuck, in waterbuck only the males have horns.

The true waterbuck all belong to one species, *K. ellipsiprymnus*. This is the largest species in the genus, 48-53 in. (120-133 cm) high and weighing 475 lb (216 kg). There are no inguinal glands (in the groin), unlike the other species, and the face glands are also absent. The coat is long and wiry, forming a mane on the neck. The general color is brownish, and there is a white ring or patch on the rump. The horns are long, simply divergent and forwardly concave— like very elongated reedbuck horns. Waterbuck are heavily built, adult males developing a thick neck and haunches.

Waterbuck are inhabitants of the African savanna. While sometimes living in fairly arid country, they normally keep down by the banks of rivers, spending the night in the thick riverine-cover.

The second subgenus, *Adenota*, contains animals with a short coat and no mane. The horns are rather short, somewhat lyrate, and face glands and inguinal glands are present. The kob, *K. kob*, is found from southwestern Kenya through the savanna zone west to Senegal and on the Bijagos Islands. A male kob stands 34-36 in. high (85-90 cm) and weighs 200-220 lb (90-100 kg); a female weighs only 137-145 lb (62-66 kg). The color varies geographically from reddish orange to nearly black, with white round the eye and base of the ears. There is a black line down the front of the fore legs, and the muzzle, lips, underside, insides of the thighs and a band above the hoofs are white. The smaller puku, *K. vardoni*, is found in the Chobe district, eastern Caprivi, the Kwango region, Zambia, Central Malawi, southern Tanzania and part of Katanga. It stands 32-35 in. (80-88 cm) high. The male weighs 150-170 lb (68-77 kg) and the female 125-140 lb (57-64 kg). The puku has no black on the fore legs. no white hoofband, and a narrower white eye-ring. The coat is longer and rougher and the inguinal glands open forwards instead of backwards as in the kob.

The third and last subgenus, *Hydrotragus*, contains two species which are both known as lechwe. They have a coarse, long coat like true waterbuck, but no mane and long slender lyrate horns with, however, a double curve, up and then back, somewhat gazellelike. There are no face glands, but rudimentary inguinal glands. The Nile lechwe or Mrs. Gray's waterbuck, *K. megaceros*, has horns with a longer, more pronounced backward curve. The adult male is nearly black with a large white patch in front of the withers (usually), joined via a white line up the back of the neck to a pair of white eye-rings. The chin, lips, belly, inner surfaces of hind legs, and a hoofring are also white. Females and young are yellow brown, with white on the head, but not on the neck or shoulders. Both sexes are 35-40 in. (88-102 cm) high. This species lives in the swamps of the Nile, Bahr-el-Ghazal and Sobat, and between the Baro and Ghilo rivers in adjacent parts of Ethiopia. Little is known of its habits or social organization. Herds are said, however, to contain about equal proportions of males and females.

The true lechwe, *K. leche*, is found from northern Botswana, north of Lake Ngami, to eastern Angola, the upper Zambesi, Kafue, Chambeshi and Luapula rivers, and the southeastern Congo to 6° S. It is 40 in. (1 m) high, with long coarse hair. The horns are shorter, and not swept back so far. The white eye-ring is present but not the white shoulder- and neck-markings.

WATER BUFFALO, *Bubalus arnee*, the largest species of Asiatic buffalo, domesticated over very large areas of the tropics and subtropics and in the Mediterranean region.

Water buffalo have horns that face backward and outward in their basal part, and gradually turn inward so that the tips face each other. They are triangular in section and irregularly ridged. Unlike true cattle, genus *Bos*, water buffalo have a straight back with no hump on the shoulders, a sparsely haired gray skin, a smooth tongue, a small scrotum and large hoofs. They are leanly built, taller than an ox of equal weight with a longer body and a low head.

Wild buffalo are found in northern India, from the Ganges and Brahmaputra plains to part of Orissa and southeastern Madhya Pradesh, and Nepal and from northern Burma into Indochina, and also in Ceylon. In most parts of their range they are slaty black, with the legs white from just above the knees and hocks to the hoofs, and a white crescent on the throat. The newborn calf is very light, almost yellowish in color. The horns either curve up in a semicircle, with the tips close together, or else spread out horizontally going slightly up and in at the tips.

Indian wild buffalo, *Bubalus arnee arnee*, are 5½-6½ ft (170-200 cm) high, and some bulls may weigh 2,600 lb (1,170 kg). The horns may be as much as 6 ft (180 cm) long around the curve, and 4½ ft (140 cm) is not unusual. The Ceylon wild buffalo, *B. a. migona*, is much smaller, only 58 in. (150 cm) high, with horns not exceeding 39 in. (1 m) in length. Similar small buffaloes are recorded from Vietnam. The buffalo found in the Mishmi Hills of Assam is said to be a separate, dun-colored form, *B. a. fulvus*. A population of wild buffaloes exists in Sarawak, but these are almost certainly feral domestic stock.

Domestic buffalo are smaller than Indian wild buffalo, and have differently shaped shorter and thicker horns, the tips of which are less inturned. They have been transported over most of the tropical and subtropical latitudes, and are the staple beasts of burden or milk and meat providers for much of the world's human population. There are about 78 million domestic buffalo in the world, of which 45 million are in India.

Wild buffalo live in tall grass jungles and reed brakes near swamps. In the southern parts of their range, they live on rather harder ground. They form small herds of 10-20, which sometimes combine to form larger grazing groups. They graze in the early morning and again in the evening, avoiding the heat of the sun, and have been known to feed at night. The members of the herd communicate by grunting; bulls occasionally bellow like a domestic bull. It is unfortunate that wild buffalo are fond of cultivated plants, and this fact, together with their competition with tame buffalo for grazing, has caused a reduction in their numbers. It is probable that there are under 2,000 remaining in the wild, of which 400 are in Kaziranga, Assam.

WATER FLEAS, tiny crustaceans deriving their common name from their method of swimming, which is basically a hop and drop using their branched antennae. To maintain its position in the water, a water flea has to swim continuously, otherwise it sinks to the bottom. Not all water fleas swim. Some have become so highly modified for life in mud that they spend all their time tunneling through the surface layers at the bottoms of pools. *Iliocryptus sordidus* is such a species. The name *sordidus* derives from the fact that when it molts it does not cast off the old carapace. As a consequence it always carries a coat of debris on its back, *sordidus* being from the Latin for wearing dirty clothes.

WATER MEASURERS, a group of small, slender bugs, often little more than $\frac{1}{3}$ in. (1 cm) long, with long, stiltlike legs, which are used to carry the insect slowly and ponderously over the surface film of stagnant ponds. At first sight, water measurers may be confused with water scorpions, but can be identified, on closer examination, by the very elongate head, which may be at least as long as the thorax. The 70 or so species are found mainly in the tropics. They feed on small insects such as mosquito larvae.

WATER SCORPIONS, *Nepa*, aquatic bugs with the front pair of legs modified for seizing prey. These look a little like a scorpion's claws, but they clasp prey with a jackknife action. The bugs are oval and flattened, brown, and look like small dead leaves, a resemblance heightened by a slender "tail" at the hind end of the body, which looks like a leaf stalk. This tail is a breathing tube that the bug pushes up through the surface film to take in air, and it may possibly have a very faint resemblance to a scorpion's tail, so assisting the derivation of the common name. Water scorpions are not uncommon among the decaying leaves around the edges of ponds and lakes. They are carnivores and stalk their prey on their second and third pairs of legs, which are long and stiltlike. There are 150 species in various parts of the world, and the largest is no more than 2 in. (5 cm) long.

Related to the water scorpions and in the same family is the water stick insect *Ranatra*, long in body and with long slender legs and a breathing tube nearly as long as the rest of the body. It looks very like the more familar stick insect that lives on land.

WATER SNAILS, aquatic or amphibious gastropod mollusks that usually have a well-developed horny or calcareous shell. Their shells are often spirally coiled like a typical snail, but one important and widespread group have such a flattened spiral to their shells that the coil is in one plane, like a wheel. These flattened water snails are called ram's horn or trumpet snails (Planorbidae).

The water flea *Daphnia*, rendered transparent by laboratory reagents for examination under a microscope. The body is enclosed in a bivalved carapace.

Water snails are herbivores, feeding on aquatic plants, algae and detritus, using their rasplike feeding apparatus, the radula. They can often be found crawling or floating upside down from the surface of the water, suspended from a mucus raft. They can also be found in moist situations on the land at the marshy junctions between water and land. They are most abundant in slow-moving rivers or streams or in lakes. Fast-moving water rarely has many snails as they become dislodged and so washed downstream. Because water snails have a calcareous shell they are mostly confined to "base-rich" (calcareous, neutral or alkaline) waters, the shell thickness being governed by the available calcium. Lakes and rivers with low calcium usually have a poor molluskan fauna; waters rich in calcium have a large number of species and individuals.

WATER SPIDER, *Argyroneta aquatica*, the only species of spider that lives more or less permanently below the surface of the water, although several spiders can run over water and some can run down plants below the surface to escape enemies or catch water insects. The water spider is small-bodied and long-legged, the front part of the body light brown with faint dark markings, the chelicerae reddish brown and the abdomen grayish, covered with short hairs. The females are usually 8-15 mm long and the males 9-12 mm, although females of up to just over an inch (28 mm) long have been recorded. The water spider ranges across temperate Europe and Asia.

Although living under water the water spider is dependent on air for breathing. It rises to the surface and entraps a bubble of air around its hairy body, then descends to its silken thimble-shaped diving bell and into this releases the bubble. This is repeated until the diving bell is filled with air.

The water spider remains in its bell during the day and entraps any water insects passing by, but at night it goes out to hunt. The male mates with the female in her bell in spring or early summer, and the female subsequently lays 50-100 eggs in a silken bag that takes up the upper half of the cavity of the bell. After the eggs hatch, the spiderlings stay in the bell for a few weeks, then leave it to live independently.

WATTLEBIRDS, a crowlike group of New Zealand forest birds, with brightly colored fleshy wattles at the corners of their bills.

The kokako, *Callaeas cinerea*, or wattled crow is a big bluish-gray bird with a black face. It completely lacks fear of man and can be watched as it bounds and glides through the forest, rarely flying any distance. The food is largely young leaves and fruit, but it also eats invertebrates. One foot is often used to hold the food while it is torn apart. The organlike song can be heard miles away and is one of the most pleasant sounds of the New Zealand forest. The nest is a massive structure of twigs, rotten wood, moss and fiber placed in a tree. The female incubates the clutch of two or three pale gray, speckled eggs for about 25 days, and both sexes help to feed the young.

The tieke, *Philesturnus carunculatus*, is a glossy

Male waterbuck of the type known as defassa.

The water spider of Europe, which spins a silken underwater diving bell for storing air.

black bird the size of a blackbird with a bright chestnut "saddle" across its back. The juvenile of the South Island race is brownish all over and was long considered to be a separate species. They were once common throughout the New Zealand forest, moving in noisy flocks, searching the ground and tree trunks for insects, berries and nectar. The Maoris regarded them as a bird of omen, guarding ancient treasures and foretelling the outcome of war. The nest is built low down in a hollow tree or a dense piece of vegetation. The two to three very pale brown eggs take about three weeks to hatch, and, though the male does not incubate, he brings most of the food for the nestlings.

WAXWINGS, three similar species of small forest birds constituting the subfamily Bombycillinae. The most widely distributed of these, *Bombycilla garrulus,* is called simply "waxwing" or, not very happily, "Bohemian waxwing." It inhabits northern latitudes in Europe, Asia and North America, and three geographical subspecies are recognized. Two further species exist: the so-called Japanese waxwing, *B. japonica,* a native of eastern Siberia and only an irregular winter visitor to Japan, and the cedar

waxwing, *B. cedrorum,* a native of northern North America and an irregular winter visitor as far south as the Caribbean.

Waxwings are about 6½-7½ in. (16-19 cm) long, with short, broad and slightly hooked bills and short, stout legs. The plumage is remarkably soft and is mostly brown, grayish-brown or reddish-brown, with a black throat, wing quills and tail quills. *B. garrulus* has a yellow wing bar, and *B. japonica* has a crimson one and also red tips to the tail feathers where the other two have yellow. In all species (but, strangely, not in all individuals), the shafts of the secondary wing feathers are prolonged in red, drop-shaped tips like sealing wax—hence the English name. A brown swept-back crest is another conspicuous characteristic.

Waxwings are birds of the northern forests, particularly coniferous forests. The nests, placed in trees, are bulky collections of twigs with a lining of moss and fibers to which grasses and feathers are added. The eggs are grayish blue, with profuse black markings. In summer the birds are largely insectivorous, but in autumn and winter berries predominate in the food, especially those of the rowan tree in the case of *B. garrulus.* The birds are capable of

survival in the extremely severe conditions of winter on or above the Arctic Circle, but many make regular migrations to the west, east or south. Every few years these take place on a greater scale and may extend to countries that are little visited in ordinary years. Such irruptions, on the part of *B. garrulus,* often reach the British Isles in the late autumn, where the birds attract attention by their ornamented plumage, by their notable gregariousness and by their lack of fear of man.

WEASEL, *Mustela nivalis,* the smallest carnivore in the world, in the same family (Mustelidae) as the stoat and very similar in appearance to it, but smaller. The weasel can usually be recognized by the uneven line separating the cream color of the ventral surface from the dark reddish upper side. There are variations in the patterning of this color demarcation line that facilitate the identification of individual weasels in the field. Also, the shorter, less fluffy tail does not have the stoat's dark tip. The fore paws may be white, and light flecks can also appear on the head or snout.

There has been great controversy about how many species of weasel exist, and now it seems that even the American "least" weasel, *Mustela rixosa,* is conspecific with the European, and that *Mustela nivalis* is the only valid species. True, the northern individuals are smaller than the southern ones in their very wide holarctic distribution. Strangely enough, where stoats and weasels are found together, weasels are always smaller than stoats, whereas in north Africa where the stoat is absent, weasels are larger, filling two roles or ecological niches at once. Whatever the length of the individual may be, a marked size difference between the sexes persists. Because of these ecological factors, a male weasel's size may vary from 6.7-11.7 in. (17-29.5 cm), and the weight ranges between 2 and 4.5 oz (60-130 gm). The female, sometimes only half the size of the male, measures 6-7.5 in. (15-22 cm) and weighs 1.5-2 oz (45-60 gm). Not at all hampered by its diminutive size, the weasel often pursues, catches and kills prey many times its own weight and size. Agile and muscular, it is constantly on the alert, often giving the impression of being fearless, even to the point of attacking a human being who has blocked the entrance to its burrow. A stoat, in the same situation, would flee or remain motionless. Perfectly adapted physically to its predatory life, there is an old saying that a weasel can squeeze down a hole the size of a wedding ring, and indeed, once the head has pushed through the body will follow down a mouse hole or through a knothole in a fence. It feeds on an eclectic diet of mice, voles, squirrels and insects. Fledglings and, more rarely, adult birds may also be taken. The legend of the weasel sucking blood from its victim may be based on the fact that when attacking a prey much larger than itself, the usually lethal neck bite must be repeated again and again, covering the weasel's snout with blood. It may, in times of plenty, kill more than it can eat, thus usefully serving as a check to the rodent population explosion but also spreading disaster in chicken coops. A pair of weasels can easily kill more than 2,000 rodents a year, which is ample compensation for the death of a few chickens. Territorial

The weasel of Europe and North America, the smallest of all carnivores.

like most other mustelids, the weasel uses the two scent glands situated below the base of its tail to mark boundary lines along branches and grass hummocks.

Two litters of four to eight cubs are born, one in the spring, one in late summer, after a six-week gestation. The cubs from the early litter may mature quickly enough to breed in their first year. The life expectancy is about six years.

WEAVERS, a large family of seed-eating birds, which includes the sparrows, best known for the pendent woven nests of some of the typical species. Their natural distribution is within the Old World, centered mainly on Africa.

In the subfamily Ploceinae, the typical main group of weavers, are the 67 species of the genus *Ploceus* and the 10 *Malimbus* species. These are sparrowlike, the females, immature birds and males out of breeding plumage all having brown plumage with blackish streaking on backs, wings, and heads, and pale buffish underparts. The males in breeding dress develop bright color, most of the *Ploceus* species being mainly yellow with variable areas of black, and *Malimbus* species being similar but patterned in red and black, a difference apparently correlated with the moist forest habitat of the latter. These

birds are normally sociable at all times and nest in colonies, sometimes of considerable size. The pendent nests are built onto the tips of twigs or palm fronds, or more rarely in lower swamp growth. The male usually makes an initial hanging ring and then builds out to form a rounded nest chamber in one direction, and in the other extends the entrance to an extent that varies considerably from one species to another but that may result in anything from a small porch to a long downward-pointing tube up which the birds will enter. During building, the male displays by clinging to the structure and hanging downward, flapping the wings and making harsh chattering noises. Interest shown by the female may be necessary to stimulate completion of the nest.

Polygamy may occur in such colonies. The egg color and pattern vary very considerably between individuals in some species, suggesting either that a closed nest is a recent evolutionary trend in this family and that selection for camouflage has ceased, permitting variation, or that there is a need for individual recognition of clutches. The young are fed extensively on insects, and one or two species also feed mainly on insects when adult. The timing of the breed-

ing season may be governed not only by the needs of the young but also by the need for a particular plant material, in a particular state, for building the complexly woven and resilient nest structures.

The social weavers of the genus *Pseudonigrita* lack the typical streaked feather pattern. The gray-headed social weaver, *P. arnaudi,* has a grayish white crown, and the black-headed social weaver, *P. cabanisi,* is white below and has a black cap. These two build nests that are round spiky masses of grass stems, with two entrances on the underside, one of which is closed before nesting begins. The nests of the second species are built more around the end of a twig, rather than across one, as in the first. Such nests are placed close together in low trees, the first species often using thorn trees inhabited by stinging ants. The social weaver, *Philetairus socius,* of southwestern Africa is an undistinguished streaky brown bird, famous for its nest. The individual nests of a flock, with entrance holes on the lower side, are built in contact with one another to form huge masses. Each mass is roofed over with a curved roofing layer, which may help to shed rain; and this roof is built first, the nests being subsequently

Taveta golden weavers, *Ploceus castaneiceps*, one of many species of sparrowlike birds of Africa.

built into the underside of it. They are not woven but consist of a tightly packed mass of grass stems, "thatched" together by over-lapping. Nests may be added around the edge later, and adjacent masses may be joined. A colony may consist of several such groups, and a single mass may contain over a hundred nests. These colonial nest masses are reused for many years. This is a species of arid grasslands where it feeds in the open, and nests may be built in scattered or solitary trees.

The males of the 15 species of true sparrow of the genus *Passer* are patterned with black, chestnut and yellow. They are birds of grass-land and arid places, taking readily to grain. The nests vary from rounded, domed structures built roughly of twigs with grass linings, to holes in trees or rocks. Displays are limited to wing fluttering and chirping at the nest site. The rock sparrows, five species of the genus *Petronia*, are dull sandy brown or gray birds, lacking conspicuous markings. They live in open country, nesting in holes and crevices.

WEEVER FISHES, marine bottom-living, poisonous fishes related to the red mullets, stargazers and jawfishes. The name "weever" derives from an old French word, *wivere*, itself deriving from a Latin root meaning poisonous, a reference to the poisonous spines found in the weevers. There are two species found off the coasts of northern Europe, the greater weever, *Trachinus draco*, and the lesser weever, *T. vipera*. The latter grows about to 6 in. (15 cm) in length and is much more abundant than the former, which reaches 20 in. (51 cm). The two

species are fairly similar in appearance. The body is a deep yellow-brown with gray-blue streaks, and the belly is a pale yellow. The most striking feature is the first dorsal fin, which contains 5-6 spiny rays and is black in color. The poison glands are at the bases of these rays. In the lesser weever there is another poison gland at the base of the spine projecting backward from the gill cover. The spine is sheathed in skin, but when this is ruptured the poison is released along the grooved spine and enters the wound. The venom, which resembles that of certain snakes, is not fatal but can be extremely painful.

The weevers frequent shallow water, burying themselves in the sand at the bottom with only the top of the head and the black dorsal fin visible. They are not infrequently caught in shrimping nets, and there is a risk of treading on them while shrimping.

WEEVILS, forming the largest family in the animal kingdom, with over 35,000 species, the vast majority of which are immediately re-cognizable by the greatly drawn-out head forming a long snout, so they are also known as snout beetles. The snout or rostrum bears on its tip the mouthparts, which are character-istically small, with short palps (sense organs) but with powerful mandibles for chewing hard vegetable matter such as seed coats or wood. Most weevils have antennae with a long first segment, the scape, followed by a variable number of smaller segments known collectively as the funiculus, the scape usually enlarged at the tip of the antenna to form a club. Such

antennae are usually geniculate, that is, they are elbowed or bent between the scape and the funiculus. If the rostrum is very long the antennae are usually attached near its mid-point, but weevils with a short rostrum have the antennae close to its tip.

Weevils are generally oval or pear-shaped with well-developed wing covers, and the body wall, which is very hard, is often tuberculated, pitted, roughened or grooved. Frequently they are scaly, and although most species have dull coloring, some rank among the most brilliantly colored of insects. Although weevils are partic-ularly well represented in the tropics, they exist wherever seed-producing plants occur. Their larvae, because they live surrounded by their food inside seeds, buds and stems, are fat, white legless grubs armed with a powerful pair of jaws. They may pupate in the food mass, but in many species the larvae drop to the ground when fully fed and pupate in the soil.

Many species are injurious to crops, and a con-siderable number are very serious pests, one of the most notorious being the cotton boll weevil, *Anthonomus grandis*, which causes damage amounting to hundreds of millions of dollars annually in the cotton belt of the United States. The females drill holes in cotton-blossom buds, laying a single egg in each. The life cycle may take only about 25 days, and in the warmer parts of its range there may be ten generations per year. The cotton weevil is exceedingly difficult to control because eggs, larvae and pupae develop inside the cotton buds and bolls, and furthermore the adults feed by boring into the plants and so are not easily affected by insecticides. The grain weevils, *Sitophilus granarius* and *S. oryzae*, are cosmopolitan pests wherever grain is stored. The female bores a small hole in a wheat grain or similar cereal and deposits a single egg inside. The larva spends its whole life inside the grain, completely eating out the kernel.

WELS, *Silurus glanis*, a large catfish found in the bigger rivers of Europe; in the upper part of the Rhine and its tributaries and in rivers to the east of the Rhine that flow into the Black, Aral, Caspian and Baltic seas. It also occurs in the Caspian and Baltic seas. It is not native to Great Britain but has been introduced into some private lakes and has flourished.

The wels is a naked catfish with a small dorsal fin just behind the head and a long anal fin. There are two barbels on the upper jaw and four on the lower. The jaws have rows of fine teeth, and there is a patch of fine teeth in the roof of the mouth. The color is variable, but is basically dark shades of blue-gray, brown or deep olive, lighter on the flanks, which are often mottled, and shading to pale gray on the belly. The fins often have reddish edges. Partial or complete albinos are fairly common.

The wels is a solitary fish, hiding in deep pools during the day and swimming into shallower water in the evening to feed on small fishes. It is a considerable predator and will eat not only fishes but also frogs, birds and small mammals. There is a record from the last century of a child being eaten by a wels, a not altogether unlikely occurrence since they can grow to about 9 ft (2.7 m) and reach a weight of 700 lb (318 kg). Records of even larger specimens may, however, be exaggerated. The wels spawns in shallow water, and the eggs are laid on plant leaves. The parents guard the eggs until the young hatch out, looking very like tadpoles. Growth is very rapid, the fishes reaching 6 lb (2.7 kg) by the third year. A fish of 180 lb (82 kg) was found to be 24 years old, indicating that some of the largest specimens may well be 90 years old.

The wels is of some economic value. The flesh is eaten, the swim bladder and bones are used for the glue, the skin is tanned into leather and the eggs provide a sort of poor man's caviar. The fish is eagerly sought after by anglers.

A blue *Eupholus* weevil of New Guinea.

WHALE SHARK, *Rhincodon typus*, the largest of all fishes but a harmless species found in the warm waters of the world. The whale shark can be immediately recognized by the lines of pale spots on a grayish body and by the wide mouth at the end of the snout (the mouth being underneath in most other sharks). The body is heavily built, with ridges down the flanks, and the snout is blunt. Like the basking shark, the whale shark feeds on small planktonic animals, straining them from the water with a fine mesh of rakers in the throat region. There are, however, numerous small teeth in the jaws, of which 10 or 15 rows are functional at any one time. The fish cruises through the water with its mouth open when feeding, although whale sharks have also been seen apparently feeding in a vertical position with their heads pointing upward. As in the basking shark, the enormous oily liver probably helps to maintain buoyancy, and these fishes often lie at the surface where they are occasionally rammed by ships. The largest recorded whale shark was 45 ft (13.5 m) in length, but there have been reports of specimens believed to be much larger than this, and a length of 60 ft (18 m) may well be reached. A Florida specimen of 40 ft (12 m) was estimated to weigh 13½ tons (about 13,000 kg). In spite of this enormous size, the whale shark shows none of the aggressive tendencies of its relatives, and men have actually clambered onto them while they lay at the surface. Unlike many sharks, which give birth to live young, the whale shark is an egg-layer, and the egg cases are commensurate with the size of the fish, measuring over 2 ft (60 cm) in length.

Fewer than 100 specimens have been caught and described by scientists, although many more have been observed, and there is still much to be learned of the biology of these huge creatures.

WHELK, a common name given to a wide variety of marine mollusks related to snails. Most of these animals have shells which at first sight are somewhat similar. They have a typical spirally coiled shell with a thick, durable calcareous wall, and the group includes the largest British operculate snails, some of which have shells up to 8 in. (20 cm) long, like the red whelk, *Neptunea antiqua*. The unusual feature of the shells of this group is the spout, or siphonal canal, at the mouth of the shell. This holds the siphon that, in the living animal, acts as an inhalant tube, allowing it to draw clean, well-oxygenated water into the mantle cavity and so over the gill. The siphon is also a sense organ.

Mediterranean species of the group (*Murex brandaris*, *M. trunculus* and *Thais haemastoma*) were used to produce the well-known dye Tyrian purple, which often colored the clothes of important and wealthy citizens of the Roman Empire. Empty whelk shells are one of the most favored homes for the hermit crab, the new tenant then enabling various sea anemones to colonize the shell.

The common whelk, *Buccinum undatum*, has a number of popular names, including the white whelk and the buckie. It is usually found at extreme low tide level, or below tide level, on all types of shore. It is a voracious carnivore feeding on crabs, worms, bivalve mollusks and fresh carrion. In the British Isles it is fished commercially on the east coast and eaten as a traditional part of the "seaside holiday" diet, although its rather tough rubbery flesh does not place it among the most edible of mollusks. However, the tradition seems to have started in Roman times, and whelks were an important dish in the feasts of the Middle Ages.

Whelks feed on living bivalves, such as mussels. Normally mussels are protected from such predation because their hinged shell (the shell is made up of two valves, hence bivalves) snaps closed when disturbed. However, whelks creep among partially open, actively feeding mussels and prevent them from closing by wedging them with the spout region of their own shell. They then feed on the soft parts of the mussels with an elongated protrusible feeding structure, the proboscis, which is a very mobile snout. This remarkable organ when fully extended has the mouth at its tip and so is ideally suited to feeding on the soft parts of animals held at some distance from its own body. While feeding, whelks secrete a large amount of saliva, which contains mucus as a lubricant and nerve poisons that can paralyze their prey. They have a much-modified radula with few teeth, and like all carnivores they have only a short gut.

WHIRLIGIG BEETLES, small oval, black, shiny beetles, less than ½ in. (1.2 cm) long as a rule, that swim rapidly in tight circles on the surfaces of pools and backwaters of streams. They can also plunge under water if danger threatens from above. Their eyes are divided into upper and lower halves, which are adapted to looking upward into the air and downward into the water. There are 400 species throughout the world.

Whirligigs live in swarms of several hundred, and although they may all be gyrating rapidly at the same time they do not collide. Their antennae, so arranged that they lie in the surface film, perceive vibrations from an approaching neighbor so that avoiding action can be taken. The antennae also detect the movements of insects that fall from the air onto the surface film and provide the whirligigs with food.

The females lay their eggs in rows on submerged water plants. The larvae, just over ½ in. (1.2 cm) long when fully grown, are slender. They swim or creep about the bottom feeding on other insects or on plants.

WHITE-EYES, tiny, thin-billed arboreal birds, usually having a ring of white around each eye. Although there are a large number of species, 85 in all, they are mostly remarkably alike and the majority are included in the genus *Zosterops*. Typically they are green on head, back, rump and tail, and gray or yellow below, sometimes with some reddish-brown on the flanks. The ring round the eye is formed by white feathers and differs in size and distinctness in various species. The white-eyes occur through Africa and the Oriental and Australasian regions.

White-eyes are about 3-5 in. (8-13 cm) long, with a slender, pointed and often slightly decurved bill. They have strong feet which allow them to cling in a variety of postures and aid their feeding. They are almost entirely arboreal but occur in low bushes and scrub in fairly open country as well as in tall trees, and they are found in such vegetation from coastal mangroves to mountain forests, but tend to inhabit the edges of denser forest rather than its interior. They are gregarious, usually living in small flocks except when breeding, and are constantly active.

The nests are small deep cups made of fine fibrous material, sometimes with moss on the exterior, and are slung in the fork of a thin twig.

The clutch consists of two or three pale blue or white eggs. They are incubated by both parents and the incubation period may be very short, as little as just under 11 days. They are cared for by both parents. The young birds may lack the white eye-rim at first.

WHITE-TAILED DEER, *Odocoileus virginianus*, the most widespread deer in the American continent, its distribution extending from Brazil and Peru in South America northward through Central America and Mexico into the United States and southern Canada. In the United States it is resident in practically every state except Alaska, and possibly Nevada and Utah, although in both wanderers may occur in the extreme north. It is present in all the adjacent provinces of southern Canada, and also Nova Scotia. It is absent from Labrador, Northwest Territories, Yukon and Newfoundland. A recent estimate has put the white-tailed deer population of the United States at over 8 million. In 1965 over 30 million hunting

White whale or beluga *Delphinapterus leucas*, a relative of the narwhal.

licenses were issued, of which a very substantial number were just for the white-tail. The race that frequents the Florida Keys—the Keys deer, *O. v. clavium*—is, however, becoming scarce. Throughout its entire range from southern Canada to northern South America no fewer than 38 subspecies are recognized, and since the range of many of these overlap, interbreeding has occurred widely. However, it can be said that the larger forms of white-tailed deer are to be found in the north and the smaller in the south. Thus the shoulder height of bucks from the most northern forms (*O. v. borealis*, *O. v. dacotensis* and *O. v. ochrourus*) will measure about 40 in. (102 cm) as compared to the small white-tailed deer, *O. v. margaritae*, from Margarita Island, Venezuela, which is only some 24 in. (61 cm) high.

The white-tailed deer is so called by reason of its longish white tail which is raised erect as the deer bounds away when alarmed, thus acting as a danger signal to other deer. This deer is often referred to as the Virginian deer, but strictly speaking this should only apply to the subspecies of that name, the typical deer, *O. v. virginianus*, from Virginia.

In summer the general color of most subspecies is one of reddish-brown, with a whitish patch on the throat and inside the ears. This coat is replaced by a more somber one of gray to grayish-brown in winter. Fawns are spotted.

WHITE WHALE, or beluga, *Delphinapterus leucas*, a toothed whale related to the narwhal and about the same size, 15 ft (5 m) long, is an arctic species largely restricted to within the Arctic Circle. It has no dorsal fin and like the narwhal has a reasonably mobile and defined neck. In this way both resemble the river dolphins, and there is a similarity also in the flippers and other skeletal features, as well as in the middle ear, which is supposedly much more primitive than in the true dolphins. Unlike the narwhal the white whale has some eight to ten teeth on each side of both upper and lower jaws, and these may be more complicated than those of most dolphins, having accessory cusps. The white color is a feature of the older animals; the young are dark gray at first, becoming mottled as they lose pigment, to become yellowish cream and eventually white by the age of four or five years.

The white whale feeds on a wide variety of foods from quite large fish to cuttlefish and Crustacea. It is reported as moving into rivers in quite large numbers, and here it may hunt in remarkably shallow water for bottom fish. White whales have always been a source of food and leather for Arctic communities, their skin being thicker than is usual in whales, and from about the 17th century many have been taken commercially. But the numbers available have fallen off and, whereas about 2,000 a year were taken by Tromsö vessels off Spitsbergen at the end of the last century, only a few now remain as they have moved farther east to breed in greater security.

WHITING, *Merlangius merlangus*, a fish related to the cod found in the eastern North Atlantic but rare in the Mediterranean. The whiting is so commonly seen on fishmonger's slabs that a color description would seem to be superfluous. In life it is yellow-brown with golden mottlings on the belly and superb pink and purple reflections playing along the flanks. The dead whiting is a pale image of the live fish. As in the cod, there are three dorsal fins and two anal fins, but there is no barbel under the chin. The lateral line series of scales are brown and curve downward halfway along the body. There is a black blotch at the bases of the pectoral fins. Whiting are slender fishes and smaller than the cod, growing to about 14 in. (36 cm) and weighing about 3 lb (1.2 kg). The largest fish caught weighed 8 lb (3½ kg).

The whiting is found in the northeastern Atlantic but wanders a little farther south than the cod. It lives in large shoals that usually swim and feed near the bottom, especially over sand or mud. They are carnivores, feeding mainly on small fishes and crustaceans and will swim nearer the surface if tempted by shoals of fishes. A female whiting lays about 200,000 eggs in spring, usually at depths of about 150 ft (50 m). During the spawning season the shoals stay together, but afterward they break up and intensive feeding begins, usually in deeper waters. In early summer the shoals begin to re-form over rich feeding grounds in moderately deep water of about 100 ft (30 m). They remain there during the summer, but as winter approaches they move into shallower waters and can be readily caught on rod and line at dusk.

WHYDAH, a name given to long-tailed, mainly black-plumaged weaverlike birds, and as a result used for species from two different subfamilies of the weavers. In one case it is used for some species of the genus *Euplectes*, inhabiting open country and using the long tails and dark color of the males for advertisement, and in the other for the parasitic widow birds of the subfamily Viduinae.

WILD BOAR, *Sus scrofa*, a large woodland-dwelling hog with long tusks and stiff, dark, gray-black or brown hair. The snout is typically piglike with a mobile disc. There are four toes on each foot, of which the middle two are used for walking. The head and body length is about 70 in. (178 cm), and the height at the shoulder is about 40 in. (100 cm), and the tail is 12 in. (30 cm) long. The weight of males varies from 150-500 lb (67-227 kg) and that of females from 80-330 lb (36-150 kg). The dental formula is: incisors $\frac{3}{3}$; canines $\frac{1}{1}$; premolars $\frac{4}{4}$; molars $\frac{3}{3} \times 2 = 44$. The upper canines turn outward and upward, and these wear against the bottom canines causing sharp edges to form.

Wild boar are found throughout the deciduous wooded areas of Europe and northern Africa and throughout southern Asia to the Malay peninsula and islands of Java, Sumatra, the Philippines, Japan and Taiwan. It is now extinct in England and southern Scandinavia, but it has been widely established by introduction into the Bismarck Archipelago, Louisade Archipelago, New Guinea, New Zealand, North America and the Solomon Islands. In the U.S.A. it is now found in Tennessee, New

Whiting with brittlestars on the rock in the background.

European wild boar, ancestor of the domestic pig.

Hampshire, California and a few other places. In Europe the rut is in winter between November and February. After a gestation period of four months, the young are born from March to May. In parts of the Mediterranean basin breeding is in spring (April-May). Litter sizes vary from five to eight. Young are born with nearly continuous light stripes on a darker brown background.

The wild boar is principally a vegetarian, eating a variety of green vegetation, acorns, berries, roots and tubers, but it will occasionally eat worms, insects, and even reptiles and birds' eggs. It is very destructive to gardens and farm crops where its habitat borders on agricultural land. It is generally regarded as a game animal and is hunted extensively, sometimes with dogs and sometimes on horseback and can be dangerous when wounded. The wild boar is one ancestral species for domestic swine. See pig, domestic.

WIREWORM, the larva of the click beetle. Wireworms are elongated and very smooth yellow-brown grubs that live in the soil and grow to a maximum length of nearly 1 in. (2.5 cm). Although different species of wireworms may be found in almost any soil, the natural habitat of the commonest species, *Agriotes obscurus*, is grassland. Permanent grassland may be very densely colonized by wireworms (up to 8 million per acre), as there is an almost inexhaustible supply of food in the roots, crowns and lower stems of grass plants. In such situations, wireworms do little real damage as grass plants quickly regenerate damaged parts or new plants take the place of plants that have been killed. However, when permanent grass is plowed for arable cultivation, the wireworms may become a very serious destructive force, especially if the first crop is a cereal. Crops such as field beans, peas and flax are particularly resistant to wireworm attack: the first two because the seeds are very large and the root growth is strong, and the latter owing to the vast numbers in a crop of flax seedlings (about 15 million per acre) and their stringy nature. In general it is found that arable land of long standing has relatively low populations of wireworms, as regular plowing and cultivation expose the grubs to birds, frost, hot sunshine and desiccating wind. The long life cycle of the wireworm of 4-5 years allows the safe use of leys (temporary grass) in a crop rotation.

WISENT, *Bison bonasus*, or European bison, similar to the American bison but now nearly extinct in its truly wild form. Since the extinction of the true aurochs, the ancestor of domestic cattle, the wisent has itself sometimes been called "aurochs." Wisent differ from the American bison, which may be called simply "bison," in numerous features. The hindquarters are higher and the shoulder less humped, so that the back is more nearly horizontal; the color is uniform brown instead of being nearly black on the fore parts; the hair on the head, limbs and tail is of the same type as on the rest of the body instead of being crisp and curled; the hair on the forehead is only 8 in. (20 cm) long compared with 20 in. (50 cm) in the bison, and is drawn forward and lies down, instead of forming a cap and overhanging the nose. The tail of the wisent is long, reaching below the hocks, and long-haired, but not tufted as in the bison, and the hoofs are broad and weakly curved, whereas in the bison they are narrow and strongly sickle-shaped, their tips overlapping.

In the wisent, as in the kouprey, the renewing of the horn is a noticeable process. The points of the new horn begin to pierce the old sheath at about four years of age, although the conspicuous fraying of the kouprey is not seen. Wisent rub their horns against tree trunks and bushes, and this helps with the shedding of the old sheath.

Formerly, wisent were spread over a great deal of central and eastern Europe, and in the latter half of the 18th century they were still to be found in the upper Irtysh and Ob region of Kazakhstan. Nothing is known about these

Asiatic wisent, but the European ones were divided into two very distinct subspecies, one in the lowlands and one in the Caucasus mountains. The lowland wisent, *Bison bonasus bonasus*, which still exists as a pure race, is the larger of the two. Old bulls may measure 6 ft (2 m) at the shoulder and 5 ft (1.5 m) at the croup, and weigh 1,870 lb (850 kg). The hump is comparatively high, so that the line of the back is curved. The head is rather large. The hoofs also are large, being 4 in. (10 cm) long. The body hair is nearly smooth, and the color is brownish with ocherous tones. The Caucasus wisent, *B. b. caucasicus*, which no longer exists in its pure form, was only 5 ft 4 in (1.6 m) at the shoulder, weighing 1,540 lb (700 kg). The hump was low, the line of the back comparatively straight and flat. The hoofs were short, only $3\frac{1}{2}$ in. (8.5 cm) long, and the body hair was rather finely curled, almost like an American bison's, and the color a warm sepia.

WOLF, *Canis lupus*, supposed ancestor of the domestic dog, ranges throughout the temperate and coniferous forests and tundra of the northern hemisphere, and is a large predator, usually weighing 60-120 lb (27-55 kg). It has a broad chest, small pointed ears and long legs adapted for a cursorial life. Although the majority are tawny with a cream chest and black markings on the shoulders, tail tip, and near the base of the tail, pure white or black individuals are occasionally found.

Wolves are pack hunters (although they can and do hunt individually) in order to take advantage of the numerous moose, deer and elk herds found in the north. Packs are large during the winter, with up to 30 members, but in the summer when there are many smaller mammals available, like snowshoe hares and beavers as well as ungulate calves, the packs often break up into smaller groups. In addition, the spring dispersal often results from the onset of the breeding season when there is increased tension.

Mating occurs between January and March, and the pups are born two months later, usually in a hole dug by the mother, or in an enlarged fox den. Both the father and other adult wolves of both sexes may help in rearing the litter by guarding the cubs if the mother joins an evening hunt and by bringing meat back to the growing litter.

Whatever their size, packs usually have an established territory traversed over a period of weeks in the search for food. The size of the territory is partly dependent upon the size of the pack, but also upon the number of prey species available. Maximum density is probably never more than one wolf per sq mile (2.7 sq km). Territories are demarcated by scent marks placed on conspicuous trees, rocks and bushes along the wolves' trails. Howling also informs the wolves of each other's location. A pack has little tolerance for strangers, and intruders are chased away and even killed if they are too persistent in efforts to join the group.

Since wolves cannot easily overcome healthy adult moose or caribou, they typically prey upon weaker members of the ungulate population, e.g., the elderly, sick and young. Before beginning a chase, they test potential victims by sham charges, and if an animal stands its ground and fights back, the wolves usually abandon it. Even if an extended chase begins, however, wolves are not always successful, and there may be several chases before a victim is finally brought down.

Within a wolf pack, there is a complex social organization, with a top male and female who are largely responsible for patrolling the territorial boundaries, settling disputes between pack members, and controlling the movements of the pack. They may be the only breeding pair. Below these two animals in status are the other wolves, usually arranged in two linear hierarchies with little interaction between the male and female rank orders. At the very bottom of the social order is often an outcast who lives on the edge of the pack and exists by eating the scraps left by the group. This social system is established and maintained by the wolf's very complicated sign language consisting of many movements of the tail, ears, mouth and body as well as vocalizations that indicate the status and mood of each pack member.

WOLF SPIDERS, hunters on the ground belonging to the spider family Lycosidae, which leap on their prey. Some are entirely vagrants; others live in deep silk-lined burrows from which they pounce on passing insects. They range in body length from $\frac{1}{4}$ in. (5.5 mm) in the genera *Pardosa* and *Pirata* to the largest species, *Lycosa ingens*, on Deserta Grande, Madeira, which reaches $1\frac{3}{4}$ in. (45 mm).

A high cephalothoracic narrowing in front bears a row of four small eyes with two large eyes above it looking forward. Farther back are two eyes which look upward.

The species of *Pardosa* are small vagrants on open ground, which carry an egg sac attached to their spinnerets until the young hatch and clamber on their mothers' backs. Species of *Pirata* are small chocolate-brown spiders that frequent damp ground and can run down plants under standing water. *Arctosa* usually live in sandy situations and are pale and mottled in color. They make burrows, as do most of those belonging to the genus *Lycosa*. The burrow-dwellers come to the entrance to sun their egg sacs.

Being comparatively long-sighted, the males resort to visual courtship and have special ornamentations that are displayed. In *Pardosa*, the palps are often conspicuous, and these are waved in harmony with the legs. In *Lycosa*, it is often the front legs that are specially enlarged or colored, and these are waved. Where the female lives in a silk-lined burrow the male drums and vibrates on the silk at its entrance.

WOLVERINE, *Gulo gulo*, or glutton, a large terrestrial mustelid looking like an oversized, heavy-bodied marten and weighing 26-64 lb (13-28 kg), found in the Old and New World tundra regions. Two caramel-colored stripes, contrasting with the long black-brown fur, run along the flanks to the rump, where they overlap. The head with small ears nearly concealed in the fur is typically mustelid. The short, muscular legs with clawed feet have haired soles, a trait this species shares with other subpolar mammals such as the polar bear. Males are usually larger than females, measuring up to 40.5 in. (113 cm) in length including the bushy tail. The male wolverine may share its vast home range, sometimes reaching over 100 sq miles (260 sq km), with two or three

A wolf spider *Geolycosa blackwalli*, of Europe.

females. By relentlessly patrolling this area, using regular trails and leaving scent and scratch marks on definite grass hummocks or tree trunks, it attempts to discourage and intimidate other males.

The wolverine's diet shows seasonal variation and covers a wide range, including berries, insects, fish, eggs, birds, lemmings, carrion and even provisions left behind in a trapper's cabin. During a blizzard, it may sleep for several days, tunneled deep under the snow, but it does not hibernate for any length of time.

The breeding season is in July, and delayed implantation takes place so that two to five cubs are born the following spring or early summer. Blind for 20 days, the young are suckled until they are ten weeks old, and are fiercely protected by the mother. The adult male will drive off his male offspring when they are two years old, and they then set out to establish a territory of their own. Sexual maturity is reached at four years. Wolverines in captivity are usually silent animals; a growl or a grating whine may sometimes be heard. Their life expectancy is approximately 15 years. Trapped intensively for their valuable pelt (used for lining parka hoods), the wolverine has now become quite rare and is found only in the upper Arctic limits of its range.

WOMBATS, heavy, stockily built, burrowing marsupials of bearlike appearance with a vestigial tail and broad head, up to about 4 ft (1.2 m) in length and weighing up to 70 lb (32 kg). They are quadrupedal with short legs and with five toes on each foot, the first toe of the hind foot lacking a claw. The wombats are usually placed in a distinct family containing but three living species. Their closest living relative appears to be the tree-dwelling koala.

The common wombat, *Vombatus hirsutus*, and the island wombat, *Vombatus ursinus*, have dark brown to almost black, coarse, thick fur. They are distinguished from the hairy-nosed wombat, *Lasiorhinus latifrons*, by having the muzzle between the nostrils covered with hairless granulated skin. The ears of common and island wombats are shorter than those of the hairy-nosed wombat. The incisors of wombats are reduced to a single pair in each jaw, separated by a long, toothless gap (diastema) from the single premolar and four molar teeth of each side of upper and lower jaws. All the teeth are rootless, but extend a long way into the bone and grow continuously throughout life from persistent pulps.

Common wombats are confined to the eucalypt forests of eastern continental Australia, and island wombats to Tasmania and Flinders Island in Bass Strait. The hairy-nosed wombat is a dry-country animal found in the drier parts of South Australia, near the lower Murray River, and on the Nullarbor Plain and in a small part of inland Queensland.

Wombats have a pouch containing but two teats, and they give birth to a single young in late autumn, which is independent of the mother by the following summer.

Wombats are nocturnal and herbivorous and feed on grasses, roots and other vegetable matter. The common wombat is found in South Australia in areas of low limestone cliffs, and the burrow is often made beneath limestone. Favored burrowing areas are often adjacent to

freshwater swamps or lakes, and the roots of swamp plants then form the staple diet. The burrow system of the hairy-nosed wombat is usually also beneath a limestone outcrop and much more extensive than that of the common wombat. Each system may have up to a dozen entrances, and the burrows connect beneath the ground.

WOODCHUCK, *Marmota monax*, one of the most familiar animals of North America, where it is popularly referred to as the "chuck." It belongs to the squirrel family and could be described as a woodland ground squirrel. Its closest relatives are the marmots, which are sometimes referred to as "rock chucks." The woodchuck is found throughout the northeastern parts of the U.S.A. and all the forested areas of Canada.

The woodchuck is a large stocky rodent, up to 2 ft (60 cm) long with a rather flat-topped head and a short but moderately bushy tail. The pelage is a rather uniform brown. It is equipped with long strong claws with which it excavates a deep burrow. The main entrance has a conspicuous mound of soil, but there are usually other less conspicuous bolt holes.

Woodchucks feed in daylight, usually in the early morning or late afternoon, and are therefore fairly often seen, even if it is only a rear view as they bolt for their burrows with surprising agility. But like their relatives, the mountain marmots, woodchucks also like to sit sunning themselves at the entrance of their burrows, where they can be observed without much difficulty. They are vegetarians, feeding mainly on green fodder, although at fattening-up time in autumn acorns and windfall apples help to lay down fat in preparation for hibernation.

A slater, *Philoscia muscorum*.

The sea slater *Ligia oceanica* among acorn barnacles.

The woodchuck is one of the most profound hibernators among the mammals of North America. It retires to the end of its burrow, which it seals off with soil before curling up for the winter. The body temperature and pulse rate fall, and the animal is torpid until spring.

Mating takes place soon after emerging from hibernation and may be preceded by a considerable amount of territorial fighting. The young usually number about four, and they are very small, naked and blind at birth. They first emerge from the burrow when about a month old, but is is another month before they become independent.

WOODCOCK, medium-sized wading birds, related to the sandpipers, 10-13 in. (25-32 cm) long, with long straight bills and gray-brown and rufous plumage. They differ from other wading birds in their solitary habits and in living in woodland and coppices. The woodcock, *Scolopax rusticola*, is found throughout Europe and northern Asia and in winter reaches north Africa and India. Some six or seven species of *Scolopax*, similar to *S. rusticola*, are found in southeast Asia on various islands, including the Celebes and New Guinea. The American woodcock, *Philohela minor*, occurs in the eastern United States from Florida to southern Canada and, in winter, west to Texas. Woodcock feed on earthworms and other insects found by probing in soft soil. The bill is sensitive at the tip, and food is located by touch. It is also slightly flexible at the tip and can be parted sufficiently to grasp anything it feels.

In the breeding season males execute a display flight known in Britain as roding, usually at dusk and dawn. The flight follows a set route and is accompanied by two special notes, one of which may be made by the wings. The young are hatched and raised by the female alone from four brown cryptically marked eggs laid in a hollow lined with dead leaves. Carrying the young in flight by the female is well documented. They are usually held between the feet and the body but have been seen on the female's back.

WOOD CREEPERS, a small, rather uniform, family of medium-sized American perching birds, sometimes called "wood hewers." There are 50 species, which range in size from 5-15 in. (13-40 cm). Generally they are olive, with rufous wings and tail and light streaks on the head and underparts. The tail feathers are usually stiffened and the legs short and powerful, for climbing trees. The bill is stout, compressed and long for probing bark. The family ranges from northwestern Mexico (Sonora) to northern Argentina, Bolivia and Peru, occurring in Trinidad and Tobago, but not in the Antilles. Probably most breed in holes, but nesting habits are generally unknown. They are solitary, arboreal birds, living in wooded areas and behaving like tree creepers or woodpeckers.

WOODLICE, also called sow bugs, are small crustaceans belonging to the Isopoda, which means that they have more or less flattened bodies and legs of more or less equal length. They are related to the sea slaters inhabiting the seashores throughout the world. Some species, known as pill bugs, can roll into a ball. The antennae of all woodlice are well developed; the antennules small and relatively insignificant. Woodlice are capable of rapid walking movements on land which are similar to the walking movements of isopods on the sea floor. They breathe through their abdominal legs (pleopods), the first pair of which are modified to long pointed cones to assist in reproduction.

Their food is mainly vegetable, but they will

eat dead animals and their own cast skins, and they are cannibalistic. They are scavengers and can also cause much damage to cultivated crops in gardens and greenhouses. They are eaten by small mammals and some birds. Some species secrete a sticky repugnant fluid that makes them distasteful to many possible predators.

The young are developed within a brood pouch. The number of eggs per brood averages 22-35, with an incubation period of 32-45 days, 60-80 in the pill bug. Usually two broods are produced each year and in captivity common woodlice have been kept alive for four to six years.

Woodlice are normally found in damp places such as under stones, among fallen leaves, under the bark of rotting trees and in crevices. Although each species may have a wide range of habitat each is more abundant in some habitats than others, and it is possible to find several kinds of woodlice in one rubbish heap. The pill bugs are more abundant in chalky districts and can tolerate drier habitats than other woodlice. One species is only found in ants' nests. Desert woodlice avoid the rigors of their environment by burrowing deeply into moist micro habitats in the desert sand. Woodlice do not hibernate; in winter they seek protection in such places as outhouses and manure heaps.

WOODPECKERS, birds specialized for obtaining their food from the trunks and branches of trees. The 210 species are all, excepting the two wrynecks, similar in appearance. They occur on all the major continents except Australasia, but are absent from Madagascar and some oceanic islands. The majority are 6-14 in. (15-35 cm) long, but the piculets are small, down to $3\frac{1}{2}$ in. (8.7 cm) long, and a few species are up to 22 in. (55 cm) long. The normal posture of these birds is upright, clinging to a vertical surface with widely straddled, short legs bearing large strong feet with sharp curved claws. The feet are zygodactyl, two toes pointing forward and upward while the outer toe is directed backward and outward. The original hind toe is reduced in size or lost altogether in some genera. Woodpeckers rarely perch in transverse postures.

Typical woodpeckers have well-developed tails, wedge-shaped at the tip, with very strong quills that protrude beyond the tip of each pointed tail feather and, when the tail is pressed against a trunk or branch, act as a prop, the spines of the quills resisting the abrasion that would destroy the weaker feather vane. Their habit is to fly to a low position on a trunk or branch and then move up it in a series of jerky hops; but woodpeckers can also hop backward down a vertical surface, or sideways across it.

The wings are rounded, and flight, although it may be swift, is usually undulating, with a swoop between each wing beat, becoming more direct in the larger species. The neck is slender but muscular, and the head relatively large. The bill is strong and straight, fairly thick at the base and tapering evenly to a sharp tip. It is used for chipping away wood, and the muscles and structure of the head and neck are adapted for driving the bill forward with considerable

The greater spotted woodpecker of Europe.

force, and absorbing the shock of the blow. The bill hacks open rotten wood and exposes the insects on which most woodpeckers feed. In addition, the tongue is long and slender, mobile and wormlike. It can be extended considerably.

Some species are soberly colored in greens, browns and grays, sometimes patterned with spots or bars, a few showing extensive yellow or red coloring, and a large number are boldly patterned in black and white, the last being barred, spotted or with large contrasting areas of black or white feathers. Many species have signal patches of red or yellow on the head, rump, or upper tail coverts; and a number possess a short crest.

The majority occurs in woodland of some kind, the larger species usually being associated with tall forest trees. The typical woodpeckers spend most of their time on the trunks and larger branches of trees, hacking away dead or rotten wood in search of burrowing insects. Smaller species may spend more time on thinner outer twigs. Some species may occur in more shrubby growth, and several woodpeckers, although typical tree birds in other respects, spend much of their time on the ground probing for ants, the green woodpecker, *Picus viridis*, of Eurasia, and the flickers, *Colaptes*, of North America, being typical examples of this. The ground woodpecker, *Geocolaptes olivaceus*, of South Africa, is a more extreme example, spending its life on the ground in open but rocky country, where it can roost among the rocks and tunnel in the ground to make a nest.

Although food is mainly insects, some species take other food as well. Larger seeds or nuts may be used. The Scandinavian population of the greater spotted woodpecker, *Dendrocopos major*, feeds extensively on the seeds of pine cones, making a hole in a tree to which these are carried, wedged and opened. The acorn woodpecker, *Melanerpes formicivorus*, of western North America, eats acorns in a similar fashion and will also store them in large numbers by fitting them into series of holes specially excavated in tree trunks. The North American sapsuckers drill horizontal rows of holes in the bark of a tree, returning repeatedly to drink the sap that oozes out.

In spite of the varied methods of feeding, all these birds are typical woodpeckers, together forming the subfamily Picinae. The tiny woodpeckers known as piculets are grouped in a small subfamily, the Picumninae. They are widely scattered, in South America, Africa and Asia. They are the smallest woodpeckers, 3-4 in. (7.5-10 cm) long, usually perching as well as clinging and lacking the spiky, stiff tail feathers. They have small, relatively weak bills that they use mainly on rotten wood.

The young are naked and blind. They have special pads on the leg joints on which they can rest and shuffle about on the hard floor of the nest chamber. Partly grown young tend to be rather noisy in the nest and to betray its presence. When they leave they may be able to fly only weakly, but provided this will carry them from tree to tree, even with loss of height, they can rely on their ability to climb rapidly to carry them out of reach of predators.

WOOD PIGEON, *Columba palumbus*, also known as the ring dove, a western palearctic

The European green woodpecker, whose call is like maniacal laughter.

Woodpeckers are much the same, except in size and color, wherever they may live. A typical example is this black-and-white South American woodpecker.

bird of both town and countryside, found in large flocks in winter. A considerable pest to farmers and therefore widely persecuted, it maintains its numbers well over most of its range. In northwest Europe it has even extended its range northward over the last 100 years. Typically a bird of woodland and farmland, it has spread to the very edges of wooded country, sometimes even beyond the treelines, when it nests on the ground, and into towns and cities. The wood pigeon is some 16 in. (41 cm) long, somewhat larger than domestic pigeons, generally bluish-gray with a mauve-pink breast and cream belly. It has a white band across the wings, which is conspicuous in flight, and a white patch on each side of the neck surrounded by an iridescent area of purple-pink and green. The tail is gray with a very dark tip. The legs are rather short, and the feet are reddish purple. Like other pigeons the wood pigeon flies strongly. When disturbed in trees or bushes it frequently "explodes" from cover with loud wing claps and crashes through twigs and foliage. Characteristically, it raises and lowers the tail again a few seconds after alighting. The display flight consists of a series of swoops in the air in which the bird rises steeply, claps its wings loudly, and then glides down on outstretched wings to repeat the process. Its most common "song" is a phrase of five notes: "coo-cooo-coo, coo-coo," a common sound of the countryside during the warm summer months, but its call during courtship display is more varied.

WOOD WASPS, large insects with conspicuous

black and yellow or metallic blue coloration, which inhabit forests. They are not true wasps but are related to sawflies. The adult females have stout ovipositors, which are used to drill through the bark of trees and into the wood, usually only trees that are past their prime. Inside the tubes thus formed, eggs are laid. Most wood wasps have an associated symbiotic fungus that females store in sacs in their abdomen, and inject into the wood during oviposition. The fungus develops in the timber, contributing to the death of the tree, and providing food for the developing larvae. The wood-wasp larvae inhabit the wood for one or more years, producing long tunnels packed tight with chewed-up fragments of wood. The larvae pupate in the tunnel, and the young adults chew their way out when they are mature. The giant wood wasp or horntail, *Sirex gigas*, of Europe, 1 in. (2.5 cm) or so long, is often mistaken for a hornet. The corresponding genus in North America is *Tremex*.

WOODWORM, the immature stage of certain wood-boring beetles found in buildings, either in furniture, structural timber or joinery, especially of *Anobium punctatum*, also known as common furniture beetle. As infestations of furniture are insignificant compared with attacks on softwood constructional timber and flooring, we might do well to follow the New Zealand example and call this $\frac{1}{6}$-in. (4-mm) long, brown beetle, the house borer.

Like many other insect pests, *Anobium punctatum* occurs in the wild, where it causes no serious harm. Holes left by the woodworm can easily be detected in the branch scars or exposed dead sapwood of trees where the bark has been removed by cattle or hedging operations. These trees are almost always hardwoods, oak and orchard trees predominating. Softwood trees, the cone-bearing species, seldom show the results of woodworm attack out of doors, contrasting strikingly with what is found inside buildings.

The larva hatches out after two or three weeks by eating its way through the bottom, pitted half, of the eggshell, thereby ingesting the yeast cells that come to play an important part in the breakdown of cellulose.

In fact, the larvae cannot survive without the yeast cells in their gut; it is an obligatory symbiosis. A New Zealand research worker, J. M. Kelsey, turned a number of eggs of this species upside down so that the young larvae ate their way out of the smooth unpitted eggshell and, therefore, did not ingest any yeast cells. They all died within a few days. From the eggshell the young larva bores directly into the wood and never leaves it.

WRASSES, a family of perchlike fishes related to the parrotfishes, typically having fairly elongated bodies with a long-based dome fin (the first part spinous) and a shorter anal fin. The lips are fleshy, and in addition to strong teeth in the jaws there is a set of powerful molar teeth in the throat, the pharyngeal teeth, the lower set being fused into a triangular plate. In size, the wrasses range from small reef fishes of 3 in. (8 cm) to large species reaching 10 ft (3 m) in length. Many of the cleaner-fishes, which remove parasites from larger fishes, are wrasses.

The tropical wrasses are usually more brightly colored than those from temperate waters, but one of the prettiest is the cuckoo wrasse, *Labrus mixtus*, of European coasts, a species which grows to about 12 in. (30 cm) in length. Like many wrasses, the colors of the males differ from those of the females. The male has bright blue on the head and back, while the flanks, belly and dorsal and anal fins are yellow. There is also a blue longitudinal stripe running across the cheek and along the flank. In the females the color is similar to that of the young,

the general background being orange to red with three dark spots on the back. These fishes commonly occur in about 100 ft (30 m) of water but will often come inshore. In northern latitudes they tend to swim singly, but form shoals farther south. They are often caught in lobster pots, which suggests that they scavenge for their food. They build a nest during the breeding season, and this is guarded by the male.

WRECKFISH, or stonebass, *Polyprion americanus*, a marine perchlike fish that derives its common name from its habit of frequenting wrecks and pieces of floating debris. It lives in the open waters of the tropical Atlantic and Mediterranean. The wreckfish has a deep and compressed body with a long dorsal fin, the soft-rayed parts of the dorsal and anal fins being prolonged into lobes. Reported to attain 6 ft (1.8 m) in length and to weigh up to 100 lb (45 kg), it is occasionally caught in British waters but is generally found in the warmer parts of the Atlantic.

WRENS, the name of several groups of small birds. The true wrens are the Troglodytidae, a family of about 60 species of small to medium-sized perching birds.

The wrens in the family Troglodytidae are mainly compact little birds, 3¾-8¾ in. (10-22 cm) long, with short to long tails which in many species are held cocked upwards. They nearly all have brown plumage, lighter below, often with stripes of black or white and dark barring on the wings and tail. The bill is rather long, slender and slightly downcurved. The legs are long with strong feet. The wings are short and rounded.

The greatest diversity of "true" wrens is in Central America. A number are familiar birds in North America and one species, the wren or winter wren, *Troglodytes troglodytes*, has reached Europe, Asia and north Africa. The majority live in thick low vegetation but some are found on bare moorland, among cacti in deserts, in rain forests, deciduous northern woodlands and in the lush vegetation of tropical swamps. As they live in the thick cover most wrens are more readily heard than seen, the loud, clear songs and loud chacking or ticking

call notes being all that prevents them from being passed unnoticed.

All the true wrens are insect-eaters, mainly picking insects from foliage and from the ground litter under vegetation. The northern cactus wren, *Campylorhynchus brunneicapillus*, of the United States often overturns stones and dried dung in its search for insects, and the American song wren, *Cyphorhinus phaeocephala*, shovels aside dead leaves. A few species sometimes take tiny reptiles, spiders and even small fish.

Most wrens build bulky, roofed nests of moss, dead leaves and other materials but a few, such as the house wren, *Troglodytes aedon*, the ochraceous wren, *T. ochraceus*, and the rufous-browed wren, *T. rufociliatus*, build cup-shaped nests inside cavities in trees, walls or rocks.

The 2-9 eggs are white with brown or lilac spots and blotches that may be either dense or sparse in different species. The eggs are incubated by the female alone, but in some species she is fed on the nest by the male. The young are fed either by the female alone or by both parents and they fledge after about 15-19 days depending on the species. After fledging the young of many species are dependent on the parents for food for a surprisingly long period after they leave the nest. In some species the young from early broods help their parents with later broods.

WRYNECKS, small slim birds about 6 in. (15 cm) long, belonging to the woodpecker family but looking more like perching birds than woodpeckers. Wrynecks belong to the subfamily Jynginae of the woodpecker family Picidae but differ from the true woodpeckers, Picinae, in having a soft tail, the feathers of which are not pointed. The plumage of the common wryneck, *Jynx torquilla*, is gray-brown above, streaked and patterned like a nightjar, and white and pale buff below with darker bars and spots. It is a tree-haunting bird with a liking for the upper branches but can often be found in low bushes or on the ground, especially during migration and in the winter quarters. It breeds in Europe, Asia and northwest Africa.

The food is insects, largely ants. The wryneck does not bore into wood like a woodpecker but uses its long tongue to pick insects from bark, leaves and crevices too deep to insert the bill. Wrynecks are territorial, but they rarely drum like woodpeckers and the territory is advertised more usually by a long, drawn-out, high-pitched cry, "ki-ki-ki...," rather like that of a kestrel. The nest site is an existing hole in a tree, a nest box or a hole in a bank or wall such as a sand martin burrow. No nesting material is used, the eggs being laid on the floor of the hole. The clutch size varies from 5-14 eggs, larger clutches probably being laid by two females. If eggs are removed as they are laid, a single female will produce up to 40 or more in a season. Incubation is carried out by both sexes and lasts 12 days. The young are fed from the bill by both parents and fledge in about 20 days. When handled, wrynecks twist the head about in a most peculiar fashion.

The European wren is a troglodyte, typically nesting in caves where it can sometimes be seen clinging to the ceiling by its feet, upside-down.

X·Y·Z

XENOPUS, a genus of amphibians with four species restricted to Africa; the clawed frog, *Xenopus laevis*, is the most widely known. This 5-in. (12.5-cm), purely aquatic frog is limited to southern Africa, where it is found in pools or streams that dry up in summer causing the animals to aestivate. Claws on the inner three toes of each foot give this frog its common name. The clawed frog is best known to doctors as the "pregnancy frog" as it was once used extensively to test suspected pregnancy in women. Other species are: *X. gilli* of South Africa, and *X. mulleri* and *X. tropicalis*, of tropical Africa.

YAK, *Bos mutus*, (= *B. grunniens*), shaggy ox of the high plateau of Tibet, distinguished by the long fringe of hair on shoulders, flanks, thighs and tail, which may sweep the ground. It has been domesticated and is the staple beast of burden and milk-producer of the central Asian highlands. Wild yaks are larger than the domestic animals, bulls sometimes standing 6 ft 8 in. (2 m) at the shoulder, and weighing 1,150 lb (525 kg). They are always black, whereas domestic yaks may be a variety of colors, from black to brown, white or piebald. The wild yak also has longer horns, which turn in at the tips. In domestic yak the tips often turn out outward, and some breeds are polled.

Wild yak are found all over the Tibetan plateau and enter Indian territory in the Changechenmo valley, Ladak. Bulls travel alone or in groups of two or three; cows and calves form big herds of any number from 20 to 200. In summer, yaks ascend to 14,000-20,000 ft (4,270-6,100 m) where there is permanent snow and the temperature can be –40°F (–5°C); the herd picks its way through the snow in single file, each animal placing its feet in the tracks of the one before. In winter, the yaks come lower, and late in the year (late autumn) the rut starts, when the bulls fight fiercely, although fatalities rarely result. The bull stays for one or two months with the cows it has gathered around it, then they separate again. The young are born the following April to June, in time for the fresh moor grass, which has been uncovered and watered by melting snow.

YAPOK, *Chironectes minimus*, an aquatic opossum closely related to the American opossums and the only aquatic marsupial. The feet are broadly webbed, and the tail, although prehensile, is much thicker than that of other opossums. Unlike most other marsupials the male, as well as the female, possesses a pouch into which the scrotum is pulled. The yapok is about 20 in. (50 cm) long, and the tail measures about 12 in. (30 cm). The underparts, inner faces of the thighs, upperparts of the hind feet and terminal portion of tail are white. In swimming, the hind feet are used alternately, the tail streaming out behind.

The yapok has a wide distribution in Central and South America, from the southernmost parts of Mexico through Nicaragua, Panama and other Central American countries to Colombia, Venezuela and Guyana and southward to northeastern Argentina. Litters of two and three are the rule, but up to five young have been found in the female's pouch. The yapok builds nests of leaves, but has also been found in holes in the banks of streams. It eats crayfish, shrimps and other crustaceans.

ZAMBEZI SHARK, *Carcharhinus leucas*, a species probably worldwide in tropical and subtropical waters but remarkable for entering estuaries and rivers and even living permanently in fresh water. This species has many different common names: in Australia it is known as the whaler shark, in Central America the Lake Nicaragua shark and in the United States the bull, cub or ground shark. It has a broad head with short rounded snout and small ears. The body is heavy, up to 10 ft (3 m) long and weighing over 400 lb (200 kg), with a prominent triangular first dorsal fin and a small second one set well back toward the tail. The teeth are triangular with saw edges. The back and flanks are light to dark gray, the underside white. Wherever it occurs the Zambezi shark has a reputation for being aggressive. It attacks large fish including other sharks and will attack human bathers and even boats. There seems to be a connection between the shark's attacks and the presence of fresh water, there being a greater incidence of attack on bathers in the freshwater Lake Nicaragua than in coastal waters. See man-eater sharks.

ZEBRA. The three species of this horselike herbivore have in common that they live in Africa south of the Sahara and have individual stripe patterns like human fingerprints. In other respects they are as different from each other as are the other equids.

The living plains zebra can be subdivided into three fairly uniform subspecies. In east Africa and down to the Zambesi is found the Boehm's zebra, or Grant's zebra, *Equus quagga boehmi*, which has a pronounced stripe pattern even down to the hoofs. The stripes are black or dark brown on a white background, although in foals they are often light reddish-brown. Shadow stripes, that is dark stripes between the regular ones, occur in all populations of this subspecies, but are more common in the southern part of its distribution. The mane is, especially in adult stallions, often straggly and sometimes completely absent.

Rhodesia, Angola, Mozambique and Transvaal is the home of Chapman's zebra, *E. quagga chapmani*, the stripe pattern of which is somewhat reduced on the legs. Shadow stripes are common, the background color is more cream than white, and the mane is well developed. Farther south and west, in southwest Africa, Botswana and Zululand, lives the Damara zebra, *E. quagga antiquorum*, the leg stripes of which are further reduced and have often completely disappeared.

The plains zebras are the most numerous of all wild equids, and their total number has been estimated at about 300,000, of which about 200,000 live in the Serengeti Mara area in east Africa. The habits of the plains zebra have been investigated in the wild. They were studied particularly in the Ngorongoro Crater in Tanzania, an area of 100 sq miles (250 sq km) of mostly open steppe country. The findings of this study of a population of about 5,000 Boehm's zebra proved to be valid for other populations of the same and of other subspecies. The animals were studied for several years, individuals being recognized by their stripe patterns.

Particularly interesting is the social organization. Plains zebra live in small units of up to 16 members. There are two types of units: the families, consisting of one stallion, one to six mares and their young; and secondly the bachelor groups of surplus stallions, including subadult ones. Adult stallions are occasionally seen solitary. Plains zebra individuals and/or groups are not territorial—that is, they do not defend a certain area against conspecifics. They live in large home ranges of 30 to well over 100 sq miles (80-250 sq km), which they share with many other zebra. The groups, especially the families, are permanent units, and their adult members stay together for years, even for the lifetime of the animals. This coherence is, however, not enforced by the stallion. Should he die or leave the family because of old age, a new stallion will take the group on as a whole. When a member of a group is lost, the other animals will search for it for hours and even for days. The integrity of the groups is maintained irrespective of the animals of different groups crowding at water holes or when fleeing. The members of a group recognize each other by their stripe patterns, voices and scent.

In the families there is a rank order, in which the stallion is the top animal and the mares hold the subordinate places in descending order of rank. This order is demonstrated when the group migrates. The top mare leads the group, followed, in single file, by the other mares according to their rank. The stallion usually brings up the rear, but it may also lead. Each mare is followed by its foals in age order, the youngest first. Zebra families and stallion groups often aggregate to form larger herds. They do so on their sleeping grounds at night, during migrations, when thousands of zebra move together, and in good grazing zones. There is, however, no organization above the level of families and stallion groups.

Two subspecies of the mountain zebra, *Equus zebra*, are known, the Cape mountain zebra, *E. z. zebra*, which was formerly common in the Cape Province, and the Hartmann mountain zebra, *E. z. hartmannae*, of southwest Africa and southern Angola. There are fewer than 100 individuals left of the Cape mountain zebra. About 70 live in the Mountain Zebra National Park near Cradock, and another small population is in a reserve near Swellendam. The Hartmann zebra are more numerous.

They live in the mountain ranges along the Namib Desert and even in the Namib itself. Their number has been estimated at 5,000-8,000 The Hartmann zebra stands about 51 in. (1.30 m) at the shoulder, the Cape mountain zebra only 48 in. (1.20 m). Apart from their stripe patterns, both mountain zebras are characterized by their dewlaps, a fold of skin under the neck. They are specialized mountain animals able to climb even steep, rocky canyons to get to water.

Their social organization corresponds even in details to that of the plains zebra: they live in permanent, nonterritorial families and stallion groups.

The Grevy's zebra, *Equus grevi*, is the largest of all wild equids, measuring over 5 ft (1.5 m) at the shoulder. It inhabits the semidesert of northern Kenya and parts of Ethiopia and Somalia.

The social organization of the Grevy's zebra is completely different from that of the other zebras, as the stallions are territorial and there are no permanent bonds between adults.

ZEBU, *Bos indicus*, the characteristic "native" cattle of Asia and Africa, easily recognized by the prominent hump over the shoulders and the large baggy dewlap under the throat. They usually have large drooping ears and prominent horns. The typical color is steel gray, with the front half of the animal often a darker smoky gray, but many different color varieties exist. The characteristic hump is a fleshy structure rising above the animal's back and supported internally by specially elongated neural spines of the thoracic vertebrae. Zebu are large, comparatively trim cattle, much less stocky than many of the western breeds, and are also noticeably longer in the face. There are over 30 recognized varieties, the largest being the Gujarot zebu of Pakistan.

Zebu do not have the high milk yield or superior meat carcasses found in the best breeds of western cattle. However, they are often used in breeding programs to upgrade the heat tolerance of European varieties, thus rendering them more suitable for hot regions. The Santa Gertrudis, for instance, results from crossing shorthorn cows with zebu bulls. Cattle varieties that contain zebu blood have been found to be resistant to drought conditions and can even suckle their calves when ordinary cattle are physiologically seriously embarrassed.

ZORILLA, *Ictonyx striatus*, or African polecat, similar in appearance to the polecat of the same family, Mustelidae, but having a black and white striped coloration reminiscent of the New World skunks. When alarmed, it erects its hair, waves its tail, hisses and sometimes feigns death. Moreover, the zorilla nearly always sprays its enemy with fluid from the anal scent glands, a secretion which has a much more foul and persistent odor than that of the African striped weasel, a species with which it is often confused. In most aspects of its ecology and behavior, the zorilla is similar to the polecats.

Group of Boehm's zebra of East Africa. The beautiful front, back and side views, illustrate the typical pattern of this subspecies.